10th International Workshop "Strong Microwaves and Terahertz Waves: Sources and Applications" 2017

EPJ Web of Conferences Volume 149 (2017)

Nizhny Novgorod, Russia
17 - 22 July 2017

Editor:

A.G. Litvak

ISBN: 978-1-5108-4640-1

Printed from e-media with permission by:

Curran Associates, Inc.
57 Morehouse Lane
Red Hook, NY 12571

Some format issues inherent in the e-media version may also appear in this print version.

Printed by Curran Associates, Inc. (2017)

For additional information, please contact EDP Sciences – Web of Conferences at the address below.

EDP Sciences – Web of Conferences
17, Avenue du Hoggar
Parc d'Activité de Courtabœuf
BP 112
F-91944 Les Ulis Cedex A
France

Phone: +33 (0) 1 69 18 75 75
Fax: +33 (0) 1 69 28 84 91

contact-edps@webofconferences.org

Additional copies of this publication are available from:

Curran Associates, Inc.
57 Morehouse Lane
Red Hook, NY 12571 USA
Phone: 845-758-0400
Fax: 845-758-2633
Email: curran@proceedings.com
Web: www.proceedings.com

TABLE OF CONTENTS

Chairman of the Workshop – Alexander Litvak, Institute of Applied Physics, Russia

Organized by
Institute of Applied Physics, Russian Academy of Sciences
http://iapras.ru/

GYCOM Ltd.
http://www.gycom.ru/

Sponsored by
Russian Foundation for Basic Research
http://www.rfbr.ru/rffi/ru/

Informational support by
Association for Microwave Power in Europe for Research and Education (AMPERE)
http://www.ampereeurope.org/

Program Committee

International Advisory Committee

Petr Bagryansky, Budker Institute of Nuclear Physics, Russia

Monica Blank, CPI, USA

Vladimir Bratman, Ariel University, Israel and Institute of Applied Physics, Russia

Gregory Denisov, Institute of Applied Physics, Russia

Gian Piero Gallerano, ENEA-Frascati, Italy

Sergey Golubev, Institute of Applied Physics, Russia

Timothy Grotjohn, Michigan State University, USA

Eugeny Gusakov, Ioffe Institute, Russia

Jay Hirshfield, Yale University, USA

Toshitaka Idehara, University of Fukui, Japan

John Jelonnek, Karlsruhe Institute of Technology, Germany

John Lohr, General Atomics, USA

Gregory Nusinovich, University of Maryland, USA

Gun-Sik Park, Seoul National University, Korea

Alan Phelps, University of Strathclyde, UK

Alexander Shkurinov, Moscow State University, Russia

Michael Tendler, KTH Royal Institute of Technology, Sweden

Manfred Thumm, Karlsruhe Institute of Technology, Germany

Anatoly Vikharev, Institute of Applied Physics, Russia

Local Program Committee
(Institute of Applied Physics, Russia)

Gregory Denisov

Naum Ginzburg

Mikhail Glyavin

Sergey Golubev

Egor Gospodchikov

Sergey Kuzikov

Alexander Shalashov

Vadim Skalyga

Anatoly Vikharev

Vladimir Zapevalov

Irina Zotova

Scientific Secretary of the Workshop – **Sergey Mishakin,** Institute of Applied Physics, Russia edp sciences

EPJ Web of Conferences

Statement of Peer review

In submitting conference proceedings to EPJ Web of Conferences, I certify to the Publisher that I adhere to the **Policy on Publishing Integrity** of the journal in order to safeguard good scientific practice in publishing.

1. All articles have been subjected to peer review administered by the proceedings editors.
2. Reviews have been conducted by expert referees, who have been requested to provide unbiased and constructive comments aimed, whenever possible, at improving the work.
3. Proceedings editors have taken all reasonable steps to ensure the quality of the materials they publish and their decision to accept or reject a paper for publication has been based only on the merits of the work and the relevance to the journal.

Title, date and place of the conference

10th International Workshop "Strong Microwaves and Terahertz Waves: Sources and Applications", July 17-22, 2017
Nizhny Novgorod, Russia.

Proceedings editor(s):

Professor A.G. Litvak

Date and editor's signature

June 30, 2017

EPJ Web of Conferences **149**, 01001 (2017) DOI: 10.1051/epjconf/201714901001
10th International Workshop 2017 "Strong Microwaves and Terahertz Waves: Sources and Applications"

New trends in gyrotron development

G.G. Denisov

Institute of Applied Physics, Russian Academy of Sciences, Nizhny Novgorod, Russia, den@appl.sci-nnov.ru,
Gycom Ltd., Nizhny Novgorod, Russia

Abstract

Gyro-devices provide the highest CW or average power of microwaves in centimeter, millimeter and sub-millimeter wavelength ranges and therefore they are very attractive as microwave sources for many of applications such as plasma fusion, radiolocation, ion sources, telecommunication, technology, spectroscopy and some other. In last years an essential progress in the device development was demonstrated. The paper presents State-of the Art in the device development, new demands in the parameters enhancement and possible ways to achieve the goals.

1. Gyro-Devices. State-of-the Art

Gyrotrons for plasma fusion installations usually operate at frequencies 40–170 GHz [1–6]. Requested output power of the tubes is about 1 MW and pulse duration is between seconds and thousands seconds (depending on plasma machine parameters). In ITER installation there will be 24 of 170 GHz gyrotron systems with 1 MW microwave power each. ITER requirements include also high efficiency of the gyrotrons over 50%, possibility of power modulation with frequency up to 5 kHz, compatibility of the gyrotron complex with ITER control system. In May, 2015 a Russian Prototype of ITER Gyrotron System was completed and its operation was demonstrated [2–5]. The system includes gyrotron oscillator, liquid-free superconducting magnet, supplementary magnets, several electric power supplies, cooling systems control and protection systems, and other auxiliary units. The gyrotron system shows reliable operation with required parameters. In October, 2015 Final Design Review Procedure for the gyrotron system was successfully passed and in 2016 fabrication of the first serial gyrotron system was completed. Megawatt power at very long pulses (300–1000 s) was also demonstrated with gyrotron at 140 GHz frequency for electron-cyclotron systems of EAST (China) and KSTAR (Korea) superconducting tokamaks. Megawatt power gyrotrons with moderate pulse duration from 2 to 10 seconds were developed for TCV, HL-2A, and ASDEX Upgrade tokamaks.

There are successful developments of megawatt gyrotrons [1, 6] by European team including industrial company Thales, by CPI (USA), by Japanese cooperation of QST/Toshiba and Tsukuba University. The point worth mentioning is that the new stellarator W7-X is equipped now with 9 gyrotrons capable to operate at 140 GHz frequency and power of 0.8–1.0 MW in 1800 second pulses.

Gyro-amplifiers are intrinsically more difficult in realization because of multi-mode microwave systems, but in recent years there is a remarkable progress in their development. At IAP/GYCOM we are developing an original concept of a gyro-TWT that is based on the use of a helically corrugated waveguide that radically changes the dispersion of the modes of a circular waveguide. The operating mode has sufficiently high and almost constant group velocity at zero axial wavenumber which enables broadband operation of the helical-waveguide gyro-TWT with minimum sensitivity to electron velocity spread. A number of experiments have proved the main theoretical predictions and advantages of gyro-TWTs of this type [7–9]. In particular the output power of about 10 kW and instant frequency band more than 2 GHz were demonstrated with CW Ka-band tubes and more than 150 kW in pulse tubes with 10% duty factor. Several directions of research on the helical-waveguide gyro-TWT are currently under development at the IAP. Among the actual topics are the following: tubes with a high (40–50 dB) gain; devices with a new microwave system for inputting and outputting the radiation to and from a tube through one window; high-power tubes operating at the third cyclotron harmonic; tubes operating at 95 GHz and 260 GHz [10].

2. New Customer Demands

Traditional goals in gyrotron development are the power increase and the frequency increase. Future plasma machines as DEMO [11] claim for EC system higher frequency (comparing with ITER) 230 GHz (170 GHz), module power increase 1.5–2.0 MW (1 MW), efficiency enhancement to value higher 60% (50%), multi-frequency operation in order to avoid wide angle scanning of wave beams in plasma. There are mutual contradictions in these listed requests, for example, the higher frequency and the higher power require bigger gyrotron cavity (higher operating mode) and this affects gyrotron efficiency. Nowadays such a combination of parameters looks unreal – similar case with ITER requirements 25 years ago.

Besides power and efficiency increase recently more requirements became more acute in aspects of oscillation spectrum control. As it was mentioned it is considered gyrotron operation at different modes with different frequencies. For one mode operation frequency stabilization is desired and frequency tuning within resonance curve is necessary. Phase and frequency locking is required to make several gyrotrons as coherent oscillators. For spectroscopy application high frequency stabilization is needed. For some cases generation of periodical sequences of ultrashort (wide spectrum) phase-coherent pulses is very useful.

3. New Approaches and Recent Results

There are new developments as phase locking of gyro-oscillator, stabilization of gyrotron frequency by the optimal reflection from a remote object, stabilization of gyrotron frequency by phase locking loop (PLL), and generation of ultra-short pulses in gyro-TWT with nonlinear absorber in the feedback loop [12–17].

A gyrotron oscillator operating efficiently at modes of both rotations was developed and tested in Proof-Of–Principle experiment. The gyrotron operation can be switched by means of a cathode coil between two modes: co- and counter- rotating ones with respect to electron rotation. The main operating modes are $TE_{\pm 28.12}$ at 170 GHz frequency. A new synthesized quasi-optical mode converter composed of 3D metal reflectors provides coupling out of the generated waves in the form of two Gaussian wave beams passing the window disk with ±10° angles with respect to the disk normal, each of them corresponds to definite direction of the mode rotation. The measured gyrotron output power (up to 2 MW), interaction efficiency (34%) and diffraction losses in the mode converter (< 2%) agree well with the design values. The new gyrotron scheme alloys principal enhancement in the device parameters – possibility of electronic switching of output wave beam direction by a cathode coil, frequency stabilization by reflection, possibility to arrange an effective scheme to input signal and to provide frequency/phase locking of a gyrotron-oscillator. The latter advance gives very promising perspectives in development of megawatt gyrotrons with higher frequency (higher operating mode) stabilized by an external signal, possibility to make several gyrotrons coherent and combine their power. The first experiments on gyrotron frequency stabilization by optimum reflection (A. Kuftin, A. Chirkov, G. Denisov, 2016) are very encouraging. For rather high voltage variation of 95–100 kV frequency of the 2 MW gyrotron was stable within 2 MHz. Frequency stabilization of moderate power 263 GHz gyrotron by PLL resulted in ultra-narrow (record) oscillation spectrum width as 1 Hz [16].

Detail simulations and preliminary experiments were made for the scheme of 30 GHz gyro-TWT with non-linear absorber in a feedback loop. For the second harmonic operation with 68 kW/10 A electron beam the output radiation of the device consists of periodic pulses with 400 kW power and 0.2 ns pulse duration (corresponding frequency bandwidth of 5 GHz) with 7 ns interval between pulses. The non-linear absorber based on cyclotron absorption at second harmonic was successfully tested.

Acknowledgements. This discussion on gyrotron development trends would be unthinkable without the work results of my colleagues and collaborators whose group is fortunately large in number.

References

1. *M.K. Thumm*, "State-of-the-Art of High Power Gyro-Devices and Free Electron Masers", KIT Scientific Publishing, 2015.

2. *A.V. Krasilnikov, et al.* Progress with the ITER Project Activity in Russia. Nuclear Fusion, 2015, 55, #10, 104007 (11 p).

3. *G.G. Denisov, A.G. Litvak, A.V. Chirkov, et al.* Development Status of Gyrotron Setup for ITER ECW System. Proceedings of IRMMW THz Conference, Hong Kong, August, 2015.

4. http://www.iterrf.ru/

5. *L. Popov, M. Agapova, Yu. Belov et al.* Super-High Power Gyrotrons for Electron-Cyclotron Plasma Heating. This conference.

6. Proceedings of 26th IAEA Fusion Energy Conf., Kyoto, Japan, October 2016, Conf. ID: 48315 (CN-234).

7. *S.V. Samsonov, I.G. Gachev, G.G. Denisov, et al.* "Ka-band Gyrotron Traveling-Wave Tubes with the Highest Continuous-Wave and Average Power", IEEE Trans. on Electron Devices, 2014, vol. 61, no.12, p. 4264-4268.

8. *S.V. Samsonov, A.A. Bogdashov, G.G. Denisov, et al.* Proof-of-Principle Experiment on High-Power Gyrotron Traveling-Wave Tube with a Microwave System for Driving and Extracting Power Through One Window. IEEE Microwave and Wireless Components Letters, 2016, vol.26, no.4, p. 288-290.

9. *S.V. Samsonov, A.A. Bogdashov, G.G. Denisov, et al.* "Recent experiments and simulations on gyro-TWTs with helically corrugated waveguides", Proc. of 41st Int. Conf. on Infrared, Millimeter and Terahertz Waves, Copenhagen, Denmark, p. H4B5, Sept. 2016.

10. *S.V.Samsonov, G.G.Denisov, I.G.Gachev et al.* Development of Gyrotron Traveling-Wave Tubes at IAP and GY-COM. This Conference.

11. Procedings of 4th IAEA DEMO Programme Workshop. 14-18 November, 2016, Karlsruhe, Germany.

12. *A.V. Chirkov, G.G. Denisov and A.N. Kuftin*. Perspective gyrotron with mode converter for co and counter-rotation operating modes. // Applied Physics Letters, **106**, 263501 (2015); doi: 10.1063/1.4923269.

13. *V.L. Bakunin, G.G Denisov, Y.V Novozhilova*. Zones of Frequency Locking by an External Signal in a Multimode Gyrotron of a Megawatt Power Level. Radiophysics and Quantum Electronics, 2016, 58 (12), 893-904.

14. *Yu.V.Novozhilova, V.L.Bakunin, A.V.Chirkov et al.* Influence of Mode Competition and External Wave Frequency Modulation on Gyrotron Frequency Locking. *This Conference.*

15. *M.Yu. Glyavin, G.G. Denisov, M.L. Kulygin, Yu.V. Novozhilova*. Stabilization of gyrotron frequency by reflection from nonresonant and resonant load. Technical Physics Lett., July 2015, V. 41, Issue 7, p. 628-631.

16. *G.G. Denisov, A.P. Fokin, M.Yu. Glyavin, et al.* High precision frequency stabilization of a 100W/263 GHz continuous wave gyrotron. Proceedings of 18th Int. Vacuum Elecronics Conf., London, 24-26 April, 2017.

17. *N.S. Ginzburg, G.G. Denisov, M.N. Vilkov, et al.* Generation of "gigantic" ultra-short microwave pulses based on passive mode-locking effect in electron oscillators with saturable absorber in the feedback loop. Physics of Plasmas, 2016, 23 (5), 050702.

Recent activities of ITER gyrotron development in QST

Y.Oda[1], R. Ikeda[1], K. Takahashi[1], K. Kajiwara[1], T. Kobayashi[1], K. Sakamoto[1], S. Moriyama[1], C. Darbos[2], and M. Henderson[2]

[1]National Institute for Quantum and Radiological Science and Technology, Naka, Japan oda.yasuhisa@qst.go.jp
[2]ITER Organization, St-Paul-lez-Durance, France

Introduction

QST is committed to the development of high power RF power sources (or gyrotrons) for plasma heating in fusion devices. Our main activity is the development of 170GHz gyrotron for electron cyclotron resonance heating and current drive (EC H&CD) system of ITER project which requires 20 MW of power injection into fusion plasma. In ITER, 24 sets of 1 MW gyrotrons are planned to be installed of which Japan procures 8 sets. QST (previous JAEA) has already demonstrated 1MW-800s-55%, 0.8 MW-1hr-57% with the $TE_{31,8}$ mode gyrotron [1] and also demonstrated >1 MW operation with $TE_{31,11}$ mode gyrotron, which can generate higher power than previous design. Now the QST-ITER gyrotron has finished its designing phase and proceeds to manufacturing and test phase. The next stage of R&D activity in QST aimed multi frequency operation including >200 GHz frequency region, for future fusion devices such as DEMO. In this paper, present status of these gyrotron developments are described.

Status of ITER gyrotron

An electron cyclotron heating and current drive system in ITER is designed to inject RF power of 20 MW to actively control plasmas. The RF power is supplied by 24 gyrotrons. QST procures to supply 8 gyrotrons which produce 170 GHz / 1 MW each in addition 1 equatorial launcher for ITER.

The specification of ITER gyrotron is 170 GHz / 1 MW output / 50% efficiency / 3600s pulse duration. The oscillation mode in the cavity is $TE_{31,11}$ mode to keep marginal heat load on the cavity surface. The internal mode converter was optimized for 170 GHz single frequency operation. The triode electron gun is utilized and the operation of high frequency power modulation up to 5 kHz is achieved by control of anode voltage.

The final design review of the gyrotron was held in 2015 and the gyrotron design was approved for manufacturing. Manufacturing of auxiliary components for gyrotron operation system, for example the super conducting magnets (SCM), started in 2015 and first two sets of SCMs were delivered to QST test stand in the beginning of 2016. In 2016, the first set of two gyrotrons was manufactured by Toshiba Electron Tube & Devices (TETD) and the tubes were delivered to QST Naka Fusion Institute. Figure 1 shows the photograph of ITER gyrotron.

The short pulse test of the first ITER gyrotron has started since 2017 April after preparation of test stand. A new anode and body power supply system, which has the same configuration with ITER high voltage power supply (HVPS) system was installed into the gyrotron test stand. The new power supply system consists of DC HVPS devices that generate the high voltage and incorporates the switching components for 5 kHz modulations as required for controlling the applied HV to the gyrotrons. The ITER gyrotron was installed into the test stand with ITER SCM and has achieved 1 ms pulse duration at 170 GHz oscillation and a single peak Gaussian profile beam.

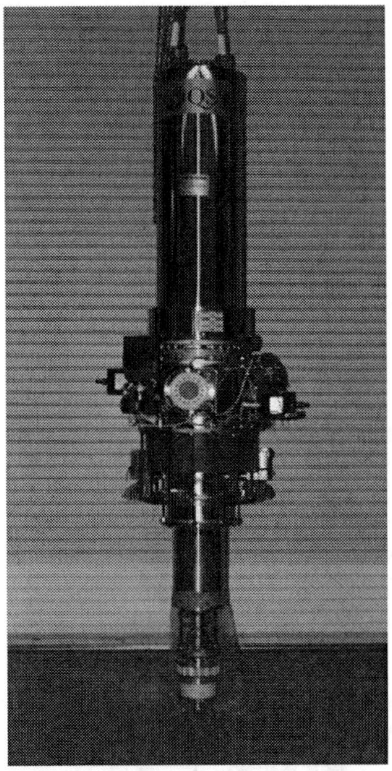

Fig. 1. Photograph of the ITER gyrotron

Development of multi frequency gyrotron

A gyrotron with $TE_{31,11}$ mode cavity is suitable for multi frequency as the radiation angle out of the internal launcher as well as beam directivity from the diamond window is compatible with the mode group of $TE_{19,7}$-$TE_{25,9}$-$TE_{31,11}$-$TE_{37,13}$-.... And, injection beam radii into the cavity resonator to excite these modes are 9.25 mm ($TE_{19,7}$), 9.19 mm ($TE_{25,9}$), 9.13 mm ($TE_{31,11}$) and 9.10 mm ($TE_{37,13}$) and are controllable using a gun magnet around the magnetron injection gun (MIG). In addition, gyrotrons with a triode MIG have capability to optimize the pitch factor, the ratio of parallel and perpendicular electron velocities, to adapt different magnetic field profile. The oscillation frequencies of $TE_{19,7}$-$TE_{25,9}$-$TE_{31,11}$-$TE_{37,13}$ are 104 GHz, 137 GHz, 170 GHz and 203 GHz, respectively and are matched for the diamond window transmission frequencies of 102 GHz, 136 GHz, 170 GHz and 204 GHz for a window thickness of 1.853 mm.

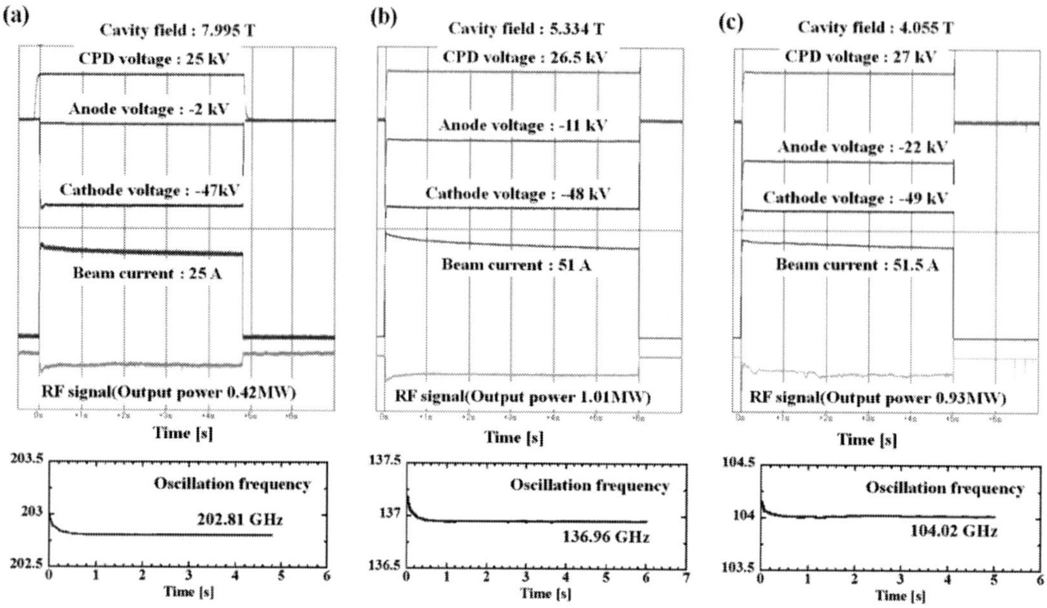

Fig. 2. Time histories of long pulse operations for (*a*) $TE_{37,13}$ mode, (*b*) $TE_{25,9}$ mode and (*c*) $TE_{19,7}$ mode

The operation of $TE_{31,11}$ mode/170 GHz has demonstrated 1.23 MW with an electrical efficiency of 47 % for 2 s, and 1.17 MW, 45 % for 5 s. For stable long pulse operations 1.02 MW/46% oscillation for 300 s was demonstrated.

The operation of the $TE_{19,7}$ mode, $TE_{25,9}$ mode, and $TE_{37,13}$ have also been demonstrated with the measured beam profile was relatively circular and centered on the output window for the other thee frequencies. The results include 1 MW output power for 2 s pulse was achieved at $TE_{19,7}$ mode (136.8 GHz); $TE_{25,9}$ mode (103.9 GHz) very short pulse of 0.3 ms of $TE_{37,13}$ (202.96 GHz) of 0.9 MW. This was achieved at beam voltage of 75 kV, beam current of 65 A, and a magnetic field of 7.94 T. And 4.8 s pulse operation has achieved the output power of 0.42 MW and an electrical efficiency of 38%. The beam voltage and beam current of this operation are 72 kV (CPD voltage 25 kV) and 25 A, respectively. 1 MW-level tests will be started soon. Figure 2 (a) shows the time history of long pulse operation with $TE_{37,13}$ mode.

The extension of pulse length for 137 GHz and 104 GHz oscillations were carried out as shown in Fig. 2 (b) and (c), which show the time history of 137 GHz and 104 GHz, respectively. Up to now, the output power of 1.01 MW for 6 s was achieved and the total efficiency of 41 % was obtained in the 137 GHz oscillation. In the 104 GHz oscillation, the output power of 1 MW/41 % for 2 s and 0.93 MW/37 % for 5 s were achieved. In these shots, adjustment of a magnetic field strength was not yet performed because the beam current was rapidly decreased by cathode cooling effect. Therefore, the improvement of the efficiency by applying the magnetic field control in longer pulse length can be expected.

References

1. *K. Sakamoto et al.*, Nat. Phys. 3, 411 (2007).
2. *R. Ikeda et al.*, Fusion Eng. Des. 96–97, 482 (2015).
3. *K. Kajiwara et al.*, Appl. Phys. Express 4, 126001 (2011).
4. *R. Ikeda et al.*, submitted to J. Infrared Millim. Teraherz Waves.

EPJ Web of Conferences **149**, 01003 (2017) DOI: 10.1051/epjconf/201714901003
10th International Workshop 2017 "Strong Microwaves and Terahertz Waves: Sources and Applications"

Electron cyclotron resonance ion sources – physics, technology and future challenges

O. Tarvainen[1], T. Kalvas[1], H. Koivisto[1], V. Skalyga[2], I. Izotov[2], D. Mansfeld[2],

[1]University of Jyväskylä, Jyväskylä, Finland, olli.tarvainen@jyu.fi
[2]Institute of Applied Physics, RAS, Nizhny Novgorod, Russian Federation

The performance of Electron Cyclotron Resonance Ion Sources (ECRIS), producing high charge state ions from a great variety of elements, has improved dramatically over the past decades, thus enabling significant advances in accelerator-based nuclear physics. The order-of-magnitude performance leaps of ECR ion sources, e.g. from 15 μA of extracted O^{6+} in 1974 [1] to 4700 μA in 2016 [2], result from improvements to the magnetic plasma confinement, increases in the microwave heating frequency and techniques to stabilize the plasma at high densities.

The design of modern ECR ion sources is based on semi-empirical scaling laws, suggesting most importantly that the extracted current at the peak of the ion charge state distribution scales with the microwave frequency squared [3], i.e.

$$I_{peak} \propto \omega_{RF}^2. \quad (1)$$

Direct comparison of 1st – 3rd generation ECR ion sources operating at frequencies below 10 GHz, at 10–18 GHz and above 18–28 GHz, respectively, is complicated by the fact that the peak of the charge state distribution also shifts with the frequency. However, it is evident that increasing the microwave frequency improves the extracted currents of highly charge ions. Several explanations for the frequency scaling have been proposed. These include for example damping of the microwave electric field at plasma densities exceeding the critical (cut-off) value and so-called RF pitch angle scattering limiting the electron density, both scaling with the microwave frequency squared.

The magnetic field of an ECRIS is a superposition of solenoid and sextupole fields resulting to a minimum-B structure serving several purposes.

(i) The field configuration provides a closed resonance surface fulfilling the ECR-condition

$$\omega_{RF} = \omega_{ce} = eB_{ECR}/\gamma m_e, \quad (2)$$

where ω_{ce} is the electron gyrofrequency, which depends on the magnetic field strength B and electron energy through the relativistic γ-factor (e and m_e are the elementary charge and electron mass, respectively). The resonant interaction with the microwave electric field assures that a sufficient number of electrons gain energies required for ionization of high charge state ions. The magnetic field fulfilling the resonance condition for non-relativistic electrons can be written conveniently as

$$B_{ECR}\,[\mathrm{T}] = f_{RF}\,[\mathrm{GHz}]/28. \quad (3)$$

(ii) The minimum-B configuration results to appropriate electron confinement. The development of ECR ion sources over the past decades has led to the formulation of the following scaling laws [4] for the magnetic field:

$$B_{inj}/B_{ECR} = 4 \quad (4)$$

$$B_{rad}/B_{ECR} = 2 \quad (5)$$

$$B_{ext} \approx 0.9B_{rad} \quad (6)$$

$$B_{min} \approx 0.4B_{rad}, \quad (7)$$

where B_{inj}, B_{rad}, B_{ext} and B_{min} are the magnetic field at the injection, radial wall of the plasma chamber, extraction and B-minimum, respectively.

(iii) The minimum-B configuration suppresses magnetohydrodynamic instabilities.

The scaling laws in Eqs. (1) and (4)–(7) set a practical limit for ECRIS technology relying on room-temperature (RT) technology if the desired power consumption for generating the magnetic field is < 1 MW. In RT-ECRISs the solenoid field is created by electromagnetic coils while the radial sextupole field is formed by an array of permanent magnets. Figure 1 shows the dependence of the maximum solenoid field at the injection (B_{inj}) and the sextupole field at the chamber wall fulfilling the scaling laws on the microwave frequency. Commonly applied frequencies are highlighted by the dots. The practical RT-limits of both field components, set by the power consumption and permanent magnet remanence / coer-civity, are depicted by horizontal lines crossing the data curves at ~18 GHz, which is deemed as the upper frequency limit for 2nd generation RT-ECRISs.

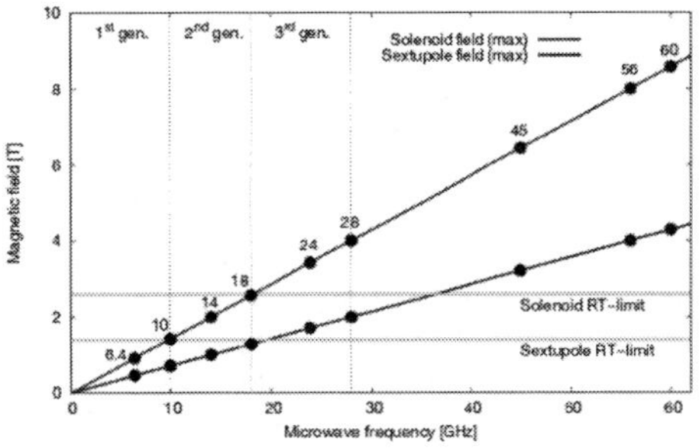

Fig. 1. Frequency scaling of the ECRIS solenoid and sextupole magnetic fields

It is evident from Fig. 1 that 3rd generation ECRISs operating at 18–28 GHz frequencies must rely on superconducting technologies. Building a 4th generation ECRIS operating at microwave frequency up to 56 GHz is feasible using state-of-the-art Nb_3Sn superconducting wire [5]. Increasing the frequency beyond 56 GHz requires either accepting a substandard field strength or R&D on innovative ECRIS concepts, e.g. the ARC-ECRIS [6] feasible up to 100 GHz.

It has been recently shown [2] that the extracted currents of high charge state oxygen ion beams extracted from the 28 GHz VENUS ECRIS increase linearly with the microwave power up to 10 kW. Since the maximum currents scale roughly as frequency squared and the extracted current per incident watt is constant, the power required to operate high frequency sources up to their full capacity should also scale as frequency squared [7], i.e.

$$P_{RF} \propto f_{RF}^{2}. \qquad (8)$$

The power scaling shown in Fig. 2 shows that the required power for 14, 18–24 and 28 GHz sources is 2–3 kW, 4–6 kW and 10 kW, respectively, which matches well with the experimental evidence. This implies that operating a 4th generation ECRIS e.g. at frequencies of 45–60 GHz would require up to 40 kW microwave power. Such high power poses a significant challenge on managing the localized heat load determined by the electron losses. Furthermore, the required combination of frequency / power therefore restricts the choice of applicable microwave generators as indicated in the figure.

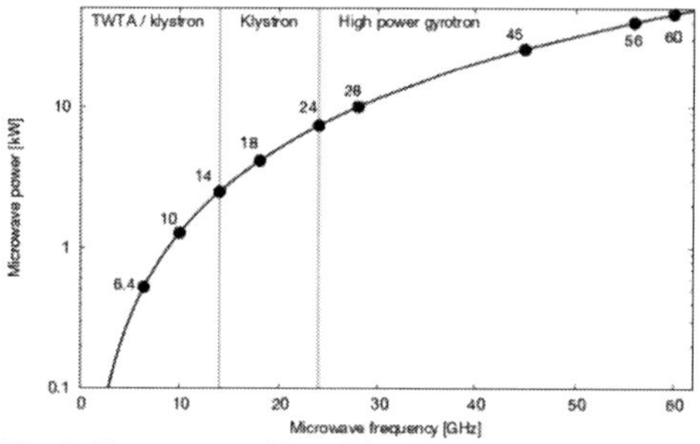

Fig. 2. Frequency scaling of the ECRIS microwave power

The ECRIS magnetic field strength at the extraction affects the ion beam quality. The emittance ε of the ion beam, describing the relative position and transverse momentum of the extracted ions at each transverse plane (*x*, *y*), scales (theoretically) with the magnetic field at the extraction as

$$\varepsilon_{x,y} \propto B_{ext}. \qquad (9)$$

The increase of the emittance with the magnetic field needs to be taken into account when matching the extracted beam into the acceptance of the subsequent accelerator. The presumable increase of the emittance at strong magnetic field poses yet another challenge for the R&D of next generation ECRISs.

The magnetic field scaling laws (4)–(6) can be understood qualitatively from the point-of-view of plasma confinement and losses preferably directed towards the extraction. The scaling of the B_{min} which can be written alternatively as

$$B_{min}/B_{ECR} \approx 0.8 \qquad (10)$$

is not evident. It has been shown [8] that when the B_{min}/B_{ECR}-ratio exceeds a certain limit, typically 0.7–0.85, the extracted currents of high charge state ions are restricted by periodic onsets of electron cyclotron instabilities. The kinetic instabilities arise from the anisotropy of the ECRIS electron velocity distribution which results from the resonant heating process affecting only the perpendicular (to B-field) velocity component of the electrons. The instabilities affect the current, longitudinal energy spread and background impurities of the extracted ion beams due to periodic loss of plasma confinement. The instability threshold can be affected by so-called two frequency heating, which allows expanding the ion source parameter space corresponding to stable plasma regime [9]. This is demonstrated in Fig. 3 showing the stabilizing effect of two frequency heating on O^{6+} beam current. Although, the physical mechanism explaining the suppression of the kinetic instabilities is unknown, it is evident that two frequency heating plays a crucial role in the future of ECRIS R&D.

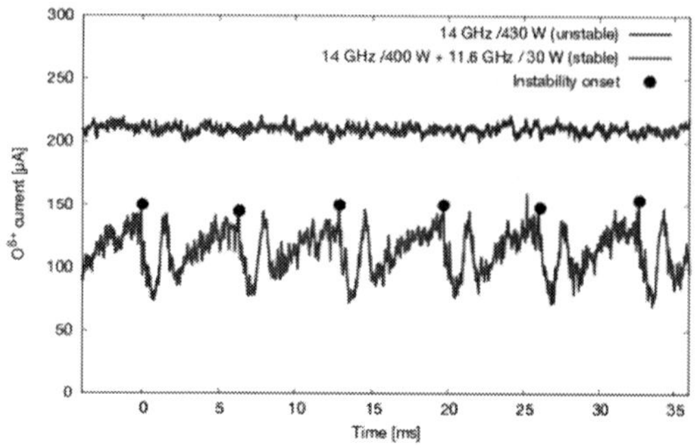

Fig. 3. Suppression of kinetic instabilities by two frequency heating. Data recorded with the JYFL 14 GHz ECRIS

It is concluded that while the performances of ECR ion sources have increased by several orders of magnitude since their conception, severe technological challenges, mainly related to the magnetic field, lie ahead if frequency scaling is pursued in the R&D of future generation machines. Further development of techniques improving the quality and temporal stability of the extracted ion beam require deeper understanding of the underlying plasma physics.

Reported work devoted to the plasma stabilization was supported by the grant of Russian Science Foundation # 16-12-10343.

References

1. *V. Bechtold, N Chan-Tung, S. Dousson, R. Geller, B. Jacquot,* and *Y. Jongen* ECR ion source for multiply-charged oxygen beams // Nucl. Instruments and Methods, 178 (1980), p. 305-308.

2. *D.Z. Xie, W. Lu, J.Y. Benitez, C.M. Lyneis,* and *D.S. Todd,* Recent Production of Intense High Charge Ion beams with VENUS // THA001 in Proc. ECRIS 2016, Busan, Korea, jacow.org

3. *R. Geller* ECRIS: The Electron Cyclotron Resonance Ion Sources // Annu. Rev. Nucl. Part. Sci. 40 (1990), p. 15.

4. *D. Hitz, A. Girard, G. Melin, S. Gammino, G. Ciavola and L. Celona* Results and interpretation of high frequency experiments at 28 GHz in ECR ion sources, future prospects // Rev. Sci. Instrum. 73, (2002), p. 509.

5. *C. Lyneis, P. Ferracin, S. Caspi, A. Hodgkinson,* and *G.L. Sabbi* Concept for a fourth generation electron cyclotron resonance ion source // Rev. Sci. Instrum. 83, 02A301 (2012).

6. *P. Suominen and F. Wenander* Electron cyclotron resonance ion sources with arc-shaped coils // Rev. Sci. Instrum. 79, 02A305 (2008).

7. *C.M. Lyneis* Scaling Laws in Electron Cyclotron Resonance Ion Sources // MOA001 in Proc. ECRIS 2016, Busan, Korea, jacow.org

8. *O. Tarvainen* et al. Beam current oscillations driven by cyclotron instabilities in a minimum-B electron cyclotron resonance ion source plasma // Plasma Sources Sci. Technol. 23 (2014) 025020.

9. V. Skalyga et al. Suppression of cyclotron instabilitity in Electron Cyclotron Resonance ion sources by two frequency heating // Phys. Plasmas 22, 083509 (2015)

Powerful neutron generators based on high current ECR ion sources with gyrotron plasma heating

V.A. Skalyga[1], S.V. Golubev[1], I.V. Izotov[1], R.L. Lapin[1], S.V. Razin[1], R.A. Shaposhnikov[1], S.V. Sidorov[1], O. Tarvainen[2]

[1]Institute of Applied Physics RAS, Nizhny Novgorod, Russian Federation, skalyga@ipfran.ru
[2]University of Jyväskylä, Jyväskylä, Finland

Introduction

Neutron sources are widely used in fundamental and applied research. Manifold applications require neutron fluxes of various intensities, energy spectra and degrees of collimation. Several types of neutron sources have been developed for the purpose including nuclear reactors, radioisotope sources, D-D and D-T generators, plasma pinch machines, accelerators etc. Each neutron source type occupies a niche fulfilling the parameters desired in a given application.

The present paper presents the results of recent investigations devoted to development of a new generation of compact D-D (D-T) neutron generators based on a high-current quasi-gasdynamic ECR ion source, which utilizes powerful gyrotron radiation of mm-waveband for plasma creation [1]. Such approach allows to increase the plasma density in the discharge significantly in comparison to conventional ECRISs, which utilize microwave radiation with frequencies on the order of 10 GHz. In experiments with a 37.5 GHz gyrotron the plasma density reaches values of $2\cdot10^{13}$ cm^{-3}. Significant increase of the plasma density leads to a change of the plasma confinement mode. A so-called quasi-gasdynamic confinement was realized in the presented experiments instead of the collisionless one, which is common for modern ECRISs. The plasma lifetime, which is much shorter than in conventional classical ECRIS, can be expressed as $\tau = \frac{L \cdot R}{2V_{is}}$, where L – magnetic trap length, R – trap mirror ratio (ratio between magnetic field in the magnetic mirror and in the trap center), V_{is} – ion sound velocity. Short plasma lifetime provides high plasma flux density from the trap. Under our experimental conditions typical plasma lifetime is on the order of 10 μs. The described peculiarities of quasi-gasdynamic ECR discharge sustained by mm-waveband radiation, namely, short lifetime and high density, provide unprecedented plasma flux densities of several eA/cm^2.

Experimental results

Experiments aimed to demonstrate the benefits of the high current gasdynamic ECR ion source as a part of D-D neutron generator were conducted at SMIS 37 facility [2], schematically depicted in Fig. 1. The plasma was created and sustained inside a d = 4 cm vacuum chamber by pulsed (1.5 ms) 37.5 GHz linearly polarized radiation with power up to 100 kW. The simple mirror magnetic trap with 25 cm length was created by pulsed solenoids, providing a mirror ratio of 5. The magnetic field strength was varied in a range of 1–4 T at mirror plugs, whereas the resonant field strength is 1.34 T for 37.5 GHz.

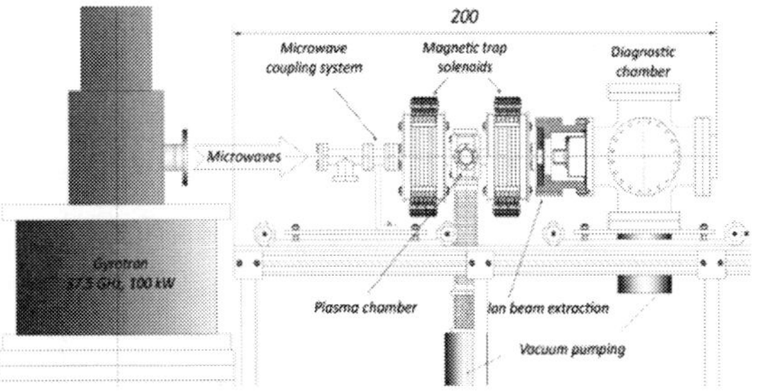

Fig. 1. SMIS 37 experimental facility

Microwave radiation was guided into the chamber quasi-optically through a special coupling system. The pulsed gas feed line was incorporated into the coupling system i.e. the neutral gas was injected axially. The ion extraction and beam formation were realized by a two-electrode (diode) system consisting of a plasma electrode and puller.

The best result was obtained with plasma electrode aperture of 10 mm. The maximum deuterium beam current was 500 emA, which corresponds to a current density of 630 emA/cm^2 at the plasma electrode. To our knowledge this is a record for ECR ion sources. Beam emittance was measured with the "pepper-pot" method and its normalized rms value was 0.07 pi·mm·mrad. Studies of the extracted charge state distribution have shown that 94% of it consists of atomic deuterons (D+) (contribution of molecular D_2^+ ions is less than 6%). For our knowledge such combination of parameters makes the obtained beam the brightest for any kind of ion sources with normalized rms brightness value of 100 A/(pi·mm·mrad)^2.

Measurement of the produced neutron flux was performed by bombarding a heavy ice target with 300 emA D^+ beam accelerated to 45 keV energy. An example of the neutron counter signal is presented in fig. 2-a and the corresponding amplitude spectrum (collected over 100 pulses) in fig. 2-b. The amplitude spectrum of the acquired signals matches the reference one, proving that the signals are generated by neutrons, not X-rays or electrical noise. The D_2O target yield was 10^9 s^{-1}. It is worth noting that the area of the target was about 1 cm^2, which corresponds to a neutron flux density of about 10^9 $cm^{-2}\cdot s^{-1}$ at the target. Obtained results allow to deduce the prospects of D-D neutron generator development using deuterium ion source based on quasi-gasdynamic ECR discharge, sustained by powerful gyrotron radiation. Commercially available TiD_2 targets are capable of generating up to 10^8 s^{-1} neutrons per 1 emA of deuterium beam bombarding the target with 100 keV energy.

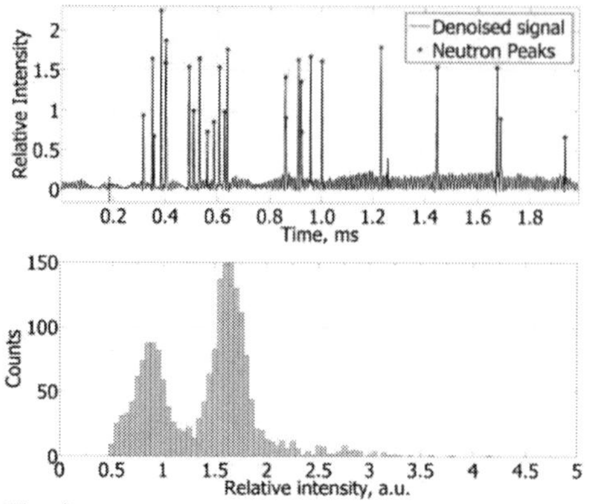

Fig. 2. (a) Single pulse waveform from neutron counters

Each peak corresponds to a single detected neutron. (b) - Amplitude spectrum of the neutron signals collected over 100 pulses. Two peaks on the spectrum correspond to different counters.

In the case of target bombardment with the demonstrated beam current (current density 630 mA/cm^2) in the present work and accelerated up to 100 keV energy the neutron flux density could reach $6\cdot10^{10}$ s$^{-1}\cdot$cm^{-2}.

Point-like neutron source

One of the applications of high density neutron fluxes is neutronography. Traditionally for this purpose paraxial neutron beams from nuclear reactors and accelerators are utilized [3]. Another approach is to create a point-like source.

The use of a high-current ECR ion source with quasi-gasdynamic plasma confinement and heating with gyrotron microwave radiation allows the formation of light ion beams with uniquely low emittance (for a given current level). Such a low emittance enables focusing of the ion beam into a small spot. By numerical modeling with open IBSimu code [4] it was shown that an ion beam with the properties listed above could be focused with a simple magnetic lens into 100 µm scale spot. Results of these calculations are shown in fig. 3.

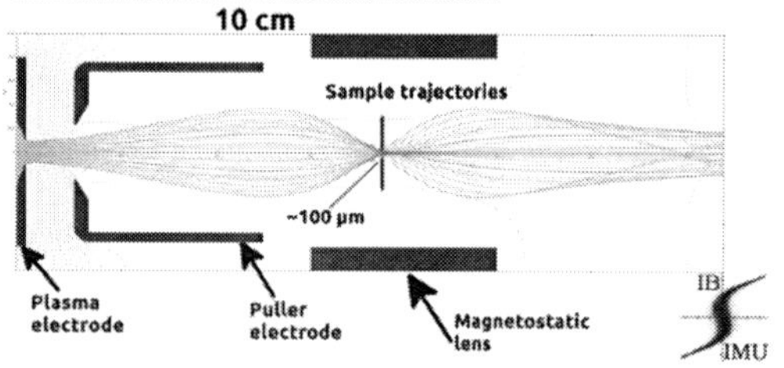

Fig. 3. Simulation of the beam focusing with IBSimu software

Estimation presented in [5] show that the point-like source based on high current ECR ion source could produce a neutron yield up to 10^{11} s^{-1} for D-target, and up to 10^{13} s^{-1} for T-target.

D-D neutron generator for BNCT

One of the most interesting and important applications of high yield neutron generators discussed above is boron-neutron capture therapy (BNCT) of oncologic diseases [6]. The main problem hindering the development of BNCT is the necessity of neutron fluxes exceeding 10^9 s$^{-1}\cdot$cm^{-2}. The only machines able to produce the required neutron flux are nuclear reactors and large-scale accelerators plagued by tremendous price and strict safety rules for protecting the staff and patients. The D-D neutron generator scheme discussed here is free of those shortcomings as it does not use radioactive isotopes including tritium and/or high energy particles, thus not requiring heavy X-Ray shielding. Moreover, the device is compact allowing it to be installed in literally any existing clinic, and has significantly lower price in comparison to the aforementioned technologies.

The described experiments were performed in pulsed mode. BNCT, however, requires continuous (CW) flux of neutrons. Thereby a CW D-D neutron generator using deuterium ion source based on CW ECR discharge in a permanent magnet trap, sustained by 28 GHz / 10 kW gyrotron is under construction at IAP RAS. Schematic view of the future facility is shown in fig. 4.

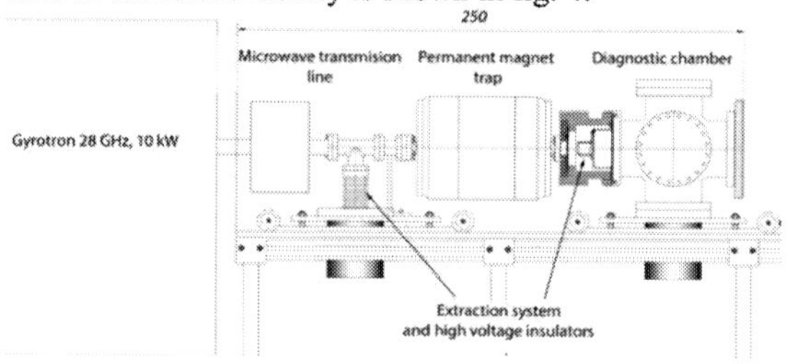

Fig.4. Scheme of the CW ECR ion source for neutron production

The first experiments on plasma confinement in the quasigasdynamic regime in CW mode have demonstrated that 24 GHz/5 kW gyrotron radiation can provide plasma flux from a trap with 1 A/cm^2 density. Thus, it is possible to obtain continuous ion beams with similar parameters as reported above.

Acknowledgments. The reported work devoted to the CW neutron production was supported by Russian Science Foundation, grant #16-12-10343. The reported work devoted to the point-like neutron source creation was supported by Russian Science Foundation, grant # 16-19-10501.

References

1. *V.Skalyga, I.Izotov, S.Golubev, A.Sidorov, S.Razin, A.Vodopyanov, O.Tarvainen, H.Koivisto, T.Kalvas.* New progress of high current gasdynamic ion source. Rev. Sci. Instrum. 87, 02A716 (2016)
2. *V. Skalyga, I. Izotov, A. Sidorov, S. Razin, V. Zorin, O. Tarvainen, H. Koivisto, T. Kalvas.* High current proton source based on ECR discharge sustained by 37.5 GHz gyrotron radiation. *JINST,* v.7, P10010 (2012)
3. *I.S. Anderson, R.L. McGreevy and H.Z. Bilheux,* Neutron imaging and applications, Springer, U.S.A. (2009)
4. *T. Kalvas, O. Tarvainen, T. Ropponen, O. Steczkiewicz, J. Arje and H. Clark,* IBSimu: " a three-dimensional simulation software for charged particle optics. Rev. Sci. Instrum. 81 (2010) 02B703
5. *S. Golubev, V. Skalyga, I. Izotov, A. Sidorov..* JINST, V. 12, 2017_JINST_12_T02003 (2017)
6. *W. Sauerwein, A. Wittig, R. Moss, and Y. Nakagawa, editors,* Neutron Capture Therapy. Springer, Berlin Heidelberg (2012).

EPJ Web of Conferences **149**, 01005 (2017) DOI: 10.1051/epjconf/201714901005
10th International Workshop 2017 "Strong Microwaves and Terahertz Waves: Sources and Applications"

ECRH system, microwave diagnostics and experimental results in the EAST tokamak

F.K. Liu, H.D. Xu, X.J. Wang, D.J. Wu, M.H. Li, J. Zhang, Y. Liu, T. Zhang, Q. Zhang, Y.P. Zhao, J.G. Li and the EAST team

Institute of Plasma Physics, Chinese Academy of Sciences, Hefei, China, mhli@ipp.ac.cn

Electron cyclotron resonance heating (ECRH), as a highly efficient and controllable heating tool, has been widely used in tokamak magnetic confinement devices [1]. Usually, the EC power can be injected as narrow Gaussian beams, which gives rise to highly localized power deposition. Consequently, ECRH and electron cyclotron current drive (ECCD) have been expolied to an ideal candidate for local transport studies and for the control of various magnetohydrodynamics (MHD) instabilities occurring in tokamak plasmas. Unlike other RF heating techniques, EC wave can propagate in vacuum and the coupling is insentitive to the parameters at plasma edge, which suggests that the antenna can be located far away from the plasma. This advantage makes ECRH an attractive scheme for a thermonuclear device like ITER, in which, the launcher cannot be close to the plasma in order to avoid the material erosion and thermal damage.

EAST is a fully superconducting tokamak with a major radius of R_0 = 1.90 m and a minor radius of a = = 0.45 m. The EC system will consist of a 4 MW continuous wave (CW) at 140 GHz (second harmonic of the extraordinary mode) for heating and current drive applications [2, 3]. In line with the long pulse operational capability of EAST, it was designed for a pulse length up to 1000 s. Benefitting from a pulse step modulator (PSM) type of power supply system it has a capability of fast modulation frequency up to 1 kHz. There are four gyrotrons produced by two different venders, namely, No1.and No.3 are provided by Russian company GYCOM, and the other two by CPI, USA. The specific parameters are shown in

Table 1. Specific parameters of gyrotrons

	No.1 and No.3	No.2 and No.4
Vender	Gycom	CPI
Type	Diode	Diode
Cavity mode	$TE_{22.8}$	$TE_{28.7}$
Max. output power	1 MW	900 KW
Pulse Length	100~1000 s	100~1000 s
Output beam	Gaussian beam	Gaussian beam
Collector type	Depressed	Depressed
Collector power limit	1.1 MW	1.8 MW
Cathode voltage	-46 kV	-59 kV
Beam current	43 A	40 A
Body voltage	+25 kV	+22 kV
Filament power	1000~1200 W	200~250 W

Table 1. The EC power will be launched from four mirrors located in the same port at low field side. In order to satify the physical need of changing power deposition in a large region, the launch angles can be continuously varied with range of over 30 degree in poloidal direction and ± 25 degree in toroidal as shown in Fig. 1. At present, No.1 and No.2 gyrotron systems have been installed and the layout is shown in Fig. 2. However, only No.1 system has been used for plasma experiments.

Figure 3 shows a typical waveform of long pulse and high electron temperature discharge with ECRH. The magnetic field on the axis was set to be 2.5 T, in order to guarantee the EC power absorbed in the central region. The EC power of 0.4 MW was injected at t = 3 s and turned off at 102 s. The central electron temperature measured by Thomson scattering was high as T_{e0} > 4.5 keV during the EC heating phase. The contribution of ECRH to electron temperature can be found by comparison between pulse #59892 and pulse #59891 in which the plasma parameters and other RF heating power were the same. As shown in Fig. 4, the electron temperature is lowered by ~ 1.2 keV in the core region (ρ < 0.25) for pulse #59891 without ECRH.

Long pulse H-mode discharge with the combination of lower hybrid (LH), ion cyclotron (IC) and EC waves heating has been obtained in EAST. Typical traces of an H-mode discharge with ECRH (t = 1.3–31.4 s) and ECRH (t > 31.4 s) are illustrated in Fig. 5. It can be seen that when the EC power of about 0.3 MW was turned off at 31.4 s, the H-mode confinement phase was still maintained. However, the plasma stored energy (W_{MHD}) was decreased gradually from 145 kJ to 103 kJ, which indecates that ECRH is very important for high performance plasma. Besides, the loop voltage (V_{loop}) was increased slightly after EC power switched off, passibly resulted from the decrease in electron temperature which affects LH current drive efficiency significantly [4].

Microwave (MW) techniques such as quasi-optical reflector / lens, heterodyne detection, and lossless transmission line, are applied widely in plasma diagnostics on EAST, including the electron cyclotron emission (ECE) temperature measurement and the reflectometry for density and density fluctuation measurement. The ECE diagnostics consists of two radiometer systems and a Michelson interferometer [5]. The typical characteristics of the diagnostics are listed in Table 2. Thanks to the good spatial and temporal resolution, it can be used for investigation on MHD instabilities in addition to providing conventional temperature profiles. A new quasi-optical antenna with better poloidal resolution has been developed recently, which can be used for T_e fluctuation measurement in next campaign. The X-mode polarized Q-, V- and W-band reflectometries allow to measure the density profile from plasma edge to core [6]. These reflectometries featured by a high time resolution of 50 μs demonstrate the ability for study of pedestal physics on EAST as shown in Fig. 6. The systems are currently being upgraded to couple/decouple Q-, V- and W-band based on quasi-optical method, and to use double-ridged horns with a broad band of 33–110 GHz for launcher and receiver.

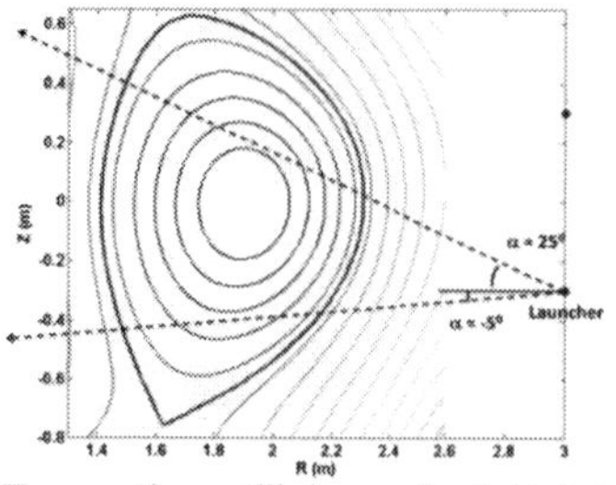

Fig. 1. EAST magnetic equilibrium and poloidal steering capability for the EC launchers at $Z = -0.3$ m. The red thick curve represents the lost closed magnetic surface

Fig. 2. The installed No.1 and No.2 gyrotron systems

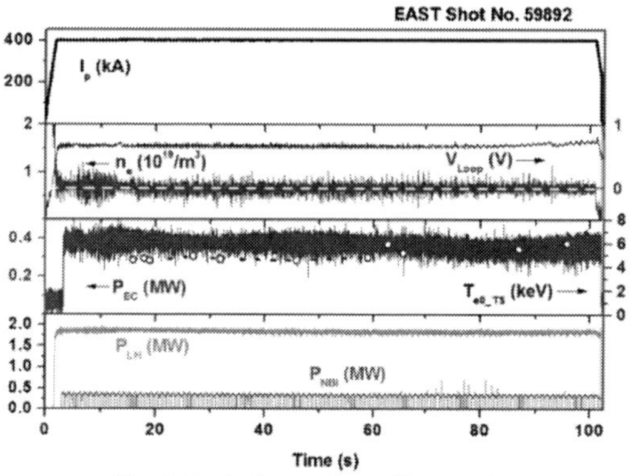

Fig. 3. Typical waveform of long pulse and high electron temperature discharge with ECRH

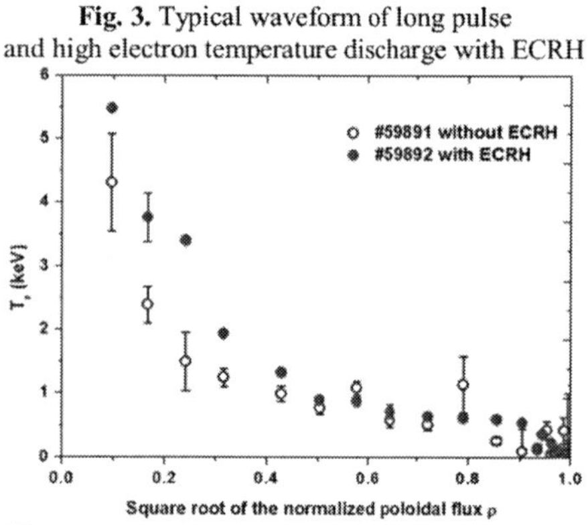

Fig. 4. Comparison of electron temperature profiles with ECRH and without it

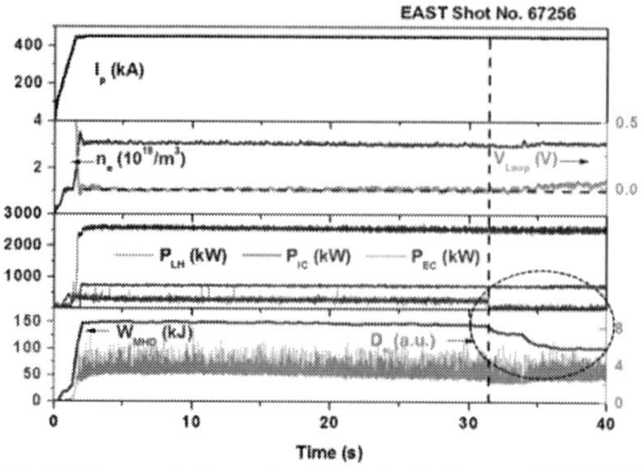

Fig. 5. Time evolutions of an H-mode discharge with ECRH (t = 1.3–31.4 s) and without ECRH (t > 31.4 s)

Table 2. Details of ECE diagnostics

Name	32-channel radiometer	16-channel radiometer	Michelson interferometer
Frequency coverage	104–168 GHz	97–120 GHz	80–500 GHz
Spectral resolution	0.5 GHz	0.2/0.5 GHz	2.8 GHz
Temporal resolution	2.5 us	2.5 us	30 ms

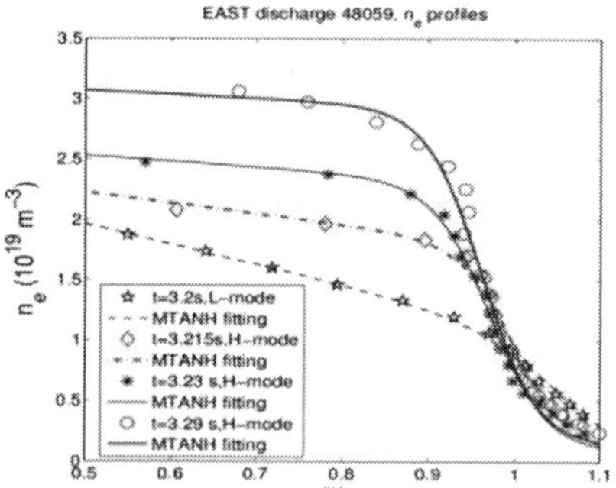

Fig. 6. An example of electron density profiles measured by fast frequency sweeping reflectometry

References

1. *ITER Physics Expert Group on Energetic Particles,* Chapter 6: Plasma auxiliary heating and current drive Nucl. Fusion 1999. V. 39, No. 12. P. 2495-2539.

2. *Wang X. J. et al.* Progress of high power and long pulse ECRH system in EAST Fusion Eng. Des. 2015. V. 96-97. P. 181-186.

3. *Xu H. D. et al.* Development and Preliminary Commissioning Results of a Long Pulse 140 GHz ECRH System on EAST Tokamak Plasma Sci. Technol. 2016. V. 18, No. 4. P. 442-448.

4. *Li M. H. et al.* Study on lower hybrid current drive efficiency at high density towards long-pulse regimes in Experimental Advanced Superconducting Tokamak Phys. Plasmas 2014. V. 21. P. 062510-1- 062510-9.

5. *Liu Y. et al.* A Michelson Interferometer for Electron Cyclotron Emission Measurements on EAST Plasma Sci. Technol. 2016. V. 18, No. 12. P. 1148-1154.

6. *Qu H. et al.* Q-Band X-Mode Reflectometry and Density Profile Reconstruction Plasma Sci. Technol. 2015. V. 17, No. 12. P. 985-990.

Update on the DIII-D ECH system: experiments, gyrotrons, advanced diagnostics, and controls

John Lohr[1], Rigoberto Brambila[1], Mirela Cengher[1], Yuri Gorelov[1], William Grosnickle[1], Charles Moeller[1], Dan Ponce[1], Antonio Torrezan[1], Lawrence Ives[2], Michael Reed[2], Monica Blank[3], Kevin Felch[3], Claudia Parisuaña[4], Alexandra LeViness[5]

[1]General Atomics, San Diego, CA USA, lohr@fusion.gat.com
[2]Calabazas Creek Research, San Mateo, CA, USA
[3]Communications and Power Industries, Palo Alto, CA, USA
[4]Pontificia Universidad Católica del Perú, Lima, Peru
[5]University of Alabama, Tuscaloosa, AL, USA

Abstract

The ECH system on DIII-D is continuing to be upgraded, while simultaneously being operated nearly daily for plasma experiments. The latest major hardware addition is a new 117.5 GHz gyrotron, which generated 1.7 MW for short pulses during factory testing. A new gyrotron control system based on Field Programmable Gate Array (FPGA) technology with very high speed system data acquisition has significantly increased the flexibility and reliability of individual gyrotron operation. We have improved the performance of the fast mirror scanning, both by increasing the scan speeds and by adding new algorithms for controlling the aiming using commands generated by the Plasma Control System (PCS). The system is used for transport studies, ELM control, current profile control, non-inductive current generation, suppression of MHD modes, startup assist, plasma density control, and other applications. A program of protective measures, which has been in place for more than two years, has eliminated damage to hardware and diagnostics caused by overdense operation. Other activities not directly related to fusion research have used the ECH system to test components, study methods for improving production of semiconductor junctions and materials, and test the feasibility of using ground based microwave systems to power satellites into orbit.

1. System Overview

With the addition of a seventh gyrotron to the six in regular operation, about 4.5 MW injected rf power will be available for experiments. Although the pulse lengths are at present limited administratively to 5 s, compatible with the 6–7 s pulse lengths of DIII-D plasmas, the gyrotrons reach thermal equilibrium after a few seconds, and longer pulses will be available when longer duration DIII-D plasmas are being used. Results of tests of the 117.5 GHz tube will determine the course to be taken with future system upgrades. One additional upgrade, that has been committed at present, will be the combining of parts from two 1 MW 110 GHz tubes, that suffered collector failures, into a single gyrotron with an improved collector made from CuCrZr material. This gyrotron should be available in 2018 and boost the injected rf power to greater than 5 MW from eight gyrotrons. The power supplies for all the gyrotrons in the system employ autotransformer/rectifiers, delivering about 90–100 kV to tetrode modulator/regulators that provide highly regulated voltage to the gyrotrons, either two in parallel or singly. The rf is carried in evacuated circular waveguides with diameter 31.75 mm, propagating the $HE_{1,1}$ mode about 90 m to the tokamak with 75–80% efficiency.

2. Control System Upgrades

Previously, the gyrotron output power, including a variety of pre-programmed modulation options, was controlled by waveform generators with good time resolution and reliability, but with limited options beyond four waveform choices. These were simple continuous, and three oscillation waveforms, sinusoidal, triangular, and rectangular. Faults from such events as cathode or body overcurrent, loss of rf generation, or failure to track the commanded waveforms, would result in the termination of operation for the faulted gyrotron for that plasma shot. In cases where two gyrotrons were powered from a single tetrode, a fault in either gyrotron would terminate output from both.

All the DIII-D gyrotrons are now controlled by FPGA based logic, which has greatly increased the flexibility of fault handling, improved fault protection and, in a major improvement, allowed restart of the gyrotrons after a fault has cleared. As an example, Fig. 1 shows a case in which a gyrotron faulted, shut off, and then successfully restarted and continued generating full power as if nothing had happened. The FPGA controller is programmed to look for inconsistencies between the applied voltages and the measured performance parameters, such as output rf power, body current and cathode current, all digitized at 2 MS/s. In this case, the system detected a very small oscillation of the body current when the body voltage was stable. This inconsistency triggered a rapid shutdown of the gyrotron within about 50 μs and a rapid restart to full output within about 50 ms. The brief fault was not a factor for the experiment and the shot, which previously would have been a failure, was in practice a success.

This capability has also allowed a modest increase in delivered rf power, since the gyrotrons can now be operated closer to the point where higher power is achieved but at risk of loss of rf generation. The tubes will now be restarted in this event.

3. ELM Control with ECCD

After Edge Localized Modes (ELMs) were recognized as a potentially serious threat to the safe operation

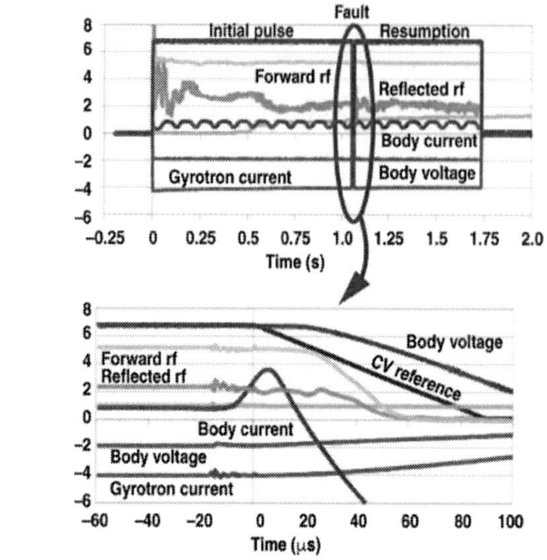

Fig. 1. In this case, an incipient body current fault was identified by the FPGA controller. Rapidly halting the shot enabled an immediate restart with nearly no lost time

of large devices such as ITER, research programs have attempted both to understand these fluctuations and to mitigate their effects. On the front lines of this effort are ECH/ECCD. The microwave systems have been used in conjunction with externally applied 3D fields to suppress or augment ELMs in studies of the ELM phenomenon. An example is shown in Fig. 2, in which ECCD in the plasma periphery from six gyrotrons, generating current in the direction opposite to the plasma current, successfully suppressed ELMs. A large number of similar experiments have studied the ability of ECH/ECCD to mitigate this instability.

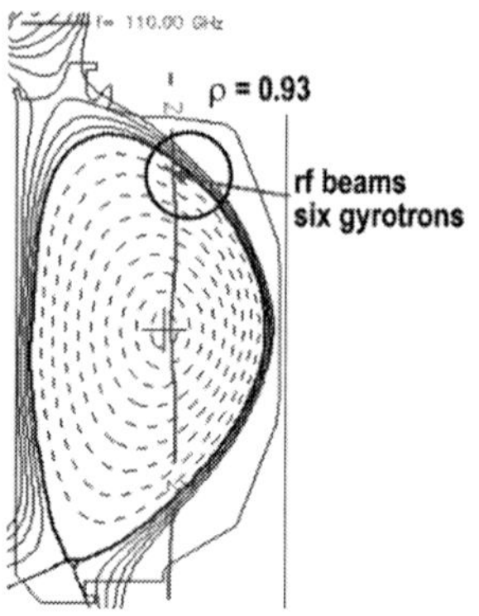

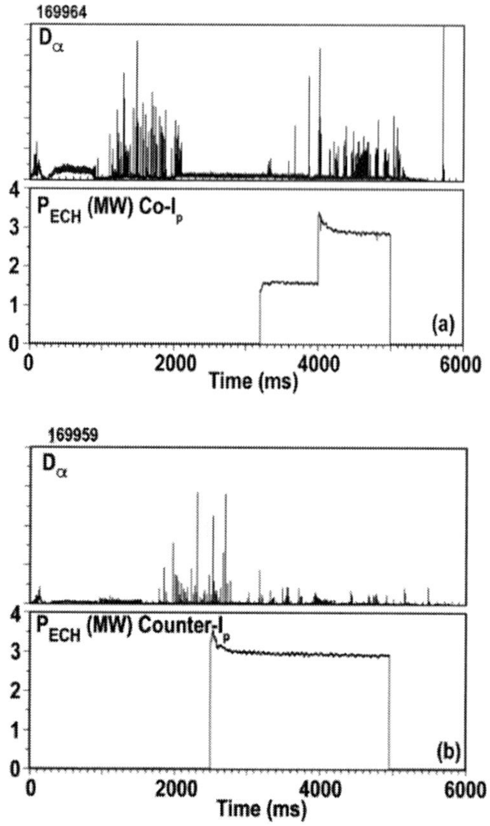

Fig. 2. This example shows the effect on ELMs of about 3 MW of electron cyclotron power deposited in the edge of the plasma beyond rho=0.9. The small amount of EC driven current was opposite to the plasma current for suppression and in the direction of the plasma current for augmentation of the ELMs. Different combinations of externally applied magnetic fields can reverse the effect of the ECH

Acknowledgment

This work was funded by the U.S. DOE under Cooperative Agreement DE-FC02-04ER54698. DIII-D data shown in this paper can be obtained in digital format by following the links at https://fusion.gat.com/global/D3D_DMP.

EPJ Web of Conferences **149**, 01007 (2017) DOI: 10.1051/epjconf/201714901007
10th International Workshop 2017 "Strong Microwaves and Terahertz Waves: Sources and Applications"

Medium size tokamak T-15MD as a base for experimental fusion research in Russian Federation

Alexander Romannikov and Fusion research Centre team

NRC "Kurchatov institute", Moscow, Russia, Romannikov_AN@nrcki.ru

History of tokamak has more than 65 years. "Kurchatov institute" is a leader in Russia of high temperature plasma researches on tokamak facilities. It was the first in the world that has built the first tokamak and it is remained the world leader in researches on tokamaks for long time. At present time, the new tokamak T-15MD is being built in NRC "Kurchatov institute".

T-15MD main parameters are:

Parameter	Value
Major radius of torus R, m	1.48
Aspect ratio	2.2
Plasma current I_p, MA	2.0
Elongation, k	1.9
Triangularity, δ	0.3–0.4
Plasma configuration	SN, DN
Discharge pulse duration, s	10
Toroidal magnetic field at plasma axis, T	2
Magnetic flux swing in central solenoid, Wb	6
Neutral beam injection power, MW	6
ECR heating power, MW	7
ICR heating power, MW	6
LH heating power, MW	4

The magnetic system with divertor configuration of T-15MD facility for confining of hot plasma is already created. The toroidal magnet system consists of 16 D-shaped coils (Fig. 1) that form an arched structure [1–3].

Fig. 1. Toroidal winding together with poloidal field coils

The level of ripples at the outboard plasma boundary is approximately 0.8%. Each coil has 50 turns that are wound by a hollow conductor made of silver-copper alloy. The toroidal winding is cooled with distilled water. Each coil has four parallel water branches. The total water mass flow rate through the winding is 43 m^3/h, with a pressure drop of 0.8 MPa.

The toroidal magnet is charged by eight thyristor convertors (22 kA, 1 kV each). The rated value of the toroidal magnetic field at the plasma axis is B_0= 2T. The current plateau duration is ten seconds.

The central solenoid (Fig. 2) consists of three separated coils. Each coil is wound with two parallel hollow conductors that have a trapezoidal form and are made of silver-copper alloy. The coils are charged by independent power supply systems.

Fig. 2. Central solenoid

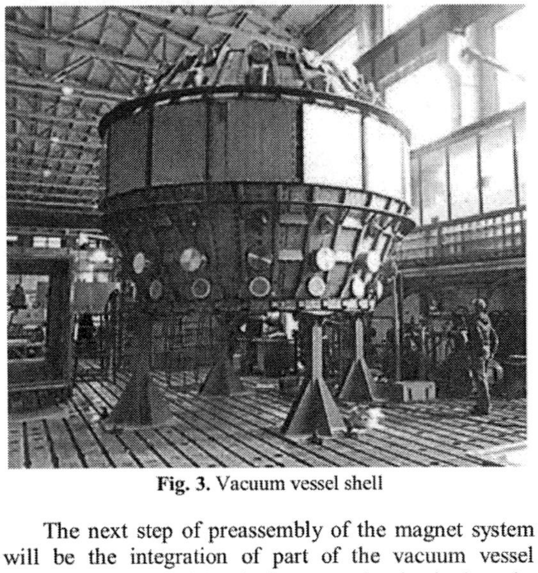

Fig. 3. Vacuum vessel shell

The next step of preassembly of the magnet system will be the integration of part of the vacuum vessel (Fig. 3) into toroidal winding. At the beginning the "cold" and "hot" vacuum tests of tokamak vacuum chamber will be carried out at the plant in Bryansk. After the preliminary assembling of the main tokamak systems will be carried out. For one vacuum vessel shell will be cut to four parts (four quarter sections). Half of the toroidal coils and two horizontal field coils will disassembled. Using the special rails two sectors of the vacuum vessel will be integrated inside the half of the toroidal winding. After finishing the preassembly of magnet system together with vacuum vessel parts, all elements will be disassembled and delivered to the Kurchatov Institute for assembly of the tokamak T-15MD in 2017–2018. The Tokamak T-15MD should begin operations in 2019.

The engineering systems [4] are important part of the overall system of the tokamak. The part of which is already ready now. They consist from:

1) High vacuum pumping system of the chamber with four turbopumps with total capacity of ~ 10 m^3/s (H_2) and two cryogenic pumps with the capacity of 4 m^3/s (H_2) each.

2) Distiller water cooling system with a pressure of 0.5 MPa and 1.0 MPa. All equipment is installed and adjusted.

3) Power supply systems. The consumed total power in an impulse with current of plasma 2 MA and additional heating of plasma of 20 MW will make 300 MBA. Reconstruction of substations No. 745 and No. 110 (110/10 kV) are planning to complete in 2017–2018.

4) Management information system [5] which is created as modular, scalable, with possibility of replication, the system focused on work of large experimental physical installation. Now the Data-processing center is created, installation of equipment of the main systems is carried out, tests and check of operability of some subsystems are carried out, commissioning for realization of physical start-up of installation in 2018–2019 is carried out.

The experimental program of T-15MD facility is based on experimental dates of T-10 tokamak achieved during a few last years. The main tasks of experimental program of T-15MD tokamak are: support of the ITER project in field of physical and technological database; development of tokamak fusion neutron source; fusion investigations in frame of "National Russian program on fusion and plasma technology" developing at present time.

More detailed of the experimental investigations program on the tokamak T-15MD includes the following tasks: the possibility of obtaining high values of beta as a way to reduce the cost of fusion reactor; research of plasma-materials interaction, including graphite, tungsten and lithium; monitoring of the plasma current and pressure profiles; research of the possibility of implementing a large value of beta and density in steady-state operations fully non-inductive current regimes; the control in real time of the stabilization, balance, heating and confinement the high temperature plasma, etc. Tokamak T-15MD will be used as a facility, which will be tested in the materials and technology of the first wall and divertor.

The installation is equipped with system of additional plasma heating and current drive (P_{sum} = 15–20 MW), which will allow one simultaneous the achievement of high temperature plasma ($T_i \sim T_e \sim$ 5–9 keV) and plasma density (ne ~ 10^{20} m^{-3}) in the discharge with pulse duration up to 30 s. This system involves the injection of neutrals (3 injector on 2 MW/75 keV H^0, each), electron-cyclotron heating (7 gyrotrons of 1.0–1.5 MW each, f =110–120 GHz with the possibility of heating at the second harmonic, and Bershtein's modes), ion cyclotron heating (3 antennas on 2 MW each, including the possibility of maintaining the current helicons), lower hybrid heating and current drive (grill with a power of 4 MW, f = 2.45 GHz).

References

1. *Bondarchuk, E.N., Azizov E.A., Alekseev A.B., et al.* Engineering Problems of Tokamak T-15 Electromagnet System Reconstruction // IEEE Transactions on Applied Superconductivity, 2012, Volume: 22 , Issue: 3, Page(s): 4201604.

2. *Khvostenko P.P., Azizov E.A., Alfimov D.E. et al.* The magnet system of the Tokamak T-15 upgrade // Fusion Engineering and Design, 2015, vol. 98-99, 1090-1093.

3. *Khvostenko P.P., Anashkin I.O., Belyakov V.A., et al.* Preassembly of the tokamak T-15MD magnet system // Fusion Engineering and Design, 2017, http://doi.org/10.1016/j.fusengdes.2017.03.064

4. *P.P. Khvostenko, I.O. Anashkin, V.A. Belyakov, et al.* Status of Tokamak T-15MD // 26th IAEA Fusion Energy Conference, 17–22 October 2016, Kyoto, Japan // Book of abstracts, FIP/P7-40, p.749; eproceedings, fec2016-preprints, preprint0289.pdf

5. *Motckin Y.Ya., Sokolov M.M., Khvostenko P.P.* Information and control system for experimental fusion device "Tokamak T-15" // MKA : BKC, 2015, №3, 29-37.

EPJ Web of Conferences **149**, 01008 (2017) DOI: 10.1051/epjconf/201714901008
10th International Workshop 2017 "Strong Microwaves and Terahertz Waves: Sources and Applications"

Development and applications of THz gyrotrons

M.Yu. Glyavin

Institute of Applied Physics of Russian academy of Sciences, Nizhny Novgorod, Russia, glyavin@appl.sci-nnov.ru

Introduction

The last decade has contributed to the rapid progress in the development of THz sources in particular gyrotrons. The CW gyrotron developed at IAP RAS provides radiation with a power up to 1 kW in the frequency range 0.26–0.52 THz and the gyrotrons with pulsed magnetic fields operates with 100 kW power level at 0.7 THz and 1 kW at 1 THz frequency.

Results

The main results of IAP RAS activity with THz band gyrotron development are summarized below.

A series of experiments on application of CW THz-radiation for spectroscopy, nano-powders manufacturing, and gas discharge initiation has been carried out [1–3]. The possibility of achieving of unique results has been demonstrated. For example, the increase of the sensitivity of a radio-acoustic detector by several orders of magnitude, several-times increase in powders formation speed has been demonstrated. For the first time, a continuous (tens of seconds) gas discharge with a characteristic size of 1 mm localized in a gas stream has been demonstrated. The duration of the discharge was limited only by overheating of quasi-optical elements of the transmission line.

A physical model of accounting for the ion compensation of the space charge of the electron beam in the transporting channel of the gyrotron, including the cavity region, has been developed. An analytic solution of the ion compensation problem in a long-pulse regime have been derived. An equilibrium concentration of ions in the gyrotron's operating space after the transient processes and the profile of the electron concentration in the transporting channel have been determined. These results allow one to estimate the change of the pitch-factor due to a partial compensation of potential degradation in the gyrotron's cavity and, thus, are important for starting currents, efficiency and mode competition analysis [4].

For solution of the mode selection problem in gyrotrons at the cyclotron harmonics, a scheme, based of decreasing of diffraction quality factors of quasi-cut-off parasitic modes through their transformation into other, traveling modes was suggested. The cavity of the gyrotron in the simplest case has the form of a quasi-regular systems with a single irregularity (local radius increase) situated at the center of the cavity. While the radius of the regular parts of the cavity corresponds to the cut-off of the operating mode being excited at the second cyclotron harmonics, the local increase of the radius corresponds to the cut-off of the same wave, but with an additional radial variation. Numerical simulation of the electrodynamic cavity of a 1.3 THz gyrotron at the second cyclotron harmonic has been preformed. Owing to the profile of the cavity, the starting current at the second cyclotron harmonic (about 10 A) becomes substantially lower than the currents of the parasitic waves at the fundamental harmonic (20–30A). These results make it attractive to experimentally realize gyrotrons with unique parameters: selective excitation of the 2nd harmonic, power of tens of kW and frequency of 1.3 THz.

An experimental investigation of a pulsed gyrotron with an axis-encircling electron beam (80 kV / 0.7 A) with a sectioned klystron-like cavity, operating at the third cyclotron harmonic at a frequency of about 1 THz has been carried out. A sectioned klystron-like cavity with a long wave-beam interaction space and a relatively low diffraction Q-factor has been suggested.

The scheme of a gyrotron with a planar configuration of the interaction space and sheet electron beam proposed by the authors of the project is of interest for the generation of sub-millimeter radiation at the harmonics of the gyro-frequency. Through the PIC simulation by the CST STUDIO SUITE code, the ability to effectively generate the 3rd harmonic in a planar gyrotron with an operating frequency of about 250 GHz has been demonstrated. The gyrotron was fed by a sheet electron beam with an energy of 80 keV, a current of 15 A and the pitch factor ~ 1. In simulations, the steady state regime establishment at the 3rd cyclotron harmonic generation was observed at the guiding magnetic field of about 3.35 T. Generation efficiency at these settings was 8% and the output power ~ 50 kW. Increasing of the power level up to 0.3 MW while maintaining a selective excitation of the operating mode at the third harmonic in this circuit can be achieved by increasing the cross-section of the system up to 10 cm (i.e., up to 100 wavelengths) with a corresponding increase in the total beam current.

The THz band pulsed gyrotron has been used successfully for initiation of gas discharges [5]. Such plasma is promising for development of a point source of multicharged ions, a source of high–energy ultraviolet (extreme ultraviolet EUV or XUV) [6] and remote detection of remote detection of concealed radioactive materials [7]. The discharge developed and was localized only in a small region of space (about 1 mm) near the gas injection hole, where the pressure was high (close to the atmospheric one) and the breakdown conditions (the "point-like" discharge) were fulfilled. The density of electrons in the discharge was measured by observing the Stark broadening of the atomic radiation line H_α of hydrogen (656.3 nm) which was present in the discharge as a minor admixture. The plasma density in the discharge was equal to about $2 \cdot 10^{16}$ cm^{-3}, which exceeds the cut-off density for a frequency of 0.67 THz used to maintain the discharge. The localized gas discharge has been successfully obtained also with 1 kW / CW / 0.26 THz gyrotron.

The same gyrotron has been used also for the high-resolution molecular spectroscopy in a gas mixture. As a result of analysis of experimental profiles of known lines of the rotational SO_2 spectrum, it was demonstrated for the first time that an increase of radiation power by about three orders of magnitude leads to a proportional increase of the RAD spectrometer sensitivity. This permitted, in particular, to observe the weak transitions predicted earlier in the SO_2 molecule.

The phase lock loop control of anode voltage has been used for frequency stabilization of 0.26 THz tube. The width of the frequency spectrum was decreased to 1 Hz measured at intermediate frequency IF = 350 MHz, which corresponds $\Delta f / f = 3{\cdot}10^{-12}$ with measurement time of a few seconds. The long-term stability was defined by reference clock ($\delta f / f \sim 10^{-9}$ for quartz clock and up to $\delta f / f \sim 10^{-12}$ for rubidium clock). Obtained spectrum width and frequency stability were previously achieved in backward-wave oscillators, utilized in spectroscopy, but with power levels of tens of mW, while stabilization system with anode voltage control in gyrotrons has no apparent limiting factors in the field of gyrotron output power – as it was demonstrated with output power of 40 W [8].

The last interesting feature is high speed production of pure nanopowders by material evaporation and condensation. In the case of evaporation by 1 kW / CW / 0.26 THz radiation, the evaporation rate was increased approximately in order in comparison with traditional heating methods.

For the future progress of gyrodevices we proposed tube with improved selection based on multi-beam scheme or non-uniform cavity profile. The optimization of the operational parameters and the experimental setup are in progress now with dual beam 0.79 THz CW gyrotron. The generation of series fundamental harmonic modes in the wide frequency range (0.2–0.4 THz) with output power of the order of 100 W is obtained and second harmonic operation with frequency closed to design one and output power about 5 W has been observed. Next step of this project will design 1 THz third harmonic gyrotron with relatively low operating voltage [9–11]. The possibility of energy recovery was investigated to grove up operation efficiency, same as was demonstrated, for example in [12].

For the excitation second harmonics at tube with pulsed coil and, correspondently, the possibility of 1.3 THz generation the scheme based on the cavity with a phase corrector is under development. The preliminary calculation shown, that the starting currents of fundamental harmonic parasitic modes can be increased in an order in contrast with traditional cylindrical cavity.

Conclusion

Thus, the series of powerful THz gyrotrons has been developed and successfully tested at the IAP RAS. The recent progress in the THz gyrotrons development opened a number of novel applications in the high-power terahertz science and technology.

Acknowledgements

The work supported by the Russian Science Foundation under the project No. 14-12-00887.

References

1. M.*Yu. Glyavin, T. Idehara, S.P. Sabchevski.* Development of THz gyrotrons at IAP RAS and FIR UF and their applications in physical research and high–power THz technologies // IEEE Trans. on Terahertz Science and Technology, 2015, V.5, No.5, P.788-797

2. *M. Glyavin, G. Denisov, V. Zapevalov, M. Koshelev, M. Tretyakov, A. Tsvetkov.* High-power terahertz sources for spectroscopy and material diagnostics // Physics-Uspekhi, 2016, V.59, No.6, P.595-604

3. *M.Yu. Glyavin, A.V. Chirkov, G.G. Denisov et al.* Experimental tests of 263 GHz gyrotron for spectroscopy applications and diagnostic of various media // Rev. Sci. Instr., 2015, V.86, No.5, P.054705

4. *A.P. Fokin, M.Yu. Glyavin and G.S. Nusinovich.* Effect of ion compensation of the beam space charge on gyrotron operation // Phys. of Plasmas, 2015, V.22, No.4, P.043119

5. *M.Yu. Glyavin, A.G. Luchinin, A.A. Bogdashov et al.* Experimental Study of the Pulsed Terahertz Gyrotron with Record-Breaking Power and Efficiency Parameters // Radioph. and Quant. Electr., 2014, V.56, No.8-9, P.497-507

6. *A.V. Vodopyanov, M.Yu. Glyavin, S.V. Golubev et al.* Plasma density in discharge sustained in inhomogeneous gas flow by high-power radiation in the terahertz frequency range // Technical Physics Letters, 2017, V.43, No.2, P.186–189

7. *G.S. Nusinovich, P. Sprangle, V.E. Semenov, D.S. Dorozhkina and M.Yu Glyavin.* On the sensitivity of THz gyrotron based systems for remote detection of concealed radioactive materials // Journal of Applied Physics, 2012, V.111, P.124912

8. *G.G. Denisov, A.P. Fokin, M.Yu. Glyavin et al.* High precision frequency stabilization of a 100W/263 GHz continuous wave gyrotron // 18th International Vacuum Electronics Conference 24 - 26 April 2017, London

9. *N.S. Ginzburg, M.Yu. Glyavin, A.M. Malkin et al.* Improvement of operation stability at high cyclotron harmonics in the double-beam THz range gyrotrons // IEEE Trans. On Plasma Science, 2016, V.44, No.8, P.1303-1309

10. *M.Yu. Glyavin, N.A. Zavolskiy, A.S. Sedov and G.S. Nusinovich.* Low-Voltage Gyrotrons // Phys. of Plasmas, 20, 033103 (2013)

11. *N.S. Ginzburg, M.Y. Glyavin, I.V. Zotova, I.V. Zheleznov, A.P. Fokin.* Optimization of terahertz range gyrotron self-excitation conditions by increasing the lifetime of cyclotron oscillators in low-voltage interaction space // Tech. Phys. Lett., 2017, V.43, No.1, P.110-113

12. *M. Glyavin, A. Kuftin, N. Venediktov, V. Zapevalov.* Experimental investigation of a 110 GHz/1 MW gyrotron with the one-step depressed collector // Int. J. of Infrared and Millimeter Waves, 1997, V.18, No.11, P.2129-2136

EPJ Web of Conferences **149**, 01009 (2017) DOI: 10.1051/epjconf/201714901009
10th International Workshop 2017 "Strong Microwaves and Terahertz Waves: Sources and Applications"

Sub-THz technology for dynamic nuclear polarization in nuclear magnetic resonance (DNP NMR): transverse confinement of microwave propagation through heterogeneous solid DNP samples

Frank Engelke[1], Armin Purea[1], Christian Reiter[1], Fabien Aussenac[2]

[1]Bruker Biospin GmbH, Rheinstetten, Germany, frank.engelke@bruker.com
[2]Bruker Biospin S.A., Wissembourg, France

Passive microwave components for DNP NMR, the resulting spatial microwave field distribution at the sample site, the sample structure, the molecular structure, and DNP spin physics are closely linked to each other. We study specific examples of low-temperature (100 K) DNP samples, characterized by a high degree of spatial heterogeneity on a sub-mm scale. We report on our work in progress investi-gating the local microwave propagation modified through transverse confinement in DNP samples with possible consequences for the DNP spin transfer mechanism.

DNP NMR [1] on solids samples, in particular, under magic-angle spinning (MAS) has become an extremely useful technique in solid-state NMR to enhance the weak NMR signal by one or two orders of magnitude [2–4]. Application examples range from the study of biomacromolecules [5–6] to material samples like zeolites or mesoporous silica with small molecules or catalytic compounds adsorbed at the inner surfaces of the mesoporous material [7–9].

The mechanism by which DNP allows to enhance the nuclear spin polarization signals is based on a coherent spin polarization transfer in a strong static external magnetic field from unpaired electron spins in paramagnetic centers (*e.g.*, radicals) to nuclear spins under irradiation of electron spin-resonant microwaves. A particular DNP mechanism working at high magnetic fields (*e.g.*, 9–18 T) requires *bi*radicals in the sample, now with two unpaired electron spins coupled by dipole-dipole interaction. Together with a nearby situated nuclear spin they form a three-spin system that gives rise to the DNP cross effect [10–11] being specifically efficient in DNP MAS NMR.

Probes for MAS NMR capable of performing DNP experiments [12] inherit various traits from the subsystems composing them, which may lead to conflicting or challenging probe hardware boundary conditions. The sample is packed into a zirconia or sapphire MAS rotor spinning at a speed of ca. 10 kHz, driven by an airbearing microturbine. Tightly wound around the MAS rotor there is an NMR rf solenoidal coil with interturn spacings such that the radially (to the rotor axis) incident HE_{11} or TEM_{00} Gaussian microwave beam can irradiate the sample inside the rotor. A schematic view is shown in Fig. 1. The MAS system and the sample are kept at temperatures around 100...200 K in order to keep the spin relaxation time T_{1e} of the unpaired electron spins in the biradicals at large enough values (micro to milliseconds) to maintain the electron spin polarization or coherences during the DNP experiment.

For a variety of DNP samples an additional complication appears whenever the sample is heterogeneous on a sub-mm scale. Examples are powders of mesoporous or nanoporous samples, layered samples of oriented biomembranes, or, *e.g.*, simply powders with a grain size distribution on the order of 100 to 400 μm, impregnated by a solvent (*e.g.*, tetra-chloroethane, TCE) and dissolved in it some biradical (AMUPol, TEKPol, and others) frozen at low temperature. It has been shown experimentally, that the strongly inhomogeneous microwave field in such samples leads to quite high DNP enhancement factors, in the case of dielectric particle powders impregnated with frozen biradical solutions one even may observe DNP enhancement factors being two- to three-fold higher [13] than those in homogeneous samples with well defined periodic microwave field distributions.

The heterogeneity of such samples strongly affects the microwave propagation, because in the case of frozen impregnated powders there is a random distribution of spatial regions with differing dielectric constants, giving rise to multiple reflections and dif- fraction of local microwave beamlets. It is well known from first principles, that for charge-free regions Gauss' law div $\mathbf{E} = 0$ applies. Close to surfaces, e.g., of the dielectric particles, and inside cavities or pores of sub-wavelength dimensions this leads to local microwave electric fields **E** and local Poynting vectors **P** that are not orthogonal anymore, because the presence of surfaces and pore walls imposes transverse confinement to the local field such that longitudinal **E** field components arise. Fig. 2 represents an example of dielectric particles (sapphire, $\varepsilon = 9.6$) embedded in frozen TCE ($\varepsilon = 2.5$) and illustrates the simulated (by CST Microwave Studio®, www.cst.com) spatial distribution of the angle between the local Poynting vector and the local electric field in a rotor axial plane parallel to the incident microwave beam. The resulting complex pattern reflects the randomness inherent in the powder distribution of sapphire particles and causes an equally randomized distribution of local microwave electric and magnetic field amplitudes and *directions* which directly relates to the performance of DNP experiments. Only the microwave magnetic field vector component orthogonal to the external static magnetic field couples to the electron spin and causes electron spin transitions that lead to saturation (equal occupation of Zeeman spin levels).

In conclusion, we suggest that the simulated complex spatial distribution of amplitudes and local angular dependence of Poynting and electric field vectors appears as a result of the transverse confinement [14–15] in strongly heterogeneous DNP samples. It gives rise, apart from the powder distribution of EPR *g* tensors, to a statistical distribution of microwave magnetic field vectors representing the randomness of particle or pore distribution in the sample. Yet we do not know the details of

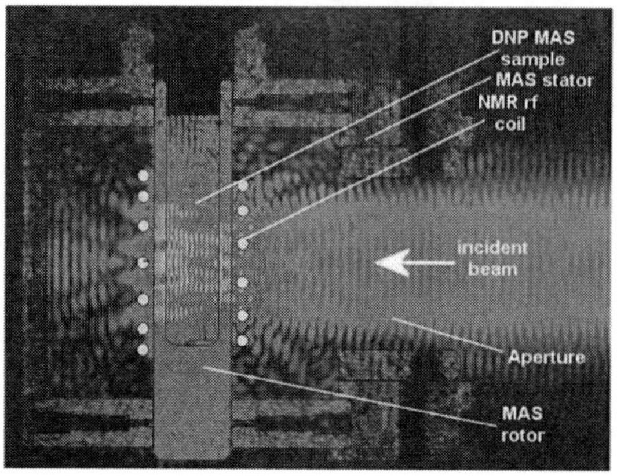

Fig. 1. Schematic view of an DNP MAS NMR setup with the rotor containing the sample (here supposed to be homogeneous), surrounded by an solenoidal coil tuned to the desired NMR frequencies, and the incident mm wave beam (here at 395 GHz) being diffracted and partly reflected by the coil, rotor, and the sample

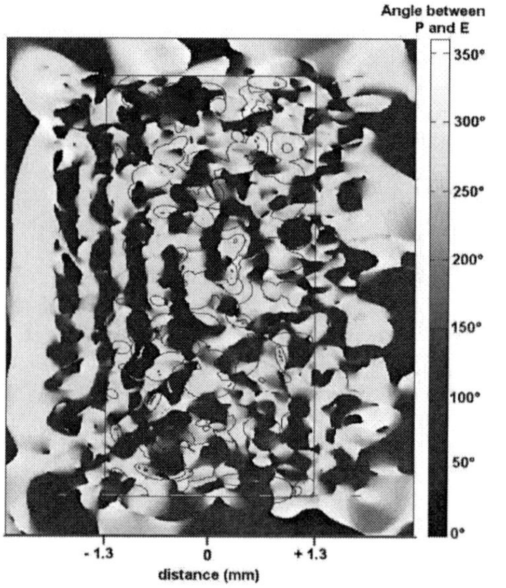

Fig. 2. Spatial distribution in a rotor axial plane of the angle between local Poyning vector **P** and local electric field vector **E** in a sample of sapphire powder (grain size distribution 100–400 μm) impregnated with tetrachloroethane / TEKPol biradical, frozen in a solid matrix at T = 100 K. The linearly polarized Gaussian microwave beam, waist 3 mm, frequency 263 GHz, is incident from the left side. In the simulation, only the rotor and the heterogeneous sample material is present. 3D voxel resolution 10 μm, 5×10^7 mesh cells

of the field distribution relevant for EPR spin excitation and DNP, but take that as a forthcoming task: to quantify the impact of the microwave field distribution on the coherent DNP spin dynamics. Further it needs to be shown in a quantitative manner that the transverse confinement in typical heterogeneous DNP samples may lead, jointly with spin diffusion in such structures, to an improvement of the enhancement factor in DNP mechanisms like the DNP cross effect.

References

1. *Carver, T.R., Slichter, C.P.* Polarization of nuclear spins in metals // Phys. Rev. 1953. V. 92. P. 212.

2. *Becerra L.R., et al.* Dynamic nuclear polarization with a cyclotron resonance maser at 5 T // Phys. Rev. Lett. 1993. V. 71. P. 3561.

3. *Gerfen,G. J., et al.* High frequency (140 GHz) dynamic nuclear polarization: polarization transfer to a solute in frozen aqueous solution // J. Chem. Phys.1995. V. 102. P. 9494.

4. *Rosay, M., et al.*, Solid-state dynamic nuclear polarization at 263 GHz: spectrometer design and experimental results // Phys. Chem. Chem. Phys. 2010. V. 12. P. 5850.

5. *Akbey, U., Oschkinat, H.* Structural biology applications of solid state MAS DNP NMR // J. Magn. Reson. 2016. V. 269. P. 213.

6. *Debelouchina, G. T., et al.* Dynamic nuclear polarization-enhanced solid-state NMR spectroscopy of GNNQQNY nanocrystals and amyloid fibrils // Phys. Chem. Chem. Phys. 2010. V. 12, N. 22. P. 5911.

7. *Lesage, A., et al.* Surface enhanced NMR spectroscopy by dynamic nuclear polarization // J. Am. Chem. Soc. 2010. V. 132, N. 44. P. 15459.

8. *Lafon, O., et al.* Mesoporous Silica Nanoparticles Loaded with Surfactant: Low Temperature Magic Angle Spinning 13C and 29Si NMR Enhanced by Dynamic Nuclear Polarization // J. Phys. Chem. C. 2013. V. 117, N. 3. P. 1375.

9. *Perras, F. A., et al.* Natural Abundance 17O DNP: Two-Dimensional and Surface-Enhanced NMR Spectroscopy // J. Am. Chem. Soc. 2015. V. 137. P. 8336.

10. *Hovav, Y., et al.* Theoretical aspects of dynamic nuclear polarization in the solid state - The cross effect // J. Magn. Reson. 2012. V. 214. P. 29.

11. *Maly, T., et al.* Dynamic nuclear polarization at high magnetic fields // J. Chem. Phys. 2008. V. 128, N. 5. P. 052211.

12. *Rosay, M., et al.* Instrumentation for solid-state dynamic nuclear polarization with magic angle spinning NMR // J. Magn. Reson. 2016. V. 264. P. 88.

13. *Kubicki, D.J., et al.* Amplifying dynamic nuclear polarization of frozen solutions by incorporating dielectric particles // J. Am. Chem. Soc. 2014. V. 136, N. 44. P. 15711.

14. *Anderson, P.W.* Absence of diffusion in certain random lattices // Phys. Rev. 1958. V. 109. P. 1492.

15. *Cherroret, N., et al.* Transverse confinement of waves in three-dimensional random media // Phys. Rev. E 2010. V. 82. P. 056603.

EPJ Web of Conferences **149**, 01010 (2017) DOI: 10.1051/epjconf/201714901010
10th International Workshop 2017 "Strong Microwaves and Terahertz Waves: Sources and Applications"

CVD diamond with boron-doped delta-layers deposited by microwave plasma

A.L. Vikharev[1], A.M. Gorbachev[1], M.A. Lobaev[1], D.B. Radishev[1], V.A. Isaev[1], S.A. Bogdanov[1], M.N. Drozdov[2], E.V. Demidov[2], E.A. Surovegina[2], V.I. Shashkin[2], P.A. Yunin[2], J.E. Butler[1]

[1] Institute of Applied Physics of the Russian Academy of Sciences, Nizhny Novgorod, Russia
[2] Institute for Physics of Microstructures of the Russian Academy of Sciences, Nizhny Novgorod, Russia

Introduction

The semiconductor single-crystal CVD diamond (obtained from the gas phase during homoepitaxial deposition) is a wide band gap semiconductor with a gap width of 5.5 eV. CVD diamond has unique characteristics – high mobility of charge carriers, high carrier saturation speed, high electric breakdown field, the greatest thermal conductivity, high radiation and chemical resistance. On a combination of properties the CVD diamond is superior to other wide band gap semiconductors and is considered a promising material for the creation of a new generation of high-power and high-frequency electronic devices. The main difficulty in realization of the potential of CVD diamond as an electronic material is the problem of creating charge carriers inside it. Compared with conventional semiconductors, dopants in diamond have deeper energy levels that significantly impede the activation of the dopant (the degree of ionization of the dopant at room temperature is less than 1%). Thus, in order to create an acceptable level of conductivity, it is necessary to increase the level of doping, but in case of boron doping this leads to a decrease of carriers (holes) mobility in diamond. To solve the problem of boron doping of CVD diamond, an approach based on delta-doping technology is known. A thin layer of diamond heavily doped with boron (having a thickness of 1–2 nm and concentration of boron atoms higher than 5×10^{20} cm^{-3}) is formed inside an undoped defect-free diamond of high quality. To achieve high electronic properties (obtaining high hole mobility and conductivity of the layer), it is necessary to realize sharp boundaries between the doped and undoped materials. Recently, this problem has been successfully solved [1, 2]. This report provides an overview of the results of studies on the growth of electronic-quality epitaxial layers of diamond, the production of heavily boron-doped layers and the study of their characteristics.

Experiments

The novel microwave plasma assisted CVD reactor for growth of nanometric boron delta-doped layers with ultra-sharp interfaces between doped/undoped materials was built in IAPRAS [1]. Fig. 1 shows a schematic of the reactor. The main features of the reactor are: 1) rapid gas switching; 2) laminar gas flow; 3) axial symmetric resonant mode – symmetric discharge; 4) slow growth of diamond 40–100 nm/h. We achieve rapid gas switching from one input gas to another by a home-made electronic switch. The residence time of the reactor is approximately 5 s.

In developed reactor the diamond deposition regimes in which one obtains thin doped delta layers with thickness of 1–2 nm with concentrations of boron about $5\cdot10^{20}$ cm^{-3} were found. Typical parameters of the delta layer under these conditions are given in Fig. 2 for the SS6-1 sample, in which the boron concentration is $4.8\cdot10^{20}$ cm^{-3} and thickness is 1 nm.

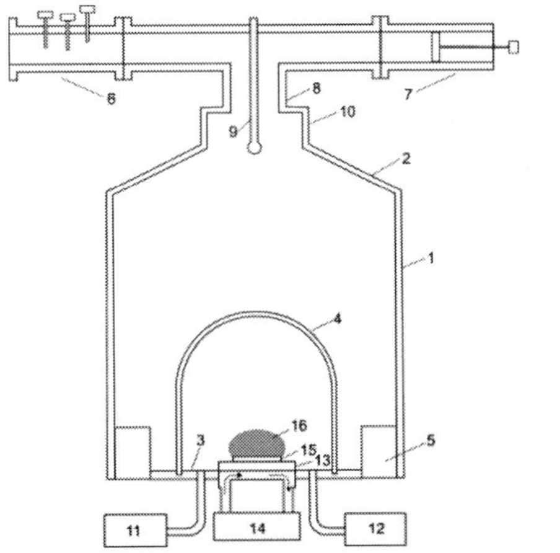

Fig. 1. Experimental setup: 1 – cylindrical cavity, 2 – circular cone, 3 – bottom cavity wall, 4 – quartz dome, 5 – step ledge, 6 – waveguide system, 7 – plunger, 8 – coaxial line, 9 – antenna, 10 – shorted cylindrical waveguide, 11 – gas supply system, 12 – gas evacuation system, 13 – substrate holder, 14 – substrate cooling system, 15 – substrate, 16 – microwave discharge

Measurement of the boron concentration in the grown samples was carried out by the secondary ion mass spectroscopy (SIMS) method using a time of flight SIMS setup (IONTOF TOF.SIMS-5). The depth of the etch craters for calculation of the analysis depth from the etching time was carried out using white light interference microscopes Talysurf CCI 2000 and Zygo NewView 7300. Sputtering was performed by Cs^+ ions with energy of 1 keV at angle of 45°. The probing was carried out by Bi^+ ions with energy of 25 keV. The output of the cluster ion C_2B^- was registered. Quantitative calibration of the mass spectrometer was performed using a test HPHT crystal implanted with boron ions with three different maximum concentrations (10^{18}, 10^{19} and 10^{20} cm^{-3}).

Capacitance–voltage measurements were performed on sample Tr06 with two 'delta layers' and an underlying heavily doped layer. Mesa structures of diameters 25 to 400 μm were formed by lithographic masking and etching down to the heavily boron doped layer. Ohmic contacts were formed after etching on the heavily doped layer p^{++} layer by deposition of Ti/Mo/Au metallization and annealing. Schottky contacts were formed on the top of the mesas with Cr/Al metallization. The profiles of holes concentration and boron concentration in the Tr06 structure are shown in Fig. 3. For the investigated boron delta layer in diamond, the activation energy was determined to be ~ 60 meV, which is less than the activation energy of boron in diamond (370 meV) and corresponds to a high boron concentration, which according to SIMS data exceeds $3\cdot10^{20}$ cm^{-3}.

For another sample, Tr12, with higher boron concentration and less thickness of delta layer the hole profile is changed. The profiles of holes concentration and boron concentration in the Tr12 structure are shown in Fig. 4. It is seen that the portion of holes outside of the delta layer sufficiently increased then it was for Tr06 sample.

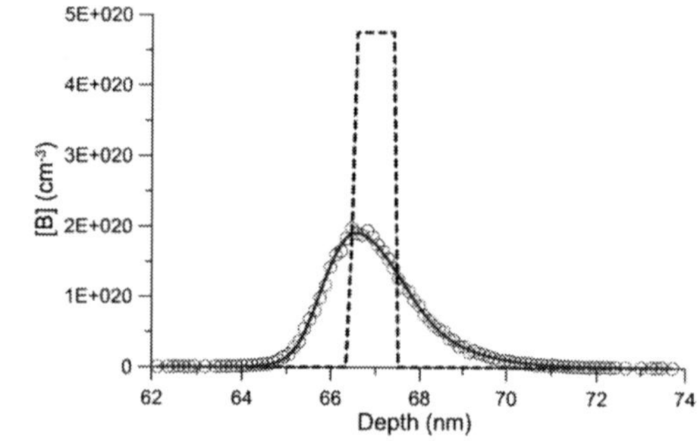

Fig. 2. Boron concentration profiles determined by SIMS (circles); boron concentration profiles recovered using depth resolution function (dashed line); convolution of recovered profile and depth resolution function

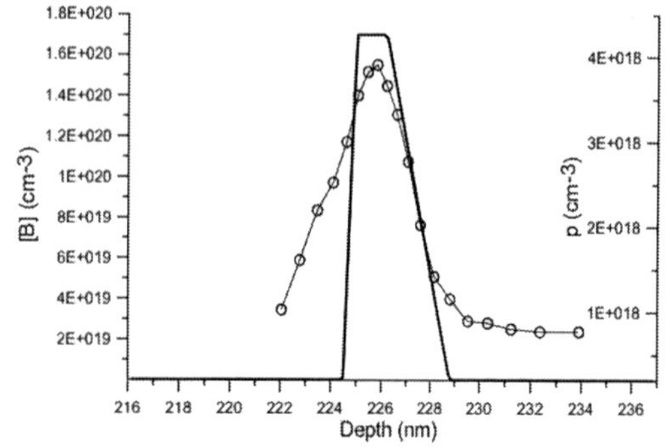

Fig. 3. Comparison of boron concentration profile recovered using analytical depth resolution function (solid line) and hole profiles extracted from C-V analysis measured at room temperature (circles), Tr06 sample

Measurements of the holes mobility in various samples were made by the Van der Pau method, and the maximum hole mobility was determined to be 300 V^2/cm·s. Because of the spatial separation of holes and dopant boron atoms, the measured holes mobility consists of an integral value from different regions of diamond – doped with low mobility and undoped with high mobility. This is confirmed by the study of the dependence of the mobility and concentration of holes on the temperature of the sample. Thus, for the first time in the epitaxial diamond films with delta layers, the effect of increase of mobility of holes was observed due to the separation with the doping boron atoms. This proves the high quality of the fabricated structures – sharpness of transitions in the doping profile, the high quality of undoped epitaxial layers, and the high concentration of boron atoms in the delta layer.

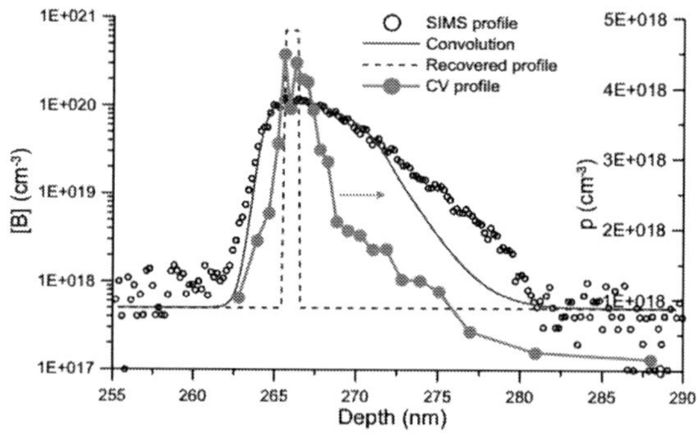

Fig. 4. Comparison of boron concentration profile recovered using analytical depth resolution function (dashed line) and hole profiles extracted from C-V analysis measured at room temperature (circles), Tr12 sample

Acknowledgments

This work was supported by Act 220 of the Russian Government (Agreement no. 14.B25.31.0021 with the host organization IAP RAS).

References

1. *A.L. Vikharev, A.M. Gorbachev, M.A. Lobaev et al.*, Novel microwave plasma-assisted CVD reactor for diamond delta doping, Phys. Status Solidi RRL, v.10, Issue 4, 2016, p. 324–327.

2. *J.E. Butler, A.L. Vikharev, A.M. Gorbachev et al.*, Nanometric diamond delta doping with boron, Phys. Status Solidi RRL, 1–6 (2016) / DOI 10.1002/pssr.201600329

EPJ Web of Conferences **149**, 01011 (2017) DOI: 10.1051/epjconf/201714901011
10th International Workshop 2017 "Strong Microwaves and Terahertz Waves: Sources and Applications"

ECRH on CFETR – physics and technology needed

Jiangang Li, F.K. Liu, Y.P. Zhao, X.J. Wang, M.H. Li, H.D. Xu
Institute of Plasma Physics, Chinese Academy of Sciences, Hefei, China (j_li@ipp.ac.cn)

The Chinese Fusion Engineering Testing Reactor (CFETR) is the next device for the Chinese magnetic confinement fusion (MCF) program which aims to bridge the gaps between the fusion experiment ITER and the demonstration reactor DEMO. CFETR will be operated in two phases: Steady-state operation and tritium self-sustainment will be the two key issues for the first phase with a modest fusion power up to 200 MW. The second phase aims for DEMO validation with a fusion power around 1 GW.

Operation scenarios have been assessed by integrated modeling based on advanced H-mode physics with high magnetic fields (up to 7 T). High frequency electron cyclotron resonance heating (170–230 GHz), lower hybrid current drive (5–7.5 GHz), off-axis negative-ion neutral beam injection will be used together with high bootstrap current for achieving steady-state advanced operation.

Fully non-inductive CFETR scenarios have been developed with a self-consistent core-pedestal-equilibrium model. Table one is the baseline scenarios based on the integrated modeling. Different modelling of steady state operation conditions by the combination of ECCD with NNBI, ECCD with high field lower hybrid wave current drive were presented in this paper.

Concept design for CFETR EC system which is 20 MW 170 GHz from mid-plan port and 230 GHz from top window depending on the operation scenarios. 750 kA–1.5 MA plasma current can be driven by chosen different location and rotation angle. Further optimization for get the best current drive condition is still under way.

Table. Baseline steady state scenarios for CFTER

	Start point	Phase I	Phase II
R_0, a (m)	6.6,1.8	6.6 , 1.8	6.6, 1.8
P_{NBI} , P_{ECH}(MW)	29.5, 1.0	35.8, 20.0	33.9, 20.1
Fusion Gain Q_{FUS}	2.9	3.0	14.9
r P_{fus} (MW)	111	169	811
B_T (T), I_p (MA)	6.0, 7.5	6.0, 7.6	6.0, 10.0
Bootstrap Fraction	60%	63.6%	84.4%
β_N	1.46	1.89	3.15
H_{98Y2}	1.3	1.3	1.3
Γ_{NW}(MW/m^2)	0.15	0.19	0.92
Diverter Loading P_{DIV}/R_0 (MW/m)	7.1	10.4	25.8

The main technologies needed to fulfill CFETR EC mission are 170 GHz /230 GHz gyrotrons, long distance transmission, power supply, control and protection system and integrated antenna.

Design of 170 GHz gyrotron is finished. Key components R&D started 1–2 years ago, such as helium-free magnet, transmission line, and gyrotron itself. The first step is to develop a gyrotron shown in Fig. 1 with 1 MW power and 5–10 s pulse length at 140 / 170 GHz within next 3–5 years. The following step is to develop the CW 1 MW gyrotron at 170 GHz and 230 GHz. The assembly and commission of the whole system for short pulse system will begin before the end of 2017. The other key components for CW operation, such as diamond window, MOU box and damping load are under design.

Fig. 1. Short pulse 1 MW gyrotron

The key sub-system is the antenna. Two options are under concept design-which based on current developments of microwave antenna for ITER ECRH system. One is top window launcher and another one is the mid-plane antenna shown in Fig. 2. Due to the limitation of space, beam combinationwill be used. 20 transmission lines which transfer 1MW power each from gyrotron are combined into the mid-plan port. By combine 5 units into one which handle 5 MW per mirror unit. Steering mechanism similar with that used in EAST will be adopted to have capacity for change the power deposition position and real time NTM suppression. All components inside port are actively cooled with full shielding function. The first steering mirrors are equipped with shutter for plasma contamination during plasma discharge cleaning.

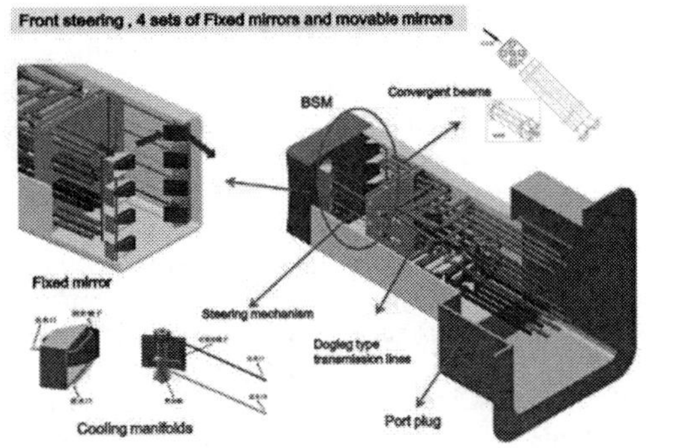

Fig. 2. Concept design for mid-plan port antenna

The detailed physical modeling, concept design, research and development activities including R&D on long transmission line, helium free magnet, gyrotron for aiming high performance steady state operation in CFETR, will be introduced in this paper.

References

1. Y. Tao Song, et. Al., IEEE TRANS. ON PLAS. SCI., VOL. 42, 495(2014)
2. B.N. Wan et. al., IEEE TRANS. ON PLAS. SCI. Vol. 42, 503(2014)
3. V. Chan, A. Costley, Nuclear Fusion, 2015, 55(2)
4. H.D. Xu, X.J. Wang, F.K. Liu et. al., Plasma Science and Technology, Vol.18, (2016) 442
5. Yunying Tang, Xiaojie WangFukun Liu, et. al., Fusion Engineering and Design, Vol. 89, Issue 6, (2014) 847–852.
6. Yunying Tang, Xiaojie WangFukun Liu, et. al., Fusion Engineering and Design 94(1), 2015
7. X.J. Wang, F.K. Liu, J.F. Shan, Fusion Engineering & Design 96-97 , 2015 , p. 181-186.

Anomalous absorption in ECRH experiments due to parametric excitation of localized UH waves

E. Gusakov, A. Popov

Ioffe Institute, St. Petersburg, Russia, evgeniy.gusakov@mail.ioffe.ru

Electron cyclotron resonance heating (ECRH) is a technique being widely utilized nowadays in magnetic fusion experiments. It is also considered for application in ITER, for plasma heating and neoclassical tearing mode control. The analysis [1-3] of different parametric decay instabilities (PDIs) potentially dangerous for mm-waves at the ECR frequencies in the presence of the monotonous plasma profiles revealed the typical microwave power threshold in the range much higher than 1 MW. This is due to huge energy loss of the interacting waves from the decay region. The typical heating system output power range available up to now, even if the gyrotrons are joined into a group, is substantially lower than the predicted PDI power threshold value. Thus, it has been taken as known until very recently that the ordinary or extraordinary electromagnetic wave propagation and absorption in the absence of the UHR being well described by linear theory is predictable in detail.

However, during the last decade many observations have demonstrated the anomalous phenomena during ECRH. Firstly, the fast ion generation and ion heating were observed at the TJ-II stellarator [4] and the TCV tokamak [5] during the 2nd harmonic X-mode ECRH pulse under conditions when energy exchange between electrons and ions should be very low. Secondly, the observations of the backscattering signal down – shifted from the pump frequency by approximately 1 GHz and correlated to the MHD mode rotation were reported [6, 7] in the 200 – 600 kW level 2nd harmonic X-mode ECRH experiment in the TEXTOR tokamak and interpreted in terms of anomalous backscattering of the pump EC wave. The anomalous scattering of the X-mode in the ASDEX-Upgrade tokamak 2nd harmonic ECRH experiment has been recently reported as well [8, 9].

The attempts of theoretical explanation of these anomalous phenomena have been undertaken during recent years [10-13]. Several theoretical models demonstrating the possibility of excitation of the low-threshold PDIs which lead to generation of the electron and ion Bernstein waves had been proposed. The cornerstone of them is taking full account of a non-monotonous plasma density profile often observed in the ECRH experiments and originated due to different physical mechanisms, among which are features of plasma confinement in the magnetic island [14] or the electron pump-out effect manifesting itself as a consequence of the anomalous convective particle flux from the ECR layer at the intensive ECRH. The poloidal magnetic field inhomogeneity is a key element of the theory as well. Based on these two elements it was shown that two- or even three-dimensional (2D or 3D) trapping of either a daughter ion Bernstein wave (IBW) or a daughter electron Bernstein wave (EBW) due to a specific multidimensional plasma inhomogeneity can be responsible for excitation of several convective and absolute PDIs possessing thresholds drastically less than those predicted by the standard model.

These theoretical results expanded significantly the standard theory which ignored peculiarities of a density profile and geometry of toroidal devices. Though the developed theoretical models demonstrated the principal possibility of the low-threshold PDI excitation in the 2nd harmonic X mode ECRH experiments at toroidal devices, they couldn't describe the main features and details of the observed anomalous effects. Predicted fast parametric instabilities were shown to be convective. The most dangerous absolute PDIs were demonstrated to have the growth rates comparable with the inverse characteristic times of the MHD and drift processes.

The relative crisis in development of the theory was, apparently, overcome in [15, 16] pointing out to the possibility that the backscattering signal generation as well as the anomalous absorption in ion channel could be a consequence of secondary nonlinear process, that accompanies a primary low – threshold absolute PDI of a different nature. A hint to this primary instability is provided experimentally in [7] demonstrating the most intensive backscattering at the plasma density in the magnetic island slightly exceeding the UHR value for half the pump frequency. In [15, 16] it was realized that at these conditions the low-threshold pump wave decay into the two upper hybrid (UH) waves is possible.

In the present paper we review recent results of the two-plasmon PDI theory development [15-24]. We introduce a model taking into account, as distinct from the standard theory [1-3], the presence of a non-monotonous density profile, which always exist on the discharge axis or may be present due to the magnetic island or the density pump-out effect. We interpret the generation of backscattering signal and the anomalous ion heating, as a result of secondary nonlinear processes that accompany a primary low–threshold two–UH–plasmon PDI driven by the pump X-mode. The threshold of the primary PDI is shown to be smaller than the one predicted in [1] due to the UH wave trapping in radial direction [15,16]. It is also shown that in the presence of the finite-size pump beam the most dangerous absolute instability could be excited. The threshold of this instability is determined in hot fusion plasma not by the UH wave damping, but by diffractive losses along the magnetic field and could be as small as several tens kW. The two-plasmon decay growth rate is very high, in the range of 10^8 s^{-1}. The primary PDI growth enhancing the UH wave fluctuations from the thermal noise level is saturated in our model due to the cascade of secondary decays of the daughter UH wave that leads to excitation of the secondary UH and ion Bernstein (IB) waves [17-20]. The threshold of this spontaneous parametric frequency down-conversion can be easily overcome for the secondary UH waves, if their radial trapping is possible. The number of the consecutive cascades is limited by the violation of the trapping condition. A set of equations describing the cascade is derived and solved numerically. The results of numerical modeling are shown to be in agreement with analytical estimations of the growth rates of the initial and secondary parametric decays and with the saturation level [19, 20]. The coupling of different daughter UH waves is responsible in the model for generation of the backscattered X wave. This mechanism appears capable of reproducing the fine details of the frequency spectrum of the anomalously reflected X wave and of the absolute value of the backscattering signal observed at TEXTOR [18]. It also predicts substantial (up to 25%) anomalous pump absorption due to the primary and secondary UH wave generation and their absorption by electrons in the vicinity of the density local maximum. The obtained anomalous absorption efficiency is large enough to consider seriously the pump wave depletion, as additional mechanism of the PDI saturation. The performed calculation results in reduction of the predicted anomalous absorption from 25% to 22% at the microwave power launched into the plasma in TEXTOR experiment [21].

It is also shown that the low-threshold absolute two-plasmon decay parametric instabilities could be excited not only in magnetic islands, but in plasma blobs or filaments elongated in the direction of magnetic field. This effect is investigated in the model experiment at the linear plasma device “Granite”, where the low threshold excitation of the two-plasmon PDI leading to anomalous absorption higher than 90% is demonstrated [22].

Based on the developed approach we analyze a more general case when only one of the UH waves is trapped in the vicinity of the density maximum whereas the second daughter wave is able to leave this region. It is shown that under these conditions excitation of the absolute two-plasmon PDI is also possible [23, 24] and can lead to substantial anomalous absorption.

Based on the proposed model one could expect a high level of excitation of UH waves at the frequency close to the half pump wave frequency and therefore substantial emission of electromagnetic waves at this frequency due to the linear wave conversion in the UH resonance. The nonlinear coupling of the daughter UH waves with the pump could lead to the measurable level of the plasma emission at the 3/2 harmonic of the pump, as it happens in the laser driven inertial fusion experiments [25]. The secondary IB waves excited in the cascade mechanism of the two-plasmon PDI saturation directly transfer the pump power to the ion component which can explain the anomalous fast ion generation at TCV and TJ-II.

It should be noted that the trapped UH waves could be excited in the experiments on the O-mode ECRH at the fundamental harmonic as well. The threshold in this case is higher (several hundred kW depending on the plasma parameters) than for the X-mode pump whereas the growth rate is large enough (in the range of 10^7 s^{-1}) to expect the non-linear saturation of the instability [26].

The authors would like to acknowledge the financial support of the Russian Science Foundation project 16-12-10043 and of the Ioffe Institute.

References

1. M. Porkolab, B.I. Cohen Nucl. Fusion 1988 **28** 239
2. Cohen B I, Cohen R H, Nevins W M and Rognlien T D 1991 Rev. Mod. Phys. 63 949
3. Litvak A G, Sergeev A M, Suvorov E V, Tokman M D and Khazanov I V 1993 Phys. Fluids B 5 4347
4. D. Rapisarda et al., Plasma Phys. Control. Fusion 2007 **49**, 309
5. A. N. Karpushov et al. Proc. of 33rd EPS Conference on Plasma Physics, 2006 **30I** P–1.152
6. E. Westerhof, S. Nielsen, J. W. Oosterbeek et al. Phys. Rev. Lett. 2009 **103** 125001
7. S.K. Nielsen, M. Salewski, E. Westerhof et al. Plasma Phys. Control.Fusion 2013 **55** 115003
8. V. Furtula et al., The Review of scientific instruments **83**, 013507 (2012);
9. S.K. Nielsen et al. 2014 In Proc. of 9th Workshop “Strong Microwaves and Terahertz Waves: Sources and Applications” July 24-30 2014 Nizhny Novgorod
10. E.Z. Gusakov and A.Yu. Popov, Phys. Rev. Lett. **105**, 115003 (2010);
11. E.Z. Gusakov and A.Yu. Popov, Nucl. Fusion **51**, 073028 (2011);
12. E. Gusakov, A. Popov, EPL **99**, 15001 (2012);
13. E.Z. Gusakov, A.Yu. Popov, A.N. Saveliev, Plasma Phys. Control. Fusion **56**, 015010 (2014)]
14. M Yu Kantor 2009 *Plasma Phys. Control. Fusion* **51** 055002
15. A.Yu. Popov, E Z Gusakov J. Exp. Theor. Phys. **120** (2015) 147
16. A.Yu. Popov, E.Z. Gusakov 2015 *Plasma Phys. Control. Fusion* **57** 025022
17. A.Yu. Popov, E Z Gusakov J. Exp. Theor. Phys. **121**, (2015), 362
18. E Z Gusakov and A Yu Popov Physics of Plasmas **23**, (2016) 082503
19. E Z Gusakov and A Yu Popov Plasma Phys. Control. Fusion **59** (2017) 025005
20. E. Z. Gusakov, A. Yu. Popov, M. A. Irzak *JETP*, **123**, (2016) 723
21. E Z Gusakov and A Yu Popov submitted to Plasma Phys. Control. Fusion
22. L.V. Simonchik et al. 2017 In Proc. of 10th Workshop “Strong Microwaves and Terahertz Waves: Sources and Applications” July 17-22 2017 Nizhny Novgorod
23. A Yu Popov and E Z Gusakov EPL, **116** (2016) 45002
24. A. Yu. Popov and E. Z. Gusakov *JETP Letters*, **105**, (2017) 78
25. E Z Gusakov Sov. Tech. Phys. Lett. **3** (1977) 504
26. E Z Gusakov, A Yu Popov, A N Saveliev and E V Sysoeva Plasma Phys. Control. Fusion, **59**, (2017) 075002

Optics for electron cyclotron resonance heating and collective Thomson scattering at the stellarator W7-X

W. Kasparek[1], V. Erckmann[2], H.P. Laqua[2], T. Stange[2], M. Weißgerber[2], C. Lechte[1], B. Plaum[1], D. Moseev[2], F. Leipold[3], M. Petelin[4], K.J. Brunner[2], H. Braune[2], S. Marsen[2], N. Schneider[2], R.C. Wolf[2], and the W7-X team[2]

[1]Institute of Interfacial Process Engineering and Plasma Technology, U. Stuttgart, Germany, kasparek@igvp.uni-stuttgart.de
[2]Max-Planck Institute of Plasma Physics (IPP), Garching and Greifswald, Germany
[3]Technical University of Denmark, Kgs. Lyngby, Denmark
[4]Inst. of Applied Physics, Russian Academy of Science, Nizhny Novgorod, Russia

In December 2015, the superconducting stellarator Wendelstein-X (W7-X) [1] became operable. During the first operational phase (OP 1.1) electron cyclotron resonance heating (ECRH) was the exclusive heating method [2], and it will remain the dominating heating system in the next operational campaign (OP 1.2). At present, 9 out of 10 1-MW gyrotrons running at 140 GHz are operable. As a versatile technique, ECRH has to drive the plasma start-up, help with wall conditioning, and provide heating and current drive, making optimum use of the spatially localized power deposition of the millimetre waves. Movable launchers allow tailoring the radial profiles of the electron temperature and the rotational transform profiles. A sophisticated control system [3] is used to manage the operation of the complete system. In addition, the system can be used for collective Thomson scattering experiments.

This paper will provide a brief review on the optical transmission system, describe its performance and first experiences, and will concentrate on new developments like in-vessel grating reflectors, Remote-steering (RS) antennas used as launchers and receivers for ECRH and collective Thomson scattering (CTS) [4], as well as possible upgrades.

System overview

Fig. 1 shows a sketch of the complete ECRH system. It consists of 10 gyrotrons with a total power close to 8 MW, CW operation. Transmission to the plasma is performed fully optically via water-cooled reflectors installed in an underground beam duct. The beams of the gyrotrons are conditioned with individual matching optics units, and transmitted in common via two multi-beam waveguides (MBWG) [5] to the experimental hall. Beam dividing optics (BDO) at the end of the MBWGs distribute the beams to 10 front-steering launchers.

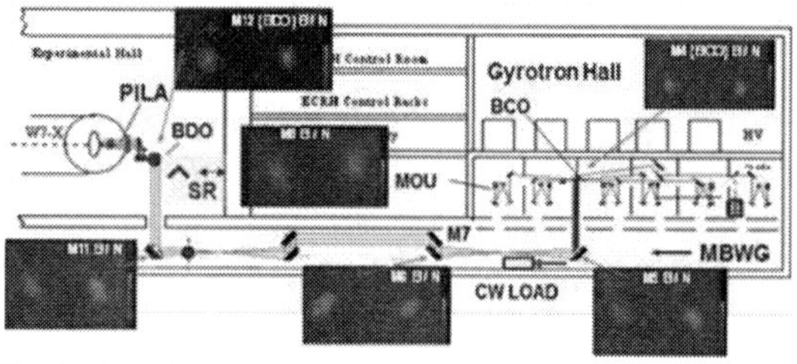

Fig. 1. Overview on the W7-X ECRH transmission system with thermal images taken on the MBWG reflectors showing beam positions relative to four markers each, as well as the beam shape on various reflectors.

In addition, two remote-steering launchers (RSL) [6] have been installed in a toroidal position where magnetic field gradient is low to study advanced current drive scenarios for example. One RSL will also be used for CTS.

Transmission characteristics

The alignment of the (in total about 150) reflectors of the transmission system and the launchers was performed by thermal imaging of the gyrotron beams on the reflector surfaces. For the MBWGs, which provide imaging from the beam combiner optics (BCO) at the input to the BDOs at the output, the alignment could be done with two beams only; all others are automatically aligned. Thermal images at various positions on the reflectors are shown in Fig. 1.

Besides ten calorimeters assigned to each gyrotron, the system employs two CW calorimetric loads installed near to the inputs of the MBWGs. By inclining one of the BCO reflectors, the respective beam can be directed into the load for power measurements. For measurements of the transmission efficiency, a retro-reflector SR can be inserted at the end of the MBWG (Fig. 1), and the backward travelling power can be directed into the load. Thus, by comparison of the direct and the power guided via the MBWG, a measurement of the transmission efficiency is possible, with the result shown in Fig. 2. From this and further measurements we conclude, that the efficiency of the MBWG itself is 0.972, and the total efficiency from gyrotron to the stellarator is >0.93.

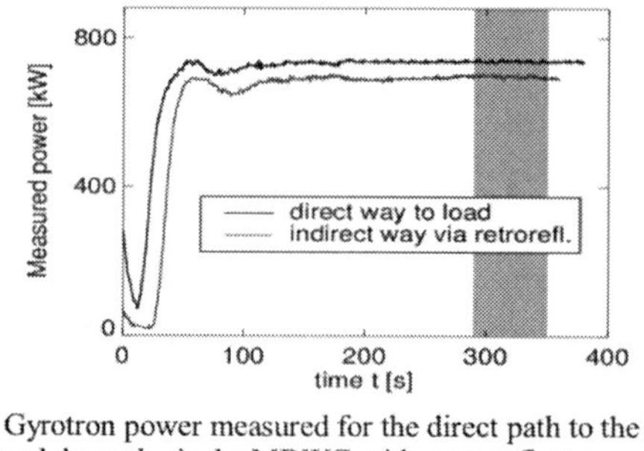

Fig. 2. Gyrotron power measured for the direct path to the load and the path via the MBWG with retro-reflectors

Main launchers and in-vessel reflectors

The ECRH power is usually injected via 10 front-steering launchers with 2-axes movable water-cooled reflectors. They are described in [1]. The calibration of the launch directions was performed with thermal targets installed in the vessel. The typical steering range is –

$8° < \varphi < +35°$ in toroidal, and $-30° < \theta < +30°$ in poloidal direction, giving large experimental flexibility.

For heating scenarios, with incomplete absorption of the ECRH beams (O2 and X3 heating), reflectors are installed on the inner wall opposite to the laun-chers, such that the remaining power is directed back to the plasma in a controlled way and further to areas which are insensitive to millimetre waves. Where necessary, the reflectors were designed as gratings with high efficiency in any polarisation [7], with the goal to keep the mode injected into the plasma.

Remote steering launchers

For investigations on trapped particles in W7-X, as demonstrator for reactor-compatible ECRH launchers, as well as for CTS experiments, two ports of W7-X have been equipped with waveguide RS launchers (RSL1 and RSL5) [6]. The remote-steering properties are based on multi-mode interference in a square waveguide leading to imaging effects: For a proper length/width of the waveguide, a microwave beam at the input of the waveguide (with a defined direction set by a mirror system outside of the plasma vacuum) will exit the waveguide (near the plasma) in the same direction.

The two RS launchers are fed via switches from the main transmission system. Main features include: (i) vacuum window at the entrance, (ii) 4.6 and 5.2 long corrugated square waveguides folded by mitre bends to fit the launchers into the W7-X structure, (iii) optimized injection of the input beam for maximum angular steering of the beam launched to the plasma. The usable steering range is at least $-15° < \varphi < +15°$. A detailed description is given in [6].

The two launchers have been installed and aligned for operation in the next campaign (OP1.2.). High-power patterns recorded in the plasma vessel for various steering angles are shown in Fig. 3. Furthermore, RSL1 can deliver the probing beam for CTS experiments [5], measuring the ion temperature profile.

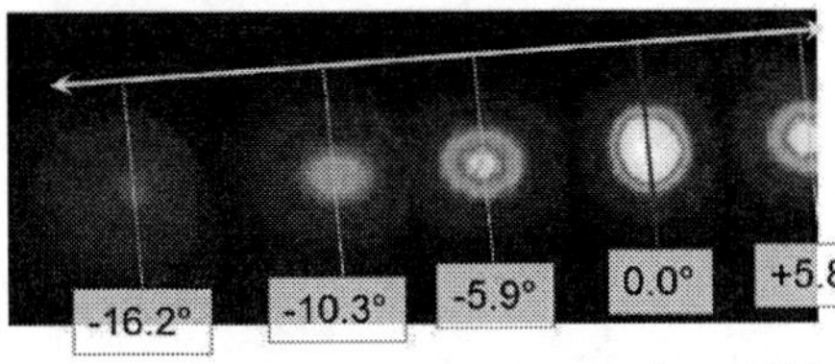

Fig. 3. High-power antenna patterns from RSL1

An optical remote-steering antenna for CTS

As receiver antenna for the CTS system, an optical remote-steering antenna was designed and built (Fig. 4). In contrast to the waveguide RSL, this antenna is broadband. Like the MBWG, it is based on the imaging properties of two confocal optics installed in series, and consists of four focusing reflectors in Z-arrangement, with surface parameter p = 0.3 [5]. Characterisation of this antenna using thermal imaging provides a beam with good quality in the angular range $-15° < phi < 15°$. Only little degradation is observed towards the limit of the steering range.

The scattered signal from this antenna is transmitted via an additional channel of the main transmission system – sharing the MWBG section with the gyrotrons – up to a spare gyrotron box, where the CTS receiver is installed.

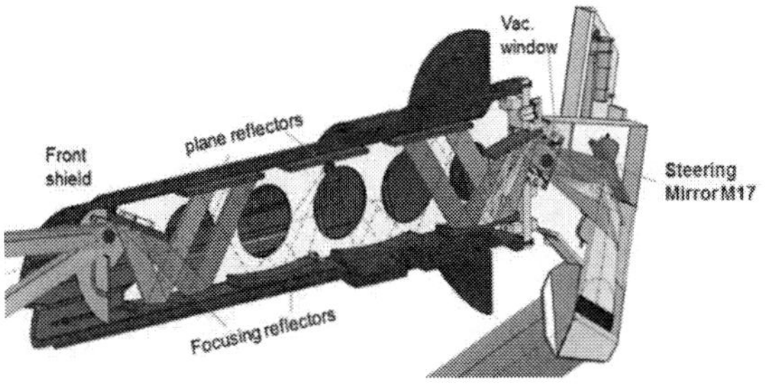

Fig. 4. Design of the optical remote-steering receiver antenna. Note the imaging of the beam pivot near the plasma to the beam pivot in the plane of the vac. window (red dots)

Resonant Diplexer as switch for ECRH and spectral/spatial filter for CTS

As a possible upgrade to ECRH and CTS, the integration of a resonant diplexer [8] is discussed. Such a device could replace the mechanical mirror switches between the main and RS launchers, and could improve the spectral purity of the gyrotron used for CTS, if necessary. A concept is based on a folded mirror resonator, which could be installed near the entrance of the MBWG or the intermediate image plane after M7 (Fig. 1).

Conclusion

The ECRH system for W7-X is characterised by high performance, and provides large experimental flexibility. The present upgrades for CTS experiments enhance the diagnostic capabilities of W7-X. An overview of past and upcoming experiments is presented in [2].

References

1. *V. Erckmann, et al.*, Electron Cyclotron Heating for W7-X: Physics and Technology // Fusion Sci. Technol. 2007. V.52, No.2. P. 291-312.

2. *J. Brunner et al.*, Continuous High-Power Microwave Heating at the W7-X Stellarator // this conference.

3. *H. Braune et al.*, Enhancements of the W7-X ECRH facility with respect to the next experiment campaign OP1.2 // this conference.

4. *D. Moseev et al.*, Design and commissioning of the collective Thomson scattering diagnostic at Wendelstein 7-X // 18th Laser-aided Plasma Diagnostics, Prague, Sept. 2017.

5. *L. Empacher and W. Kasparek*, Analysis of a Multiple-Beam Waveguide for Free-Space Transmission of Microwaves // IEEE Trans. Antennas Prop. 2001. V.49, No.3. P.483-493.

6. *B. Plaum et al*, Design of a remote-steering antenna for ECRH heating in the stellarator Wendelstein 7-X. // Fusion Engineering and Design 2015. V.96–97. P.568–572.

7. *H. Höhnle et al.*, Extension of the ECRH operational space with O2 and X3 heating at ASDEX Upgrade // Nucl. Fusion, 2011. V.51 No.8, 083013.

8. *Z. Wu et al.*, Design and Characterization of a 170-GHz Resonant Diplexer for High-Power ECRH Systems. IEEE Trans. Microwave Theory Techn. 2015. V.63, No.10. P.3537–3546.

EPJ Web of Conferences **149**, 01014 (2017) DOI: 10.1051/epjconf/201714901014
10th International Workshop 2017 "Strong Microwaves and Terahertz Waves: Sources and Applications"

Relativistic microwave oscillators with high power flux in a free space and interaction zone

M.I. Yalandin[1], M.S. Pedos[1], V.V. Rostov[2], I.V. Romanchenko[2], S.N. Rukin[1], K.A. Sharypov[1], S.A. Shunailov[1], and M.R. Ulmaskulov[1]

[1]Institute of Electrophysics, UB RAS, Ekaterinburg, Russia, yalandin@iep.uran.ru
[2]Institute of High-Current Electronics, SB RAS, Tomsk, Russia

In the recent two decades, high-power microwave (HPM) generators producing short Ka-band and X-band pulses were intensively investigated (see Refs. [1–6] and citation therein). For generation of HPM pulses with duration of units of nanosecond and less influence of microwave breakdowns of electrodynamics systems, expansion of cathode and collector plasmas, and electron beam structure distortions becomes insignificant. Though the pulse duration is short, the radiating energy rises if coherent summation of radiation from several generators is provided. When the power in each channel is high, the wave beam could be formed where power flux density is limited only by breakdown strength of the media. Using approaches of the phased arrays, the directional pattern (DP) of summarized radiation could be steered. This is achievable from pulse to pulse, as well as during a single nanosecond pulse if summarized wave beams have a shifted frequencies. The related experiments with Ka-band microwave pulses are presented below.

Super-powerful microwave pulses containing about ten oscillations could be formed in the mode of Cherenkov superradiance (SR) of electron flux propagating in elongated slow-wave structure (SWS) [1]. As a result of a single-pass, high gain amplification of initial electromagnetic (EM) perturbation, the power of SR pulse may exceed the power of the driving beam [2]. This is also because of SR pulse is shorter than the electron beam (in space) and less than its propagation time throughout the entire SWS. If the interaction with following slow wave continuous, the beam energy transfer to SR occurs due to the wave-to-particles slippage [1]. For the case of counter wave (backward synchronous harmonic) SR pulse accumulates the energy from incoming "fresh" electrons [3].

Beam-to-wave power conversion factor $K = (1–1.4)$ was obtained for relativistic superradiance Ka-band backward-wave oscillators (BWOs) [1—3]. In the recent experiments [4], voltage pulses (Fig. 1; a, b) applied to explosive electron emission (EEE) cathode formed electron beams with power of $\sim$ 2–3 GW. This was sufficient to excite a 29-GHz SR pulse (Fig. 1, c) with $K\sim1$. It was revealed that in the presence of high electric fields at the SWS wall ($E \approx 2$ MV/cm) arise breakdowns. However, rf power limitation was delayed enough to ensure SR pulse passage throughout the entire SWS. Electric field of ~1 MV/cm was achieved in quasi-stationary (QS) BWO having the SWS two times shorter than SR oscillator. This QS BWO generated 2-ns pulse with power of 0.6 GW.

Figure 2 demonstrates, that both SR and QS oscillators have (from shot to shot) generation phases linked to the voltage pulse front (Fig. 1, a) at the point of maximum derivative by time. Obviously, the faster rise time of the beam current (that correlates with the voltage front steepness), the stronger a wideband, phase-imposing seed EM perturbation is formed, when the beam is injected into SWS.

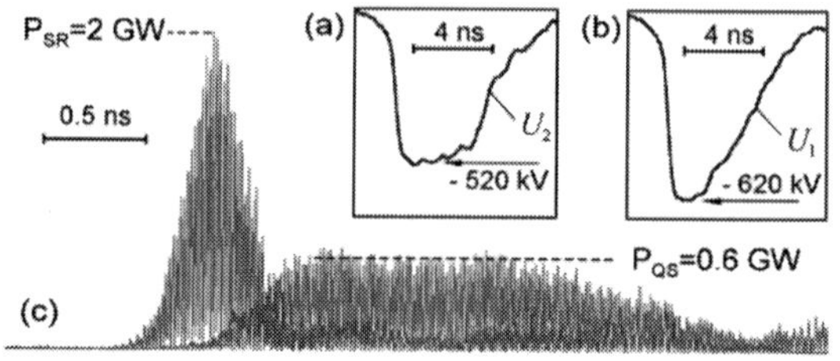

Fig. 1. (a, b) Voltage pulses fed e-beams with power of 2 GW and 3 GW, respectively. (c) Time referenced SR and QS pulses formed by BWOs in the case of the voltage (a)

One can see (Fig. 2) that SR oscillator demonstrates a fewer phase spread as compared to QS BWO. Apparently, this is the result of shortened "transient" time for a single-pass amplification of a seed EM signal by SR mechanism (Fig. 1, b) when initial current perturbation does not include information about the further voltage pulse form and/or its amplitude deviations. Besides, delayed achievement of the Cherenkov interaction condition is more probable for elongated SR-SWS where a seed EM perturbation comes back and meets the beam front with electrons, which already attained proper interaction energy.

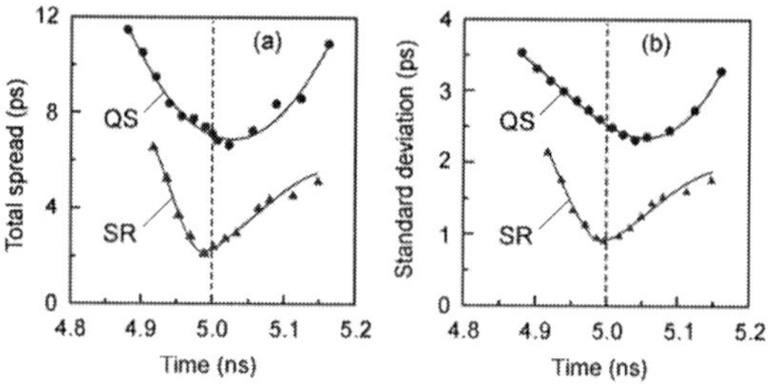

Fig. 2. Total phase spread (a) and corresponding standard deviation (b) for rf pulses of QS BWO and SR BWO depicted in Fig. 1 (c). Period of microwave oscillations is ~34 ps. Time "5 ns" corresponds to the maximum of time derivative of the voltage pulse front in Fig. 1 (a)

Stability of voltage pulses feeding EEE cathodes has less priority for the phase synchronization between several HPM generators. In the first place the pulses should be identical. This was proven (see, e.g., Ref. [5] and citation therein) when in-phase operation of multi-channel BWOs (X-band and Ka-band) was provided by a split of a single voltage pulse. Figure 3 (a) shows the radiation pattern of 2-D (2x2) array of 38-GHz, 0.6-GW SR BWOs. In the case of in-phased summation, power density in the pattern maximum achieves the value, which is equivalent to a single source with power of $0.6\times4^2\approx10\,\text{GW}$. A 200-ps width makes SR pulses insensitive to incomplete rf breakdowns of an air near vacuum windows (Fig. 3; b, c). Rearrangement of radiator's array for quasi-stationary generation (~200 MW, 3-ns FWHM) when frequency of a certain channel is shifted [6], transfers the radiation summation into waves beating, and this blurs the radiation pattern (Fig. 3; d). With that, the power flux in the main antenna direction during a single pulse varies from zero to squared maximum as much times as it defined by the frequency detuning.

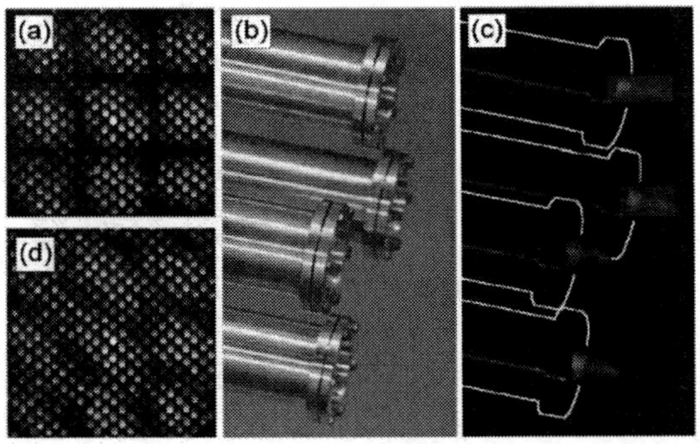

Fig. 3. DP of four-SR-BWO array reproduced at a single shot by a matrix neon-valve panel: in-phase operation (a); a frequency shift in one channel (d); (b) – Four horns; (c) – Breakdowns of the air near vacuum windows

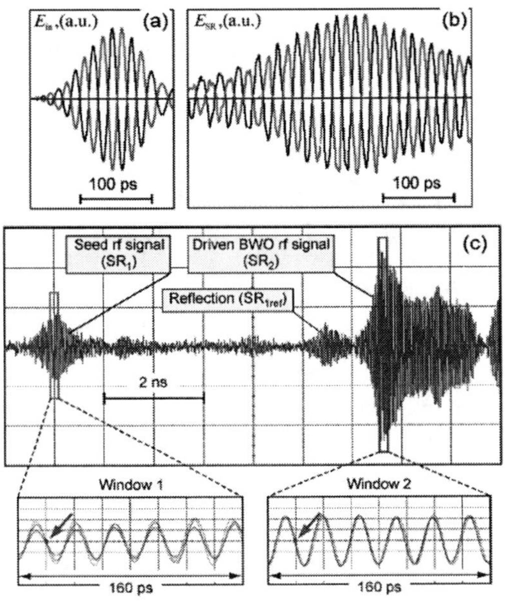

Fig. 4. (a) – Seed rf pulses (30 kW) with counter phases (0-black; π – orange) and (b) phased response of driving 400-MW SR pulse (PIC-simulation). (c) – Experiment on phase-imposed SR excitation by an external seed rf signal

Phase-imposing BWO excitation occurs when ultra-narrow external rf signal power-level overshoot the perturbation produced by the beam front. With that, excitation of auto-oscillator doesn't move to conventional amplification regime. Particle-in-cell simulation (Fig. 4; a, b) demonstrates such capability of phase imposing. Experiments were performed using two precisely synchronized electron accelerators RADAN [7] where SR BWO served as a master oscillator. Both quasi-stationary and superradiative Ka-band BWOs were explored in this regime as a driven oscillator. It was confirmed that phases of the seed and the initiated Ka-band pulses correlated within units of picoseconds on a time scale (Fig. 4; c) for the power ratio between them minimized down to -35 dB. It is important that driven oscillators were excited by seed signal earlier than in a free-running regime and at its own frequency regardless to the imposing frequency detuning up to 5% (compare windows #1 and #2 in Fig. 4, c).

In conclusion, the above described results demonstrate high radiation power density in the interaction space of an ultra-short-pulse oscillators as well as steering the pattern of coherently operating HPM generators.

The work was supported by the Russian Science Foundation (Grant No. 16-19-10312).

References

1. *Ginzburg, N.S., Zotova, I.V., Cross, A.W., Phelps, A.D.R., Yalandin, M. I., Rostov, V. V.* Generation, Amplification and Nonlinear Self-Compression of Powerful Superradiance Pulses // IEEE Trans. on Plasma Sci. 2013. V. 41, No. 4. P. 646-660.

2. *Eltchaninov, A.A., Korovin, S.D., Rostov, V.V., Pegel, I.V., Mesyats, G.A., Rukin, S.N., Shpak, V.G., Yalandin, M.I., Ginzburg, N.S.* Production of short microwave pulses with a peak power exceeding the driving electron beam power // Laser and Particle Beams. 2001. V. 21, No. 2. P. 187-196.

3. *Eltchaninov, A.A., Korovin, S.D., Mesyats, G.A., Pegel, I.V., Rostov, V.V., Shpak, V.G., Yalandin, M.I.* Review of studies of superradiative microwave generation in X band and Ka band relativistic BWOs (review) // IEEE Trans. on Plasma Sci. 2004. V. 32, No 3. P. 1093-1099.

4. *Rostov, V.V., Romanchenko, I.V., Pedos, M.S., Rukin, S.N., Sharypov, K.A., Shpak, V.G., Shunailov, S.A., Ul'masculov, M.R., Yalandin, M.I.* Superradiant Ka-band Cherenkov oscillator with 2-GW peak power // Physics of Plasmas 2016. V. 23, No. 9. P. 093103(1-4).

5. *Ginzburg, N.S., Cross, A.W., Golovanov, A.A., Phelps, A.D.R., Romanchenko, I.V., Rostov, V.V., Sharypov, K.A., Shpak, V.G., Shunailov, S.A., Ul'masculov, M.R., Yalandin, M.I., Zotova, I.V.* Coherent Summation of Emission From Relativistic Cherenkov Sources as a Way of Production of Extremely High-Intensity Microwave Pulses // IEEE Trans. on Plasma Sci. V. 44, No. 4. P. 377-385.

6. *Yalandin, M.I., Sharypov, K.A., Pedos, M.S., Romanchenko, I.V., Rostov, V.V., Rukin, S.N., Ul'maskulov, M.R., Shpak, V.G., Shunailov, S.A.* Multichannel Ka-Band Microwave Oscillator Based on Frequency-Shifted Relativistic Backward-Wave Oscillators // Radiophys. Quantum Electron. 2017. V. 59, No. 8-9. P. 629-637.

7. *Mesyats, G.A., Korovin, S.D., Rostov, V.V., Shpak, V.G., Yalandin, M.I.* The RADAN Series of Compact Pulsed Power Generators and Their Applications // Proceedings of the IEEE 2004. V. 92, No. 7. P. 1166-1179.

Generation of single and periodically repeated powerful ultrashort microwave pulses

N.S. Ginzburg, I.V. Zotova

Institute of Applied Physics RAS, Nizhny Novgorod, Russia
ginzburg@appl.sci-nnov.ru

Generation of powerful ultrashort microwave pulses (USP) is of interest for a large number of practical applications, including radars with high resolution, plasma diagnostics, spectroscopy etc. In order to achieve this purpose many methods well-known in laser physics can be implemented in microwave electronics.

In the last decade significant progress was achieved in generation of microwave USPs based on classical superradiance (SR) of extended electron bunches. Using of this effect a new type of pulse sources has been developed, which are capable of producing uniquely short (under 0.2–1 ns) electromagnetic pulses at super-high peak powers exceeding 1 GW in the millimeter waves (Ka-band) and 3 GW in the centimeter waves (X-band) with a repetition frequency up to several kilohertz [1]. A natural continuation of this research is related with the promotion of SR sources in shorter wavelengths ranges, where the oversized waveguides should be used. In the first part of this paper the results of the experimental observation of SR emission with excitation of a surface wave in the oversized periodically corrugated slow wave structures are presented. The experiments were based on a theoretical consideration in the frame of the quasi-optical approach.

A periodic train of powerful ultrashort microwave pulses with a repetition frequency of 10–100 MHz can be generated in electron oscillators with a saturable absorber installed in the feedback loop [2]. The mechanism of pulse formation in this case is similar to the passive mode-locking [3] in laser physics where such pulses are also treated as dissipative solitons. The second part of this paper is devoted to the theoretical analysis of solitons generation in a mode-locked two-section microwave oscillator consisting of Cherenkov TWT and a Kompfner absorber in the feedback loop.

Generation of Sub-THz SR Pulses Based on Excitation of Surface Waves in Oversized Waveguides

Cherenkov SR of electron bunch exciting the surface wave in an oversized corrugated cylindrical waveguide (Fig. 1a) can be considered within a quasi-optical approach [4]. In this case the radiation field near a shallow corrugation is presented as a sum of two counter-propagating *TM* polarized wave-beams:

$$H_\varphi = \mathrm{Re}\left(A_+(z,r,t)e^{i\bar{\omega}t-ikz} + A_-(z,r,t)e^{i\bar{\omega}t+ikz}\right), \quad (1)$$

propagation and mutual coupling of which is described by two non-uniform parabolic equations. The synchronous interaction of electrons with a forward partial waves leads to development a self-bunching and formation of powerful SR pulse.

Simulations show that the most optimal conditions for SR emission correspond to excitation of the backward surface wave near the Bragg frequency (π- regime, Fig.1b). For parameters of an electron bunch formed by an accelerator RADAN (electron energy of 300 keV, a total current of 2 kA, a bunch duration of 500 ps) and a corrugated waveguide with the mean radius of 3.75 mm, corrugation period of 0.825 mm, and corrugation depth of 0.36 mm the operating frequency in the resonant point is of 0.14 THz ($2r_0/\lambda \approx 3.5$). In this case the power of generated SR pulse emitted in $-z$ direction achieves ~200 MW for pulse duration of ~200 ps (Fig. 2a). As it is seen in Fig. 2b the instant spatial structure of the partial wave corresponds to formation of the evanescent surface wave with the field amplitude exponentially decaying from the corrugation.

Based on a theoretical analysis, experiments on observation of the sub-terahertz SR pulse generation were carried out in IEP RAS (Ekaterinburg). Photo of the experimental set-up is shown in Fig. 3. A typical oscilloscope trace of generated SR pulses with a duration of about 150 ps and a rise time of 100 ps reconstructed in the "power-time" coordinates is presented in Fig. 3c. Frequency measurements using a set of cut-off waveguide filters show that the pulse spectrum has a central frequency in the interval 0.13-0.15 THz. The peak power of generated SR pulses was estimated by integrating the detector signal over the directional pattern and achieved of 50-70 MW, that strongly exceeds the value obtained in the previous sub-terahertz experiments [5] with single-mode waveguides.

Generation of a Train of Dissipative Solitons in a Cherenkov Oscillator with a Kompfner Absorber in the Feedback Loop

In microwave electronic the passive mode-locking effect can be realized in a two-section oscillator which consists of an amplifying and an absorbing section Fig. 4a. As an amplifier we consider here a Cherenkov TWT which is of interest for generation of multi-gigawatt microwave pulses. A saturable absorption can be realized also based on Cherenkov TWT operating in the so-called Kompfner dip regime, when the em wave overcomes the electrons [6]. In this case both sections of an USP oscillator are described by following equations:

$$\left(\frac{\partial}{\partial z} + s\frac{\partial}{\partial \zeta}\right)\hat{A} = \frac{\omega}{c}\frac{eI_0\chi}{mc^3\gamma_0 N}\frac{1}{\pi}\int_0^{2\pi} e^{-i\theta}d\theta_0,$$
$$\frac{\partial u}{\partial z} = \frac{\omega}{c}\mathrm{Re}\left(\chi\hat{A}e^{i\theta}\right), \quad \frac{\partial\theta}{\partial z} = \frac{\omega}{c}(\Delta + \mu u), \quad (2)$$

where $\zeta = t - z/V_{\|0}$, $\hat{A} = eA/mc\omega$ is the normalized field amplitude, χ is the coupling parameter, $u = (1-\gamma/\gamma_0)$ is the relative energy variation, $\mu > 0$ is the inertial grouping parameter, $\Delta = 1 - V_{\|0}/V_{ph}$ is the initial resonance detuning, Regime of amplification realizes for $\Delta < 0$, while the regime of Kompfner dip - for $\Delta > 0$. In accordance with Fig. 4a after passing through the absorber and reflection from the partially transparent mirror, the signal returns to the input of the amplifier with a time delay $T : A(0,\zeta) = \ = RA(L_1 + L_2, \zeta - T)$, where $R < 1$ is the coefficient of reflection.

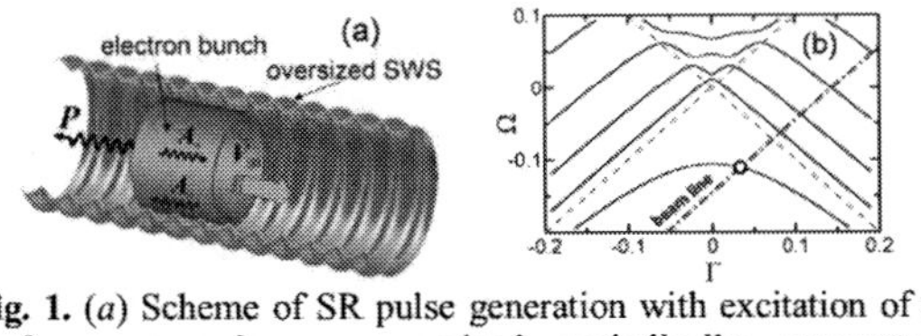

Fig. 1. (*a*) Scheme of SR pulse generation with excitation of a surface wave in an oversized periodically corrugated waveguide. (*b*) Dispersion characteristics of a corrugated waveguide and an electron beam

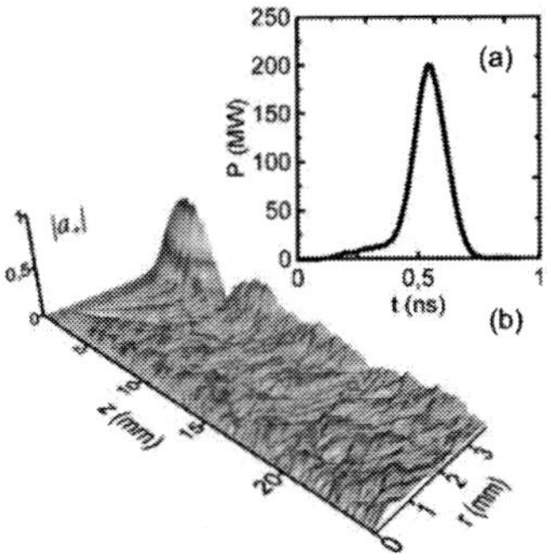

Fig. 2. SR emission with excitation of the backward surface wave: (*a*) generated SR pulse, (*b*) the structure of the forward partial wave

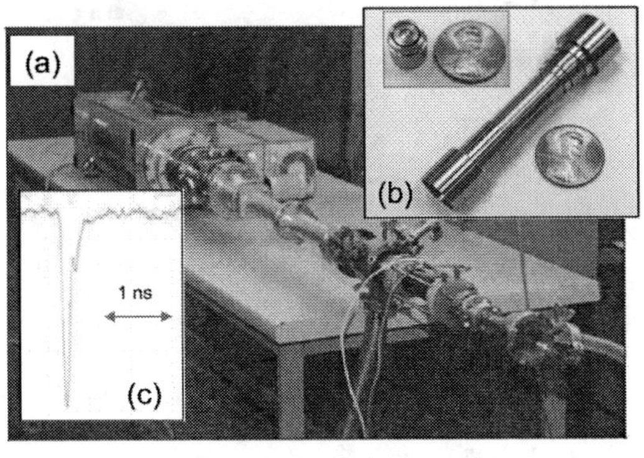

Fig. 3. Photo of the experimental set-up (*a*), a corrugated waveguide and a coaxial reflector (*b*) used for observation of superradiance with excitation of a surface wave. (*c*) Oscilloscope trace of the 0.14 THz SR pulse with duration of 150 ps and peak power up to 70 MW

The important factor for effective amplification of ultrashort em pulses is the difference of the slippage parameter $s = V_{gr}^{-1} - V_{\|0}^{-1}$ from zero. In this case due to slippage of an em pulse over electrons it can effectively accumulates the energy from different beam fractions. On the contrary, in the absorbing section for minimization of the absorber relaxation time the grazing incidence of the electron beam and the em wave is beneficial $s = 0$.

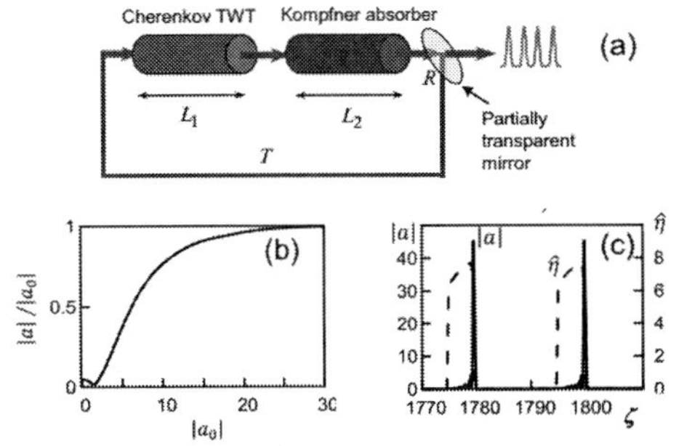

Fig. 4. (*a*) Scheme of a mode-locked microwave oscillator consisting of a Cherenkov TWT, a Kompfner absorber and a feedback loop. (*b*) Dependence of a transmission coefficient of the Kompfner absorber on the input signal amplitude demonstrating the saturation of absorption. (*c*) The output signal and instantaneous efficiency of the USP mode-locked oscillator.

Simulations were performed for following parameters of an 10 GHz amplifier: electron current of 6 kA, electron energy of 0.8 MeV, the coupling impedance of 2 Ohm. In the second section with similar parameters the Kompfner dip with saturation of absorption (Fig. 4b) can be realized by the proper choice of the period and length of a slow wave structure. Installation of such a section in the feedback loop of an electron oscillator leads to realization of passive mode-locking effect and generation of a train of ultrashort 0.8 ns pulses with peak power up to 6 GW. Based on an analogy with laser physics such pulses can be interpreted as dissipative solitons.

This work was supported by Russian Scientific Foundation (RSCF), grant No. 16-42-01078.

References

1. *Korovin, S., Eltchaninov, A., Rostov, A et al.* Generation of Cherenkov superradiance pulses with a peak power exceeding the power of driving short electron beam // Phys.Rev.E 2006. V.74 P. 016501

2. *Ginzburg, N.S., Denisov, G.G., Vilkov, M.N. et al.* Generation of "gigantic" ultra-short microwave pulses based on passive mode-locking effect in electron oscillators with saturable absorber in the feedback loop // Phys. Plasmas. 2016. V. 23. P. 050702-1-4.

3. *Haus, H.A.* Mode-Locking of Lasers // IEEE J. of Select. Topics in Quant. Electron. 2000. V. 6, No. 6. P. 1173-1185.

4. *Ginzburg, N.S., Malkin, A.M., Sergeev, A.S. et al.* Generation of Sub-Terahertz Superradiance Pulses Based on Excitation of a Surface Wave by Relativistic Electron Bunches Moving in Oversized Corrugated Waveguides // Phys. Rev. Lett. 2016. V.117 P.204801

5. *Yalandin, M.I., Shpak, V.G., Shunailov, S.A., et al* Generation of powerful subnanosecond microwave pulses in the range of 38–150 GHz// IEEE Trans. on Plasma Sci. 2000. V.28, Iss.5 P.1615

6. *Kompfner R.* On the operation of the traveling wave tube at low level // J. Brit. IRE. 1950. V. 10, No 8-9. P. 283

EPJ Web of Conferences **149**, 02001 (2017) DOI: 10.1051/epjconf/201714902001
10th International Workshop 2017 "Strong Microwaves and Terahertz Waves: Sources and Applications"

Express *in-situ* measurement of single crystal diamond growth/etching rate in microwave plasma: how to perform multiparametric kinetics study in one working day

V.G. Ralchenko[1,2,3], V.Yu. Yurov[1,2], E.V. Bushuev[1], A.P. Bolshakov[1,2], E.E. Ashkinazi[1,2], I.A. Antonova[1,2], E.V. Zavedeev[1], A.A. Khomich[1,4], V.I. Konov[1,2]

[1]General Physics Institute RAS, Moscow, Russia, vg_ralchenko@mail.ru
[2]National Research Nuclear University MEPhI, Moscow, Russia
[3]Harbin Institute of Technology, Harbin, P.R. China
[4]Institute of Radio Engineering and Electronics RAS, Fryazino, Russia

The ability to precisely control and stabilize the growth process of single crystal (SC) diamonds by a microwave plasma chemical vapor deposition (MPCVD) is of paramount importance for achieving high uniformity and quality of the produced crystals for high-tech applications. The conventional approach to experimentally optimize the process parameters relays on preparation of numerous SC samples with ex-situ evaluation of growth rate, that is a tedious and time-consuming procedure, often taking several weeks or months. Here we describe a very effective technique to measure *in situ* with high accuracy the growth rate of SC diamond using a low-coherence interferometry (LCI) [1, 2]. The key advantage of the method is a possibility to use only one substrate to deposit consecutively layer-by-layer under variable conditions and determine kinetics for each of them.

Growth

The epitaxial SC diamond films were deposited in ARDIS-100 MPCVD system [2] on (100)-oriented Ib HPHT diamond substrates in H_2-CH_4 gas mixtures in a broad range of CH_4 concentrations (1–13%) and substrate temperatures T_s (750–1150 °C).

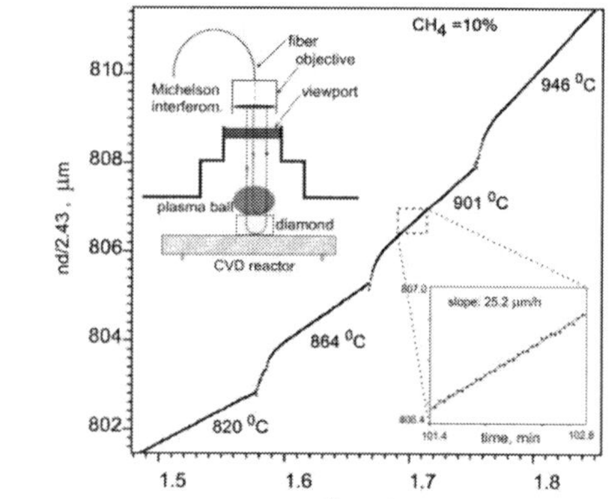

Fig. 1. Evolution of epilayer thickness with time under consecutive rise in steps of substrate temperature at 10% CH_4 content as measured with LCI. The temperature is indicated for each segment. The linear segments correspond to the growth with constant rate. Inset: the optical scheme of LCI measurement of the sample thickness

The beam of the low-coherence infrared radiation from a superluminescent diode (SLD) at λ_0 = 1560 nm wavelength with bandwidth $\Delta\lambda$ = 30 nm, was directed perpendicularly to diamond sample surface through an objective of Michelson interferometer positioned above the top quartz window of the CVD chamber. The LCI system is able to measure the temporal evolution of optical thickness $nd(t)$ of the etched sample (Fig. 1), where n is refraction index of diamond, and d is the sample thickness. Typically, less than 10 min growth run was required to get enough data points to evaluate the growth rate with high accuracy at given temperature, then switching to the next T_s value.

Rich growth kinetics is collected in a single experiment by depositing about 60 layers on diamond substrate in different regimes without switch-off the plasma. The plot for temperature dependent growth rate G(T) is shown in Fig. 2 for different CH_4 percentages. The growth rate as high as 82 µm/h is achieved by optimizing the temperature and gas composition without adding nitrogen in the gas. In certain cases, the deposition of a layer as thin as 100 nm was enough to determine the growth rate. The estimated accuracy limit of ≈ 0.4% in the G(T) measured is due the substrate temperature fluctuations and drift. The growth rate is found to follow Arrhenius dependence with activation energy E_a = 11.1 ± 1.0 kcal/mol (≈ 46 kJ/mol), in the middle of the range of 7.4–15.2 kcal/mol reported by Maeda et al. [3].

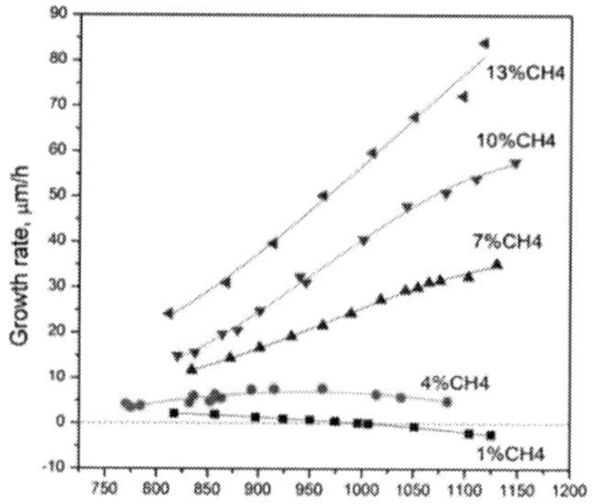

Fig. 2. Dependence of growth rate on substrate temperature G(T) at different CH_4 contents in gas mixture. Note the net etching for 1% CH_4 at T_s > 1000 K

Raman and photoluminescence (PL) spectra taken on polished cross-section of different layers, confirmed high quality of the material (Fig. 3). The diamond Raman peak width (WFHM) was essentially constant, $\Delta\nu$ = 2.4 ± 0.1 cm^{-1} through all epilayers ($\Delta\nu$ = 2.3 cm^{-1} was measured for a reference sample, natural IIa diamond), and only a small contamination of Si impurity detected by PL (SiV center at 738 nm).

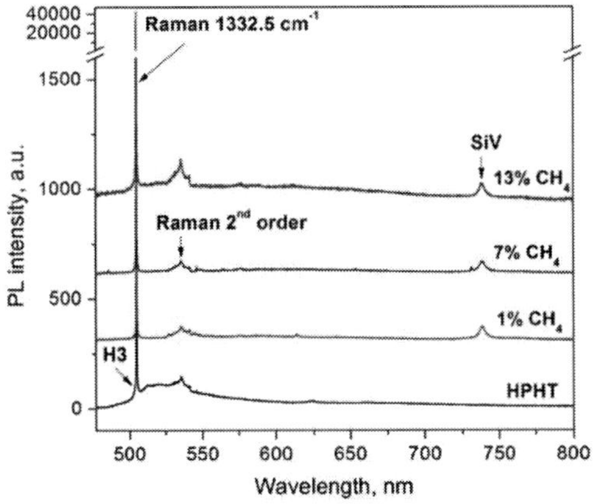

Fig. 3. PL spectra for diamond layers grown at 1%, 7% and 13% CH_4 in gas and for HPHT substrate. The spectra are zoomed to enhance the PL features

Etching in hydrogen plasma

Similarly, the etch rates in pure H_2 plasma for two sorts of synthetic diamond, (100)-oriented IIa HPHT diamond, and CVD plates, were studied *in situ* using LCI. The etching process was performed at pressure p=130 Torr, flow rate of 450 sccm, and substrate temperatures T_s = 800–1370°C, the latter being tuned by the microwave power P in the range of 2.9–4.0 kW.

Table. Diamond SC samples for etching: θ is off-angle for (100) plane, [N] is nitrogen impurity concentration.

Sample #	Size, mm	θ, degree°	[N], ppb
IIa HPHT	5×5×0.50	1.1	400
CVD	4.5×5.1×0.47	4.2	<50

The plot for etching rate ER vs T_s for the HPHT substrate, shows an unusual behavior (Fig. 4). A well defined spike in $ER(T)$ at low temperatures of 800–970°C is due to enhanced etching rate of a subsurface damaged (SSD) layer formed by polishing procedure. After the entire removal of that SSD layer, "normal" etching rate inherent to bulk material, is established. A similar picture is was observed for CVD diamond etching. The SSD layer thickness $d \approx 0.2$ μm and 0.7 μm was estimated from the $ER(T)$ plots for IIa HPHT and CVD substrate, respectively.

Assuming exponential temperature dependence for etch rate $ER(T) = A \times \exp(-E_a/RT_s)$, where E_a is activation energy, and R is universal gas constant, Arrhenius plots for the CVD and IIa HPHT samples were built based on the measured rate (Fig. 5), and the activation energy E_a of 42 ± 5 кcal/ and 32 ± 4 кcal/mol, was determined for CVD and IIa type HPHT diamond, respectively. The published data [4] for etching of Ib-HPHT diamond in H_2 MW plasma with E_a= 55.8 кcal/mol for temperatures below 950 °C are also given in Fig. 5 for comparison.

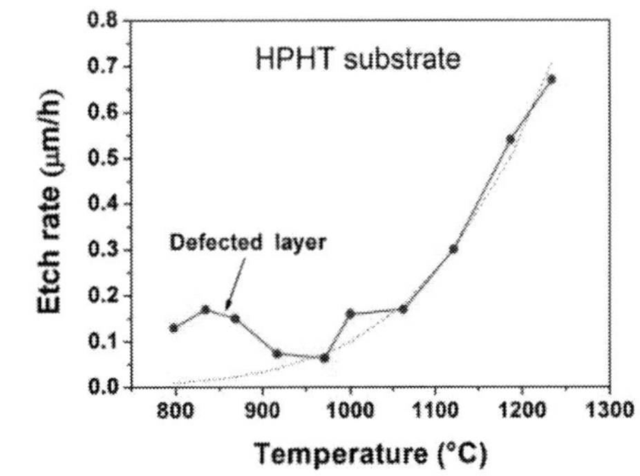

Fig. 4. Etching rates for IIa HPHT sample vs temperature. The dashed lines show Arrhenius approximations for $ER(T)$ at high temperatures (in absence of the defected layer)

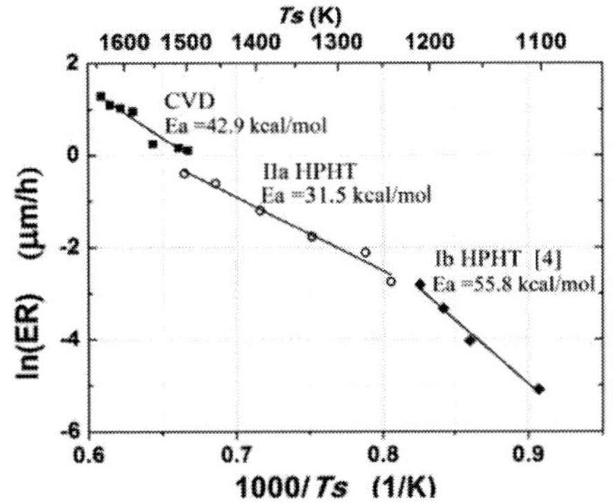

Fig. 5. Arrhenius plots for samples CVD (squares) and IIa HPHT (circles), respectively. The published data for Ib-HPHT diamond [4] (diamonds) are given for comparison

The difference in the activation energy for the CVD and HPHT diamonds can be assigned to a difference in defects abundance in these substrates. Our E_a values are the firstly reported for diamond etching in H_2 plasma at temperatures above 1000°C.

In summary, the *in situ* LCI technique dramatically reduces the time to collect kinetics data. The full cycle of the measurements, either growth or etching, can be performed during one working day.

This work was supported by the Russian Scientific Foundation, grant No. 14-12-01403-P.

References

1. *Bushuev, E.V., Yurov, V.Yu., et al.*. Synthesis of single crystal diamond by microwave plasma assisted chemical vapor deposition with in situ low coherence interferometric control of growth rate // Diam. Relat. Mater. 2016. V. 66. P 83-89.

2. *Bushuev, E.V., Yurov, V.Yu., et al.* Express *in situ* measurement of epitaxial CVD diamond film growth kinetics // Diam. Relat. Mater. 2017. V. 72, P. 61–70.

3. *Maeda, H., Ohtsubo, K. et al.* Determination of diamond [100] and [111] growth rate and formation of highly oriented diamond film by microwave plasma-assisted chemical vapor deposition // J. Mater. Res. 1995. V. 10. P. 3115-3123.

4. *Ivanov, O.A., Muchnikov, A.B. et al.* Experimental study of hydrogen plasma etching of (100) single crystal diamond in a MPACVD reactor // Mater. Lett. 2015. V. 151. P. 115-118.

Microwave discharge in liquid hydrocarbons

Yu.A. Lebedev, K.A. Averin, A.V. Tatarinov, I.L. Epstein

Topchiev Institute of Petrochemical Synthesis of the Russian Academy of Sciences (TIPS RAS), Moscow, Russia,
lebedev@ips.ac.ru

Discharges in liquids have attracted the attention of researchers and they are one of the priority areas in the study of the physics and application of low-temperature plasma [1, 2]. This is explained by both possible promising applications of such discharges in solving environmental problems, and using it to produce various gas-phase and solid products. All types of electrical gas discharges are used for these purposes, but the least explored are the microwave discharges. Current status of microwave discharges in dielectric liquids is given in [3]. This paper summarizes results of original studies of atmospheric-pressure microwave discharge in gas bubbles at the tip of the microwave antennas immersed in liquid alkanes (C_nH_{2n+2}, n = 7, 8, 10, 15, 16) differed in the boiling temperatures (from 98.42 °C to 286.79 °C). [4–10].

Experimental setup

Two experimental setups were used: the first one was based on the quarter wavelength antenna immersed in liquid hydrocarbon [4, 5, 10], and the second one was based on the coaxial line with the central conductor (antenna) immersed in liquid hydrocarbon for a particular length [6]. To obtain the maximum value of microwave field at the end of antennas it lengths were calculated from solution of Maxwell equations taking into account the geometry of discharge system and dielectric properties of hydrocarbons. This gives possibility to minimize the incident microwave power necessary to ignite and sustain the discharge. Microwave energy was delivered from a magnetron generator (frequency 2.45 GHz, microwave power below 1000 W). Discharge was non-stationary as the gas bubble, generated at the antenna end due to evaporation of liquid or by supplementary gas, is detached from the antenna end after plasma ignition and floated up. Plasma is disappeared as microwave field decreases with distance from antenna. The second reason of instability is saturation of liquid by carbonaceous solid products of plasma chemical reactions. These products effectively absorb microwaves and strongly decrease of microwave field in the antenna vicinity and thus decrease the probability of plasma ignition. This process leads to hydrocarbon heating and evaporation.

The system was oxygen free as argon gas was continuously blown through the space above the liquid surface. To simplify the discharge ignition additional argon could be introduced through the channel in antennas. The gas pressure above the liquid was close to atmospheric pressure. Microwave discharge was ignited in gas bubbles (hydrocarbons vapor or the mixture of these vapors with argon).

Plasma parameters (spectral composition, rotational and vibrational temperatures and temperature of solid products) were specified from plasma emission studied with spectrometer AvaSpec 2048 [5].

To characterize the solid phase we used a JEOL JSM-7500F high resolution field emission scanning electron microscope coupled with an INCA PentaFETx3 energy dispersive X-ray spectrometer (Oxford Instruments, UK). Raman spectra of the solid products were recorded with a RamanStation 400 (Perkin–Elmer, USA) Raman spectrometer furnished with a diode laser having λ_{ex} = 785 nm. FTIR-ATR spectra were acquired with a Tensor 27 (Bruker GmbH, Germany) FTIR spectrometer furnished with a Miracle (Pike Technologies Inc., USA) ATR attachment.

To obtain ^1H and ^{13}C NMR spectra of the solid phase, we used a Varian Unity Inova AS 500 spectrometer with probe A T3 HXY 3.2 mm was. A C/H atomic ratio in the solid phase was determined chromatographically after combustion of sample in the dynamic flash.

Sizes of residuals solid nanoparticles in a processed liquid hydrocarbons after centrifugation were studied with a Zetatrac laser analyzer (λ = 750 nm).

Processed liquid hydrocarbons were studied with GC/MS Thermo DSQ II device. The concentrated liquid sample (after evaporation) was studied with the same device and with IR-Fourier spectrometer Tensor II (Bruker GmbH, Germany).

Results and discussion

The only method for study of plasma properties in liquids is the emission spectroscopy. It was shown that emission spectra of plasma in all hydrocarbons contain only C_2 Swan bands ($d^3\Pi_g$–$a^3\Pi_u$) with rotational temperature 1000–1500 K and vibrational temperature of 7500 K. Addition of argon decreases the rotational temperature up to 600 K. Non of hydrogen atom or molecules emission was observed although hydrogen is the main gas phase product. Observed emission continuum can be attributed to the emission of solid particles produced in plasma and having the temperature of 4000 K [5].

The electron concentration in the discharge was estimated to be about 10^{14} cm^{-3} from 2D simulation of the MW discharge in liquid *n*-heptane [8, 9].

The solid carbon containing products, separated from processed liquid hydrocarbons by centrifugation (3000 rpm), are represented by particles having dimensions in the range of 100–200 nm. For all the products studied, Raman spectra have exhibited a set of necessary bands permitting to categorize these solids as a "damaged graphene", and mainly not single-layered one [4, 10].

EDS analysis of the solid samples showed presence of carbon (80–90%), oxygen (2–15%) and copper (below 2%) atoms. Source of copper is the eroded copper antenna. The carbonaceous products were kept in air atmosphere before subjected to further analysis and absorbed oxygen is showed in analysis.

NMR spectra of solid products obtained from processed hydrocarbons reveal that generation of microwave plasma in liquid alkanes leads to strong aromatization and formation of various unsaturated molecular fragments in solid products.

Initially transparent colorless liquid hydrocarbons after processing and removal of solid particles become colored transparent yellow-brown. GC/MS analysis of hydrocarbon composition before and after treatment reveals no changes in the main composition of hydrocarbons. Only insignificant amounts (<1%) of polycyclic aromatic hydrocarbons are detected in absorption and luminescence spectra of liquid hydrocarbon after processing. The question is: where the solid particles are produced – in liquid hydrocarbon from the polyaromatic compounds or in the plasma region, how it was observed in numerous papers in gas phase plasma chemistry? Analysis of our experimental results leads to the conclusion that carbonaceous nanoparticles are produced in the region of gas bubble with microwave discharge near the tip of the antenna. Then the particles are transferred into the liquid. Convective flows emerging during heating the liquid hydrocarbon spread the solid particles over the entire volume of the liquid. After prolonged treatment of a hydrocarbon with the discharge, a tree-like structure can be formed on the antenna ending [6].

Liquid hydrocarbons after centrifugation contain solid nanoparticles with sizes ranged between 1 and 3 nm.

What is the role of electron impact in the processes in discharges in liquid hydrocarbons at atmospheric pressure? It is one of the key questions in plasma chemistry.

The role of the electron impact in the decomposition of n-heptane in microwave discharge in liquid n-heptane at atmospheric pressure has been investigated on the basis of two-dimensional self-consistent model for numerical modeling of the process of formation of gas bubbles during initiation of the microwave discharge [8, 9]. The developed model has an axial symmetry. The model is based on joint solution of the Maxwell equations, Navier-Stokes equation, heat equation, continuity equations for electrons (written in the ambipolar diffusion approximation) and the n-heptane concentration (including its thermal decomposition and dissociation by electron impact) and the Boltzmann equation for free electrons of the plasma. The calculations allowed describing the dynamics of the formation of gas bubbles in the liquid, to evaluate the role of electron impact in the decomposition of *n*-heptane, and to estimate the characteristic times of various processes in the system. The results of experiments are in agreement with the simulation results.

It was shown that the electron impact dissociation can take place only in the plasma bubble in vicinity of the antenna tip where microwave field strength has maximum value. The effect of electron impact is important in the *n*-heptane decomposition at times below 10^{-3} s and is negligible at time exceeds 10^{-3} s when the gas temperature exceeds 1000 K. For larger time the thermal decomposition of hydrocarbon prevails over others and the role of electron component is only in the gas heating.

This result gives possibility to explain why we did not see hydrogen in the emission spectra of discharge? It is known the emission of hydrogen lines and bands id provides by the electron impact whereas the emission of C_2-molecules is provided by the collisions of plasma heavy particles. Measurements of emission spectra were made with time delay against the switching of the discharge and thus the short period of prevalence of the electron impact was not seen.

Decrease of the pressure should lead to increase of the role of electron impact. This is why in known experiments with lower pressures the emission of hydrogen atoms was observed.

References

1. *Bruggeman P., Leys C.* Non-thermal plasmas in and in contact with liquids. //*J. Phys. D: Appl. Phys.* 2009. V. 42, 053001.

2. *Bruggeman P. J., Kushner M. J., Locke B. R., et al.* Plasma-liquid interactions: a review and roadmap. //*Plasma Sources Sci.& Technol.* 2016. V. 25. 053002.

3. *Lebedev Yu. A.* Microwave discharges in liquid dielectrics. //Plasma Phys. Reports. 2017. V.43. P. 676-686.

4. *Lebedev Yu. A., Konstantinov V.S., Yablokov M. Yu. et al.* Microwave Plasma in Liquid *n*-Heptane: A Study of Plasma Chemical Reaction Products. // *High Energy Chem.* 2014. V. 48. P. 385–388.

5. *Lebedev Yu. A., Epstein I. L., Shakhatov V. A., et al.* Spectroscopy of Microwave Discharge in Liquid C_7–C_{16} Hydrocarbons. // *High Temperature.* 2014. V. 52. P. 319–327.

6. *Averin K.A., Lebedev Yu. A., Shakhatov V.A.*. Some results of study of a microwave discharge in liquid hydrocarbons // *Applied Physics.* 2016, V. N 2. P. 41-45 (in Russian).

7. *Tatarinov AV, Lebedev YuA, Epstein IL.* Simulation of Microwave-induced Formation of Gas Bubbles in Liquid n-Heptane. //*High Energy Chem.* 2016. V. 50. P.144-149.

8. *Lebedev Yu.A., Tatarinov A.V., Epstein I.L., and Averin K.A.* The Formation of Gas Bubbles by Processing of Liquid n-Heptane in the Microwave Discharge. //*Plasma Chemistry and Plasma Processing.* 2016. V. 36. P. 535-552.

9. *Lebedev Yu.A., Tatarinov A.V., Epstein I.L.* On the role of electron impact in an atmospheric pressure microwave discharge in liquid *n*-heptane. //*Plasma Phys. Reports.* 2017. V.43. P. 510-513.

10. *Averin K.A., Lebedev Yu. A., Shchegolikhin A.N., et. al.* Nanosize Carbon Products Formed in Microwave Discharge in Liquid Alkanes // *Plasma processes and polymers*. 2017. DOI 10.1002/ppap.20160022.

EPJ Web of Conferences **149**, 02003 (2017) DOI: 10.1051/epjconf/201714902003
10th International Workshop 2017 "Strong Microwaves and Terahertz Waves: Sources and Applications"

Influence of CVD diamond growth conditions and misorientation angle on nitrogen incorporation

M.A. Lobaev[1], A.M. Gorbachev[1], S.A. Bogdanov[1], A.L.Vikharev[1], D.B. Radishev[1], V.A. Isaev[1], V.V. Chernov[1], M.N. Drozdov[1], P.A. Yunin[1]

[1] Federal Research Center The Institute of Applied Physics of the Russian Academy of Sciences, Nizhny Novgorod, Russia, bogser@appl.sci-nnov.ru

CVD diamonds with NV-centers (nitrogen vacancy centers) are now considered as the most promising solid-state media for the realization of component base for quantum information processing due to the unique properties of NV-center electronic transitions, which allow initialization and read-out of it`s spin state [1].

The concentration and spin properties of NV-centers depend on nitrogen concentration and therefore it is important to understand the dependence of nitrogen incorporation in delta-layers on CVD growth conditions. In this work, delta-layers (1-5 nm thick) doped with nitrogen were obtained during CVD synthesis using the unique setup for delta doping created in IAP RAS [2]. It was shown that the subsequent CVD growth with and without nitrogen addition allows obtaining a two-dimensional structure of the NV centers with a given concentration and any required depth that may be determined with an accuracy of up to 1–2 nanometers. Also the influence of substrate temperature, nitrogen flow and methane content on nitrogen incorporation was investigated by the growth of nitrogen doped layers on (100) and (111)-oriented samples.

The method of delta doping in contrast to the method of ion implantation does not produce lattice defects and allows controlling the depth of the NV-center to within a few nanometers. CVD growth allows producing single-crystal diamond of high crystalline perfection with a low content of impurities that does not vary from sample to sample. Due to the absence of lattice damage, as in the case of ion implantation, NV-centers produced by delta-doping will have better properties (e.g. higher spin coherence times).

CVD diamond growth of nitrogen-doped layers was performed in the novel CVD reactor. The main features of the reactor are: 1) rapid gas switching; 2) laminar gas flow; 3) axial symmetric resonant mode – symmetric discharge; 4) slow growth of diamond (40–100 nm/h without nitrogen addition). We achieve rapid gas switching from one input gas to another by a home-made electronic switch. The residence time of our reactor is approximately 5 s. The reactor is also capable for the growth of delta-layers doped with any other impurities, such as nitrogen.

It is well-known, that incorporation of impurities during CVD diamond growth is highly dependent on the substrate orientation, and could also be affected by the misorientation angle. In order to eliminate the influence of substrate, each of the experiment was performed on one sample, varying growth conditions in a one growth process, resulting in the growth of multiple thin nitrogen-doped layers (see Fig. 1). Moreover, in this work we investigated the dependence of nitrogen incorporation efficiency on the misorientation angle using the (100)-oriented sample with multiple facets polished at different misorientation angles.

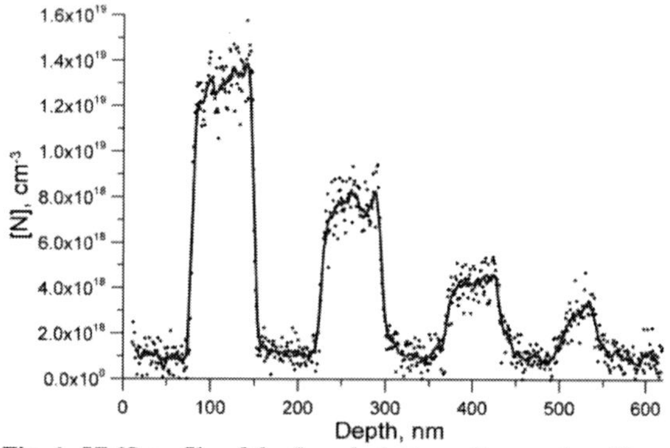

Fig. 1. SIMS profile of the Sample 1. Four nitrogen doped layers was grown at nitrogen flows 1, 2, 4 and 8 sccm. Methane flow 1.4 sccm

Our reactor is capable to vary all of the growth conditions separately during a one growth process. For the temperature variation in a one process the substrate heater was used. The growth was performed under the following conditions: pressure 40 Torr, hydrogen flow 950 sccm, methane flow 0.7–1.7 sccm. Nitrogen flow was varied from 1 to 8 sccm.

Type IIa HPHT diamond (100)-oriented substrates from New Diamond Technologies [3], 3.5×3.5×0.5 mm^3, were used for the nitrogen-doped CVD diamond layers growth. The substrates were pre-treated by ICP plasma in order to get rid of the defects introduced during the polishing procedure [4]. ICP etching up to 5 μm depth did not change the surface roughness but allowed to eliminate polishing damages.

Nitrogen incorporation during the process of CVD diamond synthesis was studied at different growth conditions: nitrogen flow, substrate temperature, methane content. High nitrogen concentration $>10^{19}$ cm^{-3} was obtained using (100)-oriented HPHT substrates. SIMS was used to measure the nitrogen concentration in the grown diamond samples. The time-of-flight secondary ion mass spectrometer TOF.SIMS-5 by ION-TOF (Münster, Germany) was used for the depth profiling and a Talysurf CCI 2000 interferometer was used to measure the depths of the etching craters for normalization of the analysis depth scale. The instrument operates in dual beam mode employing 1 keV Cs+ ions for sputtering and 25 keV Bi+ ions for probing with 45° of incidence for both ion beams. The SIMS nitrogen measurements were quantitatively calibrated using a nitrogen ion-implanted HPHT single crystal "test" diamond, having the implanted nitrogen concentration equal to 10^{20} cm^{-3}. SIMS profiles were used also for the determination of the growth rate in each of the growth regime.

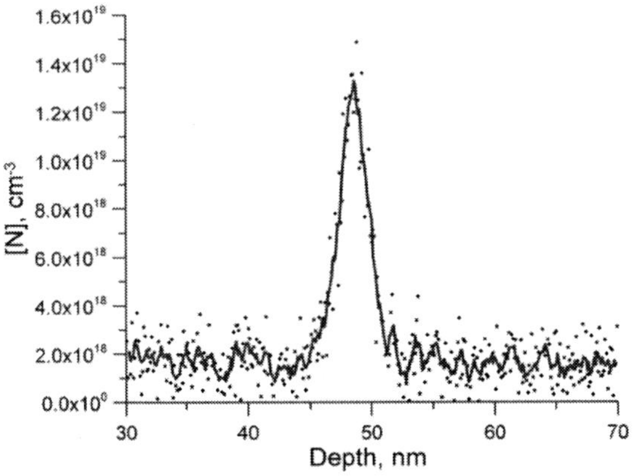

Fig. 2. SIMS profile of nitrogen doped delta-layer (thickness ~ 3 nm)

Photoluminescence of the grown samples was measured using micro-Raman system with laser excitation wavelength of 514 nm. Measurements were performed at liquid nitrogen temperature. The results of photoluminescence measurement for the Sample 1 are shown in Fig. 3. They shows the existence of both NV^0 (575 nm) and NV^- (637 nm) peaks.

In summary, we investigated nitrogen incorporation during CVD diamond growth at different growth conditions. The influence of substrate temperature, methane content and nitrogen flow on the nitrogen incorporation was extensively studied. In order to eliminate the influence of substrate, each of the experiment was performed on one sample, varying the growth conditions in the one growth process. The grown nitrogen doped layers were investigated by SIMS and photoluminescence spectroscopy. Nitrogen incorporation efficiency in our growth regime was determined as $8 \cdot 10^{-6}$.

Moreover, we succeeded in fabrication of ultra-thin nitrogen doped delta layers with peak nitrogen concentration ~10^{19} cm^{-3} and thickness ~ 3 nm (see Fig. 2). The SIMS profile of such ultra-thin nitrogen doped layer with high peak concentration is demonstrated for the first time. Such ultra-thin nitrogen-doped layers could be used for the creation of high-density NV center ensembles, which are highly desirable for ultra-sensitive magnetometry applications. Due to the absence of lattice damage, as in the case of ion implantation, NV-centers produced by delta-doping are expected to have better spin properties.

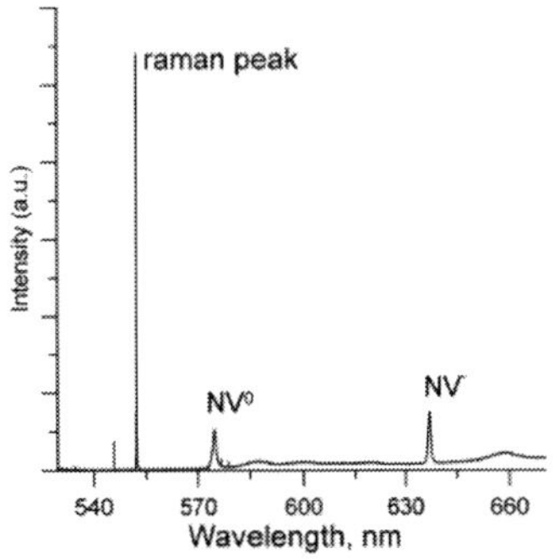

Fig. 3. Photoluminescence spectrum (77 K) of the Sample 1

Acknowledgements

The study was performed by a grant from the Russian Science Foundation (Project № 16-19-00163).

References

1. *F. Jelezko and J. Wrachtrup,* Single defect centers in diamond: a review // Phys. Status Solidi A 2006. V.203. No. 13, P. 3207-3225.

2. *A. L. Vikharev, A. M. Gorbachev, M. A. Lobaev, A. B. Muchnikov, D. B. Radishev, V. A. Isaev, V. V. Chernov, S. A. Bogdanov, M. N. Drozdov, and J. E. Butler*, Novel microwave plasma-assisted CVD reactor for diamond delta doping // Phys. Status Solidi RRL 2016. V.10. No. 4, P. 324–327.

3. ndtcompany.com

4. *A. B. Muchnikov, A. L. Vikharev, J. E. Butler, V. V. Chernov, V. A. Isaev, S. A. Bogdanov, A. I. Okhapkin, P. A. Yunin, and Y. N. Drozdov*, Homoepitaxial growth of CVD diamond after ICP pretreatment // Phys. Status Solidi A 2015. V. 212. No. 11, P. 2572-2577.

Diamond Bragg superlattice grown in microwave gas discharge for obtaining photoluminescence of single diamond color centers comprising a dense 3D ensemble

V.A. Kukushkin[1,2], M.A. Lobaev[1], D.B. Radischev[1], S.A. Bogdanov[1], M.N. Drozdov[3], V.A. Isaev[1], A.L. Vikharev[1], and A.M. Gorbachev[1,2]

[1]Institute of Applied Physics of the Russian Academy of Science, Nizhny Novgorod, Russia, vakuk@appl.sci-nnov.ru
[2]Nizhny Novgorod State University named after N.I. Lobachevsky, Nizhny Novgorod, Russia
[3]Institute for Physics of Microstructures of the Russian Academy of Sciences, Nizhny Novgorod Region, Russia

Due to its unique qualities (radiation and chemical endurance, high thermal conductivity, a very large breakdown electric field and others) chemically vapor deposited (CVD) diamond is a promising material for modern photonics and optoelectronics. Traditional optical devices on CVD diamond (e.g. waveguides and photonic crystals) are based on its post-growth treatment, for example on the creation of a periodic pattern in a CVD diamond film by electronic beam lithography, focused ion beam milling or a laser beam ablation [1]. However, such technologies have intrinsic limitations connected with difficulty to achieve a desirable spatial resolution, problems of reproducibility, defects and the roughness of the structure interfaces resulting in undesirable photon scattering.

Therefore, it seems promising to form a CVD diamond optical devices (for example, Bragg superlattices) directly during the growth process to ensure their atom-scale precision and quality. But this way encounters a difficulty that there is no heteropair for diamond (i.e. a substance with crystal lattice parameters very close to those of diamond, but a different refraction index). In principle, there exists difference of the refraction indices of diamond isotopes ^{12}C and ^{13}C, which, therefore, could be employed to form structures for CVD diamond optical devices. But this difference is very small (of order 10^{-3} and smaller in the visible and nearinfrared wavelength region). This means that for obtaining a highly reflective Bragg superlattice it is necessary to grow a very large (several thousand) number of alternating layers of ^{12}C and ^{13}C with very precisely repeated parameters, which is technologically impossible now.

So, nowadays the only way to create a highly reflective CVD diamond Bragg superlattice is the modulation doping method consisting in the growth of interchanging highly and low doped CVD diamond layers [2, 3] (please see fig. 1). The oscillations of free charge carriers (holes) in alternating electromagnetic field in the highly doped layers cause the change of the dielectric permittivity of these layers in comparison with the dielectric permittivity of the adjacent low doped layers. As a result, dielectric permittivity becomes periodically modulated in space. It is well-known that for certain wavelengths satisfying the Bragg condition electromagnetic radiation is efficiently reflected from such a superlattice even in the case of a small relative difference of the dielectric permittivities of the highly and low doped layers if the number of these layers is sufficiently large. This effect is due to coherent summation (i.e., constructive interference) of waves reflected from the consecutive boundaries of the layers.

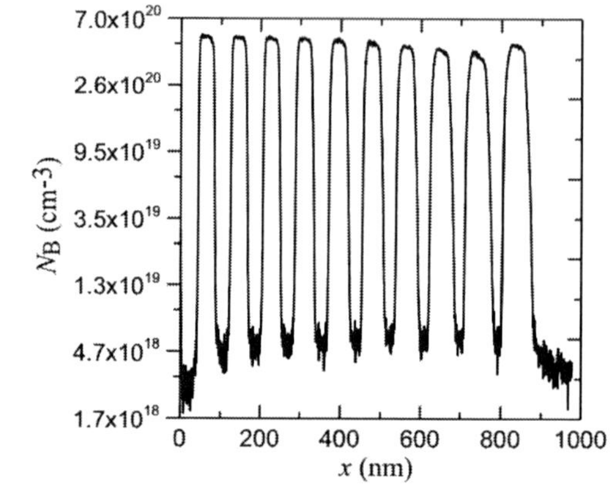

Fig 1. Secondary ion mass spectrometry (SIMS) measurements of dopant (boron) atom concentration N_B as a function of the distance x from the boundary of the CVD diamond film with air

We have grown such a CVD diamond Bragg superlattice on a home-built microwave gas discharge reactor [4]. The reactor consists of a cylindrical microwave cavity with a quartz tube placed inside on the axis of the cavity. Microwave radiation with an input power of 1.5 kW and a frequency of 2.45 GHz generates a plasma at a gas pressure 40 Torr just above a substrate holder. A non-vortex laminar gas flow with the total flow rate of 900 sccm inside the quartz tube is employed for a quick change of a gas mixture within the reactor. The low boron doped CVD diamond layers were grown in the feed gas mixture H_2+CH_4 (CH_4/H_2=0.1%). This gas mixture was quickly (in comparison with characteristic growth time) changed to H_2+CH_4+B_2H_6 (the same methane content, B_2H_6/H_2=0.1%) with an electronic switch for growing the highly boron-doped CVD diamond layers. A time value for achieving the gas mixture change in the reactor (i.e., the reactor residence time) is estimated by the Ar emission line dynamics and is approximately equal to 5 s. Due to the presence of residual boron in the reactor the low boron doped CVD diamond layers possessed some small boron concentration. Hydrogen sulfide (added to the gas mixture during the growth of the low boron doped layers) was used as a chemical getter to decrease the residual boron contamination.

Type IIa high pressure high temperature (HPHT) diamond substrates from New Diamond Technologies with dimensions 3.5 × 3.5 × 0.5 mm were used for the CVD diamond boron-doped Bragg superlattice growth.

The substrates were pre-treated by inductive coupled plasma in order to get rid of defects introduced during the polishing procedure. As a result, surface roughness $R_a \approx 0.1 \div 0.3$ nm was achieved.

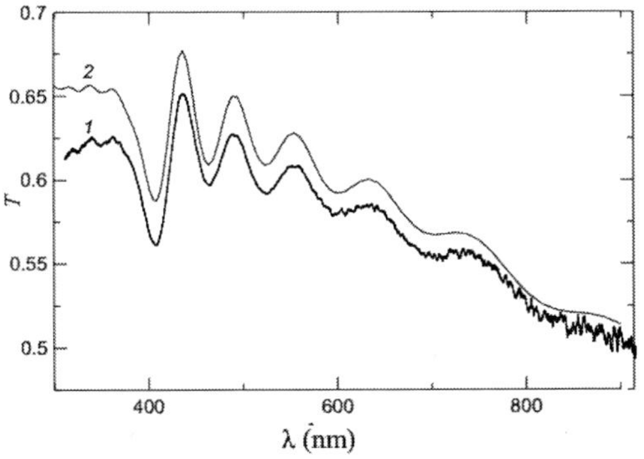

Fig. 2. Power transition coefficient T as a function of the radiation vacuum wavelength λ for normal incidence. Curve *1* – experimental data, *2* – calculation results

The vacuum wavelength dependence of a power transition coefficient T of optical and near infrared electromagnetic radiation through the diamond sample with the CVD Bragg superlattice was measured by means of Horiba Jobin Yvon FHR-1000 monochromator under the sample irradiation from the superlattice (i.e. top) side by a stabilized light source along the normal to its surface (fig. 2). The obtained experimental data were compared with the results of simulations by means of the transfer matrix method. We considered the coherence length of the used light to be larger than the superlattice thickness and, therefore, took into account the interference of fields of the waves reflected from the boundaries of the layers comprising the superlattice. In the same time, we assumed that this coherence length was smaller than the thickness of the HPHT diamond substrate (approximately 0.5 mm). So, we summed not the fields, but the intensities of the waves reflected from the interfaces the Bragg superlattice – the HPHT diamond substrate and the HPHT diamond substrate – air. Experimental curve *1* position slightly below theoretical curve *2* is explained by some additional absorption not taken into account in the calculations and connected with the sample contamination.

The Bragg CVD diamond superlattice can be used to significantly increase the spatial resolution of the color center photoluminescence collection at a fixed exciting laser beam focal volume. It will allow to collect the photoluminescence of individual color centers in their dense 3D ensembles, which is needed for quantum computing and memory applications and for measurement and imaging with high spatial resolution. The suggested method is based on the creation of a narrow waveguide for the color center photoluminescence with a small number of allowed vertical indices of guided modes. The waveguide is formed between the top surface of a CVD diamond film and an underlaid mirror – the Bragg superlattice due to the total internal reflection from the former and the Bragg reflection from the latter. Then, the frustrated total internal reflection (FTIR) method is employed to extract through the top surface of the CVD diamond film the color center photoluminescence contained only in guided modes with the same vertical index. Color centers which lie near the minima of the electric field depth distribution of these modes in the CVD diamond Bragg superlattice give small contribution to the FTIR-extracted photoluminescence and, therefore, can be disregarded. So, the exciting laser beam focal volume is divided into "dark" (i.e. lying at the minima of these modes electric field depth distribution) and "bright" (i.e. lying at the maxima of these modes electric field depth distribution) layers. In consequence, an effective (with respect to the excitation of the FTIR-extracted modes of the color center photoluminescence) exciting laser beam focal volume is comprised by only "bright" layers and, therefore, is smaller than the whole exciting laser beam focal volume. So, the maximal value of the color center 3D concentration still compatible with obtaining single color center photoluminescence becomes of order the inverse effective focal volume, which is larger than the inverse focal volume, i.e. its value in the Bragg superlattice absence.

According to the results of simulation made for a case when color centers are NV centers, the suggested method allows to increase the maximal value of the NV center concentration still compatible with obtaining single color center photoluminescence by several times at a fixed exciting laser beam focal volume. The increase is provided without the deterioration of the NV center photoluminescence collection efficiency.

The authors are grateful to P. A. Yunin for the processing of the SIMS method data. The work was supported by the Russian Science Foundation (project No. 16-19-00163).

References

1. *Tomljenovic-Hanic, S., Karle, T. J., Greentree, A. D., Gibson, B. C., Fairchild, B. A., Stacey, A., and Prawer, S.* Diamond-Based Optical Waveguides, Cavities, and Other Microstructures // *Optical Engineering of Diamond*, edited by R. P. Mildren and J. R. Rabeau. Weinheim, Germany: Wiley-VCH Verlag & Co. KGaA. 2013. P. 311–351.

2. *Fiori, A., Bousquet, J., Eon, D., Omnès, F., Bellet-Amalric, E., and Bustarret, E.* Boron-doped superlattices and Bragg mirrors in diamond // Appl. Phys. Lett. 2014. V. 105, No. 8. Art. No. 081109 (1–4).

3. *Kukushkin, V. A., Lobaev, M. A., Radischev, D. B., Bogdanov, S. A., Drozdov, M. N., Isaev V. A., Vikharev, A. L., and Gorbachev, A. M.* Bragg superlattices formed in growing chemically vapor deposited diamond // Journal of Applied Physics. 2016. V. 120, No. 22. Art. No. 224901 (1–5).

4. *Vikharev, A. L., Gorbachev, A. M., Lobaev, M. A., Muchnikov, A. B., Radishev D. B., Isaev, V. A., Chernov, V. V., Bogdanov, S. A., Drozdov, M. N., and Butler, J. E.* Novel microwave plasma-assisted CVD reactor for diamond delta doping // Physica Status Solidi Rapid Research Letters. 2016. V. 10, No. 4. P. 324–327.

Gyrotron frequency ECRIS development and the future challenges

L. Sun, H.W. Zhao, J.W. Guo

Institute of Modern Physics, CAS, Lanzhou 730000, China, sunlt@impcas.ac.cn

Gyrotron frequency amplifier has been used for the production of highly charged ion beams with high performance ECR ion sources (ECRIS) for almost 20 years. The widely used technology of microwave power coupling to the high temperature ECR plasma is still a borrowed design from the fusion machines. ECR ion source technologies and physics are advancing over the years, but the interpretation of how gyrotron frequency microwave interacts with ECR plasma and how to improve the ECRH efficiency still got little progress. Recently, experimental tests of 24 GHz with different microwave modes and waveguide openings have succeeded at IMP with some very encouraging but surprising results. This paper will discuss the international needs of gyrotron amplifiers in the ECRIS community and the highlights of ≥24 GHz microwave coupling. Typical questions and challenges in future development will be presented, especially how to meet the high performance needs of a 4th generation ECR ion source.

ECRIS concept was proposed by Geller in late 1960s [1]. ECRIS was actually developed from plasma fusion device. Plasma electrons are heated through ECR heating to high energy by coupled microwave power with the frequency in the range of 2.45 GHz to 28 GHz. The plasma is confined by a strong nested so-called mini-B magnetic field configuration (Fig. 1), which is the superposition of an axial mirror field and a radial multiple field (normally a hexapole field). Hot electrons are confined at the center of the nested field and the HCIs, which are produced by stepwise ionization while they are staying in the plasma, are trapped by the space charge established by the electrons inside the plasma. The lost ions that enter the extraction region will be accelerated by the applied HV potential to form intense mixed ion beams. HCI beams can be selected from the mixing beams with an analyzer magnet.

There are variable choices of HCI sources for heavy ion accelerators such as ECRIS, EBIS, MEVVA and LIS, and so on, however ECRIS remains an indispensable machine for worldwide heavy ion accelerator development. For high charge state intense CW or long pulse (~ms) ion beams production, ECR in source is the most powerful tool. In recent 15 years, ECRIS performance has been boosted enormously, for instance CW Xe^{30+} beam was only 10–15 eμA from a state-of-the-art ECRIS about 14 years ago, and this number has been multiplied by a factor of at least 20 with one of the 3rd generation ECRISs SECRAL [2]. The tremendous improvement is driven by the strong needs from modern advanced heavy ion accelerator projects or facilities, such as FRIB, SPIRAL2, FAIR, HIRFL/IMP, RIBF/RIKEN and so on. The application of gyrotron frequency microwave ECRH in the production of HCI beams plays significant role. A 28 GHz microwave amplifier was firstly tried in INFN/Catania with a superconducting ECRIS named SERSE [3]. Borrowed microwave coupling technique from fusion machine was used in the test. Despite the insufficient magnetic confinement to ECR plasma, the 28 GHz ECRH gave many very promising beam intensity records around 2000. The first optimum ECRIS working at gyrotron frequency is VENUS which was test at 28 GHz microwave power in 2004 [4], and therefore the ECRIS community stepped into the real 3rd G. era.

As a conventional technique, the existing 3rd G. ECRISs are all using TE_{01} mode travelling in an oversized Ø 32 mm opening circular waveguide as the gyrotron microwave power coupling scheme, which is actually a directly borrowed technique from fusion community. Ion sources working at gyrotron frequency have been tested and the capacity to produce more intense highly charged ion beams with sufficient magnetic confinement has been verified as had been predicted by the frequency scaling laws. However, at the same microwave power level or same power density level, gyrotron frequency heating is not doing as well as predicted. In most occasions, compared to 18 GHz, it is behaving just like a linear extrapolation of power effect to get more intense highly charged ion beams. At 24 GHz, SECRAL can produce the beam intensity of the ω^2 scaling, but at much higher microwave power level compared to that at 18 GHz. Similar results have also been observed with SuSI [5]. This raised the question about the coupling efficiency of gyrotron microwave power into ECRIS plasma. Therefore, studies concerning gyrotron microwave coupling and ECRH have been done in both LBNL and IMP. Different microwave injection schemes have been tried with SECRAL at 24 GHz, such as HE_{11} mode, TE_{11} mode, or TE_{01} mode with variable waveguide openings. When an Ø20 mm TE_{01} mode oversized waveguide was installed, a significant different microwave coupling behavior was observed compared to other cases. As shown in Fig. 2, the output of the ion source responded to the incident 24 GHz microwave very linearly. No saturation was found even at the power limit of the 24 GHz amplifier, i.e. 7.0 kW, which was corresponding to a power density of ~1.6 kW/l. What is more important is that the production efficiency at the same RF power level for the case of Ø 20 mm TE_{01} mode microwave is at least 30% higher than that of the Ø 32 mm TE_{01} mode or the Ø 20 mm HE_{11} mode. As is well known, for a high charge state ECR ion source, the plasma operates at a low plasma density, and therefore the plasma chamber can be regarded as a multimode cavity [1]. For a 3rd G. ECRIS, the plasma chamber diameter is typically at the dimension of Ø > 100 mm, which is much larger than that of the incident gyrotron frequency RF wavelength. For this reason, the ECRH may not be sensitive to the change of incident microwave modes. However, the microwave power coupling has consequences on how the microwave power is fed into the chamber, i.e. the microwave coupling to ECR plasma and the coupling efficiency in terms of high charge state ion beam production. The interpretation to these results still need more experimental investigations and theoretical analysis.

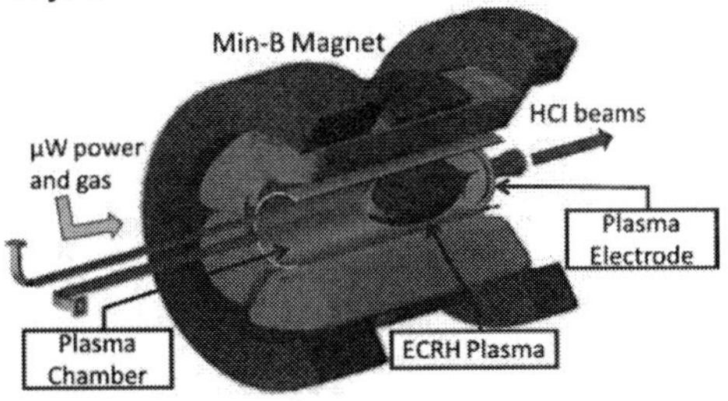

Fig. 1. Schematic plot of ECRIS

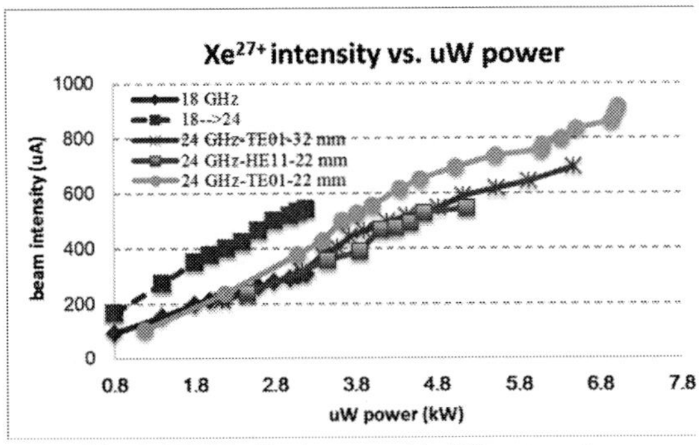

Fig. 2. ECRH effectiveness with different modes

HIAF or High Intensity heavy ion Accelerator Facility project to be launched in China, needs 50 pμA of U^{34+} from the injector ion source. The state of the art high performance ECR ion source such as VENUS can produce a beam intensity of ~11.7 pμA U^{34+} [6]which is barely 1/4 of the desired beam intensity. Obviously, the 3rd G. ECRISs still have margins for performance enhancement through technical improvements or new method applications, such as the ECRH scheme modification as presented in former paragraph. However, to have such a high gain factor, a more powerful machine is an inevitable approach. According to the history of ECRIS in the last four decades, the frequency scaling governs its development. Therefore, a 4th G. ECRIS with a performance of 4 times higher than the 3rd G. machines, is typically working at a frequency of 40~50 GHz. A 45 GHs ECRIS named FECR is under intense R&D at IMP. There are many technical and physics challenges that need long-term R&D and probably some big break-through in the development, such as Nb_3Sn magnet construction, cryogenics solution, intense beam extraction and transmission, and so on [7]. Specifically, microwave power transmission, coupling and ECRH will be critical issues for FECR. Since the wavelength of 45 GHz microwave is ~6.67 mm, quasi-optical transmission scheme is routinely utilized for high power transmission. This scheme has already been widely utilized in the fusion machines, but to accommodate with an ECRIS, the outcome is not evidenced yet. Many technical details and modifications will be made accordingly. As is demonstrated with 3rd G. ECRISs, to achieve the performance predicated by scaling laws at 24 or 28 GHz, higher power density inside the plasma chamber is needed compared to that of 18 GHz, which leads to the typical problems such as chamber cooling, bremsstrahlung radiation, cryogenic heat load and so on. For a 45 GHz ECRIS, we are inevitably encountering the same problems, or even worse. Therefore, a better understanding on how gyrotron frequency microwave heating works with a high performance ECRIS is urgent. FECR is designed to be operated at 45 GHz with a maximum input microwave power of 20 kW, i.e. a corresponding microwave power density of ~2 kW/l, which is not so challengeable, but the localized plasma loss to the chamber wall will give a >10 MW/m^2 power flux dump that will destroy an Al chamber wall easily. Consequently, it has the highest priority to improve the ECRH effectiveness at gyrotron frequency. Probably, very high power injection might be avoided so as to mitigate the other problems.

This work is supported by NSF (contract No. 11427904), MOST (contract No. 2014CB845500) and NSF (contract No. 11375244). The fruitful discussion with C. M. Lyneis, D. Z. Xie from LBNL and T. Thuillier from LPSC/Grenoble is warmly acknowledged.

References

1. *Geller R.* Electron Cyclotron Resonance Ion Sources and ECR Plasmas// Institute of Physics Publishing Bristol and Philadelphia, 1996. P. 85-92.

2. *Sun L., Guo J. W., Lu W., et al.* Advancement of highly charged ion beam production by superconducting ECR ion source SECRAL (invited)// Rev. Sci. Instrum. 2016. 87. 02A707.

3. *Gammino S., Ciavola G., Celona L., Hitz D., Girard A., Melin G.* Operation of the SERSE superconducting electron cyclotron resonance ion source at 28 GHz// Rev. Sci. Instrum. 2001. V. 72, No. 11. P.4090-4097.

4. *Leitner D., Lyneis C. M., Abbott A. R., Dwinell R. D., Collines D., Leitner M.* First Results of the Superconducting ECR Ion Source VENUS with 28 GHz//AIP Conference Proceedings, 2005. V. 749. P. 3-9.

5. *Machicoane M., Cole D., Holland K., Leitner D., Morris D., Neben D., Tobos L.* First Results at 24 GHz with the Superconducting Source for Ions (SuSI)// Proceedings of 21st Int. Workshop on ECR Ion Sources, 2014, http://www.jacow.org. P. 1-4.

6. *Benitez J., Franzen K. Y., Lyneis C., Phair L., Strohmeier M., Machicoane G., Sun L. T.* Current Development of the VENUS Ion Source in Research and Operations// Proceedings of 20th Int. Workshop on ECR Ion Sources, 2012, http://www.jacow.org. P.153-158.

7. *Sun L., Zhao H. W., Guo J. W., Yang Y., Lu W., Wu W., Ma L. Z.* Challenges for the Next Generation ECRIS // Proceedings of HIAT2015, 2015, http://www.jacow.org. P. 268-273.

ECR discharge in a single solenoid field

R.A. Shaposhnikov, S.V. Golubev, I.V. Izotov, S.V. Razin, V.A. Skalyga

Institute of Applied Physics RAS, Nizhny Novgorod, Russian Federation, shaposhnikov-roma@mail.ru

Electron cyclotron resonance (ECR) ion sources nowadays are one of the widespread systems for ion beam production in basic and applied research. One of the main directions for its development is an increase of the extracted beam current, which can be realized by production of plasma with higher density. Significant growth of plasma density in an ECR discharge is possible only with an increase of microwave radiation frequency used for the heating. In modern ECR ion sources gyrotrons are increasingly used as a source of microwave radiation. Earlier at the IAP RAS it has been demonstrated that the use of 37.5 GHz, 100 kW gyrotron radiation for plasma heating in a simple mirror magnetic trap allows to produce beams of light or multiply charged ions with record currents (up to 500 mA) [1–3]. In these studies simple mirror traps were used for plasma confinement instead of conventional "minimum – B" system due to high complexity and production cost of the last ones in case of microwave heating frequency above 30 GHz. The key feature of mentioned research was possibility demonstration for plasma creation with density up to $2\cdot10^{13}$ cm^{-3} and electron temperature about 100 eV under conditions of so-called quasigasdynamic regime of plasma confinement [4].

Such confinement regime is characterized by high density and so high electron Coulomb scattering into a loss cone that plasma losses from a trap are determined by its outgoing flux with ion sound velocity. It could be shown that in case of high enough plasma density its losses to a plasma chamber wall are always limited with ion sound velocity even without any magnetic confinement, plasma is confined mainly by electrostatic ambipolar potential. Under such conditions magnetic field of the trap is used for suppression of transversal plasma flux and gives additional geometrical factor in plasma lifetime. It is well known from fusion mirror machines studies that plasma lifetime in such systems is proportional to magnetic trap length and its mirror ratio (ratio between maximum and minimum B-field) [5].

However, the main disadvantage of a simple mirror trap is that it is favorable for development of magnetohydrodynamic instabilities, and additional methods are needed to suppress them. At the same time in IAP experiments magnetic traps with mirror ratio from 3 to 7 were used and such factor in plasma life time is not crucial. According to that it was proposed to investigate the prospects of creation a source of hydrogen ions on the basis of ECR discharge in a single solenoid sustained by powerful gyrotron radiation. Opposed to simple mirror trap this system is MHD stable and high power of microwave radiation could allow maintaining electron temperature on a sufficient level for high ionization efficiency of light gas.

The paper presents the first results on gas ECR breakdown studies in the single coil magnetic field and analysis of the prospects for the use of such plasma for ion beam production.

The first experiments were aimed to investigation of the possibility to realize the discharge in this system, determine plasma parameters and properties of the extracted ion beam. The experiments were carried out at SMIS 37 experimental facility where simple mirror magnetic trap was replaced with a single coil. Schematic view of the experimental setup (without gyrotron) is shown in fig. 1. Pulsed gyrotron radiation with a frequency of 37.5 GHz, power up to 100 kW and duration up to 1.5 ms was used for plasma ignition and heating. The magnetic field in the coil varied from 1 to 4 T (ECR for 37.5 = 1.34 T). The microwave radiation was injected into the plasma through a quasi-optical coupling system. Plasma was created in a specialized discharge chamber (working pressure from 10^{-7} to 10^{-3} Torr), the working gas (hydrogen) was fed into the discharge chamber along the system axis. Gas breakdown and plasma heating were performed under electron-cyclotron resonance conditions at the main harmonic. For beam extraction a two electrode extraction system was used. A plasma electrode with aperture diameter of 1 mm and a puller electrode with 3 mm aperture were placed in 6 mm from each other. Plasma electrode was placed outside of the coil in 12 cm from magnetic field maximum.

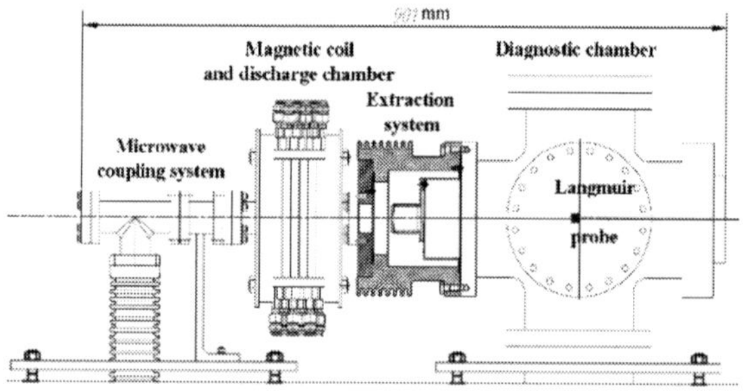

Fig. 1. Scheme of the experimental facility plasma part

At the first step experiments with a constant neutral gas injection were performed to study the possibility of the discharge ignition and to determine a threshold microwave power for it at various pressures. Breakdown curve plotted according to the experimental data is shown in fig. 2.

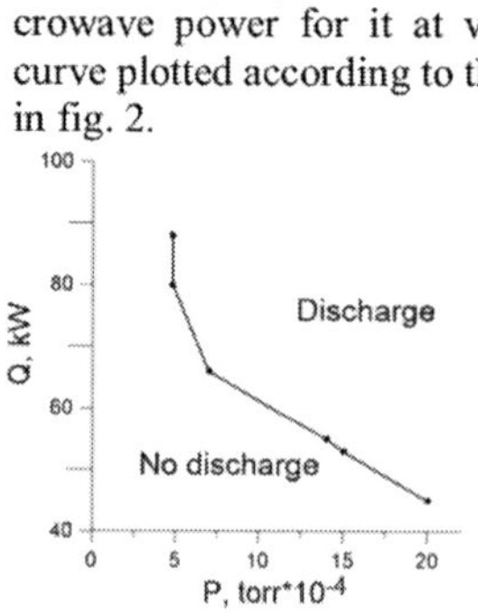

Fig. 2. Breakdown threshold curve for hydrogen

Also it was demonstrated that discharge could be realized only if maximum magnetic field in the chamber is above ECR value. It was clearly seen when the microwave pulse was shifted relatively magnetic field pulse from its maximum to the leading edge. It was done to observe breakdown behavior in case of rising in time magnetic field. The moment of discharge ignition was absolutely correlated with the time when magnetic field exceeded the ECR value.

All further investigations were performed in a pulsed mode of neutral gas injection into the system, which is standard for SMIS 37 operation. As the next step plasma parameters were studied using the regular Langmuir probe technique. The probe was placed in the diagnostic chamber on the system axis (see fig. 1) where magnetic field was weak enough (about 0.02 T). The probe current-voltage characteristics were recorded in a wide range of microwave power and gas pressure above the valve which was used for pulsed gas injection into the trap. Result of its processing for electron temperature determination is shown in fig. 3.

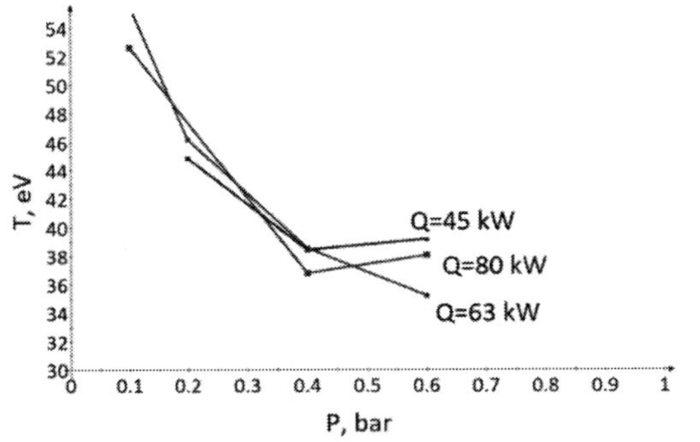

Fig. 3. Electron temperature dependence in hydrogen discharge for different microwave power and conditions of neutral gas injection

It is clearly seen that fine tuning of the gas pressure in the begging of microwave pulse by varying of the pressure above the valve and the moment of its opening could allow to obtain electron temperature up to 50 eV. Such temperature level is close to optimal for effective hydrogen ionization.

One of the main purposes of these experiments was to demonstrate a possibility of dense outgoing plasma flux production for high current ion beam formation. For these studies a two electrode extraction system was installed. Measured dependence of the extraction ion current on the voltage on extraction system is shown in fig. 4 in case of microwave power Q = 80 kW, gas pressure in the buffer chamber above the valve P = 0.18 bar.

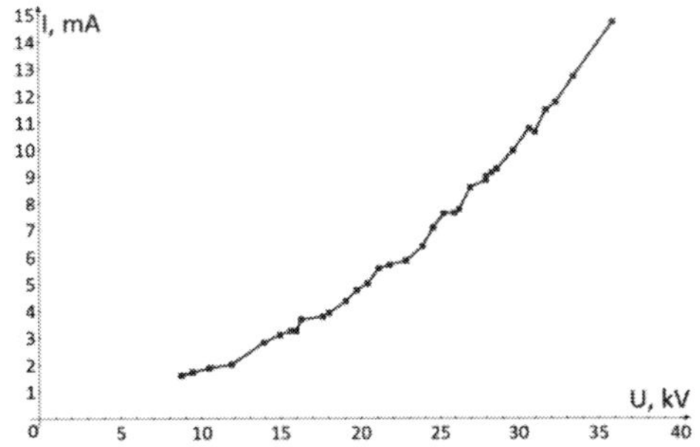

Fig. 4. Extracted ion beam current dependence on high voltage between electrodes

From figure 4 it can be seen that it was possible to obtain very high extraction current density values, 13 mA of total beam current corresponds to current density in the plasma electrode of more than 1.5 A/cm^2. It should be noted that the system parameters was not optimized, so, it might be possible to obtain better results. However, in experiments, which already have being carried out, high values of the ion current were demonstrated. This fact shows the prospects of the ion source design based on the system with one coil. High current ECR source of protons (or deuterons) with a single coil magnetic field looks interesting for applications where high ion current densities are required, such as ion beam injection to modern particle accelerators or compact powerful D-D neutron generators development.

Acknowledgments

Presented work was supported by the grant of Russian Science Foundation # 16-19-10501.

References

1. *S. Golubev, I. Izotov, S. Razin, A. Sidorov, V. Skalyga, A. Vodopyanov, V. Zorin, A. Bokhanov*. High Current ECR Source of Multicharged Ion Beams. Nuclear Instruments and Methods in Physics Research B, v. 256, p. 537 – 542 (2007).
2. *A. Sidorov, M. Dorf, A. Bokhanov, I. Izotov, S. Razin, V. Skalyga, V. Zorin, A. Balabaev, P Spädtke, J. Roßbach*. Multi-aperture ion beam extraction from gas-dynamic electron cyclotron resonance source of multicharged ions. Review of Scientific Instruments, 79, 02A317 (2008).
3. *V. Skalyga, I. Izotov, S. Razin, A. Sidorov, S. Golubev, T. Kalvas, H. Koivisto, and O. Tarvainen*. High current proton beams production at Simple Mirror Ion Source 37. Review of Scientific Instruments, v. 85, no. 2, p. 02A702-1 – 02A702-3 (2014).
4. *V. Skalyga, I.Izotov, S.Golubev, A.Sidorov, S.Razin, A.Vodopyanov, O.Tarvainen, H.Koivisto, T.Kalvas*. New progress of high current gasdynamic ion source. Review of Scientific Instruments. **87**, 02A716 (2016)
5. *V.V. Mirnov, D.D. Rutov*, Pisma JTF, v.5, p. 678 (1979).

Investigation on mm-wave sintering of metal powder compacts using in-situ dilatometry and electrical resistivity measurements

M. Mahmoud[1,2], G. Link[1], J. Jelonnek[1] and M. Thumm[1]

[1] Karlsruhe Institute of Technology, IHM, Karlsruhe, Germany, guido.link@kit.edu
[2] King Fahd University of Petroleum and Minerals (KFUPM), KSA & City for Scientific Research and Technological Applications (SRTA City), Egypt

Introduction

Sintering of metal powders is a well-established industrial process for specific applications, which allow a cheap series production of metal parts even with complex shape. The residual porosity, always present in sintered power compacts, limits their mechanical strength. On the other hand, for several products such porosity is needed. Products of that kind are for example: metal filters showing a porosity of about 50%, or self-lubricating plain bearings with a value between 5% and 30%, depending on the technological working conditions. In the case of self-lubricating plain bearings made of sintered iron or sintered brass, the open pores are filled with lubricants. Pore-free sinter metals are obtained by hot pressing or impregnation with a low-melting metal

The use of microwave technology for metal powder sintering had been reported by many researchers, demonstrating improvements in the materials properties and/or in the overall process. Although microwave are not able to penetrate into bull metals, significant penetration depth in metal power compacts has been reported by theoretical as well as experimental investigations [1, 2].

Here, the progress of microwave sintering of metal power compacts with focus on copper powders was investigated by systematic variation of process parameters. Those investigations were supported by both in-situ electrical resistivity and dilatometry measurements that give important information about the sintering progress and microstructural changes.

Experimental

In this study microwave sintering experiments on copper powder compacts were done at 30 GHz in a compact gyrotron device. Changes in sample length due to thermal expansion or sintering were measured by use of a commercial dilatometer from Linseis GmbH, Germany. This has been adapted to the mm-wave applicator and was combined with in-situ electrical resistivity measurements using the four-wire method [3]. The sintering temperatures were limited to 1000°C, ensuring sufficient temperature difference to the melting point of copper. The sintering process was performed at inert gas atmosphere such as nitrogen (N_2) and argon (Ar) as well as in a forming gas environment with 8% hydrogen (N_2/H_2).

For sample preparation commercial gas atomized spherical copper (Cu) metal powder from Alfa Aesar was utilized with a mesh size 100, equal to a maximum particle size of 149 μm (see Figure 3a). Copper compacts with different green densities were prepared using a uniaxial mechanical press. The prepared compacts were cylindrical in shape with an average diameter of 6.4 mm to exactly fit into the sample holder of the dilatometer system. The length was in the range from 8 to 11 mm.

The as-received copper powder and the microwave sintered Copper samples were characterized using X-ray diffraction (XRD), optical microscopy and scanning electron microscopy (SEM).

Results and Discussion

Systematic investigation on the sintering behavior with respect to process parameters and sintering atmosphere were performed [4]. The heating process was controlled along a preset temperature-time program using the temperature signal of an S-type sheath thermocouple, touching the sample surface.

Fig. 1 shows the in-situ resistance and dilatometry measurements of copper powder compacts with green densities of about 76% th.D., sintered at 1000°C with 10 min. soaking time using the 3 different process gases. A significant expansion behavior was observed in all the Cu compacts processed using the hydrogen containing gas (N_2/H_2). This abnormal expansion behavior was not observed when using the other two gases (N_2 & Ar). Furthermore, a sharp decrease in the electrical resistivity of the Cu compacts was observed at around ~190–210°C when using forming gas, while a gradual steady decrease on the electrical resistivity of similar samples occurred when using the other gases, reaching its minimum at around 600°C.

These different trends in the electrical resistivity could be explained by the difference in the reduction reaction of the thin copper oxide layer on the metal powder surfaces when comparing H_2 induced reaction, in case of forming gas and thermal induced reduction, in case of N_2 and Ar. Whereas in the first case water vapor is produced, in later case O_2 is formed according to the following reaction equations:

$$Cu_2O + H_2 \rightarrow 2Cu + H_2O \qquad 2\,Cu_2O \rightarrow 4\,Cu + O_2$$

$$CuO + H_2 \rightarrow Cu + H_2O \qquad 2\,CuO \rightarrow 2\,Cu + O_2$$

With progressing thermal or chemical reduction the electrical resistivity of the sample decreases due to the continuously improving electric contact between the Cu particles. In case of forming gas, when water vapor is formed and trapped in intra- and intergranular pores of the developing copper microstructure, due to increasing vapor pressure during heating, the sample expands as can be seen in Figure 1. This expansion was significant in the temperature range from 300°C to 500°C followed by a subsequent sintering stage starting at around 700°C that compensates the previous expansion and reaches a final densification comparable to the N_2 process. This unique expansion was observed for any green density, process parameter as well as for other grades of Cu powders.

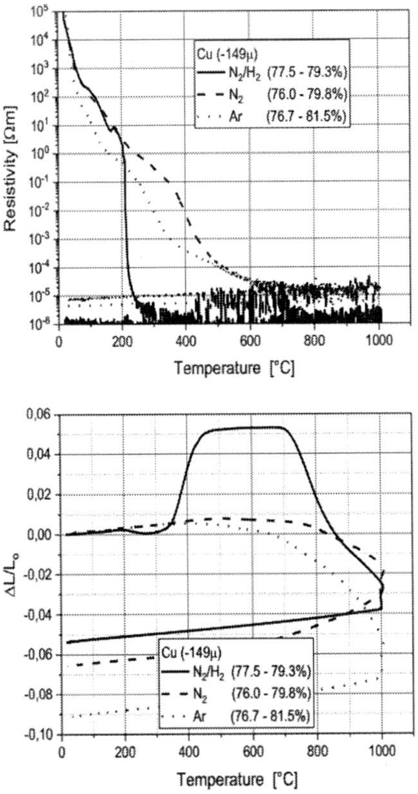

Fig. 1. Resistivity (top) and relative change in length (bottom) of Copper power compacts during sintering at 1000°C with 10 min. soak and subsequent free cooling with different process gases. Numbers in brackets reveal relative green and sintered density in % of theoretical density

This effect can even be identified by comparison of copper powder particles heated at similar conditions for 5 min at 550°C (see Fig. 2). Whereas no distortion could be observed for particles processed in Ar or N_2, significant effects are obvious in the forming gas case. The expansion effect is revealed by irregular bursts in particle surfaces, obviously caused by vapor, developing during H_2 reduction of CuO_2 and captured in closed pores till the pressure reaches the mechanical strength of those.

Comparing the sintering curves in Fig. 1 reveals the best densification behavior in Ar gas. On the other hand, the lowest resistivity could be achieved by using forming gas probably due to most complete reduction reaction. Further parameter studies revealed higher sintered densities with increasing soaking time, and higher green density, as expected. Furthermore more fine grained grades of copper particles allow highest sintered densities and in much shorter times when compared with similar coarse-grained Cu samples. Finally the MW processed Cu samples had achieved higher densities than similar ones processed using conventional heating, as in the Ar gas case.

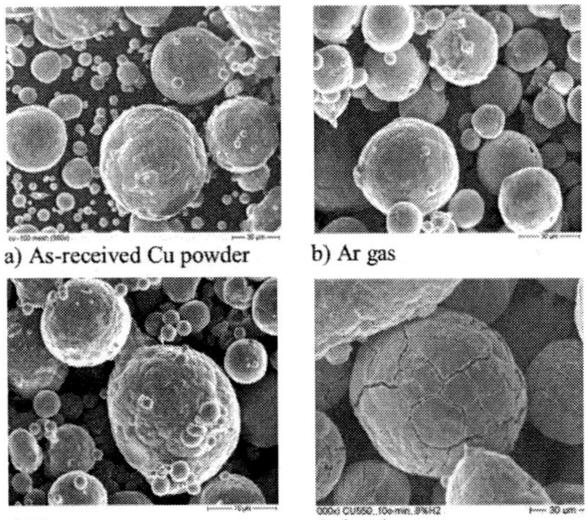

a) As-received Cu powder b) Ar gas
c) N_2 gas d) forming gas

Fig. 2. SEM of the as-received Cu powder (maximum particle size 149 μm), compared to MW treated powder at 550 °C for 5 min in different atmospheres

Conclusions

Systematic experimental studies on mm-wave sintering of copper powder compacts reveal the existence and influence of oxide layers on particle surfaces. Whereas bulk metals hardly allow microwave heating due to limited skin depth, volumetric heating of metal power compacts is possible. Native copper oxide films formed on the particle surface under ambient conditions dominate in the microwave absorption and interaction, the sintering behavior and the microstructural changes.

Acknowledgement

We acknowledge the support of this project by Deutsche Forschungsgemeinschaft (DFG) under the reference number TH 656/3-1 within the special Indo-German (DST-DFG) research program.

References

1. *Takayama S., Link G., Miksch S., Sato M., Ichikawa J. & Thumm M.*, Millimetre wave effects on sintering behaviour of metal powder compacts // Powder Metallurgy, 2006 V. 49 N. 3 P. 274-280.
2. *Rybakov K.I., Semenov V.E., Egorov S.V., Eremeev A.G., Plotnikov I.V., and Bykov Yu.V.* Microwave heating of conductive powder materials // Journal of Applied Physics 2006 V. 99, 023506
3. *Link, G.; Mahmoud, M.M.; Thumm, M.*, Dilatometric study and in situ resistivity measurements during millimeter wave sintering of metal power compacts // *Processing and Properties of Advanced Ceramics and Composites IV*. John Wiley & Sons, New Jersey. 2012. P. 145-149.
4. *Mahmoud, M.M., G. Link, and M. Thumm*, The role of the native oxide shell on the microwave sintering of copper metal powder compacts. // Journal of Alloys and Compounds, 2015. V. 627 P. 231-237.

EPJ Web of Conferences **149**, 02008 (2017) DOI: 10.1051/epjconf/201714902008
10th International Workshop 2017 "Strong Microwaves and Terahertz Waves: Sources and Applications"

Effective magnetic permeability of compacted metal powders at microwave frequencies

I.I. Volkovskaya[1,2], V.E. Semenov[1], and K.I. Rybakov[1,2]

[1] Institute of Applied Physics, Russian Academy of Sciences, Nizhny Novgorod, Russia, volkovskaya.ira@mail.ru
[2] Lobachevsky State University, Nizhny Novgorod, Russia

The calculation of the absorption of electromagnetic waves in an inhomogeneous medium consisting of many particles relies on the effective values of the dielectric permittivity and the magnetic permeability of the medium [1]. The subject of this paper is the effective magnetic permeability of an inhomogeneous medium consisting of metal particles. The response of a medium consisting of metal particles when exposed to a high-frequency magnetic field is determined by the induced magnetic moments associated with the excitation of eddy electric currents in the particles.

Until now, no method for calculating the magnetic-type losses has been proposed that adequately takes into account the mutual influence of particles when their volume concentration is significant (for example, during the sintering process). If the relative volume concentra-tion of conductive particles in a medium, η (defined as the ratio of volume of the particulate material to the total volume of the medium), is so small that each particle can be considered isolated, the effective permeability of the medium may be defined as

$$\mu_{\text{eff}} = 1 + 4\pi\eta\alpha_m, \quad (1)$$

where α_m is the complex magnetic polarizability of a spherical metal particle of radius a placed in an alternating magnetic field in vacuum [2]:

$$\alpha_m = -\frac{3}{8\pi}\left[1 - \frac{3}{(ka)^2} + \frac{3}{ka\tan(ka)}\right], \quad (2)$$

where $k = \omega\sqrt{\varepsilon}/c \approx (1+i)/\delta$, $\varepsilon = \varepsilon' + i4\pi\sigma/\omega$, σ is the conductivity of the particle material, ω is the cyclic frequency of magnetic field oscillation, $\delta = c/\sqrt{2\pi\sigma\omega}$ is the penetration depth of the magnetic field into metal (skin depth), c is the speed of light, and i is the imaginary unit.

To account for the mutual influence of particles at high values of the relative volume concentration, it is necessary to determine the magnetic dipole moment of a particle in a uniform alternating magnetic field $\mathbf{H}_0$ when the particle is surrounded by an effective medium with a magnetic permeability μ_{eff}. Similarly to [2], the magnetic field outside the spherical particle is a superposition of the external field and the dipolar field created by the particle:

$$\mathbf{H} = \mathbf{H}_0 - A\nabla\left(\frac{\mathbf{H}_0 \cdot \mathbf{r}}{r^3}\right), \quad (3)$$

where A is an arbitrary constant that is related to the magnetic dipole moment of the particle, $\mathbf{p}_m$, as

$$\mathbf{p}_m = \left(\frac{1 + 2\mu_{\text{eff}}}{3}A + \frac{\mu_{\text{eff}} - 1}{3}a^3\right)\mathbf{H}_0. \quad (4)$$

The magnetic field inside the conductive particle can be represented as

$$\mathbf{H} = B \operatorname{rot}\operatorname{rot}\left[\frac{\sin(kr)}{r}\mathbf{H}_0\right], \quad (5)$$

where B is an arbitrary constant, $k = \omega\sqrt{\varepsilon\mu}/c$, ε and μ are the dielectric permittivity and magnetic permeability of the material of the particle. A and B can be obtained from the conditions at the boundary between the spherical particle and the effective medium that surrounds it. This determines the magnetic dipole moment of the particle (4):

$$\mathbf{p}_m = -\frac{a^3}{2}\cdot\frac{\mu_{\text{eff}}\left[\xi(1+2\mu) + (ka)^2\right]}{\xi(\mu_{\text{eff}} - \mu) + (ka)^2\mu_{\text{eff}}}\mathbf{H}_\infty, \quad (6)$$

where $\xi = ka/\tan(ka) - 1$.

The effective medium can be represented as a medium filled with metallic spherical particles with magnetic dipole moments $\mathbf{p}_{mj}$ and void space. The sum of all the dipole moments of metallic spherical inclusions in the volume of the effective medium V is equal to the total dipole moment of this volume $\mathbf{p}_{m\,\text{eff}}$:

$$\mathbf{p}_{m\,\text{eff}} = \sum_j \mathbf{p}_{mj}N_j, \quad (7)$$

where N_j is the number of metallic inclusions of sort j inside the volume V of the effective medium. On the other hand, the total dipole moment of the effective medium $\mathbf{p}_{m\,\text{eff}}$ is equal to the product of the magnetization $\mathbf{M}_{\text{eff}}$ and the volume V of the medium:

$$\mathbf{p}_{m\,\text{eff}} = \mathbf{M}_{\text{eff}}V = \frac{\mu_{\text{eff}} - 1}{4\pi}\mathbf{H}_0 V. \quad (8)$$

Inserting (6) and (8) in (7), we obtain an equation for the effective magnetic permeability (the degree of which is equal to the number of phases in the system):

$$\mu_{\text{eff}} = 1 - \frac{3}{2}\sum_j \frac{\mu_{\text{eff}}\left[\xi_j\left(2\mu_j + 1\right) + (ka)^2\right]}{\xi_j\left(\mu_{\text{eff}} - \mu_j\right) + (ka)^2\mu_{\text{eff}}}\eta_j. \quad (9)$$

Since the magnetic field outside the spherical particle is calculated within a quasistatic dipolar approximation, the particle radius must be small in comparison with the wavelength in the effective medium:

$$\left|\frac{\omega}{c}\sqrt{\varepsilon_{\text{eff}}\mu_{\text{eff}}}a\right| \ll 1. \quad (10)$$

If the medium consists of metallic nonmagnetic particles with high conductivity ($4\pi\sigma/\omega \gg \varepsilon'$) and void space, Eq. (9) reduces to a quadratic equation with two parameters: the ratio of the particle radius to the skin depth a/δ and the relative volume concentration of the particles η. Similar to the previous model (1)–(2), in the model described here the dependency of the imaginary part of the effective magnetic permeability on a/δ exhibits a maximum when the particle size and the skin depth are of the same order of magnitude.

Shown in Fig. 1 is the dependency of the real (a) and imaginary (b) parts of the effective magnetic permeability, μ_{eff}, on the relative volume concentration of solid matter. In contrast to the model (1) – (2), in the proposed model the real part of μ_{eff} remains positive at high values of the particle concentration. The imaginary part exhibits a maximum at a particle concentration $\eta \approx 0.7$ (for $a/\delta \gtrsim 3$). The decrease in the imaginary part of μ_{eff} at higher volume concentration of conductive inclusions is apparently due to the influence of a dense medium on the magnetic field in it. At $a/\delta \lesssim 3$ the imaginary part of μ_{eff} increases monotonically with increasing concentration.

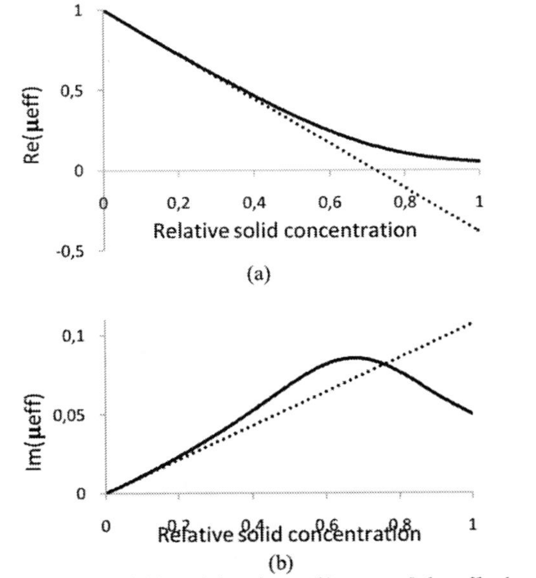

Fig. 1. The real (a) and imaginary (b) parts of the effective magnetic permeability of the metallic powder vs. the relative volume concentration of solid. The ratio of the particle radius to the skin depth is $a/\delta=20$. The solid line corresponds to the newly developed model (9), the dotted line corresponds to the previously used model (1) – (2)

As follows from Fig. 1, both models give the same result in the domain of low concentrations. With an increase in concentration, there are differences in the models in the value of μ_{eff} and in the character of the dependence of μ_{eff} on the concentration. When the volume fraction of particles tends to unity, the model (9) does not coincide with the case of a homogeneous metal, since it is assumed that the particles are electrically isolated from each other (e.g., by native oxide layers), and the eddy currents induced by the alternating magnetic field are localized inside small areas inside the medium.

Finally, it is necessary to determine the range of parameters in which the applicability condition (10) of the developed model is satisfied. The effective dielectric permittivity was calculated in the effective medium approximation that takes into account the inhomogeneous structure of the electromagnetic field inside metallic inclusions and the presence of insulating layers on metallic particles [3].

Shown in Fig. 2 is the dependency of the parameter $\left|\frac{\omega}{c}\sqrt{\varepsilon_{eff}\mu_{eff}}a\right|$ on the thickness of the insulating layer and the concentration of particles.

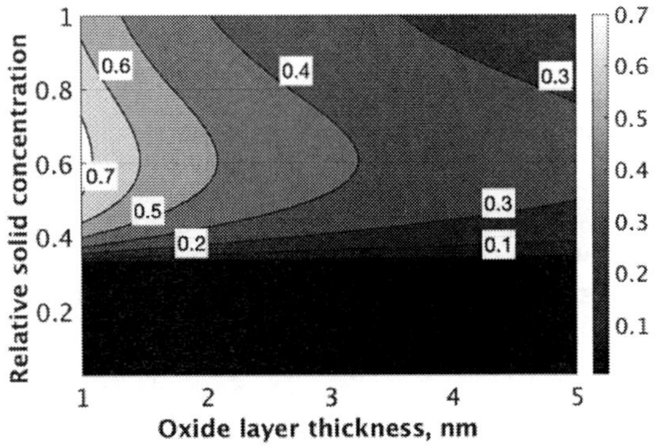

Fig. 2. Contour lines of the parameter $\left|\frac{\omega}{c}\sqrt{\varepsilon_{eff}\mu_{eff}}a\right|$ vs. the thickness of the insulating layer on particles (horizontal axis) and the relative volume concentration of particles (vertical axis). Copper powder, conductivity $\sigma = 5.96 \cdot 10^7$ S/m, particle radius $a = 10$ μm, frequency $f=\omega/2\pi = 24$ GHz

It can be seen from Fig. 2 that for the given values of the conductivity and frequency the condition (10) is satisfied when the insulating layers on particles are not too thin. In particular, it is obviously satisfied at a concentration of metallic particles below the conductivity percolation threshold (0.33), since the absolute value of the effective permittivity of the medium differs little from unity in this case.

In conclusion, a model for the calculation of the effective magnetic permittivity of powder metal compacts has been developed. The proposed model makes it possible to determine the values of the real and imaginary parts of the magnetic permeability of metal powder compacts in the microwave range. The developed model can be used to calculate microwave absorption in dense compacts of metal particles insulated by oxide layers.

This research was supported by Russian Science Foundation, grant # 17-19-01530.

References

1. *Bergman, D. J., Stroud, D.* Physical properties of macroscopically inhomogeneous media // *Solid State Physics: Advances in Research and Applications*, edited by H. Ehrenreich and D. Turnbull. New York: Academic Press. 1992. V. 46. P. 147–269.

2. *Landau, L. D., Lifshits, L. P., Pitaevskii, E. M.* Electrodynamics of Continuous Media. New York: Pergamon, 1960.

3. *Rybakov, K. I., Semenov, V. E.* Effective Microwave Dielectric Properties of Ensembles of Spherical Metal Particles // IEEE Transactions on Microwave Theory and Techniques. 2017. V. 65, No. 5. P. 1479–1487.

EPJ Web of Conferences **149**, 02009 (2017) DOI: 10.1051/epjconf/201714902009
10th International Workshop 2017 "Strong Microwaves and Terahertz Waves: Sources and Applications"

Sources of ultraviolet light based on microwave discharges

A.V. Vodopyanov

Institute of Applied Physics of Russian Academy of Sciences, Nizhny Novgorod, Russia, avod@yandex.ru

Today micro- and nano- electronics industry requires a source of extreme ultra-violet (EUV) radiation with a wavelength of 13.5 ± 1 % nm for high resolution projection lithography. The size of the emitting region must be less than 1 mm. One of the most promising sources of EUV light is considered to be a source that uses a pulsed CO_2 laser radiation focused on a specially formed stream of droplets of tin with dimensions of the order of 0.1 mm. However, along with tangible achievements in these light sources have a number of fundamental flaws that do not allow us to consider the problem of creating a EUV light source to be solved.

We propose a source of EUV radiation based on plasma heated by microwave radiation. The nonequilibrium ECR plasma is an effective source of radiation in the EUV region. Radiation of the plasma can lie in the region of 100 angstroms. This range contains the line radiation of multiply charged ions excited by electron impact.

The EUV line emission from a pulsed electron-cyclotron resonance discharge in argon, maintained by a high-power millimeter-wavelength beam in a magnetic mirror trap, was studied using a multilayer mirror EUV monochromator. The EUV spectrum was measured, and the absolute spectral intensity of emission was determined in a 6–17 nm wavelength range. The discharge can be used as an effective source of EUV light with an efficiency of the microwave to EUV light power conversion on a level of 10% and a maximum spectral power density of 7.3 kW/nm at a wavelength of 9 nm. Plasma was created and sustained by the microwave radiation with frequency of 37.5 GHz and power of about 100 kW. Scheme of the experiment shown in Fig. 1 [1].

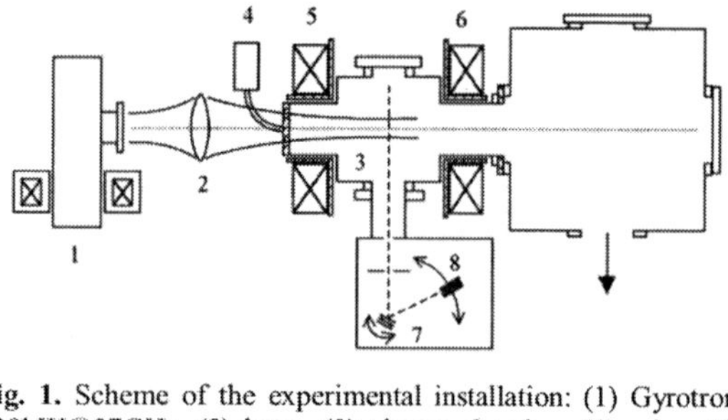

Fig. 1. Scheme of the experimental installation: (1) Gyrotron 130kW@37GHz, (2) lense, (3) plasma chamber, (4) valve, (5, 6) magnetic coils, (7) multilayer mirror, (8) EUV detector

Increase of the frequency of the heating radiation makes it possible to sustain plasma with higher density and propose the hybrid source of EUV light. Multiply charged ions are efficiently generated and excited in a discharge. Tin ions was injected into the magnetic trap from a vacuum-arc discharge and additionally stripped in the ECR plasma. Multicharged tin ions emit line radiation in the desired wavelength range. A radiation power of 50 W in a wavelength range of 13.5 nm ± 1% and an efficiency of about 1% for the conversion of the micro-wave radiation absorbed in the plasma to the extreme ultraviolet radiation were achieved in the experiments. Plasma was sustained by the microwave radiation with frequency of 75 GHz. Scheme of the experiment shown in Fig. 2 [2].

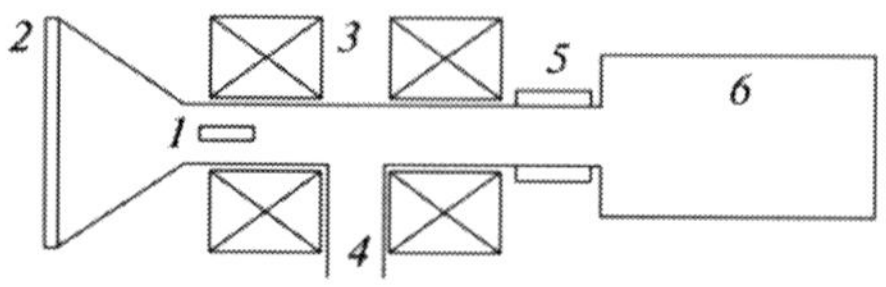

Fig. 2. Scheme of the experimental installation: (1) plasma generator, (2) microwave window, (3) magnetic coils, (4)pump-out channel, (5) magnetic cloak, (6) EUV detector or ion extractor and time-of-light analyzer of ion spectrum

A further increase in the heating radiation frequency and a transition to the terahertz range makes it possible to move close to the plasma parameters necessary to create a point-like source of extreme ultraviolet with the parameters necessary for high-resolution lithography. An increase in plasma density with increasing frequency of the heating wave to the value of 10^{15} cm^{-3} and above makes a plasma resonance heating mechanism effective with small plasma size [3]. The main idea of creating of a point-like discharge with high emissivity in the required wavelength band is the realization of a breakdown in a nonuniform gas jet with the scale of the inhomogeneity of the order of 1 mm. In this case, breakdown conditions fulfilled only in a small region of space and discharge cannot go beyond it [4]. The scheme of the experiment is presented on Fig. 3. Typical photo of the discharge is presented on Fig. 4.

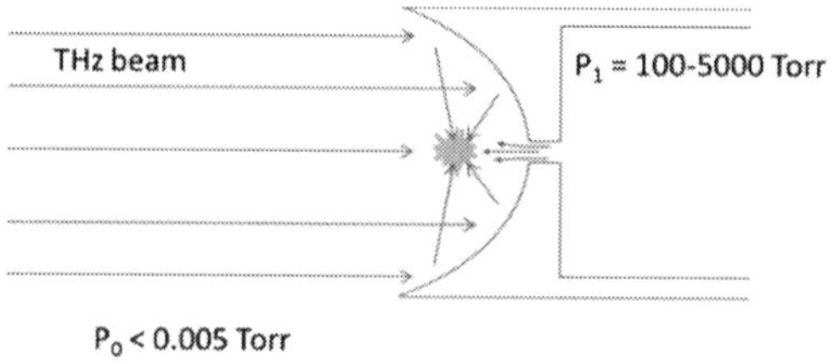

Fig. 3. Scheme of the experiment

THz radiation was coupled into the discharge vacuum chamber and focused by a parabolic mirror; THz wave power density reached 40 MW/cm^2, ensuring stable gas breakdown for the pressure values of about 20 Torr. A small nozzle (80 μm in diameter) connected to a buffer volume was incorporated in the mirror for inhomogeneous gas flow production. Neutral gas flow was $\sim 10^{20}$ particles/s at a gas pressure of 2–3 bar in buffer volume, which corresponds to ~1 bar pressure at the nozzle outlet.

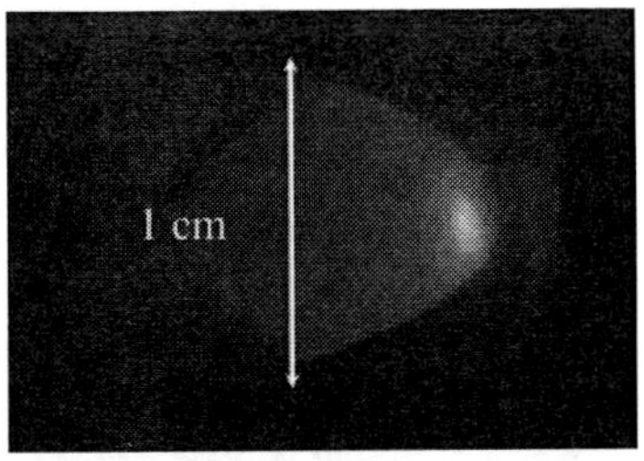

Fig. 4. Photo (in visible light) of the point-like discharge. Argon. Radiation frequency of 670 GHz

The electron concentration was determined using Stark-effect induced broadening of the H_α atomic emission line (656.3 nm) of hydrogen present in discharge as a small impurity in residual gas. The maximum observed Stark broadening of the H_α line corresponded to a plasma density on the order of 2×10^{16} cm^{-3} (see Fig. 5) which exceeded the critical value for the given frequency (670 GHz) of radiation sustaining the discharge [5].

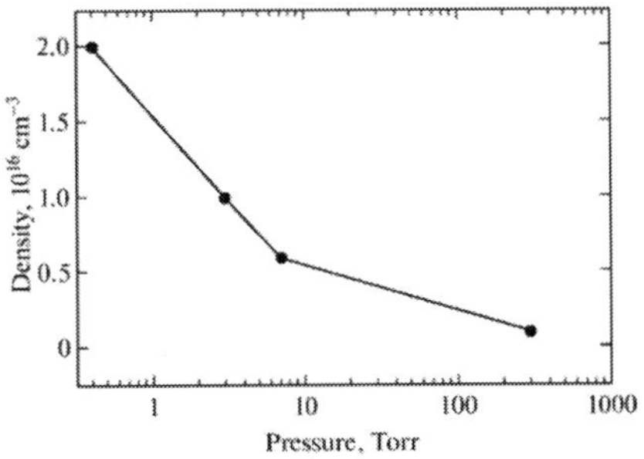

Fig. 5. Plasma density vs. background argon pressure in the discharge chamber

Output EUV radiation was measured using both photomultiplier with operating range from 112 to 400 nm and photodiode covered with different filters. Both measuring devices were connected to oscilloscope. For our experiment we used a single Mo/Zr filter with a transmission of 45 % near 13.5 nm wavelength range and almost zero in the rest range (see Fig. 6).

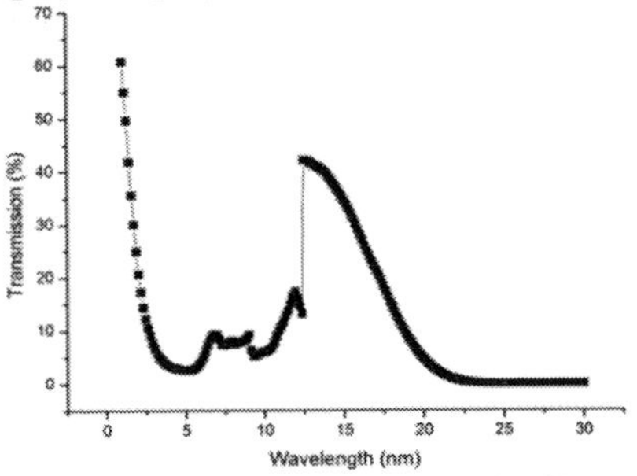

Fig. 6. Transmission spectrum of the Mo/Zr filter

Photo of a point-like xenon discharge in optimal for EUV light conversion conditions is shown in Fig. 7. In this case there is an area inside the plasma witch is slightly darker than the rest plasma's sphere. That can be caused of the fact that temperature in the center of the discharge is high enough to generate EUV photons, so plasma starts to be invisible for the camera.

Fig. 7. Photo (in visible light) of the point-like discharge in the optimal conditions. Xenon. Radiation frequency of 670 GHz

Measured in the optimal conditions EUV radiation of the point-like discharge in the wavelength range 12 - 17 nm was of 100 W/cm^3. Conversion efficiency from THz radiation to EUV emission can be raised by increasing duration and power of the heating pulse.

Acknowledgements

The work was supported by the Russian Science Foundation, project No. 14-12-00609.

References

1. *A. V. Vodopyanov, S. V Golubev, V. G. Zorin, A. Y. Kryachko, A. Y. Lopatin, V. I. Luchin, S. V Razin, and A. N. Smirnov*, "Soft x-rays generated by the electron-cyclotron resonance discharge in heavy gases sustained by a high-power microwave beam in a magnetic trap," // Tech. Phys. Lett., vol. 26, no. 12, pp. 1075–1077, 2000.

2. *A. Vodopyanov and S. Golubev, D. A. Mansfeld, A. G. Nikolaev, K. P. Savkin, N. N. Salashchenko, N. I. Chkhalo, G. Yu. Yushkov*, "Extreme-ultraviolet source based on the electron-cyclotron-resonance discharge," // JETP Lett., vol. 88, no. 2, pp. 103–106, 2008.

3. Brodskii, Yu. Ya., S. V. Golubev, V. G. Zorin, and G. M. Fraiman, "Plasma-Resonance Discharge," // Sov. Phys. JETP, vol 61 (3), 453, 1985.

4. *M. Yu. Glyavin, S. V. Golubev, I. V. Izotov, A. G. Litvak, A. G. Luchinin, S. V. Razin, A. V. Sidorov, V. A. Skalyga, and A. V. Vodopyanov*, "A point-like source of extreme ultraviolet radiation based on a discharge in a non-uniform gas flow, sustained by powerful gyrotron radiation of terahertz frequency band," // Appl. Phys. Lett., vol. 105, no. 17, 174101, 2014.

5. *A. V. Vodopyanov, M. Yu. Glyavin, S. V. Golubev, A. G. Luchinin, S. V. Razin, M. I. Safronova, A. V. Sidorov, A. P. Fokin* "Plasma Density in Discharge Sustained in Inhomogeneous Gas Flow by High-Power Radiation in the Terahertz Frequency Range"//Tech. Phys. Lett., 43, 186, 2017.

A study of RF power station for microwave rocket launch system

Y.Oda[1], T. Imai[2], K. Shimamura[2], M. Fukunari[3], H. Katsurayama[4], N. Ohnishi[5], and K. Komurasaki[6]

[1]National Institute for Quantum and Radiological Science and Technology, Naka, Japan, oda.yasuhisa@qst.go.jp
[2] Tsukuba University, Tsukuba, Japan
[3] Research Center for Development of Far-Infrared Region, University of Fukui, Fukui, Japan
[4] Yamaguchi University, Yamaguchi, Japan
[5] Tohoku University, Sendai, Japan
[6] The university of Tokyo, Tokyo, Japan

Introduction

Microwave Rocket is a space launch system which acquires propulsion energy by a high power microwave beam. Since the microwave power station for the launch system, which is the cost driver of the system, will be built on the ground and it will be used for millions of launches, Microwave Rocket is expected to realize a low cost launch system. To deliver microwave power to the vehicle traveling to the satellite orbit, the millimeter-wave band high power beam is demanded for long distance beam transmission up to 100 km and gyrotrons are expected as RF power source. Two kinds of thrust generation mechanism for Microwave Rocket are proposed, namely CW type and repetitive pulse type. CW type utilizes heat exchange system or steady plasma to heat up propellant by microwave power [1, 2]. The repetitive pulse type utilizes aerial breakdown which induces shock wave (or blast wave) causes impulsive thrust [3].

There are some studies on Microwave Rocket such as demonstrative launch of a small model [3], theoretical approach of thrust generation model by atmospheric breakdown caused by millimeter wave [4, 5], analysis of launch scenario from ground to the orbit [6], and etc. These studies presented the possible requirement of the thrust performance, the momentum coupling coefficient from microwave power to thrust should be 100–500 N/MW. When the launch of some hundreds kg of the vehicle is planned for instance, 100 MW – 1 GW of microwave power in average is required. In this report, a study of GW-class microwave power source design will be presented. We tried to design such a super-high power source by current existing technologies now utilized for the electron cyclotron heating (ECH) system for fusion devices. For example, ECH system for ITER generates 20 MW of 170 GHz RF power by clustering 24 gyrotrons [7]. With consideration of similar technology, the approach of clustering hundreds of MW-class gyrotrons [8–12] is evaluated and the necessary technologies for auxiliary system such as microwave transmission and power control are discussed.

Required system for GW-class beam station

The microwave beam station for Microwave Rocket launching system consists of RF power sources, antenna system, and transmission line as similar to RF heating system.

To generate 100 MW – 1 GW RF power which is requirement for Microwave Rocket launch, the only realistic solution for such high power system is the cluster of MW-class gyrtrons which are available technology today. Then RF power source of the beam station forms a cluster of some hundreds of gyrotron tubes. The required power level is extremely high in such application, all the gyrotron tubes of the cluster is expected to operate at its maximum operation power to minimize the number of tubes. Then RF power control device, power divider or combiner, is also another key function when the active phased array antenna system requires to control the power of element for instance.

The antenna of the beam station is another key device to realize the requirement of very long distance RF transmission. The beam propagation in free space is usually based on Gaussian beam optics and the diameter of the beam determines its performance. For 100 km transmission to provide power to the vehicle in launching trajectory, the antenna is expected to have more than 100 m in diameter. In such huge scaled system, RF transmission technology will dominate the specification of the beam station and it is required to realize higher performance.

RF power source

The RF power source of the beam station consists of the cluster of gyrotron tubes. This configuration is similar to RF power source of ECH system. One of the restrictions of gyrotron cluster forming is the minimum distance between gyrotrons caused by interaction of magnetic field from neighboring gyrotron system. Kajiwara et al. analyzed the effect of magnetic field of neabering gyrotrons and the magnetic field profile at collector is affected by the neighboring magnet [13]. This may causes the abnormal increase of the collector heat load. The analysis result finally determined the location of gyrotrons in the building. In the GW-class beam station, the minimium distance between gyrotrons is the most dominant parameter to determine the scale of the RF source and the beam station site. The power source builidng can become far larger than the huge antenna system for long distance transmission.

Transmission line

RF power generated by gryotrons is transmitted to antenna system. Since scale of the antenna system for the beam station is expected greater than 100 m, the distance of RF transmission from output window of gyrotron to antenna element is longer than 100 m or more. For such long distance transsmission technology is low-loss transmission technology is mandatory.

There is two major method for the long distance low-loss transmission, namely corrugated waveguide and quasi-optical beam transmision. The corrugated waveguide system is widely used in ECH transmission line [14]. In ITER ECH system, its RF transmission disntance and effieicency is more than 160 m and 89%, respectively [15]. A signel waveguide line can be transmit more than 1 MW of RF power and it is good point to combine the multi beam lines in limited space. On the other hand, it is mandatory to evacuate the waveguide for MW-class transmission and output window is necessary at radiation point. The quasi-optical beam transmission is also used for ECH transmission line such as in W7-X [16]. The beam transmission is able to operate in atmospheric ambient. However the power density is limited and it will be hard to combine many transmission line.

Power control system

RF power combining and dividing technology is very useful for active control of RF power. The high power RF diplexer which utilizes ring resonation is developed as fast switching device of RF beam for ECH system, such as FADIS by W. Kaparek et al. this technology can be applied for power divider and combiner of beam lines [17]. M. Petelin et al. proposed the cascated resonator which can be utilized for multiplexer [18]. Yet thre requirement of power control is not clearly defined for beam station. Indeed the avilities of these technology is promissing to realize necessary operation of beam power control without changing gyrotron output.

Antenna

The antenna system for the microwave beam station for rocket launching needs to transmit RF beam for more than 100 km distance at high power effieciency. For such long distance beam transmission, the scale of antenna system is more than 100 m in diameter [19]. By using large scle antenna, the power density of the beam near by antenna becomes some-tens W/cm^2 even for the GW-class total output power. This power density level is far lower than MW-class tranmission line. Then generally RF power in the beam line is divided to several elements at antenna.

To launch the rocket, it is necessary to track the beam to launching vehicle which flies away from the station. For this operation, the active phased array technology is expected very usuful because the active phased array antenna can control the beam direction and focusing point by controlling phase and power profile of front plane of the antenna. To realize such operation, power and phase control for high power RF is one of key technologies for develiopment of GW-class beam station system.

References

1. A. R. Bruccoleri, et al., J. Propulsion Power, 23, 4, 889 (2007).
2. G.M. Batanov, et al., Plasma Physics and Plasma Electronics, 241 (1989).
3. T. Nakagawa, et al., J. Spacecraft Rockets, 41, 151, (2004).
4. Y. Oda, et al., J. Propulsion Power, 25, 1, 118 (2009).
5. Y. Oda, et al., J. Appl. Phys., 100, 113308 (2006).
6. H. Katsurayama, et al., J. Space Tech. Sci., 20, 2, 32, (2005).
7. C. Darbos, et al., J. Infrared, Milli. Terahz, Wav.. 37, 1, 4 (2016).
8. K. Sakamoto, et al, Nature Phys., 3, 411 (2007).
9. K. Kajiwara, et al., Fusion Eng. Des., 86, 955 (2011)
10. G.G Denisov, et al., Nuclear Fusion, 48, 054007 (2008).
11. J.Lohr, et al., J. Infrared Milli. Terahz. Wav., 32, 241 (2011).
12. T.Kariya, et al., J. Infrared Milli. Terahz. Wav., 32, 241 (2011).
13. K. Kajiwara et al., Effect of the stray magnetic filed on the gyrotrons for ITER, 34th International Conference on Infrared, Millimeter, and Terahertz Waves (IRMMW 2009), Busan, Korea, 2009.
14. J. Doane and R. Olstad, Fusion Sci. Tech., 53, 39, (2008).
15. F. Gandini et al., Fusion Sci. Tech., 59, 709 (2011).
16. V. Erckmann et al., Fusion Sci. Tech., 52, 291 (2007).
17. W. Kasparek et al, Nuclear Fusion, 48 054010 (2008).
18. M. Petelin et al, Plans for quasi-optical experiments at JAEA and NIFS, Proc. 9th International Worksho on Strong Microwaves and Terahertz Waves: Sources and Applications, Nizhny Novgorodo, Russia, 2014.
19. Fukunari et al., J. Infrared, Milli. Terahz, Wav., 38, 2, 176 (2017).

Excitation of wakefields by relativistic electron bunches in the dielectric waveguide filled with radially inhomogeneous plasma

G.V. Sotnikov, P.I. Markov and I.N. Onishchenko

NSC "Kharkov Institute of Physics and Technology", Kharkov, Ukraine, sotnikov@kipt.kharkov.ua

One of shortcomings of the dielectric wakefield accelerators (DWA) limiting the effective length of an acceleration is susceptibility of drive bunches to beam breakup (BBU) instability [1]. With the purpose to suppress this instability recently it has been proposed to fill the drift channel of dielectric structure with plasma of certain density [2]. The investigations of plasma DWA (PDWA) [2] were made for dielectric wakefield structures with homogeneous radial plasma density. Such consideration is right if plasma comes to the drift channel from external source.

Other, widely way of plasma creation is capillary discharge. For this case on big times from the beginning of discharge (it is more than a half of pulse width of discharge current) steady distribution of plasma density with inhomogeneous distribution of discharge density forms [3]. At short distances from the axis a radial dependence of plasma density can be described with the parabolic dependence [3, 4]

$$n(r) \approx n_0[1+0.33(r/a)^2], \qquad (1)$$

where a is the inner radius of dielectric tube.

Investigations on excitation of wakefield in PDWA with and inhomogeneous transverse distribution of plasma density were not carried out so far. For practical application of the proposed accelerating structure it is important to know how the accelerating gradient and focusing forces acting on test bunches will change in case when plasma in the drift channel is created as a result of the capillary discharge.

Here we report the results of numerical simulation of wakefield excitation in the dielectric structure filled with plasma created in capillary discharge. Wakefield is built up at injection of electron bunch in dielectric tube with permittivity of 3.8; outside diameter of dielectric tube (is equal to the diameter of metal waveguide) is 1.2 mm, inner diameter is 1.0 mm. Energy of drive bunch electrons is 5 GeV, bunch charge is 3 nC, its length was 0.2 mm, solid bunch has diameter of 0.9 mm. For the testing the excited wakefield we used a witness bunch. Witness bunch has the same energy and sizes as drive bunch, the charge of witness bunch was less by ten times than drive bunch charge.

The drift channel (internal area of dielectric tube) is filled with plasma of on-axis density $n_0 = 4.41\times10^{14}$ cm^{-3}. Three different models of dependence of plasma density on radius were investigated:

a) homogeneous radial plasma density;

b) parabolic dependence of plasma density on radius according eq. (1);

c) the dependence obtained numerically by N.A. Bobrova [3] et al. when modeling capillary discharge (see Fig. 3 of Ref [3] or Fig. 2 of Ref. [4]). Last numerical dependence we will denote as BPM dependence.

At numerical simulation by means of the 2.5D PIC code created by us we examined a structure of wakefield and dynamics of electron bunches at their motion in the drift camera.

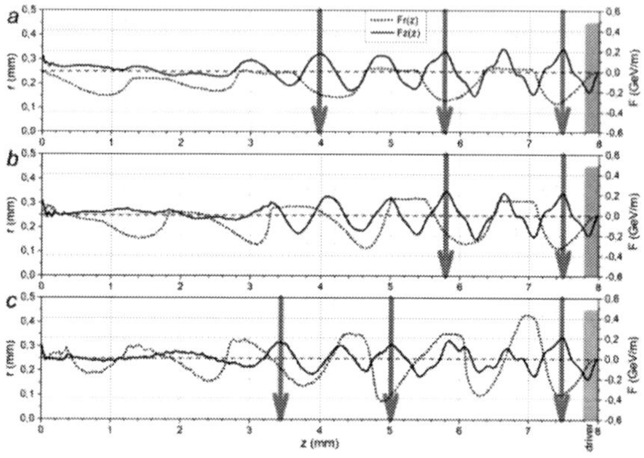

Fig. 1. Axial profiles of the transverse wakefield force (short dot line) and longitudinal wakefield force (solid line) acting on a test electron at the distance of 0.45 mm from center of the accelerating channel at time t = 26.69 ps for tree cases of plasma density distributions pointed above (a, b and c)

In Fig. 1 axial profiles of the Lorentz wakefield force acting on test electron in dielectric waveguide with plasma filling for the time t = 26.69 ps from the beginning of injection of drive bunch for different dependences of plasma density on radius are shown: a) homogeneous, b) parabolic (1) and c) the BPM dependence [3]. The grey rectangle near output end shows the location of the drive bunch electrons. Vertical arrows have noted the possible placements of the witness bunch providing not only its acceleration at a local maximum of the longitudinal force F_z, but also the transverse focusing caused by negative value of transverse force F_z in these locations.

One can conclude from Fig. 1 the amplitude (~ 200 MeV/m) and frequency (~372.2 GHz) characteristics of the longitudinal wakefield (accelerating gradient) weakly depend on what model of dependence of plasma density on radius is used in calculations. At the same time the transverse (focusing) wakefield significantly changes both amplitude, and frequency characteristics. Thus for model proposed by N.A. Bobrova (Fig. 1c) amplitudes of transverse field is approximately twice more, and the frequency by 1.4 times higher, than at homogeneous distribution (Fig. 1a). It should be taken into account at elaborating the two-beam PDWA [2].

Let's consider a dynamics of drive and witness bunches in the excited wakefields. To provide the required shift of the witness bunch relative to drive bunch (see arrows in Fig. 1) it is necessary to carry out the injection of the witness bunch with certain delay t_{del}.

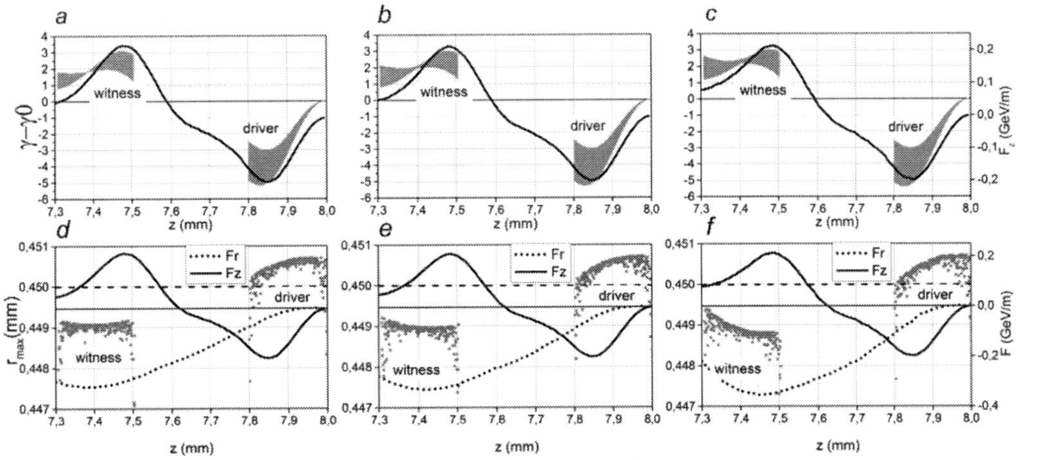

Fig. 2. a, b, c - the phase plane "energy - longitudinal coordinate" combined with dependence of longitudinal force; d, e, f - the configuration space displaying the positions of peripheral electrons of bunches combined with profiles of longitudinal (solid line) and transverse (dot line) forces for different dependences of plasma density on radius: a, d - homogeneous, b, e - parabolic, c, f - the BPM dependence [3]. The delay time t_{del}= 1.651 ps

In Fig.1 rightmost arrows for the three cases correspond the same delay time t_{del}= 1.651 ps.

At this delay it is observed both acceleration and focusing of the witness bunch for all studied dependences of plasma density on radius. The phase planes "energy – longitudinal coordinate" and the positions of peripheral electrons combined with dependences of longitudinal and transverse forces for different dependences of plasma density on radius are given in Fig. 2. The left column corresponds homogeneous plasma, middle column shows the results of the simulations for the parabolic law of plasma density (1), and the right column is for the BPM dependence of radial plasma density.

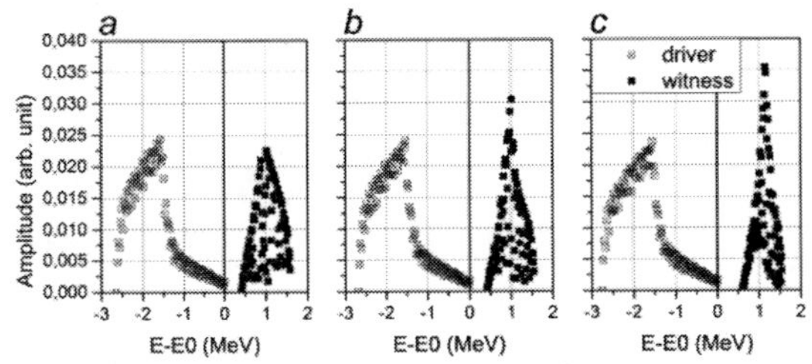

Fig. 3. Distribution functions of electrons of the drive and the witness bunches on energy at t = 26.69 ps for different dependences of plasma density on radius: a) homogeneous, b) parabolic (1), c) the BPM dependence [3], E_0 = 5 GeV. The delay of witness bunch is 1.651 ps

In Fig. 3 distribution functions of electrons of the drive and witness bunches on energy for different dependences of plasma density on radius at delay t_{del}= 1.651 ps are shown. Comparing Figs. 2 and 3 it is possible to conclude that the greatest acceleration and focusing of test bunch are observed for the dependence of plasma density obtained numerically by N.A. Bobrova et al. [3].

If we chose other delays of witness bunch pointed in Fig. 1 by arrows we can obtain either greater acceleration or stronger focusing of witness bunch than described above. But a greater acceleration provides a worse focusing and vice versa.

The results of numerical simulation by "particle in cell" method, given above, testify that inhomogeneity of plasma in transverse cross-section is noncritical and advantages of PDWA structure with homogeneous plasma remain. I.e. even if plasma in PDWA is created by capillary discharge [3, 4], opportunity to accelerate and focus test bunch remains. Moreover, the greatest acceleration and the best focusing of test bunch are observed for the BPM dependence obtained numerically in work [3].

It should be note that the drive bunch is focused both in homogeneous, and in inhomogeneous plasma.

Work supported by NAS of Ukraine program "Perspective investigations on plasma physics, controlled thermonuclear fusion and plasma technologies", Project P-1/63-2017.

References

1. *Li C., Gai W., Jing C. et al.* High gradient limits due to single bunch beam breakup in a collinear dielectric wakefield accelerator // Phys. Rev. ST Accel. Beams, 17, 091302(2014).

2. *Sotnikov G.V., Kniaziev R.R., Manuilenko O.V. et al.* High gradient limits due to single bunch beam breakup in a collinear dielectric wakefield accelerator // Nuclear Instruments and Methods in Physics Research Section A: Accelerators, Spectrometers, Detectors and Associated Equipment. 2014, V. 740. P. 124-129.

3. *Bobrova N. A., Esaulov A. A., Sakai J.-I. et al.* Simulations of a hydrogen-filled capillary discharge waveguide // Phys. Rev. E, 65, 016407 (2001).

4. *Steinhauer L.C., Kimura W.D.* Quasistatic capillary discharge plasma model // Phys. Rev. ST Accel. Beams, 9, 081301 (2006).

EPJ Web of Conferences **149**, 02012 (2017) DOI: 10.1051/epjconf/201714902012
10th International Workshop 2017 "Strong Microwaves and Terahertz Waves: Sources and Applications"

Data rates of SubTHz wireless telecommunication channels

G.M. Bubnov, I.V. Lesnov and V.F. Vdovin

Institute of Applied Physics RAS, Nizhniy Novgorod, Russia, vdovin_iap@mail.ru

The work is devoted to the actual problem of data rate of wireless telecommunication channels. Presented analysis of the telecommunication channel of subterahertz (subTHz) frequency range – as the most promising band for the implementation of wireless telecommunications for space links and terrestrial communications of high capacity.

There are some market researches, indicating a big and increasing interest to THz communications [1].

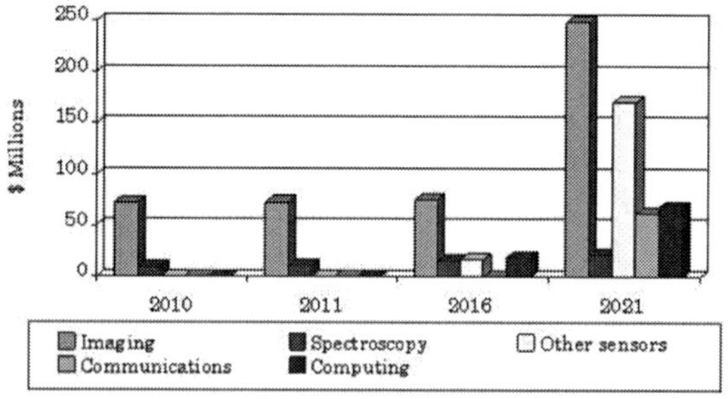

Fig. 1. Forecast of the THz telecommunication market potential [1]

Channel is considered as a combination of technical realization of high effective transponder – transmitter together with the atmospheric open line. The means to achieve a high signal/noise ratio and high data rates is usage of low-noise cryogenic (subTHz) receivers actively developing in IAP for radioastronomy and atmosphere spectroscopy [2]. The theoretical analysis of data rates for various natural and technical implementa-tions of communication channels demonstrated reasonable limits of cooling of receivers about ~10 K, providing a weighty increase channel capacity, while deeper cooling is impractical due to atmosphere restrictions in certain ranges and conditions of signal propagation, including altitude and seasonal features. The presented theoretical analysis of data rates have been confirmed by experiments related to the implementation of the telecommunication channel Earth – Space for communication with deep space missions, performed in the framework of the international project Exo-Mars, launched on March 13, 2016. There are cooled down 7 K front end of receivers with total noise less then 20 K (including 6 K noise temperature of receivers) [3]. See Fig. 2.

The calculation of the capacity of wireless telecommunications was carried out on the basis of the the Shannon – Hartley theorem, under the Dryagin – Goronina limitations [4]:

$$SNR = \frac{P_{signal}}{P_{noise}}, \qquad (1)$$

where P_{noise} – is the power of additive white Gaussian noise and P_{signal}, power of signal.

The receiver noise temperature, or T_n, in the the Shannon – Hartley theorem would be a more accurate parameter then noise factor (F). This value does not include the temperature T_o which for cooled receivers (to get extremely high sensitivity, receivers normally are incorporated into the cryogenic cooling system) may significantly differ from the standard T_o=300 K. T_n is related to F through the well-known formula:

$$F = 1 + T_n/T_o. \qquad (2)$$

A more precise formula for determining the signal-to-noise ratio of the receiver will therefore be as follows:

$$SNR = P_{rx} / k\ B\ (T_n+T_o) \qquad (3)$$

thus yielding the developed Shannon formula for the link capacity estimation:

$$C = B\ \log2\ (1+ P_{rx} / k\ B\ (T_n+T_o)). \qquad (4)$$

SNR of a total channel with atmosphere is determined by two models – free-path-lose attenuation and integral atmospheric attenuation:

$$P_{signal_{db}} = P_{tx} + G_{tx} + G_{rx} + 20\log\frac{c}{4\pi d\nu} - \alpha_h, \qquad (5)$$

where α_h – integral total atmospheric attenuation from h to ~200 km.

In our work we have analyzed capacity of channel Earth – Space depends of Equivalent receiver noise temperature, ambient temperature, altitude of a receiver location, angle of inclination relative to the zenith.

The best results are obviously for open space communications, however there are definite possibilities not

Cryo LNA in lab

TNA 1500 antenna

CryoLNA in focus TNA 1500

Fig. 2

only for Space – Earth cannels but for terrestrial too, but only for atmospheric windows 1.3 and 0.8 mm.

References

1. Gert de Lange. Valorization Seminar. TeraDec Meeting, Apr. 2012, SRON, Utrecht, the Netherlands.
2. V. F. Vdovin . and I. I. Zinchenko Modern Millimeter and Submillimeter receiver systems for Radio Astronomy. Radiophysics and Quantum Electronics, Vol. 52, No. 7, 2009, p. 461.
3. Cryogenically Cooled Low-Noise Amplifier for Radio-Astronomical Observations and Centimeter-Wave Deep-Space Communications Systems / V. F. Vdovin, V. G. Grachev, S. Yu. Dryagin, A. I. Eliseev, R. K. Kamaletdinov,D. V. Korotaev, I. V. Lesnov, M. A. Mansfeld, E. L. Pevzner, V. G. Perminov, A. M. Pilipenko, B. D. Sapozhnikov, V. P. Saurin // Astrophysical Bulletin – 2016. – V. 71. – № 1. – pp. 125-128.
4. Yu.A. Dryagin and K.A. Goronina On the Noise Parameters of Diode Mixers. Journal of Radio Electronics, № 1, January 2002.

EPJ Web of Conferences **149**, 02013 (2017) DOI: 10.1051/epjconf/201714902013
10th International Workshop 2017 "Strong Microwaves and Terahertz Waves: Sources and Applications"

Optical emission spectroscopy for diagnosis of diamond growth and etching processes in microwave plasma

V.Yu. Yurov[1,2], E.V. Bushuev[1], A.P. Bolshakov[1,2,3], I.A. Antonova[1,2], V.G. Ralchenko[1,2,3], V.I. Konov[1,2]

[1] General Physics Institute RAS, Moscow 119991, Russia, yurov6591@gmail.com
[2] National Research Nuclear University MEPhI, Moscow 115409, Russia
[3] Harbin Institute of Technology, 150001 Harbin, P.R. China

For a better control and understanding of growth process of CVD single crystal diamond in microwave (MW) plasma it is important to monitor a large number of the gas phase parameters (chemistry, temperature, concentrations, etc.). A relatively simple and non-perturbing method to measure the plasma parameters is the optical emission spectroscopy (OES) [1, 2]. The OES advantage is also in a fast response to variation of the plasma parameters in the reactor. Here we present the results of OES application to characterize the MW plasma (2.45 GHz) in course of diamond deposition in H_2-CH_4 mixtures and etching in pure H_2.

The emission from the plasma passed through a quartz window in the CVD reactor chamber and an image was formed by a lens onto a matted silica plate (OES configuration (I)). The emission collected by a 0.6 mm diameter quartz fiber was directed to a spectrometer (Ocean Optics HR 4000, wavelength range of 400–800 nm, spectral resolution 0.6 nm. The fiber tip was translated across the image of the plasma to record spatially resolved emission spectra along X or Z axes, parallel and perpendicularly to the substrate, respectively, with the istance of –25 to +25 mm along the X axis (counting from the plasma center), and from zero (the substrate position) to 50 mm along Z axis. The probed volume was a cylinder with diameter of 1.5–2 mm directed along the plasma ball diameter (Y direction), thus the spatial resolution of ~1.5 mm in the OES was obtained both for X and Z axes [2].

In another OES configuration (II) a M833 spectrometer (Solar Laser System) with high spectral resolution of 0.01 nm, equipped with a double grating with 2400 lines/mm and CCD image sensor Hamamatsu S10420-1006, was used. The optical fiber system collected the emission from a cylinder of 12 mm diameter directed horizontally along the plasma ball diameter (Y direction). The gas temperature was evaluated from OES data measured in the central region of plasma by the analysis of the (0,0) band of C_2 Swan system ($d^3\Pi_g \rightarrow a^3\Pi_u$) following the procedure described in [3, 4]. A thermal equilibrium between heavy species' translational and rotational modes, and a Boltzmann distribution of the rotational levels of the considered vibrational state are assumed.

The spectroscopy of the Swan transitions of the C_2 molecule (λ = 516 nm) with well-resolved rotational components was used to measure the gas temperature with accuracy of ±150 °C upon wide variation, 1÷13%, in methane concentration [5]. The rotational temperature T_{rot} only weakly depends on the absorbed MW power (Fig.1). It increase from ~3000 K at low (2%) CH_4 content to 4000–4500 K at higher (7–13%) CH_4 concentrations.

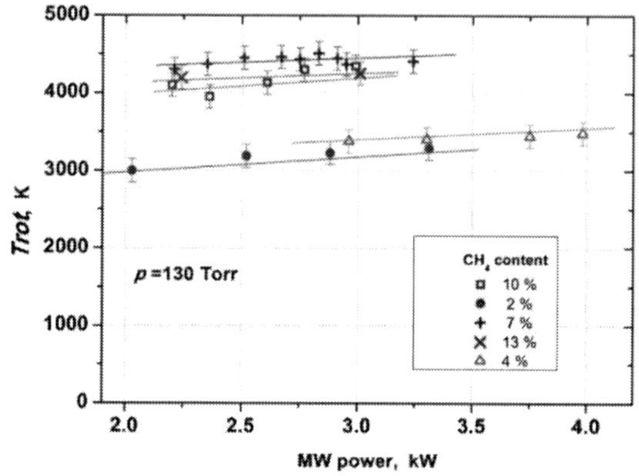

Fig. 1. Rotational temperature T_{rot} in the core of plasma cloud measured from C_2 fine structure spectrum vs MW power at CH_4-consentations: 2, 7, 4 10, 13, 10%, and pressure p = 130 Torr

The EOS data are useful for estimation of MW power density in the discharge. As the atomic hydrogen line H_α is excited mainly by electron impact, the intensity of H_α line gives an information about the plasma size [6]. With spatially resolved EOS for atomic hydrogen line H_α the intensity profiles of plasma glow were reconstructed in different cross sections. This, in turn, allows correct assessment of the MW power density, which eventually governs the rates of all chemical reactions in the plasma [1, 7].

We used two different approaches to measure H_α profiles in vertical (Z) and horizontal (X) directions of the discharge. The first one was based on spatially-resolved EOS in configuration (I) and spectrometer measured H_α line intensity. The second method implies using photographs of the plasma captured with a digital mirror camera Canon EOS 650D through a red H_α–filter passing only the H_α line emission (λ = 656.5 nm ± 6 nm). The spatial profiles of the H_α line emission along X and Z axes were determined from those images, and the shape of the plasma was very well fitted with Gaussian profiles, with further calculation of the volume and microwave power density (MWPD) values. It was important to save images as a RAW-data, that keep linearity in intensity in the photo [1]. The both approaches gave consisted results and very good fits with Gaussian profiles.

Diamond etching with atomic hydrogen in a microwave plasma is an important process in preparation of hydrophobic surfaces, highly stable interfaces modified with a variety of biomolecules, in pre-treatment of single crystal (SC) substrates in a H_2 or H_2-O_2 mixtures to eliminate defects before homoepitaxy. Due to low etching rates the etching kinetics measurements may take much time, and these data still are rather rare.

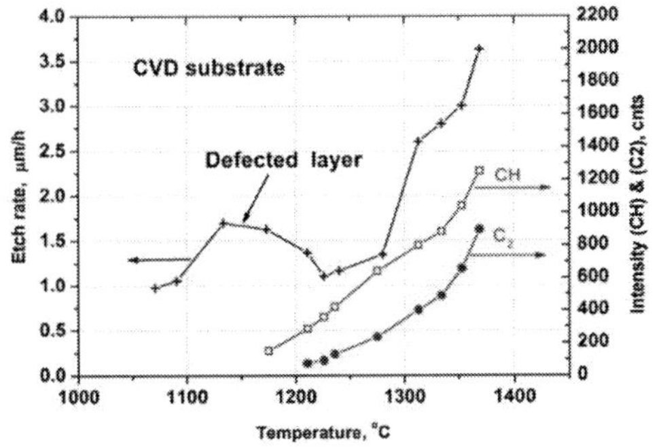

Fig. 2. Etching rate for CVD sample (crosses) in pure H_2 plasma and intensities of CH (open squares) and C_2 (filed circles) lines in OES spectra of the plasma vs substrate temperature at pressure 130 Torr

The EOS is also interesting for study diamond etching in pure hydrogen, which is relevant for substrate pretreatment in diamond growth technology. Since the etching of diamond supplies carbon species into the plasma, OES is able to monitor the emission lines from CH and C_2 radicals. High resolution OES configuration (II) was used in such experiments. Simultaneously, the etching rate of (001) plane of SC diamond was measured with a low coherent interferometry [7]. The etch rate increases with the substrate temperature T_s, however, non-monotonically (Fig. 2). The etching rate spike in the temperature range of 1170–1230 °C is caused by enhanced etch of a defective surface layer formed by the sample polishing. At higher T_s the etching rate follows Arrhenius law with activation energy of 43 ± 5 кcal/mol.

At the lower T_s of 1092 °C only the hydrogen lines H_β and H_γ are seen in OES spectra, but with the temperature increase to T_s = 1369 °C the CH line at 431 nm and the C_2 peak (516 nm) are revealed, as the supply of carbon to the plasma becomes significant. While the primary etching products of CVD sample are hydrocarbons C_xH_y, the CH and dimer C_2 radicals observed in OES, form by further reactions in the plasma.

Both I_{CH} and I_{C2} intensities monotonically increase with etch temperature, following in general the enhanced etch rate. The CH_x radicals are assumed to be primary etching products, which are then consumed in a part for conversion to a number of carbon-containing species, such as C_2H_2, CH_4, C_2, and others to reach an equilibrium in the plasma [8]. We checked that CH and C_2 lines appeared in OES spectra only in presence of a hot diamond substrate in the reactor. No lines associated with carbon-containing species were observed in the plasma optical emission spectra in absence of the substrates under identical MW power and pressure. Therefore the CH and C_2 emission we monitored was surely not due to the plasma etching of carbon contamination from the reactor chamber walls.

As the substrate temperature in the present work was tuned by variation of microwave power P, rather than by an independent heating/cooling, a question arises if the plasma chemistry (concentration of atomic H) contributes to the etching kinetics in addition to the main effect of the substrate temperature. The elementary processes and concentration of species in the plasma are determined by the microwave power density (MWPD) P/V, where V is the plasma volume. We checked how the MWPD depends on the absorbed power, using spatial profiling of H_α line intensity, and we found the MWPD to be almost constant. This suggests a stability of plasma chemistry and atomic hydrogen flux towards the diamond surface, therefore we assign the observed etch effects solely to the temperature-depended surface chemical reactions.

The measurement of OES intensities of CH and C_2 lines could be a simple and convenient qualitative indicator of etching rate of diamond in H_2 plasma, and presumably also in plasmas with more complex compositions, such as H_2-O_2 or H_2-Ar.

This work was supported by the Russian Science Foundation, grant No. 14-12-01403-P.

References

1. *M.A. Lobaev, S.A. Bogdanov, et al.* Method of power density determination in microwave discharge, sustained in hydrogen–methane gas mixture // Diam. Relat. Mater. 2016. V. 66. P. 177-182.

2. *A.P. Bolshakov, V.G. Ralchenko, et al.* High-rate growth of single crystal diamond in microwave plasma in CH4/H2 and CH4/H2/Ar gas mixtures in presence of intensive soot formation // Diam. Relat. Mater. 2016. V.62. P. 49-57.

3. *X. Duten, A. Rousseau, et al.* Rotational temperature measurements of excited and ground states of C_2 ($d^3\Pi g$ - $a^3\Pi u$) transition in a H_2/CH_4 915 MHz microwave pulsed plasma // J. Appl. Phys. 1999. V. 86. P. 5299-5301.

4. *G. Lombardi, F. Bénédic, et al.* Determination of gas temperature and C_2 absolute density in Ar/H2/CH4 microwave discharges used for nanocrystalline diamond deposition from the C_2 Mulliken system.// Plasma Sources Sci. Technol. 2004. V. 13. P. 375–386.

5. *E.V. Bushuev, V.Yu. Yurov, et al.* Express in situ measurement of epitaxial CVD diamond film growth kinetics // Diam. Relat. Mater. 2017. V. 72. P. 61–70.

6. *J. Ma, M.N.R. Ashfold, and Y. A. Mankelevich* , Validating optical emission spectroscopy as a diagnostic of microwave activated CH4/Ar/H2 plasmas used for diamond chemical vapor deposition // J. Appl. Phys. 2009. V.105. P. 043302.

7. *E.V. Bushuev, V.Yu. Yurov, et al.* Synthesis of single crystal diamond by microwave plasma assisted chemical vapor deposition with in situ low-coherence interferometric control of growth rate // Diam. Relat. Mater. 2016. V. 66. P. 83-89.

8. *J. Ma, J.C. Richley, et al.* Probing the plasma chemistry in a microwave reactor used for diamond chemical vapor deposition by cavity ring down spectroscopy //J. Appl. Phys. 2008. V. 104. P. 103305.

Dependence of boron incorporation in delta layers on CVD diamond growth process and misorientation angle

M.A. Lobaev[1], A.M. Gorbachev[1], A.L. Vikharev[1], D.B. Radishev[1], V.A. Isaev[1], S.A. Bogdanov[1], P.A. Yunin[2], M.N. Drozdov[2], J.E. Butler[1]

[1] Institute of Applied Physics of the Russian Academy of Sciences, Nizhny Novgorod, Russia
[2] Institute for Physics of Microstructures of the Russian Academy of Sciences, Nizhny Novgorod, Russia

Introduction

Diamond is a promising material for the next generation of high-power and high-frequency semiconductor devices. In terms of its physical properties: high carrier saturation velocity ($2.7 \cdot 10^7$ cm/s), high electron and hole mobility at low doping level, and record thermal conductivity, diamond significantly exceeds other semiconductor materials. Diamond is a wide-band gap semiconductor/insulator (band gap of about 5.5 eV), and partly because of this, all known dopants create the deep-lying levels with high activation energy. At room temperature only a very small fraction of the dopants are ionized. To create an acceptable level of conductivity, it is necessary to increase the level of doping, which inevitably leads to a decrease of the carrier mobility in diamond. To solve the problem of diamond doping the delta-layer technology is being developed. A nanometer thin layer of highly boron-doped diamond (the thickness of a few nanometers, $N_B > 10^{20}$ cm^{-3}) is introduced in the undoped high-quality defect-free diamond. The achievement of high carrier mobility in this region requires the realization of sharp boundaries between doped and undoped material. This has been a rather difficult experimental problem. In this paper, we present the results of investigation of the boron incorporation into diamond delta doped layers as a function of the growth conditions and misorientation angle.

Experiment

Delta layers doped with boron were grown in home-made CVD reactor, described in detail in [1]. The reactor consists of a cylindrical cavity with a quartz tube placed on its axis. Inside the tube there is a substrate holder, over which a plasma is created using a magnetron at a frequency of 2.45 GHz. Inside the quartz tube, a laminar vortex-free gas flow is maintained at a flow of 900 sccm, which makes it possible to rapidly switch the composition of the gas mixture. Non-doped diamond is grown in a gas mixture of H_2+CH_4 (CH_4/H_2 = 0.1%). To grow the delta layer, this mixture rapidly (in comparison with the characteristic growth time of the diamond) changed to a gas mixture containing boron H_2+CH_4+B_2H_6 (CH_4/H_2 = = 0.1%, B_2H_6/H_2 = 0.1%) using a gas switch. The characteristic time for changing the composition of the gas mixture was found to be about 5 seconds. Typical epitaxial diamond growth rates were 40 to 80 nm/h. Due to presence of residual boron in the reactor the undoped diamond layers had low boron concentration, typically 10^{17} cm^{-3} or less, an amount which has little impact on the carrier mobility. Addition of hydrogen sulfide to gas mixture was used as a chemical getter to decrease the residual boron contamination during growth of undoped layers [2].

For the growth of CVD diamond delta layers doped with boron, IIa type high pressure high temperature (HPHT) grown diamond substrates from New Diamond Technologies with (100) orientation and size 3.0×3.0×0.5 mm were used. These substrates are nearly strain free as observed through crossed polarizers, with very low dislocation densities (ca. 10^2 cm^2). The substrates were mechanically polished to a surface roughness (Ra) of 0.1 nm, measured with a Zygo NewView 7300 white light interferometer on an area of 0.22×0.22 mm. To remove defects introduced in the near surface region by polishing, a layer of 4–5 μm was homogeneously etched from the substrate in ICP plasma (Oxford Instruments, Plasmalab 80). As a result, defect-free substrates with an atomically smooth surface were used to grow the delta layers of the CVD diamond.

Measurement of the boron concentration in the grown samples was carried out by the secondary ion mass spectroscopy (SIMS) method using a time of flight SIMS setup (IONTOF TOF.SIMS-5). The depth of the etch craters for calculation of the analysis depth from the etching time was carried out using white light interference microscopes Talysurf CCI 2000 and Zygo NewView 7300. Sputtering was performed by Cs^+ ions with energy of 1 keV at angle of 45°. The probing was carried out by Bi^+ ions with energy of 25 keV. The output of the cluster ion C_2B^- was registered. Quantitative calibration of the mass spectrometer was performed using a test HPHT crystal implanted with boron ions with three different maximum concentrations (10^{18}, 10^{19} and 10^{20} cm^{-3}).

The influence of growth conditions of CVD diamond on the incorporation of boron into the delta layer was investigated by growing structures containing several (3–5) delta layers. One of the growth process parameters was varied when growing each layer. Thus, the dependence of the boron incorporation on one of the growth parameters can be obtained within one growth process, which excludes the influence of such factors as the miscut angle and pre-treatment of the substrate, the "history" of the reactor, etc. In the case when several substrates were used to obtain the results, the substrates were selected with similar miscut angles of about 1°.

Figure 1 shows the dependence of the boron concentration in the delta layer on the substrate temperature with the methane content in the gas mixture CH_4/H_2 = 0.07%. During the experiment, all process parameters were kept constant, and only the temperature of the substrate was varied using an external heater. It is evident from Fig. 1 that boron is best incorporated into the delta layer in CVD diamond in the 850–900 °C temperature range.

At higher temperatures, the efficiency of boron incorporation decreases.

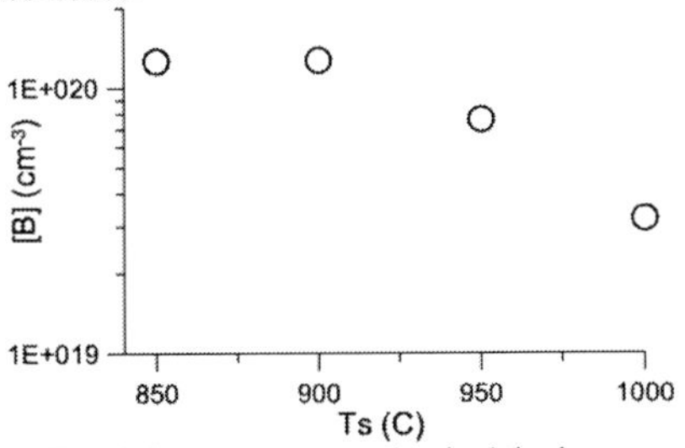

Fig. 1. Boron concentration in delta layer as function of substrate temperature

Fig. 2 shows the dependence of the boron incorporation into the delta layer on the B/C ratio in the gas mixture at a methane content of CH_4/H_2 = 0.07% (open circles) and CH_4/H_2 = 0.14% (black circles). The temperature of the substrate was kept constant at 850 °C. It is evident from Fig. 2 that with a B/C ratio in a gas mixture of less than 20000 ppm, the incorporation of boron increases linearly with an increase of the B/C ratio. At B/C ratio in the gas mixture in the range from 20000 ppm to 30000 ppm, saturation of the boron incorporation in the delta layer is observed. At a ratio values greater than 30000 ppm, the incorpo-ration of boron in the delta layer decreases. An increase in the methane content in the gas mixture from CH_4/H_2 = 0.07% to CH_4/H_2 = 0.14% leads to an increase in the incorporation of boron in the delta layer. Thus, to obtain heavily boron-doped diamond delta layers, it is necessary to select process parameters such that the B/C ratio in the gas mixture does not exceed 20000 ppm.

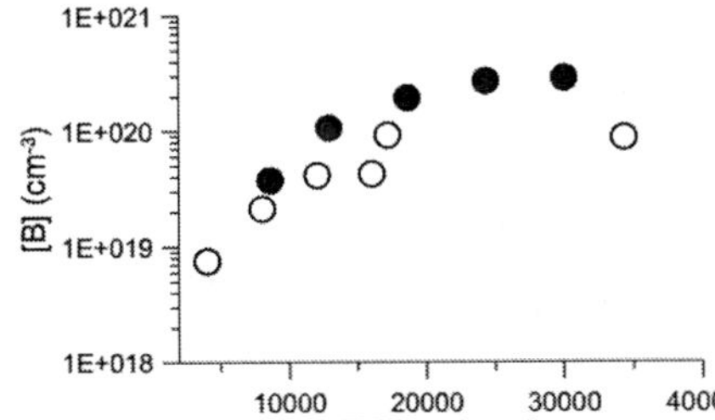

Fig. 2. Boron concentration in delta layer measured by SIMS as a function of [B]/[C]gas ratio for two different methane content CH_4/H_2 = 0.07 % (open circles) and CH_4/H_2 = 0.14 % (black circles)

It is known that the boron incorporation into diamond during the CVD growth process depends on the miscut angle of the substrate [3]. To investigate the influence of the miscut angle on the incorporation of boron in the delta layer, the sample was used which surface is divided into three sectors with different angles of surface miscut angle relative to the crystallographic direction [001]. On that substrate, a structure containing 5 delta layers was grown with different B/C ratios in the gas mixture. Fig. 3 shows the dependence of the boron concentration in the delta layer on the B/C ratio in the gas mixture for different angles of deviation of the crystallographic direction [001] from the normal to the surface. The highest efficiency of boron incorporation in the delta layer is observed for the 0° miscut angle. As can be seen from Fig. 3, the incorporation of boron in the delta layer saturates at B/C ratio in the gas mixture over 20000 ppm for all miscut angles. For miscut angles of 1° and 3°, the incorporation of boron at a large B/C ratio is almost equal.

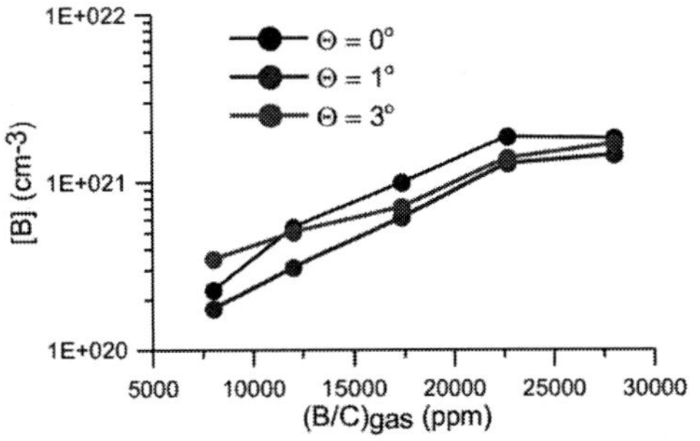

Fig. 3. Boron concentration in delta layer measured by SIMS as a function of [B]/[C]gas ratio for different miscut angles

Conclusion

Noval MPACVD reactor was used for the investigation of boron incorporation in delta layers on CVD diamond growth process and misorientation angle. As the result of experiments the optimal diamond deposition parameters at which thin doped delta layers with boron concentration (5–10) 10^{20} cm^{-3} with thickness of 1–3 nm were found. This research allowed us to obtain heavily boron doped thin layers and to implement two-dimensional hole "gas" in diamond with high mobility and hole concentrations.

Acknowledgments

This work was supported by Act 220 of the Russian Government (Agreement no. 14.B25.31.0021 with the host organization IAP RAS).

References

1. *A.L. Vikharev, A.M. Gorbachev, M.A. Lobaev, A.B. Muchnikov et al.* , Novel microwave plasma-assisted CVD reactor for diamond delta doping // Phys. Status Solidi (RRL), 2016, 10, 324.

2. *S. C. Eaton, A. B. Anderson, J. C. Angus, Y. E. Evstefeeva et al.* Co-doping of diamond with boron and sulfur // Electrochem. Solid-State Lett., 2002, 5, G65.

3. *M. Ogura, H. Kato, T. Makino, H. Okushi et al.*, Misorientation-angle dependence of boron incorporation into (001)-oriented chemical-vapor-deposited (CVD) diamond // Journal of Crystal Growth, 2011, 317, 60.

EPJ Web of Conferences **149**, 02015 (2017) DOI: 10.1051/epjconf/201714902015
10th International Workshop 2017 "Strong Microwaves and Terahertz Waves: Sources and Applications"

Cooperation and competition of solid state and vacuum microwave devices in radar applications

A.V. Galdetskiy, S.V. Sherbakov

JS Istok, Fryazino, Russia, Galdetskiy@mail.ru

In recent years significant attention is attracted to employing of GaN solid state devices in radar applications. They demonstrated impressive progress in output power and reliability. For example, single GaN transistor has achieved output power level 300 W in X-band with power added efficiency 37% [1]. Connection of such perfect performances with active electrical scanned array (AESA) concept apparently makes possible to create radars in centimeter wavebands quite competitive in power with designs based on TWT and klystrons. As result a new concepts of high power radars are discussed like Space Fence radar, which contains 36000 of transmitting modules of total power 2.69 MW (average – 0.81 MW) in S-band [2].

It could be seemed that the epoch of vacuum tubes in radars tends to end. However, comparing radars based on solid-state and vacuum amplifiers, we must take into account additional features.

As rule, AESA radars are much more expensive comparing vacuum counterparts with passive ESA.

Important radar's parameter is velocity of beam control, which influences a number of simultaneously handled targets. In radar, based on vacuum tube and passive ESA, the beam is usually controlled by ferrite phaseshifters. These unique components possess extremely small microwave losses and high power capability. However, their typical phase switching times are rather large – tens of microseconds. In AESA phase commutation is implemented by solid-state switches having switching times units and tens nanoseconds. So fundamentally solid-state AESA can process larger number of targets than traditional ESA using vacuum devices (if it is not limited by digital part of radar). Taking this fact in mind we can conclude that the opportunity of surviving of vacuum devices in competition with solid-state ones in radar applications of centimeter bands is based on development of fast phaseshifters having small microwave losses and high power capability. The most probable candidates for this role are phaseshifters based on MEMS switches (the switching time 0.2–0.5 us) [3–] and ferroelectric phaseshifters [6, 7].

Besides this very remarkable feature of AESA is the absence of frequency and directional selectivity of each single module. This means that the circuit of each module receives all interference signals from hemisphere. This implies requirement of very large linearity and dynamic range of the receiver (~120 dB) which is not so easy to realize. This means that AESA possesses increased vulnerability to countermeasures and interferences. Contrary to AESA passive phaseshifters in ESA have very large dynamic range (they handle output power of own transmitter), and active devices in receiver after summation circuits are protected by directivity of the array which possesses suppression of side lobes by ~30 dB.

The problem is closely related with protection devices of radar receiver. In AESA solid-state protection devices are not very effective due to strong limitations in volume and mass. They have rather long recovery time and significant transmission at high power levels. In contrast, single receiver in passive ESA can be very effectively protected by vacuum cyclotron device [8, 9] – another superior vacuum device for radar applications.

Nevertheless, along with competition we see also examples of good cooperation of vacuum and solid-state microwave devices. Increasing of output power of solid-state amplifier have made possible to decrease significantly requirements for gain of vacuum amplifier which results in remarkable decreasing of mass and dimensions of the tube. As example we can point out to wideband, small gain TWT (booster) - the foundation of microwave power modules (MPM) – high power amplifiers, which are fabricated not for some specific application but for common use in various systems. Such concept of component-of-the-shelf (COTS) significantly decreases expenses and terms of the development of new radar systems.

References

1. *Kikuchi, K., Nishihara, M., Yamamoto, H. et al* An X-Band 300-Watt Class High Power GaN HEMT Amplifier for Radar Applications // SEI Technical Review, 2015, 81, P. 40-44.
2. *Gallagher, J., Haimerl, J. A., Higgins, T., Gruber, M.*, Space Fence Radar Leverages Power of GaN // Microwave journal, 2016, 8, P. S6-S10.
3. *Nordquist, C.D., Dyck, W., Kraus, M. et al* A DC to 10-GHz 6-b RF MEMS time delay circuit // IEEE Microwave and Wireless Components Letters. 2006. v. 16., No 5. P. 305-307.
4. *Maciel, J.J., Slocum, J.F., Smith, J.K., et al* MEMS Electronically Steerable Antennas for Fire Control Radars // in 2007 IEEE Radar Conference, 2007. Pp. 677-682.
5. *Kosmin, D.M., Kotelnikov, I.V., Osadchy V.N., et al* Phaseshifters for communication phased arrays in 2 ÷ 30 GHz frequency bands // in Proc. Of Conf. Microwave electronics, S-Petersburg, June 4-7 20126 v.2., p. 102-105 (in Russian).
6. *Kozyrev, A.B., Osadchy, V.N., Pavlov, A.S., et al* 30 GHz steerable beam antenna based on ferroelectric phase shifters. // Proc. of Progress In Electromagnetics Research Symposium, July 5-14, 2000, v.1, P.48
7. *Kazakov, S., Shchelkunov, S., Yakovlev, V., et al*, Fast High-Power Microwave Ferroelectric Phase Shifters for Accelerator Application // in Proc. in Proc. Advanced Accelerator Concepts: 13 Workshop, 2009, P. 477-482.
8. *Vanke, V., Lopukhin, V., Savvin, V.* Supernoiseless cyclotron-wave amplifiers // Soviet Physics Uspekhi, 1970, v. 12(6), P. 743-755.
9. *Budzinsky Y.A., Kantyuk S.P.* A new class of self-protecting low-noise microwave amplifiers. in 1993 IEEE MTT-S International Microwave Symposium Digest, 1993, v.2, P. 1123-1125

Synthesis of micro- and nanostructures with controllable composition in the chain plasma-chemical reactions initiated by the radiation of a powerful gyrotron in the mixtures of metal-dielectric powders

N.N. Skvortsova[1,2], N.S. Akhmadullina[3], G.M. Batanov[1], V.D. Borzosekov[1,4], L.V. Kolik[1], E.M. Konchekov[1,4], N.K. Kharchev[1], A. A. Letunov[1], D. V. Malakhov[1,4], E.A. Obraztsova[1,2], A. E. Petrov[1,4], K.A. Sarksian[1], V.D. Stepakhin[1,4], O.N. Shishilov[5,6]

[1] Prokhorov General Physics Institute, RAS, Moscow, Russia, mukudori@mail.ru
[2] National Research Nuclear University MEPhI, Moscow, Russia
[3] A.A. Baikov Inst. of Metallurgy and Material Science, Moscow, Russia
[4] Pirogov Russian National Research Medical University, Moscow, Russia
[5] Institute of Fine Chemical Technologies, Moscow, Russia
[6] O3-Innovation, LLC, Skolkovo Innovation Center, Moscow, Russia

The report presents a new type of exothermic chain reactions initiated by powerful microwave radiation of gyrotron in mixtures of metal-dielectric powders. In this microwave-plasma-chemical process new substances are synthesized with controlled composition [1]. Chain self-sustaining reactions in which initially appearing products take part in the formation of new products are the basis of many technological processes of synthesis of substances. Chain reactions occur usually at a high rate, and often have the character of an explosion [2].

Ceramic materials are widely used in modern technology so they receive wider attention of the researchers due to their unique mechanical and physical properties. Plasma methods are among the most promising techniques for the synthesis, modification, and processing of micro- and nano-structured ceramic materials [3]. There have been many examples of nanostructured material syntheses using various discharge types: spark, arc, glow, high-frequency, microwave, and so on. However, gyrotrons, being one of the most powerful sources of microwave discharges, have not been used in the regarded technology for a long time. This report will show the synthesis of -structures with controllable composition in chain reactions in powder mixtures: Mo+B, Mo+BN, Ti+B, W+B. A+-AlN, Mg+AlN+Al_2O_3, etc.

Experiments were carried out with stand-mounted plasma-chemical complex of the MIG-3 installation with the gyrotron (frequency 75 GHz, pulse duration up to 12 ms, power up to 550 kW) [4]. A specially designed plasma-chemical reactor (Fig. 1) [5] was placed between two mirrors of the MIG-3 quasi-optical tract.

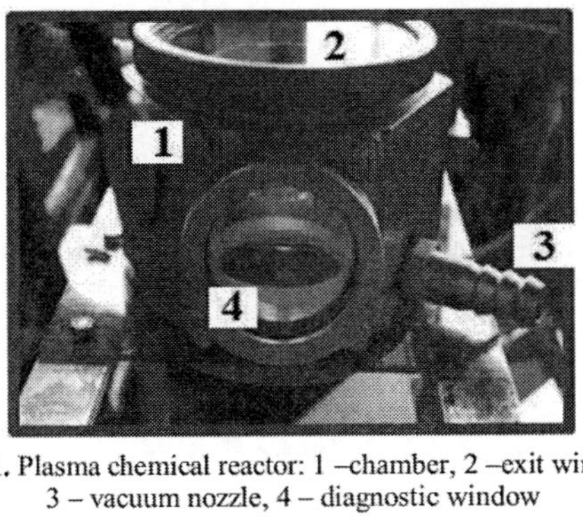

Fig. 1. Plasma chemical reactor: 1 –chamber, 2 –exit window, 3 – vacuum nozzle, 4 – diagnostic window

Microwave breakdown in the reactor was effected in the mixture of metal and dielectric powders placed on a quartz substrate with free upper surface (fig. 2). A quartz tube was set inside the reactor to collect the synthesized substances for further analysis. The evolution of direct and reflected gyrotron power during the microwave discharge was monitored using a system of microwave detectors, which were calibrated using the calorimeter [6].

The samples were prepared in the following way: 1-mm-thick layer of dielectric powder (2) (fig. 2) was put on the quartz plate (1) followed by a 0.5–0.7 mm layer of metal-dielectric powder mixture (3). The layers were some squeezed with a flat quartz plate and quartz cylinder (6) was set up. The upper surface of the powder remained open to ensure free plasma-gas-dynamic scattering of reaction products as well as movement of the heated gas into the quartz cylinder volume.

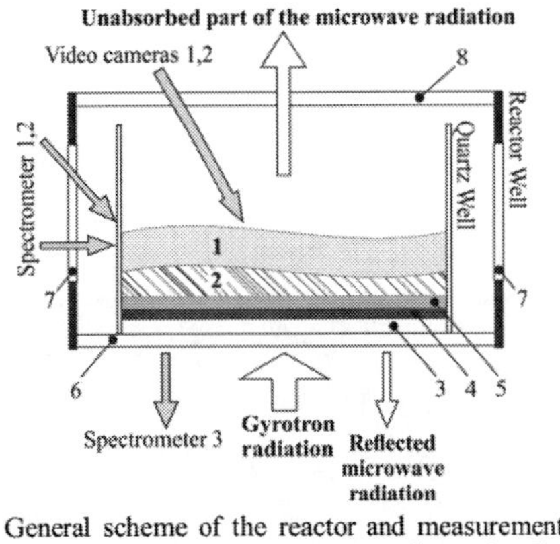

Fig. 2. General scheme of the reactor and measurements: 1 – working gas, 2 – plasma, 3, 6, 8 – quartz plates, 4 – dielectric powder layer, 5 – powder mixture of metal and dielectric, 6 – quartz cylinder (d = 70 mm, L = 350 mm), 7 – diagnostic window. The arrows on the left-hand side indicate directions along which measurements were carried out by two cameras (#1, 2) and three spectrometers (#1, 2, 3)

The gyrotron microwave radiation was injected through the bottom quartz plate. Irradiation of the samples was carried out with 5–100 single 2–12-ms pulses with intervals no less than 20 s. The gyrotron radiation power was 250–450 kW which corresponds to intensity of ~6–10 kW·cm^{-2} inside the powder mixture. The synthesized substances were deposited on both the walls of the quartz cylinder and the remaining surfaces of the reactor including the upper quartz window. The produced materials composition and structure were analyzed using Raman spectroscopy, X-ray diffraction, energy dispersive microanalysis, transmission and reflection electron microscopes.

Previously it was shown that to transform the initial mixture of powders into a new material the gyrotron discharge parameters should be chosen so that the microwave discharge was followed by an afterglow phase. Only under these conditions absorption of microwave radiation is almost complete in the powder mixture, and followed by initiation of chemical reactions of synthesis of substances. In described experiments certain threshold values of pulse duration, gyrotron power and metal content in the every powder mixture should be found and reached.

Plasma-chemical chain reactions were observed in the atmosphere of air/nitrogen for mixtures of different metals (titanium, molybdenum, aluminum, and est.) with boron, boron nitride and other dielectrics without and in the presence of small amounts of $C_2B_{10}H_{12}$, $NaBH_4$ and H_3BO_3 as hydrogen sources.

In our experiments we found all three stages of the evolution of chain process.

The first stage – initiation takes place in the upper layer of the powder during the gyrotron pulse (Coulomb explosion process) with arising its temperature up to 1 eV. Than dust particles, atoms and ions pushed out of the upper surface into the reactor.

The second stage – synthesis of complex molecules under conditions of a large temperature gradient over the height of the reactor and in the presence of ensembles of dust charged particles (dusty plasma). This stage time reaches tens of seconds. The self-oscillating character of such reactions and the increase in the yield of reaction products using a catalyst was found. Properties of ensembles of dust particles (tracks, velocities, lifetime, etc.) were measured. Photo of the ensemble of dust particles that appears in the volume of reactor 32 ms after turning of the gyrotron pulse is shown in Fig. 3.

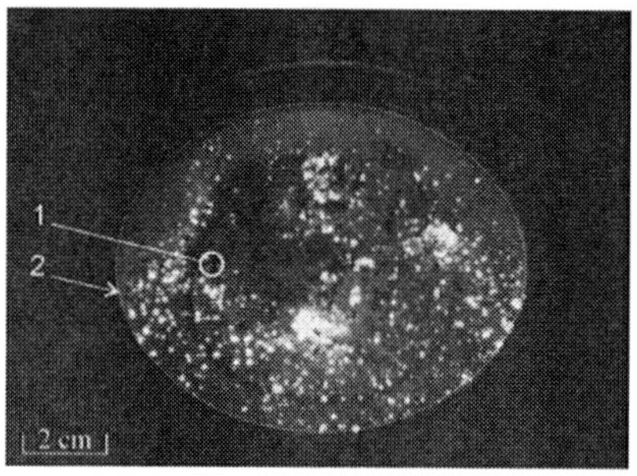

Fig. 3. Dust particles (1) in the reactor on 32 ms after switching off the 300 kW 2-ms gyrotron pulse (Ti+B powder mixture with catalyst H_3BO_3). 2 – quartz cylinder wall

The third stage – deposition of substances on the centers of crystallization with the formation of composites of micro- and nano-sizes, spraying on the walls, cooling of substances in the reactor, stopping the chain process. A typical spherical structure synthesized in the chain reactions initiated in the mixture Ti+B with catalyst H_3BO_3 is represented in Fig. 4. This spherulite consists of titanium nitride and titanium oxide and has a developed surface.

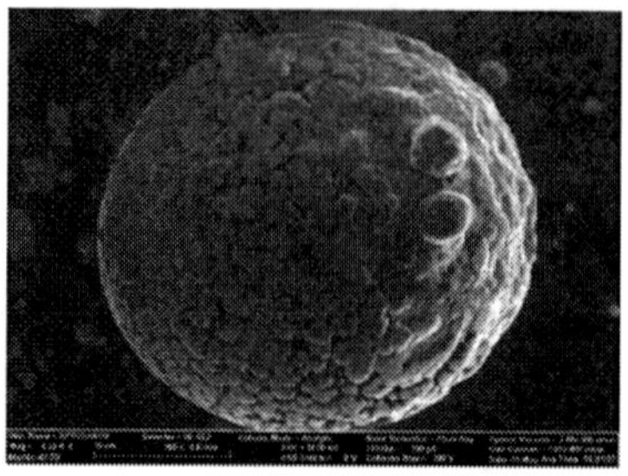

Fig. 4. Spherulite consists of titanium nitride and titanium oxide. 300 kW 2-ms gyrotron pulse

We observed plasma-chemical processes in the Al_2O_3 + AlN mixture with 5% of Mg adding in the discharge under 8 ms pulses of 350 kW. Formation of well-cut microcrystals of $MgAl_2O_4$ was confirmed by SEM and XRD.

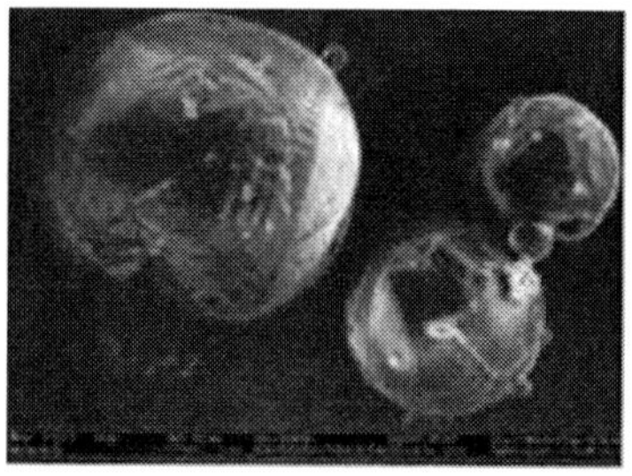

Fig. 5. Microcrystals of $MgAl_2O_4$ (powder mixture Al_2O_3 + AlN + 5% Mg)

We are grateful to Russian Foundation for Basic Research for financial support (project #17-03-00630-a).

References

1. *Batanov, G.M., Borzosekov, V.D. et al.* Microwave method for synthesis of micro- and nanostructures with controllable composition during gyrotron discharge. // J. Nanophotonics, 2016, 10(1), 012520.

2. *Semenov, N.N.* // Development of the theory of chain reactions and thermal ignition. Moscow. 1969.

3. Ivanovskii A.L. Non-carbon nanotubes: synthesis and simulation // Russ. Chem. Rev. 2002. V.71. No. 3. P. 175–194.

4. *Batanov, G.M., Belousov, V.I., Bondar, Yu.F., et al.* // A New MIG-3 Gyrotron Complex for Creation and Heating of Plasma in the L-2M Stellarator and the First Experimental Results. Plasma Physics Reports, 2013, Vol. 39, No. 13, pp. 1088–1095 2013.

5. *Skvortsova, N.N, Stepakhin, V.D., Malakhov, D.V., et al.* Relief Creation on Molybdenum Plates in Discharges Initiated by Gyrotron Radiation in Metal–Dielectric Powder Mixtures // Radiophysics and Quantum Electronics. 2016. V. 58. No. 9. P. 701–709.

6. *Ukryukov, G.V., Malakhov, D.V., Skvortsova, N.N., et al.* // Microwave absorption in the gyrotron discharge ibn a mixture of powders of metals and dielectrics in the mode of installation of exothermic chemical reactions. Engineering Physics. 2017. No. 2. P. 28–36.

EPJ Web of Conferences **149**, 02017 (2017) DOI: 10.1051/epjconf/201714902017
10th International Workshop 2017 "Strong Microwaves and Terahertz Waves: Sources and Applications"

Influence of intense coherent electromagnetic radiation on several types of radioactive decay

S.N. Andreev[2], E.V. Barmina[2], A.K. Kaminsky[1], S.N. Sedykh[1], G.A. Shafeev[2], I.A. Shcherbakov[2], A.V. Simakin[2], V.I. Stegailov[1], S.I. Tyutyunnikov[1]

[1]Joint Institute for Nuclear Research, Dubna, Russia, snsed@yandex.ru
[2]Prokhorov General Physics Institute RAS, Moscow, Russia

Optical laser radiation

Many configurations have been proposed to use high-power laser radiation for acceleration of the charged particles [1]. Required power density for such experiments varied from 10^{18} W/cm^2 up to 10^{20} W/cm^2. For example, X-rays of MeV energy have been produced by accelerated plasma electrons when laser beam with intensity of $5*10^{18}$ W/cm^2 was focused on tantalum sheet. Energy of these X-rays was high enough to initiate a set of nuclear reactions, which have been registered experimentally [2].

We investigate an influence of coherent electromagnetic radiation on different types of radioactive decay. Our experiments indicate that some nuclear reactions can be initiated even at laser beam intensity about $1*10^{12}$ W/cm^2. The key feature is the presence of metallic (chemically inactive Au as a rule) nanoparticles. In plasmon resonance regime the wave field near nanoparticles can be concentrated up to the necessary level. Layout of the typical experiment is shown on fig. 1.

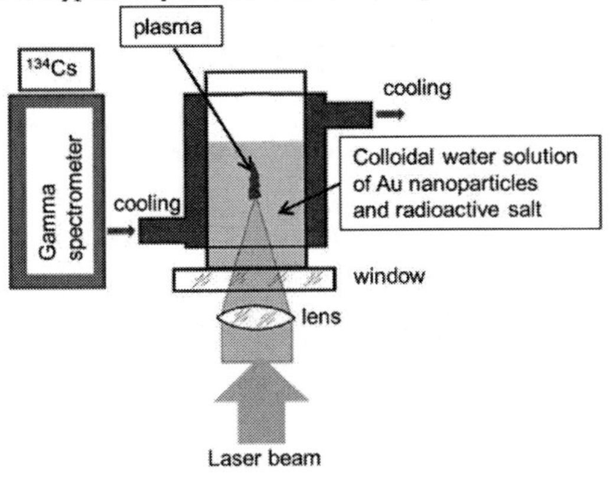

Fig. 1. Layout of the experiment on irradiation of water solution of the radioactive salt by the medium-power laser in presence of the metallic nanoparticles

We use several types of optical and near-infrared lasers – Cu vapor, femtosecond Ti:sapphire, 90-ps Nd:YAG, 350-ps Nd:YAG, 10-ns Nd:YAG, Their pulse energy varied from 0.1mJ up to 40 mJ, repetition rate varied from 10 Hz up to 20 kHz. Power density in focal region without influence of the nanoparticles varied from $1*10^{11}$ W/cm^2 up to $1*10^{12}$ W/cm^2. Nanoparticles being used in our experiments are produced by laser ablation by metal target in the water.

We have observed partial Hg transmutation into Au inside heavy water D_2O [3]. Laser irradiation in presence of Au nanoparticles initiates the decay of ^{232}Th [4]. Very exciting result was a significant (up to 50%) decrease of gamma-activity of ^{238}U and ^{235}U series after laser irradiation in presence of beryllium nanoparticles [5].

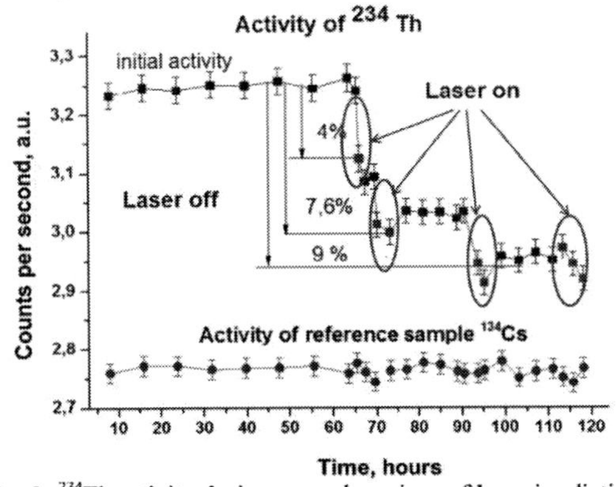

Fig. 2. ^{234}Th activity during several sessions of laser irradiation of UO_2Cl_2 water solution in presence of Au nanoparticles as the laser wave concentrators. ^{134}Cs is a reference sample

Figure 2 presents the results of long-time measurement of the ^{234}Th activity during several sessions of laser irradiation. Each session decrease the activity. Effect of each next irradiation is less than the previous one due to fragmentation of nanoparticles and their escape from the plasmon resonance.

Microwave radiation

There are some similar experiments in microwave and radiofrequency regions. 3 GHz radiation caused increasing the ^{51}Cr activity at 1% [6]. 0.07% of the ^{137}Cs activity growth was registered during irradiation of the sample by 4.1 MHz and 1.55 MHz radiation with a maximum power of 50 kW [7]. Our experiment using 30 GHz FEM shows about 1.5% variation of the ^{152}Eu activity [8]. In all these experiments there were no special measures to increase the power density.

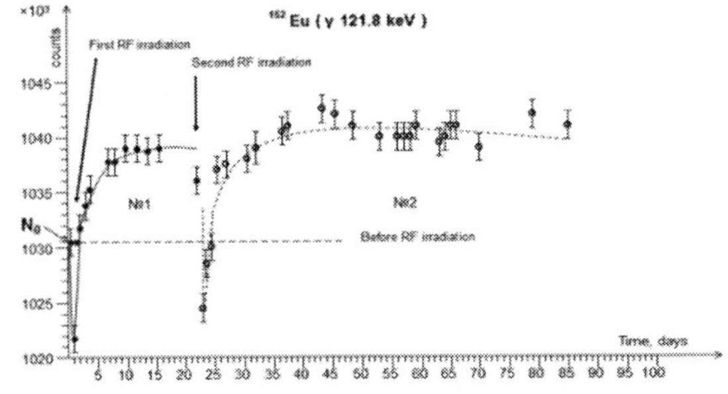

Fig. 3. Activity of the ^{152}Eu during two irradiations by 15 MW 30 GHz JINR-IAP FEM oscillator

Field concentrators in microwave experiment

We assume that all changes in the rate of the radioactive decay were caused by the charged particles accelerated in the plasma, produced in focal region in presence of field concentrators (metal nanoparticles). We have in mind that the power density about $1*10^{18}$ W/cm^2 is enough to produce some types of nuclear reaction without special field concentrators. Estimation of the maximal energy of the proton inside the electromagnetic wave with wavelength λ and intensity *I* can be done by [9]:

$$E_{\max} = 3.6 * \lambda\sqrt{I/10^{18}}, \quad (1)$$

where E_{max} is in MeV, λ in microns, *I* in W/cm^2. Indeed, the wave with length of 1 micron and intensity of $1*10^{18}$ W/cm^2 can accelerate protons up to 3.6 MeV, which is more than the threshold of many nuclear reactions. In our experiments with power densities of $1*10^{12}$ W/cm^2 the presence of field concentrators (nanoparticles) can increase the field intensity up to 10^5 and (nanoparticles dimers) to 10^6 times.

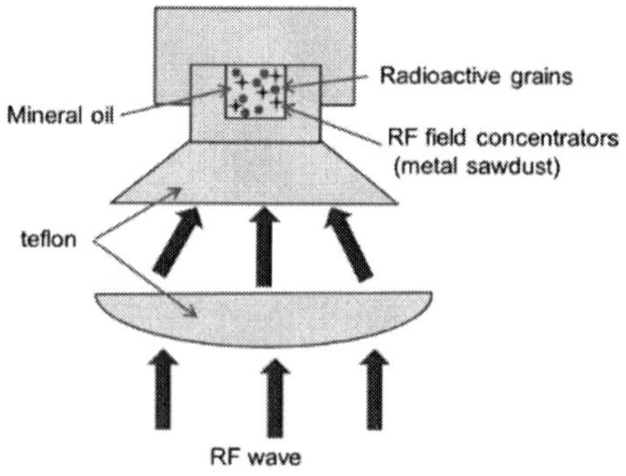

Fig. 4. Proposed layout of the experiment on the RF irradiation of the radioactive sample. Metal sawdust should be the RF field concentrators

Let us estimate the necessary rate of field concentration in microwave region. For wavelength of 1 cm (10^4 microns) the desired proton energy of 3.6 MeV requires the field intensity of $1*10^{10}$ W/cm^2. The diffraction limits the area of focal spot by 1 cm^2, so we need or 10 GW radiation source, or 10 MW radiation source and $1*10^3$ times field concentrators. For microwave region the role of the field concentrators can play metal sawdust. The layout of proposed experiment is shown on fig. 4.

References

1. *Krushelnick K., Najmudin Z. and Dangor A.E.* Particle acceleration using intense laser produced plasmas // Laser Phys. Lett. 2007. V. 4, No. 12 P. 847–862.

2. *Schwoerer H., P. Gibbon P., S. Düsterer S. et al.* MeV X rays and photoneutrons from femtosecond laser-produced plasmas // Phys. Rev. Lett. 2001. V. 86, No. 11 P. 2317–2320.

3. *Shafeev G.A., Bozon-Verduraz F. and Robert M.* Experimental evidence of transmutation of Hg into Au under laser exposure of Hg nanodrops in D_2O // Physics of Wave Phenomena 2007. V. 15, No. 13 P. 131–136.

4. *Simakin A.V. and Shafeev G.A.* Initiation of nuclear reactions under laser irradiation of metal nanoparticles in the presence of Thorium aqua ions // Physics of Wave Phenomena 2008. V. 16, No. 4 P. 268–274.

5. *Barmina E.V., Sukhov I.A., Lepekhin N.M. et al.* Application of copper vapour lasers for controlling activity of uranium isotopes // Quantum Electronics 2013. V. 4, No. 6. P. 581–596.

6. *Rybasov A.G., Ryzhkov V.A., Len' N.A. and Sulakhshin A.S.* Patent of Russian Federation No 2100858 (C1) G21F 9/00 1995 (in Russian).

7. *Reiss H.R.* Observation of the acceleration by an electromagnetic field of nuclear beta decay // A Letters Journal Exploring the Frontiers of Physics EPL 2008. V. 81 P. 42001.

8. *Golubykh S.M., Gons Z., Kalinnikov V.G. et al.* Influence of high-power pulse RF irradiation on several types of Beta-decay // Proceedings of 9th Intern. Workshop "Strong microwaves and terahertz waves: sources and applications" 2014. P. 61–62.

9. *Maksimchuk A., Flippo K., Krause H. et al.* High-energy ion generation by short laser pulses // Plasma Phys. Rep. 2004. V. 30 No 6 P. 473-495.

High-performance spectrally selective pyroelectric detection of millimeter and submillimeter waves using ultra-thin metasurface absorbers

S.A. Kuznetsov[1,2], A.V. Arzhannikov[1,3], N.A. Nikolaev[4]

[1]Novosibirsk State University, Novosibirsk, Russia, SAKuznetsov@nsm.nsu.ru
[2]Rzhanov Institute of Semiconductor Physics SB RAS, Novosibirsk Branch "TDIAM," Novosibirsk, Russia
[3]Budker Institute of Nuclear Physics SB RAS, Novosibirsk, Russia
[4]Institute of Automation and Electrometry SB RAS, Novosibirsk, Russia

Introduction

In the last decade, following a rapid progress in millimeter-wave (MMW) and submillimeter-wave (SMMW) technologies [1, 2], a lot of efforts are undertaken in device physics to improve conventional or realize novel instrumental solutions using metamaterials. Indeed, the metamaterial structures, especially the plasmonic ones, are capable of noticeably extending functional properties of MMW/SMMW devices hardly attainable with classical approaches. In particular, via tailoring the structure's absorptivity A it is possible to attain $A\approx 100\%$ when the structure's thickness d is much smaller than the peak absorption wavelength λ [3–5]. Such a solution is highly demanded in thermal detectors, e.g. pyroelectric ones, since minimization of d is essential for decreasing the absorber's thermal capacity and, therefore, for achieving high sensitivity and low response time of the thermal device. Due to the small thickness ("ultra-thinness"), such artificially engineered planar subwavelength structures are frequently referred to as metasurface absorbers (MSAs). It is worth noting that thinning the MSA inevitably leads to narrowing its fractional absorption bandwidth $\Delta\lambda/\lambda$.This conclusion stems from the fundamental Rozanov restriction on the thickness of an arbitrary metal-backed magnetodielectric absorber [6]:

$$d \geq \left|\int_0^{\infty} ln|S_{11}(\lambda)|d\lambda\right|/(2\pi^2\mu_s), \qquad (1)$$

where S_{11} is the amplitude reflection coefficient and μ_s is the relative static magnetic permeability of the slab.

The narrowband response of ultra-thin MSAs is usually adduced as their drawback incompatible with broadband applications. On the other hand, the conditions $d/\lambda << 1$ and $\Delta\lambda/\lambda << 1$can be turned to the advantage as they readily allow implementation of thermal sensors with high spectral discrimination. Arranged in compact spectrometric arrays with tailored spectral and polarization responsivities, the MSA-based detectors do not necessitate the use of external dispersive elements to measure radiation spectra and polarization, as well as to realize wavelength- and polarization-sensitive imaging [5]. Such a novel instrumental solution is commercially unavailable for MMW/SMMW bands, wherein it is relevant for numerous applications.

In this contribution we overview the results of theoretical and experimental investigations of high-performance MSAs designed for narrowband operation in the frequency range of 0.1–1 THz and integrated with compact spectrometric detectors of a pyroelectric type. The detectors were developed for spectral analysis of high-power broadband subterahertz emission from a plasma source driven by a high current relativistic electron beam at the GOL-3 plasma facility (BINP SB RAS, Novosibirsk) [7].

MSA Fundamentals

In a basic design, the MSA is realized as a single-layer frequency selective surface (FSS) of a capacitive type patterned on a thin dielectric slab with uniform back metallization ("ground plane" or GP) (Fig. 1,a). In an equivalent circuit approach [8, 9] the lossy grounded slab of a small thickness ($d/\lambda \ll 1$) is described as an effective inductor with the input impedance $Z_S \cong j\omega L_S + R_{GP}$, where $L_S = \mu\eta_0 d/c_0, \mu$ is the relative magnetic permeability of the slab, $\eta_0 \cong 377\ \Omega$ is the free spaceimpedance, c_0 is the speed of light, ω is the angular cy, R_{GP} is the lumped resistance. The MSA is interpreted as a parallel connection of the impedanceZ_S and the FSS impedance $Z_{FSS} = 1/(j\omega C_{FSS}) + j\omega L_{FSS} + R_{FSS}$ (Fig. 1, b). Since $\mathrm{Im}(Z_{GP}) > 0$ and $\mathrm{Im}(Z_{FSS}) < 0$ within the MSA band, the structure acts as a resonant LCR-circuit. By properly adjusting the equivalent circuit parameters $\{C_{FSS}, L_{FSS}, R_{FSS}, R_{GP}, L_S\}$ the MSA impedance in the resonance can be settled equating to that of free space: $(Z_{FSS}^{-1} + Z_S^{-1})^{-1} = \eta_0$. Under this condition the MSA exhibits no reflection that means total absorption of the impinging wave: $A = 1 - |S_{11}|^2 = 0$.

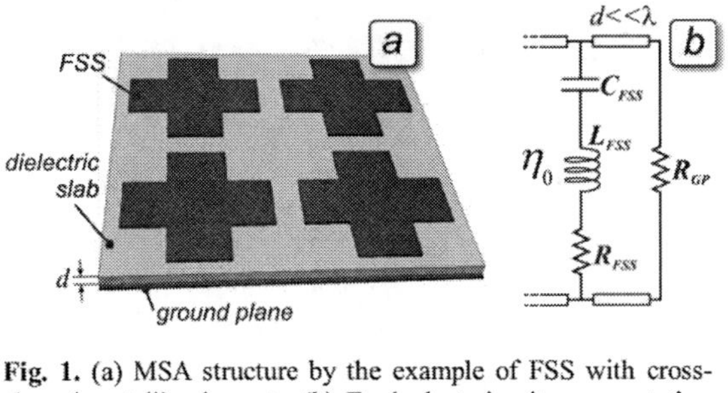

Fig. 1. (a) MSA structure by the example of FSS with cross-shaped metallic elements. (b) Equivalent circuit representation of the MSA. The resistanceR_{FSS} describes cumulative energy dissipation in the FSS and dielectric slab, while $R_{GP}$is in charge of ohmic losses in the ground plane

The fractional bandwidth $\Delta\lambda/\lambda$ and the thickness-to-wavelength ratio $d/\lambda$for the absorption resonance arecompactly expressed via the resistive parameter $r = (R_{FSS} + R_{GP})/\eta_0$:

$$\Delta\lambda/\lambda \cong 4\sqrt{r}/(1+\chi), \quad d/\lambda \cong \sqrt{r}/(2\pi\mu), \qquad (2)$$

where $\chi = [1 + 4L_{FSS}/(C_{FSS}\,\eta_0^2 r)]^{-1/2}$[9]. The relations (2) distinctly necessitates the use of low-loss materials in the MSA configuration as it allows minimizing the thickness and bandwidth of a thermal sensor due to decreasing the resistive parameter$r$. In this work, polypropylene (PP) and thermally sputtered aluminium were chosen as optimal materials for the MSA implementation (see also [8, 9]).

Design and Characteristics of MSA-Integrated Pyro-Sensors

We developed a set of 29 selective with spectrally shifted sensitivities, gradually overlapping the frequency range from 94 GHz to 1 THz. Each detector was produced via combining an ultra-thin MSA with a discrete commercial IR pyro-sensor MG33 from the Russian manufacturer "Vostok" (Fig. 2, a, b) [9]. All 29 spectrally different MSAs were implemented on basis of a PP film 15 μm thick, thus yielding ratios $d/\lambda = 1/213$–$1/20$ at the fractional absorption bandwidth $\Delta\lambda/\lambda = 3$–$6\%$. The geometry of cross-shaped polarization insensitive metallic elements was chosen in the FSS design (Fig. 1, a), while their optimal structural parameters were found via 3D full-wave simulations in ANSYS Electromagnetic Suite software. The FSS micro-patternfabrication was realized with a conventional technology of contact photolithography, which was specially adapted for a work with flexible PP substrates [10].

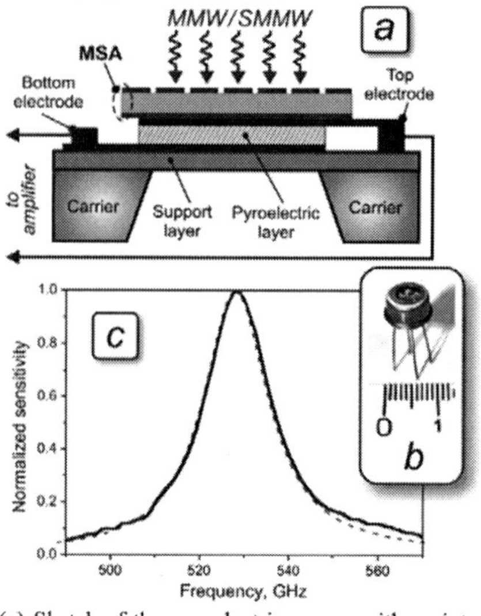

Fig. 2. (a) Sketch of the pyroelectric sensor with an integrated resonant absorber. (b) Appearance of the accomplished sensor in a standard KT-3 housing. A fragment of the ruler with 1 mm marks is shown for scale. (c) Example of spectral characteristics for the detector optimized for 530 GHz: *dashed line* – absorptivity of the free-standing MSA; *solid line* – normalized sensitivity of the MSA-integrated sensor

The basic scheme for a MSA-integrated detector is shown in Fig. 2, a. Its pyroelectric part is represented by a pyroelectric layer with top and bottom metallic electrodes, which are electrically connected with an integrated preamplifier and collect a pyroelectric charge induced in the pyro-layer. Manufactured independently, the MSA is attached directly to the top electrode through the GP layer and fixed throughout the periphery with a heat-conducting paste. Upon resonant absorption of the incoming MMW/SMMW radiation in the MSA, the induced heat is rapidly transferred to the pyroelectric and generates the wanted electric signal.

Spectral characterization of the developed MSAs and MSA-integrated sensors using the techniques of BWO-based and laser-driven CW THz spectroscopy demonstrated good concordance between simulations and measurements (Fig. 2, c). Our tests also showed that in the MMW/SMMW bands such detectors keep high values of their response speed (~2–3 ms) and sensitivity (~ 50 kV/W, NEP ~ $1.0{\cdot}10^{-9}$ W/Hz$^{1/2}$) changed insignificantly against the regime of IR detection. In tote, the proposed approach represents an effective and relatively low-cost instrumental solution in the technology of uncooled thermal sensors.

Acknowledgements

The work is partially supported by the Russian Foundation for Basic Research (Project #17-32-80039).

References

1.*Bründermann, E., Hübers, H.-W., Kimmitt, M.F.* Terahertz techniques //Springer series in optical sciences. Berlin Heidelberg: Springer-Verlag. 2012. V. 151.

2. *Song, H.-J., Nagatsuma T.* Handbook of terahertz technologies: devices and applications // Singapore: Pan Stanford Publishing Pte Ltd. 2015.

3. *Watts, C. M., Liu, X., Padilla, W. J.* Metamaterial electromagnetic wave absorbers // Adv. Mater. 2012. V. 24, No. 23.P. OP98–OP120.

4. *Ra'di, Y., Simovski, C. R., Tretyakov, S. A.* Thin perfect absorbers for electromagnetic waves: theory, design, and realizations // Phys. Rev. Applied. 2015. V. 3, No. 3.P. 037001.

5. *Ogawa, Sh., Kimata, M.* Wavelength- or polarization-selective thermal infrared detectors for multi-color or polarimetricimaging using plasmonics and metamaterials // Materials. 2017.V. 10, No. 5.P. 493.

6. *Rozanov K. N.* Ultimate thickness to bandwidth ratio of radar absorbers // IEEE Trans. Antennas and Propag. 2000. V. 48, No. 8.P. 1230–1234.

7. *Thumm, M. K. A, Arzhannikov, A. V., Astrelin, V. T. et al.* Generation of high-power sub-THz waves in magnetized turbulent electron beam plasmas // J. Infrared, Millimeter, and Terahertz Waves. 2014. V. 35, No. 1. P. 81–90.

8. *Kuznetsov, S. A., Paulish, A. G, Gelfand, A. V. et al.* Extremely thin metamaterial absorbers for subterahertz waves: from fundamentals towards applications in uncooled bolometric sensors // Proc. SPIE. 2012. V. 8423. P. 84230S.

9. *Kuznetsov, S. A., Paulish, A. G., Navarro-Cia, M., Arzhannikov, A. V.* Selective pyroelectric detection of millimetre waves using ultra-thin metasurface absorbers // Sci. Reports. 2016. V. 6. P. 21079.

10. *Navarro-Cia, M., Kuznetsov, S. A., Aznabet, M., Beruete M. et al.*A Route for Bulk Millimeter Wave and Terahertz Metamaterial Design // IEEE J. Quant. Electr.2011. V. 47, No. 3.P. 375–385.

Microwave pulse delay at propagation through the 1D electromagnetic crystals

V.S. Babitski[1], V.G. Baryshevsky[2], A.A. Gurinovich[2], E.A. Gurnevich[2], P.V. Molchanov[2], L.V. Simonchik[1], M.S. Usachonak[1] and R.F. Zuyeuski

[1]Institute of Physics of NAS of Belarus, Minsk, Belarus, l.simonchik@ifanbel.bas-net.by
[2] Institute for Nuclear Problems of BSU, Minsk, Belarus
[3]Museo Storico della Fisica e Centro di Studi e Ricerche "Enrico Fermi", Roma, Italy & CERN

Interest to electromagnetic pulses propagation in matter with the group velocity much smaller than the speed of light in vacuum increases recent decades. Effect of group velocity reduction for a wave in a spatially periodic matter (natural crystal, photonic crystal, etc.) exists for different ranges and variety of waves (X-ray, optical, THz and microwave) due to generality of laws of wave diffraction. Multiple stu-dies report significant reduction of group velocity v_{gr} of an X-ray pulse passing through a crystal, when Bragg diffraction occurs [1]. A neutron delay in crystal in conditions of diffraction was predicted and observed [2]. Slow light, which is a promising solution for time-domain processing of optical signals, is a subject for multiple studies [3]. Experimental observation of this effect for microwave radiation generated by an electron beam in the photonic crystal built of metal threads was reported in [4].

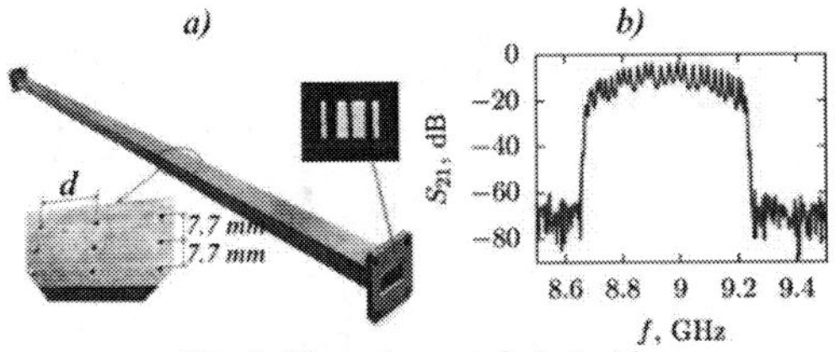

Fig. 1. Photonic crystal photo (a) and its frequency response (b)

In the present report the experiment to measure microwave pulse delay in a specially designed photonic crystal (spatially periodic structure built of metal rods inside a WR90/R100 waveguide of 1 m length) is presented. Each crystal plane comprises 3 copper rods of 1.25 mm diameter placed normally to the wider waveguide walls (Fig. 1, *a*). The crystal period d = 23.2 mm corresponds to the half-wavelength in the waveguide for 9.15 GHz frequency. The microwave field structure in the waveguide corresponds to TE_{10} mode of a rectangular waveguide. Since the delay time is proportional to the length of photonic crystal, the number of periods N was chosen to be rather high (N = 33) to make the delay time convenient for measuring. The photonic crystal frequency response is presented at Fig. 1, *b* and shows some amplitude ripples, which number is determined by the number of crystal periods.

Microwave pulse propagation in the photonic crystal was simulated by CST Microwave Studio [5] to define the incident pulse width making pulse delay effect evident. The simulation of transmitted and reflected pulses in time-domain was produced at the incident pulse duration from 4 to 30 ns. It was shown that delay effect is well resolved, when the width of the incident pulse is 10 ns and shorter. The time distance between maxima of transmitted and reflected pulses is about 23 ns (Fig. 2), that corresponds to pulse delay time τ. Both the transmitted and reflected pulses have chain of equidistant peaks with time interval between adjacent those of about 2τ.

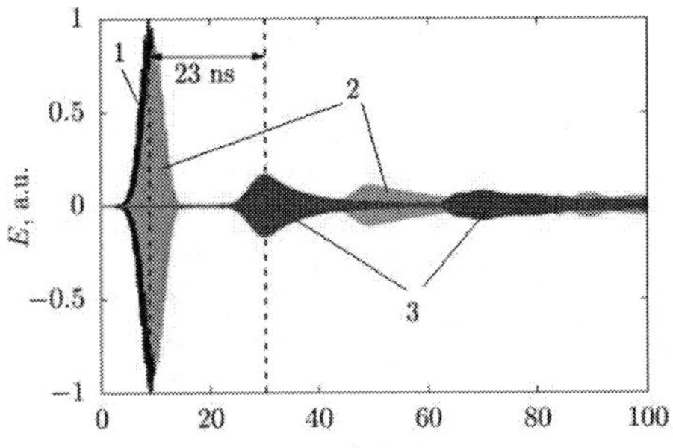

Fig. 2. Simulated electric field strength for the incident (1), reflected (2) and transmitted (3) waves

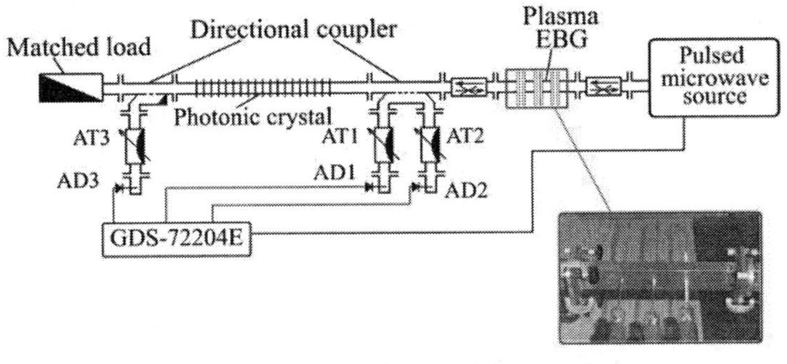

Fig. 3. The experiment layout and the setup photo

Experiment was produced using a high power magnetron oscillator, which keyword parameters are as follows: frequency about 9.15 GHz, pulse duration about 150 ns, repetition rate about 2 kHz, peak power 50 kW.

Microwave pulse shortening was produced by a one-dimensional electromagnetic band gap (EBG) plasma structure (in other words, electromagnetic crystal) formed in a WR90/R100 waveguide section by 3 discharges in the glass tubes filled by neon at pressure about 70 Torr [6]. These tubes were installed perpendicularly to the wide wall of the waveguide at 30 mm distance from each other (see Fig. 3, inset at the bottom). When the discharges with electron densities exceeding 10^{14} cm^{-3} are initiated in the tubes, the plasma electromagnetic crystal, a stop band central frequency of which is about 9.15 GHz, is formed. Discharge ignitions in the tubes take place in result of micro-wave breakdown of working gas under an effect of incident powerful microwave pulses. Incident and transmitted through the plasma electromagnetic crystal pulses are presented at Fig. 4.

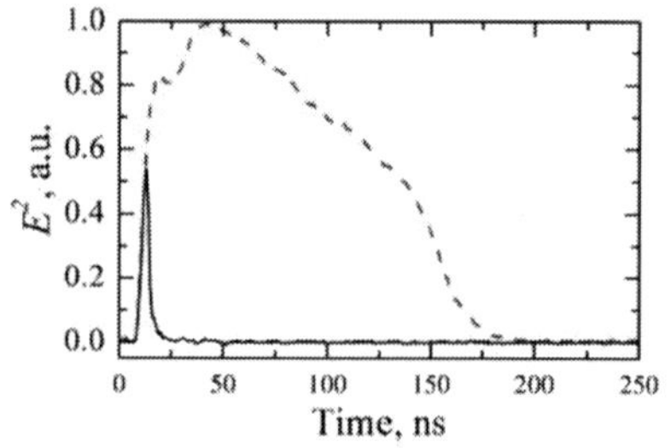

Fig. 4. Microwave pulses incident on the plasma EBG (dashed line) and passed through it (solid line)

Part of the incident to the photonic crystal microwave power was guided by an additional coupler, attenuator AT1 and waveguide-to-coaxial transition AD1 to a microwave diode and detected by oscilloscope GDS-72204E (200 MHz, 1 Gigasample/sec). Reflected and transmitted pulses were similarly detected by the same oscilloscope via channels AT2-AD2-microwave diode and AT3-AD3-microwave diode, respectively. Oscilloscope is triggered by sync pulse from the microwave source. To ensure equity of measuring arms the layout was preliminary tested with the empty waveguide of the equal length and without the photonic crystal.

Microwave pulse transmitted through the photonic crystal was normalized to the power of the incident pulse and compared with that obtained in simulation. The detected envelopes of the incident microwave pulses and the transmitted one are shown at Fig. 5 by (line+symbols) curve. The envelope shaded area in Fig. 5 shows simulated squared electric field strength (E^2) for the pulses.

Comparison of experimentally obtained and simulated pulse envelopes demonstrates good agreement in shapes of peaks and their amplitudes. The amplitude of the transmitted pulse is observed to be 30 times smaller as compared to the amplitude of the incident pulse. The observed delay time is $\tau \approx 23$ ns that is also in good agreement with simulation results. The pulse group velocity v_{gr} in the photonic crystal is $v_{gr} = L_{cr}/\tau \approx$ $\approx 3.2\times10^7$ m/s $\approx 0.11c$, where c is the speed of light in vacuum. Also as it was predicted in simulations the transmitted signal oscillations were detected experimentally. Signal peaks were separated at about 46 ns that in full accordance with the simulation results.

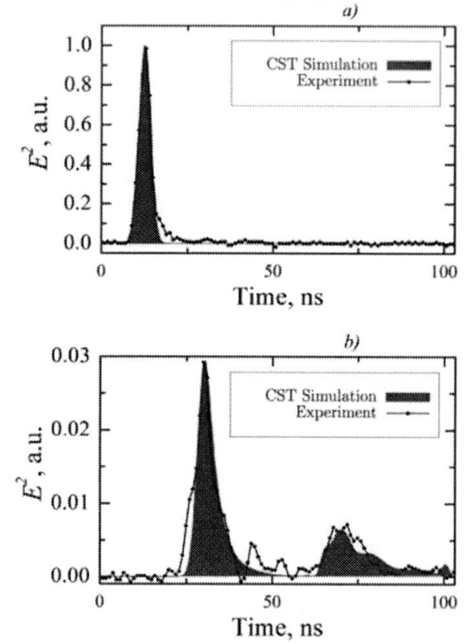

Fig. 5. (*a*) The envelope of the microwave pulse incident on the photonic crystal and (b) the microwave signal transmitted through it: experiment (line+symbols curve), simulation (shaded area)

The obtained results can be applied in high-speed information transmission systems, which use periodic crystal-like structures to control propagation of microwave radiation.

References

1. *Aleksandrov E.B., Zapasskii V.S.* // Phys. Usp. 2006. V. 49. 10671075.

2. *Voronin V.V., Lapin E.G. et al.* // JETP Letters. 2000. V. 71. P. 76.

3. Baba T. // Nature Photonics. 2008. V. 2. P. 465.

4. *Baryshevsky V.G., Gurinovich A.A.* // arXiv:1101.4162. 2011.

5. https://www.cst.com/products/cstmws

6. *Arkhipenko V.I., Callegari Th. et al.* // Journal of Applied Physics. 2014. V. 116. 123302.

Effect of methane on stability of plasma in a MW-assisted hydrogen-methane plasma

Swaminathan Prasanna, A. Michau, C. Rond, S. Farhat, K. Hassouni, A. Gicquel
LSPM, CNRS 3407, Universite Paris 13 Villetaneuse France

MW assisted Hydrogen-methane plasmas have been used extensively for growth of carbon materials such as CVD diamond, graphene and carbon nano-tubes. Graphene is usually grown at low pressures (>25 mbar) where as CVD diamond can be grown at relatively high pressures. Microwave radiation energizes the few free electrons to very high temperature, which collide with heavy species resulting in a self-sustaining plasma. In a conventional resonating MW cavity, the plasma occurs close to one of the available anti-nodes in the reactor and the substrate is placed closed to this anti-node enabling growth of diamond or graphene. The quality of the deposit depends on the coupling between microwave and the plasma which again depends on various factors such as MW power, pressure, concentration of the precursor gases etc. Understanding the coupling between the microwave and the plasma is critical for design and optimization of efficient MW reactors.

In this article, we discuss the effects of concentration of methane in a hydrogen plasma for carbon material processing in a MW assisted bell jar reactor which has been used for growing CVD diamonds and graphene at LSPM [1] (schematic shown in Fig. 1). The reactor is characterized by having two anti-nodes, one close to the substrate and the other close to the top of the bell jar reactor. At normal operating conditions, the plasma is formed close to the substrate which is desirable. However as the MW power is increased, the plasma can become unstable as two plasma zones can be formed.

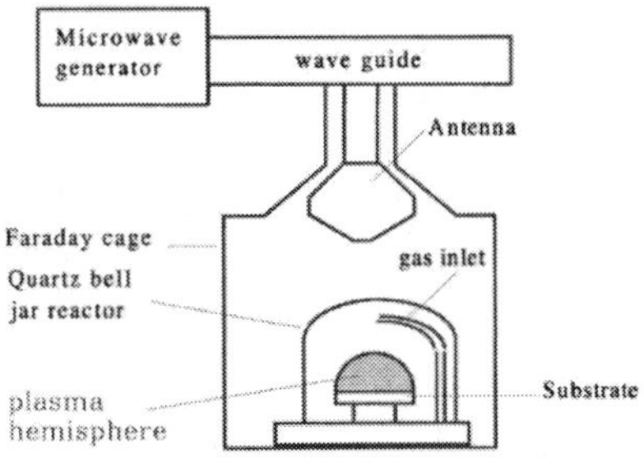

Fig. 1. Schematic of MW bell jar plasma reactor

We have observed that the addition of methane into the reactor affects the general plasma characteristics. The threshold MW power density (PMWD) at which the plasma becomes unstable have been measured for various concentration of methane in a MW bell jar reactor. This is achieved by visualizing the plasma emissive value at different conditions. Fig 2. shows the effect of concentration of methane on the threshold microwave power density. Addition of methane lowers the threshold microwave power density. Self-consistent simulations [2] have also been performed for these conditions (25 mbar). The difference between the numerical simulations and experiments may be due to uncertainty of absorbed microwave power, and also sensitivity of the numerical scheme. It has to be noted that only input power is available and there is no quantitative measurement of the MW power absorbed by the plasma. However, similar trends have been observed with experiments and numerical simulations as seen in Fig. 2.

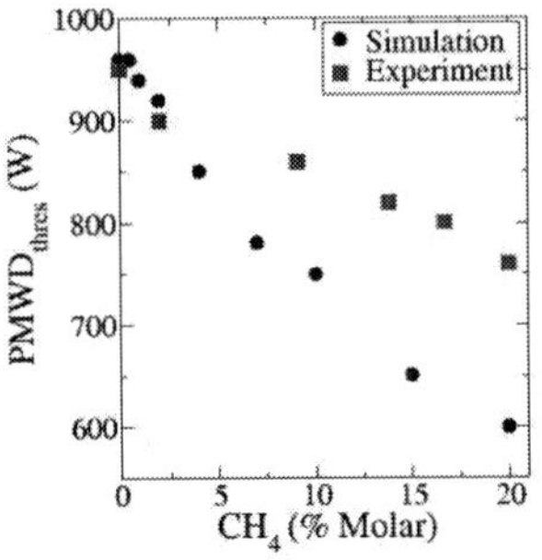

Fig. 2. Threshold Microwave power density at which plasma becomes unstable in a MW bell jar plasma reactor at 25 mbar

Similar results have been observed numerically for different reactor pressures. The present results are important because the effective range of operation of the MW reactor depends on the precursor gas concentration and pressure. The present observation can be explained by the dominant ions present in the reactor. In a pure hydrogen plasmas, the dominant ions under these conditions is H_3^+. However, with the addition of methane, the dominant ion species shifts to CH_5^+ and $C_2H_3^+$ depending on the concentration of methane. As the hydrocarbon ions are heavier compared to the hydrogen ions, the diffusion of hydrocarbon ions is much lower than the hydrogen ions. Consequently, higher ion density and thus electron density (due to electro-neutrality and ambi-polar diffusion prevailing under these conditions) is observed in the plasma core. This means the hydrocarbon plasmas observe higher power dissipation close to the substrate than the hydrogen plasmas. This has been confirmed from the numerical simulations. As a result the size of the plasma for with hydrocarbons is larger than the pure hydrogen plasmas for same MW power conditions Formation of secondary plasmas occurs when sufficient electrons are available at the second anti-node. This process is accelerated for a hydrocarbon plasma due to large size of plasma at the same input microwave power.

The present results show that the MW-plasma coupling is affected by the concentration of methane. To design efficient and optimum reactors, it will be important to understand the effects of the precursor gases on MW-plasma coupling. Further studies would be conducted to understand the MW-plasma coupling in different gas discharges at wide range of operating conditions.

Acknowledgements. This work was supported by ANR through two projects ANR-11-LABX-086/ ANR-11-IDEX-0005-02 and ANR2011BS09029.

References

1. *Gicquel, A., Hassouni, K., Farhat, S., Breton, Y., Scott, C.D., Lefebvre, M. and Pealat, M.* Spectroscopic analysis and chemical kinetics modeling of a diamond deposition plasma reactor. // Diamond and Related Materials, 1994, 3, 581-586.

2. *Prasanna, S., Rond, C., Michau, A., Hassouni, K. and Gicquel, A.* Effect of buoyancy on power deposition in microwave cavity hydrogen plasma source.// Plasma Sources Science and Technology, 2016, 25, 045017.

EPJ Web of Conferences **149**, 02021 (2017) DOI: 10.1051/epjconf/201714902021
10th International Workshop 2017 "Strong Microwaves and Terahertz Waves: Sources and Applications"

Microstructure of the microwave fast-sintered $MgAl_2O_4$ ceramics

A.A. Sorokin, S.V. Egorov, Yu.V. Bykov, A.G. Eremeev, V.V. Kholoptsev,
K.I. Rybakov, S.S. Balabanov[1], A.V. Belyaev[1]

Institute of Applied Physics, Russian Academy of Sciences, Nizhny Novgorod, Russia, asorok@appl.sci-nnov.ru
[1] G.G. Devyatykh Institute of Chemistry of High-Purity Substances, Russian Academy of Sciences, Nizhny Novgorod, Russia

Polycrystalline $MgAl_2O_4$ spinel is considered as one of the major candidates for various applications that require a combination of properties including high optical / IR transparency and excellent mechanical performance, particularly in harsh environments. Among many different techniques to produce translucent or transparent spinel, the two-stage method based on the pressureless or hot press sintering with subsequent hot isostatic pressing (HIP) proved itself the most successful. But high pressure (≥150 MPa) and high temperature (≥1600 °C) commonly used in HIP greatly limit the potential of its use.

Recently, a significant attention has been attracted to the investigation of very rapid ("flash") sintering of ceramics achieved by applying a dc or ac voltage to the sample [1]. In these experiments, many (mostly oxide) ceramics have been sintered nearly to full density in several seconds. In particular, in [2] the $MgAl_2O_4$ spinel was flash sintered to a relative density of 97.9 % at the applied DC field of 1000 V/cm. This paper reports on the fast microwave sintering of $MgAl_2O_4$ ceramics to nearly full density under high-rate heating and zero isothermal hold time.

Magnesium aluminate spinel powder was produced from isopropoxide $MgAl_2(OPr^i)_8$ by method described in detail elsewhere [3]. Specific surface area of powder was about 100 m^2/g after calcining in air at T = 900 °C. Samples with 15 mm dia and 2.5 mm thickness were uniaxially pressed at 300 MPa to a relative density of 39 % of the theoretical value. Some samples were impregnated with yttrium oxide which was introduced by dipping them into 0.5 mol/l water solution of $Y(NO_3)_3$ with consequent heating in air at 70 °C. The content of Y_2O_3 in these samples was equal to 0.8–1.0 % wt.

Samples were heated in the applicator of a gyrotron system with a microwave power up to 6 kW at a frequency of 24 GHz with a feedback computer control of the power. The details of the experimental scheme can be found in [4]. Samples were heated at 2–4 MPa pressure of the residual air at a rate of 50 °C/min up to 1250 °C and then at rates of 15...200 °C/min to the preset maximum temperature of sintering, T_{max}, in the range of 1400–1780 °C. When the preset T_{max} was reached, the microwave power was switched off automatically and a sample cooled down along with the thermal insulation.

Shown in Fig. 1 are the relative densities of the pristine and Y-doped samples heated at a rate of 100 °C/min as the function of the temperature T_{max}. The relative density of the Y-doped samples exceeded 99 % at all rates of heating to $T_{max} \geq 1500$ °C, whereas it was about 98 % for the pristine samples at $T_{max} \geq 1600$ °C. The relative densities of doped and pristine samples sintered conventionally in the similar regimes at T_{max} = 1500 °C were only 92.2 and 89.1%, respectively.

The density of Y-doped samples starts to exceed the density of pristine samples at T_{max} above 1400 °C.

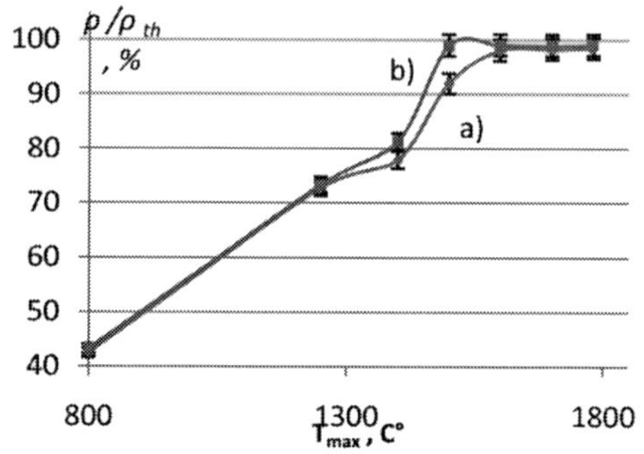

Fig. 1. Density of pristine (a) and Y-doped (b) samples heated at a rate of 100 °C/min to the temperature T_{max}

The densification enhancement of Y-doped samples is explained by the changeover to the transient liquid phase sintering, as it is clear from the data on T(t) and $P_{mw}(t)$ behavior at the final stage of heating. It can be seen that at T ≥ 1600 °C the microwave power P_{mw} required to sustain the prescribed heating rate does not increase (as it should due to the growth of the heat losses from the sample, which are proportional to T^4) and even falls down towards the end of the process (Fig. 2). The decrease in the microwave power can only be explained by a sharp increase in the microwave absorption coefficient, i.e. in the effective high-frequency conductivity of the material of the sample, which is associated with the development of the thermal instability. The onset temperature of the instability decreases with an increase in the microwave power deposited per unit volume of sample.

The mechanism of fast densification caused by the development of thermal instability has been suggested in [4, 5]. In brief, if the temperature and the density of the deposited power are high enough, the particle surface premelts at a temperature well below the melting point of the bulk of grains due to the abundance of impurities and defects in the near-surface layer. All the mechanisms inherent in liquid phase sintering become activated in the transient quasi-liquid medium, leading to an accelerated densification.

The crucial role of the liquid phase formation in the process of fast microwave $MgAl_2O_4$ ceramics sintering is confirmed by the study of the microstructure of the sintered samples. As seen from the SEM image of the unpolished surface of a sample sintered at T_{max} = 1500 °C (Fig. 3a), it has a dense and uniform microstructure with an average grain size of about 150 nm, close to the initial particle size. At higher temperature, $T_{max} \geq 1600$ °C, intense abnormal grain growth (AGG) takes place, much more pronounced in the Y-doped samples (Fig. 3b). It is well-known that AGG is typical for liquid phase sintering. The morphology of the microstructure suggests that the growth of large-size grains occurs by some collective effects rather than by the ordinary mechanism based on the grain boundary mobility of individual particles.

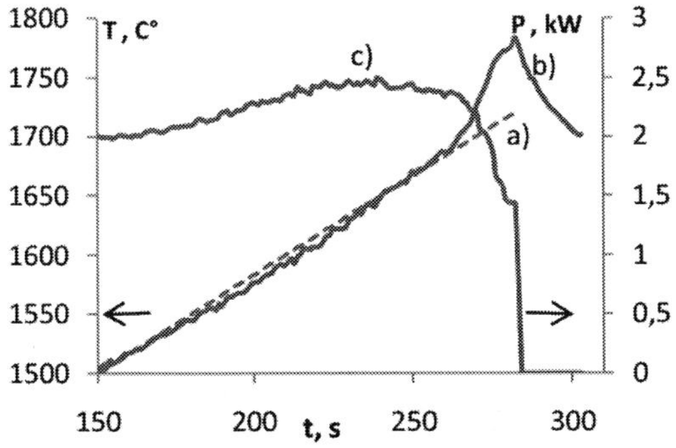

Fig. 2. Time dependence of the preset (a) and actual (b) temperatures, and microwave power (c) during heating of an Y-doped sample at a rate of 100 °C/min to T_{max} = 1780 °C

The formation of the liquid phase at the high temperature stage of sintering is confirmed also by the effect of fast segregation of Y atoms at the boundaries of the growing grains (Fig. 3c). From the pattern and scale of Y atoms distribution, seen in the SEM images as white lines along boundaries, the grain boundary mobility can be estimated, which, as it turns out, exceeds the value measured from normal grain growth in dense stoichiometric $MgAl_2O_4$ ceramics [6] by a factor of 10^2-10^3.

The X-ray phase analysis revealed small amounts (about 1–2 wt. %) of Y-Al-O compositions in the fast microwave sintered Y-doped samples: the $Y_4Al_2O_9$ phase is present in the samples heated to 1250 °C, and YAG – in the samples heated to 1780 °C. The composition $MgAl_2O_4$ – YAG has an eutectic temperature of 1775 °C at conventional heating under normal pressure. It is known that microwave heating can decrease the temperature of some phase transformations and solid state reactions [7]. One can not exclude a similar effect of the microwave heating on the reduction of the eutectic temperature of the given composition resulting in the earlier formation of the liquid phase.

The reason why the abnormal grain growth is also observed during the fast microwave heating of the pristine spinel needs further investigation.

In conclusion, we have demonstrated fast microwave sintering of $MgAl_2O_4$-ceramics to nearly full density in less than two minutes without any isothermal hold. The results of the microstructure study on the sintered samples show that the fast microwave densification of $MgAl_2O_4$-ceramics takes place by the transient liquid phase formation.

This research was supported in part by Russian Foundation for Basic Research, grant # 16-08-00736.

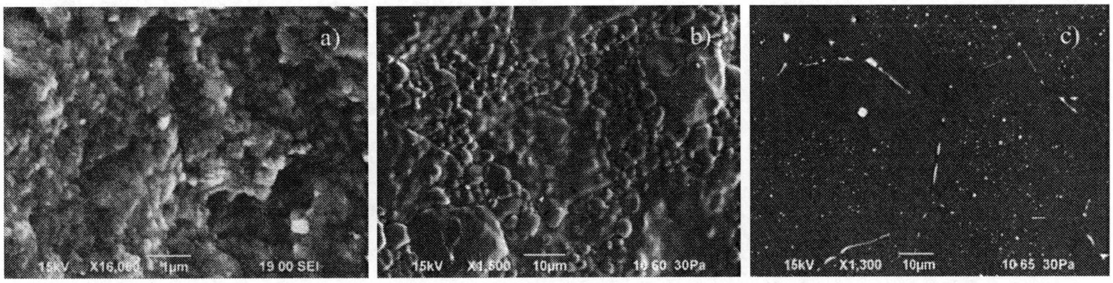

Fig. 3. Microstructure of unpolished surface of Y-doped $MgAl_2O_4$ sample sintered at 1500°C (a) and at 1780°C, unpolished (b) and polished surfaces (c)

References

1. *Yu, M., Grasso, S., Mckinnon, R.,* et al. Review of flash sintering: materials, mechanisms and modeling // Advances in Applied Ceramics, 2017. V. 116, No. 1. P. 24-60.

2. *Yoshida, H., Biswas, P., Johnson, R.* et al. Flash sintering of magnesium aluminate spinel ($MgAl_2O_4$) ceramics // *J. Amer. Ceram. Soc.* 2017. V. 100, No. 2. P. 554-562.

3. *Balabanov, S., Vaganov. V., Gavrishchuk, E.* et al. Effect of magnesium aluminum isopropoxide hydrolysis conditions on the properties of magnesium aluminate spinel powders // Inorganic Mater. 2014. V. 8, No. 8. P. 830-838.

4. *Bykov, Yu. V., Egorov, S. V., Eremeev, A. G.*, et al. Flash microwave sintering of transparent Yb:$(LaY)_2O_3$ ceramics // J. Am. Ceram. Soc. 2015. V. 96, No. P. 3518-3524.

5. *Bykov, Yu. V., Egorov, S. V., Eremeev, A. G.*, et al. On the mechanism of microwave flash sintering of ceramics // Materials 2016, V. 9, No.8. P. 684-702.

6. *Chiang, Y.-M., Kingery, W. D.* Grain-boundary migration in nonstoichiometric solid solutions of magnesium aluminate spinel: I, Grain grows study // *J. Amer. Ceram. Soc.* 1989. V. 72, No. 2. P. 271–277.

7. *Esposito, L., Piancastelli, A., Bykov, Yu.,* et al. Microwave sintering of Yb:YAG transparent laser ceramics // Opt. Materials 2013. V. 35, P. 761-765.

High rate production of nanopowders by the evaporation – condensation method using gyrotron radiation

A.V. Vodopyanov[1], A.V. Samokhin[2], N.V. Alexeev[2], M.A. Sinayskiy[2], A.I. Tsvetkov[1], D.A. Mansfeld[1], M.Yu. Glyavin[1], A.P. Fokin[1] and V. I. Malygin[1]

[1]Institute of Applied Physics of Russian Academy of Sciences, Nizhny Novgorod, Russia, avod@yandex.ru
[2]Baykov Institute of Metallurgy and Materials Science, Moscow, Russia

A new method for metal oxide nanopowder production was proposed recently by the authors [1, 2]. It is the evaporation–condensation using a focused microwave radiation. The source of microwaves was the technological gyrotron with frequency of 24 GHz and power up to 7 kW with the energy density flux of 13 kW/cm^2. Radiation was focused on a layer of powder of the treated material to ensure its evaporation, subsequent condensation of vapor in the gas stream, and deposition of particles on the water-cooled surface. A scheme of the experiment is shown in Fig. 1.

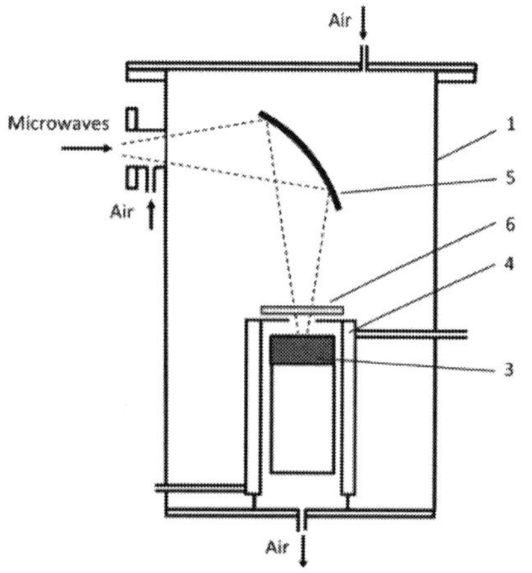

Fig. 1. Schematic view of the evaporation–condensation section: 1 – vessel, 2 – thermally insulating material, 3 – crucible with material, 4 – water-cooled collector, 5 – microwave focusing mirror, and 6 – quartz window

Deposited powders consist of particles whose sizes are in the range of 20 nm to 1 μm. The powder consists of particles which have different shapes—close to spherical shape as well as octahedral, which indicates that the mechanism of particles formation is "vapor–liquid–crystal" as well as "vapor–crystal." Fig. 2 shows examples of nanopowders obtained. The produced powders are the ensembles of particles whose sizes are in the range of 20 nm to 1 um. The WO_3 powders consist of particles having different shapes close to spherical shape as well as octahedral, which indicates that the mechanism of particles formation is vapor–liquid–crystal as well as vapor–crystal. The formation of the nanoparticles in evaporation–condensation processes occurs by vapor condensation and it is determined by temperature conditions, which in turn depends on the flux density of microwave energy and velocity distribution of the gas incident on the evaporation surface. If the particle growth is from the vapor phase at temperatures above the melting point, the particles are spherical. The deposited tin oxide powder consists of particles with sizes of 20–200 nm, preferably with equiaxed shape and a facet. The obtained zinc oxide powder has a complex morphology consisting mainly of two types of particles–faceted rods with the length of 200–300 nm and a transverse dimension of about 60 nm and nanowhiskers of approximately the same length but with a transverse dimension not greater than 20 nm. The maximum evaporation rate was 100 g/hr. Evaporation rate depending to microwave power is shown in Fig. 3.

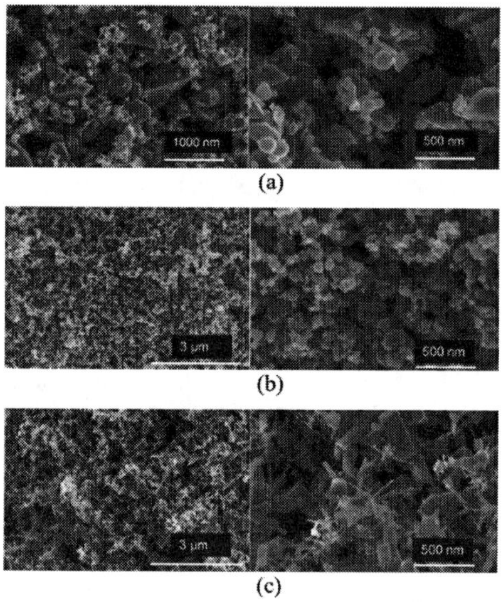

Fig. 2. Microscope images of powders.
(a) – WO_3, (b) – SnO_2, (c) – ZnO

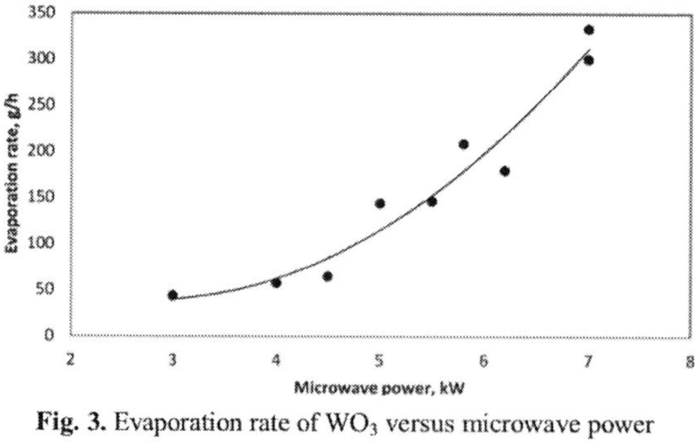

Fig. 3. Evaporation rate of WO_3 versus microwave power

This method was upgraded by the use of a subterahertz gyrotron setup with output frequency of 263 GHz and a nominal power of 1 kW as a radiation source. In this case microwave-energy flux density was of about 20 kW/cm^2. A significant increase in the substance evaporation rate compared with the previous experiments has been shown. But in this case evaporation rate is slightly decreasing with time, see Fig. 4.

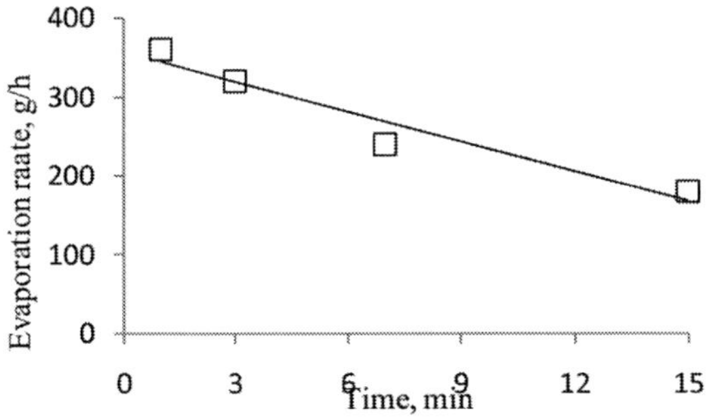

Fig. 4. Time dependence of the evaporation rate for WO_3

This is due to the formation of the increasing crater at the place of action of the microwave beam on the sample. On the one hand, this leads to a change in the flow pattern of the quenching gas in this region and, on the other hand, reduces the power density of the microwave beam on the heated surface due to its leaving of the focal waist. These drawbacks can be partly compensated by using a mechanical system of height adjustment of a crucible with the sample and upgrading the air flow system.

The experimental results demonstrated that the size of nanoparticles could be adjusted by tuning the heating power – cooling air flow ratio. Fig. 5 shows the typical dependence of the specific surface of the obtained powder on the quenching air flow rate.

A significant increase in the substance evaporation rate compared with similar experiments using a technological gyrotron setup with a frequency of 24 GHz as a radiation source was demonstrated. An increase in the radiation frequency leads to the improvement of the heating efficiency due to an increase of the absorption coefficient and provides a greater power density due to a better focusing of the radiation. The sizes of the nanoparticles obtained in the experiments range from 20 to 500 nm. The maximum rate of evaporation of tungsten trioxide WO_3 was about 360 g/h and of zinc oxide ZnO, about 50 g/h.

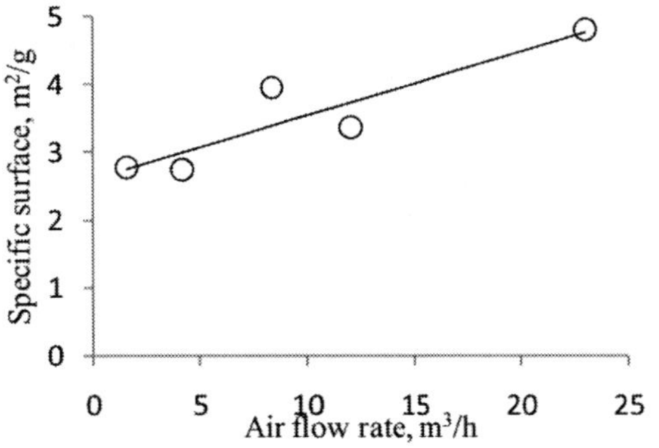

Fig. 5. Specific surface of the obtained powder on the air flow rate for tungsten trioxide WO_3

Acknowledgements

The work was supported by the Russian Science Foundation, project No. 15-12-00046.

References

1. *A. V. Samokhin et al.*, "Production of WO3 tungsten oxide nanopowders by evaporation-condensation process using focused 24-GHz microwave radiation," High Energy Chem., vol. 49, no. 4, pp. 267–272, 2015.

2. *A. V. Samokhin, N. V. Alexeev, A. V. Vodopyanov, D. A. Mansfeld, and Y. V. Tsvetkov*, "Metal Oxide Nanopowder Production by Evaporation– Condensation Using a Focused Microwave Radiation at a Frequency of 24 GHz," J. Nanotechnol. Eng. Med., vol. 6, no. 1, p. 11008, 2015.

EPJ Web of Conferences **149**, 02023 (2017) DOI: 10.1051/epjconf/201714902023
10th International Workshop 2017 "Strong Microwaves and Terahertz Waves: Sources and Applications"

High-temperature microwave pyrolysis of peat as a method to obtaining liquid and gaseous fuels

T.O. Krapivnitskaia[1,2], A.A. Bogdashov[1], A.N. Denisenko[1], M.Yu. Glyavin[1], Yu.K. Kalynov[1], S.V. Kuzikov[1], N.Yu. Peskov[1], L.L. Semenycheva[2] and A.V. Stricovskiy[1]

[1]Institute of Applied Physics RAS, Nizhny Novgorod, Russia, kto465@yandex.ru
[2]Lobachevsky State University of Nizhniy Novgorod, Nizhny Novgorod, Russia

Currently, scientific researches and engineering developments in the field of physical-chemical processing of organic raw materials, in particular, aimed on production of so-called renewable liquid and gaseous fuels, are becoming increasingly important. One of the materials that recently attracted high research interest from the point of view of the great variety of its practical applications is a peat [1].

There are various methods of the peat processing; however, the most promising is a high-temperature microwave pyrolysis. The main advantages of microwave pyrolysis over traditional thermal heating systems include the following: high processing efficiency, volume character and high thermal efficiency (up to 90 - 95%) of microwave heating, increase in the speed of chemical and physical-chemical reactions under the action of microwave radiation, possibility to achieve high temperatures and deep processing of biomaterial, high environmental safety, etc.

The installation for the high-temperature peat destruction under the action of microwave radiation was developed at the IAP RAS based on industrial 2.45 GHz / 1 kW magnetron (Fig. 1).

In the recent series of experiments, optimization of the electrodynamic system transporting radiation from the magnetron into the interaction vessel (reactor) was carried out. To reduce thermal loads on the barrier window an additional cooled waveguide section was installed for condensation of the resins products of the processing. Tuning waweguide elements were added to keep high transmission of the radiation into the reactor during radical change of the absorption capacity of peat in the process of high temperature treatment [2].

The first series of experiments on this installation was carried out. As a result of high-temperature peat pyrolysis, an oily fraction, a carbonaceous residue and a pyrolysis gas were obtained. A series of experiments was also carried out using traditional thermal heating of peat. The obtained samples were analyzed on GCMS-QP2010 Ultra-GC-mass spectrometer. Chromatograms of pyrolytic gases were compared and showed a higher yield of combustible gases when microwave exposure to samples (Fig. 2, 3).

A CH-analysis of the initial sample loaded into the reactor and a carbonaceous residue after the pyrolysis process were also carried out (Table 1).

Fig. 1. Photograph of experimental test facility for peat pyrolysis heated by microwave radiation:
1 – industrial magnetron; 2 – power supply system; 3 – air cooling system; 4 – matching wave-guide elements; 5 – barrier window; 6 – connection waveguide; 7 – operating camera; 8 – water cooling system; 9 – thermoelectric couple; 10 – pressure-gauge; 11 – the heat exchanger

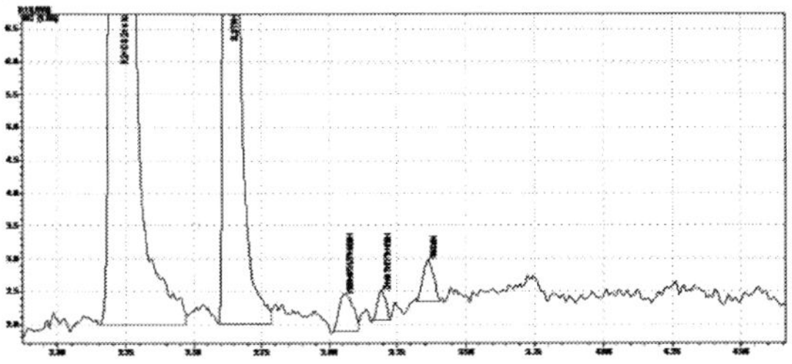

Fig. 2. Chromatogram of pyrolysis gases during thermal heating

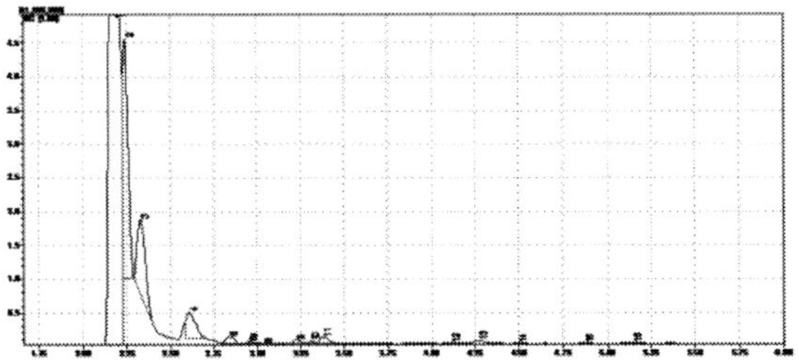

Fig. 3. Chromatogram of pyrolysis gases at microwave heating

The mass fraction of carbon in percent after processing the sample with microwave radiation was more than 2 times greater than the mass fraction of carbon in the initial sample [3]. According to the initial results of the analysis of the samples, it can be said that the proposed peat processing method is more effective than traditional heating methods and allows to significantly increase the rate of pyrolysis process passage, as well as to increase the yield of combustible gases for their further application. As further development of work on microwave pyrolysis, an increase in the efficiency of the output of combustible gases from organic fuel and, thus, the efficiency of the installation, as well as the possibility of creating a closed cycle for powering the device and its autonomous operation.

A prospective set of applications under development currently is related to RF impact on biological objects including the microwave grain crop drying in order to eliminate the fungal contamination, RF methods for lumber disinsectization and destruction of pest, etc.

Preliminary basis allows to make a conclusion on the prospective of using the microwave radiation in technological processes, in an increasingly large number of applications and on economic advantages of this approach.

Table 1. The results of CH-analysis

Name	ω (C),%	ω (H),%	ω (N),%	ω (S),%
initial sample	31	7,9	0,52	‹ 0,3
after microwave processing	64,28	5,02	1,4	‹ 0,3

References

1. *J.Yang, H.Chen, W.Zhao, J.Zhou.* TG–FTIR-MS study of pyrolysis products evolving from peat // Journal of Analytical and Applied Pyrolysis. № 117, 296-309 (2016)

2. *M.Yu.Glyavin, Y.K.Kalynov, T.O.Krapivnitskaia, et.al.* Development of microwave test facility for peat pyrolysis // Proc. of the 26th Int. Conf. "Microwave and Telecommunication Technology" (CriMiCo 2016), Sevastopol, Russia, 2016, 10, 2287-2293

3. *T.O.Krapivnitskaia, N.Yu.Peskov, L LSemenycheva, et al.,* Perspective applications of high-power microwave radiation in problems of exposure to organic materials and biological objects // Proc. of the VI All-Russian Scientific and Technical Conference "Electronics and Micro-electronics Microwave", St. Petersburg, Russia, impress.

Theory of resonant stationary discharge with multiply charged ions in plasma flow propagating in mirror magnetic trap

I.S. Abramov, E.D. Gospodchikov, A.G. Shalashov

Institute of Applied Physics RAS, Nizhny Novgorod, Russia, abramov@appl.sci-nnov.ru
Lobachevsky State University of Nizhny Novgorod, Russia

Nowadays semiconductor industry requires reliable sources of extreme ultraviolet radiation (EUV) for next generation projection lithography [1]. In practical applications the only radiation fitting EUV spectral region is line radiation of multiply charged ions of Sn or Xe. These ions have considerable number of lines within the wavelength band of 13.5±1% nm [2, 3] that corresponds to peak values of reflection coefficient for existing Mo/Si multilayer mirrors. The most advanced projects use laser produced plasma to obtain required ion charges [4, 5]. These facilities are characterized by about 4–5% conversion efficiency to 13.5±1% nm band and total output of EUV radiation up to 250 W of averaged power.

Microwave discharge seems to be a perspective alternative to laser produced plasma. Pioneer experiments performed at IAP RAS (Nizhny Novgorod, Russia) demonstrated possibility of this plasma usage for generation of EUV radiation [6]. In the experiments, Sn plasma from vacuum arc source propagates in axial vicinity of the mirror magnetic trap. The flow is heated by 75 GHz / 50 kW gyrotron. The key feature of the experiments is resonant nature of plasma heating providing effective power load directly into the electron component that results in high rates of electron impact ionization and excitation without sufficient ion heating. In the paper we propose the theory of considered discharges that is of interest in terms of further source development.

Let us consider a plasma flow along confining magnetic field $B(x)$. The effective cross-section $\sigma(x)$ of the flow is unambiguously defined from the magnetic field line freezing-in, $B\,\sigma = \text{const}$. Plasma is characterized with the densities $n_j(x)$ of ions with charge-state $Z_j = j$, the flow velocity $u\,(x)$, the electron density $n_e(x)$ followed from the quasi-neutrality condition, and the effective electron temperature T_e. Variation of the electron temperature along the flow is neglected due to high electron thermal conductivity typical for high-power microwave discharge [7]. Electron temperature considerably exceeds the temperature of ions, thus we also neglect the ion pressure as compared to the electron pressure. Discharge duration is defined by the pulse length of the supporting microwave source which scales from 100 µs to CW. It is much longer than internal times of the discharge evolution, e.g. ion fly-time through the trap, so we consider a stationary discharge. Under these assumptions, we use the following fluid equations

$$\partial_x(\sigma n_j u) = \sigma(k_{j-1} n_e n_{j-1} - k_j n_e n_j), \quad (1)$$

$$\partial_x(\sigma \Sigma M n_j u^2) = \sigma \partial_x(n_e T_e), \quad (2)$$

representing the particle balance for ions and total ion and electron momentum balance [7, 8]. Here the sum is taken over all ion species, M is the ion mass, and k_j is the ionization coefficient of j-th ion fraction. Electron temperature T_e enters this model as a parameter. The boundary conditions are defined by the initial ion beam composition at the plasma gun, the ion beam current and the presence of the transition from the subsonic flow inside the trap to the supersonic one in the expanding region providing decreasing electron density needed for thermal insulation of the discharge. Formal mathematical difficulties considering dependence of this transition location on preceding ionization dynamics may be resolved as it is proposed in [8].

Power load that is needed to support the discharge may be estimated from the energy conservation

$$P(T_e) = A\sigma n u \frac{Mu^2}{2}\bigg|_{x\to\infty} + \int \sigma p_{\text{ion}} dx + \int \sigma p_{\text{exc}} dx. \quad (3)$$

The first term represents the convective losses, here $A \sim 2.5 - 5$ depending on the details of plasma absorber. Other terms describe the volumetric losses for ionization and line excitation of ions, characterized with densities

$$p_{\text{ion}} = \sum_j E_j k_j n_e n_j, \quad p_{\text{exc}} = \sum_j \sum_l \Delta E_{jl} k^*_{jl} n_e n_j. \quad (4)$$

Here p_{ion} is the power spent to ionize a unit volume of multiply charged plasma, E_j is the ionization energy and $k_j\, n_e$ is the effective rate of electron impact ionization for j-th ion, p_{exc} is the power spent to line excitation, inner sum is over allowed transitions in the spectrum of j-th ion, ΔE_{jl} and $k^*_{jl}\, n_e$ are the transition energies and rates. In the studied conditions, the rates of non-radiative processes and the photon trapping are low: once excited, an ion spontaneously relaxes to the ground state emitting a photon that leaves the discharge volume. So, with good accuracy, p_{exc} may be interpreted as radiation losses.

In the absence of data related to an electron distribution function formed in the discharge, we use monoenergetic cross-sections averaged over Maxwellian distribution. Fitting parameters for the electron-impact ionization processes in Sn plasma, presumably used for EUV radiation sources development, are taken from [9]. For the estimation of the excitation coefficients, Bethe approximation for dipole allowed transitions is used [10],

$$k^*_{jl} = 1.6\times10^{-5}\,\frac{\text{cm}^3}{\text{s}}\left[\frac{T_e}{1\text{eV}}\right]^{-\frac{3}{2}} \frac{f_{jl}\langle g(\varepsilon_{jl})\rangle}{\varepsilon_{jl}\exp(\varepsilon_{jl})}, \quad (5)$$

where $\varepsilon_{jl} = \Delta E_{jl}/T_e$, f_{jl} is the oscillator strength of corresponding transition. Function $\langle g(\varepsilon_{jl})\rangle$ is averaged Gaunt-factor. In our model, we use Gaunt-factor in the form proposed in [11] and corrected in [12], see Fig. 1. The oscillator strengths are taken from NIST Atomic Spectra Database [13] for Sn^{+1} and, in the absence of such data that is typical for highly charged ions, calculated by Cowan's suite of codes [14]. Such calculations were performed for Sn^{2+} – Sn^{19+}. Results for EUV band are presented in Fig. 2 for several ion species. Only allowed transitions are taken into consideration in our model due to their higher probabilities and presence of numerous resonance lines ($\Delta n = 0$) in spectra of Sn ions that correspond to situation when allowed transitions dominate.

Discharge simulations within the model show relatively good agreement to available experimental data [6] realizing flow regimes with similar emitted EUV power and obtained ion charges.

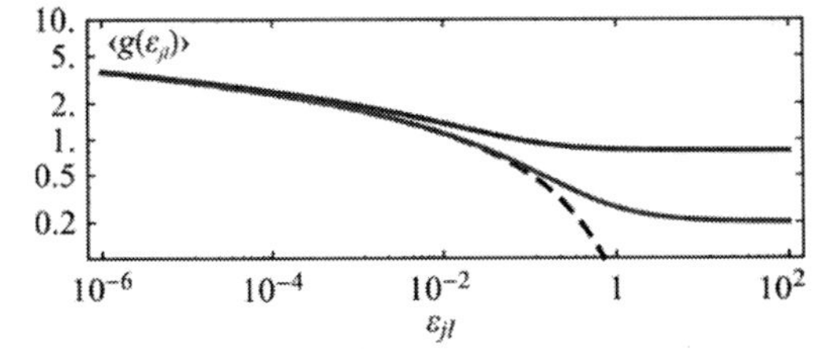

Fig. 1. Gaunt-factor $\langle g(\varepsilon_{jl})\rangle$ averaged over Maxwellian distribution of electrons: red line – as formulated by Van Regemorter [13], blue line – after correction by Sampson and Zhang [14] (used in the modeling), black dashed line – high-temperature asymptotics given by $\sqrt{3}E_1(\varepsilon_{jl})/2\pi$

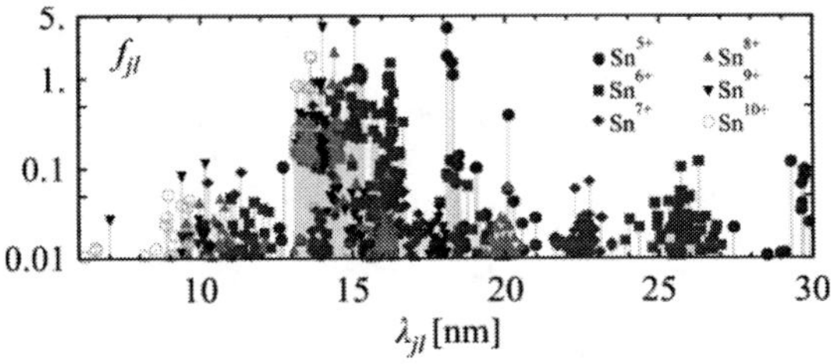

Fig. 2. Calculated oscillator strengths of allowed transitions in EUV spectral region for Sn^{+5} – Sn^{+10}. Ion species are indicated by different colors and unique geometric figure.

Modeling is performed in a wide range of plasma parameters and possible magnetic configurations taking into account physical constrains related to now available sources of microwave radiation and plasma. As a result, we find out opportunities for optimization of future EUV sources based on the discussed concept. An example of such optimization is illustrated in Fig. 3. The figure shows flow regime with an improved magnetic configuration. Such discharge emits about 40 kW of EUV radiation in 13.5±1% nm with 20% conversion efficiency. The size of EUV emitting region is about 1 cm × 0.3 cm^2 what is far from the point-like source optimal for the beam forming optics. Therefore, some additional losses in efficiency may be expected due to not ideal EUV focusing; however, this is compensated by the very high emitting efficiency and total power of EUV radiation.

As a summary, we point out that scaling of the results of existing experiments on the microwave assisted EUV radiation in Sn plasma shows good potentials for the development of an industry-ready source featured with a multi-kW level in 13.5±1% nm band. Most required hardware, such as compact high-current plasma sources and high-power high-frequency gyrotrons, are available.

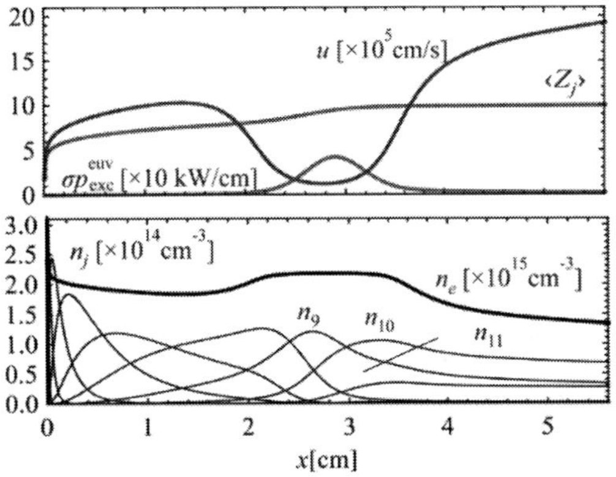

Fig. 3. Simulation of EUV-radiating discharge: variation along the trap axis of flow velocity u (red line, top panel), average ion charge $\langle Z_j\rangle$ (green line), linear power density σp_{exc} of radiation in 13.5±1% nm EUV band (violet line), ion fractions' densities n_j (thin black lines, bottom panel), and electron density n_e (thick black line). Model parameters are: trap length L = 5.6 cm, mirror ratio R = 9, initial plasma diameter d = 0.2 cm, arc current I_a = 2.9 kA, initial beam composition $n_1(0)/\Sigma n_j(0) \approx 0.325$, $n_2(0)/\Sigma n_j(0) \approx 0.567$, $n_3(0)/\Sigma n_j(0) \approx 0.108$, and other components are negligible.

The work was supported by Russian Science Foundation (grant No. 14-12-00609).

References

1. *C. Wagner, N. Harned*, Nat. Phot. 2010. V. 4. P. 24 .
2. *S. Churilov, Y. N. Joshi, J. Reader*, Optics Letters 2003. V. 28. P. 1478.
3. *J. White, P. Hayden, P. Dunne et al.* Journal of Applied Physics 2005, V 98, P. 113301.
4. *A. A. Schafgans, D. Brown, I. Fomenkov et al.* Proc. SPIE 2015. V. 9422. P. 94220B-1.
5. *H. Mizoguchi, H. Nakarai, T. Abe et al.* Proc. SPIE 2016. V. 9776. P. 97760J-1.
6. *N. I. Chkhalo, S. V. Golubev et al.* J. Micro Nanolithogr. MEMS MOEMS 2012. V. 11, P. 021123-1.
7. *I. S. Abramov, E. D. Gospodchikov, A. G. Shalashov*. Radiophysics and Quantum Electronics. V. 58. P. 914.
8. *A. G. Shalashov, I. S. Abramov, S. V. Golubev, E. D. Gospodchikov*. JETP 2016. V. 123. P. 219.
9. *A. V. Philippov, V. M. Povyshev, A. A. Sadovoy et al.* JINR Commun. 2002. V. E9-2002-5. P. 1.
10. *J. D. Huba*. NRL: Plasma formulary No. NRL/PU/6790–04-477 (2004). Eq. (4) on page 52.
11. *H. Van Regemorter*. The Astrophysics Journal 1962. V. 136. P. 906.
12. *D. H. Sampson, H. L. Zhang*. Phys. Rev. A 1992. V. 45. P. 1556.
13 *A. Kramida, Yu. Ralchenko, J. Reader, and NIST ASD Team*. NIST Atomic Spectra Database (v. 5.4). Available at http://physics.nist.gov/asd (NIST, Gaithersburg, MD, 2016).
14. *R. D. Cowan. The theory of atomic structure and spectra* (Univ.of California Press 1981)

EPJ Web of Conferences **149**, 02025 (2017) DOI: 10.1051/epjconf/201714902025
10th International Workshop 2017 "Strong Microwaves and Terahertz Waves: Sources and Applications"

Study of plasma parameters in a continuous ECR discharge sustained by 24 GHz/5 kW gyrotron radiation in a quasi-gasdynamic mode

I. Izotov, V. Skalyga, S. Golubev and A. Bokhanov

Institute of applied physics RAS, N.Novgorod, Russia, ivizot@ipfran.ru

Introduction

Recent investigations of gasdynamic ECR plasma [1] resulted in development of pulsed high current ion sources capable to deliver ion beams with 100–500 mA current and normalized RMS emittance below 0.1 pi·mm·mrad [2] thanks to heating with mm-waveband gyrotron radiation. While the pulsed operation of gasdynamic ion sources was studied in details [1, 2], CW operation of such devices is not well investigated. SMIS 24 experimental facility was installed at the IAP RAS. At SMIS 24 the plasma is heated by 24 GHz radiation of CW gyrotron and confined in a simple mirror trap similar to SMIS 37 facility [1, 2], which was used for pulsed experiments. Detailed studies of CW gasdynamic ECR plasma parameters and ion beam extraction are presented in this work.

Experimental results

The scheme of SMIS 24 facility is shown in Fig. 1. The microwave radiation is generated by CW 24 GHz 5 kW gyrotron (manufactured by GYCOM). Radiation is delivered by means of over-sized cylindrical waveguide to high-voltage DC-break. Following the DC-break the microwaves pass through a specially developed coupling system which maximizes the electric field at ECR surface and minimizes reflection.

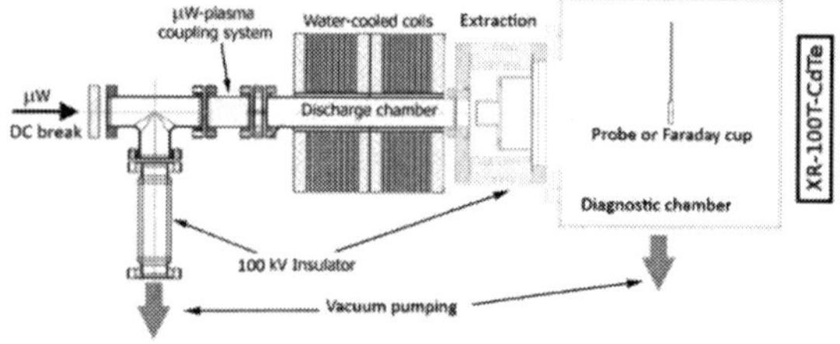

Fig. 1. SMIS 24 experimental facility

Plasma chamber size is 380 mm length and 32 mm in diameter. The magnetic field is produced by a pair of water-cooled coils of Bitter type forming a simple mirror trap. The maximum magnetic field on the axis reaches the value of 1 T. The mirror ratio (B_{max}/B_{min}) is equal to 3.5, the distance between magnetic mirrors (i.e. trap length) is 130 mm. The magnetic field distribution inside the plasma chamber is shown in figure 2.

A two-electrode extraction was used in the experiment for ion beam formation. Plasma and puller electrodes had 3 mm and 5 mm apertures, respectively. The available extraction voltage was up to 20 kV. The extraction system was placed far behind the magnetic mirror (see Fig. 2) to reduce the ion current density so that the total extracted current could meet the power supply maximum ratings.

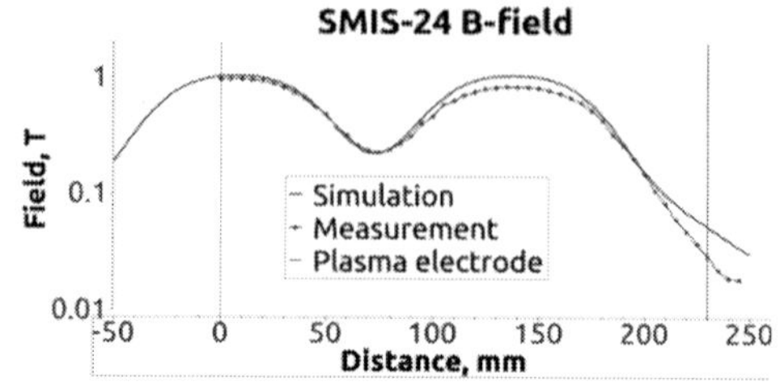

Fig. 2. Measured and simulated magnetic field strength on the axis of SMIS 24

A Langmuir probe was used for the measurement of plasma flux parameters (electron temperature and density). The probe was placed in the diagnostic chamber, where the magnetic field strength is low enough to consider the plasma unmagnetized. The Faraday cup was used to measure the total beam current. An x-ray spectrometer Amptek XR-100T-CdTe was used for plasma bremsstrahlung measurements. X-ray detector was placed on the axis of the magnetic system outside the vacuum enclosure facing towards the plasma. Helium was used in the experiments in the pressure range of 10^{-4}–$5\cdot10^{-4}$ Torr.

Electron temperature and density were estimated from Langmuir probe I-U curves. An example of typical I-U curve is shown in figure 4 in the case of 4 kW microwave power and $2.8\cdot10^{-4}$ Torr helium pressure. Curves shown: logarithm of electron current (thick solid line), derivative of electron current (dashed line), two linear fits (thin lines), estimated plasma potential is 4.1 V and indicated with vertical line. I-U curve in Fig. 3 is typical for bi-Maxwellian electron energy distribution function (EEDF). One can observe two characteristic energies of electrons, which are 17 eV and 49 eV.

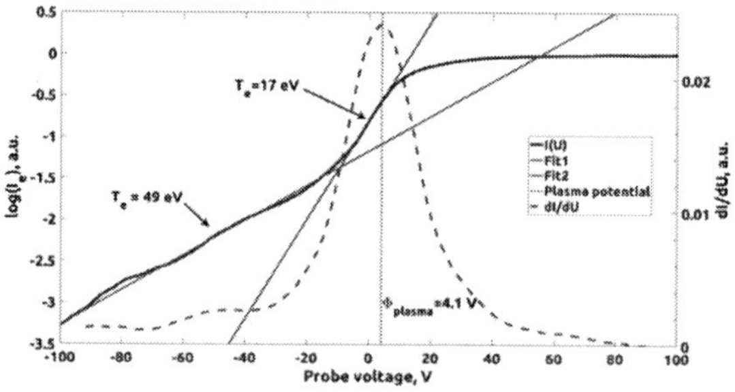

Fig. 3. A typical Langmuir probe I-U curve

The same shape of I-U curves is observed for all examined settings. It is believed that the EEDF of an ECR plasma in a magnetic trap consists of (at least) three populations [3]: (1) "cold" electrons with temperature close to the first ionization potential, (2) "warm" electrons with temperature dependent on the heating power and (3) a "hot" population creating high-energy bremsstrahlung.

Energies of "cold" and "warm" popula-tions were calculated by means of I-U curves analysis, whereas characteristic energy of the "hot" population was estimated as a spectral temperature of bremsstrahlung spectrum [4]. An example of bremsstrahlung spectrum with spectral temperature fit is demonstrated in Fig. 4 for 4 kW microwave power and $2.8 \cdot 10^{-4}$ Torr helium pressure.

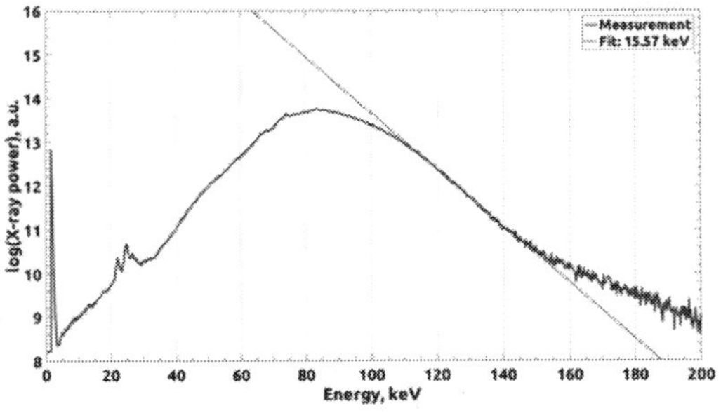

Fig. 4. Bremsstrahlung spectrum and spectral temperature fit

Measurements of characteristic electron energies are summarized in Fig. 5, where all electron "components" are plotted together.

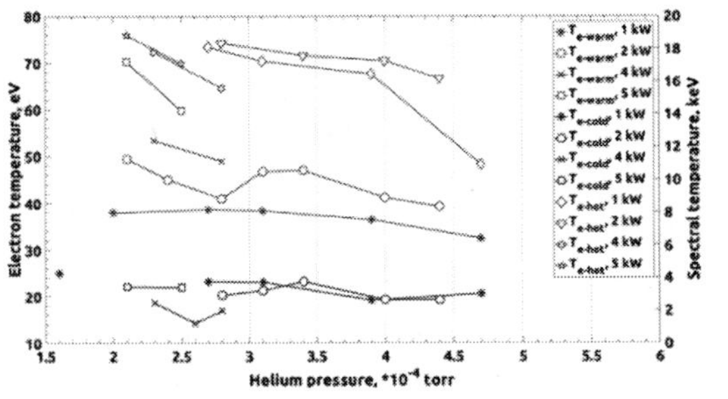

Fig. 5. Characteristic electron energies

It is of note that regardless the helium pressure and heating power characteristic energy of "cold" electrons remains in the range of 15–25 eV, which is consistent with helium ionization potential (24.47 eV). Energy of "warm" component is growing with the heating power, though dependence on pressure is unclear. "Hot" component energy seems to depend on helium pressure rather than on heating power.

All of acquired I-U curves had shown a distinct maximum of dI/dU (see Fig. 3), which enables the rough estimation of electron density (assuming Maxwellian EEDF with temperature in-between "cold" and "warm" components), as [5]:

$$I_e(U = \phi_{plasma}) = en_e S_{probe} \cdot \sqrt{eT_e/2\pi m_e}, \quad (1)$$

where n_e is the electron density, e and m_e is the electron charge and mass, S_{probe} is the current collecting surface, T_e is the electron temperature in eV.

Estimated electron density for different heating power and helium pressure is shown in Fig. 6. Assuming that the plasma flux follows magnetic field lines, the density value was corrected by the factor of 28.5, which is the ratio between magnetic field strength in the mirror point and the probe position. Given the scatter of data, no explicit dependence was found. However, it is of note that estimated plasma density is on the order of a cut-off density, which is equal to $7 \cdot 10^{12}$ cm^{-3} at 24 GHz.

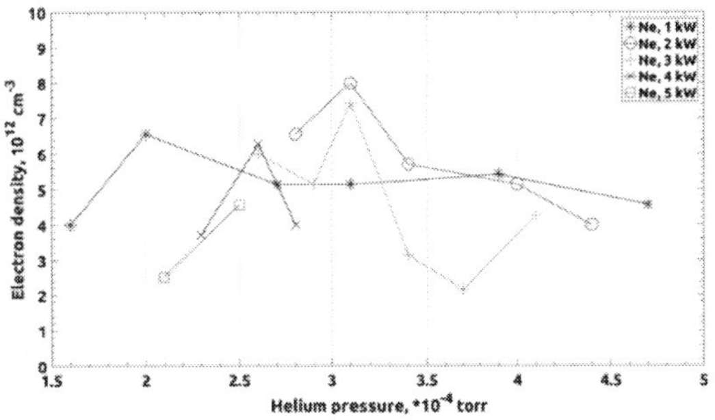

Fig. 6. Plasma density estimated for mirror point

Experiments on ion beam formation didn't show strong dependence on gas pressure and heating power. The facility demonstrated a stable long-term (>1 hour) extraction of CW ion beam with mean value on the level of 2.8 mA at 1 kW power, $4 \cdot 10^{-4}$ Torr helium pressure and 20 kV extraction voltage. Ion beam current density at extraction region was 40 mA/cm^2, which corresponds to plasma flux density of 1.13 A/cm^2 at the mirror point. The region with flux density of ~600 mA/cm^2, which is the value of plasma flux at extraction region in SMIS 37 facility [2], may be found in-between, thus suggesting the possibility of forming continuous ion beams with parameters similar to those at SMIS 37, if the extraction system is shifted towards higher flux, i.e. closer to the mirror point (see Fig. 2).

Conclusion

The experiments in gasdynamic CW ECR discharge confinement by 24 GHz gyrotron radiation demonstrated plasma parameters similar to those obtained in pulsed mode at SMIS 37 facility. Three electron components were observed – "cold" fraction with energies on the order of helium ionization potential, "warm" fraction with energies in the range of 30–70 eV, which is optimal for low charge states production, and "hot" fraction with spectral temperature in the range of 10–20 keV. Plasma density found to be on the cut-off level. A long-term reliable operation with ion current density >1 A/cm^2 (estimated) was demonstrated.

Acknowledgements

The work was supported by Russian Science Foundation, grant #16-12-10343.

References

1. *Golubev S et al.* // Nucl. Instrum. Meth. in Phys. Research B, 2007, 256, 537
2. *Skalyga V et al* // Rev Sci. Instrum., 2014, 85, 2, 02A702-1
3. *Douysset G et all* // Phys. Rev., 2000, E 61 3015
4. *Markus Hohl et al* // Plasma Sources Sci. Technol., 2005, 14 692
5. *Robert L. Merlino* // American Journal of Physics, 2007, 75.

EPJ Web of Conferences **149**, 02026 (2017) DOI: 10.1051/epjconf/201714902026
10th International Workshop 2017 "Strong Microwaves and Terahertz Waves: Sources and Applications"

First experiments on applying the gasdynamic ECR ion source for negative hydrogen ion production

R.L. Lapin[1], V.A. Skalyga[1], I.V. Izotov[1], S.V. Golubev[1], S.V. Razin[1], O. Tarvainen[2]

[1]Institute of Applied Physics of Russian Academy of Sciences, Nizhny Novgorod, Russia, lapin@ipfran.ru
[2]University of Jyväskylä, Department of Physics, Jyväskylä, Finland

Introduction

H^- ion sources are of great demand for charge exchange injection into cyclotrons and storage rings. It has been recently demonstrated that a gasdynamic ECR ion source based on ECR discharge in a simple mirror trap is very efficient for proton beam production [1–2]. Therefore, it was suggested to use the gasdynamic plasma source as the first stage of an H^- ion source based on volumetric production through dissociative electron attachment [3–4] (DEA).

The first experiments were performed in a pulsed mode with 37 GHz / up to 100 kW gyrotron radiation in a dual-trap magnetic system, consisting of two identical simple mirror traps. The schematic of the experimental facility is presented in fig. 1. The first trap was used for plasma production under ECR conditions. Dense hydrogen plasma flux (with estimated current density of about 1–5 A/cm^2) from the first trap was allowed to flow into the second trap through a perforated conducting plate. The grid placed between the traps prevents microwave leakage into the second part to avoid electron heating there. Such configuration was chosen to produce a flux of "hot" electrons (energy 50–100 eV), which can effectively ionize and excite high vibrational states of hydrogen molecules in the second trap, and to produce anions there as the result of DEA with "cold" electrons (energy < 10 eV). Thus, the plasma confined in the second trap presumably consists of two electron populations enabling the volume production of H^- ions. Such approach is similar to the one suggested in [5] where heating 2.45 GHz ECR discharge was used as a plasma cathode producing "hot" electrons. A simple two-electrode extraction system was used for the ion beam extraction. Diameters of a plasma electrode and a puller were 5 and 10 mm respectively. The plasma chamber and plasma electrode had a negative potential up to –15 kV with respect to the ground. To reflect electrons extracted from the plasma together with the negative ions two pairs of rectangular permanent magnets were installed just after the puller. Both magnet pairs produced a magnetic field transversal to the extracted beam and had opposite directions to compensate each other's influence on the ion trajectories. Coils of the dual-trap magnetic system were able to provide magnetic field maximum ranging from 1 to 2.5 T. Thus it was possible to observe the role of ECR heating in the first trap (ECR field for 37.5 GHz is 1.34 T). In the experiments two schemes of neutral gas injection were used: pulsed gas injection into the first trap and constant gas feeding into the first trap through a calibrated leak combined with pulsed gas injection into the second trap. Pulsed gas injection was realized with a controlled valve connected to a buffer vessel with pressure from 0.05 to 1 bar. The second gas-feeding scheme was used to perform experiment with significantly different neutral gas pressures in each trap as a result of the delay between the valve opening and microwave pulse beingless than the time necessary for the gas molecules to travel to the first trap with sound velocity.

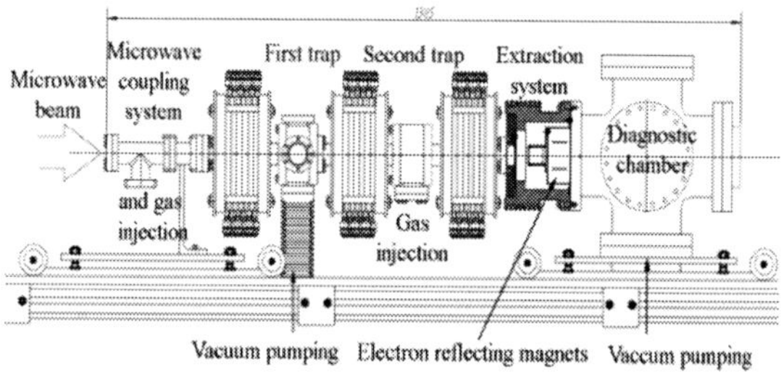

Fig. 1. Scheme of the experimental facility

The main diagnostic tools were a magnetostatic analyzer and a Faraday cup. The analyzer was used for spectrum measurements, determination of impurity level and investigation of H^- current dependence on various parameters. Transmission of the analyzer beam line is low and unknown, thus measured currents of certain ions to determine its real current in the extracted beam. The Faraday cup placed right after the extraction system and electron deflection magnets was used for total negative ion current measurements.

Measurements were done in a wide range of numerous parameters, such as gyrotron power, gas injection scheme, magnetic field strength, delay of gas pulse with respect to the microwave radiation pulse.

Experimental results

As the first step experiments with the pulsed gas injection into the first chamber without any injection into the second one were conducted. In this case the example of a typical ion beam spectrum is presented in fig. 2. It is clearly seen that a dramatic level of water-induced impurities, detrimental for H^- production, were observed. Such problem is connected to the low repetition rate (0.1 Hz) of the experimental facility and its disposal in future would probably enhance the H^- generation efficiency. Meanwhile OH^- is considered as the main negative ion and its dependencies on experimental parameters were studied together with H^-.

The dependence of H^- and OH^- currents on microwave heating power is shown in fig. 3. It was shown that, as expected, the H^- current depends linearly on the gyrotron power.

The currents of the negative ions depended on magnetic field as shown in fig.4. The H^- current was significantly higher under the condition of ECR heating, but a non-zero signal was observed even at fields below the ECR.

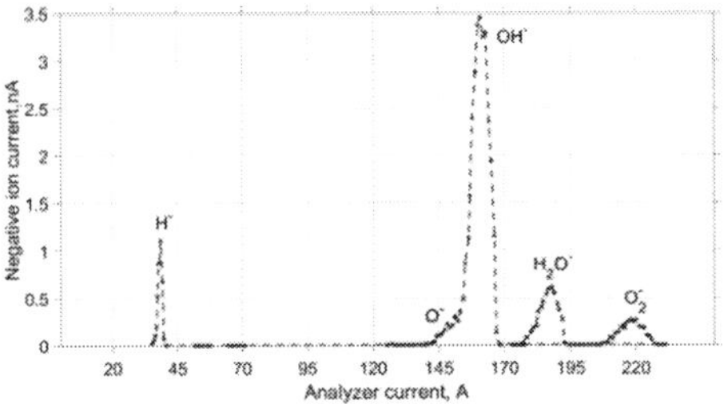

Fig. 2. Negative ion spectrum (B=1.6 T, microwave power 80 kW)

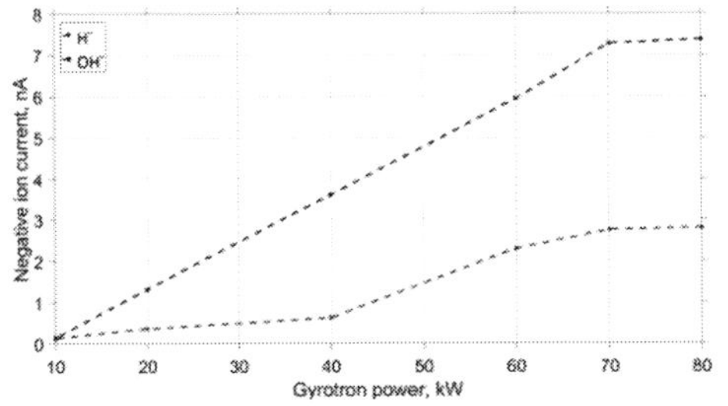

Fig. 3. The dependence of H^- and OH^- current on gyrotron power (B=1.48 T)

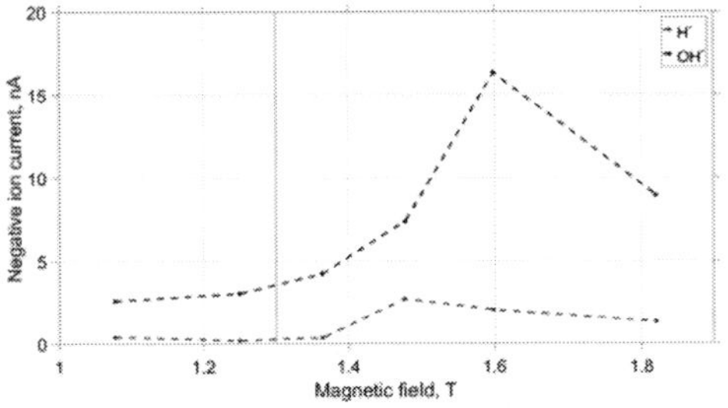

Fig. 4. The dependence of H^- and OH^- current on magnetic field ($P_{gyrotron}$=80 kW)

Data of figures 3 and 4 was collected in case of optimized neutral gas injection. The H^- current dependence on microwave pulse delay from the gas valve opening is presented in fig. 5 for different buffer vessel pressures.

Measurements of the total negative ion current with the Faraday cup have shown it at the level of 0.11 mA.

When the second scheme of neutral gas injection was tested, an increase of the total negative ion current by a factor of 15 was observed. In this case the gas pressure in the second trap was significantly increased without affecting the main discharge properties in the first trap. A typical waveform of the current is presented in fig. 6, showing a peak current of approximately 2 mA.

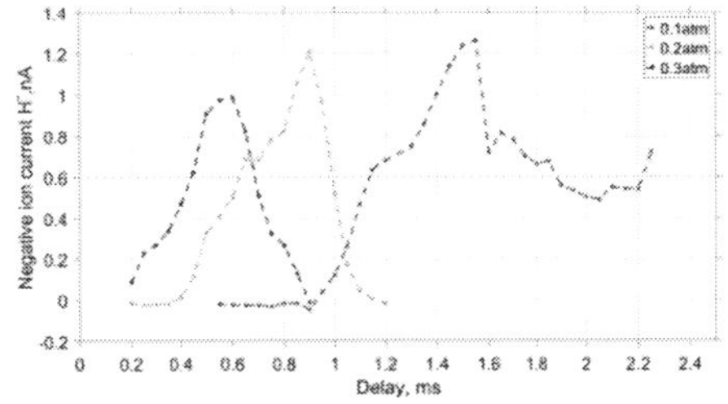

Fig. 5. The dependence of H^- current on the delay between the gas valve and microwave radiation pulses for different H_2 pressures in the buffer vessel (B=1.6 T, $P_{gyrotron}$=80 kW)

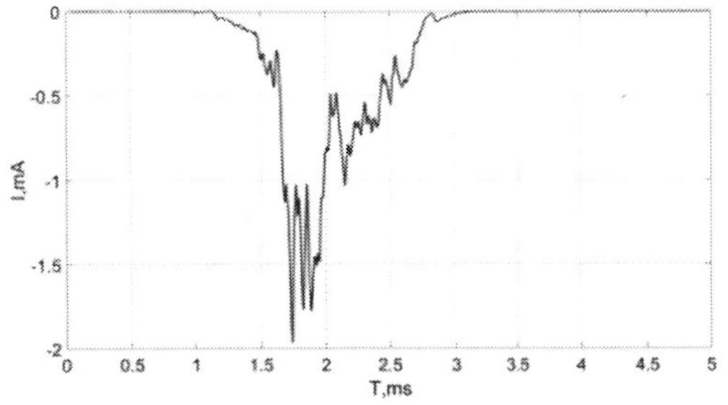

Fig. 6. Typical negative ion current waveform. Microwave power 80 kW, constant gas flow into the first trap and pulsed injection into the second one

Such increase in the total ion current could be connected with higher density of "cold" electron density induced by higher neutral gas pressure. Further experiments would be devoted to more detailed studies of this regime and better optimization of experimental conditions.

Research was carried out with the support of the Federal Agency for Scientific Organizations in the frame of state order No. 0035-2014-0026.

References

1. *Skalyga, V. A., Izotov, I. V., Razin, S.V., Sidorov, A. V., Golubev, S.V., Kalvas, T., Koivisto, H., Tarvainen, O.* High current proton beams production at Simple Mirror Ion Source 37 // Rev. Sci. Instrum. 2014. V. 85, No. 2. P. 02A702-1-3.

2. *Skalyga, V. A., Izotov, I. V., Golubev, S.V., Sidorov, A. V., Razin, S.V., Vodopyanov, A.V., Tarvainen, O., Koivisto, H., Kalvas, T.* New progress of high current gasdynamic ion source // Rev. Sci. Instrum. 2016. V. 87. P. 02A716-1-4.

3. *Bacal, M., Hamilton, G. W.* H- and D- production in plasmas // Phys. Rev. Lett. 1979. V. 42, No. 1538.

4. *Bacal, M.* Physics basis and future trends for negative ion sources / M. Bacal // Nuclear Fusion. 2008. Vol. 79, N. 2.

5. *Keller,R., Luft, P., Regis, M., Wallig, J., Monroy, M., Ratti, A., Syversrud, D., Welton, R., Anderson, D.* A Hybrid Ion-Source Concept for a Proton Driver Front-End // AIP Conf. Proceedings 773, American Inst. Of Physics, Melville, NY. 2005. P. 104–106.

New approach for a "point-like" neutron source creation based on sharp focusing of a high quality deuteron beam produced by high-current gasdynamic ECR ion source

S.V. Golubev, I.V. Izotov, R.L. Lapin, S.V. Razin, R.A. Shaposhnikov, S.V. Sidorov, V.A. Skalyga

Institute of Applied Physics of Russian academy of Sciences, Nizhny Novgorod, Russia, gol@appl.sci-nnov.ru

Introduction

Neutron tomography is one of the most exciting recent achievements of nuclear physics. It is of note that neutron tomography requires dedicated neutron source, i.e. paraxial sources with low angle spread. The only sources now able to deliver required neutron beams with sufficient intensity are nuclear reactors and large-scale accelerators in pair with collimators [1, 2]. Collimation degree is usually lies in the range of 50–300 (ratio between length and width), and the neutron flux density at collimator's end is 10^8–10^9 $s^{-1}cm^{-2}$ as a maximum.

The use of point-like neutron sources based on laser plasma induced by focusing of powerful femtosecond laser radiation onto a neutron-producing target [3, 4], for example set of deuterium clusters was discussed lately. Ions are effectively accelerated at certain conditions in such plasma, producing neutrons as a result of D-D collisions. The neutron yield of 10^6 per 1 J of laser energy has been detected at the source size of 100 μm [3, 4]. However state-of-the-art lasers having 0.1 J per pulse at 1 kHz repetition rate give average neutron yield of 10^8 s^{-1}.

The possibility of a point-like neutron source creation with the yield of 10^{10}–10^{13} s^{-1} based on high-current ion source of a new generation was discussed in [5]. It was proposed to use either D-D or D-T reactions occurring at the bombardment of D or T-loaded target with focused D ion beam. A size of the neutron emitting area would be determined by the quality of the ion beam and the efficiency of the beam focusing system (ie, the minimum size of the bombardment area). This proposal is based on recent developments [6–8] of the unique high-current ion source. Such source utilize high-power gyrotron millimeter-wave radiation for plasma production in an open magnetic trap under conditions of electron cyclotron resonance (ECR). In recent experiments it was demonstrated that this source with simple two-electrode extraction system and only 45 kV maximum voltage on it is able to produce deuterium ion beams with 400 mA current and normalized rms emittance less than 0.1 π·mm·mrad. Bombardment of a heavy ice target with described beam provided a total neutron yield of the order of 10^9 s^{-1}.

Present paper describes first experiments on proton beam focusing dedicated to the point-like neutron source development based on high current ECR ion source.

Experimental setup

Experimental studies of ion beam properties and search for optimal discharge conditions for high brightness beam formation were carried out at SMIS 37 experimental facility [6–8]. SMIS 37 provides gyrotron radiation at 37.5 GHz frequency with peak power up to 100 kW in 1.5 ms pulses for plasma heating. Microwaves were focused in the center of a discharge chamber 25 cm long and 4 cm in diameter by means of special quasi-optical microwave-to-plasma coupling system. The discharge chamber was placed in a magnetic trap of simple mirror configuration. The magnetic field in the trap was produced by means of pulsed solenoids, spaced 15 cm apart. Magnetic field in the mirror was varied from 1.4 to 4 T (ECR field for 37.5 GHz is 1.34 T). Ratio of the maximum and minimum magnetic fields of the trap was equal to 5 (i.e. B_{max}/B_{min}, mirror ratio). Deuterium or hydrogen were used as a working gas. The gas inlet into the source was realized through a gas feed line incorporated with the microwave coupling system. The ion extraction and beam formation were realized by two-electrode, i.e. single gap plasma electrode – puller electrode system. Holes in plasma electrode and puller were 5 mm and 10 mm respectively. The plasma electrode was placed as far as 10 cm downstream from the magnetic mirror to limit the extracted ion flux, which helps improving the beam transport through the puller. The maximum available extraction voltage was 50 kV. A Faraday cup was placed right behind the puller electrode to measure the total beam current passing through the extractor. Emittance of the extracted beam (all species together) was measured with "pepper-pot" method, which has been successfully tested earlier at SMIS 37 [6–8].

In the first experiment on beam focusing a simple magnetic lens with 2 T maximum field on the axis was used. It was installed just after the puller electrode. Beam current distribution and its focusing efficiency was measured using a scintillator plate for beam visualization in the focal plane of the lens.

Experimental results

The dependence of the ion beam current measured with the Faraday cup on the acceleration voltage is shown in Fig. 1 for the optimal extraction electrode configuration.

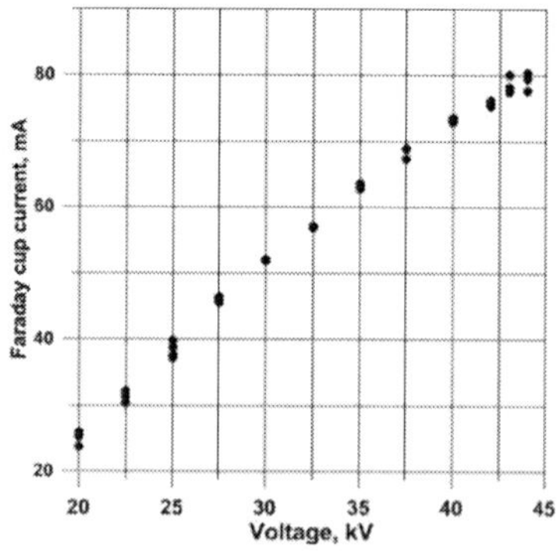

Fig. 1. Faraday cup current versus extraction voltage

For beam emittance measurement "pepper-pot" plate was placed 1 cm downstream from the puller with another 5.5 cm gap before a CsI scintillator for beam imaging. Example of the reconstructed emittance diagram is shown in fig. 2. The RMS normalized emittance appeared to be $0.06\pi\cdot$mm$\cdot$mrad.

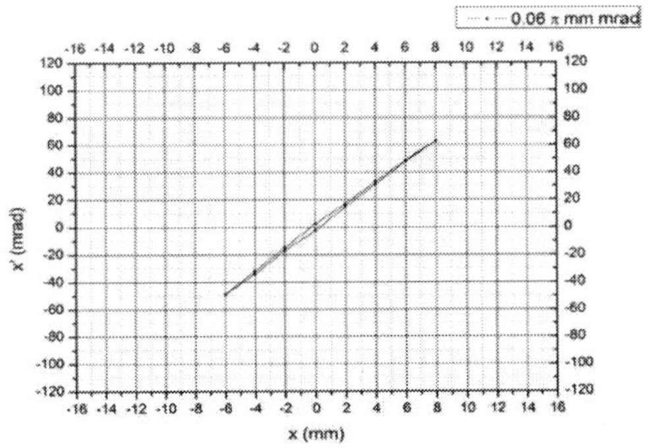

Fig. 2. Ion beam emittance diagram. Extractor: 5 mm plasma electrode aperture and 10 mm puller aperture with 11 mm gap. rms normalized emittance is be $0.06\pi\cdot$mm$\cdot$mrad

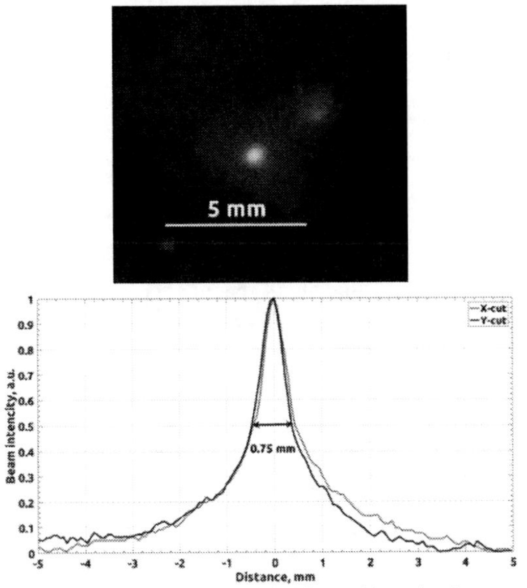

Fig. 3. Upper image: photograph of scintillator luminescence (green glow) induced by 80 mA hydrogen beam in magnetic lens focal plane (red glow corresponds to optical emission from plasma through the plasma electrode aperture); Lower image: corresponding beam space distribution in two orthogonal radial directions

Thus produced ion beam with 80 mA current and 0.06 $\pi\cdot$mm$\cdot$mrad emittance had the normalized rms brightness 22 A/($\pi\cdot$mm$\cdot$mrad)^2.

As it was mentioned above the magnetic lens was used for beam focusing as the next step. Measurements have shown that lens focus for 1.3 T field was situated approximately in 10 cm from lens center. Example of scintillator luminescence image induced by 45 keV, 80 mA hydrogen beam together with its analysis of space distribution in radial direction are presented in fig. 3.The best result for focused beam full width at half maximum was 0.75 mm.

Conclusion

The use of a high-current ECR ion source with quasi-gasdynamic plasma confinement and heating with gyrotron microwave radiation allows the formation of light ion beams with uniquely low emittance (for a given current level). Such a low emittance enables the focusing of the ion beam into the small spot. In frames of suggested approach to development of a point-like neutron source at the first experimental step-beam size at the focus of the magnetic lens less than 1 mm was demonstrated. Estimation presented in [5] show that the point-like source based on high current ECR ion source could produce the neutron yield up to 10^{11} s^{-1} for D-target, and up to 10^{13} s^{-1} for T-target. Such neutron flux corresponds to a flux density of 10^{8}-10^{10} $s^{-1}cm^{-2}$ at 10 cm distance from the source, which may be effectively utilized for neutron tomography, thus making this technique widely available.

Acknowledgments

Presented work was supported by the grant of Russian Science Foundation # 16-19-10501.

References

1. Viktor L. Aksenov, Physics-Uspekhi, V.40,N. 5(1997).
2. Ian S. Anderson, Robert L. McGreevy, Hassina Z. Bilheux. Neutron Imaging and Applications. Springer (2009).
3. T. Ditmire, et.al. Physics of Plasmas v.7, p.1993 (2000).
4. J. Davis, G. M. Petrov, and A. L. Velikovich. Physics of Plasmas 13, 064501 (2006).
5. S. Golubev, V. Skalyga, I. Izotov, A. Sidorov.Journal of Instrumentation, v. 12, 2017_JINST_12_T02003 (2017)
6. V. Skalyga, et. al.*JINST,* v.7, P10010 (2012).
7. V. Skalyga, et. al. Review of Scientific Instruments, v. 85, no. 2, p. 02A702-1 – 02A702-3 (2014).
8. V. Skalyga, I. et.al. Journal of Applied Physics v.118, p. 093301 (2015).

EPJ Web of Conferences **149**, 02028 (2017) DOI: 10.1051/epjconf/201714902028
10th International Workshop 2017 "Strong Microwaves and Terahertz Waves: Sources and Applications"

Reaching high sensitivity of radio-acoustic spectroscopy using "strong microwaves"

M.A. Koshelev, I.N. Vilkov, A.I. Tsvetkov, M.Yu. Glyavin, and M.Yu. Tretyakov

Institute of Applied Physics RAS, Nizhny Novgorod, Russia, tsvetkov@ipfran.ru

Gas molecular spectroscopy is a powerful instrument for both fundamental studies and applications, such as qualitative and quantitative gas analysis, non-invasive medicine, atmospheric remote sensing, etc. Sensitivity is one of the key parameters of any spectrometer, which determines the range of its possible applications for resolving both fundamental and applied problems. The higher the sensitivity, the higher the accuracy of measurement of the spectral line parameters and the greater the number of lines that can be observed in the experiment (the smaller the number of molecules in a gas mixture needed for their lines to appear in the spectrum) and the higher the accuracy with which the properties of the molecules can be explored.

One can recall quite a large number of currently known wideband spectrometers used for a study of the spectra of various molecules in the mm/submm wave range. They can be divided into two types according to the principle of molecular spectra registering: from variations in the characteristics of either the probing radiation (first type) or the gas being studied (second type).

For most of the mm/submm spectrometers, in which either the radiation transmitted through a cell with gas or the radiation reradiated by the gas is detected, a sensitivity close to the limit determined by fundamental physical principles is achieved (see, e.g. [1], chapter 15, p. 414). The only method that permits one to advance in solving the problem of high sensitivity achievement is known as optoacoustic (photoacoustic or radioacoustic) detection of absorption [2].

The sensitivity of any spectrometer is determined by several factors. The most crucial ones are radiation power, detection system noise, and spectral purity of the radiation. In spectrometers of the first type (e.g., classical video-spectrometer), the sensitivity increases with increasing radiation power until the detection system noise becomes surpassed by the radiation noise. Typically for mm/submm video spectrometers with a liquid helium-cooled bolometer an upper limit of radiation power is about 1 mW. Further improvement of a spectrometer sensitivity can be achieved by increasing the optical path length and by data averaging (if ambient and gas conditions are stable enough).

The situation is radically different in spectrometers of the second type (e.g., in a spectrometer with radioacoustic detection of gas absorption – RAD spectrometer). The limiting sensitivity of the spectrometer is obtained when thermal fluctuations of the membrane are defined preferably by the Brownian motion of the gas. These fluctuations do not depend on the radiation power passing through the gas cell. Meanwhile, the useful signal amplitude is directly proportional to the radiation power absorbed by the gas and, therefore, linearly increases with the power. Thus, the sensitivity of RAD method is also linearly increasing.

In this report we present our recent results on reaching the high sensitivity of RAD spectrometer using a few approaches, such as increasing radiation power, modifying the cell and measurement parameters for the noise reduction and data averaging.

Backward-wave oscillators (BWOs) are used as radiation sources in the RAD spectrometers, a series of which covers a very wide range of 35 to 1500 GHz. High stability, low level of the phase noise, and exact knowledge of the frequency are provided by the use of a phase-locked loop system of the BWO radiation referenced by a microwave frequency synthesizer signal, which is synchronized with a frequency and time standard signal. The BWO power can vary from a fraction to tens of mW (reaching more than 100 mW in the best tubes) within the operating frequency range. Larger radiation power suitable for molecular spectroscopy was not available during long time because of the limited choice of mm/submm radiation sources.

Progress in the development of high-power mm/submm radiation sources, which was achieved in recent years, made it possible to test the "power" approach of increasing the RAD sensitivity [3]. In particular, an automated facility [4] based on a gyrotron operated at a frequency of about 263 GHz in the CW regime with a radiation power up to ~1 kW was developed in IAP RAS. This gyrotron permits smooth tuning of the radiation frequency, although within small (in terms of spectroscopy standards) limits of the order of 0.2 GHz due to varying the operating voltage and temperature of the gyrotron cavity.

Experimental spectra of the SO_2 and argon mixture obtained using the RAD spectrometer for different pressures of gas in the cell at two significantly different values of radiation power are shown in Fig. 1. It is clearly demonstrated that an increase of radiation power by about three orders of magnitude leads to a proportional increase of the RAD spectrometer sensitivity. The achieved sensitivity of the spectrometer is not a limit. The spectrometer sensitivity obtained in this study is determined only by the SO_2 line saturation effect, which can be substantially reduced by proper selection of the molecule, the transition, and experimental conditions.

For reducing the influence of external acoustic and mechanical noise the cell was weighted in about ten times by small lead balls 3-mm in diameter (initial cell weight was about 2.5 kg). Analysis of experimental spectra recorded using a cell with significantly different weights shows (Fig. 2) almost sevenfold increase of the SNR of the experimental spectra.

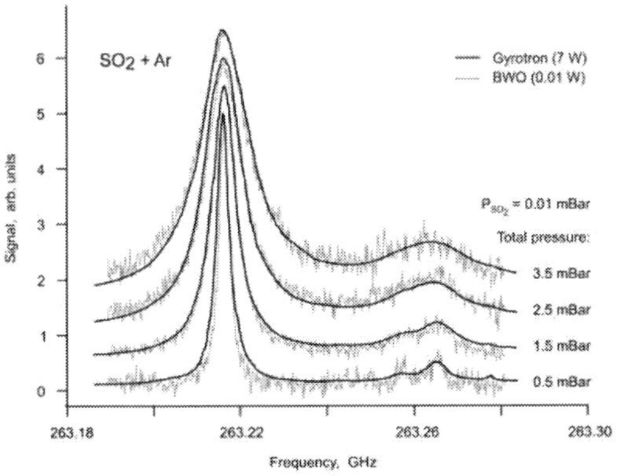

Fig. 1. An example of the experimental spectra of the SO_2 and argon mixture obtained using the RAD spectrometer for different pressures of gas in the cell at a constant level of radiation power. Partial pressure of SO_2 in the mixture is 0.01 mBar. The spectra obtained using a BWO with a radiation power of about 0.01 W (typical power of an OB-24 type tube) and using a gyrotron with output radiation power of about 7 W are shown by gray broken and black smooth lines, respectively. The synchronous detection time constant is 1 s.

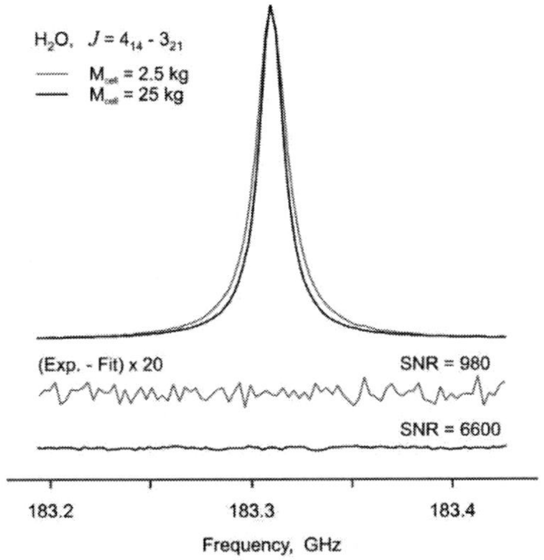

Fig. 2. Recordings of the 183 GHz water line obtained for two significantly different cell weights (2.5 kg and 25 kg). Residuals of the fit of model function to the experimental spectra zoomed in 20 times are shown in the lower part of the figure

Finally, high stability of radiation parameters (frequency and power) and experimental conditions (room and cell temperature, pressure in the cell) allowed averaging a large number of repeated experimental recordings for each chosen pressure for a multiple increasing of SNR for spectra recordings (see Fig. 3). The achieved sensitivity of the RAD spectrometer allowed the first observation of manifestation of the speed-dependence of the collision cross section of the 118-GHz oxygen fine structure line [5].

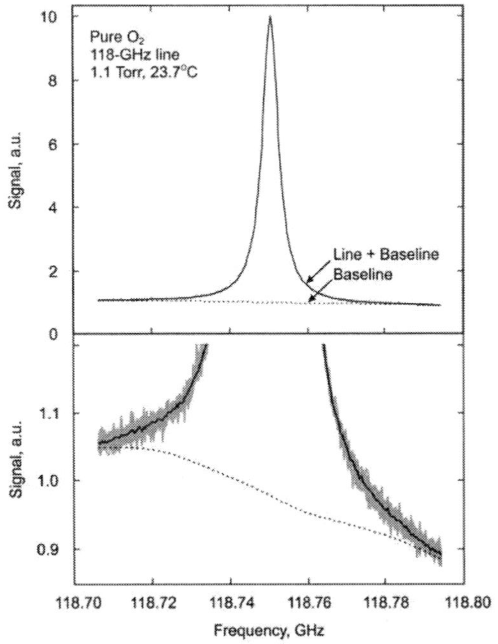

Fig. 3. Experimental spectra of pure oxygen near 118.75 GHz. The lower plot is a zoomed-in part of the upper plot. Twenty spectra recorded at 1.1 Torr of pure oxygen are shown by grey. The black curve is the result of their averaging. The dotted curve shows the baseline recorded for the same conditions with 1.1 Torr of pure nitrogen

It is worth noting that all the approaches were tested separately and being combined in one setup they can give the record sensitivity of the RAD spectrometer.

The study was supported by Russian Science Foundation (project 17-19-01602).

References

1. *Townes C. H., Schawlow A. L.* Microwave Spectroscopy, McGraw-Hill, New York-London-Toronto, 1955.
2. *Pao Y.-H.* (Ed.), Optoacoustic Spectroscopy and Detection, Academic Press, London, 1977.
3. *Koshelev M.A., Tsvetkov A.I., Morozkin M.V., Glyavin M.Yu., Tretyakov M.Yu.* Molecular Gas Spectroscopy Using Radioacoustic Detection and High-Power Coherent Subterahertz Radiation Sources // J. Molec. Spectrosc. 2017. V. 331. P 9–16.
4. *Glyavin M.Yu., Chirkov A.V., Denisov G.G., Fokin A.P., Kholoptsev V.V., Kuftin A.N., Luchinin A.G., Golubyatnikov G.Yu., Malygin V.I., Morozkin M.V., Manuilov V.N., Proyavin M.D., Sedov A.S., Sokolov E.V., Tai E.M., Tsvetkov A.I., Zapevalov V.E.* Experimental tests of 263 GHz gyrotron for spectroscopy applications and diagnostic of various media // Rev. Sci. Instr. 2015. V. 86, No. 5. P. 054705.
5. *Koshelev M.A., Delahaye T., Serov E.A., Vilkov I.N., Boulet C. and Tretyakov M.Yu.* Accurate modeling of the diagnostic 118-GHz oxygen line for remote sensing of the atmosphere // J. Quant. Spectrosc. Radiative Transfer 2017. V. 196. P. 78-86.

Study of grown single crystal diamond by optical and X-ray spectroscopy

D.B. Radishev[1], A.L. Vikharev[1], A.M. Gorbachev[1], A.B. Muchnikov[1],
P.A. Yunin[2], V.N. Amosov[3], N.B. Rodionov[3]

[1] Institute of Applied Physics of the Russian Academy of Sciences, Nizhny Novgorod, Russia
[2] Institute for Physics of Microstructures of the Russian Academy of Sciences, Nizhny Novgorod, Russia
[3] Russian State Research Center, Troitsk Institute for Innovation and Fusion Research, Troitsk, Moscow, Russia

Introduction

Single crystal diamond is a promising material for the creation of modern semiconductor electronic high-frequency and high-power devices due to its unique physical properties. Successful realization of all advantages of diamond requires high-quality ("electronic-grade") material that is the material with a lowest content of impurities and defects. Both natural diamond and a synthetic high-pressure high-temperature (HPHT) diamond do not exhibit quality necessary for modern semiconductor devices. At present the only method for obtaining artificial single crystal diamond of high quality that is reproducible from crystal to crystal is the chemical vapour deposition in plasma chemical reactors utilizing microwave discharge plasma in carbon-containing gas mixture (MPACVD method). In the process of CVD growth of diamond, the impurities concentration in the resulting material depends on the reactant gas mixture purity, also some of the defects in the HPHT substrate may be "overgrown" during deposition, resulting in the diamond with quality significantly better than seed substrate quality. In this work we investigate single crystals of CVD diamond with a low concentration of substitutional nitrogen, grown by microwave plasma-assisted chemical vapour deposition.

Experiment

CVD diamond single crystal layers were grown in a 2.45 GHz, 5 kW microwave plasma-assisted CVD reactor. Block diagram of the experimental setup is shown in Fig. 1. Reactor is based on the cylindrical cavity (1) with a diameter that allows exciting of three axially symmetric modes TM_{01n}, TM_{02n} and TM_{03n}. Upper (2) and lower (3) walls of the cavity act as mode converters. On the lower wall of the cavity, outside the quartz dome (4) limiting the gas volume, and near the side wall of the cavity, is a ledge (5) of variable height with an internal diameter selected so that the TM_{03n} mode decays in the bottom part of the cavity. As a result of superposition of three modes: TM_{01n}, TM_{02n} and decaying TM_{03n}, the distribution of the near electric field is formed so that the deposition of diamond is highly uniform: in case of polycrystalline diamond films it corresponds to 5% thickness difference at a diameter of 75 mm. Microwave power is fed into the cavity via tunable waveguide system (6) with plunger (7), connected to coaxial line (8) with antenna (9) that excites cylindrical cavity via shorted cylindrical waveguide (10) and circular cone (2). Reactor is equipped with automatic systems for supply (11) and pumping (12) of reactant gases. Substrate holder (13) is connected with water cooling system (14).

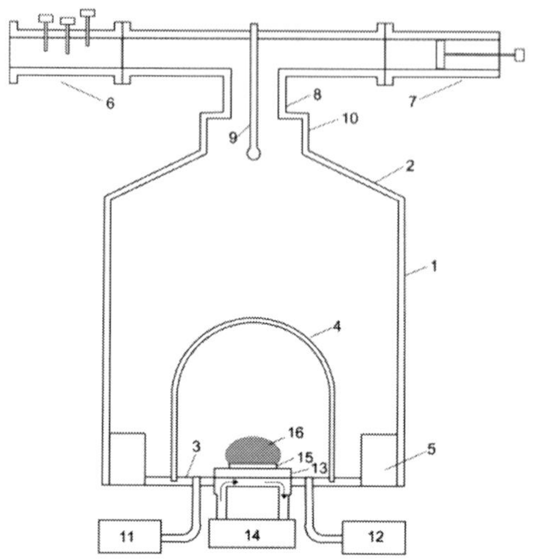

Fig. 1. Experimental setup: 1 – cylindrical cavity, 2 – circular cone, 3 – bottom cavity wall, 4 – quartz dome, 5 – step ledge, 6 – waveguide system, 7 – plunger, 8 – coaxial line, 9 – antenna, 10 – shorted cylindrical waveguide, 11 – gas supply system, 12 – gas evacuation system, 13 – substrate holder, 14 – substrate cooling system, 15 – substrate, 16 – microwave discharge

The epitaxial layers were grown on type IIa high-pressure high-temperature (HPHT) (100)-oriented single crystal seed diamonds of approximate dimensions 3×3×0.5 mm, produced by New Diamond Technology company (Sistroretsk, Russia) and the Technological Institute for Superhard and Novel Carbon Materials (Troitsk, Moscow, Russia). Six samples were used for deposition at different conditions: with varied content of methane in hydrogen-methane mixture, incident microwave power from 2.5 to 2.9 kW, at pressure 130 to 200 Torr and at substrate temperature of 950°C. Also CVD-04 and S-01 samples were grown with gas purify-

№	Sample	Layer thickness, mm	G, μm/h	Raman FWHM, cm^{-1}	Miscut angle, °	Deposition regime			Purifiers
						CH_4, %	Power, kW	P, Torr	
1	CVD-01	0.8	13.8	1.61		6%	2.7	200	-
2	CVD-02	0.85	14.5	1.53	1.83	6%	2.9	200	-
3	CVD-03	0.3		1.59	2	6%	2.8	200	-
4	CVD-04	0.25	10	1.65	0.53	6%	2.8	195	H_2
5	CVD-05	0.1	6.5	1.74	0.03	4%	2.8	150	-
6	S-01	0.25	3.8	1.56	0.28	3%	2.5	130	H_2, CH_4

ers, for hydrogen only (CVD-04) and both for hydrogen and methane (S-01). Summary of deposition regimes are presented in the Table. Samples CVD-01, CVD-02 и CVD-03 were separated (laser-cut) from the substrate, samples CVD-04 and S-01 were investigated while still on the substrate.

a)

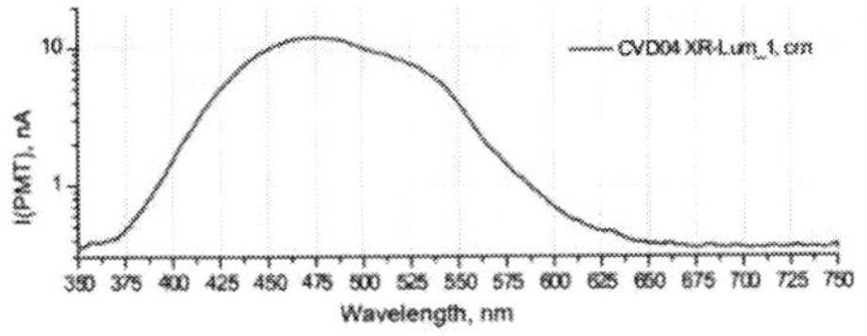

b)

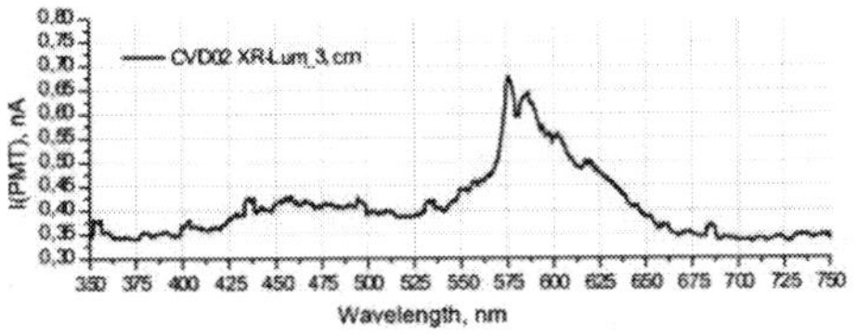

Fig. 2. XRS spectrum of: *a* – CVD-04, *b* – CVD-02

Investigations of CVD diamond layers, both separated and not separated from the substrate, were carried out using micro-Raman and photoluminescence (PL) spectroscopy methods at liquid nitrogen temperature, as well as method of X-ray luminescence spectroscopy (XRS). The most common impurity introduced into the diamond during deposition is nitrogen. It arises both from atmosphere leaks into the reactor, and from nitrogen contamination of source reactant gases. To eliminate leaks reactor was thoroughly leak-tested (N_2 leaks may also be detected by CN line emission in the optical spectrum of the discharge). To get rid of the nitrogen traces in source gases, gas purifiers were used for deposition of some of the substrates. Nitrogen incorporation into the diamond occurs mostly in form of substitutional nitrogen atoms (N_s). If there is a vacancy near the N_s atom, nitrogen-vacancy complex (N-V) may be formed. Nitrogen-vacancy complex is more energetically favorable than separated nitrogen and vacancy, also at temperatures higher than 650 °C (that is typical for CVD diamond growth condition) vacancies diffusion occurs in diamond, which leads to effective conversion of N_s atoms to (N-V) centers [1]. Conversion ratio of N_s to (N-V) almost does not depend on N_s and is about 1:300 (within the range 1:100 to 1:1000, [2]). It is hard to directly measure small concentration of N_s in diamond, but photoluminescence at liquid nitrogen temperature is known to be very sensitive method for detection of (N-V) centers and therefore can be used for N_s concentration estimation.

Raman/PL spectra of the samples grown without purifiers have pronounced luminescence lines of $(N\text{-}V)^0$ and $(N\text{-}V)^-$, along with luminescence of (N-V) centers. Width (full width at half maximum, FWHM) of Raman peak of diamond is within 1.53–1.74 cm^{-1}, that corresponds to high crystalline perfection of diamond. Raman and PL spectrum of S-01 sample, grown with both hydrogen and methane purifiers, does not have detectable lines corresponding to (N-V), $(N\text{-}V)^0$ and $(N\text{-}V)^-$ lumines-cence. According to sensitivity estimation performed earlier using substrate with thin nitrogen-doped layer with known concentration, this corresponds to N_s concentration of less than 5 ppb ($5\cdot10^{-9}$). That means that our CVD synthesis setup allows growing diamond with a low level of impurity nitrogen using source gas purifiers.

Investigation of CVD diamond samples were also made using X-ray luminescence spectroscopy – the study of spectrum of optical emission of diamond occurring under X-ray irradiation. Broad luminescence band at about 480 nm (A-band) is usually attributed to concentration of different kinds of defects that act as centers of electron-hole recombination [3]. XRS spectra of samples not detached from substrate demonstrate bright A-band luminescence associated with HPHT substrate defects (Fig. 2a). XRS spectra of samples separated from substrate shows much lower luminescence near 480 nm, but have small peak near 575 nm, possibly associated with nitrogen impurities (for sample grown without purifier, Fig. 2b). This shows that the concentration of various defects, which act as the centers of electron-hole recombination, is much lower in grown CVD layers than in HPHT substrates used. Such results demonstrate the decrease of the concentration of defects during the growth of CVD diamond and the possibility of obtaining a CVD diamond with a quality significantly superior to the quality of the HPHT substrate used.

Conclusion

Single crystal diamond is a promising material for the creation of electronic devices due to its unique physical properties, but it requires material with a low content of impurities and defects. Six experiments for deposition of CVD diamond layers were made, at different growth conditions and with or without source gases purifiers. Obtained CVD diamonds were examined using micro-Raman and PL spectroscopy, and X-ray spectroscopy. Results show that it is possible to obtain CVD diamond with no nitrogen contamination detectable with PL method and with defects concentration significantly lower than in HPHT type IIa diamond.

Acknowledgments

This work was supported by a grant from the Russian Science Foundation (Project № 17-19-01580).

References

1. Optical Engineering of Diamond, Edited by Rich P. Mildren and James R. Rabeau, 2013 Wiley-VCH Verlag & Co.

2. *A. M. Edmonds, U. F. S. D'Haenens-Johansson, R. J. Cruddace, M. E. Newton, K.-M. et al.* Production of oriented nitrogen-vacancy color centers in synthetic diamond // Phys. Rev. B, 2012, 86, 035201.

3. *Bernhard Dischler*, Handbook of Spectral Lines in Diamond, Springer-Verlag Berlin Heidelberg, 2012.

DOI: 10.1051/epjconf/201714902030

High-power microwaves against locusts and other harmful animals

V.E. Zapevalov

Institute of Applied Physics RAS, Nizhny Novgorod, Russia, zapev@appl.sci-nnov.ru

The locust is an insect that combines several types, usually collected in giant flocks, numbering several million individuals [1–3]. The damage it causes to plants, incomparable. Locusts can destroy all the plants in its path, and its invasion has always been a universal scourge. Innumerable flocks of these insects destroyed the crops, which inevitably led to mass starvation.

Locusts, acridia – several species of insects of the present locust family (Acrididae), capable of forming large flocks (up to hundreds of millions of individuals) migrating to considerable distances. A feature of biology of the locust is the presence of two phases - single and herd. Herd and single locust phases have significant differences, both in appearance and physiology, and in the nature of behavior. Locusts of a single phase usually have a protective coloration, a well-defined sexual dimorphism, and a low-active solitary lifestyle. Insects of the gregarious phase are colored more brightly and contrastively, adults are more adapted to flight. Locusts in this phase behave much more actively are formed larvae or adults. There is practically no sexual dimorphism in the herd phase. Locust has in its life cycle such stages: eggs, then larvae, the latter – imago (fig. 1). The larvae moult 5 times and after the final moult they have wings, the imago stage begins.

The hordes of locusts are able to move at a speed of 15–20 km/h and fly without interruption for up to 20 h in a row. Huge living clouds of locusts sometimes reach 10 km in width and up to 200 km in length. Two to three days of tailwind is enough for locusts to attack even remote areas of Russia.

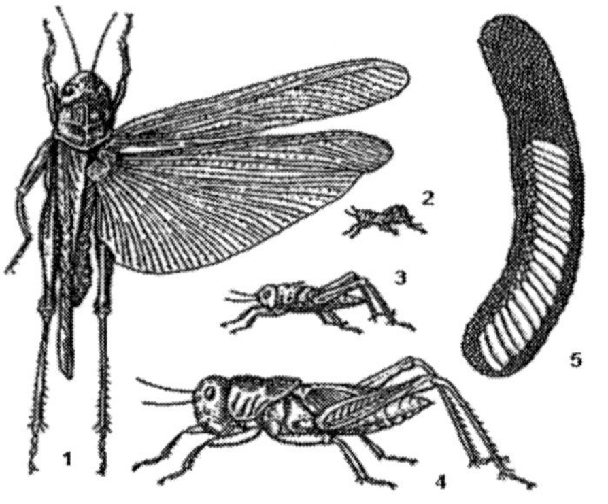

Fig. 1. Locusts: 1 – adult insect; 2–4 – larvae of the 1 st, 3 rd and 5 th instar; 5 – pod (the side wall is removed)

The fight against this evil is conducted worldwide, but almost all methods used have a strong negative impact on the environment [1-3]. Of course, in the late XX-early XXI centuries researchers have the opportunity to obtain fundamentally new data on these insects with the help of molecular-genetic, biochemical and information methods. This especially applies to the mechanisms of the transition from a single phase to a herd and back, migrations of swarms and flocks, and so on. However, these opportunities are often not implemented. This is largely due to the fact that interest in these insects (as well as research funding) falls sharply after the suppression of another outbreak, when the danger for agriculture passes. The ultimate goal in locust control is the use of preventive and proactive methods that disrupt the environment to the least possible extent. This would make agricultural production easier and more secure in the many regions where growing crops is of vital importance to the survival of the local people.

In this publication we are going to demonstrate that for control and suppression of locust (and other harmful animals) it is appropriate to use of high-power microwave systems. Important issues are the selection of the frequency range and the required power level for creating mobile microwave complexes. These conditions are determined by technical and economic factors, availability and cost of necessary sources of microwave radiation|, transportation (radiation and propagation) of electromagnetic waves, detection of insects and influence on egg capsules, kuligas (larvae), flocks (imago).

The characteristic number of individuals in the migratory flock of locusts is $N \sim 10^7$–10^9 pieces with a density of individuals n ~ 10^{-3}–10 m^{-3}. Mass of locust specimen ~ 3 g. The characteristic value of biomass in the flock is up to 3000 tons. The time of flock in flight is ~12 h per day. Locust locating by infrared radiation can be remotely conducted using thermal imagers since the temperature of the clusters is noticeably higher than the underlying surface. Radar and visual methods can also be used.

Important issue is the selection of the frequency range and required power levels for exposure to different insects. Fig. 3 shows absorption of microwaves by water (pure and with salt) as function of wavelength [4, 5]. Such absorption of microwaves by water at the body of insects (pure and mainly with salt as hemolymph) effects on egg-caps, larvae and flocks of imago. In view of the specific structure of insects (fig. 4), effects of microwaves on them are stronger than for mammals. All vital locust organs are located close to the surface [1-3], within the skin layer, in contrast to mammals (and in general vertebrates) where 83% of the energy of this radiation is absorbed by the upper layer of the skin.

For effects on eggs and larvae, it is advisable to use a high-efficient magnetrons and gyrotrons [6, 7] at the ISM frequencies ((0.915, 2.45, 5.8, 24.125 GHz etc.). Requirements for magnetron and gyrotron type devises for complexes against locust could be discussed. It is reasonable that due to technical and economical reason preferable power level now is 1-3 kW for magnetron (total power of complex near 100 kW) and order of 100 kW for gyrotron.

In adult stage (imago) locust clusters the clear advantages have gyrotron systems of mm range similar non-lethal weapons [8]. Fig 5 shows vehicle-mounted Active Denial System. This advanced system uses CPI's VGB-8095 gyrotron to generate a 95 GHz, 100 kW beam which provides a safe but effective deterrent. Similar Chinese system, known as the Poly WB-1, uses millimeter-wave beams (near 30 GHz) to scald targets from up to a kilometer away.

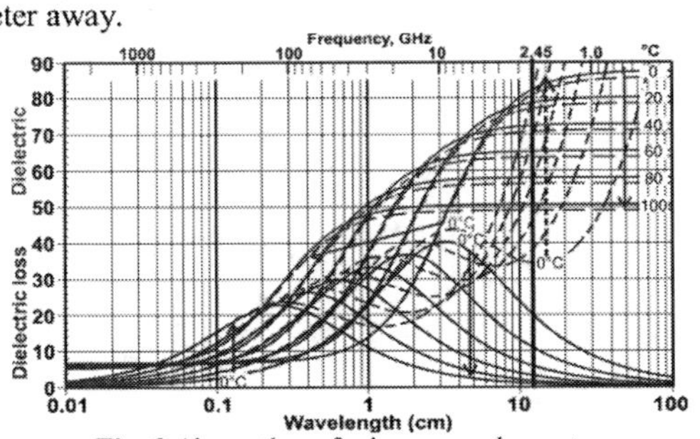

Fig. 3 Absorption of microwaves by water (pure – solid lines, and with salt – dashed lines)

It could be promising to use gyrotron working at the gyro frequency harmonics [7–10]. Fig. 6 demonstrates power and efficiency of gyrotron at the 2nd harmonics of cyclotron frequency (the TE_{03}, n=2, $\lambda_o \approx 12$ mm, $f_0 \approx 25$ GHz) vs. beam current and shows prototype of possible industrial version. Some other acceptable version of gyrotrons for such system described at [10]. For those systems RF beam control by phased array antenna it will be practically useful possibility of frequency tuning of RF source [6, 7].

For those systems RF beam control by phased array antenna it will be practically useful possibility of frequency tuning of RF source.

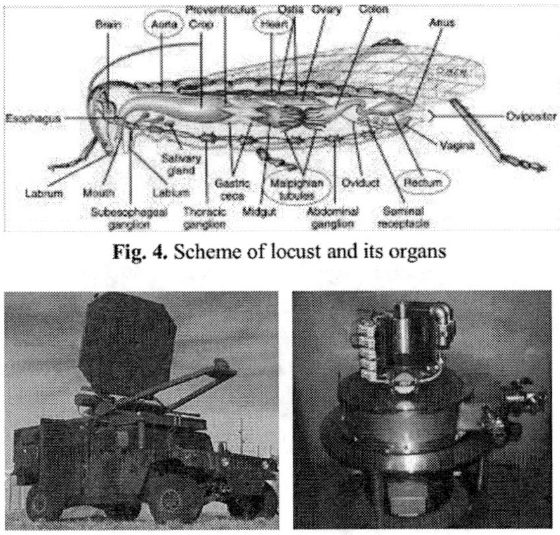

Fig. 4. Scheme of locust and its organs

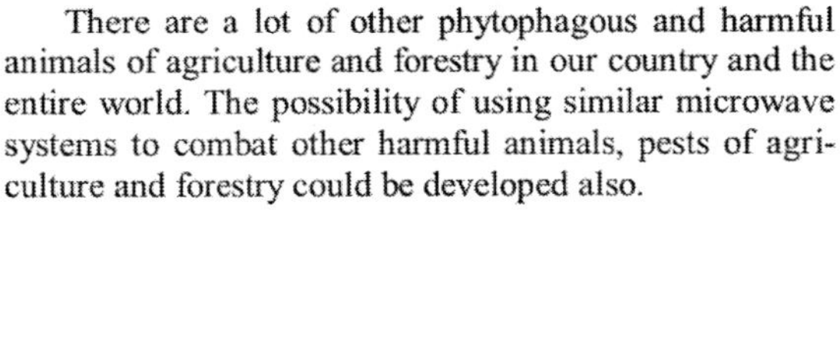

Fig. 5. Active denial system and gyrotron

There are a lot of other phytophagous and harmful animals of agriculture and forestry in our country and the entire world. The possibility of using similar microwave systems to combat other harmful animals, pests of agriculture and forestry could be developed also.

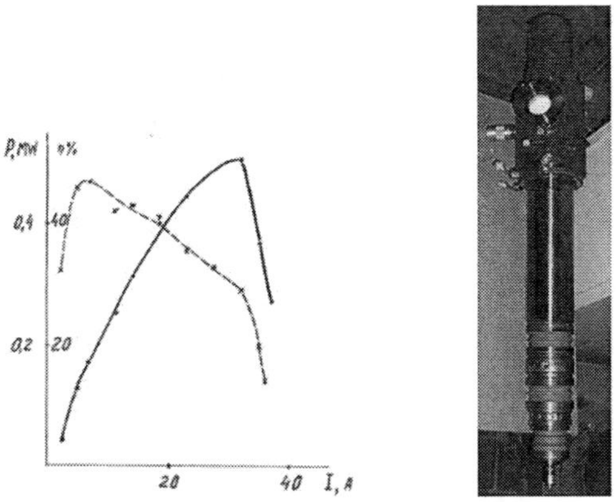

Fig. 6. Power and efficiency of 2nd harmonics gyrotron and prototype of possible industrial version

Summary

It is advisable to use high-power microwaves to control locusts and other harmful animals.

Locust locating by infrared radiation can be remotely conducted using thermal imagers because the temperature of the clusters is noticeably higher than the underlying surface.

Due to the peculiarities of the structure of insects, the effect of microwaves on them is stronger than on mammals.

To influence the laying of eggs and larvae, it is advisable to use high-efficiency magnetrons and gyrotrons at ISM frequencies. At the adult stage, gyrotron systems, analogous to non-lethal weapons, have clear advantages. The most attractive versions of magnetron and gyrotrons complexes for above mentioned tasks were pointed out.

This work was supported by the Russian Foundation for Basic Research, grant 15-42-02380.

References

1. *Resh V.H. (ed.), Carde R.T. (ed.).* Encyclopedia of Insects. Academic Press. 2003. p.666-669.

2. *A. Steedman, (ed).* 1990. Locust Handbook (3rd edition). Chatham: Natural Resources Institute. 204 pp.

3. *B.Uvarov,* 1966. Grasshoppers and locusts, V. I. Cambridge: University Press. 481 pp.

4. *J. B. Hasted,* Liquid water: Dielectric properties, in Water A comprehensive treatise, Vol 1, Ed. F. Franks (Plenum Press, New York, 1972) pp. 255-309.

5. *Sh.E. Tsimring,* Electron beams and microwave vacuum electronics, John Wiley and Sons, Inc., Hoboken, New Jersey, 2007.

6. *M.Thumm.* State-of-the-Art of High Power Gyro-Devices and Free Electron Masers. *KIT* Scientific Publishing, 2014.

7. *V.E.Zapevalov,* Evolution of gyrotrons, Radiophysics and Quantum electronics, 2011, v. LIV, №8-9, pp. 559-572.

8. *V.E.Zapevalov, S.A.Malygin, Sh.E.Tsimring,* Gyrotrons at the second harmonic of the cyclotron frequency. In the book: Gyrotrons, Gorky. IAP, 1980, pp.171-187.

9. *A.N.Kuftin, V.A.Flyagin, V.K.Lygin, et al.* 5.8-62 GHz cw gyrotrons with warm and permanent magnets for technological application. Proc. of Int. workshop "Strong Microwaves in Plasmas", 1999, N. Novgorod, 2000. pp. 671-676.

EPJ Web of Conferences **149**, 02031 (2017) DOI: 10.1051/epjconf/201714902031
10th International Workshop 2017 "Strong Microwaves and Terahertz Waves: Sources and Applications"

Gas breakdown by a focused beam of THz waves

A.V. Sidorov, S.V. Razin, A.G. Luchinin, A.I. Tsvetkov, A.P. Fokin, D.S. Sidorov, A.P. Veselov, A.V. Vodopyanov, M.Yu. Glyavin

Institute of Applied Physics of Russian academy of Sciences, Nizhny Novgorod, Russia, sevraz@appl.sci-nnov.ru

Introduction

The sources of submillimeter radiation have made significant progress recently [1-3] and so the possibility to study gas discharge phenomena in terahertz wave beams has appeared. Such gas discharge is of great interest from viewpoint of both general and applied physics.

This paper presents the results of experimental and theoretical investigations of gas breakdown thresholds for several gases (argon, krypton, xenon, nitrogen, oxygen) in a focused beam of pulsed and CW THz gyrotrons. The results of experiments for noble gases are in good agreement with theoretical calculations that were made based on a theory of breakdown of monatomic non-light gases in fields of any frequency from radio frequency to optical [4].

Experimental Setups

Two facilities with pulsed-mode [1] (0.67 THz, 20 μs, 40 kW, Fig. 1 [5]) and CW-mode [3] (0.263 THz, up to 1 kW, Fig. 2) gyrotrons were used in the experiments. Terahertz radiation was focused into a vacuum chamber where discharge occurred by means of quasioptical mirrors. The intensity of the radiation in the focal spot was 16 MW/cm^2 (rms electric field strength 80 kV/cm) for a pulsed-mode gyrotron and up to 15 kW/cm^2 (rms electric field strength 2.5 kV/cm) for a CW-mode gyrotron. Such electric field values were sufficient for the realization of a breakdown in the pressure range 1-1500 Torr.

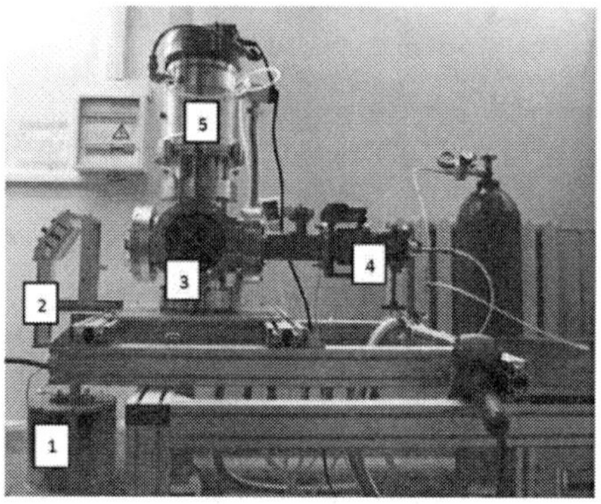

Fig. 1. Pulsed mode experimental setup (photo): 1 - pulsed gyrotron, 2 - quasioptical converter with focusing mirror, 3 - discharge chamber with optical window, 4 - gas injection system, 5 - vacuum pumping system (700 l/s).

Several photos of the discharge obtained in different working gases and at different pressures at the facilities described above are shown in Figs. 3 and 4.

Fig. 2. CW mode experimental setup (photo): 1 – gyrotron output window, 2 – the protective cover of quasioptical beam line, 3 – discharge chamber, 4 – vacuum pumping system (2000 l/s).

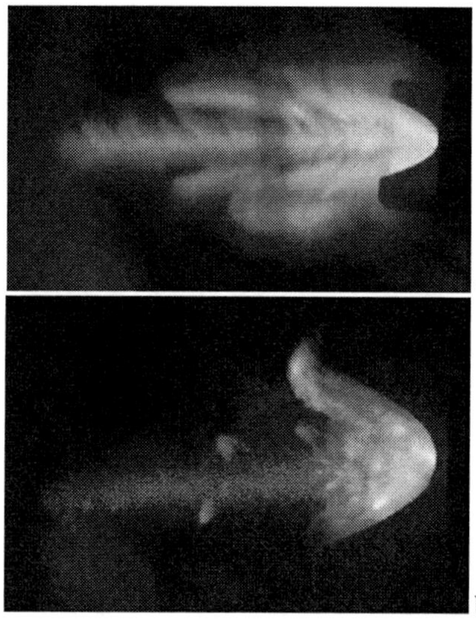

a)

b)

Fig. 3. Time-integrated photo of the emission of the pulsed discharge plasma in the visible spectral range: a) xenon, the pressure is 63 Torr; b) oxygen, the pressure is 760 Torr. The terahertz radiation is propagated from left to right.

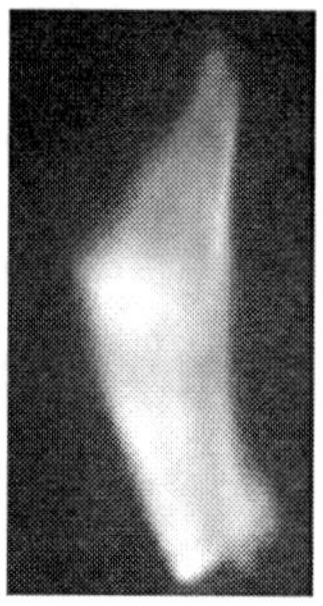

Fig. 4. Photo of the plasma glow of CW mode discharge in the visible spectral range. Argon, the pressure is 240 Torr, THz power is 950 W. The terahertz radiation propagates from top to bottom.

Experimental Results

The pressure breakdown thresholds for several gases in a focused beam of terahertz waves for a THz radiation power of 40 kW were determined during the experiments on a pulsed mode setup. The table shows the minimum gas pressures at which the discharge was observed.

Ar	Kr	Xe	N_2	O_2
8 Torr	1.5 Torr	1.5 Torr	12 Torr	18 Torr

At pressures lower than those listed in the table, the discharge could not be ignited.

Fig. 5 shows the experimental dependence of the threshold THz radiation power at which the discharge occurred, on the argon pressure for the case of the CW mode setup. According to experiments, the minimum pressure at which it was possible to ignite the discharge was about 23 Torr. The minimum of the breakdown curve is reached at pressures of 50-60 Torr.

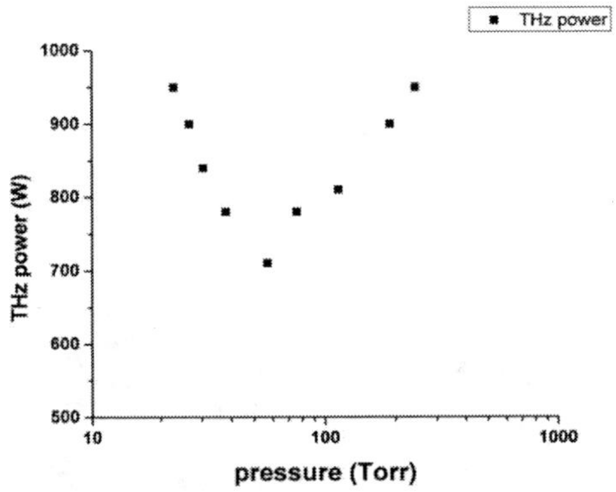

Fig. 5. Experimental threshold breakdown curve for Argon for 0.263 THz.

Discussion and Conclusion

Fig. 6 shows the calculated breakdown curve in comparison with the experimental one. The theoretical breakdown curve was calculated based on a theory of breakdown of monatomic non-light gases in electromagnetic field with frequency ranged from RF to optical [4]. The loss of electrons from the electric field area assumed to be diffusive with diffusion length Λ of 0.06 cm. The last one in our case was assumed as [6] Λ = R/2.4, where R = 1.5 mm – the radius of the wave beam in the focal spot. In this case, electron diffusion was assumed to be free.

The lower calculated curve corresponds to the diffusion length Λ = 0.6 cm. Such a value can correspond to the conditions when electron diffusion is ambipolar [6]. In this case one can see a good agreement between calculations and experimental results (black spots).

Thus, experiments were performed to determine the breakdown thresholds of various gases in focused beams of electromagnetic radiation with frequencies of 0.67 THz and 0.263 THz for the first time. Assuming that the electron losses are determined by diffusion, breakdown curves were calculated for the radiation frequency of 0.263 THz in argon. The calculations are in good agreement with experiment for the case of ambipolar diffusion.

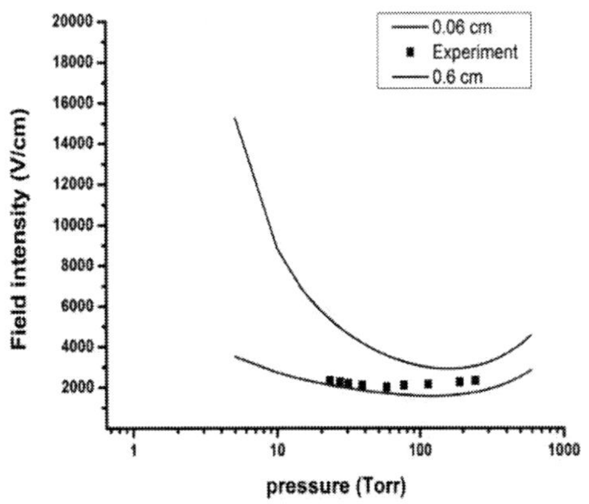

Fig. 6. Breakdown curves for 0.263 THz. Upper curve – results of calculations for Λ=0.06 cm, lower curve – Λ=0.6 cm. Black spots – experimental one.

Acknowledgements

The work was supported by the Russian Science Foundation, project No. 14-12-00609.

References

1. *M.Yu. Glyavin, A.G. Luchinin, G.S. Nusinovich, J. Rodgers, D.G. Kashyn, C.A. Romero-Talamas, and R. Pu.* A 670 GHz gyrotron with record power and efficiency. // Appl. Phys. Lett. 2012. V. 101. P. 153503.

2. *V.L.Bratman, A.A.Bogdashov, G.G.Denisov, M.Yu.Glyavin, Yu.K.Kalynov, A.G.Luchinin, V.N.Manuilov, V.E.Zapevalov, N.A.Zavolsky, V.G.Zorin.* Gyrotron development for high power THz technologies in IAP RAS. // Journal of Infrared, Millimeter, and Terahertz Waves. 2012. V. 33. No.7. P. 715.

3. *M.Yu. Glyavin, A.V. Chirkov, G.G. Denisov, A.P. Fokin, V.V. Kholoptsev, A.N. Kuftin, A.G. Luchinin, G.Yu. Golubyatnikov, V.I. Malygin, M.V. Morozkin, V.N. Manuilov, M.D. Proyavin, A.S. Sedov, E.V. Sokolov, E.M. Tai, A.I. Tsvetkov, V.E. Zapevalov.* Experimental tests of a 263 GHz gyrotron for spectroscopic applications and diagnostics of various media. // Rev. Sci. Instrum. 2015. V. 86. No. 5. P. 54705.

4. *A.I. Vyskrebentsev, Yu.P. Raizer.* A simple theory of breakdown of monatomic nonlight gases in fields of any frequency from low to optical. // Journal of Appl. Mechanics and Techn. Phys. 1973. V. 14. P. 32.

5. *M.Yu. Glyavin, S.V. Golubev, V.G. Zorin, I.V. Izotov, A.G. Litvak, A.G. Luchinin, M.V. Morozkin, S.V. Razin, A.V. Sidorov, V.A. Skalyga.* The discharge maintained by high-power terahertz radiation in a nonuniform gas flow. // Radiophysics and Quantum Electronics. 2014. V. 56. No. 8-9. P. 561.

6. *Yu.P. Raizer.* Gas Discharge Physics. // Springer, New York 1991.

EPJ Web of Conferences **149**, 02032 (2017) DOI: 10.1051/epjconf/201714902032
10th International Workshop 2017 "Strong Microwaves and Terahertz Waves: Sources and Applications"

Light emission properties of a discharge induced in a gas flow by terahertz waves in the vacuum and extreme ultraviolet range

S.V. Razin, A.V. Sidorov, A.G. Luchinin, D.S. Sidorov, S.V. Golubev, M.Yu. Glyavin, A.V. Vodopyanov

Institute of Applied Physics of Russian academy of Sciences, Nizhny Novgorod, Russia, sevraz@appl.sci-nnov.ru

Introduction

Research activity on the discharge, sustained by powerful radiation of the terahertz frequency range, has been actively conducted at the Institute of Applied Physics for a number of years [1-4]. Such a discharge is a new object of the plasma physics, and its investigations are of both fundamental and applied interest. This paper presents the results of experimental studies of the terahertz discharge plasma emission in the VUV spectral range (112 - 180 nm) and EUV spectral range (12 - 17 nm). The discharge was induced in a nonuniform gas flow (Ar, Kr, Xe) under the action of a focused beam of terahertz waves.

Experimental Setups

The experiments were carried out at two different facilities (see the block diagram in Fig. 1). The sources of terahertz radiation were gyrotrons. One gyrotron operated in a pulsed mode (setup № 1, frequency 0.67 THz, pulse duration 20 μs, power 40 kW, intensity in the beam focus 16 MW/cm^2 (for a detailed description of the setup, see [3 - 5])), another gyrotron operated in a CW mode (setup №2, frequency 0.263 THz, power 1 kW, intensity in the beam focus 15 kW/cm^2).

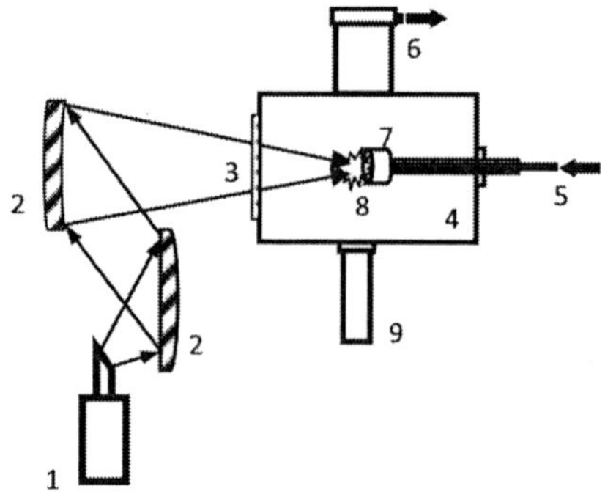

Fig. 1. Block diagram of experimental facilities. 1 – gyrotron, 2 – quasi-optical beamline, 3 – THz vacuum window of the discharge chamber, 4 – discharge chamber, 5 – working gas supply system, 6 – vacuum pumping system, 7 – nonuniform gas flow formation unit, 8 – discharge plasma, 9 – VUV photoelectron multiplier or EUV solid state detector with set of suitable filters.

The discharge was ignited in a nonuniform gas flow (Ar, Kr, Xe) under the action of a focused beam of terahertz waves. The gas flow was formed by injecting gas into the discharge chamber through a small-diameter hole (80 μm) under the pressure up to several atmospheres. An absolutely calibrated photoelectron multiplier (PMT) and a semiconductor detector in combination with a set of appropriate filters were used to measure the VUV (spectral range 112 - 180 nm) and EUV (spectral range 12 - 17 nm) emission of plasma.

Experimental Results

The performed experiments showed that the maximum power of the VUV and EUV radiation was observed in a point-like discharge with characteristic dimensions of about 1 mm [2 - 4]. In Fig. 2 for an example, a photo of such a point-like discharge realized on the above-described setup №2 with a CW mode of the gyrotron operating is given. It should be noted that earlier at the setup №1 we have already performed measurements of the VUV emission power of the point-like discharge. In particular, in the spectral range 112 - 180 nm a value of the order of 10 kW was obtained in argon [3, 4].

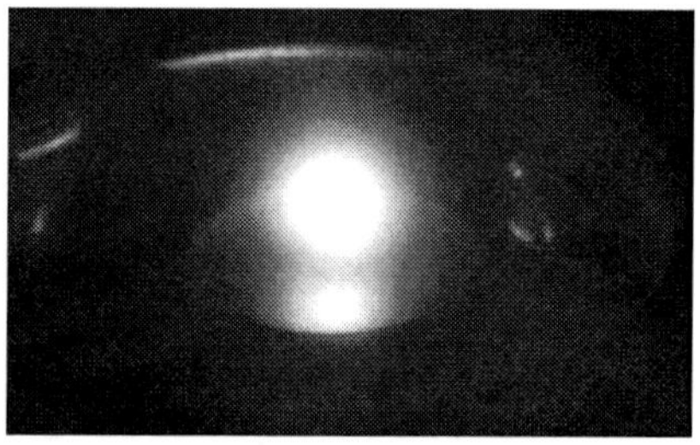

Fig. 2. A time-integrated photo of the emission of the point-like discharge plasma in a nonuniform flow of xenon in the visible spectral range. Setup № 2, the background pressure in the discharge chamber is $4{\cdot}10^{-3}$ Torr, the terahertz radiation is propagated from top to bottom.

Analysis of the experimental data showed that the most promising in terms of generation of both vacuum and extreme ultraviolet is a point-like discharge in the xenon flow. Xenon multiply charged ions emit a lot of intense lines in the VUV and EUV spectral ranges [6], in particular, in the area of 13.5 nm, which is of interest for high-resolution microlithography.

Fig. 3 shows the oscillogram of the VUV signal from the PMT obtained at setup №1 in the xenon flow. It is necessary to pay attention to the intense afterglow, the duration of which exceeds the duration of the terahertz pulse by several tens of times (it should be noted in the same way that an extremely long afterglow was observed earlier in other rare gases, in particular, in Ar [3, 4]). In addition, the intensity of the VUV radiation increased during the entire terahertz pulse. Hence it can be concluded that an increase in the gyrotron pulse duration should lead to a shift the emission spectrum of the discharge into the shorter-wave region. Probably, this should also lead to an increase in the intensity of the extreme ultraviolet.

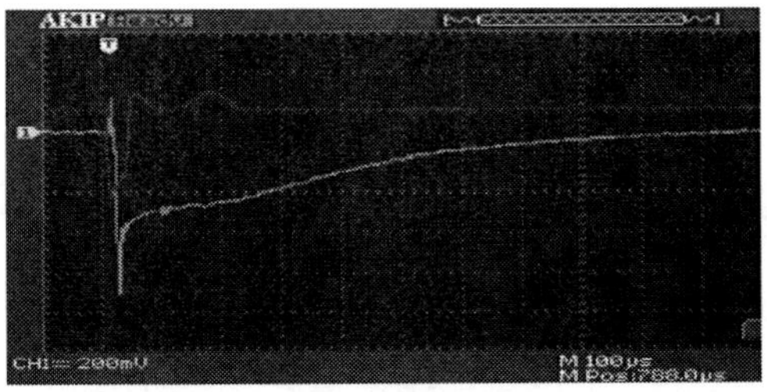

Fig. 3. Dynamics of the emission of a point-like discharge in the xenon flow in the VUV spectral region (the oscillogram of the signal from the PMT at a background pressure of $5 \cdot 10^{-3}$ Torr, lower waveform, setup №1). The upper waveform is a high-voltage pulse at the gyrotron cathode, the duration of which is approximately equal to the THz pulse duration.

According to experiments, the pressure of the gas injected into the discharge chamber has a significant effect on the plasma characteristics [5], and, consequently, on the generation of VUV and EUV emission. Fig. 4 shows the dependence of the VUV power of the plasma of a point-like discharge in the xenon flow on the pressure in the gas inlet system, obtained at setup № 2 at a maximum power of THz radiation of 1 kW. It should be noted that at the same time the vacuum evacuation rate (up to 2000 l/s) provided a low background pressure in the discharge chamber, and as a consequence, the maintenance of the point-like mode of the discharge. As can be seen from the Fig. 4, the power of the vacuum ultraviolet emitted by the plasma increases with increasing gas pressure in the gas-supply system.

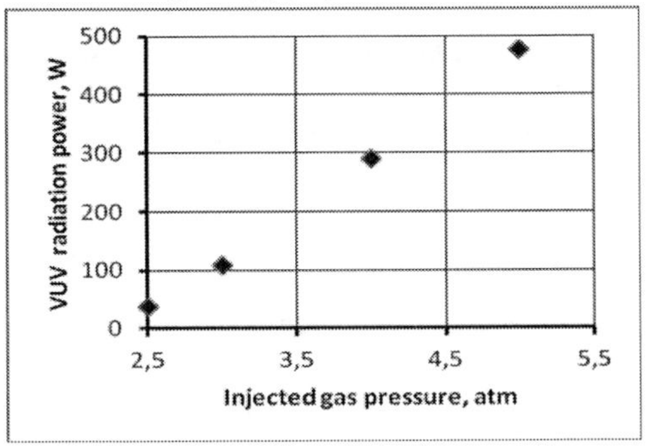

Fig. 4. The dependence of the VUV power emitted by the plasma of a point-like discharge in the xenon flow on the pressure in the gas-supply system (setup №2, power of terahertz radiation is 1 kW).

As for the EUV radiation of the THz discharges described above, the analysis of the signals from the semiconductor detector showed that specific power of the EUV radiation of the plasma in the wavelength range 12-17 nm is 0.1 W/mm^3 for set up №1, and 0.01 W/mm^3 for set up №2.

Conclusion

Thus, the performed experiments showed that the discharge plasma created by focused terahertz radiation in a nonuniform flow of rare gases emits both a vacuum (112 - 180 nm) ultraviolet and extreme ultraviolet (12 - 17 nm). In the conditions described above the power of the VUV radiation is large and reaches several tens of percent of the THz power supplied to the discharge. At the same time, the emitted EUV power is only a portion of a percent of the gyrotron radiation power. In the case of setup №1, an increase in the degree of conversion of THz radiation to an extreme ultraviolet is possible, apparently, by increasing the duration of the terahertz pulse of the gyrotron. Under the conditions of operation of the setup №2, an increase in the conversion degree is possible by increasing the THz radiation power and the pressure in the gas-supply system (in this case, it is necessary to maintain the point-like mode of the discharge, i.e., to keep the background pressure in the discharge chamber at the level of $10^{-2} - 10^{-3}$ Torr). These experiments are planned in the nearest future.

Acknowledgements

The work was supported by the Russian Science Foundation, project No. 14-12-00609.

References

1. *V.L. Bratman, I.V. Izotov, Yu.K. Kalynov, V.A. Koldanov, A.G. Litvak, S.V. Razin, A.V. Sidorov, V.A. Skalyga, and V.G. Zorin.* Features of the plasma glow in the low pressure terahertz gas discharge. // Phys. Plasmas. 2013. V. 20. P. 123512.
2. *M.Yu. Glyavin, S.V. Golubev, V.G. Zorin, I.V. Izotov, A.G. Litvak, A.G. Luchinin, M.V. Morozkin, S.V. Razin, A.V. Sidorov, and V.A. Skalyga.* The discharge maintained by high-power terahertz radiation in a nonuniform gas flow. // Radiophys. Quantum Electronics. 2014. V. 56. No 8-9. P. 561 – 565.
3. *M.Yu. Glyavin, S.V. Golubev, I.V. Izotov, A.G. Litvak, A.G. Luchinin, S.V. Razin, A.V. Sidorov, V.A. Skalyga, and A.V. Vodopyanov.* A point-like source of extreme ultraviolet radiation based on a discharge in a non-uniform gas flow, sustained by powerful gyrotron radiation of terahertz frequency band. // Appl. Phys. Letters. 2014. V. 105. P. 174101.
4. *S.V. Razin, A.V. Sidorov, S.V. Golubev, A.V. Vodopyanov, M.Yu. Glyavin, I.V. Izotov, A.G. Litvak, A.G. Luchinin, M.V. Morozkin, V.A. Skalyga.* // Proceedings of the 9-th International workshop «Strong microwaves and terahertz waves: sources and applications». Nizhny Novgorod – Perm – Nizhny Novgorod, Russia, July 24-30. 2014. P.89-90.
5. *A. V. Vodopyanov, M. Yu. Glyavin, S. V. Golubev, A. G. Luchinin, S. V. Razin, M. I. Safronova, A. V. Sidorov, and A. P. Fokin.* Plasma Density in Discharge Sustained in Inhomogeneous Gas Flow by High-Power Radiation in the Terahertz Frequency Range. // Technical Physics Letters. 2017. V. 43. No. 2. P. 186-189.
6. http://www.nist.gov/pml/data/asd.cfm

EPJ Web of Conferences **149**, 02033 (2017) DOI: 10.1051/epjconf/201714902033
10th International Workshop 2017 "Strong Microwaves and Terahertz Waves: Sources and Applications"

A possibility of remote detection of air breakdown in a focal spot of microwave beam using reflected signal

V.E. Semenov[1], E.I. Rakova[1], M.Yu. Glyavin[1], G.S. Nusinovich[2]

[1] Institute of Applied Physics, Russian Academy of Sciences, Nizhny Novgorod, Russia,
[2] IREAP, University of Maryland, College Park, Maryland 20742-3511, USA

Introduction

Propagation of high power microwave beam through atmosphere can initiate the breakdown which results in formation of plasma reflecting the incident microwave radiation. The breakdown delay time, as well as the plasma parameters, is sensitive to existence of radioactive materials which can ionize air producing thereby seed electrons in the vicinity of the breakdown prone volume [1-3]. Therefore a study of the breakdown evolution can give some particular information about the closeness of radioactive materials to the focal spot of high power microwave beam [3, 4]. The necessary information can be obtained using a reflection of the incident microwave radiation from the breakdown plasma.

In this paper the breakdown process was simulated numerically using the same model as in [2]. Calculations of the reflection evolution of the incident microwave beam were completed for the breakdown in atmospheric air. In this case the frequency of electron collisions with molecules as well as the electron diffusion coefficient and frequency of electron attachment to oxygen molecules. A dependence of frequency of the electron impact ionization on amplitude of electric field is also fixed and results of breakdown simulation are determined mainly by intensity of the incident microwave beam, microwave frequency, f, length, l_f, of the microwave beam focal region, initial distribution of plasma density in space, and intensity of the external source of ionization which is produced by cosmic rays and radioactive material in the vicinity of the focal region.

Results

Simulations were completed assuming that the amplitude of the microwave electric field in the focal point is twice the breakdown threshold. Two different frequencies f_1 = 110 GHz and f_2 = 670 GHz where considered in simulations which corresponds to the frequencies of gyrotrons used in MIT experiments [5] and in IAP-UMD experiments [6], respectively. Two different scenarios of breakdown were studied. In the first scenario, the breakdown started from well localized initial plasma that can be interpreted as a result of stochastic ionization by single particle of cosmic ray. The external ionization source was set to zero in this case. Within the second scenario, we started with the zero initial plasma density, while an action of a permanent weak source of external ionization was taken into account. In the simulations, the main attention was paid to evolution of the reflection coefficient of incident microwave beam from the breakdown plasma.

The simulation results are presented below in Figs. 1–4.

As follows from these figures, the reflection evolution depends on parameters and can be rather complicated. However the common peculiarity of this evolution is associated with a well pronounced peak of the reflection coefficient at the initial stage of breakdown. The duration of this peak is determined mainly by a value of the impact ionization frequency in the undisturbed field of microwave beam at its focal point. In case of the first breakdown scenario (localized peak of initial plasma density), the maximum value of the reflection coefficient is not very sensitive to the microwave frequency or focal length, l_f (see Fig. 1 and 2). The reflection coefficient reaches the same peak value $G \approx 0.35$ almost independently on frequency, focal length, and maximum value of initial plasma density.

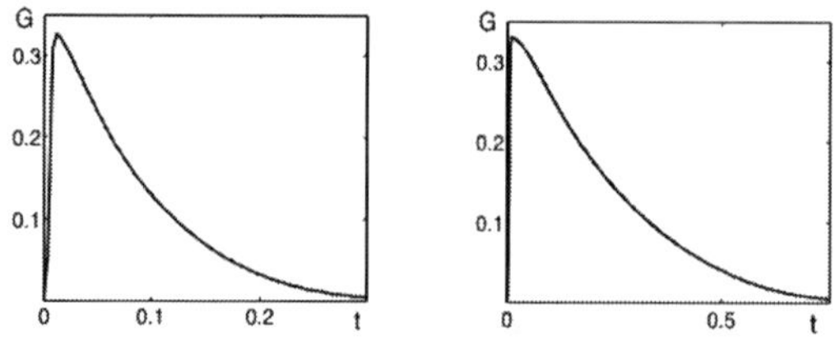

Fig. 1. The amplitude reflection coefficient, G, vs. the time t (in μs). Simulations were performed for the atmospheric air and microwave frequency f_2 = 670 GHz. The first breakdown scenario was considered (zero source of external ionization) The left and right panels present simulation results for $l_f \approx 2.3$ mm and $l_f \approx 7$ mm , respectively

In the case of the second breakdown scenario, the maximum value of reflection coefficient is sensitive mainly to a value of the parameter

$$\eta = \frac{N_s \delta}{N_c \left(1+\delta^2\right)} \cdot \exp(1/g),$$

where N_s stands for plasma density created by external ionization source in atmospheric air under equilibrium conditions (without any microwave radiation), $\delta = \nu/\omega$ is the electron collision frequency normalized to circular microwave frequency, $g = kl_f$ is the focal length normalized to vacuum wave number. When this parameter is small ($\eta << 1$), an intensity of the external ionization source does not affect the maximum value of the reflection coefficient (see Fig. 3). However, transition to the opposite limiting case ($\eta >> 1$) results in considerable decrease in the reflection coefficient (sf. Fig. 1 and 4). The further increase in the intensity of the external ionization source is accompanied only by slow logarithmic decrease in the reflection coefficient.

It should be noted that in the long time evolution of the reflection coefficient there is a qualitative difference between the cases of higher and lower microwave frequency (cf. Fig. 1 and 2). In previous studies [2, 3] it was shown that at a higher frequency the discharge evolution is associated with continuous propagation of the ionization front, whereas at a lower frequency this evolution is associated with quasi-periodic jumps of the ionization front. The continuous propagation of the ionization front is accompanied by smooth evolution of the reflection coefficient (Fig. 1), whereas the jumps of the ionization from are accompanied by oscillations in this coefficient (Fig. 2).

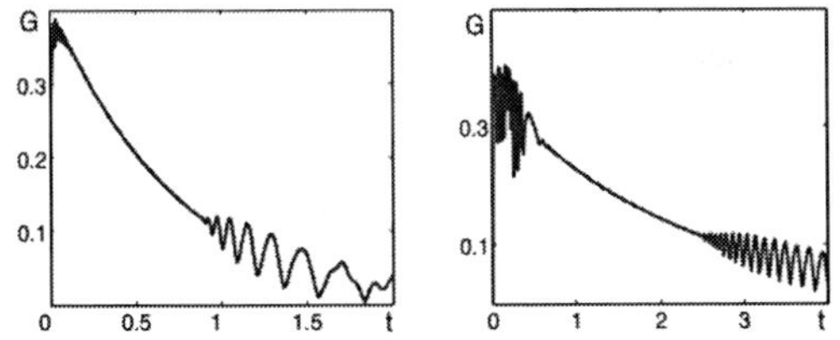

Fig. 2. The amplitude reflection coefficient, G, vs. the time the time t (in μs). Simulations were performed for the atmospheric air and microwave frequency f_l=110 GHz. The first breakdown scenario was considered. The left and right panels present simulation results for $l_f \approx 1.4$ cm and $l_f \approx 4.3$ cm, respectively

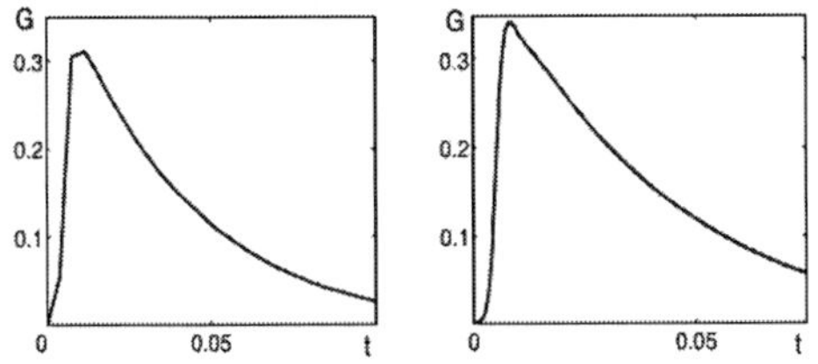

Fig. 3. The dependence of the amplitude reflection coefficient G on the time t (in μs). Simulations were performed for f = 670 GHz and $l_f \approx 0.7$ mm. The left panel presents simulation results in the case of zero external source of ionization and the right panel presents simulations results in the case of zero initial plasma density and intensity of the external ionization source which corresponds to the parameter $\eta \approx 0.01$

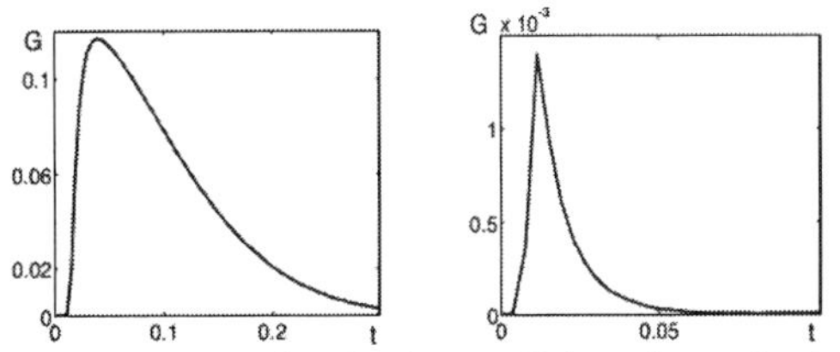

Fig. 4. The amplitude reflection coefficient, G, vs. the time t (in μs). The microwave frequency f = 670 GHz. Both left and right panels present simulation results in case of the second breakdown scenario with the same intensity of the external ionization source. However the focal lengths are different and the corresponding values of the parameter η are very different: $l_f \approx$ 2.3 mm. x (the parameter $\eta \approx 2 \cdot 10^7$) on the left panel whereas $l_f \approx 7$ mm (the parameter $\eta \approx 10^{37}$) on the right panel

Acknowledgements

The work was supported by the RFBR, project No. 17-02-00183.

References

1. *V.E. Semenov, E.I. Rakova, V.P. Tarakanov, M.Yu. Glyavin, G.S. Nusinovich* // Physics of Plasmas, 2015, **22**, 092308

2. *V.E. Semenov, E.I. Rakova, M.Yu. Glyavin, and G. S. Nusinovich* // Physics of Plasmas, 2016, **23**, 073109

3. *G.S. Nusinovich, P. Sprangle, V.E. Semenov, D.S. Dorozhkina, M.Yu. Glyavin* // J. Appl. Phys., 2012, **111**, 124912

4. *D. Kim, D. Yu, A. Sawant, M.S. Choe, I. Lee, S.G. Kim, E.M. Choi.* Remote detection of radioactive material using high-power pulsed electromagnetic radiation // Nature Communications, 2017, **8**, 15394

5. *A. Cook, M. Shapiro, R. Temkin.* Pressure dependence of plasma structure in microwave gas breakdown at 110 GHz // Applied Physics Letters 2010, 97, 1, 011504

6. *M.Yu. Glyavin, A.G. Luchinin, A.A. Bogdashov, V.N. Manuilov, M.V. Morozkin, Yu. Rodin, G.G. Denisov, D. Kashin, G. Rogers, C.A. Romero-Talamas, R. Pu, A.G. Shkvarunetz, G.S. Nusinovich.* Experimental study of the pulsed terahertz gyrotron with record-breaking power and efficiency parameters // Radiophysics and Quantum Electronics 56, 2014, 8-9, 497-507

Theory of initial stage of the breakdown in non-uniform gas flow

V.E. Semenov[1], E.I. Rakova[1], M.Yu. Glyavin[1], G.S. Nusinovich[2]

[1] Institute of Applied Physics, Russian Academy of Sciences, Nizhny Novgorod, Russia,
[2] IREAP, University of Maryland, College Park, Maryland 20742-3511, USA
sss@appl.sci-nnov.ru

In Refs. [1–3] it was suggested to use discharge in Sharply focused microwave beam to create the point-like source of extreme ultraviolet radiation. To reduce the plasma volume in experiments the discharge was ignited in gas flow from small opening into vacuum. In this case the gas density was well localized around the opening. Below the initial stage of the microwave breakdown in non-uniform gas flow was simulated numerically to understand a possibility to localize the discharge plasma and a dependence of the plasma parameters on microwave power and frequency, diameter of the opening and a value of gas pressure, p_0, at the opening cross section. For simplicity the axially symmetrical problem was considered. Within this model the sharply focused microwave beam was approximated as converging cylindrical TE wave. A dependence of the gas pressure, p, on radial coordinate was approximated by the following function

$$p(r) = \frac{p_0\, a}{\sqrt{a^2 + r^2}} \qquad (1)$$

were a stands for characteristic length of the gas density localization. It was assumed that plasma density is small compared to density of neutral molecules and therefore the ionization process does not affect the gas density distribution. The monochromatic approximation was used to describe electric field amplitude, whereas an evolution of the plasma electron density was simulated taking into account the electron impact ionization, attachment of electrons to neutral molecules and plasma diffusion. Using the normalized variables and parameters the main equations can be presented as follows:

$$\frac{1}{\rho}\frac{\partial}{\partial\rho}\left(\rho\frac{\partial F}{\partial\rho}\right) + \varepsilon F = 0 \qquad (2)$$

$$\frac{\partial n}{\partial\tau} = \frac{1}{\rho}\frac{\partial}{\partial\rho}\left(\rho\frac{1}{P}\frac{\partial n}{\partial\rho}\right) + \mu P n\left[A^{\beta}\left(\frac{1+\delta_0^2}{1+\delta_0^2 P^2}\right)^{\beta/2} - 1\right] \qquad (3)$$

Here $F(\rho,\tau)$ stands for complex amplitude of electric field, normalized to the threshold value at opening output, $\rho = \omega r/c$ is radial coordinate, normalized to the wave number of electromagnetic field in vacuum, $\varepsilon = 1 - n - i\delta n$ is complex dielectric permittivity of plasma, $n = N/N_c\left(1+\delta^2\right)$ is normalized plasma density, N_c denotes the critical plasma density, $\delta = \nu/\omega = \delta_0\, p/p_0$ denotes the normalized frequency, ν, of electron collisions with neutral molecules, $\tau = D_0(\omega/c)^2 t$ stands for the normalized time, D_0 is electron diffusion coefficient at the opening output, $P = p/p_0$ stands for the normalized gas density, $A=|F|$. Previously the very similar problem was considered in [4] where the breakdown evolution in the converging cylindrical wave taking spatially uniform gas distribution. In this case the numerical simulations shown relatively fast establishment of the stationary plasma density. For the discharge in air with relatively high density (when electron collision frequency exceeds field frequency) the stationary spatial distribution of plasma density was found to be approximately the same as predicted by V.B. Gildenburg on the base of a very simple analytical model [5]. Specifically it was found that an increase in the amplitude of the incident microwave is not accompanied by plasma density at the axis instead this increase results in an increase of the volume occupied by plasma. The plasma volume and density are important parameters when the point like source of extreme UV radiation is under discussion. Therefore a dependence of these parameters on microwave frequency and amplitude, gas non-uniformity and density was under particular interest in our simulations. The simulations were completed within wide range of the gas output pressure, p_0, length of the gas density non-uniformity, a, microwave field frequency and amplitude of the converging cylindrical wave. In all simulations it was found that stationary distribution of plasma density is established relatively fast after the breakdown. Specifically in case of relatively low output density in the gas flow (when the corresponding frequency of electron collisions is small compared to the field frequency, i.e. $\delta_0 \ll 1$) the plasma density is localized in a region which radius is small compared to electromagnetic wavelength if even a length of the gas flow non-uniformity is not so small (see Fig. 1). In this case a decrease in the amplitude of the converging cylindrical wave is accompanied by a decrease in plasma density whereas spatial distribution of plasma density remains almost unchanged (cf. Fig. 1 and 2)

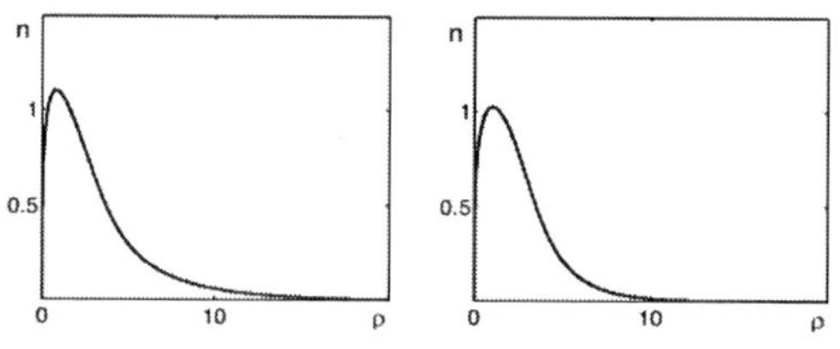

Fig. 1. Spatial distribution of stationary plasma density obtained at high value of normalized amplitude of incident electromagnetic wave (A_0=4) and relatively rarified gas density at the opening output ($\delta_0 = 0.3$). The left panel presents simulation results obtained with *ka=1*, the right panels presents simulation results with *ka=3*

On the contrary, in case of higher output density in the gas flow (when the corresponding frequency of electron collisions is big compared to the field frequency, i.e. $\delta_0 \gg 1$) an increase in the amplitude of incident electromagnetic wave is accompanied by an increase in the volume occupied by plasma which is especially pronounced at smaller length of gas density non-uniformity (see Fig. 3).

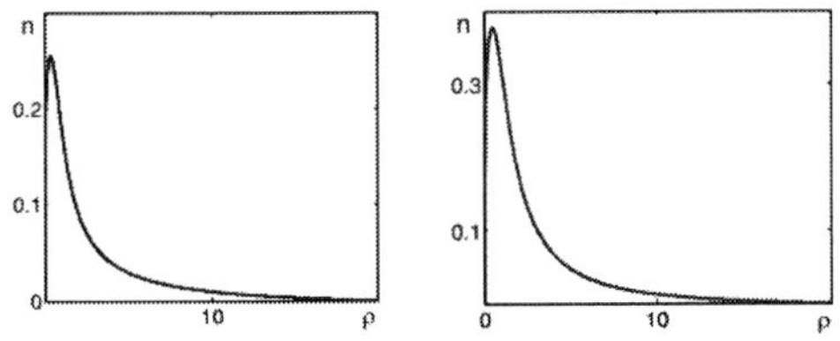

Fig. 2. Similar as in Fig 1 but calculated for less field amplitude in the incident electromagnetic wave (A_0=2)

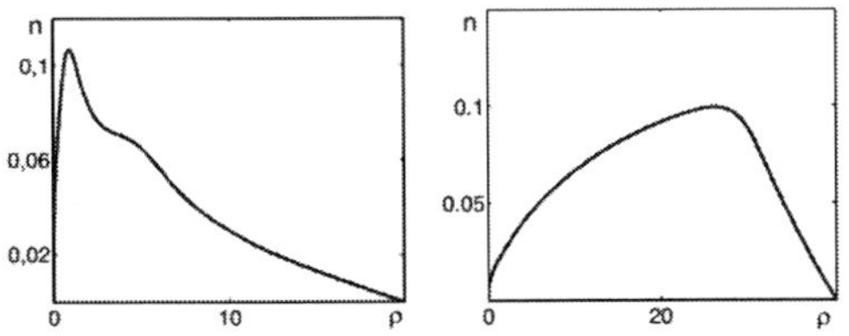

Fig. 3. Spatial distribution of stationary plasma density obtained at relatively high gas density at the opening output ($\delta_0 = 3$) and small length of gas non-uniformity *ka*=0.3. The left panel presents simulation results obtained at smaller amplitude of incident electromagnetic wave (A_0=2). The right panel presents simulation results obtained at A_0=4

On the contrary, in case of higher output density in the gas flow (when the corresponding frequency of electron collisions is big compared to the field frequency, i.e. $\delta_0 >> 1$) an increase in the amplitude of incident electromagnetic wave is accompanied by an increase in the volume occupied by plasma which is especially pronounced at smaller length of gas density non-uniformity (see Fig. 3). In the latter case an increase in the amplitude of incident wave results not only in considerable increase in plasma volume (which can exceed considerably the cube of wavelength) but also in a shift of plasma density maximum to a periphery of plasma. At higher value of *a*, one can find that plasma density distribution in space has two maxima (one in the center and another one is at the periphery of the ionized region see Figs. 4 and 5).

In this case an increase in the incident wave amplitude is accompanied mainly by an increase in the energy deposition at the periphery of ionized region which is not beneficial for creation of the point-like source of extreme ultraviolet radiation.

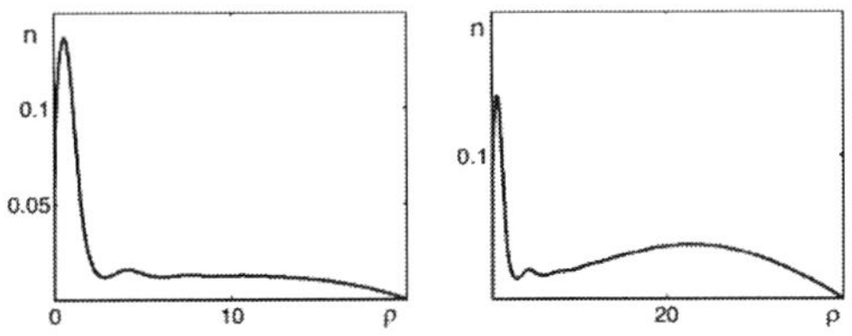

Fig. 4. Similar to Fig. 3, but calculations are completed with *ka* = 0.3

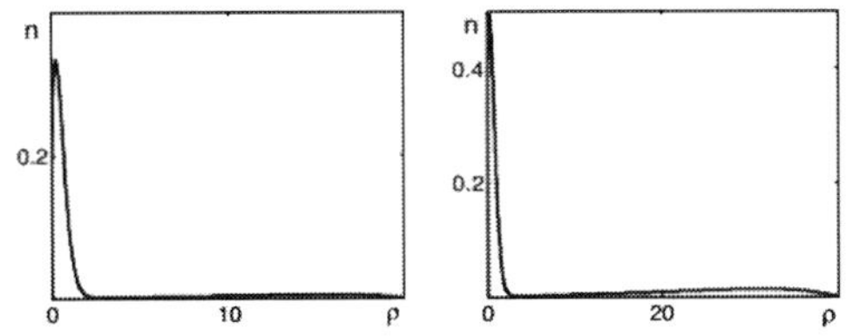

Fig. 5. Similar to Figs. 3–4, but calculations are completed with *ka* = 3

Acknowledgements

The work was supported by the Russian Science Foundation, project No. 14-12-00609.

References

1. *M.Yu. Glyavin, S.V. Golubev, V.G. Zorin, I.V. Izotov, A.G. Litvak, A.G. Luchinin, M.V. Morozkin, S.V. Razin, A.V. Sidorov, V.A. Skalyga*. The Discharge Maintained by High-Power Terahertz Radiation in a Nonuniform Gas Flow // Radiophysics and Quantum Electronics. 2014, V.56, No.8–9, P.561–565.

2. *M.Yu. Glyavin, S.V. Golubev, I.V. Izotov, A.G. Litvak, A.G. Luchinin, S.V. Razin, A.V. Sidorov, V.A. Skalyga, A.V. Vodopyanov*. A point-like source of extreme ultraviolet radiation based on a discharge in a non-uniform gas flow, sustained by powerful gyrotron radiation of terahertz frequency band // Applied Physics Letters, 2014, V.105, P. 174101.

3. *A. Sidorov, S. Razin, S. Golubev, M. Safronova, A. Fokin, A. Luchinin, A. Vodopyanov, M. Glyavin*. Measurement of plasma density in the discharge maintained in a nonuniform gas flow by a high-power terahertz-wave gyrotron // Phys. Plasmas. 2016, V.23, P.043511.

4. *I.B. Bokolishvili, A.V. Kim, G.G. Malinetsky and A.S. Petrosyan*. Numerical Simulation of a Nonequilibrium High-Frequency Discharge in a Converging Cylindrical Wave // Physica Scripta. 1990, V.42, No.4, P.442 – 448.

5. *V.B. Gil'denburg*. Electrodynamic mechanisms that limit the electron concentration in a laser spark // Sov. Phys. JETP, 1980, V.51, No.3, P.480–482.

EPJ Web of Conferences **149**, 03001 (2017) DOI: 10.1051/epjconf/201714903001
10th International Workshop 2017 "Strong Microwaves and Terahertz Waves: Sources and Applications"

Globus-M2 spherical tokamak and its mission in developing of compact fusion neutron source

V.B. Minaev[1], V.K. Gusev[1], N.V. Sakharov[1], Yu.V. Petrov[1], V.I. Varfolomeev[1], F.V. Chernyshev[1], N.N. Bakharev[1], V.V. Dyachenko[1], N.A. Khromov[1], G.S. Kurskiev[1], A.B. Mineev[2], V.A. Rozhansky[3], A.N. Saveliev[1], P.B. Shchegolev[1], I.V. Shikhovtsev[4]

[1] Ioffe Institute, St. Petersburg, Russia, Vladimir.Minaev@mail.ioffe.ru
[2] JSC «NIIEFA», St. Petersburg, Russia
[3] Peter the Great St. Petersburg Polytechnic University, St. Petersburg, Russia
[4] BINP SB RAS, Novosibirsk, Russia

Development of innovative concepts such as a tokamak based compact fusion neutron source with a nuclear blanket in a beam driven hybrid device needs research and consideration of some principal issues. The overview of the Globus-M spherical tokamak (ST) activities is presented in the report emphasizing the reasons for design and construction of novel high plasma performance machine. The Globus-M2 scientific program and the outcome of predicting simulations for the Globus-M2 plasmas are discussed.

Globus-M [1] is a low aspect ratio compact tokamak ($R = 0.36$ m, $a = 0.23$ m) with high specific ohmic and auxiliary heating power. First plasma was achieved in Globus-M in 1999. The machine has demonstrated practically all of the project objectives ever since. Target design parameters (aspect ratio – 1.5, X-point configuration, vertical elongation – 2.2, traiangularity – 0.45, average density – $1.0{\cdot}10^{20}$ m^{-3}, plasma current – 0.3 MA, auxiliary heating power – 1 MW) were achieved and some of them overcame [2, 3]. Also Globus-M has demonstrated possibility of reaching very low safety factor at the edge, which was close to 2. Only the attempt of 0.6 T magnetic field achieving in 2002 resulted in breakdown and arcing in the upper contact joint of the toroidal field coil. After repairing the toroidal field strength was limited by 0.4 T. Subsequent experiments showed that one of the main factor limiting further enhancements of plasma parameters is a relatively low toroidal magnetic field [4–6].

Globus-M2 [7] is an upgraded version of Globus-M with maximal increase of technical parameters to achieve the promising physical parameter area with sub-fusion temperature value and collisionality much less than unity. All parameters will be reached in compact geometry analogous to the Globus-M magnetic configuration and with the plasma current 0.5 MA and the toroidal magnetic field 1 T.

Magnetic field and plasma current increasing in Globus-M2 led us to complete redesign of the electromagnetic system due to new plasma equilibrium requirements and significant rise of mechanical and thermal loading with respect to Globus-M. High grade materials were used in electromagnetic system manufacturing. Inner segments of the toroidal magnetic field coils and central solenoid were made from cold extruded copper alloy hollow conductors. All poloidal magnetic field coils are water cooled too. The electromagnetic system support structure is enhanced by the upper supporting ring jointed to the lower one by four load-bearing crosspieces which reduces displacement of the toroidal field coil below 3mm under maximal load. Also novel design of coils allows toroidal field ripple minimization down to 0.4% at the plasma outer boundary. In the Globus-M2 tokamak the vacuum vessel and the in-vessel components remain the same that allows reducing of the project costs. The Globus-M2 machine at the final stage of assembling is shown in figure 1. Power supply upgrade was performed to achieve the ultimately high currents in the toroidal magnet and central solenoid too.

Fig. 1. Globus-M2 assembly

Plasma heating up to the temperatures of several keV in a compact magnetic configuration specific for the Glopbus-M2 ST requires an intensive auxiliary heating application in conditions of adequate energy confinement. Both the auxiliary heating technique development and the energy confinement improvement are in the focus of Globus-M2 research. A new neutral beam injector with 50 keV hydrogen or deuterium beam of 1 MW power will supplement the "old" one, which will be upgraded up to 40 keV energy, providing additional 1 MW power.

The ion cyclotron plasma heating scenario with light atomic mass minority was successfully tested in Globus-M experiments. New antenna, which is capable for 0.5 MW power launching into Globus-M2 plasma column, has been designed and manufactured. The antenna provides more shaped spectra of exited waves. About 70% of launched power is located in a range of parallel refractive index 10–40. With increasing of toroidal magnetic field the operating frequency is raised respectively up to 15 MHz.

In turn, that improves antenna-plasma coupling and RF power absorption significantly. Performed numerical modeling predicts essential improving of high energy ion confinement and enhancing of heating efficiency for new experimental conditions.

A problem of non-inductive plasma current start-up and sustainment is principally important for the next generation of "fusion-range" STs, as well as for the tokamak-based compact fusion neutron sources. Neutral beam injection is a well-known method for current drive (CD) complementary to bootstrap current generated by pressure gradients in hot tokamak plasmas. Numerical simulations for Globus-M2 discharges predict the achievement of conditions for better neutral beam CD efficiency and overall plasma performance. The total fraction of non-inductive current could reach 40–45% in full-power NB heated shots.

Low hybrid (LH) electromagnetic waves also demonstrated high efficiency of CD in conventional tokamaks of relatively low cost. Globus-M2 is equipped with a unique system for LHCD and will test a novel scheme for the LH wave launching with poloidal slowing down. The system consists of a klystron generator (0.5 MW, 2.45 GHz), a 10-channel power splitter and a 10-waveguide grill antenna. Launcher and feeder line construction allows to compare current drive efficiency with toroidal and poloidal slowing down of exited waves for the equal plasma conditions. Also increasing of toroidal magnetic field up to 1 T has to improve LH wave propagation significantly.

According to a preliminary modeling the increased magnetic field, plasma current and temperature in Globus-M2 will permit to drive LH current about 200 kA for the poloidal grill orientation and about 460 kA for the toroidal one. Larger driven current in the toroidal case is due to formation of a more extended quasilinear plateau in the electron distribution function compared to the poloidal grill orientation case. It should be noted that the above modeling was done neglecting the inductive electric field, which may play a noticeable role in the LHCD. Taking into account the electric field will be done in the future modeling.

Study of such complex area as scrape of layer and divertor plasma is necessary to provide high duty factor of operation and long-live plasma facing components in a compact fusion device. Comprehensive set of diagnostic tools was prepared for the experiment. Particle flux and electron temperature profiles across the lower divertor plates will be measured by means of an array of 10 embedded Langmuir probes. A new movable multi-pin Langmuir probe is installed in the equatorial midplane from the low field side. The probe allows obtaining a number of plasma parameters: ion saturation current, floating potential, electron temperature, electron density, radial electric field and Mach number. The IR camera FLIR SC7300M is planned for divertor plate thermal mesurements.

In Globus-M (0.4 T, 200 kA) D-D experiments with 26 keV beam the neutron generation rate was about 10^{10}–10^{11} n/s which corresponded to 10^{10} n/s·m^2 through the vacuum vessel surface. The NUBEAM modeling [8] shows a good agreement with neutron flux measurements on the Globus-M tokamak and predicts a neutron rate increase by two orders of magnitude for the Globus-M2 machine (1 T, 500 kA). The main reasons for that enhancement are the electron temperature growth and the substantial improvement of the fast ion confinement which allows application of a new 50 keV NB injector. The next Globus-3 machine (R/a = 0.6 m/0.36 m; B_t = 1.5 T, I_p = 1.5 MA) is currently regarded as a hydrogen prototype of compact tokamak based fusion neutron source. In the 120 keV, 4 MW D-D NBI discharges modeling predicts a neutron yield level of about 10^{15} n/s (10^{14} n/s·m^2).

Substantial improvement in plasma parameters is expected for Globus-M2: density over 10^{20} m^{-3} and temperature in keV range could be attained with high confidence. High plasma performance together with broad set of auxiliary heating equipment and comprehensive set of diagnostics provide favorable conditions for a compact fusion neutron source research and development.

References

1. *V.K. Gusev, V.E. Golant, E.Z. Gusakov, et al.* Globus-M spherical tokamak // Tech. Phys., **44** (1999) 9, 1054.
2. *V.K. Gusev, F.V. Chernyshev, V.E. Golant, et al.* Density limits and control in the Globus-M spherical tokamak // Nucl. Fusion, **46** (2006) 8, S584–S591.
3. *V.K. Gusev, S.E. Aleksandrov, V. Kh. Alimov, et al.* Overview of results obtained at the Globus-M spherical tokamak // Nucl. Fusion, 2009, **49** (10), #104021.
4. *V.K. Gusev, R.M. Aminov, A.A. Berezutskiy, et al.* Investigation of beam- and wave-plasma interactions in spherical tokamak Globus-M // Nucl. Fusion, 2011, **51** (10), #103019.
5. *V.K. Gusev, E.A. Azizov, A.B. Alekseev, et al.* Globus-M results as the basis for a compact spherical tokamak with enhanced parameters Globus-M2 // Nucl. Fusion, 2013, **53** (9), #093013.
6. *V.K. Gusev, N.N. Bakharev, V.A. Belyakov, et al.* Review of Globus-M spherical tokamak results // Nucl. Fusion, 2015, **55** (10), #104016.
7. *V.B. Minaev, V.K. Gusev, N.V. Sakharov, et al.* Spherical tokamak Globus-M2: design, integration, construction // Nucl. Fusion, 2017, **57** (6), # 066047.
8. *A. Pankin, D. McCune, R. Andre, et al.* The tokamak Monte Carlo fast ion module NUBEAM in the National Transport Code Collaboration library // Comput. Phys. Comm., 2004, **159**, 157–184.

Experimental study of Ohmic losses of polarizer mirror system

F. Leuterer[1], D. Wagner[1], J. Stober[1], W. Kasparek[2], C. Lechte[2] and ASDEX Upgrade Team[1]

[1] Max Planck Institut für Plasmaphysik, 85748 Garching, Germany
[2] Institut für Grenzflächenverfahrenstechnik und Plasmatechnologie, Universität Stuttgart, 70569 Stuttgart, Germany

Electron cyclotron heating of fusion plasmas requires to launch a high power millimeter wave beam with a well defined polarization, generally elliptical. This can be achieved with two corrugated mirrors with groove depths of $\approx \lambda/8$ and $\approx \lambda/4$. In HE11 waveguide transmission lines these polarizers are incorporated into 90° miter bends and are subject to very high power density. To study their loss behaviour we have inserted into the ASDEX Upgrade ECRH transmission line such polarizers with sinusoidal grooves of periodicity 1.07 mm, depth 0.8 mm ($\approx \lambda/4$-mirror) resp. 0.57 mm ($\approx \lambda/8$-mirror), and a thickness of 5 mm. As material we took stainless steel with higher loss and easily measureable temperature rise, measured with a Pt-100 sensor on the backside of the mirror. The measurements were done with 50 ms pulses of 375 kW at 140 GHz. Between the output of the gyrotron and the mirror under test we have a quasi-optical MOU allowing to feed the waveguide with an arbitrary polarization.

We measured the losses of such polarizers and of a plane mirror depending on the angle φ of a linearly polarized incident electric field, and on the angle α of the grooves, both with respect to the incidence plane of the miter bend. These are shown in Fig. 1. For a plane mirror and for $\varphi = 0°$ and 90° we have theoretical results

$$a_{\varphi=0°} = a_{\perp} / \cos\Theta; a_{\varphi=90°} = a_{\perp} \cdot \cos\Theta;$$

where Θ=45° is the angle of incidence in a miter bend mirror, and $a_{\perp} = 4R_S / Z_0$ is the absorption coefficient at perpendicular incidence with R_S being the surface resistance and Z_0 the free space wave impedance. From Fig. 1 top we get $a_{\perp}$ corresponding to 573 mV. This allows to normalize our measurements to $a_{\perp}$.

The measurements show a cos(2φ), resp. cos(2α) dependence of the losses and suggest an empirical relation

$$a_{norm}(\alpha,\varphi) = AE_x^2 + BE_z^2 + CH_x^2 + DH_z^2 \quad (1)$$

with E_x, E_z, H_x, H_z being the tangential electric and magnetic field components of the incident wave on the mirror surface perpendicular (x) and parallel (z) to the grooves. For a 90° miter bend these are

$$E_x = E_0 \cos\alpha \sin\varphi + 0.5\sqrt{2}E_0 \sin\alpha \cos\varphi$$
$$E_z = E_0 \sin\alpha \sin\varphi - 0.5\sqrt{2}E_0 \cos\alpha \cos\varphi$$
$$H_x = H_0 \cos\alpha \cos\varphi - 0.5\sqrt{2}H_0 \sin\alpha \sin\varphi$$
$$H_z = H_0 \sin\alpha \cos\varphi + 0.5\sqrt{2}H_0 \cos\alpha \cos\varphi$$

With measurements at $\alpha = 0°$ and 90° and $\varphi = 0°$ and 90° we can determine the coefficients A, B, C and D. The lines in Fig.1 and 3 are calculated with equ. (1) and fit very well to the measurements.

This empirical model allows to extend our AUG-polarization code to calculate for a given mirror arrangement and a given input polarization not only the output polarization, but also the loss. An example is shown in Fig. 2 displaying for a horizontal ($\varphi = 0°$) linearly polarized input electric field the output ellipticity ε, the orientation β of the ellipse, the sense of rotation r of the polarization, and the total loss as function of the mirror rotation angles $\alpha_{\lambda/8}$ and $\alpha_{\lambda/4}$. The dots show a β-scan for a constant ellipticity $\varepsilon = 0.2$ and rotation $r = +1$. Such calculations show that there are up to 4 possible settings $\alpha_{\lambda/8}$ and $\alpha_{\lambda/4}$ to realize a wanted output polatization as indicated in Fig. 2 by the black dots for an output polarization with $\beta = 30°$. These 4 settings are seen to have different loss.

To demonstrate this we selected some examples with different input and output polarizations and determined the losses for the 4 different settings. Some results are shown in Fig. 3.

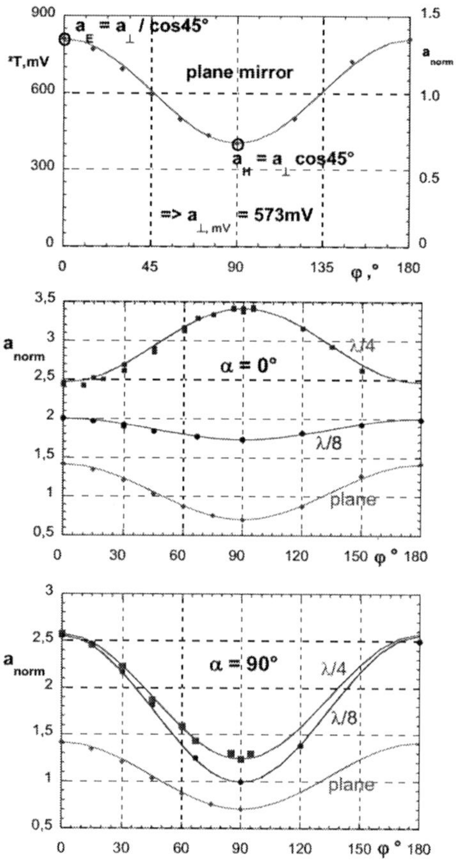

Fig. 1. measured and modelled losses of a plane, $\lambda/8$ and $\lambda/4$ mirrors as function of incidence angle φ of linear polarization at constant groove orientation α

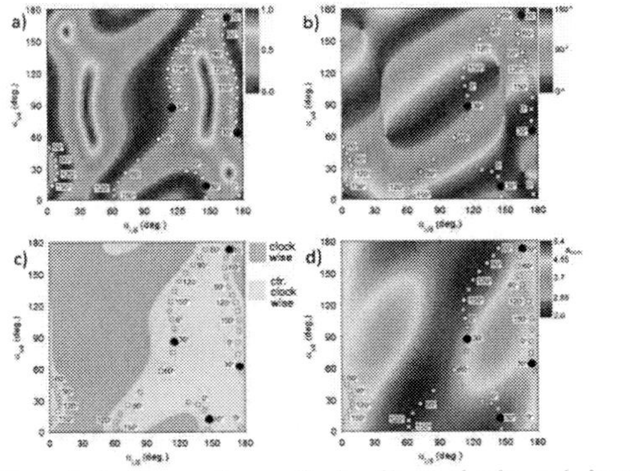

Fig. 2. polarization code result for linear horizontal input displaying a) ellipticity, b) orientation of the ellipse, c) rotation and d) loss as function of the groove angles α

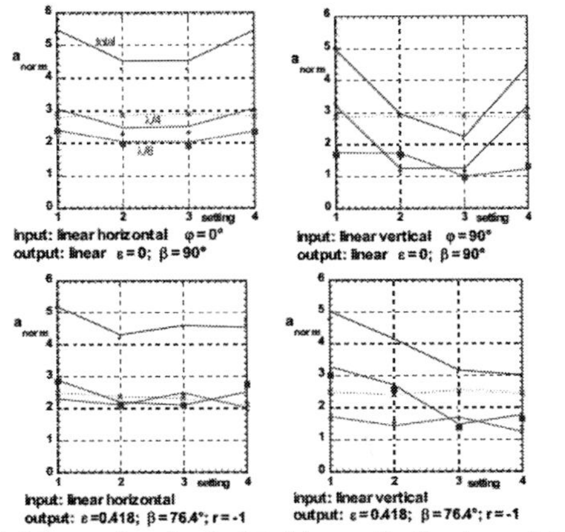

Fig. 3. Measured and calculated losses for the 4 possible settings leading to the same output polarization. Dots: total loss, squares: $\lambda/4$-mirror, rhombs: $\lambda/8$-mirror, crosses: detector signal at the end of the transmission line

We see that the loss can vary up to a factor of 2.5 depending on the setting of the individual mirrors. The constant detector signals indicate that indeed we have the same output polarization.

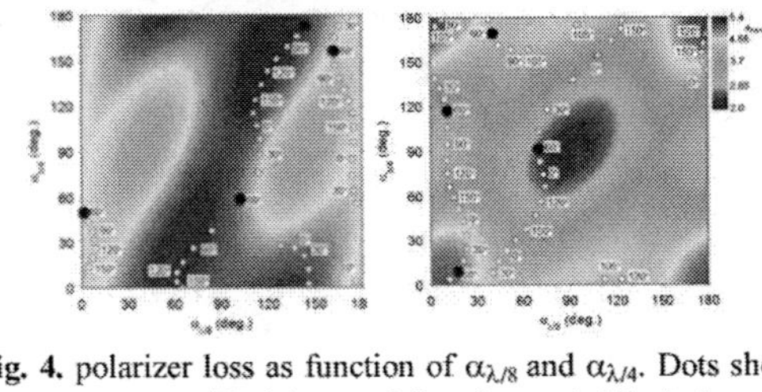

Fig. 4. polarizer loss as function of $\alpha_{\lambda/8}$ and $\alpha_{\lambda/4}$. Dots show a β-scan at const. ellipticity $\varepsilon = 0.2$ and $r = +1$. Black dots show the 4 solutions for $\beta = 60°$. Input polarization horizontal $\varphi = 0°$ (*left*) and vertical φ=90° (*right*)

The total loss, i.e. the sum of the losses of both mirrors, depends on the geometrical arrangement of the mirrors and also on the input polarization. In Fig.4 the two polarizers have the same incidence plane, but the linear input polarization is horizontal or vertical. In this example the vertical input polarization leads to much lower losses. An example for the dependence on the geometrical arrangement is shown in Fig.5 where we assumed orthogonal incidence of the two miter bends. Here too, the loss can be very different.

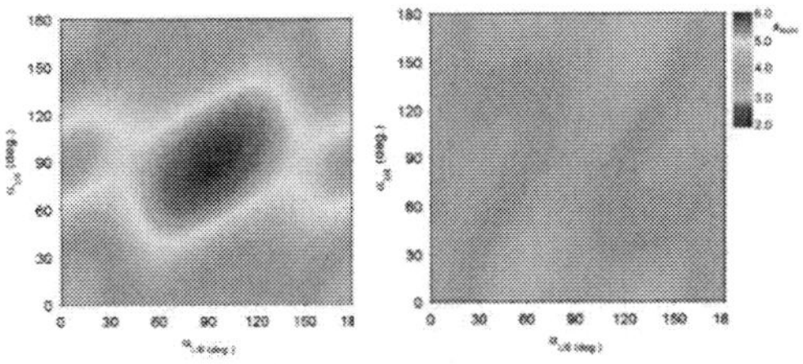

Fig. 5. polarizer loss as function of $\alpha_{\lambda/8}$ and $\alpha_{\lambda/4}$ .In this case the incidence planes of the $\lambda/4$ and the $\lambda/8$ mirrors are perpendicular to each other. Input polarization: left horizontal $\varphi = 0°$, and right vertical $\varphi = 90°$

Arbitrary polarization can also be realized with a two mirror polarizer consisting of two $\lambda/8$-mirrors instead of one $\lambda/8$ and one $\lambda/4$ mirror. We have repeated the set of measurements of Fig.3 with such a configuration. The result is shown in Fig.6. In some cases even lower loss can be realized. Such a configuration has also the advantage that only one type of grooved mirror needs to be manufactured.

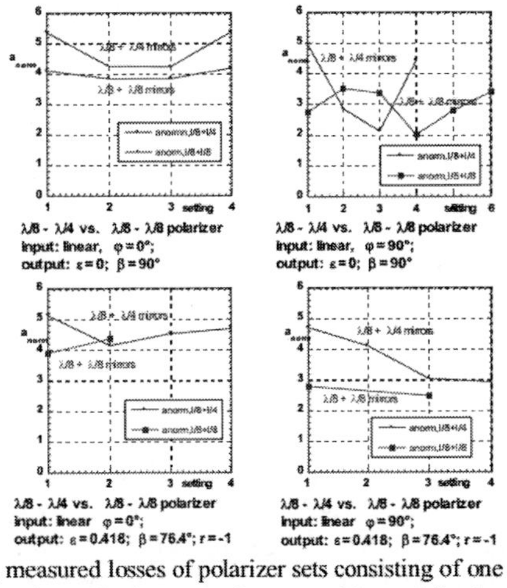

Fig. 6. measured losses of polarizer sets consisting of one $\lambda/8$ and one $\lambda/4$ mirror (dots) compared to a set with two $\lambda/8$ mirrors

In this paper we have shown that the loss of a grooved mirror as used for polarizer systems is determined by just 4 parameters *A*, *B*, *C*, *D*, which can be obtained experimentally (or numerically) and which depend only on the corrugation profile. Knowing these parameters the losses of any arrangement of a polarizer system can be calculated. There are several possible settings for one wanted output polarization. These can have differences in the loss up to a factor of 2.5. Minimum loss may enhance the power transmission capability of the waveguide line.

1. *D. Wagner, F. Leuterer,* Broadband polarizers for high power multi-frequency ECRH systems, Journal of Infrared, Millimeter and Terahertz Waves, Vol. 26 (2005), 163-172.

2. *D. Wagner, F. Leuterer et al.,* Minimization of the Ohmic loss of grooved polarizer mirrors in high power ECRH systems, Journal of Infrared, Millimeter and Terahertz Waves, Vol. 38 (2017), 191-205.

The EC-system of EU DEMO: concepts for a reactor heating system

G. Granucci[1], G. Aiello[2b], K.A. Avramidis[2a], A. Bruschi[1], G. Gantenbein[2a], S. Garavaglia[1], G. Grossetti[2b], J. Jelonnek[2a], A. Moro[1], E. Poli[5], N. Rispoli[1], D. Strauss[2b], M. Thumm[2a], I. Tigelis[4], C. Tsironis[4], T. Franke[5,6] and M.Q. Tran[3]

[1] Institute of Plasma Physics "P.Caldirola", National Research Council of Italy, Milan, Italy
[2a] IHM, [2b] IAM-AWP, Karlsruhe Institute of Technology (KIT), D-76131 Karlsruhe, Germany
[3] Swiss Plasma Center, EPFL, CH-1015 Lausanne, Switzerland
[4] NCSRD/National and Kapodistrian University of Athens, Faculty of Physics, & National Technical University of Athens, School of Electrical and Computer Engineering, Athens, (Greece)
[5] Max-Planck-Institut für Plasmaphysik, Boltzmannstr. 2, D-85748 Garching, Germany
[6] EUROfusion Consortium, Boltzmannstr. 2, D-85748 Garching, Germany

In the frame of the project for a European DEMOnstration Fusion Power Plant (FPP), the design of the auxiliary heating systems is one of the keys to heat and control the burning plasma (i.e. a plasma in reactor producing electricity to the grid from fusion reactions). The heating mix considered in EU DEMO bases on Electron Cyclotron Resonance Heating and Current Drive (ECRH&CD), Neutral Beam Injection (NBI) and Ion Cyclotron Resonance Heating (ICRH). In this first phase of the design process (called Pre-Conceptual phase) each system is under study with 50 MW as a reference power to the plasma [1]. The main tasks assigned to the EC system are reported in the Table 1, assuming the full use of the 50 MW:

Table 1. Main Physical Requirements for EC system

EC Task	Power [MW]	Localization [ρ]	Mode
Assisted Break-down	6–10	< 0.3	Heating
Ramp up and L-H transition	50	< 0.3	Heating/CD
Main heating	50	< 0.3	Heating/CD
Sawtooth control	2–10	< 0.3	CD
NTM control ($q = 2$; $q = 3/2$)	10–15	0.85; 0.75	CD
Ramp down	40	0.3–0.5	Heating

These heating tasks require different radial localization and combination of heating and non-inductive current drive (CD). A possible solution is a gyrotron capable to operate at two (or more) different frequencies, in order to cover all the physical requirements in the easiest way. Moreover in designing an EC system for DEMO, it is necessary to consider the peculiarities of a fusion power plant reactor, which are significantly different from those of any existing tokamak experiments. The power reliability and availability are major targets for the system, because in case of fault of EC system the pulse must be terminated, with a consequent unacceptable economic impact. Therefore, reliability of any each sub-system is taken into consideration when choosing the architecture of the EC system. A second objective is to minimize the recirculating power to guarantee the higher possible energy gain of the reactor with the use of a highly efficient gyrotrons [2] (above 60 %) and low losses (10%) transmission lines.

The present design considers 2 MW gyrotron units, capable to operate at 170 GHz for heating and 204 GHz for current drive. This last corresponds to a moderate upshift of 1.2 for the CD function while the two frequencies (170/204 GHz) have been selected to satisfy the λ/2-resonances of the diamond-disc window. These frequencies have been selected for the pulsed EU DEMO1 2015 baseline [3] and used as reference to calculate CD efficiency and localization. At the end of the EU DEMO Conceptual Design phase, the appropriate frequencies will be selected. Within EUROfusion research is ongoing to consider gyrotron operating frequencies of up to 240 GHz.

The transmission line (TL) is based on a quasi-optical approach, inspired to the Multi-beam (MB)TL of W7-X stellarator experiment [4], enclosing the mirrors in an evacuated pipe to make it compatible with the nuclear environment. The concept is based on a mirror confocal layout with single units composed by two mirrors forming a dog-leg and replicated few times for transmitting along straight paths (150 m as reference length).

Table 2. Transmission losses estimation of DEMO EQO for a length of 150 m

Distance between mirrors = 8 m	170 GHz	204 GHz
Ohmic Losses (H-plane - 45°)	3.2%	3.5%
Truncation (40 mirrors)	1.5%	1.4%
Mode conversion	1.5%	1.4%
Total Losses	6.2%	6.3%

One pumping system is located in each mirror unit. As reference it is considered to transmit 8 beams with one MBTL, with an overall losses estimated to be <10% (see Table 2). This choice would allow a simpler and compact solution capable to reduce the complexity of the system and to save space.

The design of the launching system is strongly limited by severe constrains imposed by the environment of a nuclear reactor as the vicinity of a hot burning plasma and under a heavy neutron bombardment. In particular, no movable parts are allowed in proximity of a burning plasma, and shielding and suitable structural component need to be included in the design [5]. A remote steering antenna, with a dogleg in the vessel port, is under study as one of the options for a launcher concept. The proper choice of steering plane and the inclination of the corrugated square waveguide are the free parameters for the required EC power localization and CD efficiency, to be combined with the engineering requirements. In Fig. 1 the result of a beam tracing performed with TORBEAM [6] is shown. Using both the frequencies the possibility to cover the assigned tasks of Table 1 is demonstrated, while the 204 GHz exhibits a higher central CD efficiency (for ST control), the 170 GHz can be efficiently used for NTM control and for central or off-axis heating.

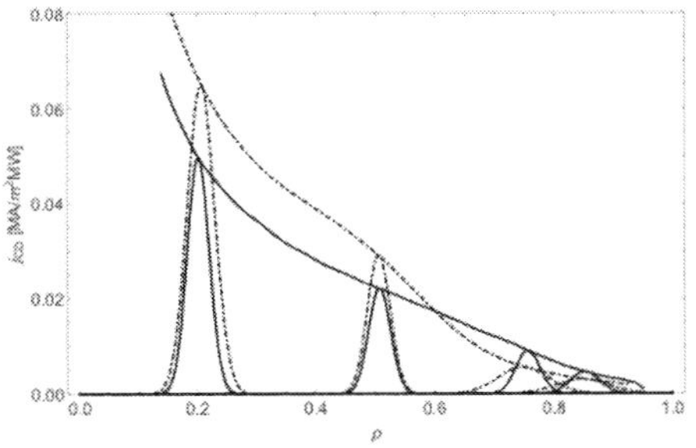

Fig. 1. With full lines (dashed lines) calculated CD profiles for 170 GHz (204 GHz) launched from equatorial plane with antenna toroidally inclined of of 17° (22°). For the different depositions a steering of ±12°is considered

A comparison of remote steering antennas and alternative simpler solutions, truncated waveguides used as launchers with step tunable gyrotron, is ongoing, on the basis of the physical requirements of Table 1. This study can be complemented considering the use of step tunable gyrotrons used in the truncated waveguide launcher concept, in order to have enough localization control for NTM stabilization. The other tasks can be satisfied increasing the number of launching points with the aid of fast microwave switches.

The reliability of the auxiliary power to be delivered, and, more in general, the RAMI of the whole EC system, is another characterizing point of the conceptual design. The maximization of reliability is obtained with appropriate system architecture, characterized by a modular design. An ideal overall reliability of 100 % to deliver 50 MW to the DEMO plasma using an EC system is possible to reach by considering a proper redundancy in the different sub units. Therefore, the system concept is based on a strong modularity which considers a cluster of 8 sources (gyrotrons) fed by one high voltage cathode power supply and individual anode power supply. The 8 beams are transmitted to the launchers with a single MBTL. The launcher of each cluster is individual and in the same port plug in structure. The best combination [7] is composed by 5 clusters of 8 units, with 7 gyrotrons operating in each cluster (and one in standby) and one full cluster in standby. Therefore only 28 lines will be on simultaneously with 12 lines off ready to substitute the unit (gyrotron or line / launcher) eventually in fault.

Acknowledgement

Parts of this work are carried out within the framework of the EUROfusion Consortium and have received funding from the Euratom research and training programme 2014-2018 under grant agreement No. 633053. The views and opinions expressed herein do not necessarily reflect those of the European Commission.

References

1. *M. Q. Tran et al. D.* EU DEMO Heating and Current Drive: Physics and Technology // 26th FEC – Kyoto (2016) FIP/7-7.
2. *J.Jelonek et al* European Research Activities Towards a Future DEMO Gyrotron // SMP, 2017, Nhizhny Novgorod.
3. *R. Wenninger et al.* The physics and technology basis entering European system code studies for DEMO // Nucl. Fusion **57**, 016011 (2017).
4. *V. Erckmann et al.* Electron cyclotron heating for W7-X: physics and technology // Fusion Sci. Technol. **52**, 291-312 (2006).
5. *G. Grossetti et al.* DEMO Port Plug design and integration studies // 26th FEC – Kyoto (2016) FIP/7-4.
6. *E. Poli et al.* TORBEAM, a beam tracing code for electron-cyclotron waves in tokamak plasmas // Comput. Phys. Comm. **136**, 90 (2001).
7. *S. Garavaglia et al.,* Preliminary conceptual design of DEMO EC system // AIP Conference Proceedings **1689**, 090009 (2015).

EPJ Web of Conferences **149**, 03004 (2017) DOI: 10.1051/epjconf/201714903004
10th International Workshop 2017 "Strong Microwaves and Terahertz Waves: Sources and Applications"

Extension of the multi-frequency ECRH system at ASDEX upgrade

D. Wagner[1], J. Stober[1], M. Kircher[1], F. Leuterer[1], F. Monaco[1], M. Münich[1], M. Schubert[1], H. Zohm[1], G. Gantenbein[2], J. Jelonnek[2], M. Thumm[2], A. Meier[3], T. Scherer[3], D. Strauss[3], W. Kasparek[4], C. Lechte[4], B.Plaum[4], A.Zach[4], A.G. Litvak[5], G.G. Denisov[5], A. Chirkov[5], V. Malygin[5], L.G. Popov[6], V.O. Nichiporenko[6], V.E. Myasnikov[6], E.M. Tai[6], E.A. Solyanova[6], S.A. Malygin[6], and ASDEX Upgrade team

[1] Max-Planck-Institut für Plasmaphysik, Boltzmannstr.2, D-85748 Garching, Germany, dietmar.wagner@ipp.mpg.de
[2]Institut für Hochleistungsimpuls- und Mikrowellentechnik,
[3]Institut für Angewandte Materialien-AWP,
Karlsruhe Institute of Technology, Kaiserstr. 12, D-76131 Karlsruhe, Germany
[4]Institut für Grenzflächenverfahrenstechnik und Plasmatechnologie, Universität Stuttgart, Pfaffenwaldring 31, D-70569 Stuttgart, Germany
[5]Institute of Applied Physics, RAS, 46 Ulyanov St., Nizhny Novgorod, 603950, Russia
[6]GYCOM Ltd, 46 Ulyanov St., Nizhny Novgorod, 603155, Russia

1. Introduction

Four two-frequency gyrotrons were in operation during the latest experimental campaign at ASDEX Upgrade (AUG), operating at 105 and 140 GHz with a pulse length of 10 s, corresponding to the maximum flat top time of AUG plasma discharges, and output powers close to 1 MW. The gyrotrons have single-stage depressed collectors and their efficiencies are around or above 50%. Recently, an old 140 GHz single-frequency ECRH system which was built in 1996 (4 gyrotrons with 500 kW output power and 2 s pulse length) has been dismantled. It is being replaced by four new 1 MW two-frequency gyrotrons using the same launching positions with slightly modified launchers [1]. The new gyrotrons will be operated in cryogen-free superconducting magnets.

2. Power Supplies

For the new gyrotrons a semiconductor-based body modulator has been developed with a maximum output voltage of 32 kV, peak and average current of 1 A and 250 mA, respectively. The voltage rise time is variable between 100 μs and 1 ms and the switch-off time is below 10 μs. The cathode voltage is provided by a tetrode-based series tube modulator with improved control system and an insulating transformer and screen grid at high potential. The output voltage can be varied during the gyrotron pulse.

3. Transmission Line

The ECRH transmission lines at AUG are based on corrugated non-evacuated HE11 waveguides with 87 mm inner diameter (Fig. 1). The new gyrotrons are placed at a linear distance of 65 m away from the tokamak, which eliminates the influence of the tokamak stray magnetic field (from which we suffered in the old ECRH installation). The total length of the transmission lines is about 90 m and the maximum number of required miter bends is between 6 and 8. The Matching Optics Units (MOUs) are equipped with a pair of phase correcting mirrors for each frequency followed by two broadband polarizer mirrors (Fig. 2).

New water loads have been developed for 1 and 10 s pulse length and 1 MW power (Fig. 3). Each MOU contains a 1 s water load which is used for system startup or intermediate conditioning. Each gyrotron can be connected to a single long-pulse load (1 MW, 10 s). Short-pulse calorimetric loads (200 ms, 1 MW) can be installed in the MOUs and are permanently mounted at the AUG ports. Online power monitoring is provided by built-in directional couplers in the second phase-correcting mirrors of the MOUs and by an improved 2-frequency miter bend integrated power monitor at the end of each transmission line.

Fig. 1. View of the HE11 waveguide transmission line at ASDEX Upgrade

Fig. 2. Gyrotron (*right*) connected to a Matching Optics Unit (*left*)

4. In-Vessel Reflectors for O2-Mode Heating

The ASDEX Upgrade Tokamak has been equipped with a solid tungsten divertor, which allows to gain experience in operation under ITER-like heat loads. Progress has been made in controlling high power ($P_{heat} \approx 20$ MW) discharges, where the power normalized to the major radius ($R_0 = 1.65$ m) is approaching ITER relevant values [2]. The potential threat resulting from the accumulation of tungsten in the plasma center can be effectively avoided, if the central heating power density is high enough. In principal, the electron cyclotron resonance heating (ECRH) system at ASDEX Upgrade (AUG) is able to fulfil this task, but it has to be operated in the O2 mode, since the plasma density in these experiments usually exceeds the X2 cut-off at the operating frequency f = 140 GHz. In the AUG case, O2 mode single pass absorption is incomplete [3]. A second pass of the unabsorbed fraction, however, improves the situation significantly [4]. Therefore it was necessary to install special reflectors into the heat shield on the central column inside the AUG vacuum vessel. It is possible to generate an efficient oblique reflection for a given wavelength and beam geometry, if an optimized grating [5] is machined into the mirror surface. In addition, for each beamline a suitable beam dump has to be selected on the low field side. This is necessary in order to spread out the unabsorbed part of the beam after the second pass, which may typically contain 10% of the initial beam power.

Fig. 3. High power water load for 1 MW, 10 s with Teflon tubing

5. Machine Protection

In case of low ECRH absorption ASDEX Upgrade is protected by a number of stray radiation detectors (sniffer probes) [1]. Typical cases for low absorption are errors in the polarization of the injected beam, plasma cutoffs or operation regimes with low mm-wave absorption [3]. As an additional passive protection, an overlapping design of the tile edges (Fig. 4) on the inner heat shield of ASDEX Upgrade increases the shielding against mm-waves and reduces the risk of severe damage. The shielding factor has been measured on a laboratory test mock-up and its worst case value of –14.4 dB (140 GHz) is considered as sufficient for protection of vital parts.

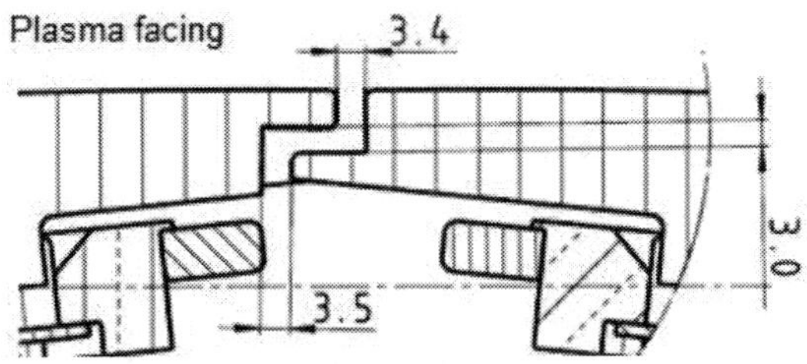

Fig. 4. Edges of heat shield tiles have been redesigned with a labyrinth-like "dog leg" structure in order to increase the shielding against millimeter waves

References

1. *Wagner, D., et al.* Status, Operation and Extension of the ECRH System at ASDEX Upgrade // Int. J. of Infrared, Millimeter and Terahertz Waves 2016. V. 37, P. 45–54.

2. *Kallenbach, A., et al.* Partial detachment of high power discharges in ASDEX Upgrade // Nuclear Fusion. 2015. V. 55. P. 053026.

3. *Erckmann, V. and U. Gasparino*, Electron cyclotron resonance heating and current drive in toroidal fusion plasmas // Plasma Phys. Control. Fusion 1994. V. 36, P. 1869.

4. *Höhnle, H., et al.*, Extension of the ECRH operational space with O2 and X3 heating schemes to control tungsten accumulation in ASDEX Upgrade // Nuclear Fusion. 2011. V. 51. P. 083013.

5. *Mangold, O., et al.*, Optimization of microwave reflection gratings for electron cyclotron resonance heating in O2 mode // Conf. Digest of the 2004 Joint 29th Int. Conf. on Infrared and Millimeter Waves and 12th Int. Conf. on Terahertz Electronics. 2004. P. 717–718. (doi:10.1109/ ICIMW.2004.1422293).

EPJ Web of Conferences **149**, 03005 (2017) DOI: 10.1051/epjconf/201714903005
10th International Workshop 2017 "Strong Microwaves and Terahertz Waves: Sources and Applications"

Electron-cyclotron waves in large-scale open traps: new questions highlighted by recent experiments

Alexander Shalashov

Institute of Applied Physics of Russian Academy of Sciences, Nizhny Novgorod, Russia, ags@appl.sci-nnov.ru
Budker Institute of Nuclear Physics, Siberian Branch of Russian Academy of Sciences, Novosibirsk, Russia

The absorption of electromagnetic waves under the electron cyclotron resonance (ECR) conditions is widely used for heating of high-temperature plasmas in tokamaks and stellarators. However, for many years the use of this method in open magnetic traps has been limited either by plasma heating in compact technological devices, such as ECR sources of multiply-charged ions and X-ray radiation [1–3], or by stabilizing of low-density plasma in fusion-relevant experiment, first of all, in GAMMA-10 at Tsukuba University [4]. This tandem mirror trap is equipped with a complicated ECRH system, including five 28 GHz high-power gyrotrons. The plug and barrier ECRH generate the axial ion-confining potential and the thermal barrier potential for electrons, respectively; thus, ECRH power is utilized to increase the confining potentials and to control of internal transport barrier formation [5]. The central-cell ECRH is aimed at increase of the bulk electron temperature, but this feature has not been demonstrated convincingly.

Until recently, the only example of the bulk electron heating in the large mirror trap was TMX-U experiment at LLNL. There the electron temperature up to 0.28 keV was obtained with ECRH at 28 GHz, but these studies were concluded soon [6]. The ECR heating of a dense plasma, comparable to those in toroidal devices, has been demonstrated in a large-scale mirror trap only in 2013. We mean successful experiments on combined plasma heating by neutral beams and 54.5 GHz fundamental harmonic X-mode ECRH performed at the Gas Dynamic Trap (GDT) in Budker Institute [7–10]. Here, a highly-localized heating of the bulk electrons leads to achieving the record for open traps electron temperature of 1 keV [8]. These studies demonstrate good prospects for the use of simple axially symmetric open traps as a powerful source neutron for fusion applications [11]. Therefore, GDT may be considered as a prototype of the next-step fusion reactors.

Note that GDT, being comparable to GAMMA-10 in dimensions and confining magnetic fields, is operated at about an order of magnitude more dense plasma, what results in very different electrodynamics conditions in these two devices. Namely, $\omega_{pe}/\omega \sim 0.8$ in GDT and $\omega_{pe}/\omega \sim 0.15$ in GAMMA-10 with ω_{pe} and ω being, correspondingly, the electron Langmuir frequency and ECRH frequency (which is close to the fundamental electron cyclotron harmonic, $\omega_{ce} \sim \omega$). In this sense, the plasma is rarefied in GAMMA-10 and relatively dense in GDT. The latter makes the refraction of radiation essential at the GDT conditions.

Implementing the efficient heating of a dense plasma at large open traps requires a revision of the prevailing ideas about the physics of cyclotron absorption [12] as well as the subsequent transport of energy and MHD stabilization of a plasma column [13]. In compact traps, millimeter wave radiation of a high-power gyrotron is usually launched from the trap end along the ambient magnetic field; such scheme allows matching the radiation and dense plasma at the plasma interface [14]. However, in a large device, this option is not possible due to obligate expander elements at the trap ends where radiation meets a cut-off. Thus, the heating radiation may be launched only through side walls of a vacuum chamber. Because of very different magnetic topology, none of the well-understood ECRH schemes developed for the toroidal plasma works well in a large open trap.

The primary ECRH scheme at GDT relies on a radiation trapping by a non-uniform plasma column [12]. This effect is caused by the dependence of the plasma refraction on the magnetic field strength. The radiation is launched obliquely through a side of a plasma column at high magnetic field (close to the magnetic mirror). As the microwave beam propagates in plasma in the direction of the trap center, the magnitude of the magnetic field decreases resulting in conditions for an internal reflection from the plasma-vacuum boundary. The plasma column acts as a kind of waveguide, heterogeneous in both transverse and longitudinal directions, whereby the radiation is delivered to the fundamental ECR surface with **k**-spectrum and polarization favorable for absorption.

This entirely new ECRH scenario substantially depends on a magnetic configuration and plasma inhomogeneity. Although successful ECRH experiments evidence for the reliability of such heating mechanism, a new electron cyclotron emission (ECE) diagnostics has been installed to improve understanding of complex wave physics in GDT [15]. The idea is to exploit the reciprocity principle: conditions favorable for the ECRH should manifest themselves as an increased ECE level of thermal electrons along the same line of sight. Thus, the new ECE diagnostics operates near the heating frequency in the geometry reversed to the ECRH one. Measured thermal emission for different discharge scenarios has essentially validated the existing physical conceptions about the microwave heating in the machine [16]. Somewhat against our expectations, it has been found that ECRH may start from weakly absorbing plasma, but during the discharge the evolution of the electron temperature leads to an eventual transition to optically thick plasma; before we believed that ECRH always starts with optically thick plasma.

Numerical modeling of the propagation and absorption of electromagnetic waves plays an important role both in preparing and understanding of the last ECRH/ECE experiments. Until recently, such modeling was only possible in the framework of the geometric optics approximation also known as ray-tracing. Combined with the Fokker-Planck quasi-linear kinetic equation for the

ECRH-driven electron distribution function, the ray-tracing works perfectly well for the low-density plasma [17]. However, for the new high-density scenario, the geometric optics may not be valid in a vicinity of wave reflection surface due to formation of caustics, and near ECR surface due to a sharp inhomogeneous damping of waves. Straightforward simulation of these effects for large devices within a complete set of Maxwell's equations is not practical, in particular, because of smallness of the wavelength. A good alternative is the consistent quasi-optical approach based on the paraxial asymptotic expansion of Maxwell's equations in an inhomogeneous dissipative media with spatial dispersion [18], recently adopted for open traps [19, 20]. New approach lets to adjust the magnetic configuration to improve the rf-power deposition profile.

Another important issue is related to fast electrons and plasma micro-instabilities generated by strong ECRH. The fraction of fast electrons in GDT is moderate compared to other devices. Indeed, an efficient power deposition into the thermal plasma component is considered as one of the major achievements of this experiment. Nevertheless, suprathermal electrons play an essential role in ECR plasma start-up recently implemented at GDT: electrons with energies about 10 keV are entirely responsible for the gas ionization and plasma pressure at initial stages of gas breakdown and seed plasma build-up [21]. Measured non-thermal ECE have unambiguously confirmed the existence of suprathermal electrons generated during the ECR heating of the main plasma [16]. Explanation of time dependence of ECE level observed in the varying magnetic field and decaying plasma is a rather challenging task. Our current hypothesis is based on the concept of stimulated micro-instabilities that cause fast losses of suprathermal electrons. Such instabilities are observed at GDT as broadband electromagnetic pulses in 5–50 GHz band and synchronized precipitations of fast electrons. Despite some similarities to kinetic EC instabilities in small traps [22], detailed explanation of such events is still a matter of our efforts.

A high-power neutron source based on GDT concept will operate at higher relative pressures β and densities [23]. In this context, ECRH at the second harmonic is attractive; however, it is efficient only in combination with heating at the first harmonic providing the initial high electron temperature. Possibilities to test such regime in GDT are discussed [24].

Another class of next-step devices, advanced reversed-field-configurations [25, 26], is characterized by even more high values of $\beta \sim 1$. For electrodynamics, this implies $\omega_{ce} << \omega_{pe}$. A convenient way to heat such overdense and weakly magnetized plasma is based on a linear transformation of high-frequency electromagnetic waves into quasi-electrostatic waves near the plasma upper-hybrid resonance. Surprisingly, the correct theory of such coupling in high-β plasma was developed just recently [27]. A rigorous approach based on exact Maxwell's equations not only correct the traditional qualitative views on wave coupling but reveals some new options for the efficient transformation of electromagnetic waves into damped quasi-electrostatic oscillations in high-β plasmas.

The author thanks to all colleagues from IAP and BINP involved in this study. The work was supported by the Russian Science Foundation (pr. 14-12-01007).

References

1. *A. V. Vodopyanov, S. V. Golubev, V. G. Zorin et al.* Tech. Phys. Lett. 2000. V. 26, No. 12. P. 1075-1077.
2. *S. V. Golubev, S. V. Razin, V. E. Semenov et al.* Rev. Sci. Instrum. 2000. V. 71, No. 2. P. 669.
3. *L. Latrasse, M. Marie-Jeanne, T. Lamy, T. Thuillier et al.* Rev. Sci. Instrum. 2010. V. 81. P. 02A324.
4. *T. Saito, K. Ishii, A. Itakura, M. Ichimura et al.* J. Plasma Fusion Res. 2005. V. 81, No. 4. P. 288.
5. *T. Cho, V. P. Pastukhov, W. Horton, T. Numakura et al.* Physics of Plasmas. 2008. V. 15. P. 056120.
6. *T. C. Simonen, R. Horton.* Nuclear Fusion. 1989. V. 29. P. 1373.
7. *P. A. Bagryansky, Yu. V. Kovalenko, V. Ya. Savkin et al.* Nuclear Fusion. 2014. V. 54. P. 082001.
8. *P. A. Bagryansky, A. G. Shalashov, E. D. Gospod-chikov et al.* Phys. Rev. Letters. 2015. V. 114. P. 205001.
9. *P. A. Bagryansky, E. D. Gospodchikov, Yu. V. Kovalenko et al.* Fusion Science and Tech. 2015. V. 68. P. 87.
10. *P. A. Bagryansky, A. V. Anikeev, G. G. Denisov et al.* Nuclear Fusion. 2015. V. 55. P. 053009.
11. *T. C. Simonen*. J. Fus. Energy. 2016. V. 35. P. 63.
12. *A. Shalashov, E. Gospodchikov, O. Smolyakova et al.* Physics of Plasmas. 2012. V. 19. P. 052503.
13. *P. A. Bagryansky, A. V. Anikeev, M. A. Anikeev et al.* AIP Conference Proceedings. 2016. V. 1771. P. 020003.
14. *E. D. Gospodchikov, O. B. Smolyakova, E. V. Suvorov.* Plasma Physics Reports. 2007. V. 33. P. 427.
15. *A. L. Solomakhin, P. A. Bagryansky, E. D. Gospodchikov, L. V. Lubyako, A. G. Shalashov, D. V. Yakovlev.* AIP Conf. Proc. 2016. V. 1771. P. 050001.
16. *A. G. Shalashov, A. L. Solomakhin, E. D. Gospodchikov et al.* Electron cyclotron emission at the fundamental harmonic in GDT magnetic mirror. arXiv:1611.09137.
17. *B. W. Stallard, Y. Matsuda, W. M. Nevins.* Nuclear Fusion. 1983. V. 23. P. 213.
18. *A. A. Balakin, E. D. Gospodchikov, A. G. Shala-shov.* JETP Letters. 2016. V. 104, No. 10. P. 690–695.
19. *A. G. Shalashov, A. A. Balakin, E.D. Gospodchikov, T. A. Khusainov.* Phys. Plasmas. 2016. V. 23. P. 112504.
20. *A. G. Shalashov, A. A. Balakin, T. A. Khusainov et al.* J. Exp. Theor. Phys. 2017. V. 124, No 2. P. 325-340.
21. *D. V. Yakovlev, A. G. Shalashov, E. D. Gospod-chikov et al.* Nuclear Fusion. 2016. V. 57. P. 016033.
22. *A. G. Shalashov, M. E. Viktorov, D. A. Mansfeld, S.V. Golubev.* Phys. Plasmas. 2017. V. 24, No 3. P. 032111.
23. *A. A. Ivanov, V. V. Prikhodko.* Plasma Phys. Control. Fusion. 2013. V. 55. P. 063001.
24. *E. D. Gospodchikov, O. B. Smolyakova, A. L. Solomakhin et al.* Electron cyclotron heating and diagnostics of plasma at second harmonic in GDT device. *This issue.*
25. *M. W. Binderbauer, T. Tajima, L. C. Steinhauer et al.* Physics of Plasmas. 2015. V. 22. P. 056110.
26. *P. A. Bagryansky, T. D. Akhmetov, I. S. Chernoshtanov et al.* AIP Conf. Proc. 2016. V. 1771. P. 030015.
27. *E. D. Gospodchikov, A. G. Kutlin, A. G. Shalashov.* Plasma Phys. Control. Fus. 2017. V. 59, No 6. P. 065003.

EPJ Web of Conferences **149**, 03006 (2017) DOI: 10.1051/epjconf/201714903006
10th International Workshop 2017 "Strong Microwaves and Terahertz Waves: Sources and Applications"

Plasma heating by microwaves in high-β devices

E.D. Gospodchikov, A.G. Shalashov, A.G. Kutlin

Institute of Applied Physics RAS, Nizhniy Novgorod, Russia, egos@appl.sci-nnov.ru
Budker Institute of Nuclear Physics SB RAS, Novosibirsk, Russia

Microwave heating of electrons under the electron cyclotron resonance conditions is one of the most efficient ways to increase the electron temperature of magnetically confined plasmas. Recent progress in plasma confinement in axially symmetric magnetic traps led to the achievement of high-β regimes, in which β-ratio between the plasma kinetic pressure and the magnetic field pressure is of the order of unity. Under these circumstances the Langmuir plasma frequency is much greater than the electron cyclotron frequency, $\omega_{pe} >> \omega_{ce}$. Therefore, the use of common schemes of electron cyclotron heating based on direct launch of electromagnetic waves from a vacuum window meets obvious difficulties because the resonance region is screened by dense plasma for all electromagnetic modes except waves propagating strictly along the external magnetic field. However, in many cases strictly longitudinal propagation is not technically possible.

One way to overcome this problem is to use the linear transformation of electromagnetic waves in quasi-electrostatic plasma oscillations in the vicinity of the plasma upper hybrid (UH) resonance

$$\omega_{UH}^2 = \omega_{pe}^2 + \omega_{ce}^2 \approx \omega_{pe}^2 .$$

Once exited, the quasi-electrostatic oscillations can be effectively damped by electrons, in particular, in overdense plasmas. Such microwave heating schemes at low cyclotron harmonics have been implemented in magnetic traps with low kinetic pressure: stellarators, classical and spherical tokamaks. The heating frequency for this systems is of the order of the electron cyclotron frequency, $\omega \sim \omega_{ce} < \omega_{pe}$. On the other hand, the opposite ordering $\omega_{ce} << \omega_{pe} \sim \omega$ is more natural for high-β devices, otherwise the linear coupling and plasma heating are localized at the very periphery of the plasma column. As the quasi-electrostatic waves are well absorbed even at high cyclotron harmonics, the key factor controlling the efficiency of the heating is the excitation efficiency of quasi-electrostatic waves by electromagnetic waves launched from vacuum.

In magnetized plasma excitation of the quasi-electrostatic oscillations is possible as a straight-forward XB tunneling of the fast extraordinary wave (X) into the electron Bernstein wave (B) in the vicinity of the UH resonance, and as so called OXB process, a consequent linear transformations of the ordinary wave (O) into the slow extraordinary wave, and then to the then Bernstein wave.

The efficiency of the XB and OX coupling (the later in major extend defines the efficiency on the entire OXB process) is traditionally evaluated with Budden–Ginzburg approach as an integral of the purely imaginary wave vector over the evanescent coupling region in the vicinity of the UH resonance [1, 2]. This approach is based entirely on the dispersion relation of waves in locally homogeneous plasma and do not take into account the peculiarities of the vector nature of Maxwell's equations in weakly inhomogeneous gyrotropic media. The resulting formulas work fairly well in case of strong magnetic field $\omega_{ce} \sim \omega$, however do not allow correct transition to the isotropic plasma, in which the efficiency of the linear coupling of electromagnetic waves to Langmuir oscillations is also known [3]. In recent papers it was shown that the existing theory does not work in the domain of finite but low magnetic fields, $\omega_{ce} << \omega$, that is of interest for high-β plasmas [4, 5].

In this work we discuss results of calculation of XB and OXB conversion efficiency for smoothly inhomogeneous weakly anisotropic magnetized plasma.

We use the following simplifications. First, we consider one-dimensional geometry in which the external magnetic field is uniform and directed along the z axis, and the plasma density changes by parabolic law along the x axis, $n_e = n_0(1 - x^2 / L^2)$. Second, we consider cold but weakly collisional plasma, thus neglecting effects of spatial dispersion due to thermal motion but introducing the effective collision rate $\nu_{eff} << \omega$. Here we rely on well-known conclusion of the equality of the weak electromagnetic energy losses due the collisional and resonant collisionless wave absorption in the vicinity of the UH resonance. In this frame, the efficiency of coupling to the quasi-electrostatic mode is calculated simply as the absorption coefficient of the electromagnetic wave. Third, we consider a plane electromagnetic wave $\propto \exp(i\omega t - i\mathbf{kr})$ launched from the vacuum at certain direction.

In case of smoothly inhomogeneous plasma $k_0 L >> 1$ transformation efficiency depends only on three combinations of physical parameters [4,5]:

$$\tilde{g} = \xi^{2/3}\omega_{ce} / \omega, \; \tilde{N}_{y,z} = \xi^{1/3} k_{y,z} / k_0,$$

where $\xi = k_0 L n_c / 2\sqrt{n_0(n_0 - n_c)}$ is the large parameter of geometrical optic, n_c is the critical plasma density. These parameters characterize magnetic field strength and direction of a wave launch correspondently.

As a reference point, the case of non-magnetized plasma $\tilde{g} = 0$ is shown in figure 1 (left panel). Here we reproduce well the analytical formula [3] for the absorption of the oblique TM wave due to coupling to the Langmuir oscillations. Non-magnetized plasma is characterized by the axially symmetric extremum $C_{TM} \approx 0.5$ reaching at $\tilde{N}_{\perp} \approx 0.66$. Absorption for the orthogonal TE wave here is negligibly small.

One can see that the dependence of the coupling efficiency over the incident direction $\tilde{N}_y$ becomes asymmetric for the TM waves. This asymmetry is already visible at extremely low magnetic fields $\tilde{g} = 0.01$ see figure 1(right panel), where the coupling efficiency increases at $\tilde{N}_y > 0$ and decreases at $\tilde{N}_y < 0$ compared with the isotropic case.

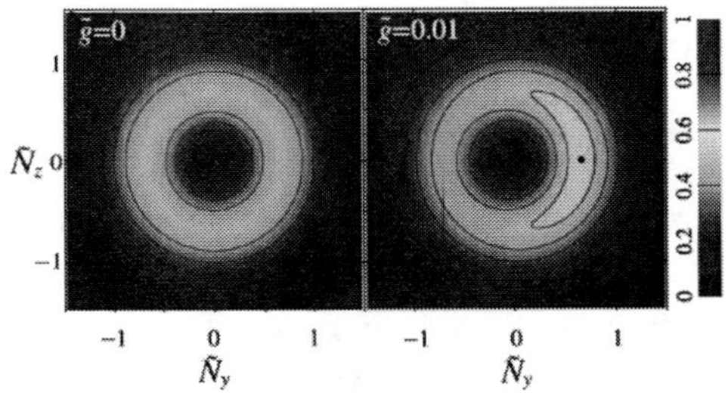

Fig. 1. Conversion efficiency for TM wave for isotropic (left panel), and weakly anisotropic (right panel) plasma

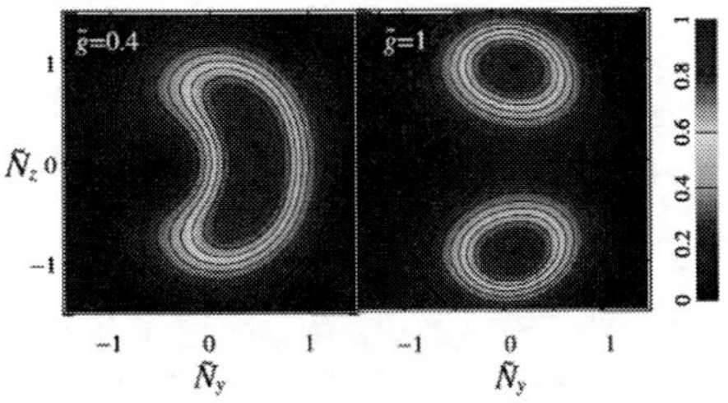

Fig. 2. Absorption coefficient for optimal polarization $C_{max} = C_{TE} + C_{TM}$ for $\tilde{g} = 0.4$ (left) and $\tilde{g} = 1$ (right)

The asymmetry becomes more pronounced with increase of the magnetic field. Efficient absorption is localized near the propagation direction perpendicular to the magnetic field, $\tilde{N}_z = 0$, but with a particular angle with respect to the density gradient, $\tilde{N}_y \approx 0.5$. At $\tilde{g} > 0.4$ the maximum saturates close to the maximum possible value $C_{TM} \approx C_{max} \approx 1$ (see figure 2). Moreover, the maximum splits into two maxima symmetric in $\tilde{N}_z$; with that bifurcation efficient absorption becomes possible in a wide range of angles to the magnetic field. At the same time, the whole region $\tilde{N}_y < 0$ corresponds to practically zero absorption. As confirmed by numerical modeling, in high magnetic fields, $\tilde{g} > 1$, the optimal polarization corresponds to the wave polarized circularly in the direction of rotation of the ion. This is exactly the O mode of a weakly magnetized plasma (see figure 3).

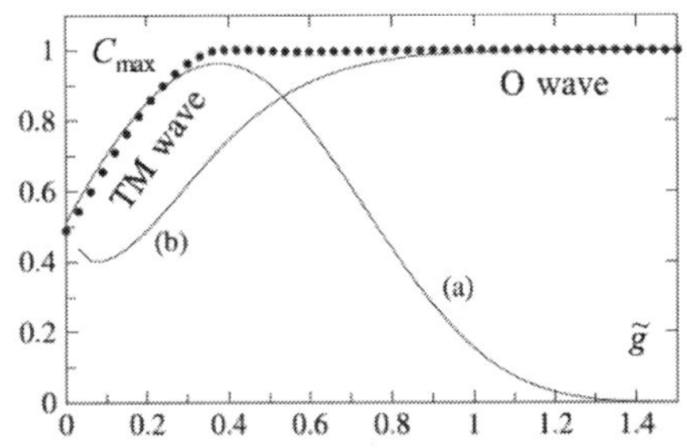

Fig. 3. Dependence of calculated absorption coefficient for optimal polarization on magnetic field strength (dotted line) versus analytical formulas for TM wave (a), and O wave (b) [5]

A straightforward approach based on the full Maxwell equations for interacting waves not only corrects the traditional theoretical views on wave coupling but shows new options for the efficient transformation of electromagnetic waves into quasi-electrostatic oscillations in weakly magnetized high-β plasmas. In our opinion, the most important, practical conclusion is demonstration of the possibility of almost total absorption for the optimally polarized plane wave by an inhomogeneous plasma slab at all magnetic fields satisfying $\omega_{ce}/\omega > 0.3\xi^{-2/3}$ At low fields, this occurs for the linearly polarized TM wave with $N_y \approx 0.5\xi^{-1/3}$ and $N_z << N_y$. With increasing magnetic field, the optimal wave becomes close to the plasma O mode propagating at $N_y << N_z$ and $N_z \approx \sqrt{\omega_{ce}/\omega}$. Another new feature outlined in this paper is pronounced asymmetry of the absorption process with respect to the plane defined by the gradient of the plasma density and the external magnetic field.

This work was supported by RFBR (project 15-02-07600).

References

1. *Budden K.G.* Radio Waves in the Ionosphere. 1966. Cambrige: Cambridge University Press
2. *Ginzburg V.L.* The Propagation of Electromagnetic Waves in Plasmas 2nd edn 1970. Oxford: Pergamon
3. *Denisov N.G.* On a Singularity of the Field of an Electromagnetic Wave Propagated in an Inhomogeneous Plasma // Sov.—JETP. 1957. V. 4. P. 544.
4. *Gospodchikov E.D., Kutlin A.G., Shalashov A.G.* Coupling electromagnetic and quasi-electrostatic waves in electron cyclotron frequency range in high-β devices // AIP Conf. Proc. 2016. V. 1771. P. 030016.
5. *Gospodchikov E.D., Kutlin A.G., Shalashov A.G.* Plasma heating and coupling of electromagnetic waves near the upper-hybrid resonance in high-β devices // Plasma Phys. Control. Fusion. 2017. V. 59. P. 065003.

3D Full-Wave modelling and mode conversion in realistic W7-X plasmas

Pavel Aleynikov, Nikolai B. Marushchenko

Max-Planck-Institut für Plasmaphysik, Greifswald, Germany, pavel.aleynikov@ipp.mpg.de

Electron Cyclotron Resonance Heating (ECRH) is the main plasma heating mechanism in Wendelstein 7-X (W7-X) Stellarator. It is provided by 10 gyrotrons at 140 GHz (corresponding to second-harmonic resonance at 2.5 T) with the power of 1 MW each.

Microwaves, launched from a vacuum, are reflected from their corresponding cut-off layers. In W7-X, the X2 and O2-modes are used in a wide range of operation scenarios. Whereas X2 has a cut-off at $1.2 \times 10^{20} m^{-3}$, the O2-mode can be used in plasmas with densities up to $2 \times 10^{20} m^{-3}$. The propagation of these modes is well tested experimentally and studied theoretically within geometrical optics (WKB).

Possible operation at higher densities would involve mode-conversion (from O- to X- to Bernstein-mode, i.e. "OXB"). The conversion of modes is outside of the applicability of WKB-theory and usually considered with the "full-wave" approach. Full-wave models are also used to study processes which involve scales comparable to the wavelength (~1mm), for example, effects of the density fluctuations.

This work reports on the development of a new 3D cold plasma full-wave code. The code utilizes the Finite Difference Time Domain (FDTD) technique [1] and has an interface with the ray-tracing code TRAVIS [2]. The computation domain is "minimized" around the WKB-trajectory obtained from the ray-tracing code; the background magnetic field is recovered from the pre-computed 3D equilibrium data. The code takes advantage of massive parallel computations with Graphics Processing Units (GPUs), which allows for up to ×100 faster calculations than on a single-CPU.

The minimization of the computation domain plays a critical role in efficient 3D computations. A so-called CPML boundary layer [3] is implemented in our full-wave code to truncate the computation domain. Note that is it common to discretize the magnetized plasma response current field (**J**) in such a way that its J_x, J_y and J_z components are co-located in space [1, 4, 5]. This facilitates the current update calculation. However, we find the stability of such a scheme to be unsatisfactory, in particular when applied to the CMPL region. Therefore we have implemented a slightly more computationally demanding but more stable scheme where **J** is discretized in the same way as the wave electric field (**E**). With this scheme, the performance of the CPML layer is found to be satisfactory.

One of the applications of our code is the calculation of the O- to-X-mode conversion efficiency in 3D plasmas.

Figure 1 shows an example of a full-wave 2D calculation of the O-X conversion process in a realistic W7-X plasma. The O-mode, launched from vacuum (bottom right corner of the left plot), propagates into the plasma and partly reflects off the corresponding cut-off layer, where a part of the O-mode is converted into a slow X-mode, which propagates toward the upper-hybrid resonance. Eventually, this X-mode will be converted to Bernstein wave. The reflected part of the O-mode has a "hole" (due to the non-uniform O-X conversion). The perpendicular component of the refractive index is computed with the use of a windowed Fourier transform and shown in Fig. 1 (right). The incoming O-mode, the reflected O-mode, and the slow-X mode (which settles near the Upper Hybrid resonance layer) can be clearly distinguished in this diagram, which appears to be in an agreement with expectations from WKB theory (i.e. dispersion curves). Note that the slow-X mode is eventually dissipated numerically without leaving the computation domain. This dissipation is of purely numerical nature. It happens when the wavelength of the mode becomes shorter than ~10 computational cells.

The W7-X plasma is strongly shaped and requires a 3D evaluation of the conversion process. In Fig. 2 an example of such a 3D calculation of the above scenario is shown (the wave is launched from the bottom of the domain). The decimal logarithm of the time-averaged E^2 is color-coded. The sequence of dark-colored points represents the corresponding WKB trajectory. It is evident from the cross-section of the reflected wave that the 3D geometry needs to be appropriately accounted for.

Thanks to its speed, the GPU-capable 3D FDTD code allows for efficient large parametric scans over a broad range of parameters.

To estimate the power conversion efficiency, we compute surface integrals of the time-averaged Poynting vector (i.e. the intensity) over every boundary of the computation domain:

$$S_i = (1/\mu_0) \iint (\langle [\mathbf{E}\times\mathbf{B}] \rangle \bullet \mathbf{n}) ds, \qquad (1)$$

where the index "i" denotes a particular boundary, **n** represents its normal vector and <...> denotes time averaging. In the above case, the OX conversion efficiency can be calculated as the difference between the intensities of the input (S_{in}) and reflected (S_{ref}) waves.

An estimate of the OX conversion efficiency is made as follows for W7-X-relevant ECRH beam parameters and plasma geometry. A Gaussian beam approximation is used as an initial condition. A beam of 1.5 cm in radius and a phase front curvature of 1 m (i.e. beam waist of radius 1.4 cm located near the O-mode cut-off) approximates the ECRH beam at the plasma entrance. The density profile is chosen arbitrarily (with normalized gradient of ~20 m^{-1} near the cutoff). The antenna launcher location was fixed, corresponding to the upper port "A, C1" in W7-X nomenclature. Then, the launcher azimuth and altitude angles were varied to find the highest OX conversion conditions. Contours of the calculated OX conversion fraction $(S_{in} + S_{ref}) / S_{in}$ are shown in Fig. 3.We find that for the given conditions (equilibrium, density profile, launcher position and beam focusing) the maximum conversion reaches 48%.

The analogous 2D calculations result in different contours of the conversion fraction (see Fig. 4). This is due to the fact that the truncation from 3D to 2D is

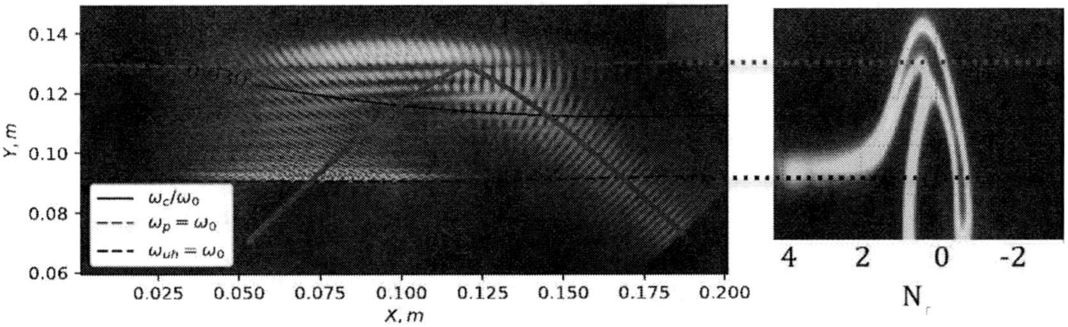

Fig. 1. An example of the calculation of the O-X mode conversion in realistic W7-X plasma. Left plot: the snapshot of the electric field amplitude (color-coded), ray tracing (solid red), antenna is in the right bottom corner. O-mode cutoff and Upper Hybrid resonance layer are plotted with the red and black dashed curves, respectively. Right plot: windowed Fourier transform of the wave field (wave refractive index), window is moved along the Y axis (the plots have common Y axis)

somewhat arbitrary in the general case: the computational domain is constructed in such a way that the distance from the ray-tracing trajectory to the computation plane is minimized. While this approximation still allows for good agreement between the ray-tracing trajectory and the full-wave solution for the reflected O-mode, the conversion fraction changes significantly (the maximum reaches 60%).

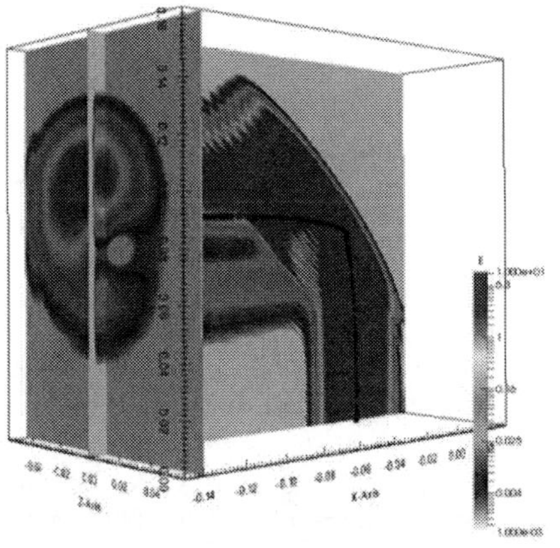

Fig. 2. 3D calculation of the scenario described in Fig. 1

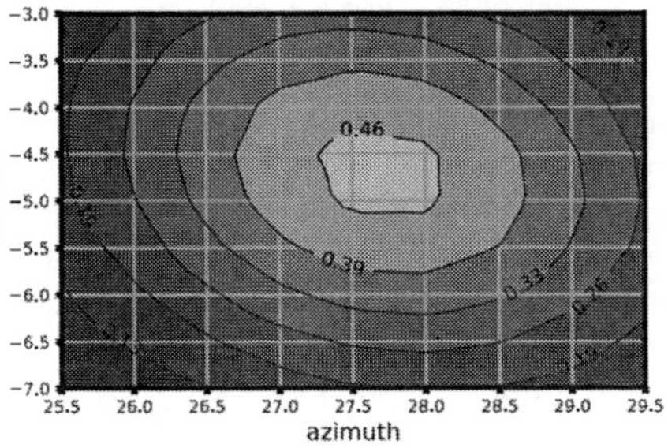

Fig. 3. Contours of the OX conversion efficiency for a given plasma parameters over varying beam aiming angles (for a fixed location of the launcher)

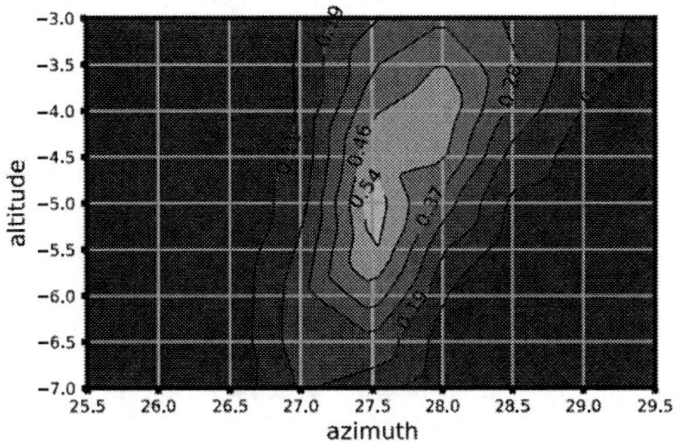

Fig. 4. Same as Fig.3 for 2D calculations

These results do not yet form a comprehensive picture of the optimal conditions for OXB heating in W7-X, but such a study is planned in future. Other future work includes: comparison of the results with analytical 3D OX conversion models [6, 7]; study of the X-O conversion for the needs of EBE diagnostics; and a study of the effects of plasma fluctuations on beam propagation.

In summary we conclude that the newly developed GPU-capable 3D FDTD code allows for a very efficient assessment of wave physics in realistic 3D plasmas.

References

1. *Inan, U., Marshall, R.* // Numerical electromagnetics: the FDTD method, Cambridge, 2011.

2. *Marushchenko N., Turkin, Y., Maassberg, H.* // Computer Physics Communications. 2014. V. 185. P. 165.

3. *Roden J, Gedney S.D.* // MICROWAVE AND OPTICAL TECHNOLOGY LETTERS. 2000. V. 27. P. 5.

4. *Kohn, A., et al.* Plasma Phys. Control. Fusion. 2008. V. 50. P. 085018.

5. *Williams, T., Kohn, A., O'Brien M., Vann R.* // Plasma Phys. Control. Fusion, 2014, V. 56 P. 075010.

6. *Gospodchikov E., Khusainov T., Shalashov A.* // Plasma Phys. Control. Fusion, 2012, V. 54 P. 045009.

7. *Popov A.* // Plasma Phys. Control. Fusion, 2011, V. 53 P. 065016.

Quasi-optical approach for inhomogeneous dissipative media with high-order spatial dispersion

A.A. Balakin[1], E.D. Gospodchikov[1,2], A.G. Shalashov[1,2]

[1] Institute of Applied Physics RAS, Nizhny Novgorod, Russia
[2] Budker Institute of Nuclear Physics SB RAS, Novosibirsk, Russia

Despite the development of codes for the electromagnetic wave propagation based on a numerical solution of the complete system of Maxwell's equations, asymptotic methods describing wave fields in smoothly inhomogeneous media with complex dielectric response still remain the main instrument of interpretation and optimization of experiments on microwave heating and current drive in magnetic confinement systems, in problems of signal generation in the Earth ionosphere and many other fields of physics and astrophysics. The main and most studied asymptotic approach is based on calculating the geometric-optics ray and absorption along them [1]. However, geometric optics can only qualitatively describe wave propagation and diffraction of the beam in more complex media with strong spatial dispersion and absorption, like, the vicinity of the electron cyclotron resonance in warm plasmas.

The quasi-optical approach is a more accurate asymptotic method, which reduce Maxwell's equations to the evolution equation for the smooth envelope of the wave beam. Such a description allows one to consistently take into account the media anisotropy, spatial inhomogeneity, dispersion, and resonance dissipation within the framework of a unified approach [2]. Its generalization to the case of anisotropic and gyrotropic media with spatial dispersion and dissipation [3] was used to simulate the propagation of quasi-optical wave beams in tokamaks and showed much higher accuracy for resonance absorption in comparison to the previously used approaches [4]. In this paper, we discuss the further generalization of the quasi-optical approach [5] and the features of its application in media of a special kind.

A quasi-optical equation [3] is the equation for the scalar complex amplitude U of the wave beam corresponding to selected electromagnetic mode in smoothly inhomogeneous media

$$H(\mathbf{r},\hat{\mathbf{p}})[U(r)] = 0. \qquad (1)$$

Here the momentum operator $\hat{\mathbf{p}} = \nabla / ik_0$ is defined as the differentiation operator in the coordinate space normalized on the vacuum wave number $k_0 = \omega/c$. The simplification of this equation is possible for paraxial wave beams (broad in the wavelength scale) with parameters varying slowly along some curve $\mathbf{r}_0$ (the reference ray). One can associate a curvilinear coordinate system $\boldsymbol{r} = (\xi,\tau)$ and canonically conjugate momenta $\hat{\mathbf{p}} = (\hat{\mathbf{q}}, \hat{p}_\tau)$ with the reference ray. Here τ is the evolutionary coordinate along the reference ray, ξ are two curvilinear coordinates orthogonal to the reference ray, $\hat{\mathbf{q}} = \nabla_\perp / ik_0$ and $\hat{p}_\tau = \partial_\tau / ik_0$. Assuming that the beam parameters change smoothly along the reference ray, we can represent the scalar amplitude U in the following form

$$U(r) = \psi(\tau,\xi) e^{ik_0 \int p_0(\tau) d\tau}, \quad |\hat{\mathbf{q}}[\psi]| << k_0 p_0 |\psi|.$$

The function $p_0(\tau)$ gives the dependence of the "carrier phase" of the wave field along the reference ray. Substituting the field in this form in Eq.(1) and expanding to the first order in powers of $\partial_\tau u$, we obtain the equation for the complex envelope [3]:

$$\hat{H}[\psi] \approx \hat{H}_0[\psi] + \frac{1}{ik_0}\left(\frac{\partial \hat{H}_0}{\partial p_0}\left[\frac{\partial \psi}{\partial \tau}\right] + \frac{1}{2}\frac{\partial^2 \hat{H}_0}{\partial p_0 \partial \tau}[\psi]\right) = 0,$$

where $\hat{H}_0 = \hat{H}(\mathbf{r}, \hat{\mathbf{q}} + p_0(\tau)\mathbf{r}_0)$ is the truncated operator with omitted derivatives with respect to τ. The last term gives the pre-exponential factor responsible for the change in intensity due to group deceleration of waves in the geometric optics limit. This term can be eliminated by replacing

$$\psi = \hat{A}[u], \quad \frac{\partial \hat{H}_0}{\partial p_0}\frac{\partial \hat{A}}{\partial \tau} + \frac{1}{2}\frac{\partial^2 \hat{H}_0}{\partial p_0 \partial \tau}\hat{A} = 0.$$

As a result, we obtain simpler equation

$$\frac{\partial u}{\partial \tau} = ik_0 \hat{H}[u], \quad \hat{H} = -\left(\frac{\partial \hat{H}_0}{\partial p_0}\hat{A}\right)^{-1} \hat{H}_0 \hat{A}. \qquad (2)$$

This equation describes the evolution of the scalar amplitude of the wave beam u along the τ-axis, taking into account the effects of diffraction, aberrations, spatial dispersion, and dissipation in the paraxial approximation. The physical meaning of the new Hamiltonian $\hat{H}$ is easy to establish if we compare the left and right sides of equation (2). This is the operator of the longitudinal wave number $\hat{p}_\tau - p_0$, expressed in terms of the transverse derivatives.

Within the approximations made, the quantity

$$\Pi_\tau = \int |u|^2 d\xi \qquad (3)$$

coincides, with an accuracy up to a dimensional coefficient, with the total energy flux in the wave beam along the reference ray. It is convenient to use the decomposition of the operator $\hat{H}$ into a Hermitian and anti-Hermitian parts

$$\hat{H} = \hat{H}_H + \hat{H}_A, \quad \begin{array}{l} \hat{H}_H = (\hat{H} + \hat{H}^+)/2 \\ \hat{H}_H = (\hat{H} - \hat{H}^+)/2i \end{array} \qquad (4)$$

Operators $\hat{H}_H, \hat{H}_A$ are Hermitian operators in the sense of the standard scalar product $\int ab^* d\xi$. As a result, the power deposition can be represented as

$$\partial_\tau \Pi_\tau = -2k_0 \int u^* \hat{H}_A[u] d\xi$$

The main difficulty in the practical use of equation (2) is the uncertainty of the exact form of the operator $\hat{H}$ for particular physical system. However, any operator can be written in the following form [6]

$$\hat{H} = \sum_{m,n} \frac{1}{n!\,m!}\left(\frac{\partial^{m+n}H}{\partial \xi^m \partial \mathbf{q}^n}\right)\frac{\xi^m \hat{\mathbf{q}}^n + \hat{\mathbf{q}}^n \xi^m}{2}. \quad (5)$$

Here $H(\tau,\xi,\mathbf{q})$ is some matrix kernel generating the linear operator. For such operator form, the real H always generates the Hermitian operator $\hat{H}$, and the imaginary operator is the anti-Hermitian one. The operator kernel $H = \Sigma H^{(n)}$ represent the sum of corrections to the Hamiltonian in powers of the characteristic inhomogeneity scale l: $H^{(n)} \sim (k_0 l)^{-n}$. In smoothly inhomogeneous media $k_0 l << 1$, therefore, as a rule, we can confine ourselves to the zero term, that is, the usual Hamiltonian expression obtained in the "locally homogeneous" approximation.

Note, the symmetry of the problem can in a number of cases impose a relation between various corrections in the representation of the operator. For example, a Hamiltonian of the form $H^{(0)} = x^2p^2$ generates the operator $\hat{H}^{(0)} = -\left(x^2\partial_{xx} + 2x\partial_x + 1\right)/k_0^2$. However, if it is known that this operator must be proportional only to the field derivatives, then the relation between the zero-order $H^{(0)}$ and the second-order $H^{(2)}$ corrections appears:

$$\left\{\begin{matrix} H^{(0)} = x^2p^2, \\ H^{(2)} = -1/k_0^2 \end{matrix}\right\} \Rightarrow \hat{H} = -\frac{1}{k_0^2}\left(x^2\partial_{xx} + 2x\partial_x\right)$$

Lets rewrite representation (5) in terms of the Fourier integral:

$$\hat{H}[u] = \iint \frac{H(\xi,\mathbf{q}) + H(\zeta,\mathbf{q})}{2} e^{ik_0\mathbf{q}(\xi-\zeta)} u(\zeta)\frac{d\zeta d\mathbf{q}}{4\pi^2k_0^2}.$$

This allows us to write an approximate solution of equation (2) in the form of an operator exponent [4,5]:

$$u(\tau+\Delta,\xi) \approx \hat{S}[u], \quad (6)$$

$$\hat{S}[u] \equiv \iint e^{ik_0\Delta\frac{H(\xi,\mathbf{q}) + H(\zeta,\mathbf{q})}{2} + ik_0\mathbf{q}(\xi-\zeta)} u(\tau,\zeta)\frac{d\zeta d\mathbf{q}}{4\pi^2k_0^2}$$

Unfortunately, the operator exponent method is justified for media without dissipation. In non-homogeneous dissipative media with spatial dispersion, the formula (6) can give a solution with exponentially growing energy, even for the case of dissipation within the framework of the geometric optic approximation (ie, for Im H>0). The formal reason is that the "dissipative" term $\hat{H}_A$ may have negative eigenvalues, since the representation (5) not guarantees the sign-definiteness of eigenvalues.

In fact, this is a known problem of correct media approximation for a physically correct description of the energy absorption of a wave beam [4]. Errors in the approximation of the Hermitian part will only perturb the wave beam phase profile, and their effect will become noticeable on the paths of many diffraction lengths. The systematic errors in the approximation of the anti-Hermitian part directly affect the amplitude of the beam. As a result, an approximate solution can, for example, exponentially increase the beam energy. This problem is especially important for microwave beams of the electron-cyclotron frequency range, when dissipation increases sharply both in the coordinate space and in the space of wave vectors.

The simplest correct approximation guaranteeing the sign-definite nature of the eigenvalues of the anti-Hermitian part of the operator is to use a model operator that can be represented as the square of some other Hermitian operator. So, we define $\hat{H}_A = \hat{G}\hat{G}$, where the new operator is constructed similarly to (6), but with kernel $G(\xi,q) \equiv \sqrt{\operatorname{Im} H(\xi,\mathbf{q})}$.

Thus, using the modified representation for the dissipative part, the quasi-optical equation (2) become

$$\frac{\partial u}{\partial \tau} = ik_0\hat{H}_H[u] - k_0\hat{G}\hat{G}[u] \quad (7)$$

To construct a numerical algorithm for solving this equation, we use the "split-step" method, which consistently takes into account the Hermitian $ik_0\hat{H}_H[u]$ and anti-Hermitian $k_0\hat{G}\hat{G}[u]$ terms

$$u(\tau+\Delta,\xi) \approx \hat{D}\hat{S}[u] \quad (8)$$

Here $\hat{S}$ is defined by the operator exponent (6) with the kernel from the real part of H. The operator $\hat{D}$ should correctly describe the weakly damped modes and provide exponential damping at large eigenvalues. One of the simplest operators satisfying the specified conditions is $\hat{D} = 1 - \hat{T}\hat{T}$ with operator $\hat{T}$ being constructed similarly to (6) for the kernel

$$T(\xi,\mathbf{q}) = \tanh\left(\sqrt{k_0\Delta}G(\xi,\mathbf{q})\right)$$

This gives us the explicit scheme (8) for the quasi-optical equation (7), which ensures a stable solution.

The technique developed in this work was used to create quasi-optical code QOOT for numerical simuation of propagation and absorption of quasi-optical microwave beams in a direct magnetic trap. Recently, the code was applied to simulate resonant electron-cyclotron plasma heating in a large-scale trap GDT in the Budker Institute [7]. It is shown that under conditions of a large installation with dense hot plasma, the quasi-optical effects of inhomogeneous dissipation can be taken into account in future optimized scenarios of ECR heating.

The work was supported by the Russian Science Foundation (pr. 14-12-01007).

References

1. Yu. A. Kravtsov and Yu. I. Orlov, *Geometrical Optics in Inhomogeneous Media*, Springer Ser. on Wave Phenomena (Springer, Berlin, Heidelberg, 1990).
2. A.V. Timofeev, Physics-Uspekhi **48**, 609 (2005); G. V. Permitin and A. I. Smirnov, JETP **82**, 395 (1996); A.A. Balakin, *et.al.*, JoPD, **40**, 4285 (2007).
3. A.A. Balakin, *et.al.*, Plas.Phys.Rep, **33**, 302 (2007); A.A. Balakin, Radiophys.Quant.Electron. **55**, 472 (2012); A.A. Balakin, Radiophys.Quant.Electron. **55**, 502 (2012).
4. A.A. Balakin, *et.al.*, Nucl.Fusion, **48** 065003 (2008); N. Bertelli, *et.al.*, Nucl.Fusion **50**, 115008 (2010).
5. A.G. Shalashov, *et.al.*, Phys.Plasmas **23**, 112504 (2016); A.G. Shalashov, *et.al.*, JETP **124**, 325 (2017).
6. A.A. Balakin, E.D. Gospodchikov, JoPB, **48**, 215701 (2015).
7. P.A. Bagryansky, *et.al.*, PRL, **114**, 205001 (2015); P.A.Bagryansky, *et.al.*, Nucl.Fusion, **55**, 053009 (2015).

EPJ Web of Conferences **149**, 03009 (2017) DOI: 10.1051/epjconf/201714903009
10th International Workshop 2017 "Strong Microwaves and Terahertz Waves: Sources and Applications"

ECRH effect on the electric potential in toroidal plasmas
(Overview of recent T-10 tokamak and TJ-II stellarator results)

A.V. Melnikov[1,2], C. Hidalgo[3], L.I. Krupnik[4], E. Ascasibar[3], A. Cappa[3], A.A. Chmyga[4], G.N. Deshko[4], M.A. Drabinskij[1,5], L.G. Eliseev[1], P.O. Khabanov[1,5], S.M. Khrebtov[4], N.K. Kharchev[1], A.D. Komarov[4], A.S. Kozachek[4], S.E. Lysenko[1], J.L. dePablos[3], V.N. Zenin[1], A.I. Zhezhera[4]

[1]National Research Centre 'Kurchatov Institute', Moscow, Russia, Melnikov_AV@nrcki.ru
[2]National Research Nuclear University MEPhI, Moscow, Russia
[3]Fusion National Laboratory, CIEMAT, Madrid, Spain
[4]Institute of Plasma Physics, KIPT, Kharkov, Ukraine
[5] Moscow Institute of Physics and Technology, Dolgoprudny, Russia

Electric field E or electric potential φ plays a key role in the transport and turbulence processes in toroidal plasmas by both mean and oscillatory components. It is believed that mean radial E_r suppresses the turbulence eddies via $E\times B$ shearing mechanism, while oscillatory E_r (zonal flows and Geodesic Acoustic Modes, GAM) provides the mechanism of turbulence self-regulation [1].

ECRH/ECCD is a powerful tool to affect and to study hot plasmas confined in closed magnetic traps, aiming for both basic plasma physics and thermonuclear fusion research [2].

Various aspects of the ECRH (using from one to three gyrotrons, operated in second-harmonic X-mode with on- and off-axis power deposition) effect on the static and oscillatory component of plasma potential were studied in two machines of similar size by Heavy Ion Beam Probing (HIBP), which is a unique diagnostics for core plasma potential, operating now on the T-10 tokamak (R=1.5m, a=0.3m, B=2.4T, P_{ECRH} = 0.4-2.4MW, f_{ECRH}=129, 144GHz) and TJ-II stellarator ($<R>$=1.5m, $<a>$=0.3m, B=1T, P_{ECRH} = 0.3-0.6MW, f_{ECRH}= 53.2GHz) [3]. Fine-focused (<1cm) and intense (150μA) Cs^+ or Tl^+ beams with energy up to 300 keV, equipped with advanced control systems provide measurements in wide density interval n_e=(0.3–5)×10^{19}m^{-3} in wide range of magnetic configurations in Ohmic and ECRH plasmas on T-10 and ECRH and NBI heated plasmas on TJ-II [4].

For any sign (positive or negative) and value of the mean electric potential of target plasmas, auxiliary ECRH pushes the potential evolution towards positive direction. So, the extra potential Δφ is always positive; it increases from 100V up to 400V with P_{ECRH}. The increase of P_{ECRH} raises the plasma electron temperature T_e, that causes the raise of the electric potential, see figs.1 and 2.

Contrary, Δφ decreases with plasma density raise. Plasma poloidal rotation, independently measured by HIBP with poloidal cross-phase of density perturbations, evolves accordingly [5].

Plasma potential profile evolution in the regimes with T_e and n_e changes is consistent with Neoclassical expectations [6].

Broadband electrostatic oscillations (up to 200 kHz) are strongly excited by ECRH in low-density plasmas of TJ-II, while for high-density plasmas, obtained in T-10 the effect found is opposite [6]. It was found in T-10 that the GAM frequency f_{GAM}

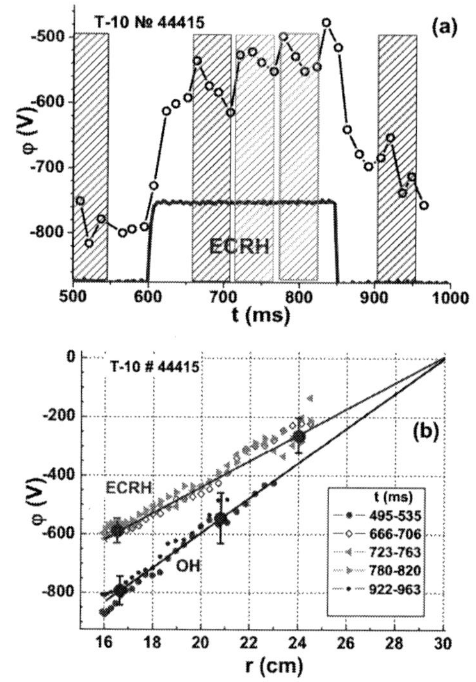

Fig. 1. Plasma potential time evolution at ρ=0.55 during an ECRH pulse (a) in ohmic (OH) discharge of the T-10 tokamak (r=16cm corresponds to ρ=0.55); Δφ (ρ=0.55)=200V. Hatched color ribbons denote the times for the profile measurements, shown in (b). The estimated error bars are in red for ECRH and in blue for OH.

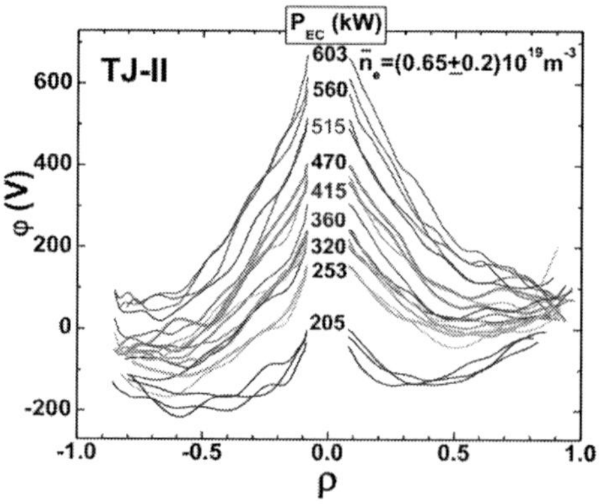

Fig. 2. Plasma potential profile dependence on P_{ECRH} in nearly on-axis ECR heating experiments in the TJ-II stellarator. Profiles are measured from Low Field Side (LFS) to High Field Side (HFS) several times during one shot.

increases with ECRH in accordance to the theoretically expected dependence on the electron temperature for the turbulence driven $f_{GAM} \sim T_e^{1/2}$ [7]. Figure 3 shows the overall T_e dependence of f_{GAM} on T_e, that was obtained in a wide range of tokamak regimes with ohmic and auxiliary ECR heating. The results show a consistency of the experimental data with the theoretical prediction for GAM, excited at r=21cm (ρ=0.7), over the whole operational limit of the machine, in which T_e varied by a factor of 4.

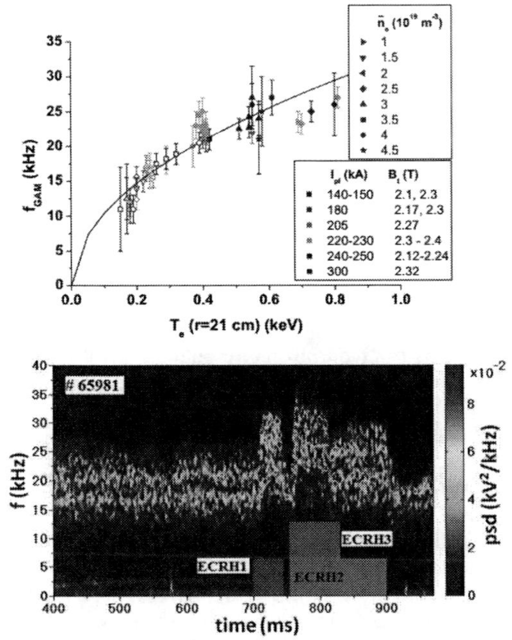

Fig. 3. ECRH effect on the GAM, as studied in the T-10 tokamak by HIBP. Upper figure: f_{GAM} dependence on T_e on OH (open symbols) and ECRH (closed symbols) discharges. Lower figure: power spectrogram of the plasma potential shows f_{GAM} time evolution in the shot with auxiliary ECRH pulses. Port-through power $P_{ECRH1} \sim$ 0.65MW, $P_{ECRH2} \sim$ 0.85MW, $P_{ECRH3} \sim$ 0.55MW. GAM is accompanied by a higher frequency satellite.

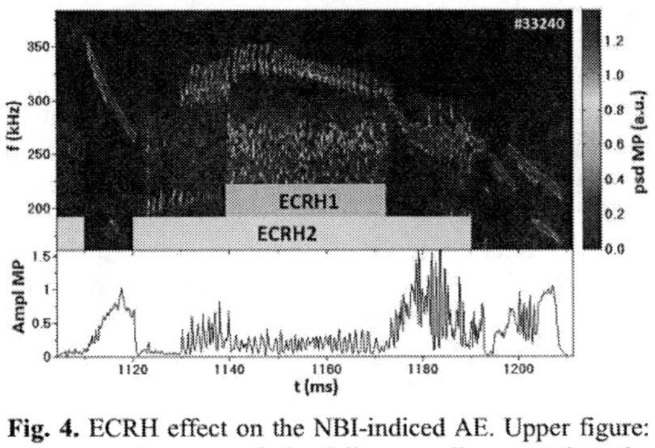

Fig. 4. ECRH effect on the NBI-indiced AE. Upper figure: power spectrogram of the Mirnov coil, presenting the transition of AE from steady frequency form in NBI phase (no ECRH) to chirping form due to ECRH/CD (ECRH2-on). An application of the second gyrotron (ECRH1) causes the suppression of the mode amplitude, as shown in lower figure.

The NBI-excited Alfven Eigenmodes (AEs) are affected strongly by ECRH: the steady frequency AEs transform to the chirping modes with frequencies up to 400 kHz [8]. A further increase of the P_{ECRH} from 0.3MW (one gyrotron on) to 0.6MW (two gyrotrons on) causes a mitigation of the AE intensity, as presented in figure 4.

In the low-density plasmas of TJ-II strong ECRH produces supra-thermal (ST) electrons, exciting electrostatic ST-modes [9].

Dual HIBP in TJ-II, consisting of two identical HIBPs located ¼ torus apart, provides measurements of stable long-range potential correlations (LRC), resembling spatially localized low-frequency zonal flows (<30 kHz) in the core of ECRH plasmas [10]. Unless LRC are also observed in the NBI-heated plasma in some conditions, their intensity in the ECRH plasma is much stronger.

Finally, various aspects of the ECRH effect on the plasma mean electric potential, broadband electrostatic turbulence, including turbulent particle flux and plasma turbulence rotation, and quasicoherent modes, including GAMs, AEs, ST-modes, will be summarized.

Acknowledgement

The work of Kurchatov team was funded by Russian Scientific Foundation, project 14-22-00193. T-10 experiments were supported by the State Corporation Rosatom contract No H.4x.241.9B.17.

References

1. *Diamond, P.H. et al.* Zonal flows in plasma—a review // Plasma Phys. Control. Fusion. 2005. V. 47. P. R35-R161.
2. *Melnikov, A.V.* Applied and fundamental aspects of fusion science// Nature physics. 2016. V. 12. P. 386-390.
3. *Melnikov, A.V. et al.* Plasma potential and turbulence dynamics in toroidal devices (survey of T-10 and TJ-II experiments) // Nucl. Fusion. 2011. V. 51. P. 083043.
4. *Melnikov, A.V. et al.* Heavy ion beam probing—diagnostics to study potential and turbulence in toroidal plasmas// Nucl. Fusion. 2017. V. 57. P. 072004.
5. *Eliseev, L.G. et al.* Two point correlation technique for the measurements of poloidal plasma rotation by Heavy ion beam probe // Plasma Fusion Res. 2012. V. 7. P. 2402064.
6. *Melnikov, A.V.* et al. Electric potential dynamics in OH and ECRH plasmas in the T-10 tokamak // Nucl. Fusion. 2013. V. 53. P. 093019.
7. *Vershkov, V.A. et al.* Density fluctuations as an intrinsic mechanism of pressure profile formation // Nucl. Fusion. 2015. V. 55. P. 063014.
8. *Melnikov, A.V.* et al. The features of the global GAM in OH and ECRH plasmas in the T-10 tokamak // Nucl. Fusion. 2015. V. 55. P. 063001.
9. *Nagaoka, K.et al.* Mitigation of NBI-driven Alfvén eigenmodes by electron cyclotron heating in the TJ-II stellarator // Nucl. Fusion. 2013. V. 53. P. 072004.
10. *Melnikov, A.V. et al.* Quasi-coherent electrostatic mode in ECRH plasmas on TJ-II. // Plasma Fusion Res. 2011. V. 6. P. 2402030.
11. *Hidalgo, C. et al.* On the influence of ECRH on neoclassical and anomalous mechanisms using a dual Heavy ion beam probe diagnostic in the TJ-II stellarator // 26-th IAEA Fusion Energy Conference. 17–22 October 2016, Kyoto, Japan. Rep. EX/P7-44.

EPJ Web of Conferences **149**, 03010 (2017) DOI: 10.1051/epjconf/201714903010
10th International Workshop 2017 "Strong Microwaves and Terahertz Waves: Sources and Applications"

Observation of ion cyclotron emission from ohmically and NBI heated plasmas in TUMAN-3M tokamak

S.V. Lebedev, L.G. Askinazi, A.A. Belokurov, D.B. Gin, V.A. Kornev, A.A. Shabelsky, A.E. Shevelev, A.S. Tukachinsky, N.A. Zhubr

[1]Ioffe Institute, St. Petersburg, Russia, e-mail: sergei.lebedev@mail.ioffe.ru

Ion Cyclotron Emission (ICE) is a fundamental phenomenon observed in space and laboratory magnetized plasmas. ICE exhibit itself as magnetoacoustic instability which develops at frequencies close to ion cyclotron resonance and its harmonics. In laboratory experiments ICE was found to be driven by resonant interaction with high energy ion population produced by fusion products or injected beam ions [1, 2, 3]. Nonmonotonicity or anisotropy of ion distribution function in velocity or physical space are considered as the energy sources for the instability [4].

ICE observation in ohmic plasmas

In the TUMAN-3M (compact circular tokamak) ICE has been found for the first time in Ohmic regime in the absence of high energy ion population. Frequency of the first harmonic emission was found to be close to IC resonance in the vicinity of pick-up magnetic probe location. Spectra span up to ninth harmonic in deuterium and up to seventh in hydrogen.

Figure 1 presents typical example of ICE spectrum obtained in hydrogen plasma in a shot with the following parameters: $B_T = 0.8$ T, $I_p = 150$ kA, $n_e = 1.6\cdot 10^{19}$ m^{-3}, $T_i(0) = 0.2$ keV. First peak in the spectrum has frequency of 14.7 MHz corresponding to hydrogen cyclotron resonance frequency at probe location. Measurements of the ICE with probe located at low field side of the torus showed reduction of first harmonic frequency in accordance with local ion cyclotron resonance at probe position. Replacement of working gas with deuterium resulted in corresponding reduction of ICE frequencies measured by the both probes.

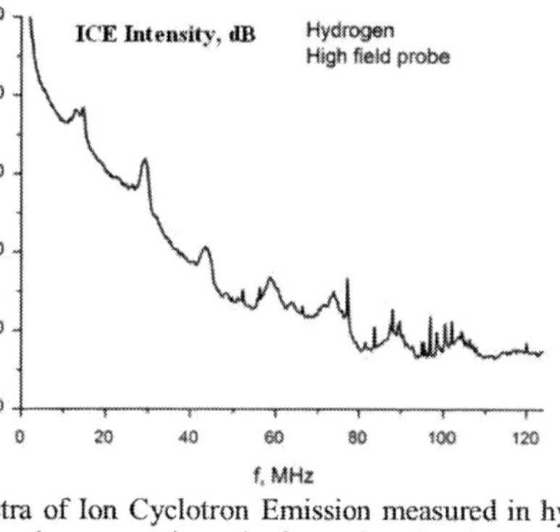

Fig. 1. Spectra of Ion Cyclotron Emission measured in hydrogen plasmas using magnetic probe located at $R = 0.42$ m (high field side of the torus)

Ion-cyclotron drift instability (ICDI) [5] is considered as a possible candidate for explanation of the observed emission. According to [5] the following criterion should be fulfilled in order to allow ICDI to develop:

$$\rho_i/a \geq 2(m_e/m_i)^{1/2}, \quad (1)$$

here ρ_i – ion Larmor radius, a– radial scale of density/temperature gradient, m_e and m_i – electron and ion masses, respectively.

Estimations have shown the above criterion can be fulfilled in the experimental conditions of TUMAN-3M tokamak. Strong background turbulence in the spectral range of the observed IC harmonics (8÷120 MHz) has been detected indicating existence of energy reservoir for ICDC excitation.

ICE in NBI heated plasmas

ICE at frequency corresponding to the deuterium cyclotron resonance near magnetic axis of the tokamak was observed during high energy NB injection of hydrogen/deuterium mixture in hydrogen target plasmas. In the described experiments beam energy E_0 was set to 16 keV and beamline of the injector was aligned in co-current tangential direction with minimal distance between beamline and torus major axis $R_{tan} = 0.42$ m. The emission was detected using pick-up magnetic probes located in both high and low field sides of the torus as seen on figures 2 and 3. The experiments were performed with the neutral beam containing mixture of deuterium and hydrogen atoms with approximate ratio of 60/40%, correspondingly.

It is noticed that replacement of working gas hydrogen with deuterium without changes in beam mixture inverts situation – ICE frequency changes from central deuterium cyclotron resonance to central hydrogen cyclotron resonance. In later case ICE frequency doubles up to $f \approx 12.8$ MHz.

The phenomenon could be understood in the framework of conjecture "beam-minority-ICE". We assume that beam containing mixture of hydrogen and deuterium ions excites the both hydrogen and deuterium central resonance waves. Depending on target plasma working gas one of the two waves is absorbed inside and cannot be detected by probes located outside plasma. In hydrogen target plasma hydrogen central ICE is absorbed whereas deuterium ICE can escape from the plasma and could be detected. Vice versa: in deuterium target plasma the deuterium ICE is absorbed but hydrogen emission leaves plasma and could be detected outside.

The hypothesis can explain decay of the ICE before NBI termination. As it is seen on figures 2 and 3 the intensity of the "6.4 MHz frequency" decreases towards the end of NBI pulse and disappears at 77th ms whereas NBI ends at 80th ms. It might be a consequence of accumulation of deuterium (injected by NBI) in target plasma which results in absorption of "beam-minority-emission".

Other features of the observed IC emission are existence of higher harmonics and fine structure of the first harmonic. Hardware available for measurements of ICE on TUMAN-3M allowed observation of second and third harmonics. These harmonics have larger width and lower intensity compared to first harmonic.

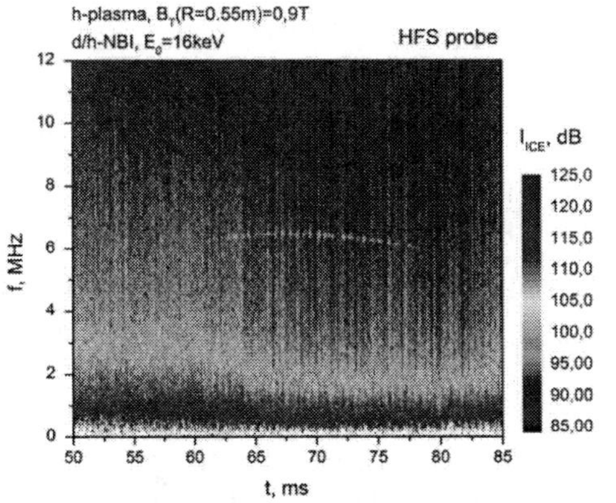

Fig. 2. Spectrogram of plasma emission measured by magnetic probe located at $R = 0.42$ m (high field side of the torus). NBI starts at 60th and ends 80th ms. Trace with frequency of $f \approx 6.4$ MHz existing during period of injection represents Ion Cyclotron Emission excited by deuterium energetic ions at $R = 0.58$ m (close to magnetic axis $R_0 = 0.53$ m)

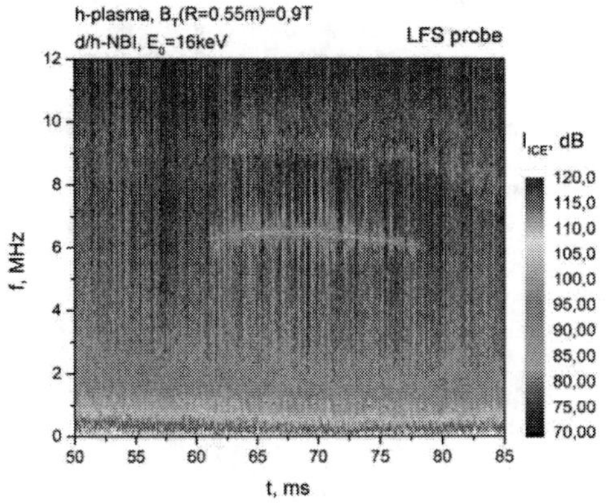

Fig. 3. Spectrogram of plasma emission measured by magnetic probe located at $R = 0.77$ m (low field side of the torus) in the same shot as presented on figure 2

It is worth to note splitting of the ICE frequency. Hydrogen minority emission reveals structure of two and in some cases of three thin lines. Observed separation of triplet/doublet lines is in the range of 100–300 kHz. Each component of the fine structure has narrow width of less than 50 kHz. Larmor radius of the energetic hydrogen ions is in the range of 0.7 cm which provides variation of cyclotron frequency of ~ 300 kHz on this scale. Thus narrow width strongly suggests the observed ICE is not a simple cyclotron emission but rather wave excited in the plasma by energetic ions. A cavity responsible for the splitting and formation of separate narrow lines should exist. For the time of writing this paper the reason of appearance of the fine structure is not clear and needs further studies.

As it was noted above frequency of ICE is close to central cyclotron resonance frequency of either hydrogen or deuterium. In order to establish existence in this area a large amount of energetic ions with high perpendicular energy a calculation of their trajectories in the geometry of TUMAN-3M NBI experiment was performed. In the calculations Monte-Carlo method for generation of ionization points and description of particle density on various trajectories was used. The calculation showed that indeed highest density of energetic ions exists in the proximity of tokamak magnetic axis. For the given amount of test particles the magnitude of $(n_i V_\perp)$ in hydrogen is by a factor of 3 higher compared to deuterium. This conclusion correlates with observation of stronger intensity of beam-minority-ICE at hydrogen cyclotron frequency than at deuterium cyclotron frequency.

Results of the calculations can be used for interpretation of splitting. According to the calculations energetic ions arising in the vicinity of magnetic axis produces regions with strong gradient of $(n_i V_\perp)$. The regions of the strong gradient are separated by distance of the order of 1 cm which can explain separation of lines by 300 kHz in fine ICE structure.

In conclusion of this section we summarize new finding made in this study. Conjecture of beam-minority-ICE is proposed for interpretation of transition between emissions at hydrogen and deuterium cyclotron frequencies with change of working gas in the target plasma. Fine structure of ICE was observed for the first time.

Acknowledgement. The study was supported by Russian Science Foundation (Project # 16-12-10285) and by Ioffe Institute.

References

1. *Bhadra, D. K., Chiu, S. C., Buchenauer, D., Hwang, D.* "Electromagnetic emission from a neutral-beam-injected plasma" // Nucl. Fusion. 1986. V. 26. P. 201-209.

2. *Cottrell, G. A., Bhatnagar, V. P., Da Costa, O., et al.* "Ion cyclotron emission measurements during JET deuterium-tritium experiments" // Nucl. Fusion. 1993. V. 33. P. 1365-1387.

3. *Cauffman, S., Majeski R.* "Ion cyclotron emission on the Tokamak Fusion Test Reactor" // Rev. Sci. Instrum. 1995. V. 66. P. 817-819.

4. *Coppi, B., Cowley, S., Kulsrud, R., Detragiache, P., Pegoraro. F.* "High-energy components and collective modes in thermonuclear plasmas" // Phys. Fluids. 1986. V. 29. P. 4060-4072.

5. *Mikhailovsky, A. B.* "Ion-cyclotron drift instability in a finite-plasma" // Nucl. Fusion. 1971. V. 11. P. 323-327.

Continuous high power microwave heating at the W7-X stellarator

K.J. Brunner[1], H. Braune[1], V. Erckmann[1], W. Kasparek[1], H.P. Laqua[1], S. Marsen[1], D. Moseev[1], B. Plaum[2], T. Stange[1], M. Weißgerber[1] and the entire W7-X team

[1]Max-Planck-Institute for Plasma Physics, Greifswald/Garching, Germany, k.j.brunner@ipp.mpg.de
[2]IGVP Stuttgart University, Stuttgart, Germany

Introduction

The W7-X advanced stellarator is a fusion experiment relying heavily on electron cyclotron resonance heating (ECRH). It uses a quasi-optical transmission line able to transmit up to 12 gyrotrons from a remote gyrotron assembly to the plasma vessel [1]. This method of transmission delivers high power densities and low losses at a moderate price tag. The microwave power is eventually coupled into the plasma using movable front steering mirrors, therefore providing a semi-arbitrary choice for the angle of incidence. In addition the incident polarization and ellipticity can be freely set via steerable polarizers in the transmission line. In addition to the front-steering mirrors the system can divert power to a remote steering launcher via a movable mirror [3].

The first operation phase (OP1.1) was conducted with 6 gyrotrons operating at 140 GHz and supplying a total of 4.3 MW of continuous heating power to the plasma. This setup already proved the capabilities of the system during the first operation phase, yielding outstanding plasma parameters, which will be briefly presented in the first section.

For the second operation phase (OP1.2) envisaged to start in September 2017, four additional gyrotrons have been installed, which will give access to up to 8.5 MW of heating power. This opens up new operation regimes, which also come with new challenges regarding heating efficiency and machine safety. This paper focuses on the developments to the heating system from a plasma physics perspective and will only briefly summarize the hardware developments in section 3. For a more thorough description of the W7-X ECRH hardware developments the reader is kindly referred to the corresponding publications also published as part of these proceedings [2, 3].

Main findings of OP1.1

The application of ECRH in fusion plasmas extends beyond plain bulk plasma heating, even though this is its main purpose. Examples for other applications wall conditioning, plasma profile shaping, instability control and electron cyclotron current drive (ECCD). During OP1.1 many of these aspects were already tested.

One of the outstanding results of this operation phase was the very high central electron temperature approaching 10 keV. These temperatures are achieved using 2nd harmonic X-mode heating (X2), which has a very narrow power deposition profile in the plasma core. The density in these discharges remained at a few $10^{19}\,m^{-3}$, which also resulted in moderate ion temperatures at roughly 2 keV due to the low collisional coupling between the electrons and ions. An example of such a discharge is seen in figure 1, which shows shot number 20160301.014. The core electron cyclotron emission (ECE) temperature can be seen in the middle plot.

A second prominent result from OP1.1 are the ECCD-induced core temperature crashes observed. The shot shown in figure 1 depicts a shot exhibiting this phenomenon as well. The top graph shows injected ECRH power per gyrotron totalling about 2 MW. The integrated toroidal plasma current measured by Rogowski coils is plotted at the bottom. The current drive was chosen to increase the rotational transform (Co-ECCD). Total X2 absorption is reached within a few tens of milliseconds of this shot and stays this way throughout. After 150 ms the then quasi-steady core electron temperature T_e is suddenly dropping sharply and recovers. The onset of this phenomenon coincides with the rise in measured I_{tor}. These sawtooth-instability-like crashes are likely the result of the iota profile changing, thus resulting in the emerging of instabilities at rational flux surfaces [4].

Developments for OP1.2

As has been mentioned before this paper will not elaborate on the ECRH hardware developments. In summary there will be a significant increase in available heating power up to 8.5 MW, which is supplied by 10 gyrotrons for up to 30 min. In addition the fueling capabilities of W7-X have been extended by a pellet injector for "deep fueling". The combination of higher fueling and increased heating power is expected to give access to plasma densities in the $10^{20}\,m^{-3}$ range. This however poses the problem that the X2 heating scheme, which has been predominantly employed during OP1.1, will suffer from significant beam refraction effects as the X2 cut-off density of $1.2\cdot10^{20}\,m^{-3}$ is approached. For this reason the OP1.2 phase of the W7-X experiment will focus on optimizing the 2nd harmonic ordinary mode (O2) heating scheme, the cut-off density of which is twice that of the X2 scheme ($2.4\cdot10^{20}\,m^{-3}$).

This heating scheme however has a lower single pass absorption efficiency. Where X2 heating easily reaches 100% even at low temperatures, the O2 absorption efficiency, which scales with T_e^2, will barely exceed 60% single-pass absorption below a core electron temperature of 2 keV. For this reason a set of Tungsten-coated (TZM) holographic tiles have been installed on the wall opposite to the ECRH launchers. These have been designed to refract about 90% of the incident ECRH power back through the magnetic axis. This setup is complemented by a non-holographic polished steel liner on the outboard side, which serves the same purpose. The plasma center exhibits the highest electron temperature and therefore the total absorbed power fraction is expected to increase from 60% to well above 80% as indicated by figure 2. This will be verified in OP 1.2.

The TZM tiles are an integral part of the efficient multi-pass O2 heating scheme and as such it is crucial, that the ECRH heating beam is well aligned on the tile's refraction surface.

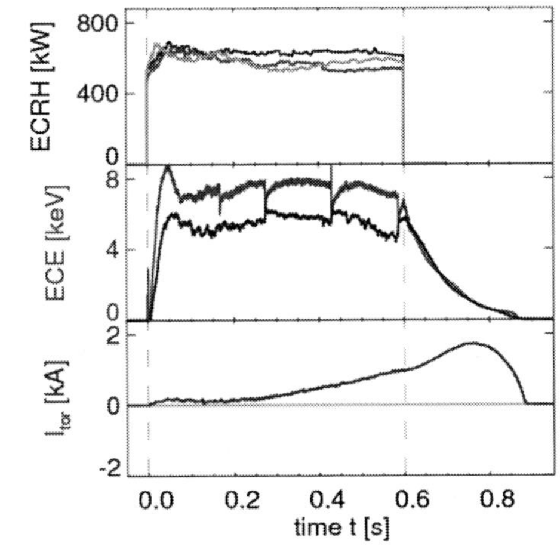

Fig. 1. W7-X shot number 20160301.014. The top shows the inserted ECRH power for each gyrotron totalling roughly 2 MW. Below are the radiative temperature measured by ECE for the central channel (red) and the channel totalling r/a = 0.4. The measured toroidal plasma current is at the bottom. The crashes of the central electron temperature as a result of Co-ECCD indicates the presence of current driven plasma instabilities (adapted from [5])

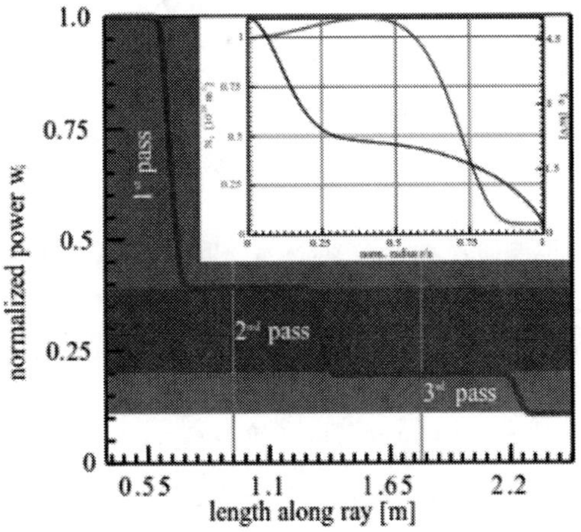

Fig. 2.TRAVIS simulation of ECRH absorption efficiency for a triple-pass through the W7-X plasma under optimum conditions. In the first pass only 60% of the ECRH power are absorbed, which can be increased significantly by adding the second and third pass. The assumed plasma parameter profiles can be seen in the inlay

However, even though the O2 beam is not expected refract a lot at the medium plasma densities expected during OP1.2, the amount of refraction suffered is still sufficient to lose about 50% of the beam at the tile. For this reason it is necessary to implement a feedback targeting system for the launcher mirrors. This will initially be implemented in a feed-forward scheme using the TRAVIS code [6]. In addition a set RF probing holes have been added to the TZM tiles, which will be used to measure the shine-through power of the ECRH beam incident on the tile. Since there is generally more than 1 probing hole, one can use this diagnostic as a feedback system controlling the beam position on the tile. This will be tested during OP1.2. To complement this diagnostic the infrared (IR) camera system observing the inboard plasma wall from the ECRH launchers has been improved to be able to detect up to 15 μm IR radiation, making the ECRH beam visible on the bare reflector tiles.

Summary

The ECRH system on W7-X delivered outstanding plasma performance during the initial experimental campaign OP1.1. For the next operation phase the system has been further expanded, giving access to even higher performance plasma regimes, which at the same time pose significant challenges for the operation of the heating system in particular the development of a safe high power O2 heating scheme at densities well above the X2 cut-off.

Acknowledgements

This work has been carried out within the framework of the EUROfusion Consortium and has received funding from the Euratom research and training programme 2014–2018 under grant agreement No 633053. The views and opinions expressed herein do not necessarily reflect those of the European Commission.

References

1. *Erckmann, V. et al.* Electron Cyclotron Heating for W7-X: Physics and Technology // Fusion Sci. Technol. 2007. V. 52 No. 2 P. 291-312.
2. *Braune, H.* Title to be added // Proc. of the SMP. 2017, P. t.b.d.
3. *Kasparek, W.* Continuous High Power Microwave Heating At The W7-X Stellarator // Proc. of the SMP 2017, P. t.b.d.
4. *Estrada, T. et al.* Transient behaviour in the plasma core of TJ-II stellarator and its relation with rational surfaces // Pl. Phys. & Contr. Fus., 2002, V. 44, P. 1615-1624.
5. *Stange, T. et al.* Advanced electron cyclotron heating and current drive experiments on the stellarator Wendelstein 7-X // Proc. of the *RFFPC*, 2017, Inv-04.
6. *Marushchenko, N.; Turkin, Y. & Maassberg, H.* Ray-tracing code TRAVIS for ECR heating, EC current drive and ECE diagnostic // Comp. Ph. Comm., 2014, V. 185, P. 165-176.

EPJ Web of Conferences **149**, 03012 (2017) DOI: 10.1051/epjconf/201714903012
10th International Workshop 2017 "Strong Microwaves and Terahertz Waves: Sources and Applications"

Isotopic effect in experiments on lower hybrid current drive in the FT-2 tokamak

S.I. Lashkul, A.Yu. Stepanov, A.B. Altukhov, A.D. Gurchenko, E.Z. Gusakov, V.V. Dyachenko, L.A. Esipov, M.A. Irzak, M.Yu. Kantor, D.V. Kouprienko, A.N. Saveliev and S.V. Shatalin*

Ioffe Institute, Politekhnicheskaya 26, St.Petersburg, Russia
*St.Petersburg State Polytekhnical University, Politekhnicheskaya 29, St.Petersburg, Russia

Result of the LHCD efficiency study

The most efficient method for sustaining the quasi-steady plasma current by means of lower hybrid (LH) waves can be implemented only at relatively low plasma densities not exceeding a certain density limit, n_{DL}. The existence of the density limit for LH current drive (LHCD) was attributed to various mechanisms: linear absorption by ions, which increases as the density tends to the resonance LH value; collision losses; scattering by drift turbulence; parametric instabilities; etc. The density limit effect has been studied and discussed for several past decades; however, it has not received comprehensive physical explanation [1].

In the present work the main attention is paid to investigation of current drive by lower hybrid waves on the FT-2 (R = 0.55 m, a = 0.08 m, $B_T \leq 3$ T, I_p = 19÷40 kA, f_0 = 920 MHz) tokamak. The results of experiments carried out at the FT-2 tokamak allow us to conclude that the most probable reason for LHCD termination in both hydrogen and deuterium plasmas at a relatively low plasma current of I = 22 kA and, accordingly, a low electron temperature is an additional reduction in T_e at the periphery of the plasma column during the LH pulse [2]. This results in a lower threshold for the parametric decay ($\sim T_e/n_e$) of the pumping wave. The parametric decay satellites, the frequencies of which are down-shifted, ($f_{sat} = f_0 - kf_{Ci}$), (k = 1, 2, 3…), are slowed down more than the pumping wave. Therefore, the plasma density, at which linear conversion of slowed down satellite waves occurs, is lower than that for the pumping wave; hence, the LH wave after parametric decay can be absorbed by ions even at the plasma periphery, without penetrating into the plasma column. At a higher plasma current (I_{OH} = 32 kA) and higher electron temperature (T_{e0} = 600 eV), the density at which the LHCD in hydrogen plasma terminates is close to the resonance value ($n_{DL}^H \approx n_{LH}^H \approx 3.5\ 10^{19}\text{m}^{-3}$). After the plasma density reaches this value, the interaction of the LH wave with electrons is replaced with direct absorption by ions, when $\omega_0 \sim \omega_{pl,i}$. As a result of the experimental study of the influence of the plasma isotopic composition on the LHCD efficiency, it is established that the efficiency η_{CD}^D in deuterium plasma on FT-2 is higher than in hydrogen plasma η_{CD}^H [2]. As it is seen from Fig. 1 (I_{OH} = 32 kA), within the density range from $\langle n_e \rangle$ = = 10^{19} m^{-3} to $\langle n_e \rangle$ = 2.5 10^{19} m^{-3} in deuterium, we have $\eta_{CD}^D \approx 0.4$ A m^{-2} W^{-1}, whereas in hydrogen plasma, this value is appreciably smaller, $\eta_{CD} \approx 0.3$ A m^{-2} W^{-1}. The resonance value n_{LH} of the density for deuterium at $N_\parallel \approx 2$ is substantially higher, $n_{LH}^D \approx 10^{20}$ m$^{-3} \geq n_{DL}^D \approx 4.5\ 10^{19}\text{m}^{-3}$. So, the obtained value $n_{DL}^D \approx 4.5\ 10^{19}\text{m}^{-3}$ is smaller by more than one-half of the n_{LH}^D. Nevertheless, in contrast to hydrogen plasma, the density at which the sharp increase in the high-energy charge-exchange (CX) atomic flux (Fast Neutral (FN) with E_{CX} = 1575 eV) $\langle n_e \rangle_{FN}^D \approx 5\ 10^{19}\text{m}^{-3}$, takes place proved to be higher than n_{DL}^D. There is an appreciable gap between the values of n_{DL}^D and $\langle n_e \rangle_{FN}^D$. Apparently, the main reason for LHCD termination in this case is the parametric decay of the pumping wave. The experimental results confirm that parametric processes intensify with increasing density during the RF pulse (see Fig. 3 in [2]). Nevertheless, we cannot exclude the influence of a substantial slowing down (up to $N_\parallel \approx$ 10) of the pumping wave as it propagates into the plasma column. The experimental evidence of influence of isotopic effect on characteristics of the ion heating observed at $\langle n_e \rangle$ rise and after LHCD termination, which demonstrates Fig. 2 for H/D plasmas.

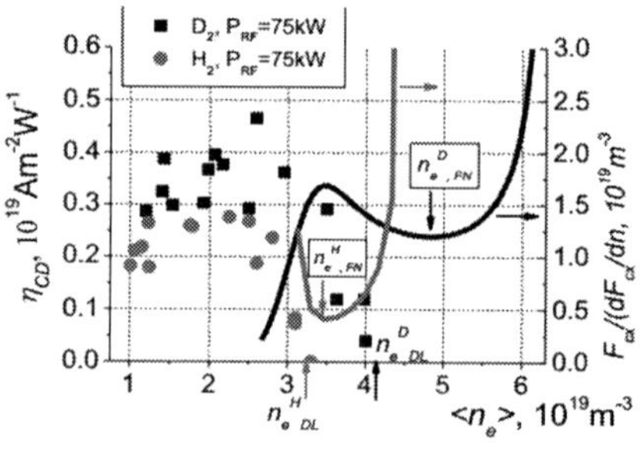

Fig. 1. Experimental values of the LHCD efficiency $\eta_{CD} = I_{RF}^N \langle n_e \rangle$ and the quantity $F_{CX}/(dF_{CX}/dn_e)$ vs. average densities of hydrogen plasma (gray circles and curve) and deuterium plasma (black squares and curve). The minimum values of $n_{e\ FN}^D$ and $n_{e\ FN}^H$ are marked with arrows

A significant part of the LHCD experimental studies on FT-2 were performed at relatively low densities of $\langle n_e \rangle$ = 10^{19} m^{-3}, when, the parametric decay of the pumping wave is absent and the highest LHCD efficiency is observed. It is established that, at relatively large values of the plasma current (I_{OH} = 32 ÷ 35 kA, $P_{RF} \approx$ 100 kW, $T_{e0} \approx$ 550 ÷ 600 eV), the quantities η_{CD} and I_{RF}^N and their dependences on the plasma parameters are close to those obtained at large tokamaks [2].

Parametric processes intensify with increasing plasma density

Parametric process on the periphery of the plasma column are considered as the main mechanism of growth $N_\parallel$ resulting in a significant reduction of LHCD efficiency and depletion of the pump wave power due to the absorption of ions in the peripheral plasma [3, 4]. The

EPJ Web of Conferences **149**, 03012 (2017) DOI: 10.1051/epjconf/201714903012
10th International Workshop 2017 "Strong Microwaves and Terahertz Waves: Sources and Applications"

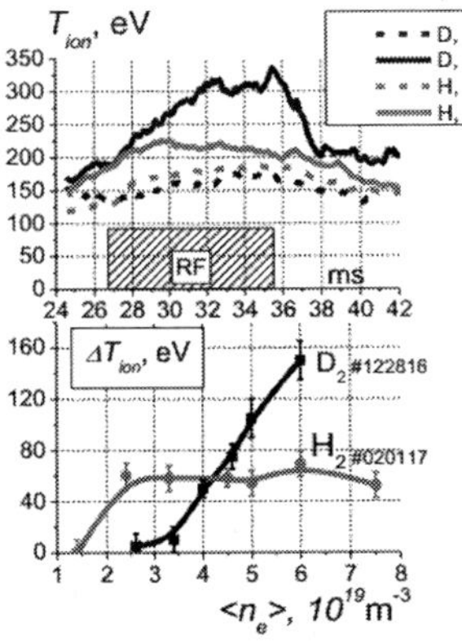

Fig. 2. (*a*) Experimental values of the $T_{ion}(t)$ at $P_{RF} = 90$ kW for H/D plasmas with initial density $<n_e>$ increase. (b) Dependence of the increment of the ion temperature vs $<n_e>$ of H/D plasmas, $\Delta T_{ion}(<n_e>)$. $\Delta E_{CX} = (364–1058)$ eV is energy neutral particle spectral interval used for T_{ion} definition

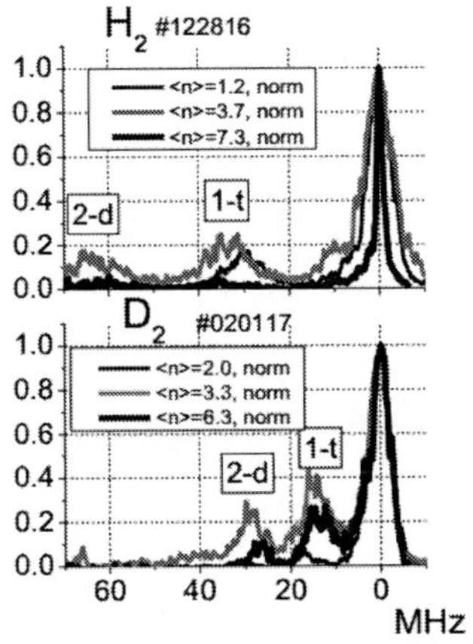

Fig. 3. The normalized intensity of peaks of the LHW pump wave spectra; 1-t and 2-d IC satellites measured by RF probe in H/D plasmas at different densities, <n>, 10^{19}m^{-3}

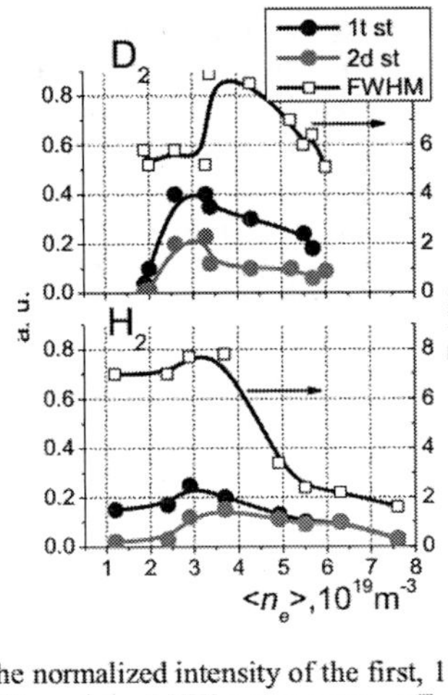

Fig. 4. The normalized intensity of the first, 1-t, and the second, 2-d, satellites of the LHW pump wave. ΔF – frequency width (FWHM) ΔF, MHz of the broadened pump wave

spectral broadening of the pump wave at development of PI is due to the excitation by the pump wave ion-sound (IS) and ion-cyclotron (IC) quasimodes.

To assess the impact of isotopic effect on the development of parametric processes on the tokamak FT-2 a comparative measurements were carried out using the movable RF probe-antenna placed in the poloidal limiter shadow at the poloidal angle $\theta = 310°$, which corresponds to the ICR in the plane of a large radius R = 0.6 m. Normalized spectra of the LH pump wave measured by the RF probe at different densities of the H/D plasma are shown in Fig. 3. It is noted that the increase in density results in a significant change in the intensity of IC satellites as well as the frequency width (FWHM) ΔF, MHz of the broadened pump wave. As shown in Fig. 4, maximum values for these dependencies as a function of $<n_e>$ are achieved with plasma densities of $<n_e> \sim 3 \div 4\ 10^{19}$ m^{-3}. Under these conditions, according to Fig. 1, the efficiency generation of LHCD is reduced almost to 0.

Characteristically, those IC satellites in deuterium plasma appear at a density higher, than the density in hydrogen plasma. With increasing density its level is approximately 2 times higher than the level of the same dependencies measured in hydrogen plasma. This fact, in particular, may indicate a significantly larger pump wave depletion effect, when large fraction of LHW power falls in sidebands. Shift down frequencies ($f_{sat} = f_0 - kf_{Ci}$), (k = 1, 2, 3...), at substantial rise of $N_{\|}$ for f_{sat} could result in additional heating of the ions on periphery of discharge. Most of the LHW power of f_0 in the course of propagation along the magnetic field lines with the gradual penetration in the plasma column reaches it central regions. During such penetration of LHW the effect of spectral broadening ΔF (Fig. 4) and additional slowdown result in decrease of the efficiency generation of the LHCD. With increasing density, more effective additional heating of the deuteron are observed Fig. 1, 2.

These and other experimental data presented in the report allows the use of isotope effect to conduct a comparative analysis for obtaining more information about the mechanisms affecting the efficiency of generation of LHCD and ion heating.

This work is executed at partial financial support by Russian Science Foundation grant 17-12-01110

References

1. *M. Goniche, L. Amicucci, Y. Baranov et al.* Lower hybrid current drive for the steady-state scenario. // Plasma Phys. Controlled Fusion 52, 124031 (2010).

2. *S.I. Lashkul, A.B. Altukhov, A.D. Gurchenko et al.* Isotopic Effect in Experiments on Lower Hybrid Current Drive in the FT-2 Tokamak. // Plasma Physics Reports, Plasma Physics Reports, 2015, Vol. 41, No. 12, pp. 990–996.

3. *R. Cesario, L. Amicucci, A. Cardinali et al.* Spectral broadening of parametric instability in lower hybrid current drive at a high density. // Nucl. Fusion 54 (2014) 043002 (18pp).

4. *M. Goniche1, V. Basiuk1, J. Decker et al.* Lower hybrid current drive at high density on Tore Supra. // Nucl. Fusion **53** (2013) 033010 (9pp).

EPJ Web of Conferences **149**, 03013 (2017) DOI: 10.1051/epjconf/201714903013
10th International Workshop 2017 "Strong Microwaves and Terahertz Waves: Sources and Applications"

Decay of the X-mode into two upper-hybrid plasmons in the plasma filament. Experimental modeling and theoretical description

L.V. Simonchik[1], A.B. Altukhov[2], V.I. Arkhipenko[1], A.D. Gurchenko[2], E.Z. Gusakov[2], A.Y. Popov[2] and M.S. Usachonak[1]

[1]Institute of Physics of NAS of Belarus, Minsk, Belarus, l.simonchik@ifanbel.bas-net.by
[2] Ioffe Institute, St-Petersburg, Russia

The propagation of electron-cyclotron (EC) waves and their absorption in plasma until recently were believed to be well described in the framework of the linear theory and therefore predictable in detail. A theoretical analysis of such nonlinear phenomena as parametric decay instabilities (PDI) of the EC waves, performed under the assumption of monotonic plasma density profile, predicts an extremely high threshold for their occurrence (more than 1 GW for backscattering instabilities) [1]. The reason for such high threshold of parametric instability is the large convective losses of the daughter waves from the region of parametric decay both in the direction of plasma inhomogeneity and of the magnetic field. In the recent decade, however, there has been accumulated a large number of observations in the ECR plasma heating experiments in tokamaks and stellarators that do not fit into a simple linear picture. The most distinct among them is the anomalous backscattering phenomena studied in detail in the second harmonic X-mode heating experiments at the Textor tokamak [2]. In these experiments it was shown that the radiation temperature of the observed frequency downshifted backscattering signal is thousand times larger than the electron temperature, and its amplitude is modulated at the magnetic island rotation frequency. According to [2], the highest level of anomalous backscattering is achieved when the plasma density in the island is equal to the upper hybrid density for the frequency equal to the half value of the pump frequency. Based on the latter observation the theoretical model [3] was proposed recently allowing to explain the effect of anomalous backscattering as a result of the two upper-hybrid (UH) plasmons PDI possessing very low threshold due to trapping of nonlinearly excited plasmons in the vicinity of the density maximum that accompanies the island. The theory [3] also predicts substantial (up to 25%) anomalous absorption due to this process.

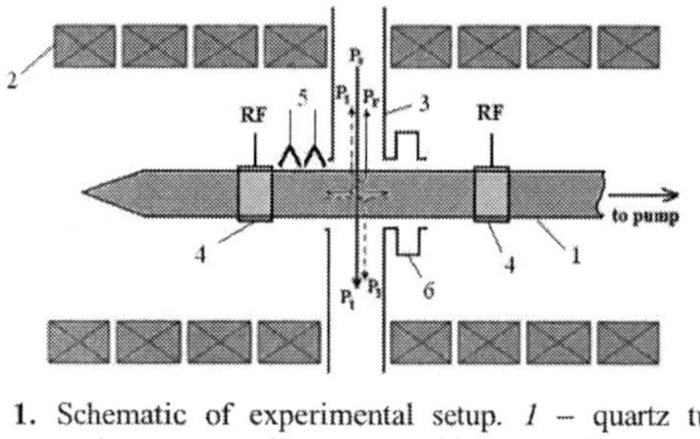

Fig. 1. Schematic of experimental setup. *1* – quartz tube; *2* – magnetic system; *3* – waveguide; *4* – RF antenna; *5* – microwave antenna; *6* – microwave cavity

In the present paper the low-threshold two-plasmon PDI leading to excitation of the trapped UH waves is modelled experimentally in the laboratory plasma. The decay occurs in a plasma filament extended in the direction of the magnetic field and produced by high-frequency discharge in long quartz tube with an inner diameter of 23.5 mm filled with argon at about 2 Pa pressure. Quartz tube passes through the waveguide (72 × 34 mm^2) in parallel to the wide wall (see Fig. 1). The RF power of about 100 W at frequency of 27 MHz is supplied to the ring electrodes placed outside of the tube and disposed on both sides of the waveguide at a distance of 300 mm.

At the maximal RF power the volume averaged plasma density measured using the cavity diagnostics is about 10^{10} cm^{-3} changing by 15–20% at the variation of the magnetic field from 0 to 450 Oe. The radial distribution of the plasma luminosity is approximated by expression $(1-r^2)^{1.6}$. Assuming it to be proportional to the plasma density we get the maximal density in the discharge about 2×10^{10} cm^{-3}.

The X-mode microwave pulses (up to 160 W) at frequency f_0 = 2.35 GHz, substantially higher than the EC and UH values, are incident onto the plasma along the waveguide. The temporal behaviors of the transmitted and reflected microwave signal, as well as the plasma luminosity are monitored in the experiment at the different pump power and magnetic field.

The theoretical model of the low-threshold two-plasmon decay proposed in [3] consists of two parts shown schematically in Fig. 2. Firstly, it is a model of the PDI linear stage describing the absolute instability leading to extra-ordinary t_X-wave decay into two UH plasmons $l_{UH}(f_1)$ and $l_{UH}(f_2)$, which are exponentially growing from the thermal noise level. Secondly, it is a model of the absolute instability nonlinear saturation as a result of cascade of several low-threshold decays in which UH waves l_{UH}' trapped in plasma and low frequency waves (ion sound l_{IS} in the case of our experiment) are excited. The saturation level determines the efficiency of the anomalous absorption.

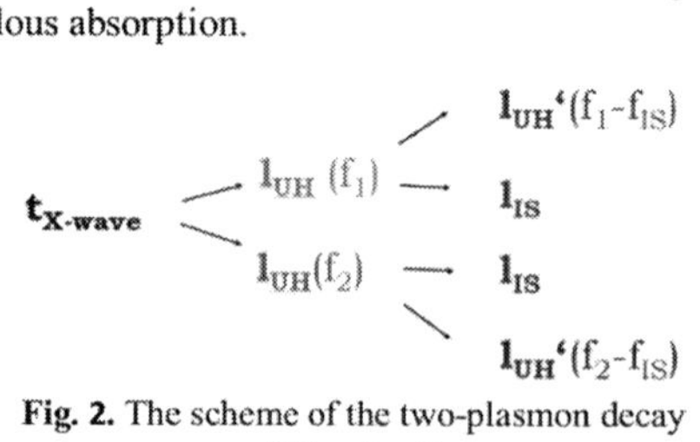

Fig. 2. The scheme of the two-plasmon decay and it's saturation

The observations of the microwave anomalous absorption at the pump frequency higher than the second electron cyclotron harmonic are performed at the power of 160 W and magnetic field smaller than 42 mT. At a small plasma density no distortions of the microwave pulse transmitted through the plasma or reflected from it are observed (Fig. 3*a*).

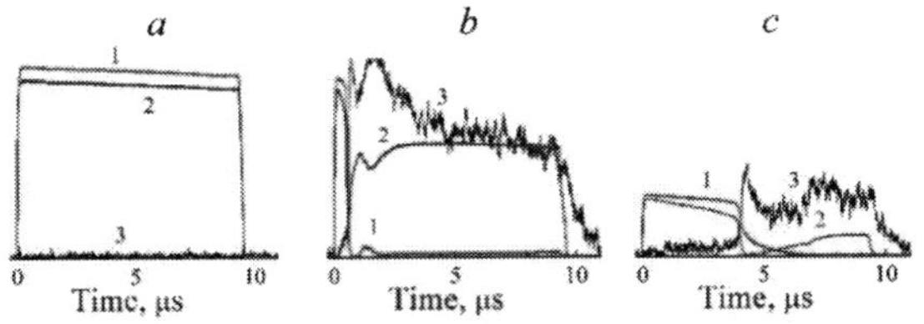

Fig. 3. Waveforms of transmitted (1), reflected (2) pulses and of the plasma luminosity (3). *a* –small plasma density, *b* – P_0 = 160 W, *c* – P_0 = 30 W

However at plasma density exceeding a critical value a fast decrease of both transmitted and reflected power is observed indicating the turning on of the strong anomalous absorption (see Fig. 3*b*). This effect is accompanied by a sharp growth of the plasma luminosity also shown in this figure. The time delay of the anomalous phenomena appearance is dependent on the pump power, as it is seen in Fig. 3*b*–*c*.

The critical density needed for switching on of the anomalous phenomena is growing with decreasing magnetic field, as it is shown by circles in Fig. 4. As it is seen there, this dependence is close to the theoretical UHR density dependence on the magnetic field plotted in the figure for the half pump frequency

$$n_e = \frac{m}{4\pi e^2}\left(\pi^2 f^2 - \left(\frac{eB}{m_e c}\right)^2\right) \quad (1)$$

The dependence of the two-plasmon instability growth rate on the pump power are studying as well. The measurements are performed for two magnetic fields 350 and 400 Oe, but at the constant averaged plasma density of about 10^{10} см$^{-3}$ and Ar pressure of about 1.5 Pa. The results for the inverse time of the anomalous absorption turning on shown in Fig. 5 appear to be in a qualitative agreement to the theory predictions for the growth rate increase with increasing pump power and for the PDI power threshold decrease with growing magnetic field. The threshold increases while the gas pressure decreases.

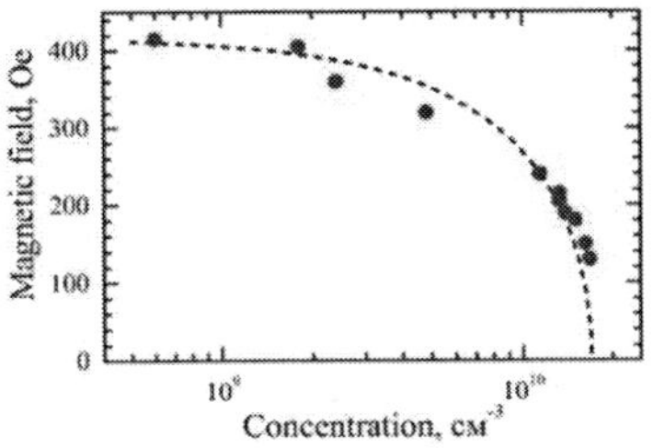

Fig. 4. Dependence of the density critical for onset of the anomalous absorption on the magnetic field. Broken line – UHR condition for 1.175 GHz; P_0 = 160 W

Using the spatially resolved plasma luminosity measurements we study the localization of the anomalous absorption, which appears to be non-central in qualitative agreement with the theoretical expectations based on the UH eigen mode localization. The radial position of the plasma radiation drastic growth was shown to be dependent on the magnetic field.

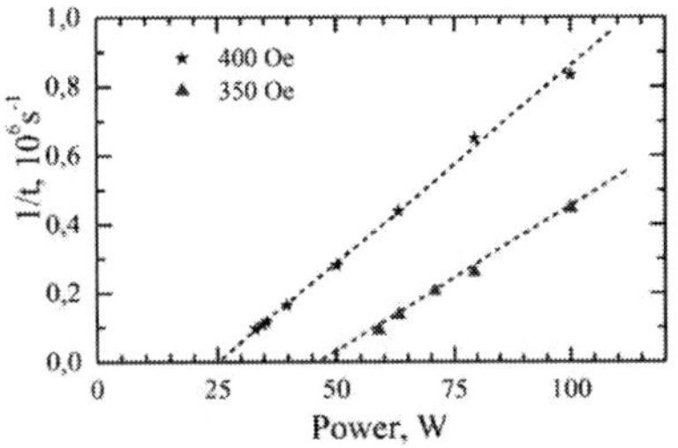

Fig. 5. Dependence of the inverse time of the anomalous absorption onset on the pump power for different magnetic fields

The low frequency waves excited in plasma as a result of PDI saturation are investigated using the UH resonance backscattering diagnostics utilizing the frequency range of 1.1–1.2 GHz. The temporal variation of the BS spectrum is demonstrated in Fig. 6. The white line on the top of the figure shows the waveform of the transmitted signal. The most intensive BS signal is observed in the period of the strongest suppression of the transmitted signal, thus in the period of the strong anomalous absorption onset. During this period the BS signal is enhanced by two orders of magnitude. The frequency shift of the BS signal is close to 2–3 MHz, which corresponds to the ion acoustic frequency range along with the PDI saturation mechanism discussed in theory.

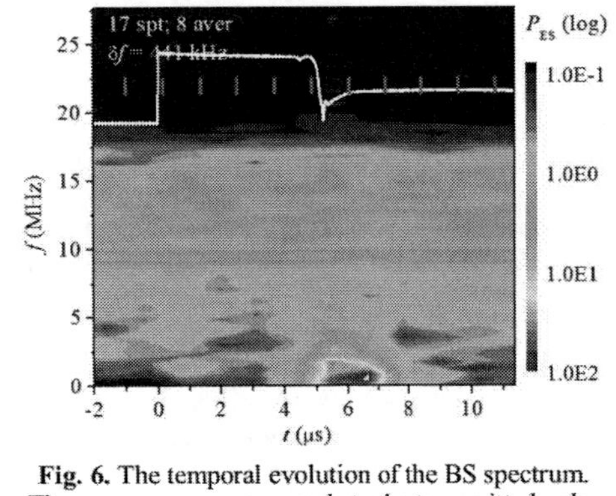

Fig. 6. The temporal evolution of the BS spectrum. The upper curve corresponds to the transmitted pulse

The work was performed under support of the BRFBR grant F16R-095 and RSF grant 16-12-10043.

References

1. *Porkolab M. and Cohen B.I.* // Nucl. Fusion. 1988. V. 28, No. 2. P. 239-254.
2. *Nielsen S.K., Salewski M. et al.* // Plasma Phys. Control. Fusion. 2013. V. 55. 115003.
3. *Gusakov E.Z. and Popov A.Yu.* // Physics of Plasmas. 2016. V. 23. 082503.

EPJ Web of Conferences **149**, 03014 (2017) DOI: 10.1051/epjconf/201714903014
10th International Workshop 2017 "Strong Microwaves and Terahertz Waves: Sources and Applications"

Kinetic instabilities in non-equilibrium plasma: a review of observations

D.A. Mansfeld

Institute of applied physics, Nizhny Novgorod, Russia, mda1981@appl.sci-nnov.ru

Resonant interaction of electromagnetic waves and energetic electrons plays an important role in the dynamics of magnetoactive plasma confined in space and laboratory magnetic traps. One of the most intriguing manifestations of such interaction is the generation of powerful electromagnetic radiation and bursts of energetic electrons from the magnetic trap due to kinetic instabilities. Kinetic instabilities are caused by the presence of positive gradients in the velocity distribution of resonant particles due to different acceleration mechanisms in space (betatron acceleration, acceleration as a result of plasma-wave turbulence and magnetic reconnection) and laboratory plasma (cyclotron heating, magnetic compression, beam heating).

Understanding of the fundamental mechanisms of generation of waves in all frequency ranges and plasma parameters is important to build a complete model of the electromagnetic background not only in the Earth's magnetosphere, but also for other planets. Field experiments, primarily satellite measurements are subject to significant restrictions because of the local nature and complexity of the measurement of spatial and temporal separation of dependencies, especially significant for non-stationary processes. In view of the universality of the physical mechanisms of generation of radiation, significant aspects of natural systems can be reproduced in the laboratory in the open magnetic configurations with magnetic mirrors. The laboratory simulation is valuable due to the fact that, firstly, in the laboratory it is possible to change the parameters of the plasma in a controlled manner, and secondly, to provide multiple repeatability. The fundamental goal of the study is to define universal mechanisms responsible for the formation of temporal regimes and dynamic spectra of stimulated emission of unstable plasma, determination of plasma emission propagation properties, and also to study the characteristics of the generated radiation impact on the plasma confinement in the trap, and the distribution of particles leaving the trap.

The opportunity to recreate different conditions for excitation and amplification of waves in plasma in a single electron cyclotron resonance (ECR) discharge pulse has been demonstrated by IAPRAS scientists, which have the priority results in this area [1–3]. The highly non-equilibrium ECR discharge plasma was created and sustained in the mirror magnetic trap by the pulsed gyrotron microwave radiation at a frequency of 37.5 GHz at power level up to 100 kW. At different stage of the discharge different conditions for excitation of waves are realized which are defined by the ratio fpe/fce (fpe – electron plasma frequency, fce – electron cyclotron frequency): (1) rarefied plasma at initial discharge stage (fpe << fce), when density of hot electrons exceeds density of cold fraction; (2) dense plasma at developed discharge stage (fpe>fce), when density of hot particles is much less than density of cold plasma; (3) at plasma decay stage (fpe<<fce), when density of hot and cold plasma fraction became the same order of magnitude (at a certain decay stage). For example, the plasma electromagnetic activity, which is realized at the advanced stage of the ECR discharge in axially symmetric magnetic trap was studied in [1], and the quasiperiodic pulsed bursts of energetic electrons at the stage of plasma decay were first observed in [3]. The results of these experiments were explained the model of the cyclotron maser [2]. The observed in experiments generation of electromagnetic radiation bursts have much in common with the natural sources of coherent cyclotron radiation - the so-called space cyclotron masers defining population energetic particles of the radiation belts, and as well as responsible for the formation of electromagnetic radiation the Earth's magnetosphere in ELF – VLF range, Jovian decametric emission and some types of solar activity. [4, 5].

In the last decade the interest in this study was renewed due the most powerful digital oscilloscopes with a broad band (up to 60 GHz) and high-speed ADC (up to 160 GS/s), which in real time measure the electric field of the wave. With their help, for the first time we have the opportunity to register with the necessary temporal resolution low-power broadband electromagnetic radiation emerging from the plasma, to determine its dynamic spectrum allocated to the background of strong interference. Measuring electromagnetic field with high temporal resolution allowed to observe various dynamic spectra of electromagnetic radiation related to at least five types of kinetic instabilities [6]. The detailed analysis of electromagnetic spectra of kinetic instabilities in non-equilibrium plasma can give a valuable diagnostic information on the plasma parameters (plasma density, the characteristic energy of the emitting particles, etc.) [7].

Studying of these mechanisms in the laboratory traps has its own value because magnetic traps filled with hot non-equilibrium plasmas have a lot of practical (technology) applications in thermonuclear fusion devices, ECR multicharged ion sources, which used in accelerators of heavy particles, UV and X-ray sources, thermonuclear neutrons sources, etc. The problem is especially relevant in connection with the use of high-power microwave sources – gyrotrons, application of which for increasing of the energy input into the plasma, can achieve not only the record values of main component plasma density and temperature, but also significantly increase the anisotropy of the velocity distribution function of «warm» and «hot» electronic components. Thus there are various kinds of plasma instabilities primarily kinetic nature that with increasing density and characteristic energy of non-equilibrium particles begin to play a crucial role in both plasma confinement and energy transfer in the balance between plasma components. Significant losses of charged particles from the trap as a result of instability leads to the generation of bremsstrahlung X-ray radiation, which is absorbed by the coolant of the magnetic systems (cryomagnet) and increases the thermal load. For example, the ECR ion sources powerful emission of high-energy electrons from the traps can lead to disturbance of the ambipolar potential and undesirable oscillations of the currents of multiply charged ions [8]. In this sense, the cyclotron instability, which is inherent with ECR heating systems, in natural way (and sometimes very significantly!) limits the achievement of the record parameters of the source. Therefore, the problem of achieving the record parameter settings (mostly, for sources of multiply charged ions) is directly related to the controlling of the dynamics of energetic electrons fraction during ECR heating. The problem of formation of suprathermal electrons is also important in understanding plasma processes in thermonuclear fusion devices. For example, despite low density the fraction of hot electrons plays an essential role in ECR plasma startup at the axially symmetric magnetic mirror device gas dynamic trap (GDT) (Budker Institute, Novosibirsk) [9]. Also, intense bursts of coherent microwave radiation were detected in plasma of Mega-Amp Spherical Tokamak (MAST) during edge localized mode (ELM) activity [10]. The most obvious reason of this phenomena is the electron acceleration by inductive electric fields during ELM activity.

Study of the formation mechanism of plasma emissions, may explain from the unified point of view a number of important phenomena occurring in the magnetospheres of Earth and other planets. For the laboratory plasma (ECR ion sources, ECR heating setups for thermonuclear fusion) the stabilization of kinetic plasma instabilities looks especially promising, since the electronic fraction with a non-equilibrium velocity distribution is naturally formed during the ECR plasma heating.

The author thanks to all colleagues from Institute of Applied Physics, Budker Institute and Jyvaskyla University involved in this study.

References

1. *A.V. Vodopyanov, S.V. Golubev, A.G. Demekhov, V.G. Zorin, D.A. Mansfeld, S.V. Razin, V.Yu. Trakhtengerts.* Laboratory modeling of nonstationary processes in space cyclotron masers: First results and prospects. // Plasma Physics Reports 2005, Volume 31, Issue 11, pp 927–937.

2. *A.G. Shalashov, A.V. Vodopyanov, S.V. Golubev, A.G. Demekhov, V.G. Zorin, D.A. Mansfeld, S.V. Razin.* Maser based on cyclotron resonance in a decaying plasma. // JETP Letters. 2006, v.84, is.6, 314-319.

3. *A.V. Vodopyanov, S.V. Golubev, A.G. Demekhov, V.G. Zorin, D.A. Mansfeld, S.V. Razin, A.G. Shalashov.* Observation of pulsed fast electron precipitations and the cyclotron generation mechanism of burst activity in a decaying ECR discharge plasma. // JETP. 2007, v. 104, is.2, p.296-306.

4. *P.A. Bespalov and V.Yu Trakhtengerts.* in Reviews of Plasma Physics, edited by Leontovich M. A., Vol. 10 (Consultants Bureau, New York) 1986.

5. *V.Y. Trakhtengerts and M.J. Rycroft.* Whistler and Alfven Mode Cyclotron Masers in Space // *(Cambridge University Press, New York)* 2008.

6. *M. Viktorov, D. Mansfeld, S. Golubev.* Laboratory study of kinetic instabilities in a nonequilibrium mirror-confined plasma. // Europhysics Letters. 2015, 109 p.65002.

7. *M.E. Viktorov, S.V. Golubev, E.D. Gospodchikov, I.V. Izotov, D.A. Mansfeld, A.G. Shalashov.* On the Mechanism of Energetic Electron Losses from the Magnetic Mirror Trap at the ECR Discharge Startup // Radiophysics and Quantum Electronics. 2013. V.56(4). P.216-227.

8. *O. Tarvainen, I. Izotov, D. Mansfeld, V. Skalyga, S. Golubev, T. Kalvas, H. Koivisto, J. Komppula, R. Kronholm, J. Laulainen, V. Toivanen.* Beam current oscillations driven by cyclotron instabilities in a minimum-B electron cyclotron resonance ion source plasma. // Plasma Sources Sci. Technol. 2014 23 025020

9. *D.V. Yakovlev, A. G. Shalashov, E. D. Gospodchikov A.L. Solomakhin, V.Ya. Savkin, P.A. Bagryansky.* Electron cyclotron plasma startup in the GDT experiment // Nuclear Fusion. 2016. V. 57. P. 016033.

10. *S.J. Freethy, K. G. McClements, S. C. Chapman, R. O. Dendy, W. N. Lai, S.J. P. Pamela, V. F. Shevchenko, R. G. L. Vann.* Electron Kinetics Inferred from Observations of Microwave Bursts During Edge Localized Modes in the Mega-Amp Spherical Tokamak. // Phys. Rev. Lett. 114, 125004.

Laboratory experiments simulating electron cyclotron masers in space

K. Ronald[1], D.C. Speirs[1], M. King[1], T. Heelis[1], S.L. McConville[1], K.M. Gillespie[1], R. Bingham[1,2], C.W. Robertson[1], A.W. Cross[1], A.D.R. Phelps[1]

[1]Department of Physics and SUPA, University of Strathclyde, Glasgow G4 0NG, Scotland, UK
a.d.r.phelps@strath.ac.uk
[2]Rutherford Appleton Laboratory, Chilton, Didcot, Oxon, England, OX11 0QX, England, UK

The study of kinetic instabilities of the electron cyclotron family driven by anisotropic velocity distributions is important, as these instabilities occur in many situations and in particular there are many examples of their occurrence in space plasmas.

There is a well-established and very successful programme of research on this topic at IAP, Nizhny Novgorod [1]. Some of the series of publications from the IAP group are cited in a helpful recent publication by Shalashov et al. [2]. At the University of Strathclyde in Glasgow there has also been continuing interest in this subject for several years [3–11]. In this paper some of the results from laboratory experiments carried out at the University of Strathclyde are presented reproducing non-thermal cyclotron radio emission mechanisms associated with the Earth's auroral region, known as Auroral Kilometric Radiation (AKR). Similar mechanisms are thought to be responsible for some of the cyclotron emissions observed from more distant astrophysical environments.

These experiments were designed and carried out to benchmark PiC simulations of the non-linear, plasma mediated, beam wave interactions associated with the generation of Auroral Kilometric Radiation. A plasma column (20 cm to 1 m in length) was required inside a cylindrical waveguide having a diameter ~5 cm. This tightly constrained the dimensions allowed for the plasma source. To meet this constraint, two Penning-discharge configurations were designed. The electrons in a Penning discharge are insulated from the cylindrical anode (the waveguide) as it is linked by a magnetic field whilst the ends of the anode cylinder are closed by cathode electrodes, providing extended electron path lengths, allowing the discharge to be maintained at low pressures (~10^{-3} mB). When a bias of a few kV was applied to the anode, the discharge ignited, establishing the plasma column. Up to a few 10's mA were provided to the discharge in either a partially-modulating or continuous mode.

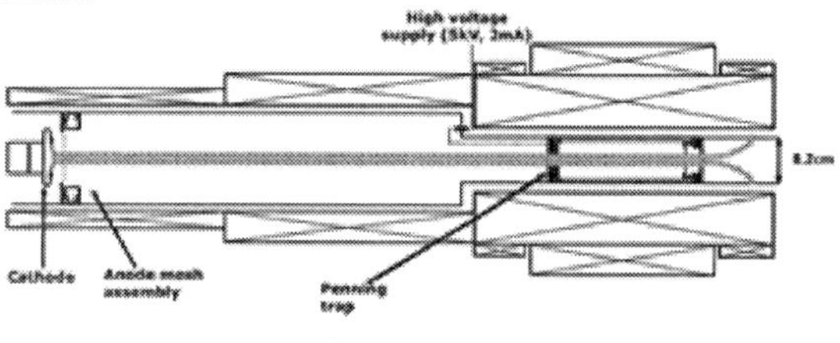

Fig. 1. Drawing of the low current trap and its location in the apparatus

The plasma density at the end of the column was inferred by observing the spectrum of plasma oscillations using a spectrum analyser connected to a small electric dipole antenna polarised along the bias magnetic field, thus avoiding problematic interpretation of the Langmuir probe IV trace caused by the strong magnetic field. The plasma density inferred at the end of the trap was found to vary with discharge current: 1×10^{15} m^{-3} for ~3 mA discharge current in a 20cm-long trap and 1×10^{16} m^{-3} for a 20 mA current in a 1m-long trap.

To support this estimate of the plasma density at the end of the 1m long plasma, and to determine the density averaged along the plasma column, an interferometric technique was implemented using a chirping microwave signal around 9.5 GHz and simultaneously measuring the phase shifts in the R and L circularly polarised TE_{11} modes using a vector network analyser over a wide range of currents. At 20 mA the average density was estimated to be approaching ~4×10^{16} m^{-3}.

Numerical PiC simulations carried out in previous research have helped to confirm the instability leading to the emissions is driven by a horseshoe electron velocity distribution.

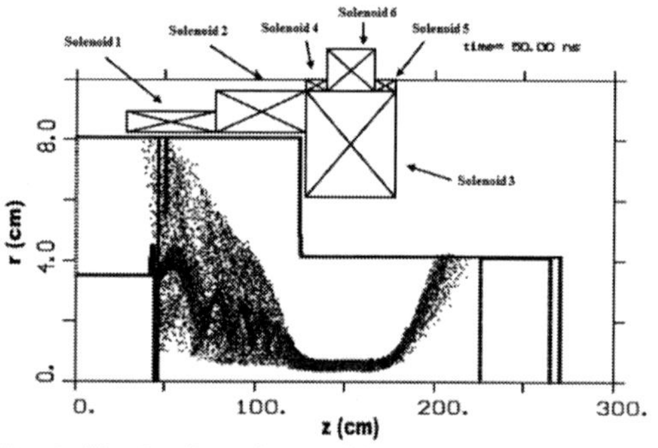

Fig. 2. The Configuration used in the KARAT PiC simulations. Note that the axial scale and the radial scale shown here are intentionally different

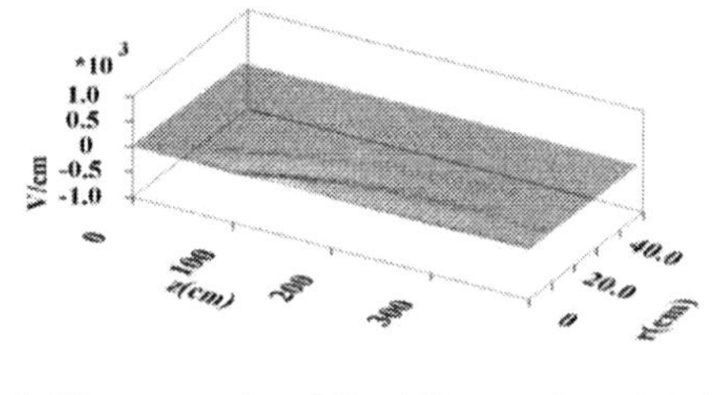

Fig. 3. 3D contour plot of E_θ within an unbounded simulation geometry at time t = 200 ns. Backward wave propagation is evident

The PiC simulations that were undertaken using software, including KARAT, have reproduced the EM wave emission observed in the experiment using a bounded geometry as shown in Fig. 2 and have also been able to simulate the wave production in an unbound environment [10] representative of the magnetosphere, as shown in Fig. 3. Some of these experiments and simulations have revealed that the instability has a significant negative axial wavenumber (backward wave component) [10, 11]. The backward wave character of the emission is evidenced by an oblique wave front propagation angle with respect to the axis of the system. The magnitude of the negative axial wavenumber reduces with increasing energy spread.

Acknowledgements

The UK Engineering and Physical Research Council (EPSRC) is thanked for supporting this research. David Barclay provided valuable technical assistance.

References

1. *Bespalov, P.A. and Trakhtengerts, V.Yu.* in Reviews of Plasma Physics, edited by M. A. Leontovich. New York: Consultants Bureau. 1986. V.10; *Trakhtengerts, V.Yu. and Rycroft, M.J.* Whistler and Alfven Mode Cyclotron Masers in Space. New York: Cambridge University Press. 2008.

2. *Shalashov, A.G., Viktorov, M.E., Mansfeld, D.A. and Golubev, S.V.* Kinetic instabilities in a mirror-confined plasma sustained by high-power microwave radiation // Phys. Plasmas. 2017. V. 24, 032111. P. 1-13.

3. *Speirs, D.C., Vorgul, I., Ronald, K., Bingham, R., Cairns, R.A., Phelps, A.D.R., Kellett, B.J., Cross, A.W., Whyte, C.G., and Robertson, C.* A laboratory experiment to investigate auroral kilometric radiation emission mechanisms // J. Plasma Phys. 2005. V. 71. No.5 . P. 665-674.

4. *Ronald, K., Speirs, D.C., McConville, S.L., Phelps, A.D.R., Robertson, C.W., Whyte, C.G., He, W., Gillespie, K.M., Cross, A.W., and Bingham, R.* Radio frequency resonator structure and diagnostic measurements for a laboratory simulation of Auroral Kilometric Radiation // Phys. Plasmas. 2008. V. 15. No. 5. 056503. P. 1-8.

5. *Speirs, D.C., McConville, S.L., Gillespie, K.M., Ronald, K., Phelps, A.D.R., Cross, A.W., Bingham, R., Robertson, C.W., Whyte, C.G., Vorgul, I., Cairns, R.A., and Kellett, B.J.* Numerical simulation of auroral cyclotron maser processes // Plasma Phys. Control. Fusion. 2008. V. 50. No. 7. 074011. P. 1-15.

6. *Gillespie, K.M., Speirs, D.C., Ronald, K., McConville, S. L., Phelps, A.D.R., Bingham, R., Cross, A.W., Robertson, C.W., Whyte, C.G., He, W., Vorgul, I., Cairns, R.A., and Kellett, B.J.* 3D PiC code simulations for a laboratory experimental investigation of Auroral Kilometric Radiation mechanisms // Plasma Phys. Control. Fusion. 2008. V. 50. No. 12. 124038. P. 1-11.

7. *Ronald, K., Speirs, D.C., McConville, S.L., Gillespie, K.M., Phelps, A.D.R., Bingham, R., Vorgul, I., Cairns, R.A., Cross, A.W., Robertson, C.W., Whyte, C.G., He, W., and Kellett, B.J.* Auroral magnetospheric cyclotron emission processes: numerical and experimental simulations // Plasma Phys. Control. Fusion. 2011. V. 53. No. 7. 074015. P. 1-11.

8. *Vorgul, I., Kellett, B.J., Cairns, R.A., Bingham, R., Ronald, K., Speirs, D.C., McConville, S.L., Gillespie, K.M., and Phelps, A.D.R.* Cyclotron maser emission: Stars, planets, and laboratory // Phys. Plasmas. 2011. V. 18, No. 5. 056501. P. 1-7.

9. *Bingham, R., Speirs, D.C., Kellett, B.J., Vorgul, I., McConville, S.L., Cairns, R.A., Cross, A.W., Phelps, A.D.R., and Ronald, K.* Laboratory astrophysics: Investigation of planetary and astrophysical maser emission // Space Science Reviews. 2013. V. 178. No. 2-4. P. 695-713.

10. *Speirs, D.C., McConville, S.L., Gillespie, K.M., Phelps, A.D.R., and Ronald, K.* Numerical simulations of unbounded cyclotron-maser emissions // J. Plasma Phys. 2013. V.79. No. 6. P. 999-1001.

11. *Speirs, D.C., Bingham, R., Cairns, R.A., Vorgul, I., Kellett, B.J., Phelps, A.D.R., and Ronald, K.* Backward Wave Cyclotron-Maser Emission in the Auroral Magnetosphere // Phys. Rev. Lett. 2014. V. 113. No. 15. 155002. P. 1-5.

EPJ Web of Conferences **149**, 03016 (2017) DOI: 10.1051/epjconf/201714903016
10th International Workshop 2017 "Strong Microwaves and Terahertz Waves: Sources and Applications"

Observation of multiple chirping events in electron cyclotron emission of non-equilibrium mirror-confined plasma

M.E. Viktorov[1,2], A.G. Shalashov[1,2], D.A. Mansfeld[1] and S.V. Golubev[1]

[1]Institute of Applied Physics Russian Academy of Sciences, Nizhny Novgorod, Russia, mikhail.viktorov@appl.sci-nnov.ru
[2]Lobachevsky State University of Nizhny Novgorod, Nizhny Novgorod, Russia

Chirping frequency patterns have been observed in the electron cyclotron emission from a strongly non-equilibrium mirror-confined plasma created by powerful microwave radiation of gyrotron (37.5 GHz, 80 kW) under electron cyclotron resonance (ECR) conditions [1]. An example of multiple chirping events in the spectrum of the electron cyclotron plasma emission is shown in fig. 1. Such patterns are typical for the formation of nonlinear phase-space structures in a proximity of the wave-particle resonances of a kinetically unstable plasma, also known as the "holes and clumps" mechanism [2, 3].

This mechanism has been extensively examined to study and interpret the Alfven wave turbulence driven by high-energy beam ions or alpha particles in toroidal magnetic traps which leads to the observation of the frequency sweeping events [4, 5]. The frequency sweeping events of energetic particle-driven Alfven instabilities were commonly observed on DIII-D, JT-60U, ASDEX-Upgrade, MAST, NSTX, START, and LHD machines with NBI heating. Here, we report the first laboratory observations of potentially the same mechanism acting in a much higher frequency domain. The focus of this work is devoted to the study of peculiar time-frequency characteristics of the electromagnetic radiation, which has been made possible only recently with the advent of methods for measuring the electromagnetic field with high temporal resolution.

In the experiments, ambient plasma and a non-equilibrium component of energetic electrons are created under ECR conditions in the simple mirror trap. The hot electron fraction with a loss-cone distribution function is a matter of kinetic instabilities of various electromagnetic modes which are reviewed in [6]. External microwave heating operates for 1 ms. Excitation of multiple chirping wave packets is observed after ECR heating switch-off with a delay from 0.1 to 1 ms. The frequency band of the observed microwave emission is below the electron cyclotron frequency f_{ce0} in the trap center. The dependency of emission frequency band on the ambient magnetic field variations indicates the cyclotron nature of excited waves.

The microwave emission is observed only in a few narrow frequency bands that are independent of all experimental conditions. Isolated narrow frequency bands may be explained by excitation of individual modes of a bounded plasma in a discharge chamber [1]. When the magnetic field decreases such that f_{ce0} becomes lower than the corresponding frequency band, the microwave emission in this particular band is not detected anymore. Simultaneously, the microwave emission at lower frequency bands may still be detected or even triggered on in a new band.

Within each frequency band the dynamic spectrum of emission is a set of highly chirped radiation bursts with both increasing and decreasing frequencies, see fig. 1. Frequency variation rate is in the range 20–40 MHz/μs. The bandwidth of this narrow-band emission is about $2\times10^{-3} f_{ce0}$. Wave packets are grouped in batches with the number of bursts in packs of about 10. The duration of narrow-band spike reaches about 10 μs. Bundles of these pulses occur quasi-periodically with a period of 10–20 μs. Sometimes very long series of pulses with a duration up to 1 ms are observed.

The fine structure of the emission spectrum in a particular frequency band is changing with the ambient magnetic field decrease. This effect is clearly seen in fig. 2 for the case of a long series of chirping events. When f_{ce0} is significantly higher than emission frequency, we observe the predominance of rising tones with a sweeping frequency that saturates at more or less constant level for all neighboring pulses (fig. 2a). With the decrease of f_{ce0}, falling tones appear in the spectrum (fig. 2b). Finally, when f_{ce0} is close to the frequency of the microwave emission, a narrowband continuous signal without pronounced chirping patterns is observed (fig. 2c). Thus, the dynamic bandwidth of the emission is reduced with the decrease of difference between the electron cyclotron frequency and the emission frequency.

The "holes and clumps" mechanism, also known as Berk-Breizman model [2], seems to be the most suitable to explain the transients with periodic frequency sweeps observed in our experiment. As the mode grows, most of the particles respond adiabatically to the wave, and only a small group of resonant particles mix and cause local flattening of the distribution function in phase space within or near the separatrices formed by the waves. However, when linear dissipation from a background plasma is present, the saturated plateau state becomes unstable and the mode tends to grow explosively. That results in the formation and subsequent evolution of long-living, as compared to the linear growth, structures in the particle distribution, so-called holes (a depletion of particles) and clumps (an excess of particles). These structures represent nonlinear waves of Bernstein-Greene-Kruskal type whose frequencies are slightly up- and down-shifted with respect to that of the initial instability. Their subsequent convective motion in phase space is synchronized to the change in wave frequency, thus leading to complex chirping patterns in dynamical spectra of unstable waves [3]. A distinctive feature of the instability under investigation is the significant effect of electron collisions providing a dissipation channel for electromagnetic energy, which leads to the formation of characteristic chirped elements in the emission spectra.

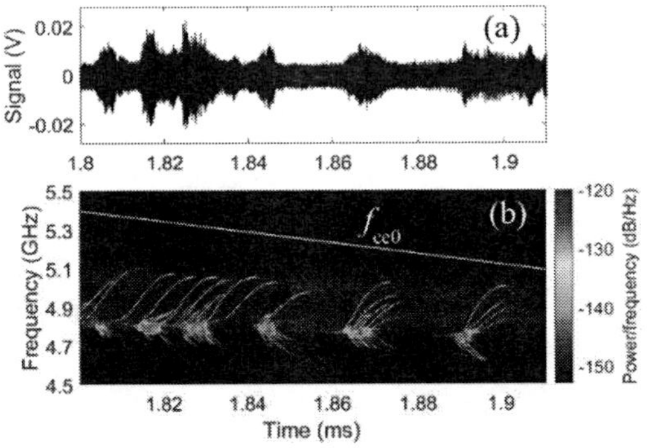

Fig. 1. (*a*) Waveform of the wave electric field oscillations and (*b*) its spectrogram. The white line shows the value of electron cyclotron frequency in the trap center on the axis. The ECR heating is operating during 0–1 ms (not shown). Adopted from [1]

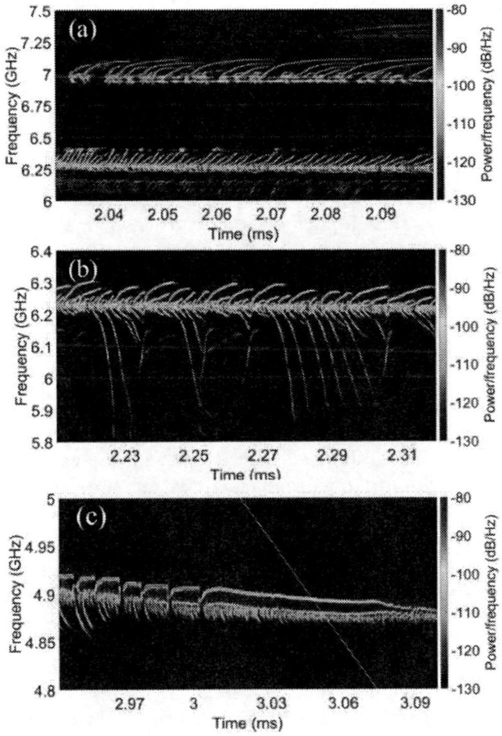

Fig. 2. Different regimes of chirping events in plasma emission spectrum observed during plasma decay. Panels (*a*) – (*c*) correspond to different sequential time slots. The white line shows the value of electron cyclotron frequency in the trap center on the axis. The ECR heating is operating during 0–1 ms (not shown)

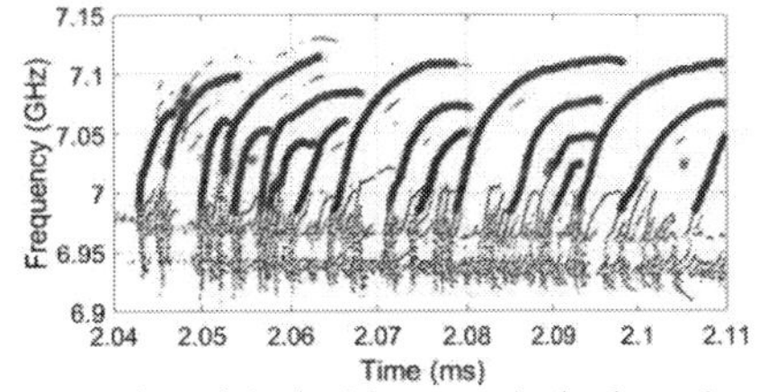

Fig. 3. Selection of single rising tones in the dynamic spectrum shown in fig. 2a (upper part)

For a very rough estimate, one can isolate the single spectral component (see fig. 3) and model the evolution in time of the chirping event. In the collisionless limit for resonant particles, a bounce average of Maxwell-Vlasov kinetic equations yields the frequency shift $\delta\omega$ of an unstable mode as [2]

$$\delta\omega = \frac{16\sqrt{2}}{3\sqrt{3}\pi^2}\gamma_L\sqrt{\nu t}. \quad (1)$$

This expression is valid for the regime where resonant particles are deeply trapped in the wave potential and perturbation of the passing particle is negligible. Fitting the measured frequency sweeps $\delta\omega = \sqrt{At}$ by constant A and assuming near-threshold condition $\gamma_L \approx \nu$, one may estimate the upper boundary for the dissipation ν or, equivalently, the lower boundary for the growth rate γ_L compatible with (1). In the experiments we obtain typical values $A = (1–3) \times 10^{21}$ s^{-3} resulting in $\gamma_L = (1.7–2.5) \times 10^7$ s^{-1}, what is in reasonable agreement with independent estimates. More details may be found in [1].

The work has been supported by RFBR (grants No. 16-02-00625, 16-32-60056).

References

1. *Viktorov, M.E., Shalashov, A.G., Mansfeld, D.A., Golubev, S.V.* Observation of quasi-periodic frequency sweeping in electron cyclotron emission of nonequilibrium mirror-confined plasma // EPL. 2016. V. 116. P. 55001.

2. *Berk, H.L., Breizman, B.N., Petviashvili, N.V.* Spontaneous hole-clump pair creation in weakly unstable plasmas // Phys. Lett. A. 1997. V. 234. P. 213.

3. *Lilley, M.K., Breizman, B.N., Sharapov, S.E.* Effect of dynamical friction on nonlinear energetic particle modes // Phys. Plasmas. 2010. V. 17. P. 092305.

4. *Breizman, B.N., Sharapov, S.E.* Major minority: energetic particles in fusion plasmas // Plasma Phys. Control. Fusion. 2011. V. 53. P. 054001.

5. *Sharapov, S.E., et al.* Energetic particle instabilities in fusion plasmas // Nucl. Fusion. 2013. V. 53. P. 104022.

6. *Shalashov, A.G., Viktorov, M.E., Mansfeld, D.A., Golubev, S.V.* Kinetic instabilities in a mirror-confined plasma sustained by high-power microwave radiation // Phys. Plasmas. 2017. V. 24. P. 032111.

EPJ Web of Conferences 149, 03017 (2017) DOI: 10.1051/epjconf/201714903017
10th International Workshop 2017 "Strong Microwaves and Terahertz Waves: Sources and Applications"

Fast events detection with the CTS diagnostic on FTU and plans for improvement

A. Bruschi[1], E. Alessi[1], B. Baiocchi[1], W. Bin[1], O. D'Arcangelo[2], F. Fanale[1], L. Figini[1], C. Galperti[3], S. Garavaglia[1], G. Gittini[1], G. Granucci[1], G. Grosso[1], L. Lubyako[4], C. Mazzotta[2], V. Mellera[1], A. Moro[1], F. Orsitto[5], F. Pallotta[1], G. Rocchi[2], U. Tartari[1], O. Tudisco[2]

[1]Istituto di Fisica del Plasma – Consiglio Nazionale delle Ricerche, Milano, Italy, bruschi@ifp.cnr.it
[2]ENEA C.R. Frascati, via E. Fermi 45, 00044 Frascati (Roma), Italy
[3]Swiss Plasma Center, Ecole Polytéchnique Fédérale de Lausanne, Lausanne, Switzerland
[4]Institute of Applied Physics, 46 Ulyanov st., Nizhny Novgorod, 603950, Russia
[5]Consorzio CREATE, Università degli Studi di Napoli, Napoli, Italy

The capability of the CTS diagnostic to detect emissions at frequencies close to the injected EC power can be used to study the occurrence of parametric decay instabilities (PDI) excited by the ECH power. In this way, from the detection of back-scattered power in the case of 2nd harmonic X-mode injection [1, 2], the sources of the signals were well explained by waves generated by PDI mechanisms [3]. On FTU, during the work with the CTS diagnostics focused on studying the effects of possible PDI occurrence on EC beams (under Enabling Research project ER-ENEA-06), other still unexplained fast signals were detected in the spectra, in experiments with injection of EC power in O-mode at the fundamental EC harmonic [4]. The possibility to analyze data with a high time resolution is crucial for the study of these fast events, and the use of a fast rate (12 GS/s max) digitizer and of the FFT of the sampled intermediate frequency signal (taken at the output of the 140 GHz heterodyne radiometer front-end) over different time-scales is mandatory [5]. In this paper the observations obtained with new experiments are compared to the previous ones. The design of the new receiving line equipped with a RS antenna and the status of the work are presented.

Experiment preparation

On the diagnostic side an effort is ongoing to exclude any parasitic effect of gyrotron spectrum, of high-frequency front-end components, of local breakdown in the antenna and other disturbances. The gyrotron frequency spectrum, previously obtained from a lossy pick-up line 50 m long used to measure the gyrotron main line frequency vs time (without using the notch filter), was now obtained with a direct pick-up from the stray radiation in the launcher region while shooting the gyrotron on the local dummy load. In this way a higher signal is collected and a sensitive spectrum of the gyrotron radiation is obtained, using the notch filter to suppress the main line. The resulting spectra from the gyrotron often include spurious lines that appear also in some CTS spectra from plasma, as shown in Fig. 1. The lines of this type have always the same behavior, starting from the beginning of the gyrotron pulse and lasting tens or hundreds of ms with a frequency decreasing with oscillations around an average speed of 2 GHz/s (Fig. 1 top right).

Whether these are due to gyrotron spurious mode or to a signal treatment in the HF diagnostic section is under evaluation. The level of the spurious line, not observed in the 2015 shots, is estimated at least 50 dB below the main gyrotron line.

The calibration of spectra received from the plasma is now performed comparing the spectra with the ones obtained during Ohmic pulses with the EC resonance inside the plasma (where the emission is only thermal ECE) and subtracting the noise spectra obtained during the same shot with the aid of a chopper/modulator working at around 10 kHz. The spectra obtained during the two chopping phases (on-off) are evaluated by proper data analysis, summed separately and then the noise part is subtracted. Examples of such spectra are shown in Fig. 2. Local breakdown in the antenna was monitored recording the light and the stray radiation level received at the back of the launching/receiving antenna, using a remote photomultiplier connected to an optical fiber as light detector and a mm-wave detector installed on a horn antenna looking into the port 8, from where the probe is launched.

Analysis of experimental observations

The analysis work has been focused on the correlation of spectral features with other plasma diagnostic observations. One of the features reported in [4] was studied in more detail, showing correlation with the appearance of a growing m:n = 2:1 magnetic island at 4.7 T central magnetic field, in asymmetric configuration (see [4]) at line-integrated density of around 1×10^{20} m^{-3}. In Fig. 2 the spectrum of shot #39848 is shown in correlation with a Mirnov Coil signal and an MHD frequency spectrum. Sudden bursts of emissions detected by CTS at frequencies multiple of the ion cyclotron frequency above and below the gyrotron frequency (at 0.8 GHz in the spectrum) show a fast time evolution (at microsecond scale) which differs with respect to phenomena observed in [1, 2]. Simultaneous variations of the stray radiation level in the vessel close to the launching port have been recorded.

New experiments were performed in 2016 at 4.7 T with two channels set to receive in parallel the two polarizations [6], both with frequency band enlarged to 4.2 GHz around the local oscillator frequency. In asymmetric configuration, emissions of another type described in [4] were observed and analyzed.

They are reported in Fig. 3, where the two channels of the receiver 1 (top, with chopping, detecting X-polarized wave), receiver 2 (without chopping, mostly O-wave) and Mirnov Coil signal are shown. Even if the appearance of these emissions is linked to the mode growth, the timing is not linked in phase with island rotation, and the duration is much less that the rotation period by at least one order of magnitude.

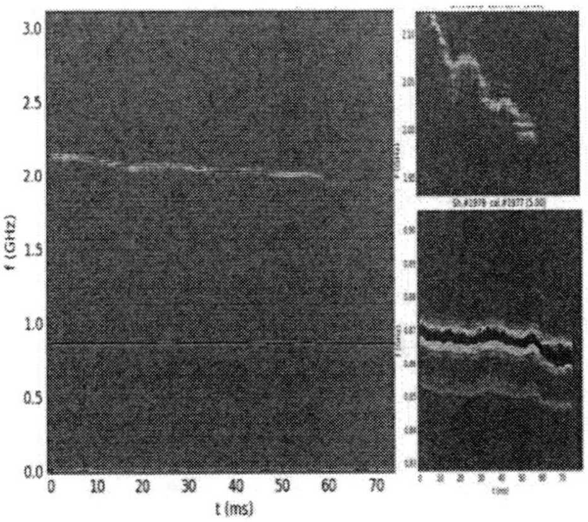

Fig. 1. Left: Spectrum obtained from the gyrotron stray power, with a pick-up at the EC launcher. The main gyrotron line which is stable in frequency at 0.9 GHz from the Local Oscillator is accompanied by a spurious line variable in frequency around 1.2 GHz above the gyrotron line. On the right a zoom in frequency of the two lines during the gyrotron pulse (the gyrotron line at the bottom)

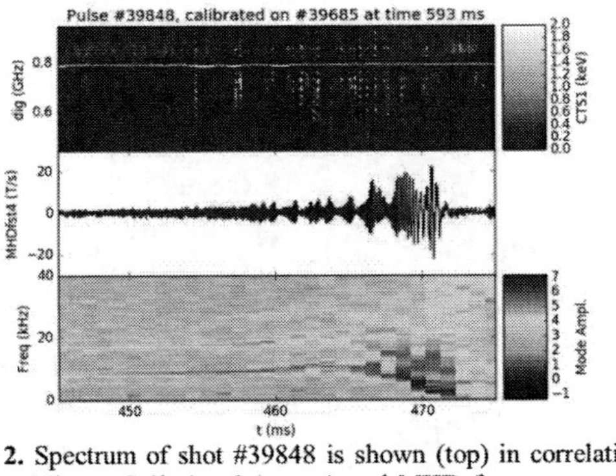

Fig. 2. Spectrum of shot #39848 is shown (top) in correlation with a Mirnov Coil signal (center) and MHD frequency spectrum (bottom). Bursts of emissions detected by CTS at frequencies multiple of the ion cyclotron frequency above and below the gyrotron frequency (at 0.8 GHz in the spectrum)

Diagnostic improvements

In order to exclude in the spectrum signatures of the gyrotron stray (by comparison with the gyrotron spectrum in the same shot) and to verify the position of the emissions (comparing with the spectra received from a different line of sight), a new receiving line is being designed, that collects radiation from the plasma through a square corrugated antenna placed in the equatorial plane of the EC port n. 8, the same of the presently used EC and CTS antenna. A mirror steering mechanism is being designed to perform steering of the received beam from outside, without moving parts into the vessel, using the well-known Remote Steering waveguide properties, to have an independent viewing line that can intersect or not the probe beam. A switch will allow also a direct reception of the stray from the gyrotron line at the port level. The path of the receiving line combines reduced-section square corrugated waveguides, 88.9 mm circular corrugated waveguides or, in case of need, also a definite-length smooth circular waveguide and quasi-optical sections with mirrors instead of mitre bends.

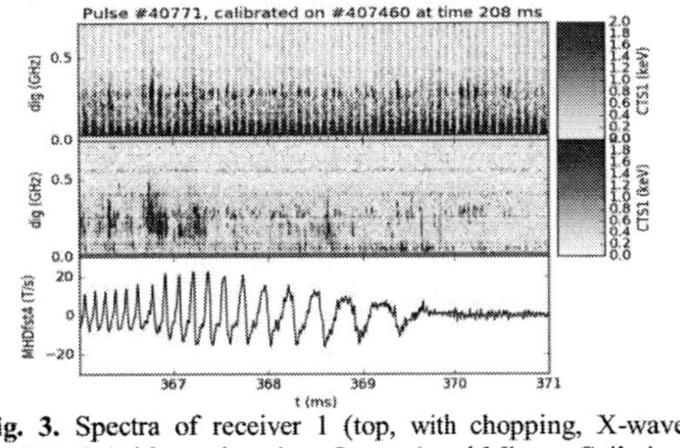

Fig. 3. Spectra of receiver 1 (top, with chopping, X-wave), receiver 2 (without chopping, O-wave) and Mirnov Coil signal for shot #40771

Acknowledgement

This work has been carried out within the framework of the EUROfusion Consortium and has received funding from the Euratom research and training programme 2014–2018 under grant agreement No 633053. The views and opinions expressed herein do not necessarily reflect those of the European Commission.

References

1. *Westerhof, E. et al.* Strong Scattering of High Power Millimeter Waves in Tokamak Plasmas with Tearing Modes // *Phys. Rev. Lett.* 2009. V. 103, 125001.
2. *Nielsen S.K. et al.* Experimental characterization of anomalous strong scattering of mm-waves in TEXTOR plasmas with rotating islands // *Plasma Phys. Control. Fusion* 2013. V. 55, 115003.
3. *Gusakov, E.Z., Popov, A.Yu.* Theory of anomalous backscattering in second harmonic X-mode ECRH experiments // *Phys. Plasmas* 2016. V. 23, 082503.
4. *Bruschi, A. et al.,* Observation of short time-scale spectral emissions at millimeter wavelengths with the new CTS diagnostic on the FTU tokamak // *Nucl. Fusion* 2017. V. 57, 076004.
5. *Bin, W. et al.,* The upgraded Collective Thomson Scattering diagnostics of FTU // *Fusion. Eng. Des.* 2015. V. 96-97, P. 733-737.
6. *Bin, W. et al.,* Advances in the FTU collective Thomson scattering system // *Rev. Sci. Instrum.* 2016. V. 87, 11E507.

EPJ Web of Conferences **149**, 03018 (2017) DOI: 10.1051/epjconf/201714903018
10th International Workshop 2017 "Strong Microwaves and Terahertz Waves: Sources and Applications"

Filters for diagnostic of Doppler reflectometry on the L-2M stellarator for operation under conditions of high ECR heating power

D.V. Malakhov[1,2,3], A.A. Kharchevsky[1,3], N.N. Bogachev[1,2,3], N.N. Skvortsova[1,4]

[1]Prokhorov General Physics Institute, Moscow, Russia, 89199945791@mail.ru
[2]Pirogov Russian National Research Medical University, Moscow, Russia
[3]Moscow Technological University (MIREA), Moscow, Russia
[4]National Research Nuclear University MEPhI, Moscow, Russia

The study of the properties of edge plasma, in particular, the plasma rotation speed, in toroidal traps is associated with the search for ways to increase the efficiency and reliability of the thermonuclear reactor. One of the methods for measuring the rate of poloidal plasma rotation in tokamaks and stellarators is Doppler reflectometry (DR). The Doppler reflectometer is designed to work as part of the experimental L-2M device to measure the rates of its poloidal rotation, as well as to measure the spatial-frequency spectra of plasma density fluctuations [1].

Diagnostics of high-temperature plasma in magnetic confinement systems is associated with a variety of problems and requires atypical engineering solutions. The powerful sources of microwave radiation (gyrotrons) used to heat the plasma [2] interfere with the detection of low-power diagnostic signals. To isolate such signals, it is necessary to provide high suppression at the ECR frequency of heating 75 GHz and to provide low attenuation at the diagnosis frequency of 30–40 GHz. The created system for filtering electromagnetic radiation consists of a band-stop filter based on a Fabry-Perot resonator and three, pin-shaped waveguide filters [3]. The CAD EMPro Keysight Technologies in a three-dimensional geometry of band-stop filter was represented as a few pairs of contiguous mica plates arranged successively at a distance from each other. Simulation of a microwave filter based on a Fabry-Perot resonator for microwave diagnostics of Doppler reflectometry, made it possible to construct an experimental filter sample with characteristics close to the given ones. The report presents the results of the development, modeling and experimental study of a filter based on a Fabry-Perot (FP) resonator. The introduction of the filter system made it possible to successfully carry out experimental measurements of the spectra of DR in the L-2M stellarator with an ECR heating power density of up to 3.4 MW/m^3. The aim of the research was to create conditions for the stable operation of DR in conditions of powerful heating ECR radiation. To reduce the effect of an external signal with a frequency of 75.3 GHz and noise in the range 60–80 GHz, it was necessary to create a new filtering system. In the filtration system we used band-stop filters. The task of these filters is to provide large attenuation (suppression) in a certain frequency range Δf near the central frequency f_0.

The system of filters for diagnostics of DR consists of three pin-waveguide filters and filter based on Fabry-Perot resonator made of mica plates. In view of the design features of the diagnostics (in particular, the overall dimensions of the waveguide paths) the installation of additional pin filters, was not possible. It was required to design and manufacture a compact band-stop filter with a central suppression frequency of 75 GHz. A filter was proposed from the sequence of Fabry-Perot resonators. Millimeter and submillimeter filters based on the Fabry-Perot resonator are actively used in various diagnostics and radio astronomy. We used Fabry-Perot resonators with flat round mica plates (dielectric permittivity of mica $\varepsilon = 6$) with diameter D = 80 mm and thickness S = 0.1 mm.

The 3D-model (XYZ coordinate system) of the resonator filter was created in the computer-aided design system Electro-Magnetic Professional (CAD EMPro) from Keysight Technologies (formerly Agilent Technologies) on the fig. 1.

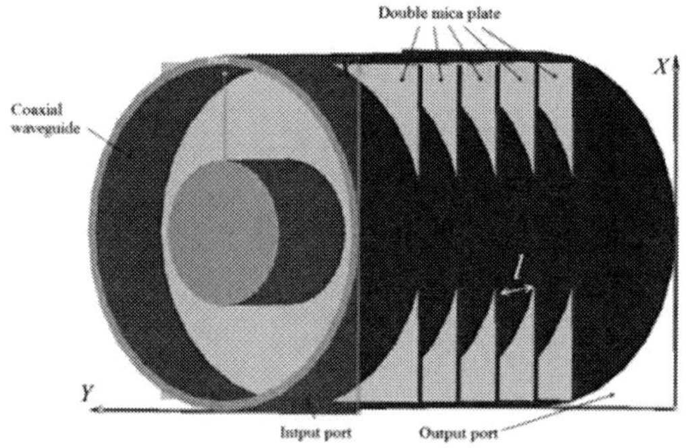

Fig. 1. The 3D-model of the filter based on a Fabry-Perot resonator

The band-stop filter with Fabri-Perot resonators is represented as a pair of mica plates in contact, located in succession at a distance l from each other. The plates are placed in a coaxial waveguide perpendicular to its axis to ensure the propagation of the TEM wave. The calculation was carried out by the Finite Element Method (FEM) in the Agilent FEM Simulator block [4], the boundary conditions at the edges of the counting area are given by Perfect Matching Layer (PML). Waveguide ports are on both sides of the filter. Test measurements showed an improvement in signal/noise ratio.

The filter characteristics were measured using an MVNA-8-350 vector network analyzer from AB Millimetre with a measurement step of 10 kHz.

The level of loses (or transmission coefficient S21) of the waveguide-pin filters measured in the ranges 29–41 GHz (operation frequencies of DR) and 62.5–100 GHz. The transmission coefficients of the pin filter do not exceed 0.5 dB in the operating frequency range of the Doppler reflectometer and for the ECR frequency of the heating radiation (75.3 GHz) more than 30 dB.

Taking into account the fact that three waveguide filters are used in the filtration system one common and one in each detector channel, the suppression of the heating frequency of 75.3 GHz in the waveguide section of the filtration system should be about 60 dB.

As can be seen on fig. 2 the level of noises in diagnostic signals were much reduced when FP filter was installed. This reduction of microwave heating in diagnostic channels allows to reliably measurement characteristics of plasmas fluctuations.

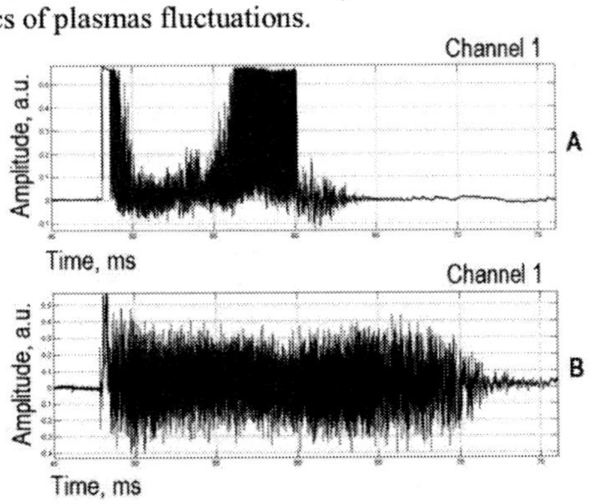

Fig. 2. Signals of Doppler reflectometry system (A) without FP filter; (B) – with installation of FP filter

As a result of the calculations, the transmission coefficient was obtained for the Fabry-Perot filter model. The results for the optimized model are shown in figure 3 in the form of a black solid curve in the range 40–80 GHz. As can be seen, at a frequency of 75.3 GHz, the attenuation value S21 = –21.9 dB was obtained.

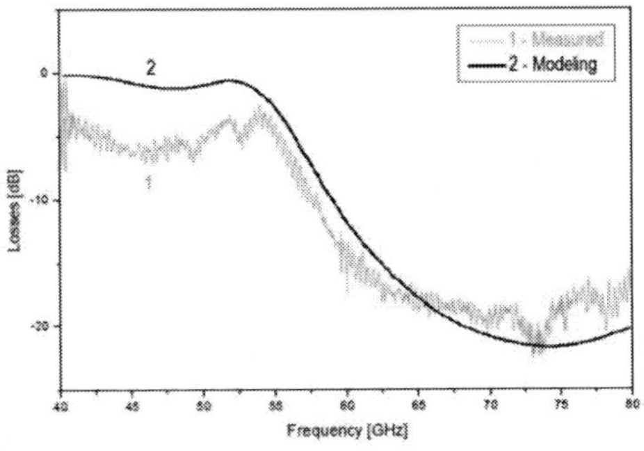

Fig. 3. The level of losses in the filter based on the Fabry-Perot resonator. Graph 1 – Measured value of filter attenuation, Graph 2 – Calculated value of attenuation of the filter model

It is worth noting that the calculated attenuation in the optimized filter model at 35 and 40 GHz does not exceed –0.9 dB. Besides on the figure 3 shows the measured transmission coefficient of a filter based on a Fabry-Perot resonator. Due to the specifics of the analyzer device, data can be removed in discrete spectral ranges. It should be taken into account that all measurements were carried out with the continuous generation of low-power radiation in MVNA-8-350 vector network analyzer of the order of hundreds of mW.

The developed and created filter, as well as the antenna system, can be used for microwave diagnostics (scattering, interferometry, reflectometry) on toroidal installations (tokamaks and stellarators) with electron cyclotron heating.

Optimization of algorithms for data analysis and the development of new software represented in [5].

The algorithm for analyzing the DR data includes the procedure for preparing data, analyzing data, visualizing and storing the results of the analysis. As part of optimizing the diagnosis of DR, the hardware and software parts were upgraded. As part of the hardware, a transition was made to the allotted ADC (2 channels, 100 MHz, 12 bit). The data transfer from the ADC is done via Ethernet over fiber optic communication. Data is stored on a NAS (Network Attached Storage) with access rights. Calculations are performed by operators on personal computers (Matlab) and on a computer server (Python, CUDA). The preprocessing procedure includes: converting the output format, normalizing the signal, searching for features in the time series. The procedure for analyzing data in general form is described in [1]. An optimized algorithm for analyzing data and searching for the velocity of poloidal rotation of a plasma column includes methods for automatically calculating a set of parameters. Such parameters are: the evolution of statistical moments, the search for the maximum Doppler shift (different algorithms), spectral analysis (Fourier and wavelet) and other parameters of signals that carry a physical meaning. Using these parameters allows you to work with such a concept as feature. After feature extraction we can talk about the application of methods of machine learning for the analysis of plasma density fluctuations.

This work was supported in part by the support of the Russian President's grant for young scientists #MK-5298.2016.8

References

1. *Pshenichnikov, A.A., Kolik, L.V., Malyh, N.I., Petrov, A.E. et al.*, The use of Doppler reflectometry in the L-2M stellarator // Plasma Physics, 2005. Vol 31. № 7. Pp. 604-611.

2. *Batanov, G.M., Harchev, N.K., Skvortsova, N.N., Sarksian, K.A. et al.*, New gyrotron complex MiG-3 for creating and plasma heating in the L-2M stellarator and the results of the first experiments // Applied Physics. 2012. № 6. Pp. 79-87.

3. *Kharchevskii, A.A., Bogachev, N.N., Malakhov, D.V.*, Optimization of doppler reflectometry system for operation in high energy input in the l-2m stellarator // Engineering Physics, №6, 2016, Pp.59-65

4. Keysight Technologies. About EMPro. URL: http://www.keysight.com/en/pc-1297143/EMPro.

5. https://github.com/malakhovdmitry/l2m

EPJ Web of Conferences **149**, 03019 (2017) DOI: 10.1051/epjconf/201714903019
10th International Workshop 2017 "Strong Microwaves and Terahertz Waves: Sources and Applications"

Interferometer system for Keda Torus eXperiment using terahertz solid-state diode sources

Wenzhe Mao[1], Jinlin Xie[1], Weixing Ding[1, 2], Hong Li[1], Tao Lan[1], Adi Liu[1], Wandong Liu[1]

[1]School of Physics, University of Science and Technology of China, Hefei, Anhui 230026, China, jlxie@ustc.edu.cn
[2]University of California-Los Angeles, P.O.Box 957099, Los Angeles, California 90095-7099, USA

A one chord interferometer system based on terahertz solid state source has been first applied on a reversed field pinch (RFP) device, Keda Torus eXperiment (KTX) [1]. The compact and lightweight platform made of non-metal materials has been used to support the system with stability by eliminating the vibration sources [2]. The whole system works continuously and steadily in the rigorous electromagnetic environment during the experiment. The line averaged electron density has been measured, ranging from $10^{18}m^{-3}$ to $10^{19}m^{-3}$ in three different operation conditions on KTX. The spectrum of the averaged electron density fluctuation is also analyzed and compared with the magnetic signal to show a mode rotation with frequency around 4kHz.

As shown in fig. 1, KTX is a middle size reversed field pinch device with a major radius 1.4 m and a minor radius 0.4 m. The whole interferometer is placed very close to KTX device to make it a compact system. The ohmic heating coils with air core are adopted, so the stray magnetic field around the machine is strong. Between the yellow ohmic field coils, the maximum stray field is around 200 Gauss. On the other hand, the high loop voltage during the plasma current ramp phase means the whole system is surrounded by high stray electric field. The terahertz solid state source is protected by 5 mm thick aluminum boxes, which are covered by 1 mm thick permalloy shell for low frequency magnetic field shielding. Two sources are used to form a heterodyne interferometer [3]. One is fixed frequency 650 GHz and the other is tunable from 630 GHz to 655 GHz. The output power of each source is about 2 mW and the signals are received by mixers with sensitivity about 750 mW/mV. Epoxy resin are used as the major material of the platform to avoid eddy current and to make the system electronically isolated. The total epoxy resin boards weigh about
200 kg. Dominate vibration effects caused by eddy current have been suppressed to the same level as background phase noise, which is about 0.05pi, corresponding to a line integral density measurement accuracy about $10^{17}m^{-2}$. The spatial resolution is 3cm in the radial direction, which is constraint by the probing beam width through plasma.

After the installation of the interferometer system, efforts were spent on the signal optimization further. The first problem is the refraction effect on signal amplitude received at detectors. As shown in fig. 2, the IF amplitude detected at the mixer almost dropped to zero during the discharge, while the reference signal was not affected at all. This phenomenon indicates that the probing beam deviated from the receive port of the Schottky diode mixer during the discharge, which is probably caused by beam refraction through plasma. The width of the mixer input port is about 4 mm. The KTX vacuum port is narrow with a diameter only 35 mm. Those restrictions result in strict condition of beam alignment. After carefully aligning the centers of probing beam and the vacuum window, the 'zero amplitude' phenomenon disappeared.

The second problem is the vibration problem.

As phase measurement system, the interferometer can provide an accurate estimation of the vibration level. Fig. 3 show the phase shift caused by system vibration during the discharge without plasma. The gray shading area is the time period of plasma pulse. During this period the vibration amplitude is less than 0.1π, which is consistent with previous bench test results [2]. By analyzing the spectra of the phase vibration signals, it is inferred that the initial vibration between 12 ms and 30 ms is the vibration of the resin board close to the vacuum window between two toroidal coils. And the late low frequency is the vibration of the support pillar. Since there is no direct contact between the whole system and the KTX machine, the vibration is caused by the sound wave propagation. In a word, the density measurement is not effected by the current vibration level. When the plasma pulse length is optimized to about 100ms after the coming KTX upgrade, this problem require to be take into consideration. Another pillar may need to strength the support structure.

Line integrated density is provided by the phase difference between the reference IF signal and the measurement IF signal. The intermediate frequency is set at 2.3 MHz, corresponding to a time resolution about 0.4 μs. Here a digital phase comparator technique [4] is used instead of analog phase comparator. KTX is now in conditioning phase, and could be operated in three different mode: Short pulse Tokamak discharge with high current (TS), Long pulse Tokamak discharge with low current (TL) and RFP discharge. In TS mode the maximum plasma current can research 200 kA. However since the maximum toroidal field of KTX is only 0.2 Tesla, this low q configuration is unstable and the pulse duration is only about 9 ms. In TL mode the maximum plasma current can research 50 kA. The pulse duration can research 25 ms, which is already beyond the shell penetration time. The minimum edge q is around 2.8 in this configuration, which is more stable than that in TS mode. The maximum plasma current in RFP mode is about 70 kA and the discharge time is about 5 ms. In RFP configuration, multi-tearing mode tent to resonant at different radial positions and overlap with each other [5]. That deteriorates the confinement and increases the plasma resistivity.

Figure 4 shows the waveforms of plasma current and center chord averaged electron density in TS, TL and RFP discharge mode. The density fluctuation amplitudes are larger in TS and RFP compared with that in TL condition. This is consistent with the stability condition mentioned above. In TS and RFP mode, plasma is more likely to scrape the wall, which results in severe neutral recycling. Another possibility is that the particle flux induced from poor confinement hit the wall and cause impurity sputtering out. The density operation region is also provided by the interferometer as shown in fig 5. The chord averaged density range from $0.5\times10^{19}\mathrm{m}^{-3}$ to $3\times10^{19}\mathrm{m}^{-3}$.

Figure 6 shows the comparison between spectrum of line electron density fluctuation and spectrum of edge magnetic fluctuation field. It is clearly shown in fig. 6a) the amplitude of fluctuation is about 0.05×10^{19} m^{-3}. Both spectrum of density fluctuation and magnetic fluctuation indicate that a mode with frequency around 4 kHz exists during the whole discharge time, and the mode should be electromagnetic. It also can be inferred that the poloidal mode numbers should be odd because the line integrated diagnostics cannot detect the fluctuations with even mode number. This is consistent with the mode number calculation based on poloidal mirnov array signal using SVD method. Above all, all results show the successful application of terahertz solid source based interferometer system to measure weak density fluctuation on toroidal magnetic confinement fusion device.

This work is supported by the Fundamental Research Funds for the Central Universities and the National Magnetic Confinement Fusion Energy Program of China under Contract Nos. 2014GB109002. This work is also supported by the National Natural Science Foundation of China 11605199.

Fig. 1. Interferometer system arrangement on KTX device

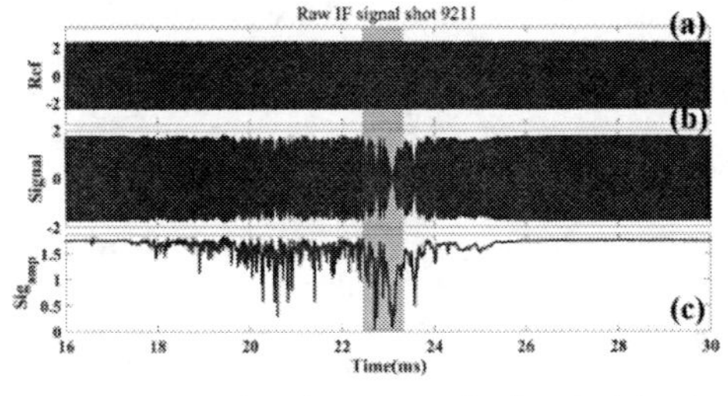

Fig. 2. Raw intermediate frequency (IF) signals: (a) reference signal, (b) probing beam signal and (c) envelop curve of diagnostic signal

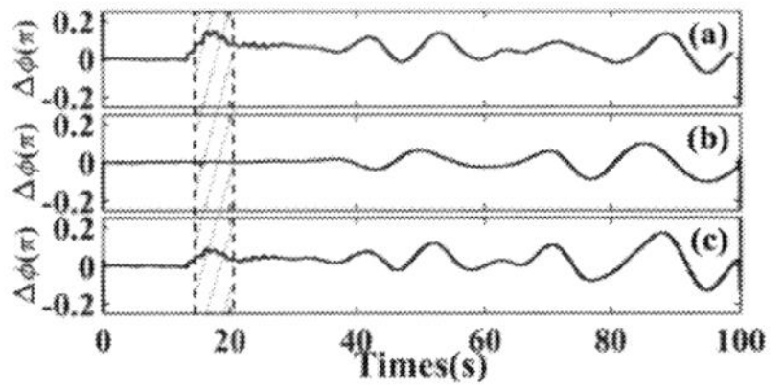

Fig. 3. Phase vibration in: a) toroidal field coils discharge only, b) Ohmic field coils discharge only, and c) Ohmic field coils and toroidal field coils discharge together

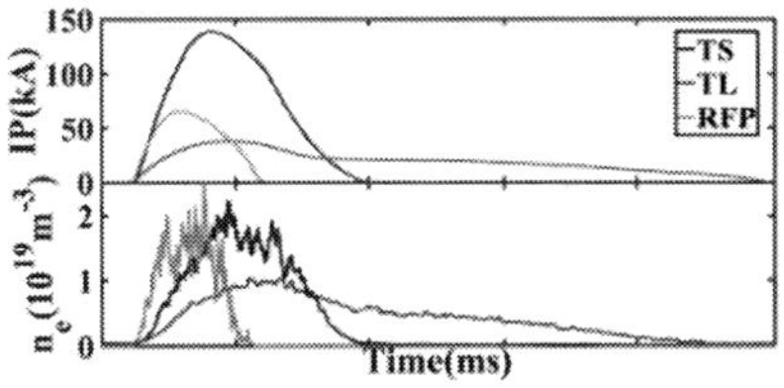

Fig. 4. Waveforms of plasma current and center chord averaged electron density in TS, TL and RFP condition

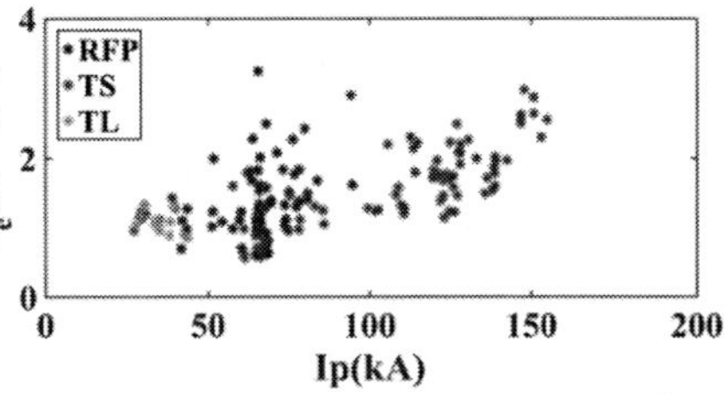

Fig. 5. Maximum plasma current versus corresponding chord averaged electron density

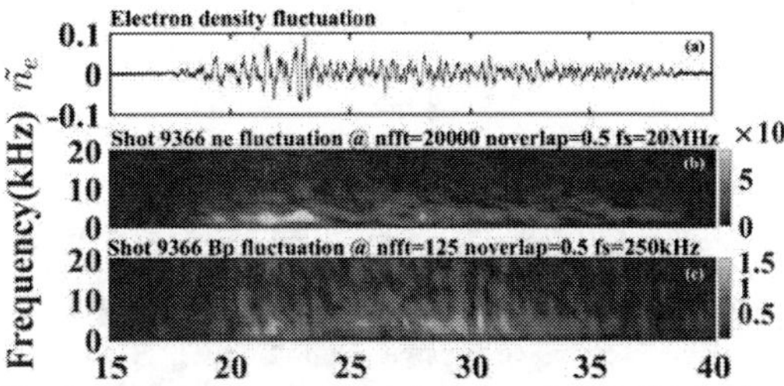

Fig. 6. a) Central chord averaged electron density fluctuation, b) Spectrum of averaged electron density fluctuation, c) Spectrum of mirnov probe

References

1. *W. Liu et al.,* Plasma Phys. Control. Fusion. 56, 094009 (2014).
2. *Mao W et al.* Rev. Sci. Instrum. 87, 11E122(2016).
3. *J.L. Xie et al.* Rev. Sci. Instrum. 85, 11D828(2014).
4. *Y. Jiang et al.* Rev. Sci. Instrum. 68, 902(1997).
5. *P. Franz et al.* Phys. Rev. Lett. 92, 125001.(2004).

EPJ Web of Conferences **149**, 03020 (2017) DOI: 10.1051/epjconf/201714903020
10th International Workshop 2017 "Strong Microwaves and Terahertz Waves: Sources and Applications"

Parametric decay instability near the upper hybrid resonance and anomalous mm-wave scattering in tokamak and stellarator plasmas

S.K. Hansen[1,2], S.K. Nielsen[2], M. Salewski[2], M. Stejner[2], J. Stober[1], and the ASDEX Upgrade team[1]

[1]Max-Planck-Institut für Plasmaphysik, 85748 Garching bei München, Germany, Soeren.Kjer.Hansen@ipp.mpg.de
[2]Department of Physics, Technical University of Denmark, 2800 Kgs. Lyngby, Denmark

Introduction and Background

The parametric decay instability (PDI) is results from a three-wave interaction in which an incident (pump) wave decays into two daughter waves. PDIs may occur in certain media under the influence of radiation with a sufficiently large amplitude, e.g. in tokamak and stellarator plasmas heated or diagnosed by mm-waves from a gyrotron. Energy and momentum conservation in the three-wave process impose the selection rules that the sum of the daughter wave frequencies, ω, and wave vectors, $\mathbf{k}$, should equal those of the pump wave,

$$\omega_0 = \omega_1 + \omega_2, \quad \mathbf{k}_0 = \mathbf{k}_1 + \mathbf{k}_2, \tag{1}$$

where subscripts 0, 1 and 2 refer to the pump wave, and the low- and high-frequency daughter waves, respectively; up-shifted high-frequency daughter waves, satisfying $\omega_3 = \omega_0 + \omega_1$, $\mathbf{k}_3 = \mathbf{k}_0 + \mathbf{k}_1$, also exist due to scattering of the pump wave by the low-frequency daughter waves. Thus, PDIs result in frequency-shifted radiation which interferes with mm-wave plasma diagnostics such as electron cyclotron emission (ECE) or collective Thomson scattering (CTS). In this work we focus on parametric decay of an electromagnetic pump wave into two electrostatic daughter waves, particularly an X-mode pump wave decaying into a high-frequency warm upper hybrid (UH) wave and a low-frequency warm lower hybrid (LH) daughter wave near the upper hybrid resonance (UHR) of the pump wave. The main motivation is observations of strong PDI-like anomalous sidebands during CTS experiments at the ASDEX Upgrade tokamak [1], shown in Fig. 1, and the LHD stellarator [2, 3]. Furthermore, similar observations have been made during 1st harmonic electron cyclotron resonance heating (ECRH) experiments at the Versator II, FT-1 and TCA tokamaks [4–6], as well as at the W VII-A stellarator [7, 8].

Results and Conclusions

For the PDI in which an electromagnetic X-mode pump wave decays into a warm UH and a warm LH wave near the UHR of the pump wave, we have generalized the results of [4, 9, 10]; the details are given in [11]. The dispersion relations describing the warm LH and warm UH waves in a simple plasma are given by [10, 11], for vanishing $\mathbf{k}_\parallel$ (with respect to the equilibrium magnetic field $\mathbf{B}$) they become,

$$\omega_1^2 = \omega_{LH}^2(1 + A_1 b_{e1}), \tag{2}$$

$$\omega_2^2 = \omega_{UH}^2 - \omega_{pe}^2 A_2 b_{e2}, \tag{3}$$

where $\omega_{UH}^2 = \omega_{pe}^2 + \omega_{ce}^2$ is the square of the UH frequency, $\omega_{LH}^2 = \omega_{ce}^2\omega_{pi}^2/\omega_{UH}^2$ is the square of the LH frequency, $b_{e1,2} = m_e T_e k_{1,2}^2/(e^2B^2)$ is the square of the wave number normalised by the inverse square of the thermal electron Larmor radius, $A_1 = 3\{\omega_{pe}^2/(4\omega_{UH}^2) + [T_i/(Z_iT_e)]\omega_{UH}^2/\omega_{pe}^2\}$ and $A_2 = 1/[1 - \omega_{pe}^2/(3\omega_{ce}^2)]$ (Z_i is the ion charge number and $T_{e,i}$ are given in energy units). From Eq. (3) and $k_2 > 0$, $\omega_2 < \omega_{UH}$ for $\omega_{pe}^2 < 3\omega_{ce}^2$ and $\omega_2 > \omega_{UH}$ for $\omega_{pe}^2 > 3\omega_{ce}^2$, indicating that propagating warm UH waves only exist on the same side of the UHR as the propagating X-mode, for which $\omega_2 < \omega_0 < \omega_{UH}$, when $\omega_{pe}^2 < 3\omega_{ce}^2$, generally limiting the occurrence of this kind of PDI to underdense and weakly overdense plasmas, e.g. found in tokamaks and some stellarator experiments. We additionally note that, from Eqs. (1) and (2), $\omega_1 > \omega_{LH}$, $\omega_2 = \omega_0 - \omega_1 < \omega_0 - \omega_{LH}$ and $\omega_3 = \omega_0 + \omega_1 > \omega_0 + \omega_{LH}$, consistent with the locations of the peaks observed in Fig. 1. In the dipole limit, $\mathbf{k}_0 \approx \mathbf{0}$ or $b_{e1} \approx b_{e2}$, it is possible to derive analytical electric field amplitude thresholds for the above PDI, as done in [9, 10, 11].

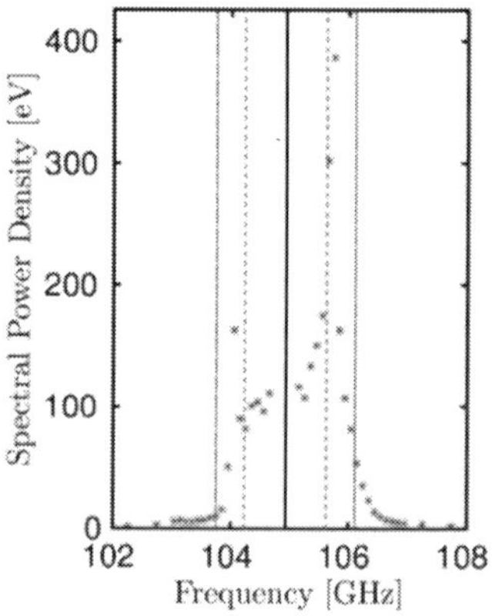

Fig. 1. CTS spectral power density recorded at ASDEX Upgrade in eV. Anomalous peaks shifted in frequency relative to the gyrotron radiation (full line) are observed; the frequency shifts slightly exceed the LH frequency at the UHR (dashed lines), but fall short of the frequency shifts predicted by dipole PDI theory (dotted lines)

The above theory may be used to explain the anomalous sidebands observed during CTS experiments at the ASDEX Upgrade tokamak [1, 10, 11], seen in Fig. 1. In these experiments, depicted in Fig. 2, 105 GHz radiation from a gyrotron is launched from the low-field side of the torus in O-mode and reflected by the high-field side vessel wall, after which part of the reflected radiation may re-enter the plasma in X-mode and ultimately reach the UHR where the PDI may occur. Ray tracing indicates the correlation between the anomalous sidebands in the CTS spectrum and a non-negligible amount of gyrotron radiation reaching the UHR in X-mode implied by the above mechanism [11]. We estimate the electric field amplitude of the X-mode radiation using simple ray and beam tracing techniques away from the UHR which are coupled to a 1D full-wave model close to the UHR [10, 11]. This allows us to convert the analytical pump electric field amplitude thresholds to a CTS gyrotron power threshold necessary to excite the PDI in ASDEX Upgrade. For the experiment under consideration, the CTS gyrotron power threshold is found to be on the order of 100 kW close to the UHR of the pump wave which is similar to the gyrotron power of 200 kW used in the experiment, indicating that the anomalous sidebands may be explained by the PDI. We note that the frequency shift predicted by the dipole theory overestimates the experimental one observed in Fig. 1 which may be understood if the warm UH wave is assumed to be forward scattered by the pump wave, following a line of inquiry similar to that of [12].

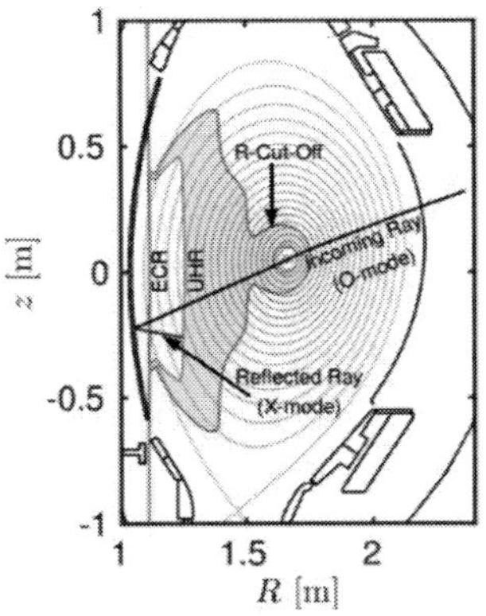

Fig. 2. CTS geometry in ASDEX Upgrade. 105 GHz O-mode gyrotron radiation is injected from the low-field side and reflected off the high-field side wall, after which part of the radiation may re-enter the plasma in X-mode and reach the UHR, where a PDI may occur due to strong field enhancement, provided that absorption at the ECR is not excessive. X-mode radiation cannot reach the UHR if launched from the low-field side since no propagating X-mode exists in the shaded area

Acknowledgements. This work was supported by a research grant (15483) from VILLUM FONDEN.

References

1. *Nielsen, S.K., Michelsen, P.K., Hansen, S.K., Kors-holm, S.B., Leipold, F., Rasmussen, J., Salewski, M., Schubert, M., Stejner, M., Stober, J., Wagner, D., The ASDEX Upgrade Team* Recent development of collective Thomson scattering for magnetically confined fusion plasmas // Phys. Scr. 2017. V. 92. 024001.

2. *Nishiura, M., Kubo, S., Tanaka, K., Seki, R., Ogasawara, S., Shimozuma, T., Okada, K., Kobayashi, S., Mutoh, T., Kawahata, K., Watari, T., LHD Experiment Group, Saito, T., Tatematsu, Y., Korsholm, S. B., Salewski, M.* Spectrum response and analysis of 77 GHz band collective Thomson scattering diagnostic for bulk and fast ions in LHD plasmas // Nucl. Fusion. 2014. V. 54. 023006.

3. *Kubo, S., Nishiura, M., Tanaka, K., Moseev, D., Ogasawara, S., Shimozuma, T., Yoshimura, Y., Igami, H., Takahashi, H., Tsujimura, T. I., Makino, M., LHD Experiment Group* Scattering volume in the collective Thomson scattering measurement using high power gyrotron in the LHD // JINST. 2016. V. 11. C06005.

4. *McDermott, F.S., Bekefi, G., Hackett, K.E., Levine, J.S., Porkolab, M.* Observation of the parametric decay instability during electron cyclotron resonance heating on the Versator II tokamak // Phys. Fluids. 1982. V. 25, No. 9. P. 1488–1490.

5. *Bulyginsky, D.G., Gusev, V.K., Djachenko, V.V., Irzak, M.A., Kantor, M.Yu., Larionov, M.M., Levin, L.S., Serebreny, G.A., Shustova, N.V.* ECR-heating of plasma in FT-1 tokamak and its influence on the ion component // *Proc. 11th European Conf. Controlled Fusion and Plasma Physics (Aachen, 1984).* Mulhouse: European Physical Society. 1984. V. 1. P. 457–460.

6. *Pietrzyk, Z.A., Pochelon, A., Behn, R., Bondeson, A., Dutch, M., Goodman, T.P., Tran, M.Q., Whaley, D.R.* Electron cyclotron resonance heating on the TCA tokamak // Nucl. Fusion. 1993. V. 33, No. 2. P. 197–209.

7. *Wilhelm, R., Erckmann, V., Janzen, G., Kasparek, W., Müller, G., Raüchle, E., Schüller, P. G., Schwörer, Thumm, M., W VII-A Team* Electron cyclotron resonance heating and confinement in the W VII-A stellarator // Plasma Phys. Control. Fusion. 1984. V. 26, No. 12A. P. 1433–1444.

8. *Erckmann, V., Janzen, G., Kasparek, W., Müller, G., Räuchle, E., Schüller, P. G., Schwörer, K., Thumm, M., Wilhelm, R., W VII-A Team* Wave absorption and transport studies on ECR-heated stellarator plasmas in W VII-A // *Proc. 10th Int. Conf. Plasma Physics and Controlled Nuclear Fusion Research (London, 1984).* Vienna: International Atomic Energy Agency. 1985. V. 2. P. 419–428.

9. *Porkolab, M.* Parametric decay instability in ECR heated plasmas // *Proc. 2nd Workshop Hot Electron Ring Physics (San Diego, 1981)*, edited by N. A. Uckan. Alexandria: National Technical Information Service. 1982. V. 1. P. 237–246.

10. *Hansen, S.K.* Parametric decay instability near the upper hybrid resonance and anomalous scattering from tokamak plasmas // M.Sc. Thesis. Kgs. Lyngby: Technical University of Denmark. 2016.

11. *Hansen, S.K., Nielsen, S.K., Salewski, M., Stejner, M., Stober, J., the ASDEX Upgrade team* Parametric decay instability near the upper hybrid resonance in magnetically confined fusion plasmas // Submitted to: Plasma Phys. Control. Fusion.

12. *Dodin, I.Y. and Arefiev, A.V.* Parametric decay of plasma waves near the upper-hybrid resonance // Phys. Plasmas. 2017. V. 24. 032119.

EPJ Web of Conferences **149**, 03021 (2017) DOI: 10.1051/epjconf/201714903021
10th International Workshop 2017 "Strong Microwaves and Terahertz Waves: Sources and Applications"

Status and design of ECRH/CD system of the upgrade of the tokamak T-15

I.N. Roy, I.O. Anashkin, N.A. Kirneva, D.A. Kislov, P.P. Khvostenko, A.V. Melnikov, A.N. Romannikov
National Research Centre 'Kurchatov Institute', Moscow, Russia, Roy_IN@nrcki.ru

The new tokamak T-15MD is under construction in National Research Center «Kurchatov Institute». The tokamak has the following parameters: R = 1.48 m, a = 0.67 m (aspect ratio A = 2.2), B = 2.0 T, I_{pl} = 2.0 MA. The poloidal system is capable of realizing the divertor both with single null and double null magnetic configuration and plasma shape with elongation k_{95} = 1.7–1.9 and triangularity δ_{95} = 0.3–0.4. Pulse duration Δt = 10 s, plasma density, $0.5\cdot10^{20}$ m^{-3}, plasma temperature $T_i(0)$, $T_e(0)$, 4–5 keV, normalized beta β_N = 1.5.

Preliminary transport analysis with ASTRA code predicts the effective electron and ion heating and transition from L to H-mode in Ohmic and auxiliary heated regimes. One of the main tasks of experimental study program on T-15U tokamak is the obtaining of physical and technological data for fusion neutron source creation.

T-15MD installation will be equipped with the auxiliary plasma heating and current drive systems (P_{aux} > 14 MW). The NBI system will consist of three hydrogen NB injectors 6 MW total power at pulse duration up to 30 s. The ECRH and CD system will consist of eight gyrotrons more then 8 MW total RF launched power at pulse duration up to 30 s. The LHH and CD system will be able to launch 4 MW RF power with 30 s duration. The ion-cyclotron heating system will consist of 3 generators, 30÷65 MHz, 2.5 MW, which will allow to inject 6 MW power with pulse duration up to 10 s.

Up to date the components of the magnetic system and vacuum chamber of T-15MD are manufactured. Next two years plan is to perform a test assembly at the factory site, disassemble, deliver it to the Kurchatov Institute site, perform the final assembly and connect to all the systems (electrical, vacuum, water-cooling, etc.).

At the same time, the design of the ECR system should be performed.

The gyrotron auxiliary heating and current drive system should allow to provide: ECRH; ECCD; Control of MHD instabilities by ECRH; ECRH pre ionization; Electron Bernstein Wave Heating.

To fulfill the above tasks, was selected the system of two with a frequency of 82.6 GHz and six gyrotrons with a frequency of 105 GHz. The power generated by each of the gyrotrons must be at least 1 MW, the preferred value is 1.5 MW in Gaussian beam. Pulse duration from 20 to 30 s. Efficiency not less 50% (gyrotrons equipped with depressed collectors).

The magnetic systems will consist of: the main cryogenic magnets (liquid Helium free, with remotely located cryocompressor), cathode coils, collector sweeping coils.

A new location was chosen for the new ECR system (Fig. 1.). Thus, to minimize the effect of magnetic fields of the tokamak. Calculations showed that at the location of the gyrotrons the radial component of the magnetic field in all regimes does not exceed 1.0 G.

The evacuated corrugated waveguide transmission line (∅63.5 mm) will transport the RF power. The length of the transmission lines will be from 27 to 42 m. They will also contain:

- Matching Optic Unit,
- RF wave polarizer providing linear or elliptical wave polarization rotation,
- waveguide switch, directing the power to the calorimetric load or to the tokamak.
- short pulse (up to 0.1 s) calorimetric measuring load,
- second waveguide switch, directing the power to the terminal load or to the tokamak
- full pulse terminal load,
- waveguide mechanical break (the tokamak will move 15 mm in radial and vertical directions due to thermal expansion),
- waveguide DC break (1 kV),
- RF vacuum valve,
- RF launcher. Two fully articulating launchers for ECRH/CD at weak field side will operate with four RF beams each. Two launchers at strong field side will operate with one RF beam each.

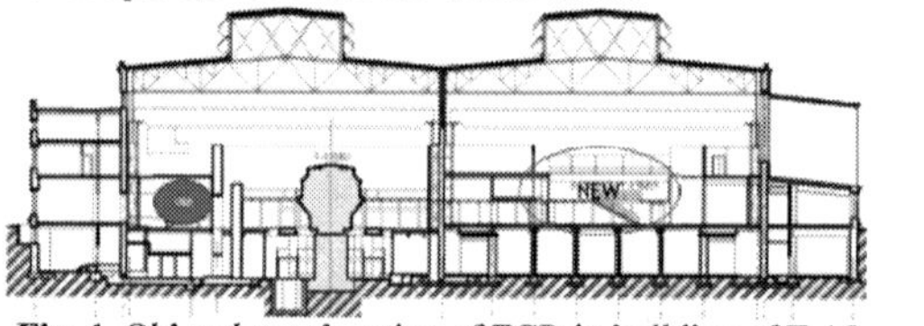

Fig. 1. Old and new location of ECR in building of T-15

The system will be equipped with the low-voltage power supplies necessary for the operation of the gyrotron: superconducting magnet power supply, collector sweep coil power supply, collector dc coil power supply, ion pump power supply, cathode filament power supply, gun coil power supply.

The water cooling system of the gyrotron and the critical elements of the transmission line should be provided.

All the equipment must be remotely controlled. Hardware-software system will be created for monitoring, control, protection and data acquisition.

Currently Kurchatov Institute, there are only two high voltage power supplies can operate with parameters required for T-15MD gyrotrons. Updating or creation of the new main high voltage power supplies is necessary to fulfill the tasks of the ECRH/CD system

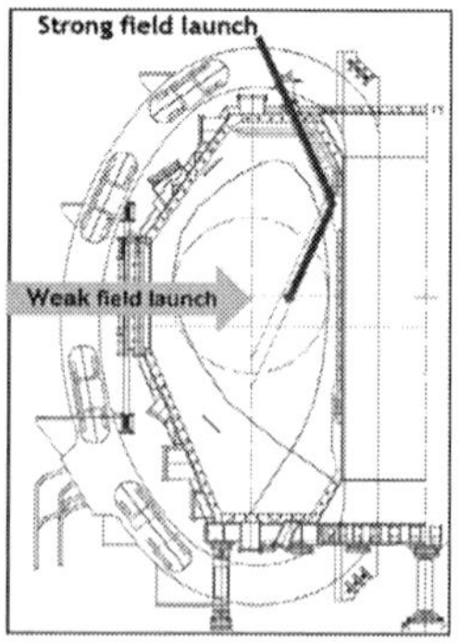

Fig. 2. Vertical cross-section of the T-15MD.

This work was partially supported by ROSATOM State Corporation, contract H.4x.241.9Б.17.1011.

EPJ Web of Conferences **149**, 03022 (2017) DOI: 10.1051/epjconf/201714903022
10th International Workshop 2017 "Strong Microwaves and Terahertz Waves: Sources and Applications"

Quasi-optical approach to reconstruction of plasma fluctuations using amplitude distribution of transmitted microwave beam

E.D. Gospodchikov, D.I. Sobolev, T.A. Khusainov, A.A. Balakin, A.G. Shalashov

Institute of Applied Physics of RAS, Nizhny Novgorod, Russia, egos@appl.sci-nnov.ru

Introduction

Despite considerable progress in the theoretical understanding of confined plasmas, there is still a need for a detailed description of the fundamental physical parameters playing a role in the processes governing equilibrium, stability, and transport. Turbulence properties are some of the most important of such parameters.

Two basic approaches are used for remote sensing of the density fluctuation by electromagnetic beams. The first one is based on the microwave reflectometry which basically allows the reconstruction of the turbulence spectrum by the spectrum of the scattered signal in the far zone [1]. The second one is phase contrast imaging (PCI), basing on reconstruction of the turbulence distribution by phase distortion of passing laser beams in geometrical optics zone [2]. In this work we present a theoretical approach to the reconstruction of inhomogeneous plasma properties by amplitude distribution of microwave beam measured after pass through the plasma in intermediate i.e. Fresnel zone. First examples of reconstructions by this method are also presented. There are two promising advantages of this approach. First, the measurement of amplitude distribution in the beam transmitted through the plasma can be done easier than phase measurements required for PCI, because it does not require a matrix of separate receivers for each transverse point hence potentially has better spatial resolution. Second, amplitude perturbations in the Fresnel zone are of the first order of phase velocity perturbations, and the technique might allow detection fluctuations with low amplitude leading to weak and/or low angle scattering.

The discussed method is based on the quasi-optical approach to modeling the propagation of wave beams developed at IAP RAS, which allows taking into account the simultaneous effects of inhomogeneity and spatial dispersion of the medium, diffraction, aberration and dissipation of the wave beam. Based on this approach, a numerical code LAQO was developed. The code is able to model the propagation and the electron cyclotron absorption of wave beams in weakly relativistic plasma in electromagnetically large devices (with typical size of few thousands of wavelengths) in a reasonable time. Calculations using LAQO proved the importance of taking into account these effects to describe correctly the RF power deposition profile in tokamak plasmas [3, 4] and in open magnetic trap [5, 6]. A small solution time for the direct problem of wave beam propagation allows the use of LAQO to solve the inverse problem, such as reconstruction of distributions of fluctuating plasma parameters using the amplitude profile of the transmitted beam. To solve the inverse problem we adopt the iterative method for the synthesis of the surface of the passive elements of microwave transmission systems developed in IAP RAS, which allows one to achieve extremely high computational precision using a several iterations of direct problem calculations [7]. Note that this technique requires knowledge of phase distribution for the incident wave beam only, and for the transmitted wave beam only intensity distribution is needed.

Reconstruction algorithm

As the first step we present the results of reconstruction of the parameters of single perturbation of refraction coefficient in homogeneous isotropic media $\delta n = \delta_0 \exp\left(-(\vec{r}-\vec{r}_0)^2 / a^2\right)$. This simplest plasma density perturbation is characterized by amplitude δ_0, position $\vec{r}_0$ and width a.

The reconstruction routine uses the following algorithm. Given the amplitude and phase of the wave beam (A_{in}, ϕ_{in}) specified in start plane, a "measureable" amplitude distribution A_{out} in measurement plane is calculated for a wave beam transmitted through the media with refraction coefficient perturbation. Then the same calculation is done for the unperturbed media, and output amplitude and phase distributions $\left(A^i_{out}, \phi^i_{out}\right)$ are obtained. After that, a substitution $A^i_{out} \to A_{out}$ is done, and the beam propagation is calculated in reverse direction. The amplitude distribution obtained in the latter calculation A^i_{in} is substituted with the one taken from the incident beam, and the difference of phase distributions $\phi^i_{in} - \phi_{in}$ is stored as the phase corrector. This sequence is repeated until the amplitude distribution calculated with phase corrector and "measured" amplitude distribution converges with enough precision. Resulting phase corrector is converted into refraction coefficient perturbation. Note that distributions along and transversely to the beam are reconstructed essentially differently. Due to the fact that the phase corrector is 2-dimensional phase perturbation, the distribution along the beam is reconstructed ambiguously. Therefore we have to use an additional scan over the position of the phase corrector for reconstruction of the longitudinal distribution of fluctuation.

Transverse distribution of refraction coefficient

Reconstructed refraction coefficient with parameters $\delta_0 = 0.05$, $a = 1\ cm$ is presented in figure 1. The wavelength was taken $\lambda = 0.5\ cm$, quasioptical beam size at input $a_{beam} = 3\ cm$, the length of propagation track is $l = 30\ cm$. Refraction coefficient perturbation is located on the beam axis.

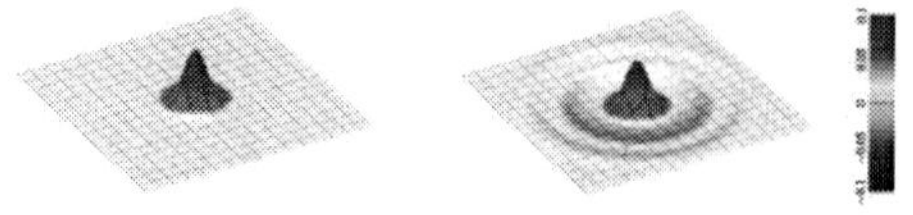

Fig. 1. Reconstruction of undisplaced perturbation. Right panel – distribution of refraction coefficient perturbation; left panel – reconstructed distribution of refraction coefficient perturbation

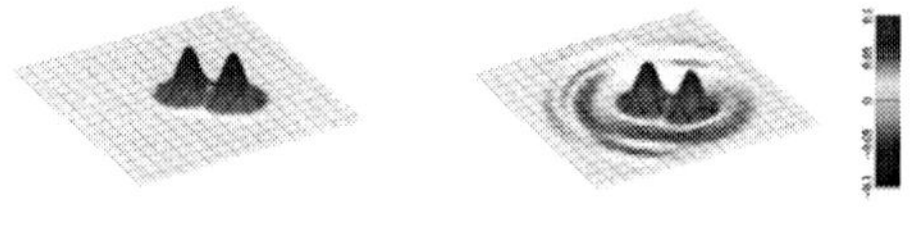

Fig. 2. Reconstruction of refraction coefficient perturbation with two maximums. Right panel – distribution of refraction coefficient perturbation; left panel – reconstructed distribution of refraction coefficient perturbation

One can see that position, amplitude and shape of perturbation in transverse plane are reconstructed with high precision, and resolution is less than the beam aperture. To illustrate this point, the reconstruction of "two-hump" perturbation is presented in figure 2.

As one can see, it is possible to reconstruct the fluctuation parameters with precision comparable to the wavelength.

Longitudinal position of perturbation

Proposed reconstruction procedure result in 2-dimensional phase corrector, which can be converted into refraction coefficient perturbation distribution with longitudinal uncertainty. It is unavoidable limitation because the original inverse problem is formally incorrect. Our calculations, however, give some hope that the longitudinal position can be reconstructed with precision about the wave beam waist. To illustrate this, we made calculations with the same parameters as presented in figure 1, but with a shifted position of the phase corrector along the beam path back and forth from the position of the perturbation. The results of modeling for phase corrector shifted relatively right position are presented in figure 3. Comparing figures 1 and 3 one can see that "inaccuracy" of the phase corrector position leads to the emergence of a strong "Fresnel lens" in the reconstructed phase corrector. Therefore, the condition of minimization of this "lens" allows reconstruction of longitudinal position of perturbation.

This method does not give the accuracy of the reconstruction of the longitudinal coordinate comparable to the accuracy of the reconstruction of the transverse distribution, but it allows to improve accuracy of reconstruction.

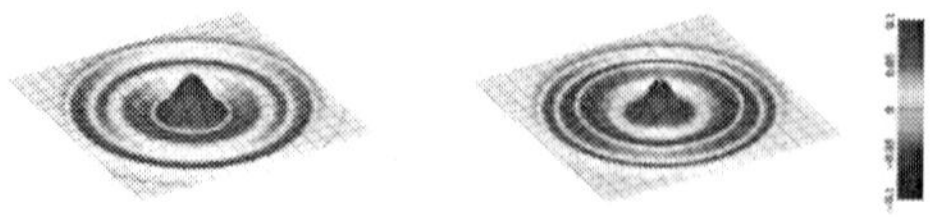

Fig. 3. Reconstruction results with different positions of phase corrector along the beam path. Right panel – before right position; left panel – after right position

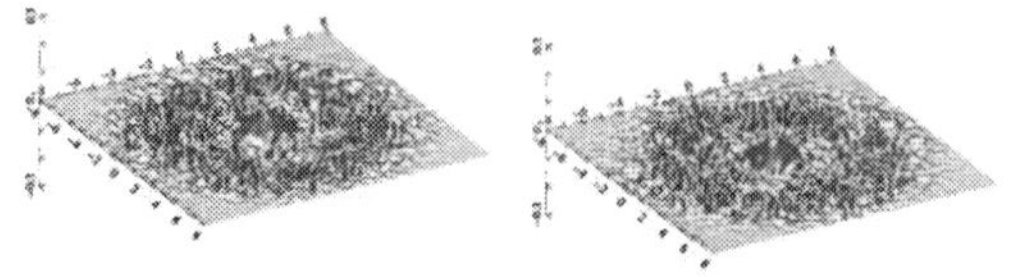

Fig. 4. Reconstruction of perturbation using noised transmitted beam distribution. Right panel after 10 procedure iterations.; left panel – 40 iteration

Transmitted beam with noised intensity distribution

Intensity distribution can be measured with limited accuracy. We demonstrated possibility of fluctuation reconstruction using noised (by 10% incoherent noise) "measured" distribution (figure 4).

This work was supported by RFBR grant № 15-02-07600.

References

1. *Skvortsova N.N., Akulina D.K., Batanov G.M., Kharchev N.K., Kolik L.V., Kovrizhnykh L.M., Letunov A.A., Logvinenko V.P., Malakhov D.V., Petrov A.E.* Effect of ECRH regime on characteristics of short-wave turbulence in plasma of the L-2M stellarator // Plasma Phys. Control. Fusion. 2010. V. 52. P. 055008.
2. *Marinoni A., Coda S., Chavan R., and Pochon G.* Design of a tangential phase contrast imaging diagnostic for the TCV tokamak // Rev. of Scientific Instr. 2006. V.77. P. 10E929.
3. *Bertelli N., Balakin A.A., Westerhof E. and Buyanova M.N.* ECCD calculations in ITER by means of the quasi-optical code// Nucl. Fusion, 2010, V. 50, P. 115008.
4. *Maj O., Balakin A.A., Poli E.* Effects of aberration on paraxial wave beams: beam tracing versus quasi-optical solutions // Plasma Phys. Controll. Fusion. 2010. V. 52, P. 085006.
5. *Shalashov A. G., Balakin A. A., Gospodchikov E. D., and Khusainov T.A.* Quasi-optical theory of microwave plasma heating in open magnetic trap// Physics of Plasmas 2016, V. 23, P. 112504.
6. *Shalashov A.G. Balakin A.A. Khusainov T.A. Gospodchikov E. D. Solomakhin A.L.* Quasi-optical simulation of the electron cyclotron plasma heating in a mirror magnetic trap // JETP, 2017. Vol. 124, No 2, P 325.
7. *Sobolev D.I., Denisov G.G.* Principles of Synthesis of Multimode Waveguide Units // IEEE Transactions on Plasma Science, 2010, V.38, No.10, P. 2825.

EPJ Web of Conferences 149, 03023 (2017) DOI: 10.1051/epjconf/201714903023
10th International Workshop 2017 "Strong Microwaves and Terahertz Waves: Sources and Applications"

Electron cyclotron heating and diagnostics of plasma at the second harmonic in the GDT device

E.D. Gospodchikov[1,2], O.B. Smolyakova[1,2], A.L. Solomakhin[2], A.G. Shalashov[1,2]

[1] Institute of Applied Physics RAS, Nizhniy Novgorod, Russia, egos@appl.sci-nnov.ru
[2] Budker Institute of Nuclear Physics SB RAS, Novosibirsk, Russia

Introduction

The electron cyclotron resonance heating (ECRH), being a standard technique for tokamaks and stellarators, has also good potentials for fusion-relevant experiments in large-scale open traps. In particular, efficient ECRH has been demonstrated recently in the mirror device GDT (gas-dynamic trap) in the Budker Institute [1–3]. In reported experiments, a combined plasma heating by neutral beams and microwave radiation results in the record electron temperatures up to 1 keV at density about 10^{13} cm^{-3} in the regime with a narrow rf-power deposition profile and up to 350 eV in the regime with a broad power deposition profile [1]. Success in ECRH experiments at GDT is convincing enough to consider the prospects of simple axially symmetric open magnetic traps as a high-power neutron source for fusion applications, including material testing, a hybrid fusion-fission reactor and nuclear waste processing [4].

Presently, the ECRH system in GDT is operating with the extraordinary mode at the fundamental cyclotron harmonic (X1). Although successful ECRH experiments at GDT evidence for the reliability of such heating mechanism, it may face essential difficulties in the next-step devices operating at higher β [5]. First difficulty is related to the very complicate plasma dielectric response near the first cyclotron harmonic that results in strong refraction [6]. Although this feature is used in the present ECRH scheme and direct ECE measurements validate our understanding of the involved physics [7, 8], next-step devices would probably require simpler solutions. Other weakness of the X1 scheme is related to the accessibility condition $\omega > \omega_{pe}$ for microwaves. In open magnetic configurations, this condition requires the launching system to be placed near a magnetic mirror, what is often technically not possible. This will be critical for the next generation devices which operate at higher electron densities.

Heating with the extraordinary mode at the second harmonic (X2) is free of the mentioned drawbacks; however it requires high enough electron temperature in order to be effective. In this context, ECRH at the second harmonic is possible in combination with heating at the first harmonic providing the initial high electron temperature. In this paper we discuss physical basics and peculiarities of regime. We also discuss our plans for experimental validation of combined X1+X2 ECRH regime in GDT by measuring the electron cyclotron emission at the second harmonic (X2 ECE).

ECRH at the second harmonic

Basics physic of electron cyclotron heating at the second harmonic is well known [9]. In tokamak conditions, effective absorption requires quasi-transverse to the external magnetic field launch of a microwave beam with extraordinary wave polarization. Essentially the same condition remains in force in an open trap. However, here a new physics appear related to the variation of a strength of the magnetic field along its direction (presumably $\mathbf{B} \parallel \nabla B$ in open traps while $\mathbf{B} \perp \nabla B$ in toroidal systems). On one hand, one can assume that quasi-transverse direction corresponds to approximately constant magnetic field along propagation path of the microwave beam. But the longitudinal wave vector is not constant due to ∇B-inhomo-geneity, thus a strong influence of the refraction is expected in an open magnetic configuration. In some sense this situation is similar to the "top launch" scenario for ECRH in tokamaks [10].

The resonance broadening in X2 ECRH is presumably defined by the relativistic electron mass,

$$|\omega - 2\omega_{ce}| / \omega \sim T_e / m_e c^2 \sim 10^{-3}.$$

Having in mind that the electron temperature in open trap is noticeably lower than in modern toroidal systems, one can expect an extremely narrow relativistic resonance region in the real space. Because of this feature, a strictly transverse launch of a wave beam is ineffective. This is illustrated in figure 1. One can see that transverse propagation corresponds to high value of optical thickness for very small fraction of rays contributing the microwave beam of finite aperture. In this situation most part of the wave beam will not be absorbed, see 1(a). With more oblique propagation, the peak optical thickness reduces, but the fraction of absorbed rays is increased. The optimal total absorption is realized at the propagation angle for which the spread of the starting positions of the absorbed rays is close to the beam width, see 1(c).

In GDT conditions this corresponds to the launching angle $\theta = 80°$ (angle between the wave vector and the magnetic field) for straight ray approximation and $\theta = 85°$ for ray-tracing modeling. This launching direction is preferable for X2 ECE and ECRH experiments. For the optimal launching angle one can expect absorption efficiency 60–70% for $T_e = 500$ eV and 10–20% for $T_e = 150$ eV.

Second harmonic ECE of bulk and hot electrons

As a first step aimed at experimental validation of combined X1+X2 ECRH regime in GDT we are going to measure the electron cyclotron emission in the extraordinary mode at the second harmonic (X2 ECE), what is obviously less expensive than implementing the X2 ECRH. The idea of the experiment is to exploit the reciprocity principle and Kirchhoff's law that link the absorption and the spontaneous plasma emission at thermal equilibrium. According to this idea, finding the X2 ECE with the effective radiation temperature comparable with the electron temperature would indicate in favor of efficient X2 ECRH in the geometry reversed to the ECE one. Such measurements will be performed with a new ECE diagnostics operating at 75 GHz (under construction now) in high-temperature ($T_e \sim 500$ eV) discharges supported by 54.5 GHz ECRH at the fundamental harmonic. In addition, new ECE diagnostics will provide data on the line-integrated electron pressure in discharge with a high resolution in time.

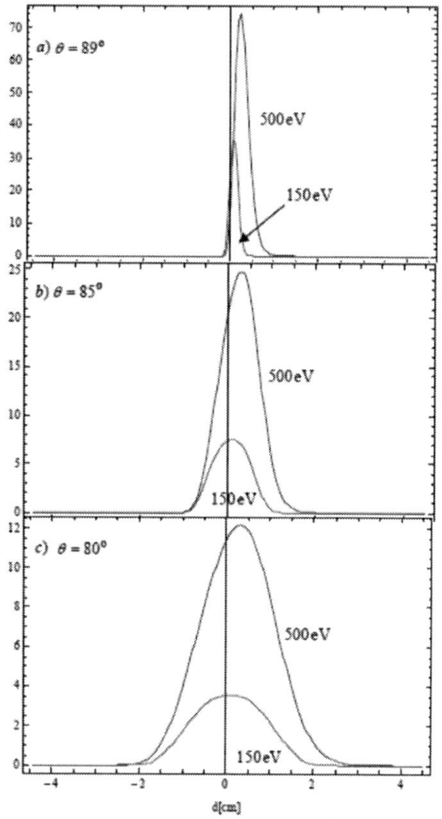

Fig. 1. Dependence of optical thickness $\tau = \int \mathrm{Im}\, k\, dl$ on transverse position d of geometric-optics ray for electron temperatures $T_e = 150$ eV and $T_e = 500$ eV. Here d is shift from the central ray of microwave beam (crossing the resonance surface at a trap axis)

The similar idea was used previously to confirm our understanding of X1 ECRH at GDT with the ECE operating in the vicinity of the ECRH frequency 54.5 GHz [7, 8]. Besides, measured by this diagnostics non-thermal ECE have unambiguously confirmed the existence of suprathermal electrons in GDT with energies about 10 keV. These electrons are generated by the intense microwave field during the ECR discharge start-up and heating of the main plasma [11]. Presence of suprathermal electrons should be taken into account in the analysis ECE at the second harmonic. Figure 2 illustrates possibility for separation of signal from hot (10 keV) and thermal ($T_e = 500$ eV) electrons exploiting the relativistic shift of the resonance frequency. Here expected radiation temperatures are calculated for the antenna which is now installed in GDT chamber.

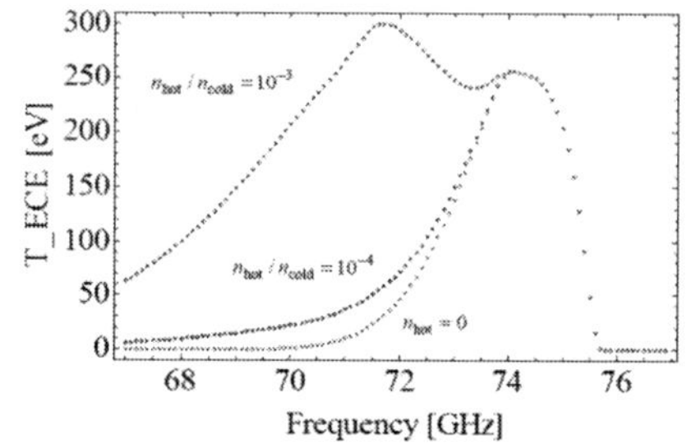

Fig. 2. Radiation temperature for different hot electron densities for 2X ECE experiments preparing at GDT

In conclusion we note that for relatively warm plasmas recently obtained at GDT ($T_e = 500$ eV) one can expect an efficient electron cyclotron heating at the second harmonic. For checking this concept we install new ECE system at 75 GHz. receiving system for registration of. Observation of near blackbody plasma emission at the second electron cyclotron harmonic should demonstrate perspective of similar heating scenarios in open magnetic traps.

The work was supported by the Russian Science Foundation (pr. 14-12-01007).

References

1. *P.A. Bagryansky, A.G. Shalashov, E.D. Gospodchikov et al.* Phys. Rev. Letters. 2015. V. 114. P. 205001.
2. *P.A. Bagryansky, E.D. Gospodchikov, Yu.V. Kovalenko et al.* Fusion Science and Tech. 2015. V. 68. P. 87.
3. *P.A. Bagryansky, A.V. Anikeev, G.G. Denisov et al.* Nuclear Fusion. 2015. V. 55. P. 053009.
4. *A.A. Ivanov, V.V. Prikhodko.* Plasma Phys. Control. .Fusion. 2013. V. 55. P. 063001.
5. *V.V. Postupaev, A.V. Burdakov, A.A. Ivanov* Fusion Science and Tech. 2015. V.68. No.1. P. 92.
6. *A. Shalashov, E. Gospodchikov, O. Smolyakova et al.* Physics of Plasmas. 2012. V. 19. P. 052503
7. *A.L. Solomakhin et al.* AIP Conf. Proc. 2016. V. 1771. P. 050001.
8. *A.G. Shalashov et al.* Electron cyclotron emission at the fundamental harmonic in GDT magnetic mirror. arXiv:1611.09137.
9. *M. Bornatici.* Plasma Phys. 1982. V.24. P.629.
10. *J.-P. Hogge, S. Alberti, L. Porte and G. Arnoux.* Nuclear Fusion. 2003. V. 43, No.11. P. 1353.
11. *D.V. Yakovlev, A.G. Shalashov, E.D. Gospodchikov et al.* Nuclear Fusion. 2016. V. 57. P. 016033

Crossed-field flows

Y.Y. Lau

University of Michigan, Ann Arbor, MI 48109, USA, yylau@umich.edu

This paper surveys crossed-field flows in conventional, inverted, and planar magnetron geometries, with and without a slow wave structure (SWS) on the anode.

An electron rotating under a general combination of radial electric field and axial magnetic field may exhibit negative, positive, or infinite mass behavior, depending on the orientation and magnitude of the electric field relative to the axial magnetic field [1]. These properties are confirmed in particle-in-cell simulations [2] on a thin electron layer whose collective interactions are shown to dominate the diocotron instability [1, 2].

The Brillouin flow exhibits a strong velocity shear, and therefore the diocotron-like instability is considered to be a dominant instability [3] in the Brillouin flow. For a smooth bore magnetron, the negative (positive) mass effect in the inverted (conventional) magnetron geometry was shown to increase (decrease) the severity of the diocotron instability [4] in the Brillouin flow. Including the SWS on the anode, the resonant interaction of the vacuum circuit mode and the corresponding smooth-bore diocotron-like mode is the dominant cause for instability [5]. This resonant interaction is far more important than the intrinsic negative (positive) mass property of electrons in the inverted (conventional) magnetron geometry. This resonant interaction severely restricts the wavenumber for instability to the narrow range in which the cold tube frequency of the SWS is within a few percent of the corresponding smooth bore diocotron-like mode in the Brillouin flow.

Since the Brillouin flow, instead of the multi-stream cycloidal flow, is the preferred state in crossed-field devices [6, 7], the Bumeman-Hartree (B-H) condition was critically examined using the Brillouin flow model [8]. It is found that the traditional, cycloidal orbit model [9] and the Brillouin flow model [8] yield the same result for the Buneman-Hartree condition only in the limit of a planar magnetron. The B-H condition derived from the Brillouin flow model shows a better match to simulation and experiment of relativistic magnetrons than the single particle model B-H condition [10].

This work was supported by Air Force Office of Scientific Research.

References

1. *Chernin, D., Lau, Y.Y.* Stability of laminar electron layers // Phys. Fluids. 1984. V. 27, No. 9, P. 2319–2331.

2. *French, A.M., Hoff, B.W., Lau, Y.Y., Gilgenbach, R.M.* Negative, positive, and infinite mass properties of a rotating electron beam // Appl. Phys. Lett. 2010. V. 97, No. 11, P. 111501.

3. *Davidson, R., Chan, H., Chen, C., and Lund, S.* Equilibrium and stability properties of intense non-neutral electron flow // Rev. Mod. Phys. 1991. V. 63, No. 2, P. 341–374.

4. *Simon, D.H., Lau, Y.Y., Greening, G., Wong, P., Hoff, B.W., Gilgenbach, R.M.* Stability of Brillouin flow in planar, conventional, and inverted magnetrons // Phys. Plasmas. 2015. V. 22, No. 8, P. 082104.

5. *Simon, D.H., Lau, Y. Y., Greening, G., Wong, P., Hoff, B. W., Gilgenbach, R. M.* Stability of Brillouin flow in the presence of slow-wave structure // Phys. Plasmas. 2016. V. 23, No. 9, P. 092101.

6. *Palevsky, A., Bekefi, G., Drobot, A. T.* Numerical simulation of oscillating magnetrons // J. Appl. Phys. 1981. V. 52, No. 8, P. 4938–4941.

7. *Christenson, P.J., Chernin, D.P., Garner, A.L., Lau, Y. Y.* Resistive destabilization of cycloidal electron flows and universality of (near-) Brillouin flow in a crossed-field gap // Phys. Plasmas. 1996. V. 3, No. 12, P. 4455-4462.

8. *Lau, Y.Y., Luginsland, J.W., Cartwright, K.L., Simon, D.H., Tang, W., Hoff, B.W., Gilgenbach, R.M.* A re-examination of the Buneman–Hartree condition in a cylindrical smooth-bore relativistic magnetron // Phys. Plasmas. 2010. V. 17, No. 3, P. 033102.

9. *Walker, R.L.* in Microwave magnetrons, Edited by G. B. Collins. New York: McGraw-Hill. 1948. P. 227.

10. *Simon, D.H.* Doctoral dissertation, University of Michigan, Ann Arbor, 2016.

Development of gyrotron traveling-wave tubes at IAP and GYCOM

S.V. Samsonov[1], G.G. Denisov[1,2], I.G. Gachev[1], A.A. Bogdashov[1], S.V. Mishakin[1], V.N. Manuilov[1,2], V.I. Belousov[1], D.I. Sobolev[1], E.V. Sokolov[1,3], E.A. Soluyanova[1,3], E.M. Tai[1,3]

[1] Institute of Applied Physics, Russian Academy of Sciences, Nizhny Novgorod, Russia, samsonov@appl.sci-nnov.ru
[2] Nizhny Novgorod State University, Russia
[3] Gycom Ltd., Nizhny Novgorod, Russia

Abstract

A brief review of research activity on gyrotron traveling-wave tube amplifiers (gyro-TWTs) with helically corrugated waveguides developing by the Institute of Applied Physics and GYCOM (Nizhny Novgorod, Russia) will be presented.

Introduction

The gyrotron traveling wave tube (gyro-TWT) is known as a broad-frequency-band variety of gyrotron-type amplifiers having potential for production of the highest average or continuous-wave (CW) power in the millimeter wavelength range [1] and therefore it is attractive as a microwave source for a number of applications such as radars, telecommunication and some other.

Since 1996 we are developing a concept of a gyro-TWT that is based on the use of a helically corrugated waveguide that radically changes the dispersion of the modes of a circular waveguide [2]. The operating helical-waveguide eigenmode has sufficiently high and almost constant group velocity at zero axial wavenumber which enables broadband operation of the helical-waveguide gyro-TWT with minimum sensitivity to electron velocity spread [3]. A number of experiments have proved the main theoretical predictions and advantages of gyro-TWTs of this type [4–6].

1. Ka-band application-oriented Gyro-TWTs

Since 2009 eleven tubes were manufactured by IAP and GYCOM Ltd., tested and shipped to customers. Parameters of the experimental prototypes for those tubes are summarized in the Table.

Main experimental parameters of Ka-band application-oriented gyro-TWTs

Parameter	Pulsed tube	CW tube
Accelerating voltage (kV)	70	40
Retarding voltage (kV)	25	20
Beam current (A)	10	1.5
Interaction B-field (T)	0.68	0.65
Magnet power consumption (kW)	25	22
Maximum output power (kW):		
pulsed	160	7.7
average	10	7.7
-1dB bandwidth (GHz)	2.4	2.1
-3dB bandwidth (GHz)	>2.7	2.6
Saturated gain (dB):		
at the band center	23	26
at the band sides	20	19
Pulse duration	100 us	hours

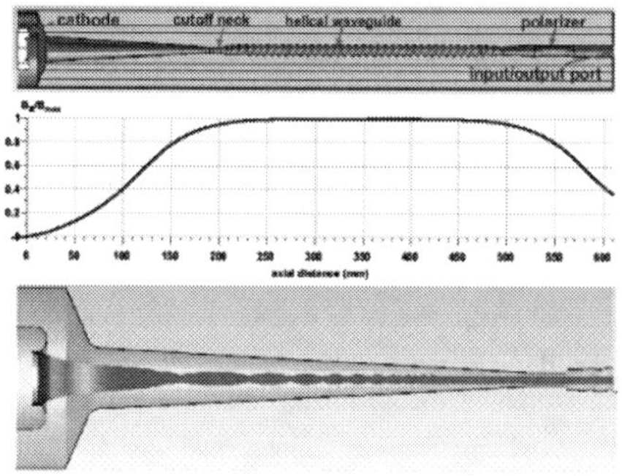

Fig. 1. CST model for simulation of a gyro-TWT from cathode to almost collector: whole geometry, B-field axial distribution and the gun part with particles' trajectories

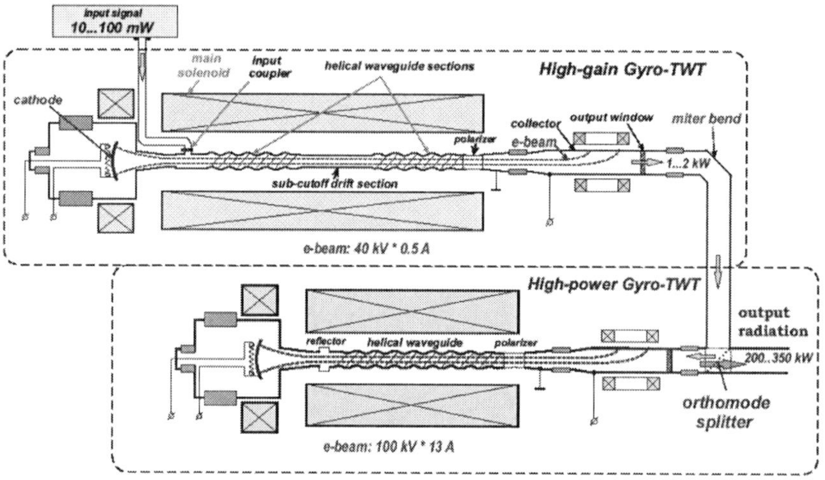

Fig. 2. Schematic view of cascade of two helically corrugated waveguide (HCW) gyro-TWTs

Due to their operation at the second cyclotron harmonic these Ka-band gyro-TWTs used relatively low magnetic fields which were generated by non-superconducting magnets with reasonable power consumption. Since no highly absorbing materials were used in the interaction circuits, these amplifiers possessed relatively low gain to avoid the reflection-induced oscillation therefore mid-power preamplifiers were needed to ensure the drive-saturated regimes.

Recently a new microwave system for feeding and extracting the radiation to and from a gyro-TWT through one oversized window was suggested and experimentally verified [7, 8] which along with significant simplification of the circuit allows principal breakthrough in the power for short millimeter waves. An implementation of this method in Ka-band can make production costs of the gyro-TWTs comparable with those for technological gyrotrons with powers of tens kilowatts.

Computer modeling of almost whole device (Fig.1: only a large-volume collector part is not included) ensures high reliability of the design and predicted behavior of the system at various experimental conditions. Basing on this modeling and previous experience a "technological" CW gyro-TWT capable for amplification of a 20-W 30-GHz drive signal to 10 kW with 1 GHz -3-dB instantaneous bandwidth or 5 kW with 2 GHz bandwidth was designed with moderate requirements to the power supplies (25 kV/2 A for the electron beam and 15 kW for the magnet).

2. Actual trends

The actual topics of research on the gyro-TWT are as follows: a cascade of two tubes ensuring the highest gain and output power in W band (Fig. 2) [9]; various microwave systems for inputting and outputting the radiation to and from a tube through one window [7, 8, 10]; tubes operating at the third cyclotron harmonic [11]; tubes prospective for DNP-NMR spectroscopy (260 GHz) [12]; tubes generating periodical sequences of ultrashort phase-coherent pulses [13].

Acknowledgements. Works mentioned in this review were supported by Federal Agency of Scientific Organizations under project 0035-2014-0023 (Sect.I), the Russian Science Foundation under Grant 16-19-10332 (Sect.II), Russian Academy of Science under Program 0035-2015-0007 (Sect.II), the Russian Science Foundation under Grant 16-42-01078 (Sect.II).

References

1. *Thumm, M.K.* State-of-the-Art of High Power Gyro-Devices and Free Electron Masers, Update 2015 // KIT Scientific Reports 7717. KIT Scientific Publishing.

2. *Denisov, G.G. and Cooke, S.J.* New microwave system for a gyro-TWT // Digest 21st Int. Conf. on Infrared and Millim. Waves, 1996, P.AT2.

3. *Denisov, G.G., Bratman, V.L., Phelps, A.D.R., Samsonov, S.V.* Gyro-TWT with a Helical Operating Waveguide: New Possibilities to Enhance Efficiency and Frequency Bandwidth // IEEE Trans. on Plasma Sci. 1998. V. 26, No. 3. P. 508-518.

4. *Bratman, V.L., Cross, A.W., Denisov, G.G., He, W., Phelps, A.D.R., Ronald, K., Samsonov, S.V., Whyte, C.G. and Young, A.R.* High-Gain Wide-Band Gyrotron Traveling Wave Amplifier with a Helically Corrugated Waveguide // Phys. Rev. Letts. 2000. V. 84, No. 12. P. 2746-2749.

5. *Bratman, V.L., Denisov, G.G., Samsonov, S.V., Cross, A.W., Phelps, A.D.R., He, W.* High-efficiency wideband gyro-TWTs and gyro-BWOs with helically corrugated waveguides // Radiophys. Quantum Electron., 2007. V. 50, No. 2. P. 95-107.

6. *Samsonov, S.V., Gachev, I.G., Denisov, G.G., Bogdashov, A.A., Mishakin, S.V., Fiks, A.S., Soluyanova, E.A., Tai, E.M., Dominyuk, Y.V., Levitan, B.A., Murzin, V.N.* Ka-band Gyrotron Traveling-Wave Tubes with the Highest Continuous-Wave and Average Power // IEEE Trans. Electr. Dev. 2014. V. 61, No. 12. P. 4264-4268.

7. *Denisov, G.G., Samsonov, S.V., Mishakin, S.V., Bogdashov, A.A.* Microwave system for feeding and extracting power to and from a gyrotron traveling-wave tube through one window // IEEE Electron Device Letters. 2014. V. 35, No.7. P. 789-791.

8. *Samsonov, S.V., Bogdashov, A.A., Denisov, G.G., Gachev, I.G., Mishakin, S.V.* Proof-of-Principle Experiment on High-Power Gyrotron Traveling-Wave Tube with a Microwave System for Driving and Extracting Power Through One Window // IEEE Microwave and Wireless Components Letters. 2016. V. 26, No. 4. P. 288-290.

9. *Samsonov, S.V., Bogdashov, A.A., Denisov, G.G., Gachev, I.G., Mishakin, S.V.* Cascade of Two W-Band Helical-Waveguide Gyro-TWTs with High Gain and Output Power: Concept and Modeling // IEEE Trans. on El. Dev., 2017. V.64, No.3. P. 1297-1301.

10. *Mishakin, S.V., Samsonov, S.V., Sobolev, D.I., Gashturi, A.P., and Denisov, G.G.* Radiation Input/Output System of a Broadband W-band Gyrotron Traveling-Wave Amplifier // *Proc. of 41st Int. Conf. on Infrared, Millimeter and Terahertz Waves*, Copenhagen, Denmark, 2016, art. no. 7758386.

11. *Mishakin, S.V., Samsonov, S.V., and Denisov, G.G.* A Helical-Waveguide Gyro-TWT at the Third Cyclotron Harmonic // IEEE Trans. on Electron Devices. 2015. V.62, No.10. P. 3387-3392.

12. *Mishakin, S.V., Bogdashov, A.A., Denisov, G.G., Samsonov, S.V., Kalynov, Yu.K., and Manuilov, V.N.* Second-harmonic 260-GHz gyrotron traveling-wave amplifier with helically corrugated waveguide // *Proc. of 9th Int. Workshop "Strong Microwaves and Terahertz Waves: Sources and Applications"*, Nizhny Novgorod – Perm – Nizhny Novgorod, Russia, July 24-30, 2014, P. 231-232.

13. *Ginzburg, N.S., Denisov, G.G., Vilkov, M.N., Sergeev, A.S., Zotova, I.V., Samsonov, S.V., Mishakin, S.V.* Generation of trains of ultrashort microwave pulses by two coupled helical gyro-TWTs operating in regimes of amplification and nonlinear absorption // Physics of Plasmas. 2017. V.24, No.2. art. no. 023103.

Non-canonical gyrotrons

V.E. Zapevalov

Institute of Applied Physics RAS, Nizhny Novgorod, Russia, zapev@appl.sci-nnov.ru

Scientific research in the 60s of the last century led to the creation of gyrotron and a number of other gyro-devices, the principle of the its actions are based on the synchronous interaction of the curved electron beam and an electromagnetic wave: the combined effect of relativistic effects and non-homogeneity of high-frequency field on the trajectories of electrons in a magnetic field leads to the intensive cyclotron radiation at the gyrofrequency and its harmonics [1]. The canonical gyrotron version includes adiabatic magnetron-injection electron gun (MIG), forming a tubular helical electron beam (HEB), oversize, weakly irregular cylindrical cavity with a diffraction radiation output, the output window of rf radiation and electron beam collector. Design of powerful gyrotron as a rule, further includes a built-in quasi-optical mode converter and depressed collector for recovery of the residual energy of the electron beam. The axial symmetry of canonical gyrotron and absence of its small-scale elements are favorable for its industrial development. For small and medium-power gyrotron for technological applications and spectroscopy efficiency increasing and frequency tuning are desirable.

Over the past few years in a number of leading countries created a high level of high-performance powerful gyrotrons from centimeter to submillimeter range [1–3]. Several fundamental scientific fields are formed, the successful development of which is directly caused by the presence of these sources (plasma heating in controlled thermonuclear fusion, development of technology of new materials, spectroscopy (EPR and dynamic nuclear polarization of NMR) and others. Each of the applications dictates own direction of development of gyro-devices, which are analyzed in this paper. For the gyrotron of low power for spectroscopy the customers have high demands for long-term stability and frequency tuning, increase of efficiency. However the complexity of the task is exacerbated by the problem of mode competition and high ohmic losses, especially in gyrotrons operating in the sub millimeter wave range on harmonics of the cyclotron frequency [2–5].

Due to a various reasons during the process of achieving of the maximal energetical characteristics (output power, efficiency, pulse energy), a number of gyro devices, other than the canonical gyrotron version, fizzled out and now were on the sidelines of progress. Such non-canonical devices have partitioned active medium, or (and) the space of interaction; non-cylindrical (ribbon or thin axial) helical electron beam (HEB); oversized strongly non-cylindrical or coaxial, quasi-optical, echelette cavities, etc. However, in the canonical gyrotron are insufficient selective properties at work on the gyro frequency harmonics and very limited possibilities for frequency tuning. The report analyzes the perspectives of non-canonical gyrotron in terms of frequency tuning and improved mode selection working on the gyro frequency harmonics.

Separation of an electron flow for several partial beams give us the possibility to improve mod selection, assisting for operating mode or suppressing parasitic modes. Fig. 1 shows photo of the electron gun for double beam second harmonic gyrotron and experimental dependencies of output power and efficiency on total beam current. Another versions of multi-beams and multi-barrel gyrotrons are discussed also.

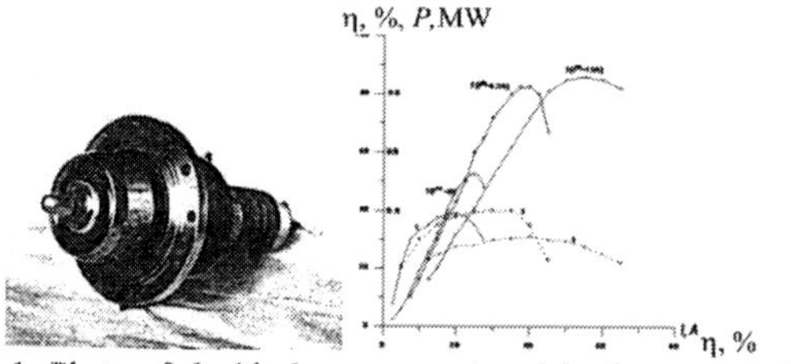

Fig. 1. Photo of double beam magnetron injection gum and dependencies of output power and efficiency on total beam current with different current of additional electron beam

Longitudinal sectioning of operation space is an effective way to ensure the selection and efficiency in microwave devices. The gyro-devices partitioning interaction space represented gyro-klystron, double-cavity gyrotron, gyrotrons with coupled cavities (CC), gyro-multiplier, etc. In connection with the mode selection problem it may be interesting to use the coupled cavity with mode conversion (CCMC) [5], shown schematically in Fig. 2.

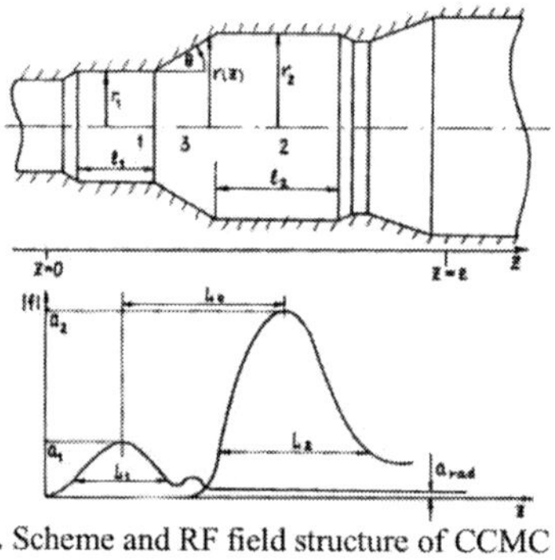

Fig. 2. Scheme and RF field structure of CCMC

The mutual modes conversion in both cavities plays a major role. Resonance Q-enhancement of the normal mode and the amplitude ratio |a1| / |a2| in CCMC creates basis for high mode selection in these systems and the reduction of radiation on the first mode provides high purity of the output-mode radiation. CCMC used in gyrotrons at the 1–3 harmonics over a broad frequency range, power and pulse duration.

Typical values of efficiency of gyrotrons with CCMC at least do not less than for single cavity gyrotron. These data indicate the feasibility of high-performance of CCMC-type gyrotron. Multi CCMC had even higher selectivity properties. As an example, Fig. 3 shows the profile and distribution of the RF field for multi-CCMC gyrotron of average power level with a frequency of 420 GHz at the third harmonic of the gyrofrequency and results of calculation of the output power, efficiency and ohmical load. Disadvantages of CCMC include the need for high-precision manufacturing. Possibility of frequency tuning for these systems is practically absent.

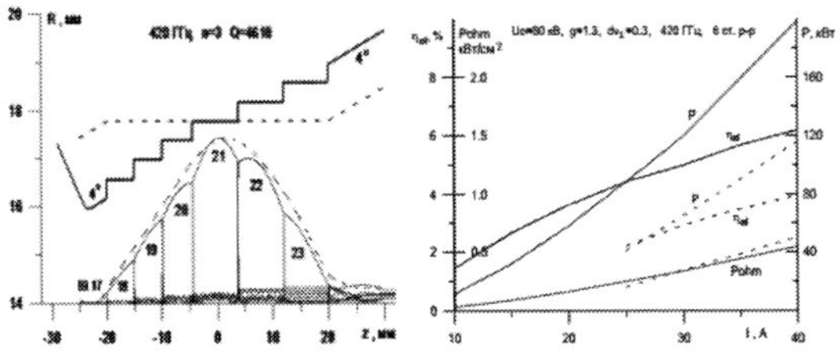

Fig. 3. Profile and RF field distribution of the multi-CCMC average power gyrotron and calculation data of the power, efficiency and ohmical load

Axially symmetric echelette cavity [6] on a some point of view is the development of multi-cavity systems [5] and has similar advantages and disadvantages. It has high selective properties and gyrotron in which such cavity was used in the experiments showed good output characteristics. Eshelette type gyrotron has great potential, but its proper numerical simulation and manufacturing techniques encounter serious problems. The complex structure of the output radiation requires the development of a new version of the QO converter. For some applications may be interested in two-mirror version eshelette gyrotron [6] shown in Fig. 4.

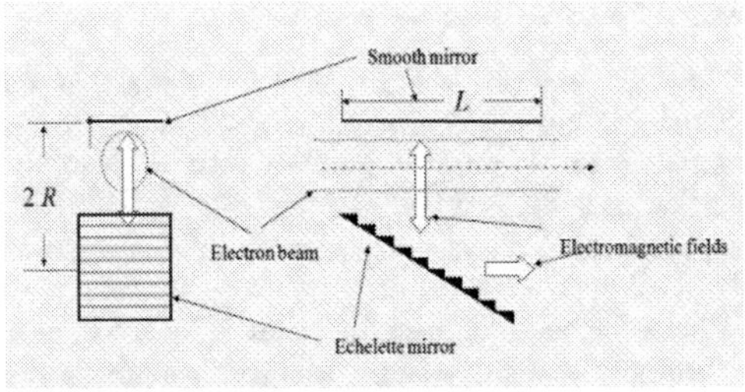

Fig. 4. Two-mirror version eshelette gyrotron

In this gyrotron version there is some possibility of frequency tuning and simplified radiation output system, but using a cylindrical beam inevitably lead to reduction of efficiency. In fact, this scheme is the development of gyrotron with quasi-optical cavity [2, 5], with the additional selection by using eshelette mirror.

Coaxial cavity draw attention of developers of gyro-devices, since its application is promising for achieving of high power levels which favor by high selectivity and absence of voltage depression of electron beam, the possibility of implementing the natural recovery scheme and frequency tuning [3]. Additional features are provided by selection of profiling the inner wall of the cavity and the coaxial insert. The achieved power levels of coaxial gyrotron exceed 2 MW at frequencies up to 170 GHz. Using a coaxial gyrotron are complicated due to technological problems associated with mounting, alignment and cooling of coaxial insert.

Great prospects for work on the cyclotron frequency harmonics have a gyrotron in a thin axial beam, often referred to as large-orbit gyrotron (LOG). In experiments with LOG with permanent magnet (field of about 1 T) was obtained [7] on these harmonics/ modes / frequency / power / efficiency: n = 3 / TE_{311} /84.88 GHz / 2.5 kW / / 6.25%; n = 4 / TE_{411} / 112.7 GHz / 0.47 kW / 0.96%. There was also observed oscillation at the 5th harmonic with frequency 138 GHz.

The LOG with pulse magnet [7] 2 kW power have been achieved at a frequency of around 0.6 THz (n = 2), 2 kW and 0.3 kW at a frequency of approximately 1 THz (n = 3) on the 2nd and 3rd harmonics of the cyclotron frequency respectively.

Summary

The report analyzes the perspective of non-canonical version of gyrotron in terms of possibility of frequency tuning and improved mode selection for operation on the gyrofrequency harmonics.

Theoretical and experimental results for millimeter and sub-millimeter range gyrotrons with different types of cavities (CCMC, coaxial, quasi-optical, echelette, etc.) are presented and discussed.

The most attractive versions of non-canonical gyrotrons were pointed out.

This work was supported by the Russian Foundation for Basic Research, grant 15-42-02380.

References

1. *G.S. Nusinovich, M.K.A. Thumm, M.I. Petelin.* The Gyrotron at 50: Historical Overview. J Infrared Millimeter Terahz Waves (2014) 35:325–381
2. *G. Nusinovich.* Introduction to the physics of gyrotrons, The Johns Hopkins University Press, Baltimore and London, 2004.
3. *M. Thumm.* State-of-the-Art of High Power Gyro-Devices and Free Electron Masers. *KIT* Scientific Publishing, 2013.
4. *V.E. Zapevalov.* Evolution of gyrotrons, Radiophysics and Quantum electronics, 2011, v. LIV, №8-9, pp. 559-572.
5. *Sh.E. Tsimring.* Electron beams and microwave vacuum electronics, John Wiley and Sons, Inc., Hoboken, New Jersey, 2007.
6. *E. Jerby, A. Kesar, M. Korol et al.* Cyclotron resonance maser array, IEEE Trans. Plasma Sceince, Vol. 27, pp. 445-455, 1999.
7. *V.I. Belousov, et al.* Research of the gyrotron with the echelette resonator. 9th Int. Workshop Strong Microwaves and Terahertz Waves: Sources and Applications. Nizhny Novgorod – Perm – Nizhny Novgorod, Russia July 24 – 30, 2014 p.166-167.
8. *V. Bratman, M. Glyavin, T. Idehara, et al.* Review of Sub-Terahertz and Terahertz Gyrodevices at IAP RAS and FIR FU. Int. Journal IEEE Trans. on Plasma Science, vol. 37, issue 1, pp. 36-43, (2009)

EPJ Web of Conferences **149**, 04004 (2017) DOI: 10.1051/epjconf/201714904004
10th International Workshop 2017 "Strong Microwaves and Terahertz Waves: Sources and Applications"

Development of advanced electron optical systems for novel gyrotrons

V.N. Manuilov[1,3], G.G. Sominskii[2] and M.Yu. Glyavin[1]

[1] Lobachevsky State University, Nizhny Novgorod, Russia, manuilov@rf.unn.ru
[2] Peter the Great St. Petersburg Polytechnic University, St.Petersburg, Russia
[3] Institute of Applied Physics, Nizhny Novgorod, Russia

Introduction

At present the application area of gyrotrons becomes every year wider and covers a lot of new directions of activity such as microwave and CVD technologies, ECRH systems, biology, spectroscopy and diagnostic of various media, hidden objects and defects identification and some others. The number of applications gradually grows and it leads to new, more stringent requirements to the power range, efficiency and frequency stability of gyrotrons. Of course, the mentioned requirements change, to some extent, the traditional scheme of the gyrotron and consequently lead to new versions and trends in the development of corresponding helical electron beam formation systems and types of the gyrotron collectors. Below the main directions in the development of new gyrotron electron optic systems is briefly described.

Results

The classical scheme of the gyrotron electron-optic system (EOS) includes the adiabatic magnetron injection gun (MIG) and cylindrical or conical collector to dissipate the rest energy of electrons after interaction with RF field in the cavity. The last one may apply the one-stage depressed collector scheme to increase the total efficiency of the gyrotron. The main modern demands to the new generation of gyrotrons are:

1) to reduce velocity spread in HEBs, especially in millimeter and submillimeter wave range;

2) to diminish the influence of reflected from magnetic mirror electrons;

3) to make the EOS less sensitive to the thermal expansion caused by high cathode temperature;

4) to make the device inertia-free, i.e. diminish the time of the switch on/off process;

5) to improve the mode competition problem and due to that reach 1 THz benchmark in CW operation regime;

6) to increase the total gyrotron efficiency and decrease the temperature load of the collector.

First two problems seem to be solved by turning to the non-adiabatic systems [1] or/and changing the regime of temperature limitation of emission to the space charge limiting regime. Two projects of the non-adiabatic EOS for 28 GHz / CW technological gyrotron and megawatt power level 170 GHz / CW are considered (see fig.1, 2). Optimization of the magnetic systems and the electrode shape is performed. The possibility to use the space charge limiting regime is examined.

Next two problems (3 and 4) can be solved, in particular, by applying the non-adiabatic systems [1] jointly with field emitters [2] instead of traditional hot cathodes. Such approach is illustrated by example of the gyrotron with an operating frequency of 263 GHz, an accelerating voltage of 15 kV and output power some tens watt aimed to use for the spectroscopy applications (fig. 3). Simulation shows that for optimized version of the electron gun the pitch-factor and velocity spread remains close to 1.2 and 30% respectively in the range of currents 0–50 mA that is quite admissible to achieve a typical efficiency of a few percent for such gyrotrons.

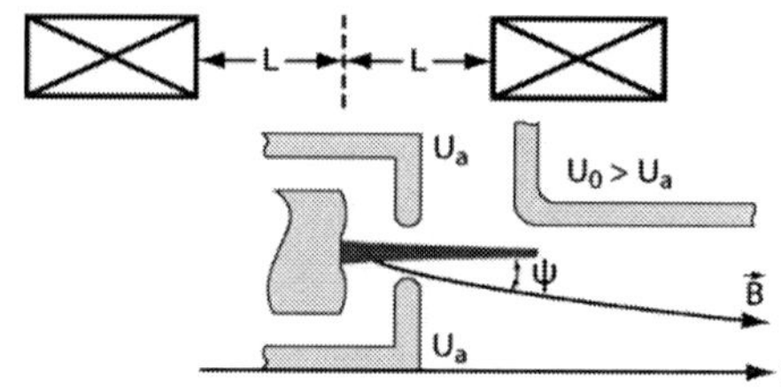

Fig. 1. The principal scheme of the HEB formation region in non-adiabatic EOS. Cathode counterrunning coils to adjust the pitch-factor value are shown

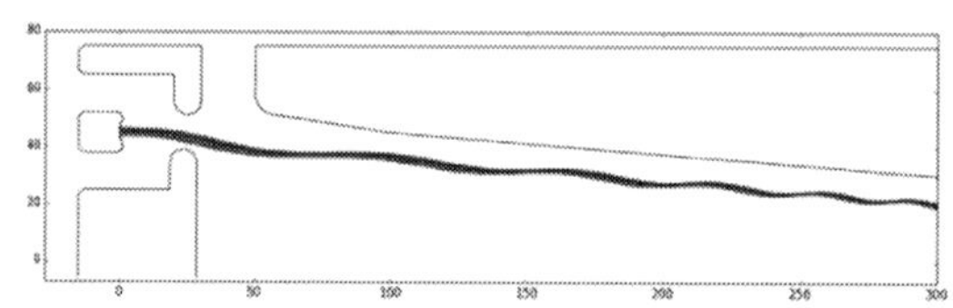

Fig. 2. The configuration of the electrodes in the triode non-adiabatic electron-optical gun for 170 GHz / 1 MW gyrotron (all dimensions are in mm)

The only way to reach 1 THz CW operation and at the same time to keep reasonable cost of the magnetic system and the gyrotron itself is to use high harmonic operation. One the effective ways to solve mode competition problem in such case is to use multi-beam gyrotrons [3] (item 5 in the list presented above). In the report the possible schemes of the multi-beam EOS are presented. The example of the two-beam adiabatic formation system with both generating beams for 790 GHz / CW second harmonic gyrotron is described and optimized (fig. 4). For a more flexible control of the parameters of the partial HEBs a triode MIG with an anode voltage of $U_a \approx 10$ kV has been envisaged. It has been assumed that both partial beams are laminar (with an incidence angle of the magnetic field on the emitting surface $\varphi \approx 25°$) in order to reduce the influence of the space–charge field and have equal pitch factors $g_{1,2} \approx 1.3$. The currents of the two beams relate as 1:2. Aiming a sufficient longevity of the cathode the current density extracted from the emitter is limited to 2 A/cm^2.

The total gyrotron based installation efficiency is extremely important for industrial CW gyrotrons. The total efficiency depends on the interaction one, efficiency of energy recovery and power consumption aimed to produce the operating magnetic field (item 6). In the report the idea to develop the gyrotron EOS with hot magnet surrounding by ferromagnetic screens is described (fig. 5). For fixed value of the operating magnetic field it allowed to decrease the power of the main coil in two times approximately. Optimization of the coils and screens geometry allowed to provide the inhomogeneity of the axial magnetic field in the cavity less than 1%. Rapid drop of the magnetic field outside the screens breaks the adiabatic conditions for magnetic field in the formation region. It makes difficult to use well known and reliable MIG for HEB generation. To keep the adiabaticity of the field the additional small power cathode coil is included into the scheme of EOS (fig. 5). Optimized MIG geometry provides high pitch-factor g > 1.6 and at the same time modest velocity spread in wide range of the operating regimes.

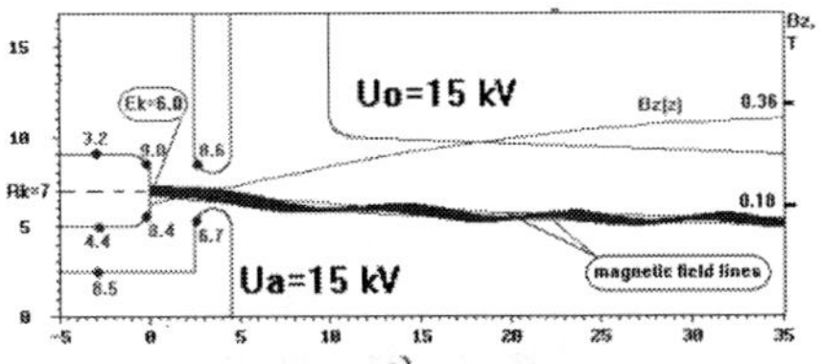

a)

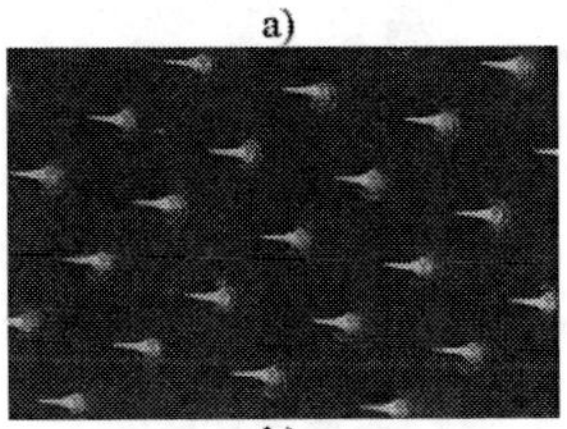
b)

Fig. 3. Scheme of the primary formation of the beam in a non-adiabatic EOS (a). Distributions of the axial magnetic field B(z) (right axis, data are in T) is presented. Dots and corresponding numbers show the electric field value (kV/mm) at the most critical points. Electric field Ek at the emitter center is also specified. All dimensions are in mm. The multi-pin stricture of the cathode (b) is shown

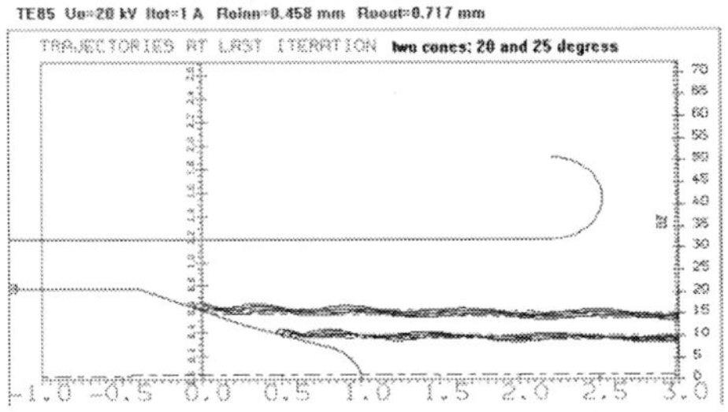

Fig. 4. The electrode profile for double-beam MIG with two generating beams

Rapid decrease of the magnetic field intensity outside the ferromagnetic screens allows also apply the new scheme of the collector with inertial expansion of the HEB and so to decrease the power density on the collector. The results of numerical simulation of such scheme are briefly discussed.

Conventional powerful gyrotron scheme includes also the one-stage CPD collector. But the corresponding gyrotron efficiency usually limited by the value 50–55%. Further increase of the CPD collector efficiency can be reached by turning to two-stage scheme. Here the main and very difficult problem is the space separation of the electrons with different energies. Some ways to solve it (collectors with crossed electric and magnetic fields, with reverse of the magnetic field) are presented in the rest part of the report.

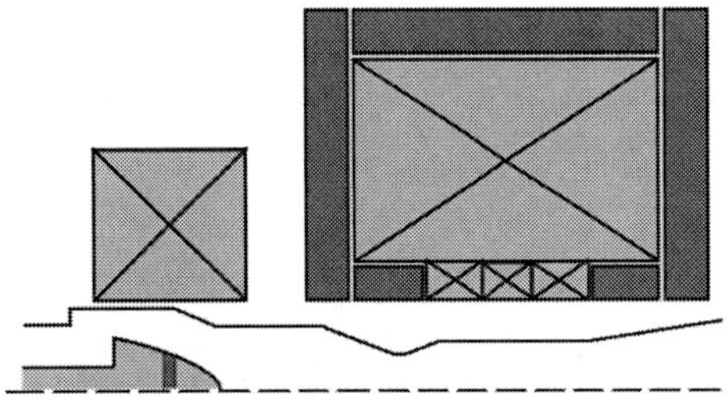

Fig. 5. Scheme of the electron optic system with ferromagnetic screens and magnetron-injection gun

Conclusion

Some new helical electron beam formation systems as well as the new concepts of collectors for gyrotrons are described and examined. It is shown that they open new promising possibilities in the development of new generation of gyrotrons with improved efficiency and increased frequency and stability.

Acknowledgements

The work was supported by the Russian Science Foundation, project No. 14-12-00887.

References

1. *A.L. Goldenberg, V.N. Manuilov and M.Yu. Glyavin.* Electron-Optical System of a High-Power Gyrotron with Non-adiabatic Electron Gun // Radiophysics and Quantum Electronics, vol.48, No 6, 2005, pp. 461-465

2. *G.G. Sominskii, T.A. Tumareva, E.P. Taradaev, M.V. Mishin, A.N. Stepanova.* Multitip semiconductor field emitters with new-type bilayer protecting coatings. // Tech. Phys., 60(1), 133-136 (2015).

3. *V.E. Zapevalov, Sh.E. Tsimring.* Multibeam gyrotrons. // Radiophys. Quantum Electron. 33, 11, 954-960 (1990).

EPJ Web of Conferences **149**, 04005 (2017) DOI: 10.1051/epjconf/201714904005
10th International Workshop 2017 "Strong Microwaves and Terahertz Waves: Sources and Applications"

Comparison between Controlled Non-Adiabatic and *E*×*B* Concepts for Gyrotron Multistage Depressed Collectors

Chuanren Wu, Ioannis Gr. Pagonakis, Stefan Illy, Gerd Gantenbein, Manfred Thumm, and John Jelonnek

Karlsruhe Institute of Technology, Karlsruhe, Germany, chuanren.wu@kit.edu

Introduction

Gyrotrons are used as RF sources for ECH systems in nuclear fusion experiments. To obtain sufficient energy-gain factor, the future DEMOnstration fusion power plant demands highly efficient gyrotrons. The interaction between electron beam and microwave in a fusion gyrotron may have in the optimal case approximately 35 % efficiency ($\eta_{\rm int}$). A large fraction of the input energy still remains in the spent electron beam. Up to $\eta_{\rm col} \approx 60$ % of the spent beam energy can be recovered by a Single-stage Depressed Collector (SDC), which leads to an overall gyrotron efficiency of about 50 % according to

$$\eta_{\rm total} = \frac{\eta_{\rm int}\,\eta_{\rm RF}}{1-(1-\eta_{\rm int})\,\eta_{\rm col}} \quad , \qquad (1)$$

where $\eta_{\rm RF} \approx 90$ % is assumed to be the ratio between the output and the generated RF power. To achieve 60 % of gyrotron overall efficiency, the collector efficiency $\eta_{\rm col}$ is required to be higher than 74 %. For this purpose, Multistage Depressed Collectors (MDC) should be introduced. In an MDC the spent electron beam will be sorted. Proper deceleration voltages (depression voltages) will be applied to each energy stage. Moreover, in order to reduce the thermal loading on the electrodes, the electron beam have to be eventually spread or swept on a large area.

Designing an MDC for fusion gyrotrons is non-trivial. Three major reasons are listed here. Firstly, fusion gyrotrons works under high magnetic field, which has a dependence on the cyclotron frequency as

$$\omega_{\rm c} = \frac{e\,B_0}{\gamma\,m_0} \quad \text{, with} \quad \gamma = \frac{1}{\sqrt{1-(v/c)^2}} \qquad (2)$$

For example, the Wendelstein 7-X 140 GHz gyrotron requires more than 5 T field and the 170 GHz ITER gyrotron requires almost 7 T. At the collector entrance, the magnetic field from the gyrotron super-conducting coil is reduced, but still can be higher than 100 mT. The magnetic confinement in the collector prohibits the electron beam to be sorted or spread, since the prerequisites for the repulsion or the universal beam spread curve is not fulfilled. Secondly, electron beams in gyrotrons are annular. The beam thickness is small compared to the electron guiding center radii, while the magnetic flux enclosed in the hollow electron beam keeps invariant in a symmetric collector. As a consequence, during the unwinding of the magnetic field mentioned in the previous point, the expansion of the enclosed hollow region inside the beam is more eager than the increment of beam thickness itself. This would results in a large MDC geometry. Thirdly, due to the presence of the magnetic field, the trajectories of secondary electrons are harder to be controlled. The magnetic field may guide secondary as well as reflected electrons back to the interaction region.

Nevertheless, there are still two concepts for gyrotron MDCs. In the first one, the gyrotron magnetic field is unwound in a special way utilizing well-controlled non-adiabatic transitions [1], whereas in the second concept, the electrons are sorted by the plasma ***E*×*B*** drift [2]. In the present work, these two concepts will be discussed and compared with each other.

Concept I: non-adiabatic demagnetization

An intuitive solution to the issue with strong gyrotron magnetic field is to unwind it from 100 mT until a low value, typically around 10 mT [1]. If this process is adiabatic, a huge part of the electron transversal motion can be converted to the longitudinal one, however an adiabatic demagnetization will take place in a long distance, hence the collector geometry will be extremely long. Another side effect of an adiabatic magnetic field transition is that the electron beam is still bundled as before and weakly confined; mixed with the applied electric field, the electron trajectories are difficult to be manipulated. This observation is also in agreement with [3]. On the other hand, a non-adiabatic reduction of the magnetic field will shrink the transition and thus will reduce the length of the MDC significantly, for the price that the electron transversal velocity will not be completely converted, so that the optimal depression voltages may not be applied. As a trade-off, the unwinding of the magnetic field should be slightly non-adiabatic.

A side effect of the non-adiabatic transition is that it may spread the electron beam over the space by influencing the ratio of transversal / longitudinal velocities of each electron individually. Furthermore, a well-designed non-adiabatic transition may spatially modulate the electron trajectories according to their energy, which can be used as an energy sorting mechanism for MDC [4].

Since the system is axis symmetric, it can be modelled and simulated in 2D. Moreover, a so-called "effective potential" (strictly to say, the energy) φ has been proposed in [1], which considers both electric and magnetic fields:

$$\varphi = \frac{1}{2\,m}\left(\frac{P_\theta - q\,r\,A_\theta}{r}\right)^2 + q\,V \quad , \qquad (3)$$

P_θ and A_θ are the azimuthal components of canonical momentum and magnetic vector potential, respectively; whereas V is the electric potential. This is an elegant way to describe the system. However, to the knowledge of the authors, the φ serves practically as a post-design diagnostic and explanation model rather than a design tool, since the synthesis of the field can still be tricky.

As shown above, the first issue mentioned in the introduction can be solved through the non-adiabatic transition. In order to target the second issue, extra coils or ferromagnetic poles are inserted coaxially. The inserted magnets "pull" the electron beam to a smaller radius; while in addition, they can further reduce the off-axial magnetic field. There is no fundamental solutions to reduce the effects of secondary electrons except tuning the fields, looking for better materials or grooving the impacting area of electrodes. Typically 8 % – 10 % of η_{col} reduction by secondary electrons can be expected.

The maximal possible number of stages depends on how wide the electron beam can be spread in space. Broadly spreading the electron beam requires a weak magnetic field (≪ 10 mT). In this case, the MDC may be sensitive to external magnetic fields and beam parameters such as beam current or misalignment. In the ideal case, a two-stage collector might have enough performance to boost the η_{total} till 60 % [4], while practically more stages might be necessary.

Concept II: *E*×*B* drift

A more promising MDC concept is the one using *E*×*B* drift. Its principle is shown in Fig. 1. The drift velocity is independent of particle mass, charge, energy and even the motion direction. In this model, each electron have an initial velocity component perpendicular to the direction of drift. The value of an electron's initial velocity depends on its initial energy. Since the same ***E***×***B*** drift is applied on each electron, the one with a higher initial velocity will travel further during the deceleration by the electric field until it lands on any electrode. The principle may allow arbitrary number of stages flexibly, as the electrons will always go to the desired electrodes, no matter how far the distances between stages are [5], which is in the previous concept not the case. This drift velocity also prevents secondary electrons to be reflected back to the interaction region, since most electrons are drifted back to the surface, where they were generated. In the worst case, a small fraction of electrons will be reflected only one stage back. This means that if there were more stages, the energy loss from those inter-stage secondary electrons would also be smaller.

There are several ways to obtain the ***E***×***B*** drift, it can be either an azimuthal ***E*** field with an axial ***B*** field [2, 6], or the contrary [7], or even some ideas out of the box that the annular beam can be transformed to one or more sheet beams [5, 8]. Realistic simulations e.g. [6]

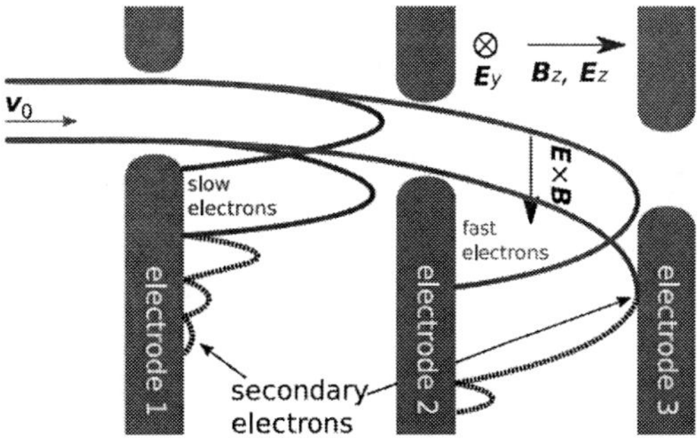

Fig. 1. Principle of the concept based on ***E***×***B*** drift

show that MDCs based on this concept can increase gyrotron efficiencies above 60 % while the efficiency reduction by secondary electrons is very low.

Conclusion

The non-adiabatic demagnetization concept and the *E*×*B* drift concept for gyrotron MDC are compared. The comparison is concluded in the following table. For high-frequency gyrotrons the concept based on the *E*×*B* drift has more advantages and will be preferred.

Acknowledgement

This work has been carried out within the framework of the EUROfusion Consortium and has received funding from the Euratom research and training program 2014–2018 under grant agreement No. 633053. The views and opinions expressed herein do not necessarily reflect those of the European Commission.

References

1. *Singh, A, Rajpatirana, S., Men, Y., Granatstein, V. L., Ives, R. L., Antolak, A.* IEEE Trans. Plasma Sci., vol. 27, no. 2, pp. 490-502, 1999.
2. *Pagonakis, I. G., Hogge, J. P., Alberti, S., Avramides, K. A., Vomvoridis, J. L.* IEEE Trans. on Plasma Sci., vol. 36, no. 2, pp. 469-480, 2008
3. *Read, M., Lawson, W., Dudas, A. J., Singh, A.* IEEE Trans. Electron Devices, vol. 37, no. 6, pp. 1579-1589, 1990.
4. *Wu, C., Pagonakis, I.Gr., Illy, S., Thumm, M., Gantenbein, G., Jelonnek, J.* German Microwave Conf., pp. 365-368, 2016.
5. *Pagonakis; I. Gr., Wu, C., Illy, S., Jelonnek, J.* Phys. of Plasmas, 23, 043114, 2016
6. *Wu, C., Pagonakis, I. Gr., Gantenbein, G., Illy, S., Thumm, M., Jelonnek, J.* Int. Vacuum Electronics Conf. 2017
7. *Louksha, O. I., Trofimov, P.A.* International Vacuum Electronics Conf. 2017
8. *Wu, C., Pagonakis, I. Gr., Gantenbein, G., Illy, S., Thumm, M., Jelonnek, J.* Phys. of Plasmas 24, 043102, 2017

	Non-Adiabatic Concept	***E*×*B* Drift Concept**
Magnetic Field	Non-adiabatic unwound to weak magnetic field typically ≤ 10 mT	Gently unwound to moderate magnetic field typically 30–60 mT for fusion gyrotrons
More Stages	Hard	Straightforward
Secondary Electrons	Not easy to handle May reduce η a lot (typically 8–10%)	The same drift is applied also on secondary electrons Small influence on efficiency (1 % if optimized [6])
Tolerances	Might be sensitive to electron beam parameters and external magnetic fields	May tolerate weak external magnetic fields and slightly misaligned beam due to the moderate ***B*** field
Design Approaches	Rotationally symmetric (2D description) Effective potential aided design and optimization	Not axisymmetric (3D description) Azimuthal ***E*** field [2, 6] or ***B*** field [7] Transformation to sheet beam [5, 8]

EPJ Web of Conferences **149**, 04006 (2017) DOI: 10.1051/epjconf/201714904006
10th International Workshop 2017 "Strong Microwaves and Terahertz Waves: Sources and Applications"

Generation of rogue waves in gyrotrons with high excess over the threshold

I.V. Zotova[1], N.S. Ginzburg[1], M.V. Morozkin, R.M. Rozental[1], A.S. Sergeev[1], A.E. Fedotov[1], V.P. Tarakanov[2]

[1]Institute of Applied Physics RAS, N.Novgorod, Russia, zotova@appl.sci-nnov.ru
[2]Moscow Engineering Physics Institute, Moscow, Russia

Many physical systems exhibit behavior associated with the emergence of the so-called rogue waves which represent dramatically high-amplitude events that occur with low probability but much more frequently than expected in ordinary wave statistics [1]. For example, in laser physics and optics rogue waves arise in the process of supercontinuum generation and represent pulses of ultra broadband light with extremely high peak power [2]. In this paper within the framework of the average approach and in direct 3D PIC (particle-in-cell) simulations we demonstrate that rogue waves can also appear in gyrotrons operating with high excess over the threshold. The mechanism of extreme spikes formation is related to simultaneous cyclotron interaction of a gyrating electron beam with forward and backward waves near the waveguide cutoff frequency as well as with the longitudinal deceleration of electrons. From the practical point of view, experimental observation of the rogue waves generation in gyrotrons can be interesting as a method for production of high-power millimeter-wave radiation with an ultra broadband spectrum. The required exceeding over the threshold can be achieved for routine parameters of the driving electron beams formed by the magnetron-injection guns due to operation at the lowest modes of microwave cavities.

Mechanism of Rogue Waves Generation

For theoretical analysis of rogue waves generation in gyrotrons we have generalized the self-consistent time-domain model of low-Q gyrotrons [3] in the frame of which the evolution of the field amplitude is described by a non-uniform parabolic equation. In the novel model, first of all an inclination of the beam line with respect to the dispersion characteristic of the waveguide mode is taken into account [4], because the maximum spectrum bandwidth of chaotic generation, including rogue waves, is determined by the difference between resonances with backward and forward waves (points *1* and *2* in Fig. 1 a). The second critical factor which should be included into consideration is the change of the longitudinal momentum of electrons. This effect is not significant for a steady-state generation regime and, as a rule, is ignored in the gyrotron theory. However, in the process of formation of giant ultrashort pulses with a very steep profile strong transverse magnetic fields are initiated even near the cut-off frequency, which, in turn, leads to transformation of the translational energy of electrons to the energy of their orbital rotation and then in the energy of an electromagnetic pulse. Correspondingly, the generalized system of equations has the following form:

$$
\begin{aligned}
& i\frac{\partial^2 a}{\partial Z^2}+\frac{\partial a}{\partial \tau}+\delta(Z)a=iI_0J,\\
& \frac{\partial \hat{p}_\perp}{\partial Z}+\frac{g_0^2}{4}\frac{\partial \hat{p}_\perp}{\partial \tau}+i\frac{\hat{p}_\perp}{\hat{p}_\parallel}\left(\Delta-1+|\hat{p}_\perp|^2+\frac{\hat{p}_\parallel^2-1}{g_0^2}\right)=i\frac{a}{\hat{p}_\parallel}, \qquad (1)\\
& \frac{\partial \hat{p}_\parallel}{\partial Z}+\frac{g_0^2}{4}\frac{\partial \hat{p}_\parallel}{\partial \tau}=-g_0^2\frac{\beta_{\perp 0}^2}{2}\mathrm{Re}\left(\frac{\hat{p}_\perp^*}{\hat{p}_\parallel}\frac{\partial a}{\partial Z}\right).
\end{aligned}
$$

where the most normalized variables are the same as in the previous gyrotron model [3], $\hat{p}_\parallel = p_\parallel / m v_{\parallel 0}\gamma_0$ is the normalized longitudinal momentum, $g_0 = \beta_{\perp 0}/\beta_{\parallel 0}$ is the initial pitch-factor, the function $\delta(Z)$ describes the profile of a gyrotron resonator. We assume that at the input cross section the electrons are uniformly distributed over cyclotron rotation phases $\hat{p}_\perp(Z=0)=e^{i\theta_0}$, $\theta_0 \in [0,2\pi)$ and have the same longitudinal momenta $\hat{p}_\parallel(Z=0)=1$. At the system input and output we apply the well-known reflectionless boundary condition [3].

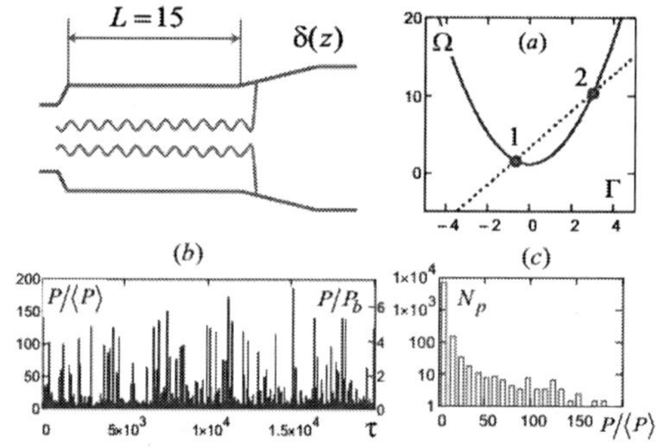

Fig. 1. Gyrotron resonator profile and dispersion diagram (*a*) of the waveguide mode and the electron beam line for the rogue wave generation (*b*) with the pulse-height distribution (*c*).

In Fig. 1 b, c the results of simulations are presented for $L=15$, $g_0=1.3$, $\beta_{\perp 0}=0.2$, $\Delta=-0.9$ and $I_0=3$. It is seen that radiation represents the sporadic formation of giant pulses with the ratio of the peak power to the average level of $P_{peak}/\langle P\rangle \sim 150 \div 200$. The corresponding pulse-height histograms have a typical for rogue waves [2] long-tail profile, with the extreme events occurring frequently than expected in the ordinary wave statistics.

The mechanism of giant pulse formation in gyrotrons includes two stages that are illustrated by Fig. 2. At the first stage, electromagnetic radiation is associated with excitation of the backward wave (Fig. 2 a) having a fairly narrow spectrum, corresponding to the low resonance point *1*. Upon reflection from the cathode neck, this radiation is partially absorbed by the electron beam, that leads to a dramatic increase in the transverse energy of electrons up to $g \sim 3.5 \div 4$. These electrons produce the forward propagating pulse, which is effectively amplified in the process of its propagation (slippage [5]) along the beam due to the difference between the group velocity of the wave and the translational velocity of particles. This process is accompanied by the pulse front steepening and excitation of a strong transverse magnetic field $H_\perp \sim \partial a/\partial Z$. It leads to transformation of the longitudinal to the transverse momenta of particles (Fig. 2 b), that, in turn, provides an additional source of rotational energy for amplification of the forward propagating ultrashort electromagnetic pulse. The appearance of such pulses leads to a significant widening of the output radiation spectrum corresponding to the frequency difference at the resonance points *1* and *2*. It is important that the peak power of generated pulses strongly exceeds not only the mean radiation level but also the kinetic power of the driving electron beam P_b (conversion coefficient of $P/P_b \approx 6 \div 9$).

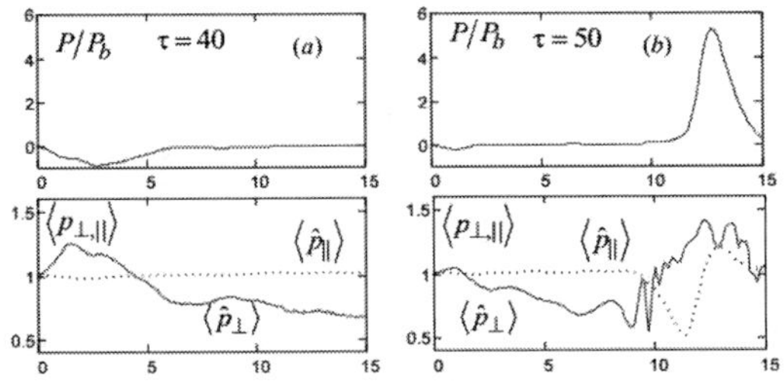

Fig. 2. Two stages of the rogue wave formation in a gyrotron: (*a*) excitation of the backward wave pumping the transverse energy of later injected particles and (*b*) deceleration of electrons with transformation of translational energy to the energy of orbital rotation

Results of 3D PIC Simulations

The possibility of the rogue waves generation in gyrotrons was confirmed by direct 3D KARAT simulations, in which we considered excitation of the TE_{11} mode with the cutoff frequency of 35 GHz by an electron beam with parameters typical for millimeter waveband gyrotrons: particles energy of 20 keV, an electron current of 2 A, and a pitch-factor of 1.3. In this case, corresponding normalized parameters are basically the same as those used above (see Fig. 1). Results of the PIC simulations are presented in Fig. 3. It is seen that the gyrotron radiation is characterized by appearance of high-amplitude ultrashort pulses with durations of 0.2 to 0.25 ns and the peak power of 200 kW, which corresponds to the conversion factor of ~5. The ratio of the pulse peak power to the mean radiation power reaches 125, that is in a good agreement with results obtained on the basis of Eqs. (1). It should be noted that strong axial deceleration of electrons was also observed in PIC simulations. From the practical point of view, experimental observation of the rogue waves generation in gyrotrons can be interesting as a method for production of high-power millimeter-wave radiation with an ultrawide spectrum, which, according to our PIC simulations, amounts to 10% at the level of –10 dB (Fig. 3 c).

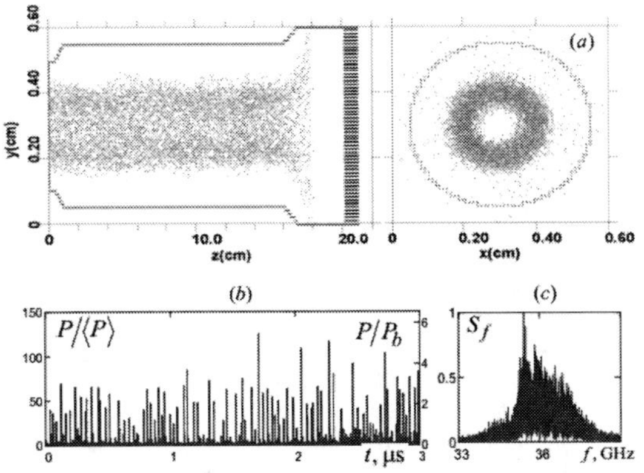

Fig. 3. Results of PIC simulations: (*a*) geometry of an interaction space, (*b*) the time dependence of the output power and typical temporal profile of a generated spike; (*c*) the broadband radiation spectrum

This work was supported by the Russian Foundation for Basic Research (RFBR) (No. 17-08-01077)

References

1. *Onorato, M., Residori, S., Bortolozzo, U., Montina, A., Arecchi, F.* Rogue waves and their generating mechanisms in different physical contexts // Phys. Reports 2013. V. 528, No. 3. P. 47–89.

2. *Solli, D.R., Ropers, C., Koonath, P., Jalali, B.* Optical rogue waves // *Nature* 2007. V. 450, P. 1054–1057

3. *Ginzburg, N.S., Nusinovich, G.S., Zavolsky, N.A.* Theory of non-stationary processes in gyrotrons with low Q resonators // Int. J. of Electronics 1986. V. 61, P. 881-894.

4. *Ginzburg, N.S., Sergeev, A.S., Zotova, I.V.* Time-domain self-consistent theory of frequency-locking regimes in gyrotrons with low-Q resonators // Phys. Plasmas 2015. V. 22, P. 033101.

5. *Zhang, T-B., Marshall, T.C.* Possibility of generating a high-power self-similar radiation pulse from a free electron laser. // Phys. Rev. Lett. 1995. V.74, Iss.6. P. 916–919.

European research activities towards a future DEMO gyrotron

J. Jelonnek[1,2], G. Aiello[3], S. Alberti[4], K. Avramidis[1], A. Bertinetti[11], A. Bruschi[5], J. Chelis[6], T. Franke[7,8], G. Gantenbein[1], S. Garavaglia[5], G. Granucci[5], G. Grossetti[3], S. Illy[1], Z.C. Ioannidis[1], J. Jin[1], P. Kalaria[1], G.P. Latsas[6], H. Laqua[9], A. Leggieri[10], F. Legrand[10], A. Marek[1], I.Gr. Pagonakis[1], D. Peponis[6], L. Savoldi[11], T. Rzesnicki[1], S. Ruess[1,2], T. Ruess[1], T. Scherer[3], M. Schmid[1], D. Strauss[3], I. Tigelis[6], M. Thumm[1,2], M.Q. Tran[4], F. Wilde[1,9], C. Wu[1], R. Zanino[11], A. Zein[1]

[1]IHM, [2]IHE, [3]IAM-AWP, Karlsruhe Institute of Technology (KIT), D-76131 Karlsruhe, Germany
[4]Swiss Plasma Center (SPC), EPFL, CH-1015 Lausanne, Switzerland
[5]Institute of Plasma Physics "P.Caldirola", National Research Council of Italy, Milan, Italy
[6]NCSRD/National and Kapodistrian University of Athens, Faculty of Physics, Athens, Greece
[7]EUROfusion Consortium, Boltzmannstr. 2, D-85748 Garching, Germany
[8]Max-Planck-Institut für Plasmaphysik, Boltzmannstr. 2, D-85748 Garching, Germany
[9]Max Planck Institut für Plasmaphysik, Teilinstitut Greifswald, Wendelsteinstrasse 1, D-17491 Greifswald, Germany
[10]Thales Electron Devices, 2 rue Marcel Dassault, Vélizy-Villacoublay, F-78141, France
[11] NEMO group, Dipartimento Energia, Politecnico di Torino, 24, c. Duca degli Abruzzi, I-10129 Torino, Italy

Coordinated by the EUROfusion Consortium, several European research institutes are working on fusion technologies towards options for a European DEMOnstration Fusion Power Plant (FPP), as a single step between ITER and a commercial FPP, to deliver net electricity by mid of this century. One of the focus areas is the research on a proper Electron Cyclotron Resonance Heating (ECRH) and Current Drive (ECCD) system for which the fusion gyrotron is one of its major key components [1].

A future FPP will probably require an ECCD operating frequency ranging from 170 GHz up to 240 GHz depending on the DEMO baseline. An RF output power of significantly higher than 1 MW (target: 2 MW) and a total gyrotron efficiency better than 60% are required. Multi-purpose operation at multiples of the $\lambda/2$-resonance frequency of the vacuum window of the gyrotron, hence in leaps of about 30 GHz (e. g. 170 / 204 / 238 GHz) needs to be considered for plasma start-up, heating and current drive. Optionally for possible steering of the absorption layer the gyrotron shall allow a fast frequency tuning in steps of around 2–3 GHz. The R&D work within the EUROfusion work package "WP HCD EC Gyrotron R&D and Advanced Developments (AD)" is focusing on all of the named targets.

Verification of the coaxial-cavity technology

The coaxial-cavity gyrotron is a promising technology for future multi-MW fusion gyrotrons [2]. In [3] a world record RF output power of 2.2 MW at short pulses (ms-range) was demonstrated. Nevertheless, the 2 MW coaxial-cavity technology, already considered for the first installation in ITER earlier [4], is still lacking its proof-of-concept regarding long-pulse operation. Major concerns are the proper alignment and thermal loading of the cavity wall and its inner conductor as well as the thermal loading of the collector. Its feasibility shall be finally demonstrated by upgrading the existing KIT 2 MW 170 GHz short-pulse pre-prototype to pulse lengths up to 1 s [5]. In parallel, work is ongoing in the field of advanced cooling concepts [6, 7]. Additionally, two new coaxial-cavity Magnetron Injection Guns (MIGs) are under manufacturing. The first is employing an advanced emitter technology whose major element is a new non-emissive coating. That will significantly reduce the velocity spread of the electrons at the emitter [8]. Secondly, a newly designed Inverse Magnetron Injection Gun (IMIG) will allow for a significant larger emitter radius and therefore increased output power at operating frequencies significantly above 200 GHz by keeping the same or even smaller size of the bore hole of the gyrotron SC magnet [9].

Studies towards a 240 GHz gyrotron

A frequency up to 240 GHz was selected for the theoretical research work towards a future FPP, considering the requirements for "multi-purpose" and "fast frequency step-tunable" operation at high-field tokamaks and for a wide range of RF beam steering. The coaxial-cavity gyrotron technology, and, as a possible fallback solution, the conventional hollow-cavity gyrotron are under investigation. Both technologies were studied regarding to the maximum achievable output power versus efficiency and stability in operation due to tolerances. A generic design strategy was developed to find the optimum operating mode for the two different cavity topologies [10]. Operating scenarios close to 2 MW for the coaxial-cavity technology and around 1.5 MW for the conventional cavity technology have been found in the theoretical analyses (ref. [11–14]).

Advances in Window Technologies

Fundamental for frequency-step tunable operation is the utilization of a proper broadband window technology. At KIT the CVD-Diamond Brewster-angle window is the favorite. The successful operation of a step-frequency tunable 1 MW short-pulse (ms) gyrotron with a synthetically manufactured diamond Brewster angle output window was demonstrated earlier [15]. Nevertheless, for future DEMO gyrotrons the development of diamond discs of larger size for higher power capabilities, advanced cooling and brazing technologies are mandatorily required and pushed forward at KIT [16, 17].

Towards a total efficiency of higher than 60%

A DEMO gyrotron will require a total efficiency of above 60% to minimize the electrical power requirements, and, ultimately the recirculating power in the balance of plant. Considering an interaction efficiency of typical 35% between the electron beam and the microwave, a large fraction of the input energy remains in the spent electron beam. Up to 60% of the spent beam energy might be recovered by a single-stage depressed collector (SDC) which will lead to an overall gyrotron efficiency of about 50% in theoretically best case. An overall efficiency of higher than 60% requires the use of advanced multi-stage depressed collectors (MDCs). Two concepts are under investigation [18] at KIT. In the first one, the gyrotron magnetic field is unwound utilizing well-controlled non-adiabatic transitions, whereas in the second one, the electrons are sorted by an E × B drift [19–22].

Intelligent control systems for gyrotrons

In future, intelligent control systems shall allow higher output powers of gyrotrons by reducing the necessary safety margins, hence allowing for an operation close to the stability limits [23]. Firstly applied to the W7-X gyrotrons, a new control concept is under investigation. The idea is to detect the stray radiation caused by the excitation of parasitic modes at the stability limits and to use that as an indicator for the operational stability of the W7-X gyrotrons. A feedback control system is under development.

Advanced gyrotron tests environments

In 2015, the final design review and procurement of the High-Voltage Power Supply (HV PS) for the new KIT FULGOR gyrotron teststand started [24]. FULGOR will allow CW operation of gyrotrons with a required input power of 5 (2nd phase: 10) MW DC. A 10.5 T SC magnet will allow the operation of gyrotrons up to 240 GHz. That includes the upgrade of all the measurements systems, which have been developed for verification of the frequency spectrum, calorimetry and quasi-optical transmission [25, 26].

Acknowledgement

This work was carried out within the framework of the EUROfusion Consortium and have received funding from the Euratom research and training programme 2014–2018 under grant agreement No. 633053. The views and opinions expressed herein do not necessarily reflect those of the European Commission.

References

1. *G. Granucci et al.* The EC-system of EU DEMO: concepts for a reactor heating system // SMP, 2017, Nhizhny-Novgorod.

2. *B. Piosczyk et al.* 165-GHz coaxial cavity gyrotron // IEEE Trans. Plasma Science, 2004, Vol. 32, No. 3, 413-417.

3. *T. Rzesnicki, et al.* 2.2-MW Record Power of the 170-GHz European Pre-Prototype Coaxial-Cavity Gyrotron for ITER // IEEE Trans. Plasma Science, 2010, Vol. 38, No. 6, pp. 1141-1149.

4. *J.-P. Hogge et al* The European 2MW, 170GHz coaxial cavity gyrotron for ITER // Joint 32nd IRMMW/15th Int. Conf. on THz Electronics, 2007, Cardiff.

5. *S. Ruess et al.* Design and Manufacturing Process for the KIT 2-MW 170-GHz Coaxial-Cavity Longer-Pulse Gyrotron // SMP, 2017, Nhizhny-Novgorod.

6. *Bertinetti et al.* Multi-physics analysis of a 1 MW gyrotron cavity cooled by mini-channels // presented at SOFT, 2016, P1.029, accepted for publication in Fus. Eng. Des., 2017.

7. *K. Avramidis et al.* Numerical studies on the influence of cavity thermal expansion on the performance of a high-power gyrotron // IVEC, London, 2017.

8. *I. Pagonakis et al.* Influence of emitter ring manufacturing tolerances on electron beam quality of high power gyrotrons, // Physics of Plasmas, 2016, 23, 083103, doi: 10.1063/1.4959113.

9. *S. Ruess et al.* An Inverse Magnetron Injection Gun for the KIT 2-MW Coaxial-Cavity Gyrotron // IEEE Trans. Electron Devices, 2016, Vol. 63, pp. 2104-2109.

10. *J. Franck, et.al.* A generic mode selection strategy for high-order mode gyrotrons operating at multiple frequencies // Nucl. Fusion, 2014, Vol. 55, No. 1.

11. *J. Franck, et. al.* Direct voltage depression calculation of arbitrary electron beams in misaligned coaxial gyrotron cavities // IEEE Trans. on Electron Devices, 2016, Vol. 63, No. 9.

12. *J. Franck, et.al.* Insert misalignment in coaxial gyrotrons: physical effects and numerical treatment // 5th ITG Int. Vacuum Electronics Workshop, 8 – 9, Sept. 2016, Bad Honnef, Germany.

13. *P. Kalaria, et. al.* RF behavior and launcher design for a fast frequency step-tunable 236 GHz gyrotron for DEMO // Journal of RF-Engineering and Telecommunications - Frequenz, 2016, Vol. 71, No. 3-4, pp. 161-171.

14. *P. Kalaria, et. al.* Systematic cavity design approach for a multi-frequency gyrotron for DEMO and study of its RF behavior // Physics of Plasmas, 2016, Vol. 23, No. 9.

15. *G. Gantenbein et al.* First operation of a step-frequency tunable 1-MW gyrotron with a diamond Brewster angle output window // IEEE Trans. Electron Devices, 2014, Vol.61, No. 6, pp.1806-1811.

16. *G. Aiello et al.* CVD diamond Brewster window: feasibility study by FEM analysis // Proc. Workshop on Electron Cyclotron Emission and Electron Cyclotron Resonance Heating, Deurne, 2012.

17. *G. Aiello et al.* Cooling concepts for the CVD diamond Brewster-angle window // IRMMW-THz 2017.

18. *C. Wu et al.* Comparison between Controlled Non-Adiabatic and ExB Concepts // SMP, 2017, Nhizhny-Novgorod.

19. *I. Pagonakis et al.* A New concept for the collection of an electron beam configured by an externally applied axial magnetic field // IEEE Trans. Plasma Sci., 2008, vol. 36, no. 2, pp. 469-480.

20. *I.Gr. Pagonakis et al.* Multistage depressed collector conceptual design for thin magnetically confined electron beams // Physics of Plasmas, 2016, 23, 043114.

21. *C. Wu et al.* Conceptual designs of E x B multistage depressed collectors for gyrotrons // Physics of Plasmas, 2017, Vol. 24, No. 4, 043102.

22. *C. Wu et al.* Novel multistage depressed collector for high power fusion gyrotrons based on an E×B Drift concept // IVEC, 2017, London.

23. *F. Wilde et al.* Measurements of satellite modes in 140 GHz Wendelstein 7-X gyrotrons: an approach to an electronic stability control // IVEC, 2017, London.

24. *M. Schmid et al.* The 10 MW EPSM modulator and other key components for the KIT gyrotron test facility FULGOR // Fus. Eng. Des., 2017, https://doi.org/10.1016/j.fusengdes.2017.02.035.

25. *M. Losert et al.* RF Beam Measurements of Quasi-Optical Mode Converters in the mW Range // IEEE Trans. Plasma Sci., 2013, Vol. 41, No. 3, pp. 628-632.

26. *A. Schlaich et al.* Time-dependent spectrum analysis of high power gyrotrons // KIT Sci. Publishing, 2015, IHM.

EPJ Web of Conferences **149**, 04008 (2017) DOI: 10.1051/epjconf/201714904008
10th International Workshop 2017 "Strong Microwaves and Terahertz Waves: Sources and Applications"

The CARM beam-wave interaction and cavity design

E. Di Palma[1], G. Dattoli[1], S. Ceccuzzi[1], E. Sabia[1], G. L. Ravera[1], I. Spassovsky[1], F. Mirizzi[2], A. Doria[1], A.A. Tuccillo[1], G.P. Gallerano[1] and the ENEA CARM Task Force

[1] ENEA Fusion Department, C.R. Frascati, Via E. Fermi 45, 00044 Frascati, Roma, Italy
[2]Consorzio CREATE, Via Claudio 21, 80125, Napoli, Italy

Introduction

A major objective to be pursued in the design of future fusion reactors is the decrease of re-circulated power, which draws a special attention to the wall-plug efficiency of heating and current drive systems. The conceptual study for a self-sustainable DEMO-nstration reactor foresees for the current drive systems an efficiency larger than 30% [1]. This represents a critical requirement to develop the future electron cyclotron systems for next-generation fusion devices.

Furthermore, a microwave tube satisfying DEMO requirements, i.e. able to provide large continuous-wave power (~1 MW) at high frequency (~250 GHz), needs intensive R&D. Nowadays, the scientific community is devoting a special effort to extend above 200 GHz the operational frequency range of the gyrotron, the most mature technology in the field of mm-wave tubes.

In this frame a research and development program has been undertaken at ENEA Frascati Research Center, aimed at realizing a microwave tube based on a Cyclotron Auto-Resonance Maser (CARM) oscillator [2], characterized by a high value of the frequency Doppler up-shift allowing a consistent reduction of the static magnetic field in the interaction cavity. The price to be paid is the necessity of exploiting high quality electron beams, with velocity spread below 0.5%, in order to ensure appropriate mode selection, adequate wave-beam power transfer with the required efficiency. The design of the CARM cavity, composed by a short smooth cylindrical waveguide section delimited by two Bragg reflectors, is a challenging task, aimed at realizing the conditions to select the right operational mode and to limit the growth of the competing modes.

The radio frequency circuit design must be supported by a numerical model allowing to accurately simulate the intra-cavity beam wave interaction. The large length of this cavity (~80 cm) compared to the wavelength under investigation impairs the effectiveness of a classical particle in cell code (PIC), based on the discretization of the Maxwell equations, due to the huge amount of computer memory and CPU time required by the resulting mesh sizes.

In parallel to the development of the full numerical analysis, we have undertaken the cavity design by the use of "home-made" 1D/3D code, based on the self-consistent procedure discussed by several authors [3–5] and on the "universal" scaling formulae that describe the CARM performances including inhomogeneous broadening effects and the beam current, as it happens in the case of Undulator FEL [6]. An effective analysis tool in term of analytical formulae and computer simulation has been accordingly obtained.

The benchmarking of both the numerical code and semi-analytical formula with a test case (coming from an experimental CARM amplifier test facility developed at the Massachusetts Institute of Technology) at low frequency (~35 GHz) reported in ref. [5] has been discussed in ref. [7].

The agreement between the procedure, we have envisaged, and the experimental results, has provided some confidence on the reliability of the method. Therefore the application of this procedure to the preliminary design of the ENEA 250-GHz CARM resonant cavity is presented. The analysis allows the understanding of the conditions to be fulfilled by the cavity Q-factor, by the length of the smooth cylindrical waveguide section and by the beam parameters, in order to achieve a convenient compromise among the CARM required efficiency and a relatively low value of the starting current.

Results

The beam-wave interaction in a cold cavity for a CARM oscillator opens three kinds of problems as highlighted in Fig. 1:

a) the suppression of the beam-wave intersection for the operating mode at low frequency;

b) the suppression of the excitation of the competitor mode near cut-off;

c) the suppression of the modes competition near the Doppler up-shift interaction.

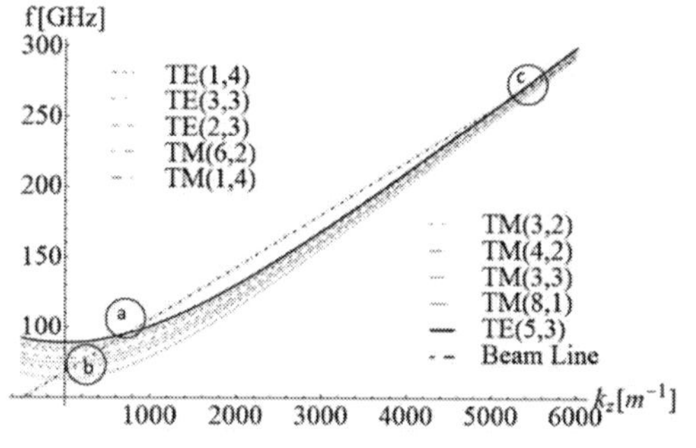

Fig. 1. Brillouin diagram of a particle beam (γ = 2.17, $v_\perp/v_\parallel$ = = 0.53) interaction with modes of a cylindrical cold cavity with radius 0.75 cm

The solution for the point a), widely discussed in ref. [7], is given by adjusting the design parameters in such a way that the starting current [4] of the near cutoff mode (intersection point a in Fig. 1) is greatly enhanced with respect to the CARM interaction. The underlying optimization is sketched in the block diagram in Fig. 2. The search for an optimum interaction length is based on a compromise between gain and efficiency for a given output CARM oscillator power.

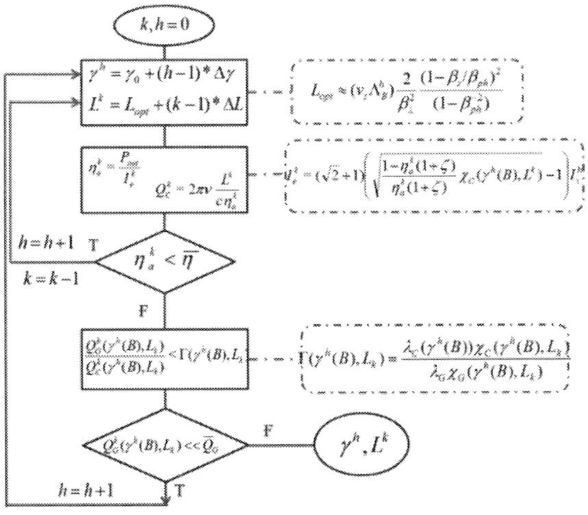

Fig. 2. Block diagram describing the procedure allowing to fix the beam energy and the smooth cylindrical cavity section length limiting the growing of the lower beam intersection with the operating mode

Major attention must be paid to the point b) for which the increase of the starting current is not sufficient for suppressing the modes growth excited by the beam near the cut-off where TM modes play a crucial role too.

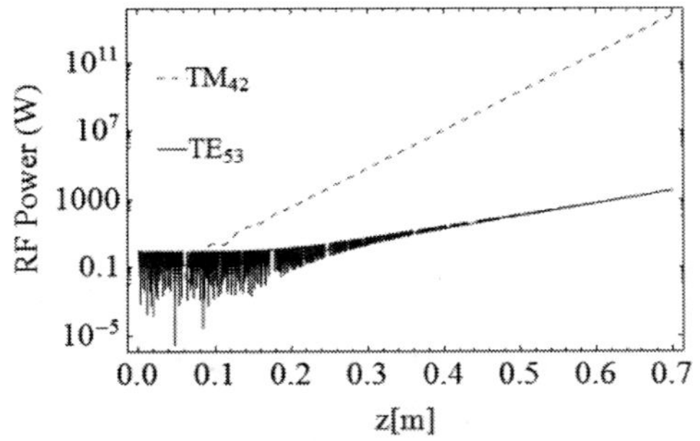

Fig. 3. The linear growth signal for an electron beam (with $\gamma = 2.17$, $v_\perp/v_\parallel = 0.53$, I = 8 A) interacting with the modes of a cylindrical cavity of radius R = 7.5 cm

In Fig. 3 it has been put in evidence how large is the linear growth signal for the TM_{42} at 70 GHz (near cut-off) with respect to the operating mode TE_{53} at 259 GHz. The starting current for TM modes is minimized at the cut-off, since the group velocity of the wave is well separated from the particle velocity, thus increasing the beam-wave energy exchange. Opposite condition occurs at CARM resonance implying an infinity value for the starting current. Due to the low value of the starting current and the high growth factor, the only way to suppress the TM mode is by using of a slotted cavity, suppressing the mode rotation, but causing a reduction of 30% of the system overall efficiency. The mode competition at CARM interaction (see point c in Fig. 1) can be kept under control by operating with an electron beam with a moderate velocity spread (lower than 0.5 %).

This statement is easily understood, by noting that the dependence of the CARM frequency on the longitudinal particle beam velocity v_z (for fixed γ beam energy) can be derived by coupling the dispersion relation of the beam-wave interaction with that of the propagating mode in a cylindrical cavity, which yields

$$\omega = \frac{-\beta_z^{-2}\Omega/\gamma - \sqrt{\omega_{cutoff}^2(1-\beta_z^{-2}) + \beta_z^{-2}\Omega^2/\gamma^2}}{1-\beta_z^{-2}} \quad (1)$$

being Ω the cyclotron frequency determined by the axial magnetic field, ω_{cutoff} the cut-off frequency of the propagating mode and $\beta_z = v_z/c$ the normalized longitudinal velocity of the particle beam. In Fig.4 we have plotted Eq. 1 vs. v_z for a beam with $\gamma = 2.17$ and $\alpha = v_\perp/v_\parallel = 0.53$ interacting with the modes supported by a cylindrical cavity of radius 7.5 mm. The plot shows that, for an electron beam within the quoted r.m.s velocity limits, the simultaneous locking of different modes does not occur. A thoroughly discussion of the ENEA CARM oscillator design criteria, along the previously sketched lines, will therefore follow.

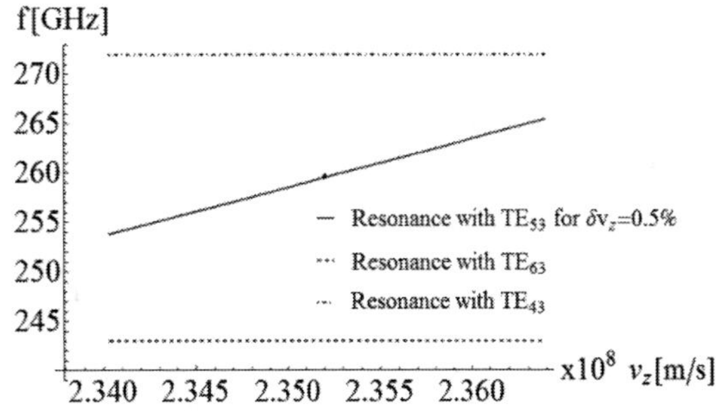

Fig. 4. The resonance frequency range for a beam line with an energy spread (0.5%) interacting with the TE_{53} mode, well separated from the resonance with the competitors modes TE_{63} (dotted line) and TE_{43} (dot-dashed line)

References

1. *Zhom,H.* On the minimum size of DEMO // Fusion Sci. Technol. 2010, V. 58 P. 613-624.

2. *Artioli M. et al.*, on behalf of CARM TEAM, A 250 GHz Radio Frequency CARM Source for Plasma Fusion, Conceptual Design Report // edited by ENEA, online at: http://www.enea.it/it/pubblicazioni/edizionienea/2016/conceptual-design-report (ISBN: 978-88-8286-339-5)

3. *Petelin, M.I.* On the theory of ultrarelativistic cyclotron self-resonance masers // Izvestiya VUZ. Radiofizika 1974, V. 17(6) P. 902-908.

4. *Bratman, V.L., Ginzburg, N.S., Nusinovich, G.S. Petelin, M.I. and Strelkov, P.S.* Relativistic gyrotrons and cyclotron autoresonance maser // Int. J. Electronics 1981, V. 51(4) P. 541-567.

5. *Chen, C. and Wurtele, J.S.* Linear and nonlinear theory of cyclotron autoresonance masers with multiple waveguide modes // Phys. Fluids B 1991, V. 3(8) P. 2133–2148.

6. *Dattoli, G., Ottaviani, P.L., Pagnutti, S.* Booklet for FEL design: a collection of pratical Formulae // edited by ENEA-Edizioni scientifiche, October 2007.

7. *Di Palma, E., Dattoli, G., Sabia, E., Sabchevski, S., Spassovsky, I.* Beam-wave interaction from FEL to CARM and associated scaling laws, to be submitted.

EPJ Web of Conferences **149**, 04009 (2017) DOI: 10.1051/epjconf/201714904009
10th International Workshop 2017 "Strong Microwaves and Terahertz Waves: Sources and Applications"

Comparison of reflector concepts for a 250 GHz CARM cavity

S. Ceccuzzi, G. Dattoli, E. Di Palma, A. Doria, G.P. Gallerano, G.L. Ravera, I. Spassovsky, A.A. Tuccillo, ENEA CARM Team*

ENEA Fusion Department, Frascati, Italy, silvio.ceccuzzi@enea.it

Introduction

The state-of-the-art on efficient high-power continuous-wave (CW) sources of coherent mm-wave radiation cannot meet the challenging future demands that mostly come from diagnostic as well as heating and current drive systems in thermonuclear fusion research. Intensive R&D is required to attain a cost-effective vacuum tube suitable to work with reactor-relevant tokamaks, even pushing forward with gyrotron technology, which is the most mature nowadays [1]. In parallel with gyrotron development, it can be thus useful to consider other source concepts like the Cyclotron Auto-Resonance Maser (CARM), theoretically promising for mm waves [2].

The study of a 250 GHz CARM for nuclear fusion was undertaken [3], aiming at a CW power of 1 MW to be pursued through intermediate steps of lower power and shorter pulse length. Some advantages of this source compared to gyrotrons are lower static magnetic field in the interaction cavity and lower ohmic losses per unit length. Such strengths are subordinate to the fulfillment of severe requirements, e.g., in terms of quality of the electron beam, stability of the high voltage modulator, and mechanical precision of the resonant cavity. Concerning the latter, the upstream mirror represents the most critical element from the realization viewpoint due to its around 800 ripples if based on the traditional Bragg reflector.

To identify the most convenient design, different types of mirror have been assessed. With respect to a recent comparison between traditional and advanced Bragg designs [4], reflectors based on tapered mean radius, trapped mode and discrete mirror are also considered here.

Comparison of reflectors

The reflector input is a circular waveguide with diameter larger than or equal to 15 mm to keep the power density for the operational TE_{53} mode below the safety threshold. Different mirror concepts have been designed maximizing the reflectivity at 250 GHz for the TE_{53} mode and constraining the transmitted power (i.e., the one flowing toward the gun) to be lower than 0.5%. For reflectors based on rippled-wall waveguides, step-type ripples have been considered. Moreover the depth of the first ripples at each side of the mirror has been tapered to reduce the level of side lobes. In the presence of multiple designs with similar performance for the same reflector type, shorter geometries have been preferred.

The types of mirror considered for this study are depicted in Fig. 1. The first one (Fig. 1a) is the traditional distributed reflector [5], based on the coupling between the forward and backward waves of the same mode. This coupling takes place by providing the waveguide wall with a small periodic perturbation that satisfies the Bragg condition, i.e., with period equal to $\lambda_g/2$, being λ_g the mode wavelength in the unperturbed waveguide. The second type of mirror (Fig. 1b) is the advanced Bragg reflector [6], i.e., a rippled-wall waveguide with the perturbation period equal to λ_g. Such condition leads to a three-wave coupling between the partial waves of the operational mode and a quasi-cutoff mode that is the $TE_{5,11}$ if the waveguide diameter is increased to 15.17 mm. The third option (Fig. 1c) is a traditional mirror having a section with tapered mean radius on the gun side. This configuration is mostly interesting because it allows a larger flexibility in tapering the radius of the annular electron beam that will pass through the resonator. The forth type of mirror (Fig. 1d) is the trapped-mode reflector [7], comprised of a taper and a backward mode converter. The TE_{53} moves to a rippled-wall section with a larger diameter of 17 mm, where it is counter coupled to the $TE_{5,12}$, being the period of the corrugations equal to $(1/\lambda_{TE5,3} + 1/\lambda_{TE5,12})^{-1}$. This high-order mode propagates back and, when passing through the down-taper, it is totally reflected because it becomes evanescent. A return trip takes place across the rippled-wall section with the $TE_{5,12}$ losing its energy in favor of the backward propagating TE_{53} mode. The last configuration (Fig. 1e) is a discrete mirror, made up of a step variation of the waveguide diameter after an adiabatic transition to a larger cross-section. If the waveguide cross-section is expanded sufficiently, its abrupt reduction can entirely intercept the non-zero transversal pattern of the operational mode and reflect it.

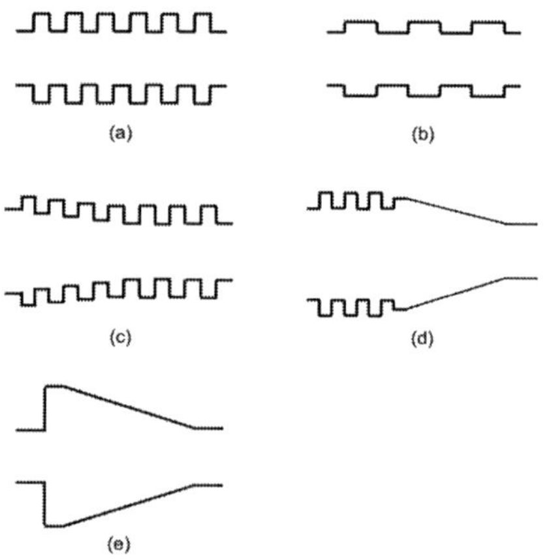

Fig. 1. Mirrors: (a) traditional Bragg, (b) advanced Bragg, (c) like (a) + rippled-wall section with tapered mean radius, (d) trapped mode reflector, (e) discrete reflector + taper

* See *Artioli M. et al.* Conceptual design report. A 250 GHz radio frequency CARM source for plasma fusion, edited by ENEA. 2016. ISBN: 978-88-8286-339-5.

An effective mode-matching code has been specifically developed to optimize the design of these oversized structures and provide a computation of losses according to perturbation method and attenuation constants. The reflectivity of the TE_{53} mode for each optimized mirror is shown in Fig. 2. These performances have been obtained with reflector lengths of around 0.51 m, 0.25 m, 0.7 m, 0.77 m and 0.6 m for the configurations of Fig. 1 from (a) to (e).

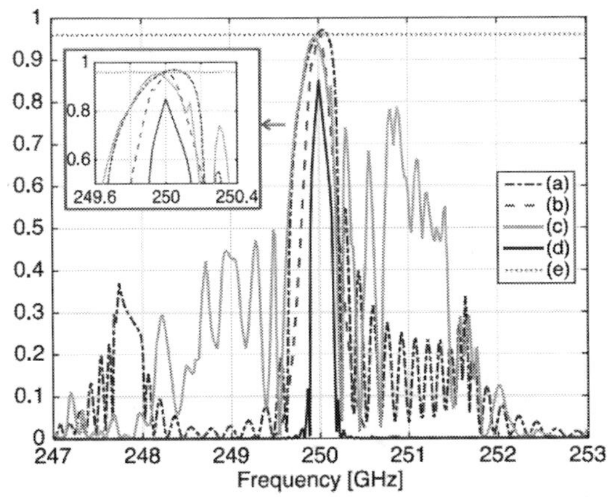

Fig. 2. Reflectivity of the optimized reflectors of Fig. 1

The discrete mirror has no variation in the plotted frequency range and its selectivity with respect to non-working modes is also extremely poor. For a taper length of 600 mm, acceptable reflectivity is obtained narrowing the waveguide down to a diameter not larger than 10 mm. This is a rather low value that conflicts with the transport of the electron beam and complicates the associated magnet system. The requirements of the latter can be relaxed using the Bragg reflector with up-tapered mean diameter. Apart from the very large side wiggles of this solution, the major drawback is the need of a longer device because the coupling between forward and reflected waves becomes less effective as the waveguide radius increases. Side wiggles can be eliminated choosing the trapped mode reflector, whose main advantage consists in a reflectivity lobe that is very steep. Nevertheless this solution is accompanied with high ohmic losses because it relies on a mode that becomes evanescent in the tapered section. Losses of the order of 14% have been computed for a rippled-wall section with a diameter of 17 mm; a larger cross-section would lead to a slightly lower dissipation and a longer device. The most attractive concept looks the advanced Bragg mirror that achieves a reflectivity of around 96.5% with the shortest length. Nevertheless its performance is very sensitive to a deviation of the geometrical dimensions from the nominal design values. In particular a variation of ±1 μm in the ripple period only produces a shift of a few hundreds of MHz in the reflectivity curves of other rippled-wall reflectors, whereas in the advanced mirror it noticeably impairs the frequency response of the structure. Finally the traditional Bragg mirror achieves a good compromise between selectivity performance, total length, ohmic dissipation and resilience to manufacturing errors.

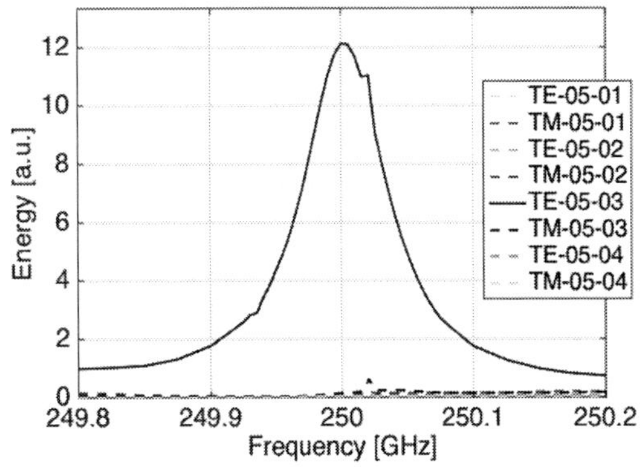

Fig. 3. Behaviour of a resonator with $Q \approx 3000$

A resonant cavity has been thus designed relying on the traditional Bragg concept. The downstream mirror has been designed to achieve a quality factor of around 3000, considering a smooth-wall section of 10 cm. The final resonator, whose predicted behaviour is shown in Fig. 3, has a total length of 0.63 m.

References

1. *Jelonnek, J., et al.* Design considerations for future DEMO gyrotrons: a review on related gyrotron activities within EUROfusion // 29th Symposium on Fusion Technology (SOFT), Praha, CZ, Sept. 5–9, 2016, P. 72.

2. *Bratman, V., Denisov, G., Ginzburg, N., Petelin, M.* FEL's with Bragg reflection resonators: Cyclotron autoresonance masers versus ubitrons // IEEE J. Quantum Elect. 1983. V. 19, No. 3. P. 282–296.

3. *Mirizzi, F., et al.* A high frequency, high power CARM proposal for the DEMO ECRH system // Fusion Eng. Des. 2015. V. 96-97. P. 393-402.

4. *Ceccuzzi, S., et al.* Traditional vs. advanced Bragg reflectors for oversized circular waveguide // Fusion Eng. Des. 2017. In press. doi:10.1016/j.fusengdes.2017.02.059.

5. *Chong, C. K., et al.* Bragg reflectors // IEEE Trans. Plasma Sci. 1992. V. 20, No. 3. P. 043504.

6. *Ginzburg, N. S., et al.* Tunable terahertz band planar bragg reflectors // Appl. Phys. Lett. 2009. V. 95, No. 4. P. 043504.

7. *Wagner, D., et al.* Sub-thz notch filters based on photonic bandgaps in overmoded waveguides // 40th Int. Conf. on Infrared, Millimeter, and Terahertz waves (IRMMW-THz), Hong Kong, CN, Aug. 23–28, 2015.

EPJ Web of Conferences **149**, 04010 (2017) DOI: 10.1051/epjconf/201714904010
10th International Workshop 2017 "Strong Microwaves and Terahertz Waves: Sources and Applications"

Plasma relativistic microwave amplifier

P.S. Strelkov[1], I.E. Ivanov[1] and V.P. Tarakanov[2,3]

[1] Prohorov General Physics Institute of Moscow, Russia, strelkov@fpl.gpi.ru
[2]Joint Institute for High Temperatures of Moscow, Russia
[3]National Research Nuclear University "MEPhI,"Moscow, Russia

Plasma relativistic microwave amplifier is investigated in General Physics Institute during ten years The new results received in 2016 are presented in this report [1,2]. The plasma relativistic microwave amplifier uses an electron beam with electron energy of 500 *keV*, beam current – 2 *kA*, and cathode voltage pulse duration of 500 *ns*. At the input of the amplifier is fed a microwave pulse with a power of about 50 *kW* from one of the three magnetrons with frequencies of 2.4 *GHz*, 2.7 *GHz* and 3.1 *GHz*. The energy of the microwave pulse at the output from the horn was measured by calorimeter with a diameter of 50 cm. The receiving antenna in the form of a conductor piece is installed on the axis of the calorimeter. The microwave signal from antenna by a coaxial cable, attenuated by 63 dB, was fed to the oscilloscope with a bandwidth of up to 4 GHz. Thus, the energy of the microwave pulse and the dependence of the electric field strength of the microwave radiation on time on the microwave beam axis in each pulse were measured. The rotation of the antenna made it possible to determine the degree of the microwave radiation polarization.

If the electric field structure at different points of the microwave beam cross section is constant in a series of microwave pulses, then the quantity $\int U^2\,dt$ and the total energy of the microwave pulse W must be proportional to each other. Here U is the amplitude of the signal from the antenna. Figure 1 shows the dependence of the quantity $\int U^2\,dt$ and the energy of the microwave pulse (W) on the plasma density

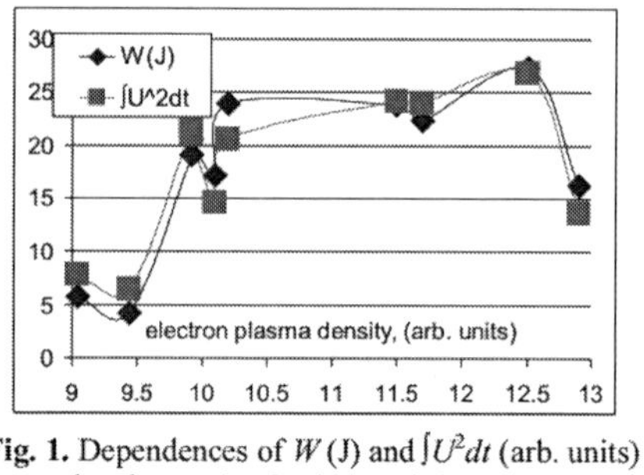

Fig. 1. Dependences of W (J) and $\int U^2 dt$ (arb. units) on the plasma density (arb. units) at 2.7 GHz

This graph demonstrates the proportionality of the quantities $\int U^2\,dt$ and W when the plasma density varies from 9 to 13 (arb. units). Moreover, the constancy of the structure of the microwave beam was confirmed by measurements [1].

The thin dielectric film with metal microparticles pasted onto it was installed at different distances from the horn. The microwave discharge occurred on this film, Fig. 2. A grid with a 5 cm cell size is superimposed on the film. It can be seen that the size of the glowing spot increases with increasing distance between the horn and the film. The electric field structure of the microwave beam cross section was calculated using the "KARAT" code. On the photos of the glow of the screen, on the left, the glowing threads are visible.

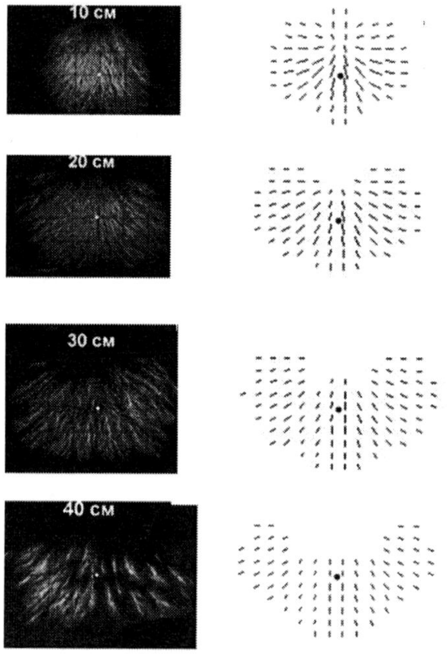

Fig. 2. Photos of the film glow, at different distances from the horn – on the left, the result of computer modeling – on the right. The dots represent the axis of the microwave beam. The length of the segments on the computer image is proportional to the electric field strength at a given point

One can see a good coincidence of the electric field structure in the experiment and in the calculation, and also the coincidence of the divergence of the beams. To measure the total energy of the microwave pulse, it is necessary to arrange the calorimeter close to the emitting horn aperture, as that the horn aperture diameter is smaller than the calorimeter diameter. At a distance of less than 70 cm, breakdown was observed on antenna. Therefore, measurements were made at a distance of 70 cm. At a distance of 70 cm the diameter of the microwave beam exceeded the diameter of the calorimeter. The divergence coefficient of the microwave beam was determined from the results of a numerical experiment. The data on the divergence of the microwave beam and Fig. 1 make it possible to calculate the total microwave radiation power according to the formula

$$P_f(t) = k\,W\,U^2_f(t)/2\int U^2(t)dt \qquad (1)$$

Here $P_f(t)$ and $U^2_f(t)$ the radiation power and the square of the output signal amplitude in the frequency band $f_0 \pm 15$ MHz, where f_0 is the frequency of the input signal.

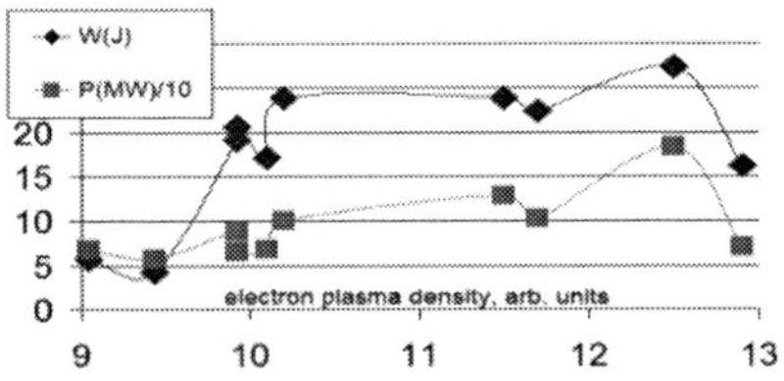

Fig. 3. Dependence of the microwave pulse energy and power on the plasma density, frequency 2.7 GHz

Figure 3 shows the dependence of the power and energy of the microwave pulse on the plasma density. The radiation power exceeds 100 MW.

The following picture were observed. The microwave pulse duration at small plasma density values is small and the maximum power is observed in the microwave pulse beginning. With an increase in the plasma density, the duration of the microwave pulse rises and exceeds the duration of the plateau of the cathode voltage pulse. The maximum microwave power is observed at the end of the microwave pulse.

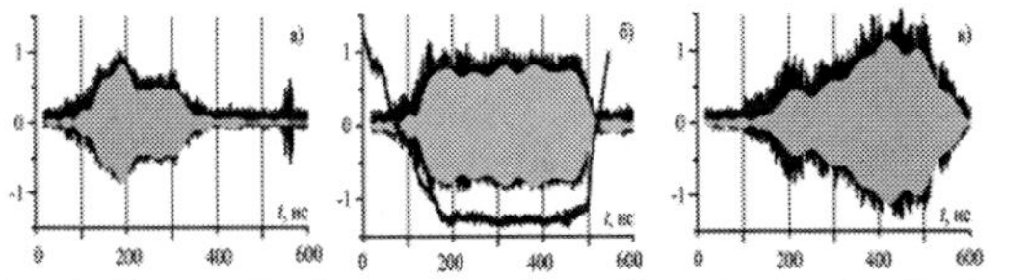

Fig. 4. Shape of microwave pulses. The plasma density increases from left to right. The accelerator voltage pulse on the cathode is shown on the middle figure. Gray color - the signal is passed through the filter 2710 ± 15 MHz from left to right. The frequency is equal to 2.7 GHz

This picture is observed for all three frequencies of the input signal.

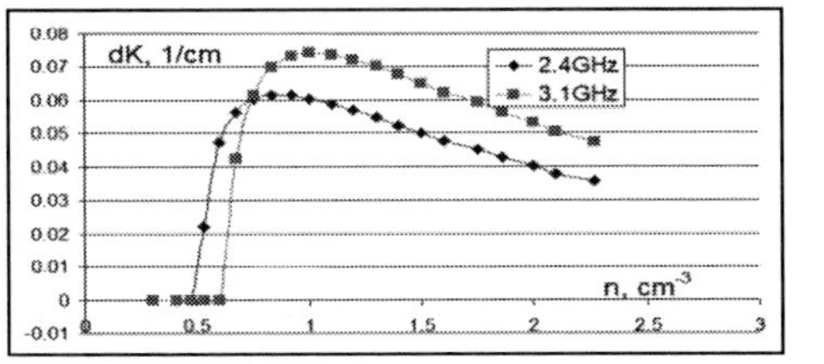

Fig. 5. Dependence of the linear amplification coefficient of the plasma wave electric field dK on the plasma density n at 2.4 and 3.1 GHz

The amplification coefficient decreases sharply when the plasma density decreases from the optimum value and decreases more smoothly when the plasma density exceeds the optimal value. In this way, one can expect different dependence of the microwave pulse amplitude on time for different values of the initial plasma density. If we assume that the initial value of the plasma density is less than the optimum value or slightly exceeds it, then a decrease in the plasma density during the current pulse of the REB should lead to a rapid decreasing of microwave power. For this reason the short microwave pulse is observed. If the initial plasma density value significantly exceeds the optimum value, then its decrease during the current pulse of the REB should lead to an observed increase in microwave pulse duration and increase in amplitude at the end of the microwave pulse. In addition, it was found that the output signal frequency is less than the frequency of the input signal by 10 MHz at a low plasma density and by 3 MHz at high plasma density. This phenomenon can also be explained by the assumption of a decrease in the plasma density during the microwave pulse. As the plasma density decreases, the wave vector of the plasma wave increases, which leads to an additional phase shift of the plasma wave at the plasma length. As a result, the output radiation frequency decreases in comparison with the input signal frequency.

The plasma density change during the mi-crowave pulse was recorded in the numerical simulation of the process of a plasma wave amplification by means of the "KARAT" code. The relativistic electron beam and the plasma were modeled by the particle in cell method.

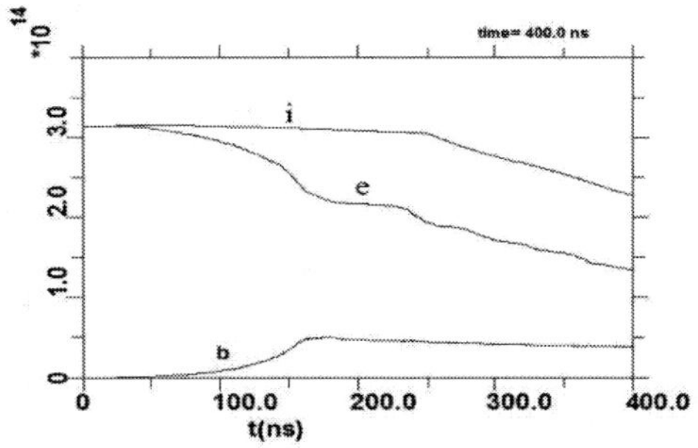

Fig. 6. Dependence of the total particles number on time: i – ions, e – plasma electrons, b – beam electrons

It can be seen that the total number of plasma electrons decreases twofold in 400 ns. Thus, the duration of the microwave pulse is limited by the decay of the plasma, which arises when the electron beam is injected into the plasma.

The microwave pulses parameters are presented in Table 1.

Table 1. Microwave pulses parameters. The percents (%) are power fraction in the frequency band $f_0 \pm 15$ MHz

(GHz)	W (J)	P (MW)	%
2.4	27	110	91
2.7	22	130	85
3.1	20	100	94

References

1 *I.S. Alekseev, I.E. Ivanov, P.S. Strelkov, V.P. Tarakanov, and D.K. Ulyanov.* Visualization of the Microwave Beam Generated by a Plasma Relativistic Microwave Amplifier// Plasma Physics Reports, 2017, Vol. 43, No. 3, pp. 340–345

2. *I.N. Kartashov, M.V. Kuzelev, P.S. Strelkov, V.P. Tarakanov.* Influence of plasma instability on the spectrum and shape of microwave pulses of a plasma relativistic microwave amplifier // article sent to the journal.

EPJ Web of Conferences **149**, 04011 (2017) DOI: 10.1051/epjconf/201714904011
10th International Workshop 2017 "Strong Microwaves and Terahertz Waves: Sources and Applications"

Development of powerful Ka-band FEM-amplifiers with broad frequency tuning

A.K. Kaminsky[1], I.V. Bandurkin[2], D.E. Donets[1], S.V. Kuzikov[2], N.Yu. Peskov[2], A.V. Savilov[2], S.N. Sedykh[1], A.A. Vikharev[2]

[1]Joint Institute for Nuclear Research, Dubna, Russia, alim_kaminsky@mail.ru
[2]Institute of Applied Physics RAS, Nizhny Novgorod, Russia

Introduction

During the several preceding years, an oscillator scheme of a free electron maser (FEM) has been realized in the joint experiments between JINR (Dubna) and IAP RAS (Nizhny Novgorod) [1]. The project was aimed at accelerating applications, and the FEM operates in a frequency range of 30 GHz and generates 20 MW / 200 ns pulses with a spectrum width of 6–7 MHz, which is close to the theoretical limit. The high output power and stability of a narrow-band operation regime allowed this JINR-IAP FEM to be used in a number of actual demand applications, including testing the components of high-gradient accelerators of the next generation [2].

However, many potential applications of powerful microwave sources require an RF-radiation phase to be controlled. Moreover, some applications need a broad instantaneous amplification band or a fast frequency tuning. In particular, for accelerating applications, precise matching of the radiation frequency with the frequency of high-gradient structure should be provided. For this purpose, the FEM-amplifier project is developed based on LIU-3000.

FEM in a grazing incident regime

Two regimes of Ka-band wave amplification have been simulated. In the first one, the electrons move along the uniform helical wiggler with the axial velocity equal to the group velocity of the operating wave (grazing regime) (Fig. 1). Provided the FEM resonance condition $\omega = (2\pi/d_w + h)v_{\parallel}$ is fulfilled, an efficient electron-wave coupling in the wide band of the operating frequencies takes place. In simulations, the possibility to provide a wide instant amplification band and, thus, fast control over the operation frequency, as well as a high efficiency and a high output power, were demonstrated. For the driving power of 4–5 kW, the simulations predicted an amplification of up to 35 dB in the range of 2–32 GHz saturating at approximately 1 m of the interaction space.

Experimental study of this scheme of the FEM-amplifier was performed at the LIU-3000 accelerator. A magnetron tunable in the range of 29.5–30.5 GHz was used to supply an input signal of about 4 kW power at the entrance of the interaction space. Circular-polarization of the driving wave-beam was provided by the section of elliptical waveguide.

In the first series of the experiments, a "regular" helical wiggler of 6-cm period and transverse field of 0.1–1.1 T was used to pump the operating transverse velocity into the electron beam focused by an axial magnetic field of around 0.15 T in a reversed guide field regime. To achieve a high efficiency of the FEM operation, an improved construction of pulsed helical wiggler was developed. The regular part of the wiggler represents a double bifilar spiral with winding currents of the same direction spaced apart to an angle of 60° (Fig. 2). Such optimization of the winding geometry ensures a high transverse homogeneity of the field inside the wiggler compared with the single bifilar wiggler. The wiggler entrance was slowly up-tapered over the initial six periods in order to provide the excitation of operating transverse oscillations in the beam, and down-tapered over the last three periods for a better beam transportation at the wiggler output.

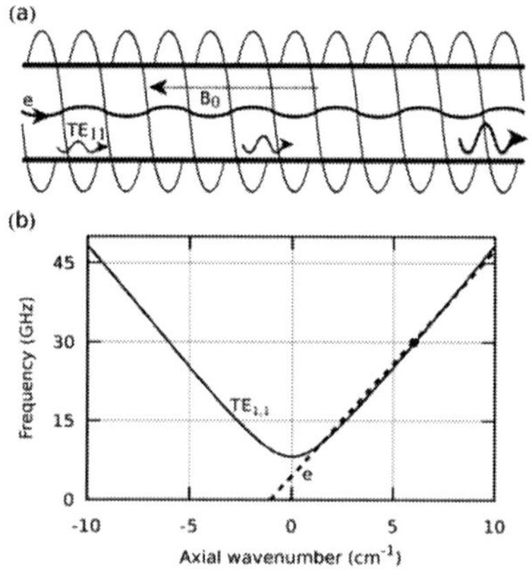

Fig. 1. (a) Schematic of the FEM amplifier with a helical wiggler and reverse guide field and (b) dispersion diagram for the 30-GHz JINR-IAP FEM-amplifier operated in the grazing regime. Solid and dashed lines show the dispersion of the $TE_{1,1}$ wave and the electron beam, respectively

During the experiments, the stable regime of input signal amplification was observed in the designed range of parameters (Fig. 3). The FEM was tuned to the maximum output power by changing the wiggler and solenoid fields, as well as by small variation of the particle energy. In optimal regime ($B_0 \approx -1.5$ kG, $B_w \approx 1$–1.2 kG), according to the calorimeter measurements, the radiation power achieved 17–20 MW at the operating frequency band around 30 GHz, in good agreement with the simulations. It should be noted that in the experiments, an about 10–15% current loss was observed in the region where the electron beam formed by the accelerator was injected into the wiggler. As a result, the beam current in the interaction space was of the level taken in the simulations. The output pulse length was near 200 ns, which is close to the total duration of the beam pulse.

The spectrum width of the amplified signal amounted to 6–7 MHz, which is close to the natural spectrum width for the observed pulse duration. For the fixed beam and electron-optical system parameters, this level of output power was observed within the whole band of the magnetron frequency tuning of ±0.5 GHz, which models the instant amplification band [5]. At the same time, an increase in the wiggler field up to 1.3 kG resulted in self-excitation of the system at parasitic frequencies.

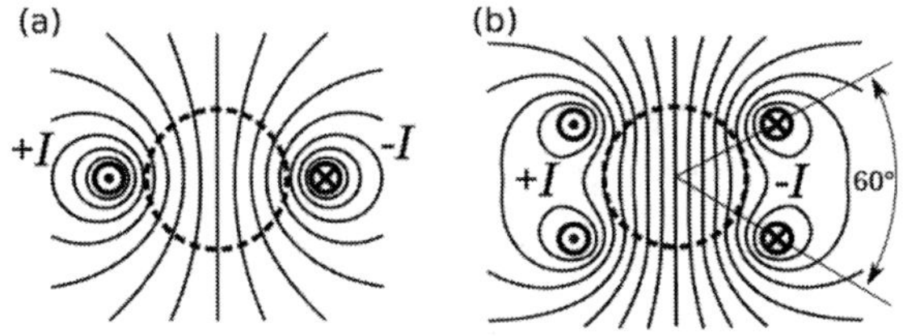

Fig. 2. Transverse magnetic field distribution inside a bifilar (*a*) and double bifilar (*b*) undulators

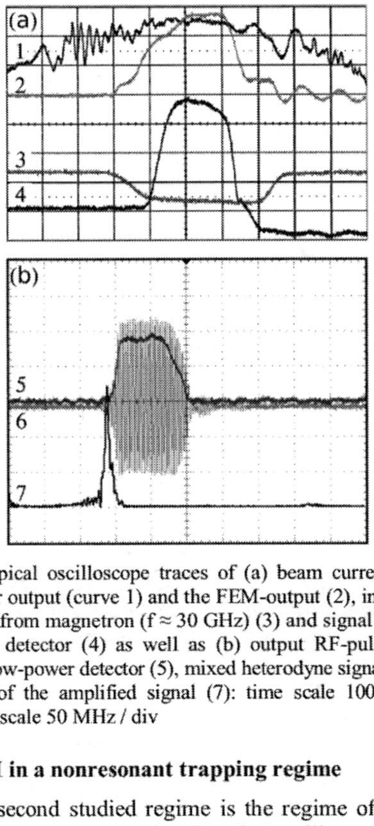

Fig. 3. Typical oscilloscope traces of (a) beam current at the accelerator output (curve 1) and the FEM-output (2), input driving signal from magnetron (f ≈ 30 GHz) (3) and signal from the RF-power detector (4) as well as (b) output RF-pulse signal from the low-power detector (5), mixed heterodyne signal (6) and spectrum of the amplified signal (7): time scale 100 ns / div., frequency scale 50 MHz / div

FEM in a nonresonant trapping regime

The second studied regime is the regime of the so-called non-resonant trapping [3, 4]. It utilizes a tapered wiggler (Fig. 4), in which the electrons are far from the resonance, $\omega > (2\pi/d_{w0} + h)v_{\|}$, at the very beginning (line 1 in Fig. 4b). While the wiggler period d_w decreases along the interaction space (line 2 in Fig. 4b), the particles get trapped by the operating wave in a proximity of an inner resonance point and after that are efficiently decelerated. Though this regime requires twice as long interaction (compared to the regime of grazing), in simulations it provides a substantially higher amplification of up to 40 MW and extremely wide amplification band up to 50%, i.e. all over Ka-band.

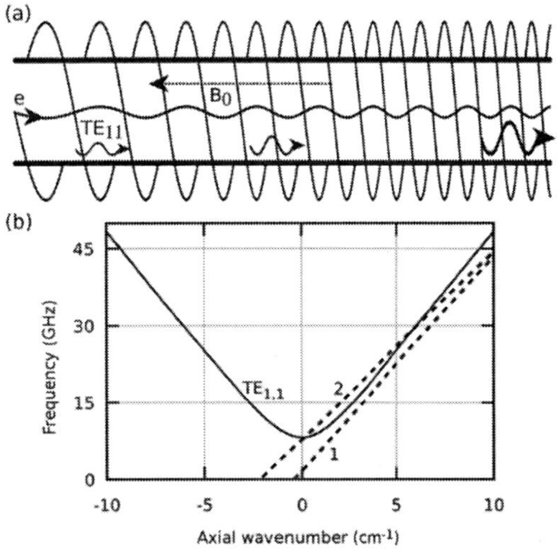

Fig. 4. (a) Schematic of the FEM amplifier with a profiled helical wiggler and (b) dispersion diagram for the 30-GHz JINR-IAP FEM-amplifier operated in the regime of nonresonant trapping. Solid and dashed lines show the dispersion of the $TE_{1,1}$ wave and the electron beam, respectively

To provide the operation of the FEM-amplifier in the non-resonant trapping regime, a helical wiggler having the period profiled from 6.2 cm to 3.6 cm and the length of up to 2 m was designed. The undulator has been manufactured, and the magnetic field distribution measurements demonstrated a good coincidence with the results of electron-optical 3D simulations. Experimental studies of FEM-amplifier of the novel type are currently in progress.

Acknowledgments

This work is partially supported by the Russian Government project # 0035-2014-0012.

References

1. *A.K. Kaminsky, E.A. Perelstein, S.N. Sedykh, e.a.* Demonstration of powerful 30-GHz FEM operation with resonant load // Tech. Phys. Lett. 2010. V. 36 No. 5 P. 37.
2. *N.S. Ginzburg, I.I. Golubev, A.K. Kaminsky, e.a.* Experiment on pulse heating and surface degradation of a copper cavity powered by powerful 30 GHz free electron maser // Phys. Rev. Accel. Beams 2011. V. 14. No. 4. P. 041002.
3. *A.V. Savilov, I.V. Bandurkin, N.Yu. Peskov.* Regime of non-resonant trapping in an FEM-amplifier // Nucl. Instr. and Meth. in Phys. Res. A, 2003, 507, 158-161.
4. *I.V. Bandurkin, A.K. Kaminsky, E.A. Perelstein, e.a.* Modeling of a High-Power Wideband Free-Electron Maser Amplifier with an Operating Frequency of 30 GHz to be Used in Particle Acceleration Experiments // Radiophys. and Quant. Electr. 2016. V. 58. No. 8. P. 607.
5. *N.Yu. Peskov, I.V. Bandurkin, D.E. Donetc, e.a.* Modeling of a High-Power Wideband Free-Electron Maser Amplifier with an Operating Frequency of 30 GHz to be Used in Particle Acceleration Experiments // Appl. Phys. Lett. 2017. V. 110. P. 013501.

Powerful narrow-band relativistic masers with Bragg resonators operating from mm to sub-mm wavelength band: resent results and prospects

N.Yu. Peskov[1], N.S. Ginzburg[1], A.K. Kaminsky[2], S.N. Sedykh[2] and V.Yu. Zaslavsky[1]

[1]Institute of Applied Physics RAS, Nizhny Novgorod, Russia, peskov@appl.sci-nnov.ru
[2]Joint Institute for Nuclear Research, Dubna, Russia

Introduction

By now, powerful narrow-band free-electron maser (FEM) operating at mm wavelength band has been realized in collaboration between JINR (Dubna) and IAP RAS (N.Novgorod) [1]. The radiation parameters achieved allows JINR-IAP FEM to be used in several applications including testing of components for high-gradient accelerators, biology-medical studies, physics of nanoparticles, etc [2]. The aim of present work is development of narrow-band relativistic masers operating up to sub-mm wavelength bands at MW/multi-MW power level for potential applications.

As an FEM microwave system capable to provide stable narrow-band operation in strongly oversize interaction space, we proposed high-selective Bragg resonators of the novel type based on the so-called advanced Bragg structures with the feedback loop involving propagating and quasi-cutoff waves [3, 4]. To decrease Ohmic losses associated with excitation of cut-off mode it is attractive to use two-mirror resonator scheme with the up-stream advanced Bragg reflector and conventional Bragg reflector with rather small reflectivity as a down-stream reflector (Fig. 1).

FEM-oscillators operating from Ka- to W-band

The JINR-IAP FEM is driven by induction linac LIU-3000 0.8 MeV / 200 A / 250 ns /1 pulse/s (JINR). A reversed guide magnetic field regime is used for the FEM operation. This regime possesses low sensitivity to the initial beam spread and, as a result, is optimal for high-efficiency FEM operation. For operation from Ka- to W-bands, helical wigglers having periods from 6 cm to 3 cm correspondingly were constructed. Enhance in amplitude of the transverse magnetic field in short-period wigglers alongside with refining its transverse homogeneity was achieved by optimization of the currents distribution in the wiggler winding. To improve the quality of helical electron beam formation at short wavelengths, the slowly up-tapered wiggler entrance was optimized as well [5].

For operation at 30 GHz, advanced Bragg resonator was constructed with the oversized parameter $\O/\lambda \sim 2$ and included advanced Bragg reflector with a feedback loop formed by two counter-propagating $TE_{1,1}$ waves and cutoff $TE_{1,2}$ wave (Fig. 2). Resonator for V-band (60 GHz) was composed with advanced up-stream Bragg reflector operating with feedback loop $TE_{1,1} \leftrightarrow TM_{1,2}$-cutoff $\leftrightarrow TE_{1,1}$ at the oversize parameter $\O/\lambda \sim 3$ (Fig. 4). For W-band (80 GHz) advanced Bragg reflectors based on excitation of the feedback cutoff wave of $TE_{1,5}$ – type were designed to provide an effective FEM operation at oversize parameter $\O/\lambda \sim 5$ (Fig. 5). In accordance with the 3D simulations, effective narrow-band reflections were demonstrated in "cold" test of Bragg structures of the novel type. In all resonators (manufactured for operation from Ka- to W-band) the reflection band in advanced Bragg structures was measured 0.5–0.7 GHz with maximum power reflection up to 80–90%, while the conventional structures had ~60% reflection in much broader band of 2.5–3 GHz.

In the proof-of-principle experiments at LIU-3000 a narrow-band operation of novel scheme of FEM-oscillators was obtained under design parameters. At Ka-band, stable single-mode operation at the frequency of 30.2 GHz was observed in accordance with simulations. The output power amounted up to 20 MW (efficiency about 15–20%) with the spectrum width of 6–7 MHz (measured by heterodyne technique) close to the theoretical limit (Fig. 3).

In the experiments at V- and W-band, the radiation spectrum was measured by means of cut-off filters set with the accuracy of about 1 GHz. Both FEM demonstrated the oscillation frequency belonging to the designed feedback loop of the hybrid resonator in the vicinity of 59 GHz and 80 GHz correspondingly.

In the resonators of optimal geometry the output power (measured by calorimeter) in both frequency regions amounted to 5–7 MW when the beam current was about 70 A (Fig. 6), which corresponded to the electron efficiency of up to 10–12%.

Project of sub-mm CARM-oscillator

Further increase of the radiation frequency in FEM is restricted by the wiggler period. As a potential candidate to advance into sub-mm wavelengths an another type relativistic maser, namely, cyclotron autoresonance maser (CARM) is attractive. Project of 250 GHz MW-power CARM operating in quasi-cw regime is under development at ENEA Frascati [6] aimed on compact plasma fusion reactors with high magnetic field. Modelling short-pulse CARM experiment is under development at LIU-3000 facility.

For this project we consider cylindrical cavity with diameter $\O/\lambda \sim 7$–10 that seems sufficient for the intense beam transportation and heating demand during RF-pulse for designed powers. In specified waveguide a $TE_{1,2}$ wave was chosen to provide needed electron bunching parameter $\mu = 1 - \beta_{ph}^{-2} \approx 0.1$ and effective interaction with the beam having oscillating velocity $\beta_\perp \sim 0.4 - 0.5$ at the axial magnetic field of about 5–6 T. According to 3D simulations using code CST "Microwave Studio", advanced Bragg structure of given transverse size and about 3–5 cm long is able to ensure needed selectivity and high reflection at designed frequency band (Fig. 7).

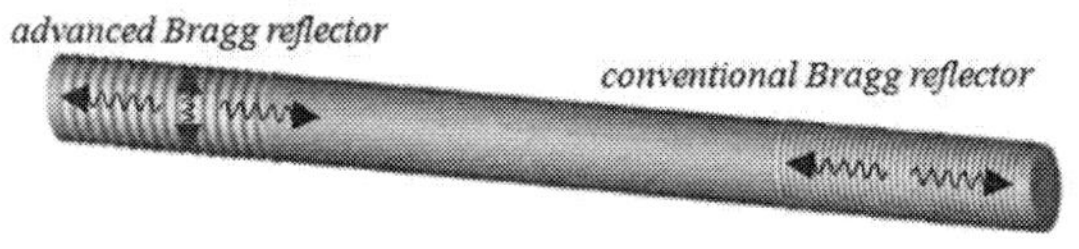

Fig. 1. Scheme of hybrid two-mirror resonator based on advanced and conventional Bragg reflectors. The partial wave-fluxes in both reflectors are shown by the arrows

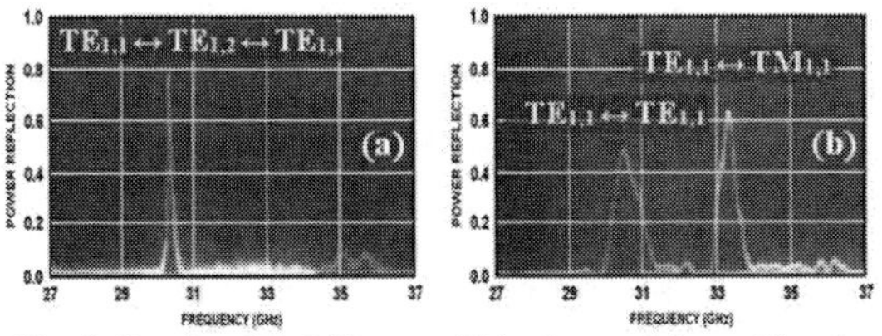

Fig. 2. Results of "cold" tests of (*a*) advanced (l_{adv} = 15 cm) and (*b*) conventional (l_{con} = 8 cm) Ka-band Bragg reflectors

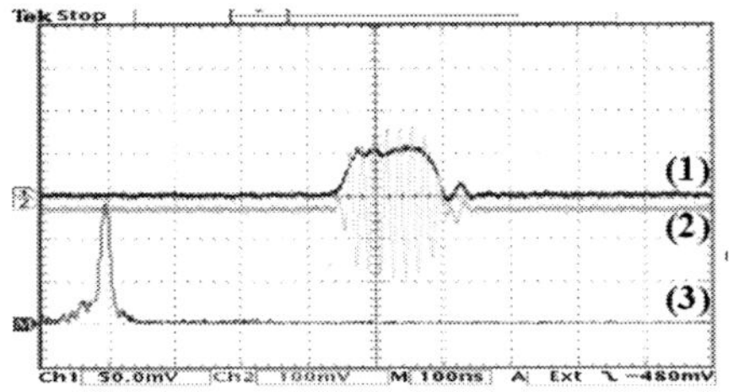

Fig. 3. Results of JINR-IAP FEM experiments in Ka-band. Typical oscilloscope traces of (1) RF-pulse (100 ns / div.), (2) heterodyne beating signal and (3) frequency spectrum (50 MHz / div.)

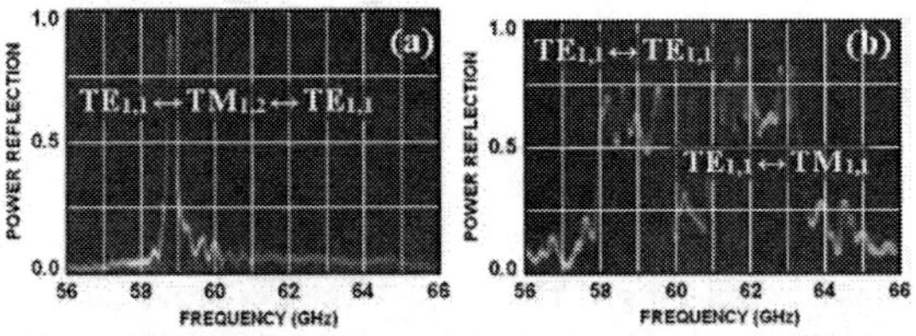

Fig. 4. Results of "cold" tests of (*a*) advanced (l_{adv} = 12 cm) and (*b*) conventional (l_{con} = 8 cm) V-band Bragg reflectors

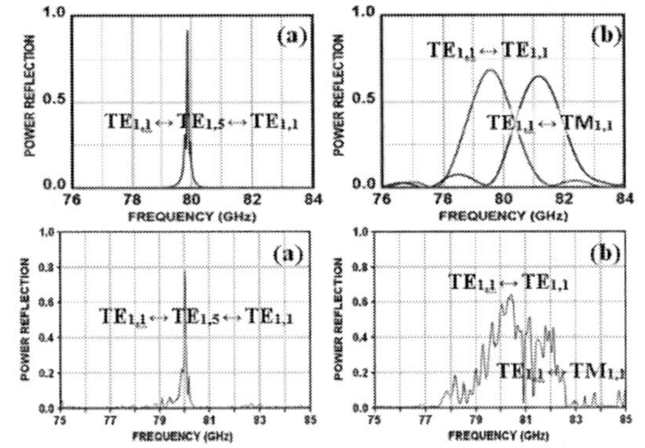

Fig. 5. Results of 3D simulations (top) and "cold" tests of Bragg structures of different types at W-band. Frequency dependence of reflection of (*a*) advanced (l_{adv} = 12 cm) and (*b*) conventional (l_{con} = 8 cm) reflectors

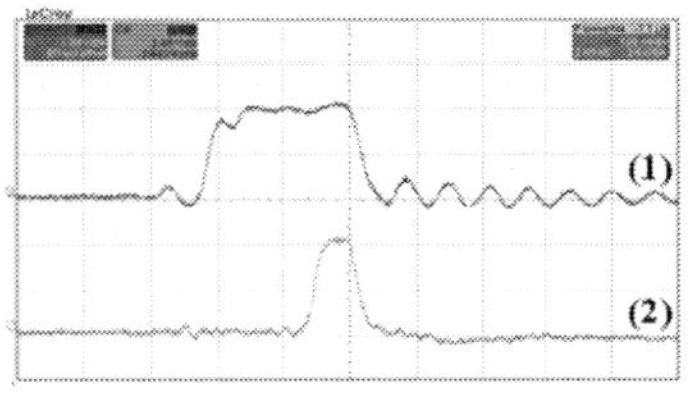

Fig. 6. Results of the JINR-IAP FEM experiments in W-band. Typical oscilloscope traces of (1) beam current and (2) output RF-pulse (100 ns / div.)

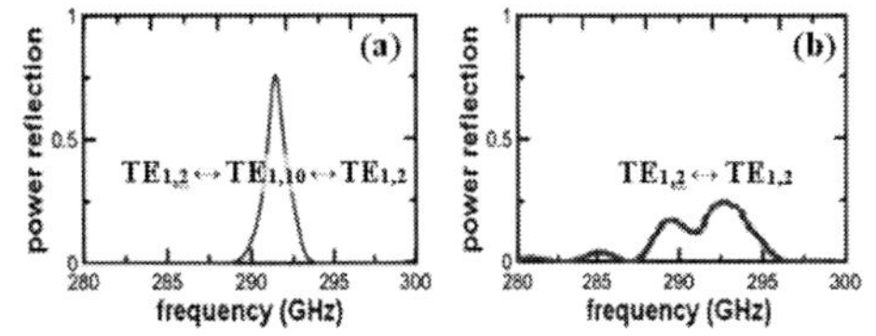

Fig. 7. Results of 3D simulations of (*a*) advanced (Ø/λ ~ 10, l_{adv} = 5 cm) and (*b*) conventional (Ø/λ ~ 7, (l_{con} = 3 cm) Bragg reflectors in 300 GHz frequency range

Theoretical analysis carried out proved operability of a narrow-band CARM operation for parameters given above. To suppress self-excitation at "parasitic" low-frequency near-cutoff modes periodical slits in the cavity were considered [7]. For designed parameters simulations demonstrated establishment of a narrow-band oscillation regime with MW-power level and efficiency of 25–30% in the cavity of 6–8 cm long and initial velocity spread in the beam did not exceeding 1–2% [8].

This work is partially supported by the Russian Science Foundation (grant #17-19-01605).

References

1. *A.K. Kaminsky, E.A. Perelstein, S.N. Sedykh, e.a.* Demonstration of powerful 30-GHz FEM operation with resonant load // Tech. Phys. Lett. 2010. V.36 No.5 P.37.

2. *N.S. Ginzburg, I.I. Golubev, A.K. Kaminsky, e.a.* Experiment on pulse heating and surface degradation of a copper cavity powered by powerful 30 GHz free electron maser // Phys. Rev. Accel. Beams 2011. V.14. No.4. P.041002.

3. *N.S. Ginzburg, A.M. Malkin, N.Yu. Peskov, e.a.* Improving selectivity of free electron maser with 1D Bragg resonator using coupling of propagating and trapped waves // Phys. Rev. Accel. Beams 2005. V.8. P.040705.

4. *N.S. Ginzburg, V.Y. Zaslavsky, I.V. Zotova, e.a.* Terahertz Free-Electron Lasers with Bragg structures based on coupling between travelling and quasi-cutoff wases // JETP Lett. 2010. V.91. No.6. P.266.

5. *N.Yu. Peskov, N.S. Ginzburg, A.K. Kaminsky, e.a.* Powerful FEM-oscillators with advanced Bragg resonators operating in a single mode regime from Ka- to W-band // Proc. of the 41th Int. Conf. on IR, MM and THz Waves, Copenhagen, Denmark, 2016, p.F2E.01.

6. *F. Mirizzi, I. Spassovsky, S. Ceccuzzi, e.a.* A high frequency, high power CARM proposal for the DEMO ECRH system // Fusion Eng. and Design 2015. V.96-97. P.538.

7. *N.S. Ginzburg, N.Yu. Peskov, A.S. Sergeev, V.Y. Zaslavsky.* Simulations of powerful relativistic masers of submillimeter wavelength band for plasma application // Proc. of the 9th Int. Workshop "Strong Microwaves and Terahertz Waves: Sources and Applications, N.Novgorod - Perm, 2014. P.239.

8. *N.S. Ginzburg, A.M. Malkin, N.Yu. Peskov, e.a.* The use of advanced Bragg structures in powerful submillimeter cyclotron autoresonance masers // Radiophys. and Quant. Electron. 2016. V.59. No.12. P.1137.

Surface-wave Bragg resonators for terahertz frequency range

A.M. Malkin, N.S. Ginzburg, S.E. Fil'chenkov, A.S. Sergeev, and V.Yu. Zaslavsky

Institute of Applied Physics RAS, Nizhny Novgorod, Russia, malkin@appl.sci-nnov.ru

Surface electromagnetic waves confined at the interface of metal and dielectric have been widely studied [1] since they possess a number of unique features making them useful for various applications, including antenna applications Raman scattering, etc. Since the evanescent waves are decelerated, Cherenkov devices based on interaction of electron beams with surface modes of the corrugated structures is of special interest in the short wavelength radiation bands, including the terahertz band [2]. These devices allow to solve the problem of both longitudinal and transverse mode selection in oversized electrodynamical systems.

In this paper, we consider a periodically corrugated section of a metallic plate as a surface-wave resonator (see Fig. 1). For the limiting case of shallow corrugation, we obtain explicitly the equations of wave coupling with Ohmic losses taken into account and derive an integral equation describing the eigenmodes. Based on this equation, we find the optimum values of the corrugation depth providing the maximum Q factor of the fundamental mode.

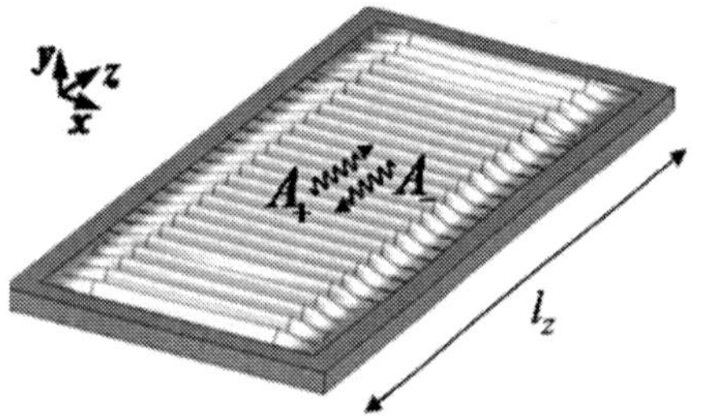

Fig. 1. Surface-wave resonator

In the case when the amplitude of sinusoidal corrugation $b(z) = (b_0/2)\cos\bar{h}z$ is small in the scale of wavelength and of the structure period d the electromagnetic field of the surface mode existing near the 1D corrugated metal plate can be presented as a sum of two quasi-optical wavebeams with slowly varying amplitudes $A_\pm$ which can be characterized by the magnetic field

$$H_x = \mathrm{Re}\left[A_+(z,y,t)e^{i(\omega_0 t - h_0 z)} + A_-(z,y,t)e^{i(\omega t + h_0 z)}\right],$$

where $h_0 = \bar{h}/2 = \pi/d$, $\omega_0 = ch_0$ is the Bragg frequency. Propagation of electromagnetic field in the vicinity of the Bragg resonance can be described by a system of coupled parabolic wave-beam equations which take into account the diffraction in y direction [2],

$$\pm\frac{\partial A_\pm}{\partial z} + \frac{\partial A_\pm}{c\partial t} + \sigma\delta(y)A_\pm + \frac{i}{\bar{h}}\frac{\partial^2 A_\pm}{\partial y^2} = i\alpha\delta(y)A_\mp \quad (1)$$

At the edges of the corrugation we set non-reflection boundary conditions: $A_+(z=0)=0,\ A_-(z=l_z)=0$. Here $\alpha = \bar{h}b_0/8$ is the coupling parameter, $\delta(y)$ is the delta function. Parameter $\sigma = \sqrt{i/2kd_{skin}}$ describes the Ohmic losses in the metal corresponding to the Leontovich boundary condition, d_{skin} is the skin depth of the metal.

Equations (1) applied to a system infinite in z direction lead to the existence of a surface wave $A_\pm \;\; e^{i\Gamma z - g_\pm y + i\Omega t}$ with a dispersion law which in the lossless case can be put down as

$$\left(\frac{\Omega}{c}\right)^2 - \Gamma^2 = \bar{h}^2\alpha^4 \quad (2)$$

Dispersion characteristic corresponding to (2) is shown in the lower part of Fig. 2.

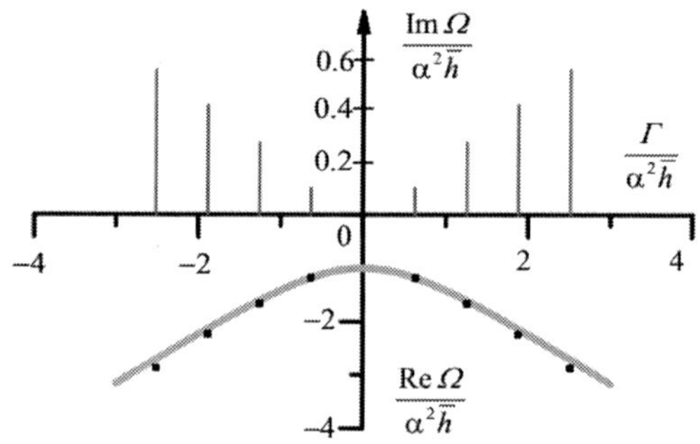

Fig. 2. Dispersion curve of the surface wave. Dots mark the surface-wave resonator modes at $\alpha^2\bar{h}l_z = 5$, lines in the upper part indicate the losses of these modes

In the case when the length l_z of the structure is finite, a set of longitudinal modes can be found. Seeking modes of (1) at an eigenfrequency of Ω, and using the Fourier transform, one can obtain an integral equation equivalent to system (1):

$$\hat{A}_+(0,z) = \\ = \sqrt{\frac{\bar{h}}{\pi i}}\int_0^z \frac{\pm\alpha(z')\hat{A}_+(0,l_z - z') + i\sigma\hat{A}_+(0,z')}{\sqrt{z-z'}}e^{-i\Omega(z-z')}dz'. \quad (3)$$

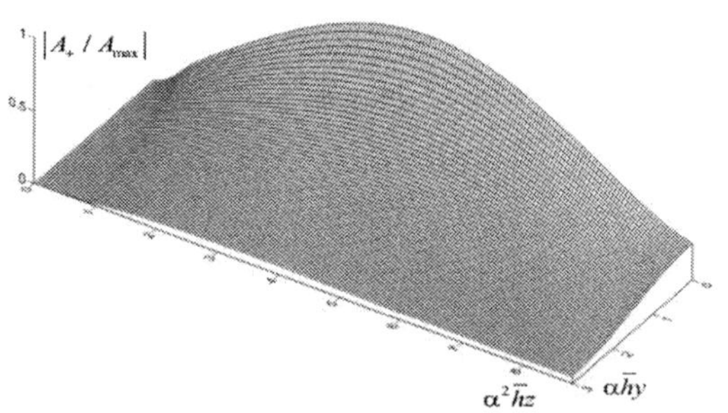

Fig. 3. Fundamental mode structure at $\alpha^2\bar{h}l_z = 8.8$

Equation (3) possesses an infinite discrete eigenmode spectrum. Frequencies and spatial profiles of eigenmodes can be found numerically. Real parts of the eigenfrequencies of the first several modes are marked with dots close to the dispersion in lower part of Fig.2 for $\alpha^2 \bar{h} l_z = 5$ while the imaginary parts of these modes are shown as vertical lines in the upper part of the same figure. It can be seen that the $\mathrm{Im}\,\Omega$ increases almost linearly. Quality factor $Q = \omega_0 (2\,\mathrm{Im}\,\Omega)^{-1}$ of the fundamental mode is about 3 times higher than that the next (second mode). It demonstrates the high selective properties of surface wave resonator. Spatial distribution of the A_+ field for the lowest (fundamental) mode is shown in Fig. 3.

Making use of the fact that according to the exact solution of (3) the longitudinal field structure of the mode tends to sine shape with increasing parameter $\alpha^2 \bar{h} l_z$ (see Fig. 3), an asymptotic formulas for Q factors can be obtained in the following form:

$$Q = \frac{1}{Q_{diff}^{-1} + Q_{Ohm}^{-1}}, \qquad (4)$$

$$Q_{diff} = \frac{\alpha^4 \bar{h}^3 l_z^3}{5\pi^2}; \; Q_{Ohm} = \frac{2}{\alpha \,\mathrm{Re}\,\sigma}.$$

Note that the diffraction Q factor increases with the length of corrugated section while the Ohmic Q remains constant. At small l_z, the diffraction quality prevails, $Q \approx Q_{diff} << Q_{Ohm}$. With increase of the corrugation length, these two terms become comparable and the total quality factor tends to saturation at $Q \approx Q_{ohm}$. These dependences are shown in Fig.4, based both on the asymptotic formulas (dashed line) and on the exact solution of the integral equation (3) for three values of σ/α parameter.

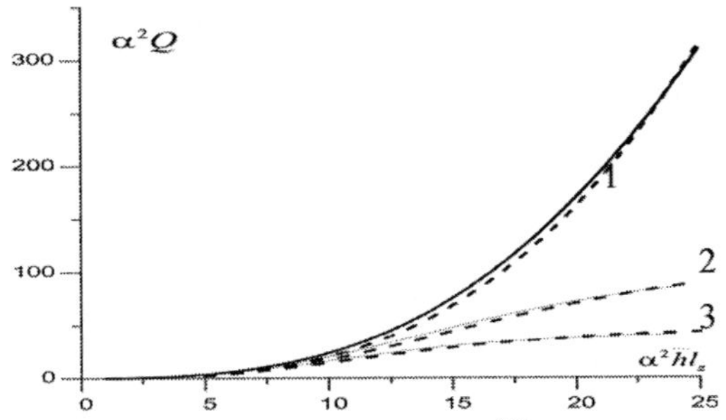

Fig. 4. Dependence of Q-factor on the $\alpha^2 \bar{h} l_z$ parameter, 1) $\sigma/\alpha = 0$, 2) $\sigma/\alpha = 0.1$, 3) $\sigma/\alpha = 0.25$

It is interesting to note that the two terms in (4) demonstrate different behavior at varying parameter α that is proportional to the corrugation depth b_0. While the diffraction term rises as $Q_{diff} \sim \alpha^4$, the Ohmic losses decrease as Q_{Ohm} α^{-1}. This can be explained in a simple way: while the wavebeam coupling rises, the modes become more confined to the corrugation thus reducing the diffraction losses; at the same time, the less the effective 'volume' of the mode is, the higher is the influence of the Ohmic losses. Total Q factor would have a maximum in its α dependence for each L. This maximum can be found from expressions (4) as

$$Q_{\max} = \left(\frac{2^4}{5^6 \pi^2} \frac{l_z^3}{\bar{h} d_{skin}^4} \right)^{1/5} \approx 0.16 \frac{l_z^{3/5}}{\bar{h}^{1/5} d_{skin}^{4/5}}$$

reached at

$$\alpha_{opt} = \left(\frac{10\pi^2}{\bar{h}^4 l_z^3 d_{skin}} \right)^{1/5} \approx \frac{3.97}{\bar{h}^{4/5} l_z^{3/5} d_{skin}^{1/5}}. \qquad (5)$$

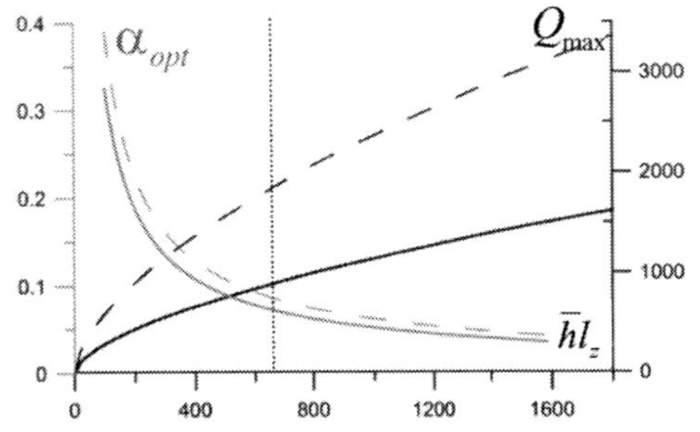

Fig. 5. Optimal α (gray) and maximum Q (black) and vs length of 1 THz resonator. Solid lines correspond to copper at 300 K, dashed lines correspond to copper at 77 K

The optimum Q and α values vs. length of the resonator are depicted in Fig.5 for a structure with a Bragg frequency of about 1 THz (d = 150 μm) made of copper at normal (solid lines, $d_{skin} \approx 0.065$ μm) and liquid nitrogen (dashed lines, $d_{skin} \approx 0.025$ μm) temperatures. According to (5), in this parameter area the small corrugation approximation is validated for structures of several dozen periods or longer. For instance, for a structure with a length of 100 periods ($\bar{h} l_z \approx 630$, see dotted line in Fig. 5), a corrugation depth of 1/10 wavelength provides an optimal Q factor of about 1000 for copper at normal temperature. For a copper structure at nitrogen temperature Q factor can reach 2000 for a slightly deeper corrugation.

Note in conclusion that similar analysis of surface-wave resonators with 2D corrugation can also be undertaken. Such structures provide mode selection in all three dimensions [3]. Using the time-domain simulation we can find the fundamental mode, its spatial profile, eigenfrequency, and Q factor.

This work was supported by RFBR grant No. 16-02-00890.

References

1. *W. L. Barnes, A. Dereux, and T.W. Ebbesen.* Surface plasmon subwavelength optics // Nature, v. 424, pp. 824–830, 2003.

2. *N.S. Ginzburg, A.M. Malkin, A.S. Sergeev, V.Yu. Zaslavsky*, Quasi-optical theory of relativistic submillimeter surface-wave oscillators // Appl. Phys. Lett. 2011, v. **99**, p. 121505.

3. *N.S. Ginzburg, A.M. Malkin, A.S. Sergeev, V.Yu. Zaslavsky.* Powerful surface-wave oscillators with two-dimensional periodic structures Appl. Phys. Lett. 2012, v. **100**, p. 143510.

EPJ Web of Conferences **149**, 04014 (2017) DOI: 10.1051/epjconf/201714904014
10th International Workshop 2017 "Strong Microwaves and Terahertz Waves: Sources and Applications"

The gyrotron – a natural source of high-power orbital angular momentum millimeter-wave beams

M. Thumm[1], A. Sawant[2], M.S. Choe[3], E.M. Choi[2,3]

[1]Karlsruhe Institute of Technology (KIT), IHM and IHE, Karlsruhe, Germany, manfred.thumm@kit.edu
[2]Department of Electrical Eng., Ulsan National Inst. of Science and Technology (UNIST), Ulsan, South Korea [3]Department of Physics, Ulsan National Institute of Science and Technology (UNIST), Ulsan, South Korea

Abstract

Orbital angular momentum (OAM) of electromagnetic-wave beams provides further diversity to multiplexing in wireless communication. The present report shows that higher-order mode gyrotrons are natural sources of high-power OAM millimeter (mm) wave beams. The well-defined OAM of their rotating cavity modes operating at near cutoff frequency has been derived by photonic and electromagnetic wave approaches.

Introduction

Orbital Angular Momentum (OAM) [1] carried by light beams (vortex beams) was discovered earlier and has been widely employed in many applications such as optical tweezers, optical drives of micro-machines, atom trapping, and optical communication. OAM provides an additional dimension (diversity) to multiplexing techniques which can be employed in addition to conventional multiplexing methods to achieve higher data rates. OAM beams have been thoroughly studied and used in the optical regime [2] but in the mm-wave region, they are still under investigation. In this frequency band, there are difficulties associated with beam-splitting and beam-combining processes as well as with the use of spiral phase plates for generation of OAM, since the wavelength is much larger compared to those at optical frequencies.

Here we introduce the natural generation of high-power OAM modes excited by gyrotrons with axial output of the rotating higher-order transverse electric mode $TE_{m,p}$ where m and p are the azimuthal and radial mode index, respectively. Such output radiation could be used for long-range wireless communication. To achieve high frequency stability and a narrow line width, a phase locking system can be utilized to control the electron beam energy. The long-term stability can be guaranteed by a reference clock. A line width as narrow as 1 Hz has been obtained [3].

OAM of Rotating Gyrotron Modes

Helically propagating rotating $TE_{m,p}$ modes in a circular waveguide can be decomposed into a series of cylindrical plane waves, each propagating at the Brillouin angle θ_B relative to the waveguide axis. The angular momentum possessed by such a ray (which is tangential to the caustic of the mode) can be expressed as

$$\vec{L} = L\hat{z} = \vec{r} \times \vec{p} = R_c \hat{r} \times p_\Phi \hat{\Phi} = R_c p_\Phi \hat{z}$$

here $R_c = (m/X_{m,p})R_0$ is the caustic radius of the operating mode, where $X_{m,p}$ is the pth root of the derivative of the Bessel function J_m, R_0 is the waveguide radius and where

$$p_\Phi = p\sin\theta_B = \hbar k \sin\theta_B\,,\ \sin\theta_B = \frac{k_\perp}{k} \text{ and } k_\perp = \frac{\chi_{m,p}}{R_0}.$$

$$\text{Therefore, } \vec{L} = R_c p_\Phi \hat{z} = \frac{m}{\chi_{m,p}} R_0 \hbar k \frac{\chi_{m,p}}{kR_0} = m\hbar\hat{z}$$

The ratio between the total angular momentum L_N and total energy W_N of N photons is given by

$$\frac{\vec{L}_N}{W_N} = \frac{Nm\hbar}{N\hbar\omega} = \frac{m}{\omega}$$

where ω is the angular frequency of the operating cavity mode which is close to the cutoff frequency $\omega_{cut} = cX_{m,p}/R_0$ in the cylindrical cavity section. Here c is the velocity of light in vacuum. Therefore follows $m/\omega = R_c/c$, which means that the OAM is proportional to the absolute caustic radius.

This is the result of the photonic approach to derive the OAM. The same value follows from the electromagnetic wave approach for the total OAM of a gyrotron cavity mode within a given waveguide volume per total energy of the electromagnetic wave in the same volume [4].

At a given frequency the OAM increases with increasing azimuthal mode number m, whereas for fixed m it decreases with increasing frequency.

Modes that have a linear polarization may possess a so-called Spin Angular Momentum (SAM) if their right-hand and left-hand rotation components are different. Table 1 summarizes the SAM and OAM features of different types of TE modes.

Table 1. Intensity on axis I_0, possible spin angular momentum (SAM) and orbital angular momentum (OAM) of gyrotron TE modes in circular waveguides

Mode Family	I_0	SAM (Polarization)	OAM
$TE_{0,p}$	no	no	no
$TE_{1,p}$	yes	yes	no
$TE_{m,p}$ (m >1)	no	no	yes

Gyrotrons with OAM mm-Wave Output Beams

UNIST has been operating a 65 kW W-band gyrotron with diode-type magnetron injection gun (MIG), $TE_{6,2}$ cavity mode and quasi-optical output in the linearly polarized fundamental Gaussian $TEM_{0,0}$ mode [5]. In case that the quasi-optical output mode converter would be removed and exchanged with a mode-conserving non-linear diameter up-taper, one would have the essential feature of a OAM gyrotron with axial mm-wave output configuration where the cavity mode is directly emitted through the dielectric tube window into free space. Of course, propagation of higher-order OAM beams generated by gyrotrons in free space results in a large diffraction angle, which however, can be compensated by the waveguide diameter taper section that is added after the gyrotron cavity to reduce the Brillouin angle θ_B of the output wave and by employing focusing lenses to reduce the diffraction. Provided that the gyrotron would also be equipped with a triode-type MIG, the tube could be electronically switch between co- and counter-rotating cavity modes (right- and left hand OAM) and between two cavity modes with different magnitude of OAM (see table 2).

Table 2. Possible operating cavity modes with different OAM of the 65 kW UNIST W-band gyrotron

Cavity Mode	$X_{m,n}$	Frequency (GHz)	Relative Caustic Radius $m/X_{m,n}$
$TE_{6,2}$	11.734936	94.8	0.51
$TE_{10,1}$	11.770877	95.1	0.85

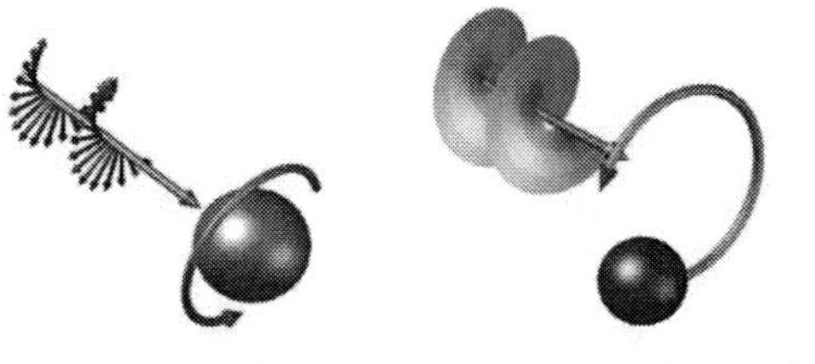

Fig. 1. Spin angular momentum (SAM) (left), where the electric field is twisting, and orbital angular momentum (OAM) (right), where the phase does the twisting [4]

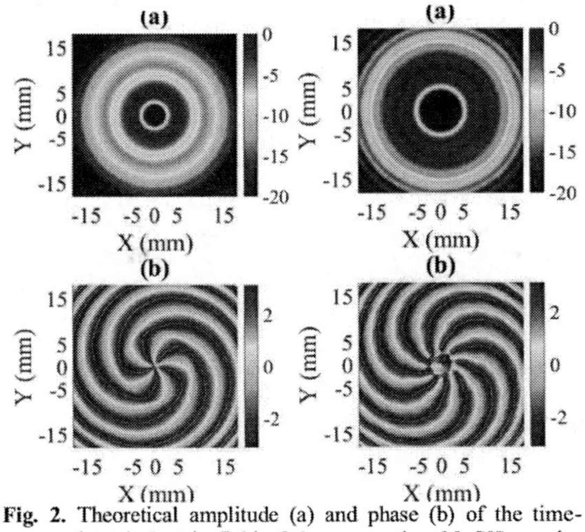

Fig. 2. Theoretical amplitude (a) and phase (b) of the time-averaged total electric field of the co-rotating 95 GHz modes $TE_{6,2}$ (left) and $TE_{10,1}$ (right) (in 10 mm distance from a 20 mm diameter aperture)

Figure 2 shows the analytically calculated amplitude and phase of the time-averaged total electric field of the co-rotating 95 GHz $TE_{6,2}$ and $TE_{10,1}$ mode, respectively. The phase patterns show m-1 spirals.

Several Ku/Ka-band 10-to-30-kW-class gyrotrons for technological applications operate in a low-order OAM cavity mode (e.g. $TE_{2,1}$, $TE_{2,2}$, $TE_{3,2}$, or $TE_{2,3}$) which also serves as axial output mode [6].

In the early 1990-ties the Research Center Karlsruhe (FZK, now KIT) performed very successful experiments with an axial output 140 GHz, 0.5 MW $TE_{10,4}$-mode gyrotron [7] having an electronic efficiency of approximately 33%. The tube was equipped with a non-linear up-taper from the 16.22 mm diameter cavity to the 70 mm diameter output waveguide with a frequency tunable double-disk, FC75 face cooled sapphire window. The high-power measurement of the output mode content using a 70 mm diameter wavenumber spectrometer [8] resulted in a purity of the co-rotating $TE_{10,4}$ output mode better than 98%.

At those times, such 0.5–1.0 MW-class short-pulse gyrotrons employing asymmetric cavity modes with non-linear up-tapers and axial output through the collector also were built by several industrial manufacturers [6], e.g.: 8 GHz, $TE_{5,1}$ and 110 GHz, $TE_{6,4}$ (THOMSON), 84 GHz, 110 GHz and 140 GHz, $TE_{15,2}$ (VARIAN), 110 GHz, $TE_{12,2}$ and $TE_{22,2}$ (VARIAN), 120 GHz, $TE_{12,2}$ and $TE_{22,12}$ (JAERI, TOSHIBA).

References

1. *Allen, L., Beijersbergen, M.W., Spreeuw, R.J.C., Woerdman, J.P.* Orbital angular momentum of light and the transformation of Laguerre-Gaussian laser modes// Phys. Rev. A 1992. V. 435, P. 8185-8189.

2. *Wang, J., Yang, J.-Y., Fazal, I.M., Ahmed, N., Yan, Y., Huang, H., Ren, Y., Yue, Y., Dolinar, S., Tur, M., Willner, A.E.* Terabit free-space data transmission employing orbital angular momentum multiplexing // Nat. Photon. 2012. V. 6, P. 488.

3. *Fokin, A.P., Denisov, G.G., Glyavin, M.Yu., Golubiatnikov, G.Yu., Lubyako, L.V., Morozkin, M.V., Movschevich, B. Z., Tsvetkov, A.I.* High precision frequency stabilization of a 100 W/263 GHz continuous wave gyrotron // Proc. 18th IEEE Int. Vacuum Electronics Conference (IVEC 2017), London, UK, 2017, GII-4.

4. *Sawant, A., Choe, M.S., Thumm, M., Choi, E.M.* Orbital angular momentum (OAM) of rotating modes driven by electrons in electron cyclotron masers // Accepted for publication in Scientific Reports – Nature, 2017

5. *Kim, S.G., Sawant, A., Lee, I., Kim, D., Choe, M., Won, J.-H., Kim, J., So, J., Jang, W., Choi, E.M.* System development and performance testing of a W-band gyrotron // J. Infrared Milli Terahz Waves 2016. V. 37, No. 3. P. 209-229.

6. *Thumm, M.* State-of-the-art of high power gyro-devices and free electron masers // KIT Scientific Reports 2016, V. 7717, KIT Scientific Publishing, Karlsruhe, Germany

7. *Piosczyk, B., Kuntze, M., Borie, E., Dammertz, G., Dumbrajs, O., Gantenbein, G., Möbius, A., Nickel, H.-U., Thumm, M.* Development of high-power 140 GHz gyrotrons at KfK for applications in fusion // Fusion Technology 1992, eds. C. Ferro, M. Gasparotto, H. Knoepfel, Elsevier Science Publishers B.V. 1993 P. 618-622.

8. *Kasparek, W., Müller, G.* The wavenumber spectrumeter – an alternative to the directional coupler for multimode analysis in oversized waveguides // Int. J. Electronics 1988 V.64 P.5-20.

EPJ Web of Conferences **149**, 04015 (2017) DOI: 10.1051/epjconf/201714904015
10th International Workshop 2017 "Strong Microwaves and Terahertz Waves: Sources and Applications"

Design and manufacturing process for the KIT 2-MW 170-GHz coaxial-cavity longer-pulse gyrotron

S. Ruess[1,2], G. Gantenbein[1], A. Heinzel[1], S. Illy[1], I. Gr. Pagonakis[1], T. Rzesnicki[1], M. Thumm[1,2], J. Weggen[1], A. Weisenburger[1], and J. Jelonnek[1,2]

[1]IHM, [2]IHE, Karlsruhe Institute of Technology (KIT), 76131 Karlsruhe, Germany
Sebastian.Ruess@kit.edu

Today's performance requirements for future DEMO gyrotrons are an operating frequency between 170 GHz up to 240 GHz and an output power of significantly larger than 1 MW (today's target: 2 MW). A total gyrotron efficiency of better than 60 % must be achieved. Multi-purpose/multi-frequency operation and frequency step-tunability are required also. It has been shown earlier, that the coaxial-cavity technology is a promising candidate [1]. In [2] a world record RF output power of 2.2 MW has been reported for short-pulses (in the range of few milliseconds). Nevertheless, it has to be proven that the coaxial-cavity technology can be used for long-pulse operation. That shall be achieved by upgrading the existing 2 MW 170 GHz short-pulse prototype used for the demonstration of the 2.2 MW record power. Additionally, a new Inverse Magnetron Injection Gun (IMIG) shall allow a significant larger emitter radius and therefore increased output power at operating frequencies significantly above 200 GHz by keeping the same or even smaller diameter of the warm bore hole of the superconducting gyrotron magnet. Both, the IMIG as well as the longer pulse gyrotron will show the way towards higher output power at higher operating frequencies of gyrotrons together with a more robust design and construction of that kind of tubes.

Heading towards a longer-pulse tube

Fig. 1 shows the major upgraded parts of the existing KIT modular coaxial-cavity tube [3,4]. In detail, for the longer-pulse tube a new modular system concept for cooling the beam tunnel, the cavity and the launcher has been developed. Fully independent water cycles for each of the named gyrotron components have been designed and verified with the multi-physics software Comsol. This concept allows the monitoring of the internal losses in each gyrotron component and of the final energy balance of the tube during long pulse operation. It has to be mentioned, that the longer-pulse 2 MW gyrotron is currently under manufacturing in the KIT workshop and the first operation is planned for mid-2017.

Introduction to the advanced IMIG

The presented Inverse Magnetron Injection gun as well as the longer pulse gyrotron will be a major step towards higher output power. Due to the larger effective emitter radius the IMIG can be operated at higher output power levels at operating frequencies above 200 GHz by keeping the same or even smaller diameter of the warm bore hole of the superconducting magnet. At these very high frequencies, magnetic field strengths of above 10 T have to be considered for the magnet. Furthermore, the advanced IMIG design is of modular type. It allows the implementation and test of new emitter technologies with significantly increased current density. The proposed design of the IMIG [1] matches the design and the nominal operating parameters of the European 170-GHz, 2-MW coaxial-cavity gyrotron, originally intended for the first installation at ITER [2]. The IMIG can also be operated in the KIT 2-MW coaxial longer-pulse gyrotron [5], as well as in conventional gyrotrons [6] and is compatible with future fusion gyrotrons at 204 GHz and multi-MW output power range.

Realization of vacuum-tight connections

One of the most critical issues in a gyrotron are the high-vacuum compatible brazed and soldered joints. OM (optical microscopy) (see Fig. 3) and SEM (Scanning Electron Microscope) cross sections and measurements of different brazed and soldered joints with variable material compositions have shown excellent vacuum compatible connections with a He leakage rate of <10-10 mbar l/s. In Fig. 3, cross sections of different brazed and soldered joints with variable material compositions are shown. The two at the top in Fig. 3 show a soldered joint with *Glidcop* and stainless steel. It can be very well seen that the solder flowed into the gap due to the capillary forces and diffuses into the Glidcop and stainless steel. In addition, excellent solder joints between CVD diamond and copper were also achieved.

Experimental Measurements

One of the main advantages of the IMIG is the optimized cooling concept of the cathode, especially for high thermally loaded components in the neighboring emitter regions and at the anode. Therefore, the thermal expansion in the IMIG could be significantly reduced.

The corresponding simulation results were already presented in [5]. In order to verify the thermomechanical simulation results a measurement setup was designed and manufactured. By using an IR-camera the temperature distribution at the neighboring region of the emitter will be measured and compared with the simulation results to verify the assumptions for the thermomechanical simulations. In the case of the IMIG, the cathode can be used as a vacuum chamber, which is closed on the top by a ZnS CG window. The measurement system is closed with a blind flange and intake socket for continuous vacuum pumping at its bottom. At the same time, the measurement system contributes also to the calibration of the filament power versus the emitter temperature.

It shall be pointed out that the IMIG is the first multi-MW Magnetron Injection Gun, which is in-house manufactured at KIT. The emitter ring is produced by 3M, Ceradyne Inc. Measurements of the already produced components indicate excellent surface conditions and small tolerances (< 20 µm) which permits a homogenous electric field at the emitter region and therefore a high quality electron beam.

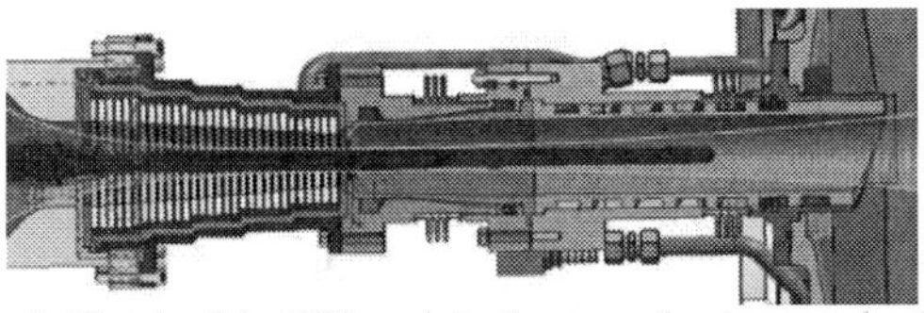

Fig. 1. Sketch of the KIT modular longer-pulse (expected pulse lengths 100 ms – 1 s) gyrotron pre-prototype. Major parts are the beam tunnel (green), the cavity (orange) and the launcher (blue)

Fig. 2. Assembled Inverse Magnetron Injection Gun during the bake out process

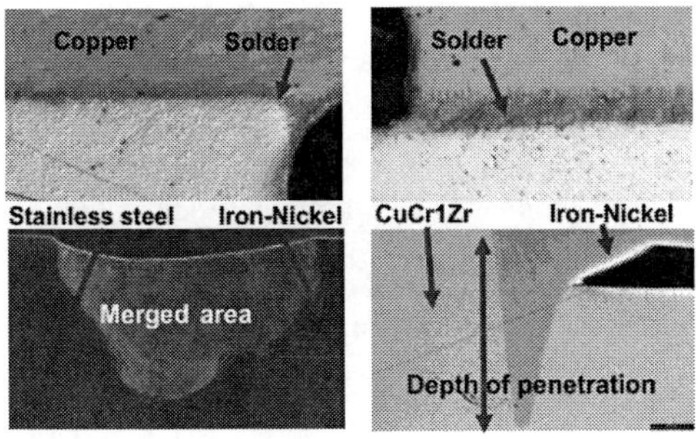

Fig. 3. Cross sections of brazed and soldered joints (OM and SEM)

Summary and Outlook

The already successfully manufactured parts of the KIT longer-pulse coaxial-cavity gyrotron are presented. In particular the construction of an innovative IMIG is discussed. The IMIG is prepared for the use with the KIT 2 MW / 170 GHz coaxial-cavity gyrotron. The IMIG permits an optimized cooling concept of thermally loaded components and allows for a given warm bore hole radius of the gyrotron magnet a significantly larger emitter ring compared to "conventional" MIGs. In addition, the IMIG is completely in-house manufactured with excellent surface quality and small tolerances. Due to the optimized construction of the beam-forming components, the IMIG was easily re-machined after the implementation of the ceramics, which results in smaller tolerances. To verify the thermomechanical simulation results, a comparison of the temperature distribution of the IMIG between the experimental results and the simulation results will be made by the use of a measurement test facility in mid-2017.

Acknowledgements

Part of this work is carried out within the framework of the EUROfusion Consortium and is receiving funding from the Euratom research and training programme 2014-2018 under grant agreement No 633053. The views and opinions expressed herein do not necessarily reflect those of the European Commission.

References

1. *B. Piosczyk, et al.* "A 2-MW, 170-GHz Coaxial Cavity Gyrotron". IEEE Transaction on Electron Devices, (32):413-417, June 2004.
2. *T. Rzesnicki, et al.* IEEE Trans. Plasma Science, Vol. 38, No. 6, pp. 1141-1149, 2010.
3. *S. Ruess, et al.* "Experimental results and outlook of the 2 MW 170 GHz coaxial-cavity gyrotron towards long pulse operation", German Microwave Conference (GeMiC) 2016, Bochum, Germany, pp. 369-372,2016.
4. *S. Ruess et al.* „Coaxial Gyrotron Development: From ITER towards DEMO ", Proc. European Microwave Conference (EuMC) 2017, Nuernberg, Germany, 2017.
5. *S. Ruess, et al.* „An Inverse Magnetron Injection Gun for the KIT 2-MW Coaxial-Cavity Gyrotron", IEEE Transactions on Electron Devices, (63), p. 2104-2109, 2016, DOI:10.1109/TED.2016.2540298.
6. *S.Ruess, et al.* Proposal of an Inverse Magnetron Injection Gun for Future Hollow-Cylindrical-Cavity High Power Gyrotrons" Proc. IRMMW 2016, Copenhagen, Denmark, September 2016, W4C.

Simulation of electromagnetic fields scattered from arbitrary shaped electric conductors

A. Marek[1], K.A. Avramidis[1], S.M. Copplestone[5], N.S. Ginzburg[4], S. Illy[1], J. Jelonnek[1,2], J. Jin[1], S.V. Mishakin[4], A.-S. Müller[3], P. Ortwein[5], M. Thumm[1,2]

[1]IHM, [2]IHE, [3]IBPT, Karlsruhe Institute of Technoloy (KIT), Karlsruhe, Germany, alexander.marek@kit.edu
[4]Institute of Applied Physics, Russian Academy of Sciences (IAP- RAS), Nizhny Novgorod, Russia
[5]Institute of Aerodynamics and Gas Dynamics, University of Stuttgart, Stuttgart, Germany

Introduction

An advanced tool (KarLESSS) for the simulation of full 3D electromagnetic scattering problems with arbitrarily shaped, large perfect conductors is under development. High accuracy in the simulation is achieved by applying a full-wave simulation method based on the electric field integral equation (EFIE) [1] which is solved numerically using a Galerkin-type method of moments (MoM) [2]. The tool was originally developed for the analysis of quasi-optical mode converters in high power gyrotrons. In this paper, additional areas of application of KarLESSS will be demonstrated. After a brief introduction of the theory in the first section, simulation results of a quasi-optical input-output system for an amplifier operating in the millimeter-wave range are shown. Afterwards, cold simulations of a helically corrugated resonator of a gyro-amplifier [3] are presented.

Theory

The components of the electric field are described with the EFIE as follows:

$$\frac{-i}{\omega\mu}\boldsymbol{n}(\boldsymbol{r})\times \boldsymbol{E}_i =$$

$$= n(\boldsymbol{r})\times \iint_S d\boldsymbol{r}' G(\boldsymbol{r},\boldsymbol{r}')[\boldsymbol{J}(\boldsymbol{r}') + \frac{1}{k^2}\nabla'\nabla'\cdot \boldsymbol{J}(\boldsymbol{r}')] \qquad (1)$$

where $\boldsymbol{J}$ is the unknown electrical surface current, $\boldsymbol{E}_i$ is the known incident field and G is the free space Green's function. In the MoM approach, the currents are expanded into a sum of weighted basis functions f. Applying this expansion into (1), using the linearity of the equation and taking an inner product of the equation with the basis functions leads to a solution for which the problem can be formulated as a matrix equation.

For a 3D problem it is known to approximate a surface by a series of triangles. Therefore, basis functions for a surface divided into such a series of triangles are required. The RWG basis functions [4] are the lowest order set of functions fulfilling this requirement. To improve the accuracy and performance, higher order basis functions [5] are used in KarLESSS. In addition, surfaces are approximated by curved triangular patches. The curved triangles are parametrized with a local B-Splines interpolation [6].

The calculation of the impedance matrix with the MoM results in a dense matrix. This is caused by the non-local kernel of the EFIE, the Green's function. Consequently, the memory requirements and calculation time for the impedance matrix are of the order $O(N^2)$, where N denotes the number of unknowns. Therefore, it is a challenge to solve the EFIE for highly over-sized scatterers such as quasi-optical systems of high-power gyrotrons. To solve this problem, a speed-up of the calculation of scattering and radiation problems based on the adaptive cross approximation (ACA) was first proposed in [7]. With the ACA, the impedance matrix can be compressed in a purely algebraic manner. No information about the integral kernel is required and the ACA can be used for different sets of basis functions as well as for different Green's functions. The premise of the ACA is that interactions of well-separated parts of the geometry result in a rank-deficient impedance matrix. Therefore, the dense impedance matrix consists of many numerically rank-deficient sub-blocks. These sub-blocks of low rank can be compressed efficiently with the ACA:

$$Z^{m\times n} \approx U^{m\times r} V^{r\times n} \qquad (2)$$

where r is the rank of the approximate matrix. If r is sufficiently small, the approximated matrix requires less memory and can be calculated faster than the full matrix. The advantage of this algorithm compared to other compression algorithms is that it requires only a partial knowledge of the original matrix. This is achieved by an iterative calculation of the approximated matrix (see [7] for details). Therefore, the memory requirements as well as the required calculation time can be reduced. In addition, the calculation of matrix-vector products with the approximated matrix is faster than for the uncompressed matrix. That accelerates the iterative solver (based on a generalized minimal residual method) which is used for solving the system of linear equations.

Quasi-Optical Input-Output System

A problem very similar to the simulation of the quasi-optical systems used in high-power gyrotrons, is the simulation of an input-output system for a gyro-amplifier(see Fig. 1 a)). The system is designed for a center frequency of 260 GHz and consists of parabolic and sinusoidal corrugated mirrors. Shown is the simulation of the output channel. For the simulation, a linear polarized $HE_{1,1}$-beam is radiated from the position of the gyro-amplifier. After it is reflected on the mirrors 1, 2 and 3, the field distribution is analyzed at the plane of the output-port (see Fig. 2 b)).

In addition to the separate simulation of the input and output channels, a simultaneous simulation of both channels was performed. The resulting surface currents on the mirrors of the output channel are shown in Fig. 1 b).

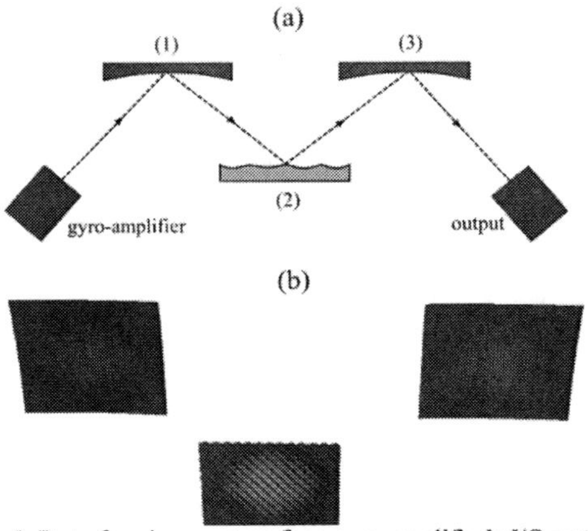

Fig. 1. Part of a mirror system for a gyro-amplifier's I/O system. a) Shematic draw. b) Calculated currents on the mirrors (a.u.)

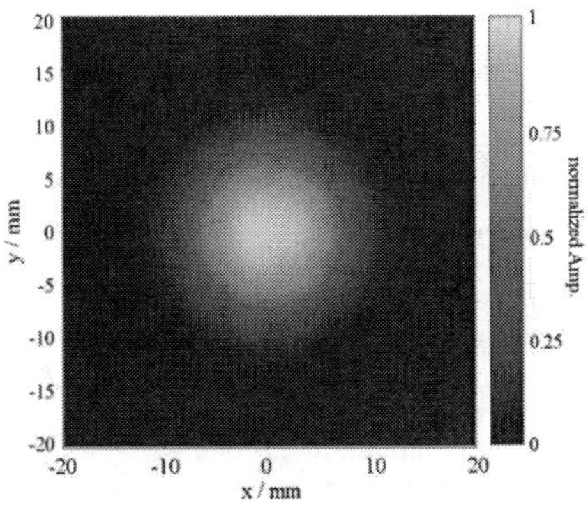

Fig. 2. Simulated field distribution for a frequency of 260 GHz at the position of the output port

Helically Corrugated Resonator

To show the flexibility of KarLESSS, field simulations, without an electron beam, of the interaction circuit of a helically corrugated gyro-amplifier [3] at 260 GHz are performed. The profile of the wall is described by:

$$r(\varphi, z) = r_0 + l\sin(\frac{2\pi}{d}z - 3\varphi) \quad (3)$$

where r_0 defines the average of the waveguide which is about 1 mm. The depth of the corrugation l is about 0.1 mm. The period of the corrugation d is also about 1 mm. At the begin and at the end of the circuit, tapers with a length of 3 periods transform the mode of a circular waveguide into eigen modes of the corrugated waveguide. In total, the structure has a length of 54 periods. The threefold right-handed corrugation couples the right-handed $TE_{2,1}$-mode with the forward-propagating left-handed $TE_{-1,1}$-input mode, whereas a right-handed $TE_{1,1}$-input mode should be almost unperturbed. In Fig. 3 the field distributions before (1st column), inside (2nd column) and after the regions with corrugations (3rd column) are shown. The upper row shows the simulated field distributions for a $TE_{-1,1}$-mode as input and in the lower row for a $TE_{1,1}$-input mode. As expected, the field distribution of the $TE_{1,1}$-mode is almost unperturbed while the $TE_{-1,1}$-mode is significantly modified. In both cases, the mode is nicely back-transformed to a cylindrical waveguide mode after the corrugations.

In the presentation, further cold analyses of the helically corrugated resonator, i.e. a simulation of the dispersion relation, will be presented.

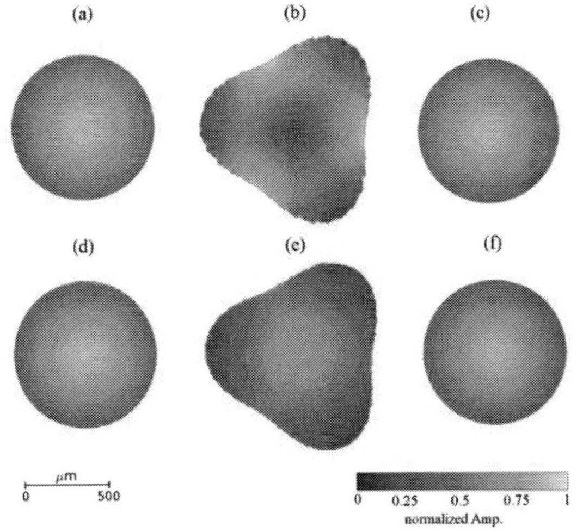

Fig. 3. Simulated field distributions for a left-handed $TE_{-1,1}$ (a, b, c) and a right-handed $TE_{1,1}$ input mode (d, e, f). Left column: entry of resonator, Center column: in the resonator, Right column: end of resonator

Acknowledgements

The research is supported by the joint RSF-DFG project (Je 711/1-1 – 16-42-01078) "Generation of Ultrashort Pulses in Millimeter and Submillimeter Bands for Spectroscopy and Diagnostic of Various Media Based on Passive Mode-locking in Electronic Devices with Nonlinear Cyclotron Absorber in the Feedback Loop".

References

1. *Neilson, J.M.; Bunger, R.* "Surface integral equation analysis of quasi-optical launchers" // IEEE Transactions on Plasma Science, 2002, vol. 30, no. 3, pp. 794-799.

2. *Harrington, R.F.* "The method of moments in electromagnetics Journal of Electromagnetic Waves and Applications" // Taylor & Francis, 1987, 1, 181-200

3. *Denisov G.G. et al.* "Gyro-TWT with a helical operating waveguide: new possibilities to enhance efficiency and frequency bandwidth" // IEEE Transactions on Plasma Science, vol. 26, no. 3, pp. 508-518, Jun 1998.

4. *Rao, S.M.; Wilton, D.R. & Glisson, A.W.* "Electromagnetic scattering by surfaces of arbitrary shape" // IEEE Transactions on Antennas and Propagation, 1982, AP-30, 409-418

5. *Cai, W.* "High-Order Mixed Current Basis Functions for Electromagnetic Scattering of Curved Surfaces" // J. Sci. Comput., 1999, 14, 73-105

6. *Piegl, L.A. & Tiller, W.* "The NURBS book" (2. ed.) // Springer, 1997, I-XIV, 1-646

7. *Zhao, K.; Vouvakis, M.N. & Lee, J.-F.* "The adaptive cross approximation algorithm for accelerated method of moments computations of EMC problems" // IEEE/EMC Transactions, 2005, 47, 763-773

EPJ Web of Conferences **149**, 04017 (2017) DOI: 10.1051/epjconf/201714904017
10th International Workshop 2017 "Strong Microwaves and Terahertz Waves: Sources and Applications"

Polarization-dependent TE_{11}-to-TE_{11}/TE_{01} waveguide mode converter for transmission line mode switching

D.I. Sobolev, G.G. Denisov, A.G. Eremeev, V.V. Holoptsev, A.I. Tsvetkov

Institute of Applied Physics, Nizhny Novgorod, Russia, nrtnm@mail.ru

Introduction

Performance of microwave setups, such as ECR ion sources, may be affected by the selected type of the operating mode of microwave transmission line. The change of operating mode, if possible, is usually time-consuming process with substitution of some components and significant realignment. However, if we consider different modes of the same regular waveguide then mode switching can be made much easier and faster with the appropriate mode converters. In this work, we present the synthesized converter for switching between TE_{01} and TE_{11} modes of hollow smooth circular metal waveguide. Both modes are widely used, and in some cases it is hard to decide which one is preferable without experiments.

Synthesis of waveguide surface

Let us briefly discuss other elements of the transmission line first. Operating mode of the 45 GHz 20 kW CW gyrotron is converted into big converging Gaussian wave beam by output converter and matching optics unit. After several meters of insulating air gap, the beam is collected by the next converging mirrors and directed into synthesized waveguide taper. The taper converts Gaussian wave beam into linearly polarized TE_{11} mode of the circular waveguide with 22 mm diameter (see later). The next mode converter can conserve this mode or convert it into TE_{01}, being turned 90 degrees around its longitudinal axis. After the converter, there is one more taper to 32.6 mm waveguide diameter.

It is known that any non-symmetric mode may propagate differently in a serpentine converter, depending on its polarization relative to waveguide bend plane [1, 2]. This effect, controlled by the waveguide synthesizing code, allowed us to achieve high efficiency in both orientations.

The iterative synthesis method is based on the previous work made by authors [3, 4]. At the beginning of every iteration step the waveguide (Fig. 1) consists of the given input cross-section S_1, given output cross-section S_2 and initial conducting wall surface S_0. If we consider only one orientation, then conversion efficiency can be is increased after iteration by calculating two field distributions inside the waveguide. The first one is calculated using the given $\mathbf{E}_1$ and $\mathbf{H}_1$ fields at the input as a boundary condition, the second one by using the spatially reversed desired $\mathbf{E}_2$ and $\mathbf{H}_2$ fields at the output. Small and smooth deformation l orthogonal to waveguide wall affects the efficiency (mutual power) P according to formulas

$$\Delta P = \int_{S_0} lF dS, \quad F = -ik\left(\mathbf{H}_{\tau 1}\mathbf{H}_{\tau -2} + \mathbf{E}_{n1}\mathbf{E}_{n-2}\right) \quad (1)$$

where τ and n means tangential and normal field components, respectively. We can minimize the average square of deformation (i.e. norm) using real and imaginary parts of (1) as equality constraints with given ΔP, and in this simple case the deformation takes the form

$$l = \alpha \operatorname{Re} F + \beta \operatorname{Im} F \quad (2)$$

where constants α and β are expressed by ΔP and surface integrals of function F [4].

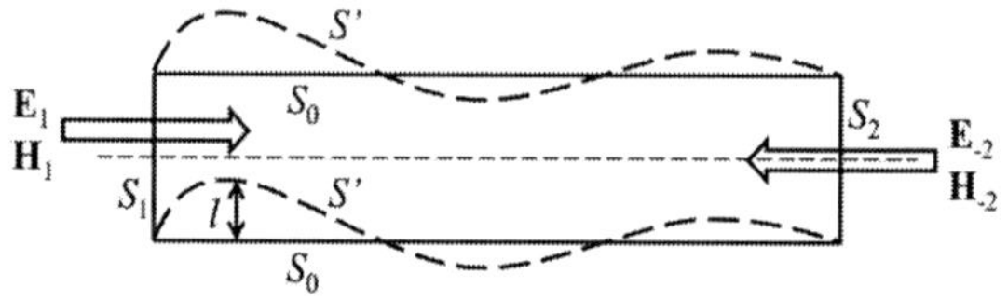

Fig. 1. Waveguide synthesis notations: two given waveguide cross-sections S_1 and S_2 (input and output, respectively), the waveguide surface before iteration S_0 and the surface after iteration S', the deformation value l at some point of the surface, the specified input fields $\mathbf{E}_1$ and $\mathbf{H}_1$, and the reversed desired output fields $\mathbf{E}_{-2}$ and $\mathbf{H}_{-2}$

To increase efficiency for both polarizations, two pairs of electromagnetic fields are calculated during the iteration and used in minimization. We used coupled-mode equations [1, 2] for field analysis, therefore we had to also add the condition of coaxiality of input and output cross-sections to allow rotation of the mode converter without realignment. This condition made the system solvable only numerically, which led to a slight decrease of the calculation speed.

Synthesized converter

We chose the waveguide diameter equal to 22 mm. It is high enough to pass the continuous power without arcing and overheating (with water cooling) and low enough to ensure adequate converter length. The length is 65 cm, which is only 10% more than then length of synthesized TE_{11}-TE_{01} converter with the same diameter and efficiency. After synthesis, the converter shape was tested numerically in CST Microwave Studio. Peak efficiency for both polarizations was well over 99.5%, but FDTD solver showed frequency shift for narrowband TE_{11}-TE_{01} conversion. However, few points calculated by frequency-domain solver showed good compliance with synthesis. It turned out that FDTD frequency shift depends on the mesh size. Power regression based on calculations with different mesh sizes suggests that the frequency shift tends to zero as mesh size approaches zero. Results for minimum cell size limited by the amount of available RAM are shown on Fig. 2.

One can see that efficiency in both orientations exceeds 99.5% in 400 MHz frequency band, which should be enough to cover the waveguide manufacturing tolerance and difference between actual and calculated gyrotron frequency. Fields in longitudinal cross-section in both polarizations are shown on Fig. 3a.

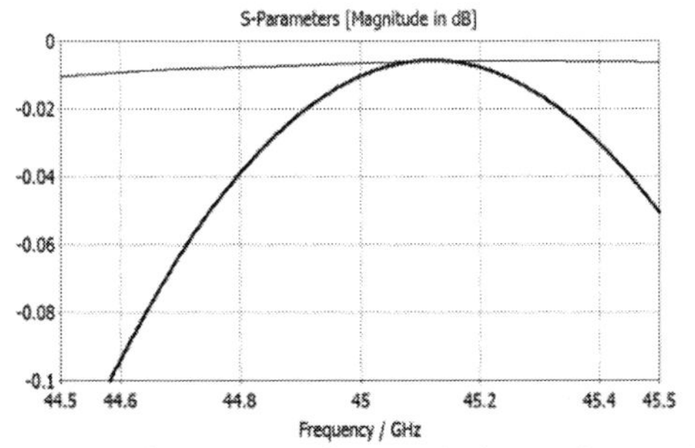

Fig. 2. CST Microwave Studio numerical test of conversion efficiency (FDTD solver): bold line corresponds to TE_{11}–TE_{01} conversion, thin line – to conservation of TE_{11} mode

The synthesized shape was split in two parts for manufacture and machined from copper. High-power measurements were done as a part of the complete setup. Output modes were observed using thermal paper and infrared imager. Thermal images presented in Fig. 3b illustrate overall efficiency of the transmission line, not only the mode converter. In spite of this, the level of spurious modes is very low.

Acknowledgements. Authors thank Dr. S. Mishakin for valuable discussions.

References

1. *Kerzhentseva, N.P.* Propagation of electromagnetic waves in bent circular waveguides // Radiotekhn. Elecktron. 1958. V. 3, No. 5. P. 649–659 [in Russian].

2. *Doane, J.L.* Propagation and mode coupling in corrugated and smooth-wall circular waveguides // *Infrared and Millimeter Waves V13: Millimeter Components and Techniques, Part IV,* edited by Kenneth J. Button. Orlando, FL: Academic Press. 1985. P. 123-170.

3. *Sobolev, D.I., Denisov, G.G.* Method for synthesis of wideband multimode waveguide elements // Radiophys. Quantum Electron. 2007. V. 50. No. 9. P. 720-725.

4. *Sobolev, D.I., Denisov, G.G.* Principles of Synthesis of Multimode Waveguide Units // IEEE Trans. Plasma Sci. 2010. V. 38. No. 10. P. 2825 – 2830.

Fig. 3. a) Fields in longitudinal cross-section calculated by CST Microwave Studio (longitudinal coordinate compressed for better viewing), b) High-power thermal images of fields at the output cross-section. Top figures correspond to conservation of TE_{11} mode, bottom – to TE_{11}-TE_{01} conversion

EPJ Web of Conferences **149**, 04018 (2017) DOI: 10.1051/epjconf/201714904018
10th International Workshop 2017 "Strong Microwaves and Terahertz Waves: Sources and Applications"

Review of the gyrotron theory

Gregory S. Nusinovich

Institute for Research in Electronics and Applied Physics, University of Maryland, College Park, MD 20742-3511, USA, gregoryn@umd.edu

This review is focused on presenting results of the studies of beam-wave interaction in high-power gyrotrons performed during last 15 years by the presenter and his co-authors. All results are gathered in three groups that can tentatively be entitled: "Megawatts and above", "Terahertz and beyond" and "Miscellaneous". All studies were performed using either non-stationary, self-consistent code MAGY [1], or self-consistent, non-stationary, multimode code based on equations formulated in Ref. [2], or using quasi-linear approach, in which the dependence of the electron current density on the intensity of the RF field is described by polynomials (see, e.g., [3]).

Megawatts and above. In this section the attention is paid, first, to the issue of mode interaction in MW-class gyrotrons operating in very-high-order modes. Results of MAGY simulations [4] illustrated by Fig. 1 demonstrate that the most dangerous are the modes with a certain frequency separation from the operating one. Since the most efficient is the operation in the regime of hard self-excitation, the most dangerous are the parasitic modes with a lower frequency located in the region of soft self-excitation, where start currents are small.

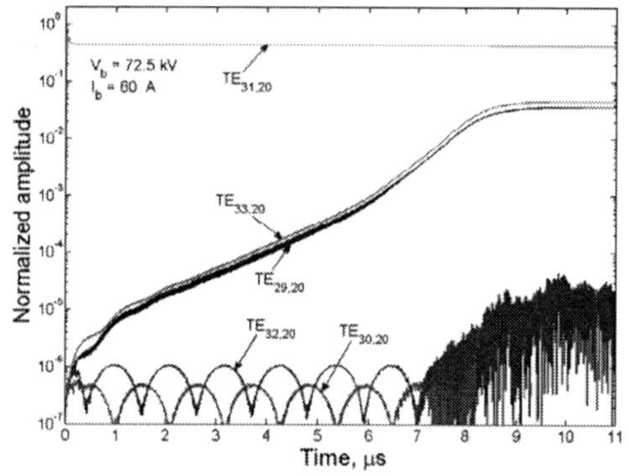

Fig. 1. Excitation of a pair of distant sidebands followed by excitation of a pair of close sidebands in a 170 GHz gyrotron operating in the $TE_{31,20}$-mode

The second issue requiring a careful consideration is the start-up scenario, i.e. analysis of methods of initial excitation of the desired mode followed by driving this mode into the region of the most efficient operation. As a rule, it requires passing through the region of soft self-excitation into the region of hard self-excitation. Such scenario in long-pulse and CW regimes can be realized by varying the external magnetic field, the cathode and mod-anode voltages and the beam current. (Original concept was formulated in Ref. [5]; the review of later studies was given in Ref. [6].) An example of passing through these regions is shown in Fig. 2.

Fast voltage rise. Typically, the cavity fill time is much shorter than the voltage rise time. Therefore it is always assumed that the amplitude of RF oscillations follows a slow voltage rise [5, 6]. As shown in Fig. 2, in this case the voltage rise in gyrotrons with diode-type guns will cause, first, oscillations of the parasitic mode with a lower frequency that can be dangerous for diagnostic channels used in plasma installations where the gyrotrons deliver the power. As shown in [7], the excitation of such parasitic modes can be avoided when the voltage rise is fast enough (Fig. 3).

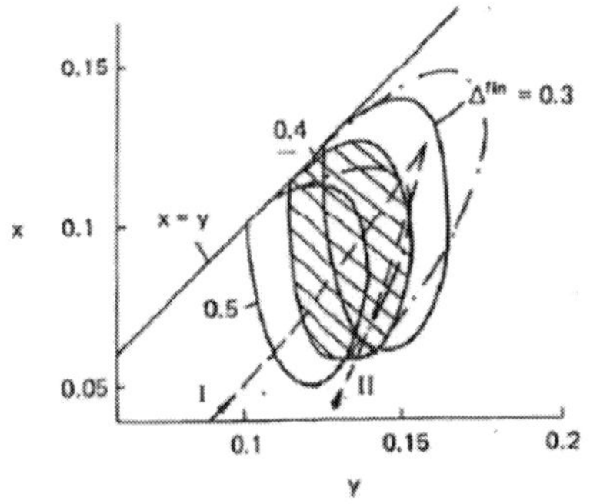

Fig. 2. Horizontal axis shows the total electron velocity squared proportional to the beam voltage; the vertical axis shows the orbital electron velocity squared (this velocity is proportional to the mod-anode voltage). The zone of self-excitation of the operating mode (dashed) is surrounded by similar zones of the high-frequency (lower voltages) and low-frequency (higher voltages) parasitic modes. Gyrotron trace I during the voltage rise corresponds to the diode-type electron gun; the trace II shows the case when during the voltage rise the difference between the beam and mod-anode voltages remains constant

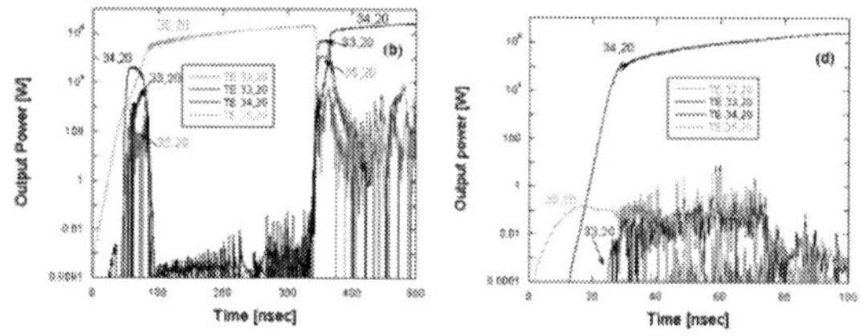

Fig. 3. Excitation of a 170 GHz FZK coaxial gyrotron. The voltage rise from 50 kV to 78 kV is equal to 500 ns in the left figure and 100 ns for the right figure

Slow processes. There are known, at least, three processes with typical times much greater than milliseconds: ion neutralization of DC space charge fields of electron beams, thermal expansion of the cavity due to the ohmic losses and, sometimes, beam current reduction due to emission cooling of the cathode. Some possibilities of mitigating these effects were analyzed in Refs. [8, 9] (see also references therein).

Terahertz and beyond. In this section we discuss, first, the role of such factors in electron beams that become more pronounced with the wavelength shortening, viz., the azimuthal nonuniformity of electron emission and the finite thickness of electron beams. The section also includes consideration of some possibilities of operating at high cyclotron harmonics allowing to increase the wave frequency at given magnetic fields.

Azimuthal nonuniformity. The azimuthal nonuniformity of electron emission reduces the efficiency of gyrotrons and also affects the mode interaction. As known (see, e.g. [3] and references therein) the interaction between two modes is purely amplitude when their frequency separation exceeds the width of their resonance curves ($|\omega_2 - \omega_1| > \omega / Q$) and their azimuthal indices are different ($m_2 \neq m_1$); similar conditions can also be written for three-wave interaction.

In the case of azimuthally nonuniform emission the electron current density can be represented as superposition of azimuthal harmonics; thus, certain harmonics can play a role in changing the nature of interaction between the modes. These issues were analyzed in Refs. [10, 11].

Beam thickness. The classical theory of gyrotrons is based on the assumption that an electron beam has no spread in radii of electron guiding centers (a thin annular electron beam). The finite spread causes efficiency degradation and may also affect the mode interaction in soft and hard excitation regimes [12]. For a 670 GHz gyrotron developed jointly by IREAP and IAP these effects were studied in Ref. [13].

A certain contradiction between requirements to the electron spread in guiding center radii and in velocities should be mentioned. For reducing the velocity spread it is beneficial to form quasi-laminar electron beams. However, the radial spread in such beams is relatively large. The trade-off between these requirements, which allows one to optimize the device efficiency, was studied in Ref. [14].

Operation at cyclotron harmonics. Gyrotron operation at cyclotron harmonics has a long history. In recent years, some issues important for operation at cyclotron harmonics were studied in [15–18].

Miscellaneous. This part of the presentation will be devoted to consideration of such effects as after-cavity interaction, efficiency and start currents in gyrotrons with tapered resonators, formation of the axial field structure in open resonators, effect of reflections, mode switching [19] and dynamics and stochasticity in gyrotrons with overlapping of cyclotron resonances [20].

References

1. *M. Botton, T. M. Antonsen, Jr., B. Levush et al.*. MAGY: A time-dependent code for simulation of slow and fast wave microwave sources// IEEE Trans. Plasma Sci., 1998,V. 26, No. 3, 882-892.

2. *Zavolsky N. A., Nusinovich G. S., Pavelyev A. B.*, Stability of single-mode oscillations and nonstationary processes in gyrotrons with oversized low-Q resonators // Gyrotrons, edited by V.A. Flyagin, G.S. Nusinovich, A.L. Goldenberg, V.K. Yulpatov, and N. A. Gorodetskaya, IAP AS USSR, 1989, P. 84-112.

3. *G.S. Nusinovich*, Mode interaction in gyrotrons//Int. J. Electronics, 1981, V. 51, No. 4, 457-474.

4. *O.V. Sinitsyn, G.S. Nusinovich and T.M. Antonsen, Jr.*, Stability of Gyrotron Operation in Very High-Order Modes// Phys. Plasmas, 2012, V. 19, 063114.

5. *G.S. Nusinovich*, Methods of voltage feeds for a pulsed gyromonotron which ensure high-efficiency in a single-mode operation //Elektron. Teknh., Elektron. SVCh, 1974, No. 3, 44-49.

6. *G.S. Nusinovich, O.V. Sinitsyn, L. Velikovich et al.*, Startup Scenarios in High-Power Gyrotrons// IEEE Trans. Plasma Sci., 2004, V. 32, No. 3, 841-852.

7. *G.S. Nusinovich, A.N. Vlasov, T.M. Antonsen, Jr., et al.*, Excitation of parasitic modes in gyrotrons with fast voltage rise//Phys. Plasmas, 2008, V.15, 103101.

8. *G.S. Nusinovich, O.V. Sinitsyn, T.M. Antonsen, Jr. et al.*, Slow processes in startup scenarios of long-pulse gyrotrons// Phys. Plasmas, 2006, V. 13, 083106.

9. *A. Fokin, M.Yu. Glyavin, and G.S. Nusinovich*, Effect of ion compensation of the beam space charge on gyrotron operation// Phys. Plasmas, 2015, V. 22, 043119.

10. *G.S. Nusinovich and M. Botton*, Quasilinear theory of mode interaction in gyrotrons with azimuthally inhomogeneous electron emission// Phys. Plasmas, 2001, V. 8, No. 3, 1029-1036.

11. *G.S. Nusinovich, A.N. Vlasov, M. Botton et al.*, Effect of the azimuthal inhomogeneity of electron emission on gyrotron operation// Ibid., No. 7, pp. 3473-3479.

12. *G.S. Nusinovich, O.V. Sinitsyn, M. Yeddulla et al.*, Effect of the radial thickness of electron beams on mode coupling and stability in gyrotrons// Phys. Plasmas, 2003, V. 10, No. 8, 3335-3343.

13. *R. Pu, G.S. Nusinovich, O.V. Sinitsyn, and T.M. Antonsen, Jr. et al.*, Effect of the thickness of electron beams on the gyrotron efficiency// Phys. Plasmas, 2010, V. 17, 083105; also, *R. Pu, G.S. Nusinovich, O.V. Sinitsyn, and T.M. Antonsen, Jr.*, Numerical study of efficiency for a 670 GHz gyrotron// Phys. Plasmas, 2011, V. 18, 023107.

14. *O. Dumbrajs and G.S. Nusinovich*, On optimization of sub-THz gyrotron parameters// Phys. Plasmas, 2012, V. 19, 103112.

15. *G.S. Nusinovich, D.G. Kashyn, and T.M. Antonsen, Jr.*, Harmonic gyrotrons operating in high-order symmetric modes// Appl. Phys. Lett., 2015, V. 106, 013502.

16. *G.S. Nusinovich, R. Pu, and V.L. Granatstein*, Suppression and nonlinear excitation of parasitic modes in harmonic gyrotrons// Appl. Phys. Lett., 2015, V. 107, 013501.

17. *M.Yu. Glyavin, A.G. Luchinin, V.N. Manuilov and G.S. Nusinovich*, Design os a dub-THz, Third-Harmonic, Continuous-Wave Gyrotron// IEEE Trans. Plasma Sci., 2008, V. 36, NO. 3, 591-596.

18. *M.Yu. Glyavin, N.A. Zavolsky, A.S. Sedov, and G.S. Nusinovich*, Low-Voltage Gyrotrons// Phys. Plasmas, 2013, V. 20, 033103.

19. *G.S. Nusinovich,O.V. Sinitsyn, and T.M. Antonsen, Jr.*, Mode switching in a gyrotron with azimuthally corrugated wall// Phys. Rev. Lett., 2007, V. 98, 205101.

20. *G.S. Nusinovich, R. Ngogang, T.M. Antonsen, Jr., and V.L. Granatstein*. Overlapping and stochasticity of electron trajectories in cyclotron masers// Phys. Rev. Lett., 2004, V. 93, 055101.

EPJ Web of Conferences **149**, 04019 (2017) DOI: 10.1051/epjconf/201714904019
10th International Workshop 2017 "Strong Microwaves and Terahertz Waves: Sources and Applications"

Simulation of gyrotrons using the high-order particle-in-cell code PICLas

S. M. Copplestone[1], P. Ortwein[1], C.-D. Munz[1], K. A. Avramidis[2] and J. Jelonnek[2]

[1]Institute of Aerodynamics and Gas Dynamics, University of Stuttgart, Stuttgart, Germany, copplestone@iag.uni-stuttgart.de
[2]Institute for Pulsed Power and Microwave Technologies, Karlsruhe Institute of Technology, Karlsruhe, Germany

Maxwell-Vlasov modelling of gyrotron devices offer deep insight into the relevant physics at nowadays moderate computational costs. The particle-in-cell (PIC) method with its long history [1,2] is a prominent example of kinetic methods for tackling such problems. Advantages of high-order Discontinuous Galerkin Methods [3] in combination with PIC schemes have since been developed [4] and are applied within the state-of-the-art code PICLas [5]. Recent improvements within this framework include high-order geometry approximation using curvilinear body-fitted instead of piecewise linear meshes and non-conforming interfaces with hanging nodes. Accurate PIC simulations of gyrotrons with open boundaries additionally require sophisticated modelling in the vicinity of open boundary conditions in order to prevent spurious reflections from corrupting the solution within the physical domain [6]. The simulation of a test-case involving a model 140 GHz $TE_{0,3}$ gyrotron [7] with and without the occurrence of dynamic after cavity interaction (ACI) is investigated regarding numerical settings and the identification of resonant dynamic ACI modes in the tapered part of the gyrotron geometry is shown via dynamic mode decomposition (DMD) [8]. It is shown that an inappropriate choice of specific parameters can lead to unfavourable results. For a more realistic application, a 140 GHz $TE_{28,8}$ gyrotron is simulated and the results are discussed.

Full-wave PIC simulations of gyrotron devices are conducted using PICLas, a code which combines high-order spatial and temporal discretisations by applying the discontinuous Galerkin spectral element approach in combination with curvilinear boundary handling. In PICLas, Maxwell's equations in purely hyperbolic form [9] are solved and the hyperbolic correction factor is set to unity throughout this work. Possible sources of numerical reflection at open boundaries, when particles or electromagnetic waves interact with the domain boundaries, are accommodated by absorbing layers, so-called perfectly matched layers (PML) [6].

The academic test-case of a model 140 GHz $TE_{0,3}$ gyrotron with two different external magnetic field profiles with either a constant or a spatially varying distribution is considered. For details on the parameters of the setup see [7]. Depending on the applied magnetic field, the occurrence of dynamic ACI is encountered on which, additionally, the applied boundary conditions take effect. For the PIC setup of the model gyrotron, a hexahedral curvilinear mesh with a geometrical order O(5) containing 2080 cells is used for a total length of the model of 120 mm. The boundary conditions at the entry and exit of the gyrotron are modelled open (first order absorbing) and the surface of the gyrotron tubing via a perfectly electric conducting (PEC) condition. The spatial and temporal order are O(6) and O(4), respectively. Time integration is done explicitly with a low storage Runge-Kutta scheme. The current beam is modelled by 85000 particles with a weighting factor of 10^5. Particle-grid coupling is carried out by using a polynomial shape function [4] and the spatial spread of each particle can be adjusted. The simulations are performed on 60 CPU, which uses ≈ 50 CPUh for 1 ns simulation time. The output power reaches 9.2 kW (averaged over several cycles) and is calculated by integration of the Poynting vector over a cross section of the gyrotron at 120 mm. PMLs with one extra layer of cells are installed at the entry and exit planes of the gyrotron. The additional cell layers either continue the taper angle of the outer radius in order to prevent a kink in the geometry or they have a cylindrical shape with a constant radius. The results are post-processed by means of DMD, for details on the algorithm see [11], for which 563 complete 3D solutions of the electric field are used, from 109.1 ns to 109.923381 ns with a sampling time-step of 0.2381 ps. The frequencies and growth rates found are shown in Fig. 1 for a non-uniform (case A) and uniform (case B) external magnetic field B_z=5.561 T with a cylindrical PML region where the damping is set to zero (small damping values yield no significant differences). The DMD reveals frequencies ω_i and growth rates ω_r of the fundamental mode at ≈ 140 GHz and its harmonics.

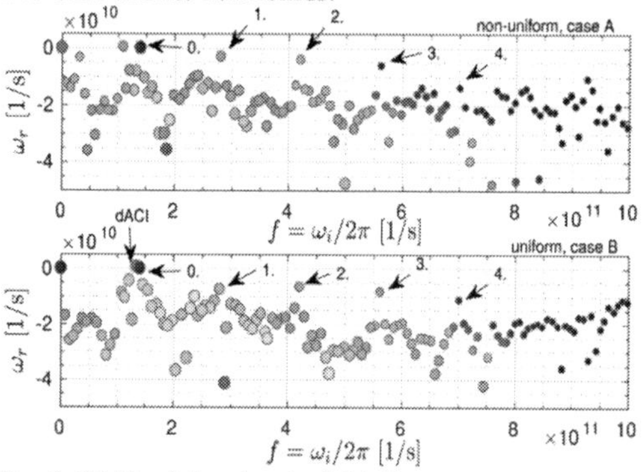

Fig. 1. DMD of the electric field: Frequencies and growth rates for the non-uniform (case A, top) and uniform (case B, bottom) external magnetic field. Note that the colour and size of the dots correspond to the Euclidean norm, correlating with the energy content, of each mode.

In case A, the fundamental frequency at 140.30 GHz is visible and no dynamic ACI is encountered. Note that the mode visible at 109 GHz does not represent any physically relevant structures. Case B shows two distinct frequencies near 140 GHz, the fundamental frequency at 140.30 GHz and the dynamic

ACI at 134.34 GHz. Additionally, the mode pattern of the DMD data of case B is depicted in Fig. 2 for the fundamental and dynamic ACI frequencies. The position in the tapered part of the gyrotron, where the dynamic ACI is excited and propagated along the gyrotron axis is clearly visible. Both the fundamental and dynamic ACI modes show a $TE_{0,3}$ structure.

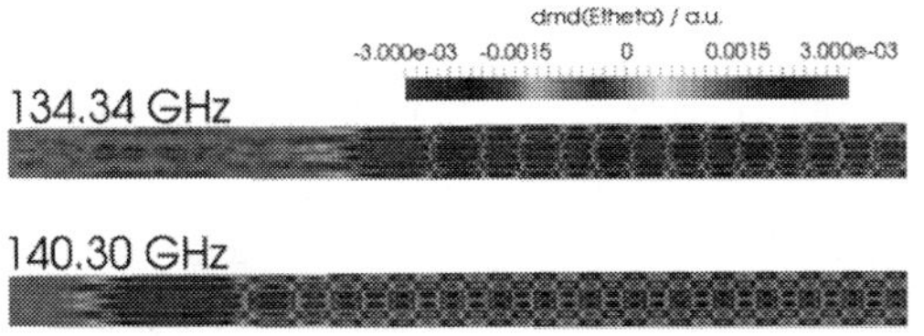

Fig. 2. DMD of the electric field in the axial x-y-plane of the gyrotron showing the resonant fundamental (bottom) and dynamic ACI mode (top) for case B (cylindrical PML).

To investigate the influence of reflection at open boundaries, a parameter variation of the damping factor ζ and PML outer radius tapering is conducted. The results are summarized in the table below. For low damping factors ($\zeta=5\cdot10^9$) and a tapered PML, the dynamic ACI frequency and output power slightly change as compared to the cylindrical PML geometry with no or small damping factors. This changes for larger damping values ($\zeta=50\cdot10^9$), where the presence of a dynamic ACI is not observed and the frequency of the fundamental mode slightly decreases. For very high damping factors ($\zeta=500\cdot10^9$, $\zeta=5000\cdot10^9$), the dynamic ACI frequency reduces by up to ≈ 2 GHz and for a cylindrical PML a shift from $TE_{0,3}$ to $TE_{2,3}$ structure is observed for the dynamic ACI mode. This may be linked to the heavy change in reflection properties at the exit plane of the gyrotron, which changes the backward scattering and excitation of certain modes in the uptaper of the gyrotron.

PML damping factor ζ, DMD frequency f, mode pattern and output Power for different cases. Note that all results refer to a uniform external magnetic field B_z.

ζ [10^9]	f [GHz]	TE Mode	P_{out} [kW]	PML radius	ACI
5	140.253	0,3	9.4	tapered	yes
	134.036	0,3			
50	140.035	0,3	9.7	tapered	no
500	140.366	0,3	10.4	tapered	yes
	133.473	0,3			
5000	140.30	0,3	10.0	const.	yes
	132.31	3,2			

In general, the results agree very well with findings reported in [7] for low and moderate damping factors in the PML. However, for larger values, we find increasing differences. As shown in [6], too high damping values attenuate the electromagnetic waves inside the PML too fast, hence, increasing the numerical reflection because the PML then functions comparable to a PEC surface, in which no fields exist due to the infinite conduction of the material.

The physical as well as the PIC parameters of the $TE_{28,8}$ gyrotron are chosen according to the case presented in [10] with a slightly increased length of the geometry due to two additional cell layers of PML regions and the entry and exit. The mesh has a length of 121 mm in z-direction, the largest diameter being 43.9 mm, and consists of 3472 hexahedral curvilinear mesh cells of geometrical order O(7). The spatial and temporal order of are O(10) and O(4), respectively. During stationary operation, there are ≈400000 particles with a weighting factor of 10^6. The external magnetic field of 5.587 T and electric current of 44 A of the electron hollow beam for soft excitation operation are used. At this point, the hard excitation parameters are used in order to shift into operation mode of the gyrotron. For this setup we observe the stable operation of $TE_{27,8}$ mode, which is in constrast to [10]. We also conducted different simulations with different grids and resolutions and observed the same result. We link these observations to the different numerical model with less reflection at the input and output planes due to the PML regions.

References

1. *C. K. Birdsall and A. B. Langdon*. Plasma Physics via Computer Simulation. Hilger, 1991.
2. *R. W. Hockney and J. W. Eastwood*. Computer Simulation Using Particles. Taylor & Francis, 1988.
3. *J. S. Hesthaven and T. Warburton*. Nodal Discontinuous Galerkin Methods: Algorithms, Analysis, and Applications. Texts in Applied Mathematics. Springer, 2008.
4. *G. B. Jacobs and J. S. Hesthaven*. High-order Nodal Discontinuous Galerkin Particle-in-cell Method on Unstructured Grids. In: J. Comput. Phys. 214.1 (2006), pp. 96–121.
5. *C.-D. Munz et al.* Coupled Particle-In-Cell and Direct Simulation Monte Carlo method for simulating reactive plasma flows. Comptes Rendus Mecanique 342.10-11 (2014), pp. 662–670.
6. *S. M. Copplestone et al.* Complex-Frequency Shifted PMLs for Maxwell's Equations With Hyperbolic Divergence Cleaning and Their Application in Particle-in-Cell Codes. In: IEEE Transactions on Plasma Science 45.1 (2017), pp. 2–14.
7. *K. A. Avramidis et al.* A comparative study on the modeling of dynamic after-cavity interaction in gyrotrons. In: Physics of Plasmas 22.5 (2015), p. 053106.
8. *P. J. Schmid*. Dynamic mode decomposition of numerical and experimental data, Journal of Fluid Mechanics 656 (2010), pp. 5–28.
9. *C.-D. Munz et al.* A three-dimensional finite-volume solver for the Maxwell equations with divergence cleaning on unstructured meshes. Computer Physics Communications, 130.1–2 (2000), pp. 83–117.
10. *A. Stock*. A High-Order Particle-in-Cell Method for Low Density Plasma Flow and the Simulation of Gyrotron Resonator Devices. PhD thesis, University of Stuttgart, 2013.
11. *O. Schmidt*. Numerical investigations of instability and transition in streamwise corner-flows. PhD thesis, University of Stuttgart, 2014.

EPJ Web of Conferences **149**, 04020 (2017) DOI: 10.1051/epjconf/201714904020
10th International Workshop 2017 "Strong Microwaves and Terahertz Waves: Sources and Applications"

Benchmarking a high-order particle-in-cell code for the simulation of a gyrotron traveling-wave tube

P. Ortwein[1], S.M. Copplestone[1], C.-D. Munz[1], A. Marek[2] and J. Jelonnek[2]

[1]Institute of Aerodynamics and Gas Dynamics, University of Stuttgart, Stuttgart, Germany, ortwein@iag.uni-stuttgart.de
[2]Institute for Pulsed Power and Microwave Technologies, Karlsruhe Institute of Technology, Karlsruhe, Germany

Introduction

In recent years, several high-order methods have been proposed for the simulation of Maxwell's equations. Discontinuous Galerkin (DG) schemes are a combination of an element local finite element formulation in which the solution is represented by a polynomial and a finite volume method which couples adjacent cells over the cells surfaces. This results in a highly parallelizable scheme which offers excellent dissipation-dispersion properties. Consequently, a high-order DG scheme has been applied within a particle-in-cell solver [1], replacing the standard Yee scheme. This development has been continued, leading to a three-dimensional full-wave particle-in-cell solver named PICLas, which has been benchmarked against a 30 GHz gyrotron operating in a $TE_{2,3}$ mode [2]. In contrast to finite difference methods, which use equidistant grids and a second order boundary approximation, PICLas uses body-fitted, curvilinear unstructured hexahedral meshes of arbitrary polynomial order. A crucial step to simulate complex geometries, like the considered gyro-TWT [3, 4], is a high quality mesh. A typical DG grid consists of coarse elements when high-order elements are used. By increasing the polynomial degree, less than one DG element per wavelength may be required for an accurate solution. This is especially challenging for the considered gyro-TWT, because its diameter approximates the wavelength. In order to represent the geometry, several elements in radial direction may be required, thus restricting the DG scheme to a lower order. We describe the utilized mesh and report preliminary results for which a homogeneous guiding magnetic field has been used.

Mesh Generation

The outer radius of the corrugation of the gyro-TWT [3] is given by

$$r(z,\varphi)=r_0 + \tilde{r}\cos(m\varphi\text{-}2\pi/d\ z), \qquad (1)$$

with r_0 = 4 mm being the mean radius, $\tilde{r}$ = 0.74 mm the amplitude, m = 3 the azimuthal number and d = 9.2 mm the axial periodic length. The corrugations are 15d long and the transition from the corrugation to two circular waveguides consists of two linear ramps of length 2d. Additionally, perfectly matched layers (PML) [2] are added at the output with a length of d.

We use our open-source mesh generator HOPR [5] to generate body-fitted, non-equidistant curved meshes of fifth order. A quality condition like the scaled Jacobian has to be fulfilled [6]. The scaled Jacobian is the smallest Jacobian of an element divided by its largest Jacobian. A value of unity indicates a cubical element and for smaller values the mesh quality decreases. A negative value indicates an invalid element. The computational grid-lines are twisted to follow the helical structure and to improve the scaled Jacobians. A picture of the mesh is shown in Fig. 1. The most inner elements are cubical. The adjacent elements are curved to blend linearly from a circle to the three-fold corrugation.

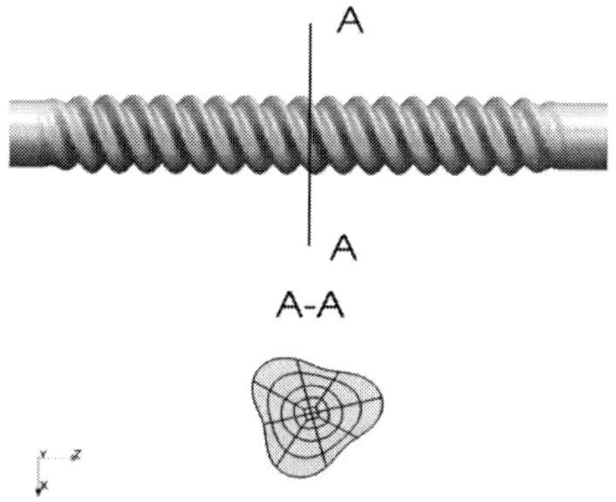

Fig. 1. The picture depicts geometry and is scaled in axial direction. The slice depicts the grid lines in the x-y plane in the center of the corrugations

Fig. 2 depicts the smallest found scaled Jacobians of the considered mesh in the range between 0.25 and 0.4. The small values indicate strong deformation of the elements. The lowest values are found at the inflection point of the corrugations between adjacent periods and at the interface between the cubical and curvilinear elements. In these regions, however, the scaled Jacobians are still sufficient to ensure a high quality solution.

Simulation Results

The simulations are conducted with our in-house PIC code PICLas. In contrast to the setup reported in [3] we simulate a simplified amplifier setup consisting of perfectly electric conducting boundaries and a homogeneous and constant guiding magnetic field. The solution is represented by a sixth order polynomial in 6336 elements. Additionally, we apply a complex frequency-shift PML at the end of the geometry with $\zeta = 350\cdot 10^9$ and $\alpha = 35\cdot 10^9$. The $TE_{1,1}$ mode is inserted at the input plane (z = 0 mm) with a power of 2 kW. The electrons are inserted with a beam current of 10 A, an energy of 70 keV and a pitch factor of 1.2. The particle weight is chosen to 10^7, resulting in about 15000 particles within the domain. A decreased particle weight of 10^6 leads to similar results, signifying that the value of 10^7 is sufficient.

We observe a high dependency regarding the simulation parameters, guiding magnetic field, beam current and voltage. The highest output power is encountered for B_z = 0.67 T. This corresponds to an output power of 100 kW and an amplification of ~17 dB. In the literature, a magnetic field of $B_z = 0.663$ T for maximum amplification of 17 dB has been found [3]. Using this guiding magnetic field, we found an amplification of ~15.6 dB.

In order to save computational resources, we considered a non-conforming mesh with reduced resolution away from the particles location. The resulting mesh consists of 3556 elements (reduction $\approx$ 50%) and no reduction in solution quality is observed.

Fig. 3 depicts the instantaneous field and particle distribution of the upper half of the gyro-TWT. The E_y-field is shown in the x-z-plane. Additionally, for each particle its Lorentz factor is shown. Inside of the corrugations, a strong bunching of the particles to the TE-mode is observed.

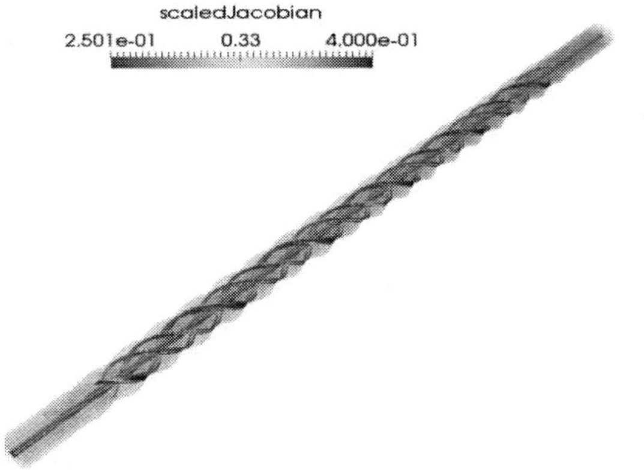

Fig. 2. The body-fitted mesh of the simulated gyro-TWT. The scaled Jacobians are depicted in the range between 0.25 and 0.4

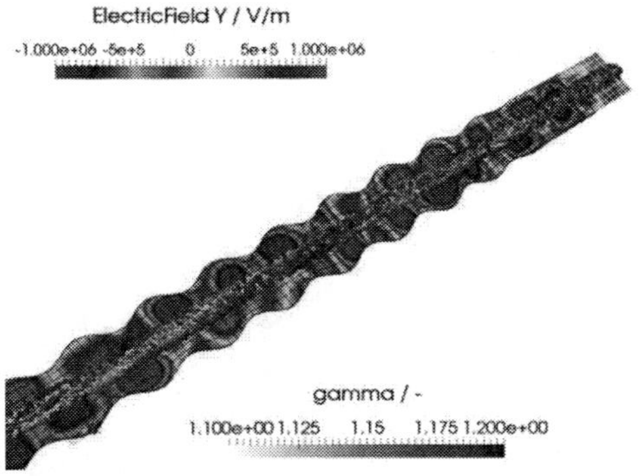

Fig. 3. Instantaneous electric field and particle distribution. Especially, a strong bunching of the particles within the corrugations is observed

Finally, a dynamic mode decomposition (DMD) [7] of the simulation results (conform mesh) is performed in a post-processing step. The DMD gives the growth and decay rates ω_r of the most dominant eigenvalues ω_i. Additionally, no a priori knowledge of the encountered modes is required. In Fig. 4 the results of the DMD is shown for 1200 snapshots with a sampling step of $\Delta t = 10^{-14}$ starting at 40 ns. The Euclidean norm of the eigenvalues is color coded, red indicating the highest amplitude. The signals at 1 and 4 are damped and match a frequency of 21.2 GHz and 42.1 GHz, a $TE_{1,1}$ and $TE_{0,1}$ mode inside the corrugations, respectively. The second and third peaks are found at 34.9 and 34.5 GHz correspond to a $TE_{1,2}$ and $TE_{1,1}$ mode.

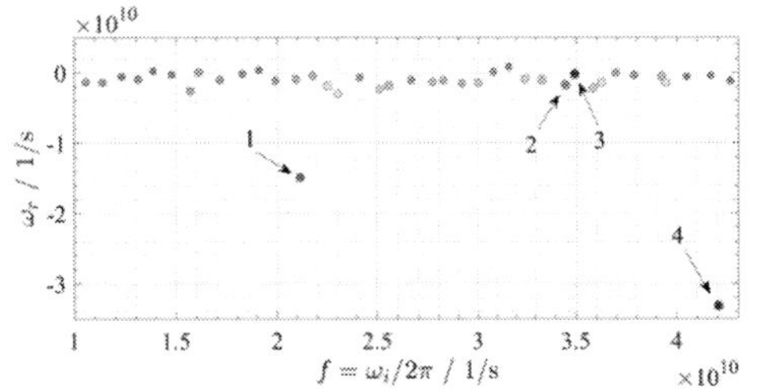

Fig. 4. Frequencies ω_i and growth rate found in a DMD. The signals with strongest amplitude are 1–4, while 2–4 are close to the inserted TE-mode at 35 GHz. Please note, that the color and size of the dots depend on the Euclidean norm

Summary

We present the first simulations of a gyro-TWT by a DG based PIC scheme with a body-fitted mesh. A high sensitivity of the operational parameters were encountered for all simulations. Here, we find a maximum amplification of the input signal by 17 dB for a constant guiding magnetic field of $B_z = 0.67$ T. By switching to a non-conforming mesh, we could reproduce the results by saving 50% of the elements. This research was supported by the DFG grants MU 1319/21-2, SFB7163 B08 and joint RSF-DFG 16-42-01078–Je 711/1-1. The authors thanks the HLRS Stuttgart for granting the computational resources.

References

1. *G.B. Jacobs, J.S. Hesthaven.* High-order Nodal Discontinuous Galerkin Particle-in-cell Method on Unstructured Grids // J. Comput. Phys., 2006, 214, P. 96-121.

2. *S.M. Copplestone, P. Ortwein, C.-D. Munz.* Complex-Frequency Shifted PMLs for Maxwell's Equations With Hyperbolic Divergence Cleaning and Their Application in Particle-in-Cell Codes // IEEE Trans. Plasma Sci., 2017, 45(1), 2-14

3. *N.S. Ginzburg, R. M. Rozental, A. S. Sergeev, I. V. Zotova and V. P. Tarakanov.* Chaotic millimeter wave generation in a helical-waveguide gyro-TWT with delayed feedback // Phys. Plasmas, 2016, 23(10).

4. *S.V. Samsonov, I.G. Gachev, G.G. Denisov, A.A. Bogdashov, S.V. Mishakin, A.S. Fiks, E.A. Soluyanova, E.M. Tai, Y.V. Dominyuk, B.A. Levitan and V.N. Murzin.* Ka-Band Gyrotron Traveling-Wave Tubes With the Highest Continuous-Wave and Average Power // IEEE Transactions on Electron Devices, 61.12, 2014, pp. 4264-4267.

5. *F. Hindenlang, T. Bolemann and C.-D. Munz.* IDIHOM: Industrialization of High-Order Methods – A Top-Down-Approach // Springer, Berlin, 2015, p.133-152.

6. *P.M. Knupp.* Algebraic Mesh Quality Metrics // SIAM J. of Scientific Computing, 23.1. 2001, p. 193-218.

7. *P.J. Schmid.* Dynamic mode decomposition of numerical and experimental data // Journal of Fluid Mechanics 656, 2010, pp. 5–28.

EPJ Web of Conferences **149**, 04021 (2017) DOI: 10.1051/epjconf/201714904021
10th International Workshop 2017 "Strong Microwaves and Terahertz Waves: Sources and Applications"

Influence of mode competition and external wave frequency modulation on gyrotron frequency locking

Yu.V. Novozhilova, V.L. Bakunin, A.V. Chirkov, Yu.M. Guznov, G.G. Denisov, A.P. Fokin, A.S. Shevchenko, N.I. Zaitsev, S.A. Zapevalov

IAP RAS, Nizhny Novgorod, Russia, julia.novozhilova2009@yandex.ru

1. Introduction

For a number of practical applications, such as plasma diagnostics or high-resolution spectroscopy, and in prospect for development of the coherently radiating gyrotron complex for large CTS setups, it is important to have a source of radiation with a highly stable frequency. Locking of gyrotron oscillation frequency by an external monochromatic wave seems to be a promising way for obtaining such source. Although in the recent decades influence of the external monochromatic signal on the gyrotron operation was studied in a number of publications [1–6], specific designs of megawatt power level gyrotrons have not been discussed. Study of frequency stabilization by a wave coming into the cavity from the external tract became of particular interest since the development of the quasi-optical converter which allows to transform a large part of wave coming from the output section into the gyrotron operating mode at the Institute of Applied Physics RAS (IAP RAS) [7]. The aim of this paper is to determine the optimal parameters of the gyrotron and of the external signal for efficient single-mode generation in the multimode gyrotron at a stable frequency. Accordingly, we consider a number of problems of multimode gyrotron frequency locking by an external signal.

2. Influence of the mode competition on the locking regime in the multimode gyrotron

We study frequency locking of a multimode gyrotron by an external monochromatic signal with a frequency close to the eigenfrequency of the operating mode. We use the approximation of fixed longitudinal structure of the field, which is valid at sufficiently high Q-factor of gyrotron cavity. This approximation allows us to study competition of several modes, including non-equidistant modes with frequency detuning comparable to the cyclotron absorption linewidth. Locking was simulated based on the example of a megawatt gyrotron prototype for ITER developed at the IAP RAS operating at TE28.12 mode at a frequency of 170 GHz with a quality factor of 1370; accelerating voltage at the end of switching stage was about 90–100 keV with a current of 50–60 A. Scenario of electron beam parameters switching was taken similar to the real one, which is used to achieve high efficiency.

Locking zones, i.e., regions where the operating mode is excited at the frequency of the external signal, were determined on the planes of various parameters: current and detuning $\Delta=\left(2/\beta_\perp^2\right)\left(1-\omega_H/\omega_0\right)$ (dimensionless detuning between gyro-frequency and eigen frequency of the operating mode and current), as well as detuning Δ and frequency of the external signal (Fig. 1, 2). We show that the width of the locking zones and maximum achievable current in the multimode gyrotron demonstrate non-monotonous dependence on the the external signal power, which can be explained by nonlinear mode competition (Fig. 1b). When the operating mode frequency of a multimode gyrotron is locked, much higher efficiency can be achieved and the frequency tuning band increases by several times in comparison to a free gyrotron (Fig. 1, 2).

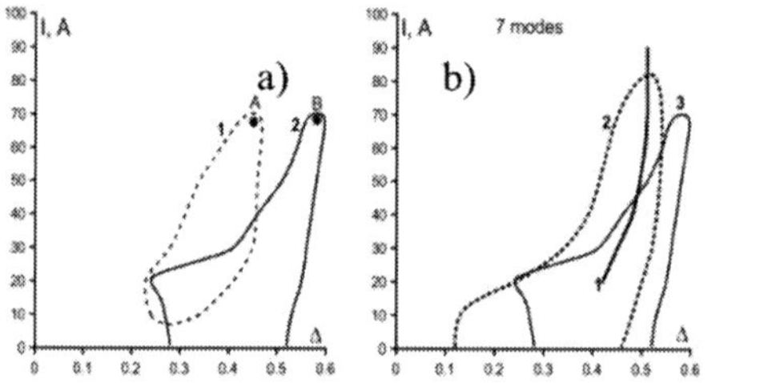

Fig. 1. Region of single-frequency generation at $TE_{28.12}$ operating mode in free gyrotron (curve 1, Fig. 1a) and locking zones (curve 2, Fig. 1a and Fig. 1b) in 7-mode gyrotron (parasitic modes $TE_{26.12}$, $TE_{27.12}$, $TE_{29.12}$, $TE_{30.12}$, $TE_{-25.13}$, $TE_{-26.13}$). 1a) external signal power is 5% of radiation power, maximum efficiency is achieved for free gyrotron at point A (63%) and for locked gyrotron at B (82%). Locking zones in Fig.1b correspond to different values of external signal power: 1) 0.01% of radiation power, 2) 0.5%, 3) 5%

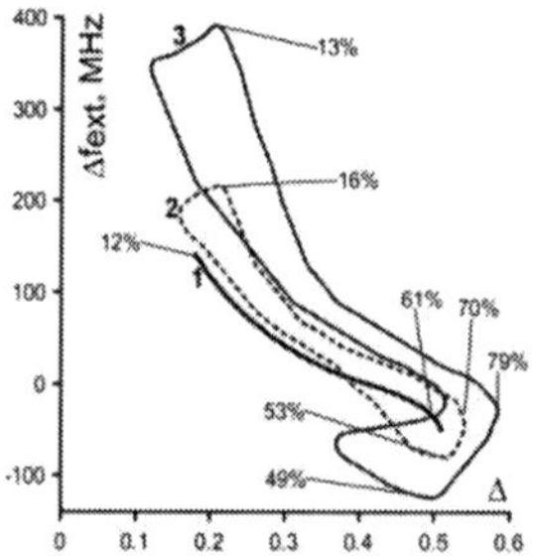

Fig. 2. Locking zones of 7-mode gyrotron with different level of external signal power at I = 50A: 1) free gyrotron, 2) external power is 1% of radiation power, 3) 5%

3. Frequency locking in the multimode gyrotron by a signal with varying frequency

We study the frequency locking of the multimode gyrotron by an external signal with a variable frequency close to the operating mode frequency. The radiation frequency follows the frequency of the external signal when the difference between the external signal frequency and free oscillations frequency does not exceed the locking bandwidth and the locking time is smaller than the characteristic time of external signal frequency variations. When the first condition is violated, beats between free oscillations and external signal occur leading to decay into neighboring modes (fig. 3). If the locking time is longer than the characteristic time of external signal frequency variations, the radiation frequency does not track the variations of external signal frequency and oscillates in an essentially smaller interval.

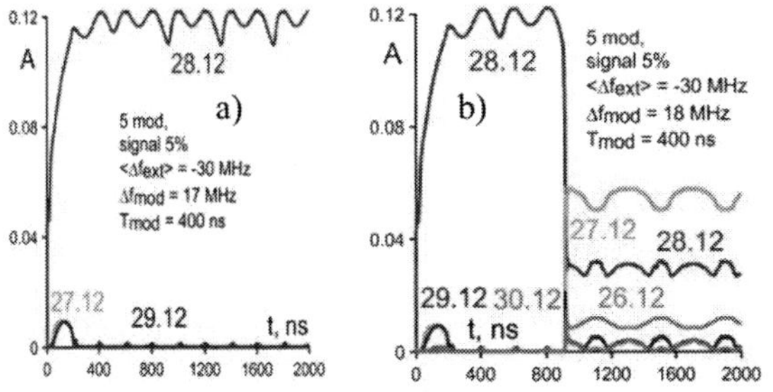

Fig. 3. Oscillograms of mode amplitudes in multimode gyrotron at a frequency of 170 GHz, $I = 60$ A, $\Delta = 0.58$, under influence of a signal with a frequency harmonically varying in time; $\langle \Delta f_{ext} \rangle$ is the difference between mean external frequency and 'cold' frequency of the operating mode, T_{mod} is the period and Δf_{mod} is the amplitude of external frequency modulation; locking bandwidth is 35 MHz; a) locking of operating mode frequency at $\Delta f_{mod} = 17$ MHz; b) decay of operating mode into parasitic ones at $\Delta f_{mod} = 18$ MHz

4. Experimental results

Locking of the gyrotron frequency by the signal of a magnetron was studied experimentally. In these experiments, the gyrotron represented an output section of a gyroklystron operating as an autooscillator at the frequency 35 GHz. The magnetron signal came into the gyrotron cavity through synthesized mode converter [7] (Fig. 4). Such a scheme permits to lock the frequency of megawatt power gyrotron unlike to the schemes used in previous experiments for low-power gyrotron with direct input of an external signal [3, 4]. Dependence of the locking bandwidth on the power of the magnetron signal was determined (Fig. 5). Experimental data are in a good agreement with the results of numerical simulations.

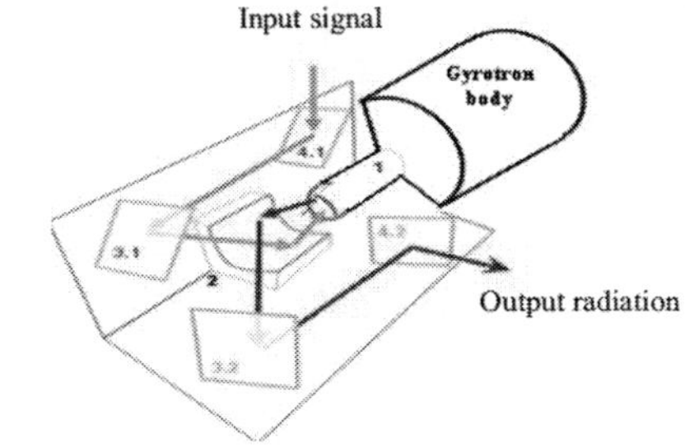

Fig. 4. Experimental setup scheme, 1 – launcher, 2 – quasi-parabolic mirror; 3.1, 4.1 – matching input mirrors; 3.2, 4.2 – matching output mirrors

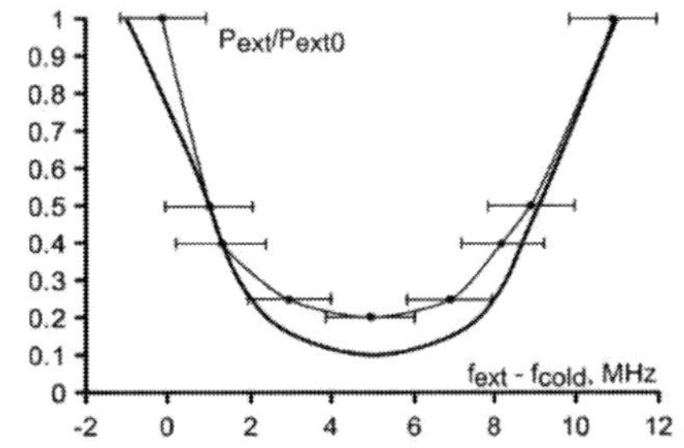

Fig. 5. Boundaries of locking band depending on the ratio between power of the magnetron signal and its maximum power. Thick curve corresponds to simulation, thin one corresponds to experiment. Maximum magnetron power level was 1/36 of gyrotron radiation power (900 kW)

5. Conclusion and acknowledgments

Influence of external monochromatic or narrow-band wave is a promising method for frequency stabilization of gyrotron.

This work was supported by RFBR grant N 15-02-01798A. We are grateful to A. N. Kuftin for useful discussions.

References

1. *Ergakov V.S., Moiseev M.A.* To the the theory of synchronization of CRM-monotron // Journal of Communications Technology and Electronics. 1978. N 12. P.2591-2599.

2. *Nusinovich G.S.* Introduction to the Physics of Gyrotrons. Baltimore: The Johns Hopkins University Press. 2004. 335p.

3. *Read M.E., Seeley R.,.Manheimer W.M.* Observations of phase locking in a single-cavity gyrotron oscillator // IEEE Trans. On Plasma Science. 1985. V. PS-13, N.6. P.398-403.

4. *Guo H., Hoppe D.J., Rodgers J., Perez R.M., Tate J.P., Conroy B.L., Granatstein V.L., Bhanii A.M., Latham P.E., Nusinovich G.S.* Phase-Locking of a Second-Harmonic Gyrotron Oscillator Using a Quasi-Optical Circulator to Separate Injection and Output Signals // IEEE Trans. On Plasma Sci. 1995. V.23, N.5. P.822-832.

5. *Ginzburg N.S., Glyavin M.Yu., Zavol'skii N.A., Zapevalov V.E., Moiseev M.A., Novozhilova Yu.V.* A Proposal to Use Reflection with Delay for Achieving the Self-Modulation and Stochastic Regimes in Millimeter-Wave Gyrotrons // Technical Physics Letters. 1998. V. 24, Issue: 6. P. 436-438

6. *Dumbrajs O., Idehara T., Watanabe S., Kimura A., Sasagawa H., Agusu L., Mitsudo S., Piosczyk B.* Reflections in Gyrotrons With Axial Output. // IEEE Transactions on Plasma Science. 2004.V. 32, Issue 3. P.899-902.

7. *Chirkov A.V., Denisov G.G., and Kuftin A.N.* Perspective gyrotron with mode converter for co- and counter-rotation operating modes. // Applied Physics Letters 106, 263501 (2015); doi: 10.1063/1.4923269.

EPJ Web of Conferences **149**, 04022 (2017) DOI: 10.1051/epjconf/201714904022
10th International Workshop 2017 "Strong Microwaves and Terahertz Waves: Sources and Applications"

High precision frequency stabilization of a 263 GHz continuous wave gyrotron

G.G. Denisov, A.P. Fokin, M.Yu. Glyavin, G.Yu. Golubiatnikov, L.V. Lubyako, M.V. Morozkin, B.Z. Movschevich, A.I. Tsvetkov

Institute of Applied Physics of the Russian Academy of Sciences, Nizhny Novgorod, Russia, ap.fokin@mail.ru

Millimeter and submillimeter wave gyrotrons with medium (kW level) power got their applications as sources for dynamic nuclear polarization/nuclear magnetic resonance (DNP/NMR) spectrometry [1, 2] and various media diagnostics [3]. One of the limiting factors for gyrotrons applications is the relatively low frequency stability and wide spectrum of the output radiation. The solution of this problem can increase the sensitivity and resolution in spectroscopy and open new prospects, such as the coherent summation of radiation of multiple gyrotrons or possibilities of suppressing hydrodynamic instabilities in plasma. The known experiments on frequency stabilization of gyrotrons utilize phase lock control to one of the electrodes of the magnetron injection gun, show that the relative half width of the frequency spectrum $\Delta f/f$ and the frequency fluctuations $\delta f/f$ can be lower than $1\cdot10^{-9}$ and $3\cdot10^{-10}$ respectively [4, 5].

Experimental setup. In our experiment, we applied the phase lock loop (PLL) against the reference generator in order to control the gyrotron anode voltage. Selection of anode voltage alteration as a way of frequency control and stabilization is caused by small capacity of anode relative to other electrodes and low anode current, so there is no need in complex and expensive power supplies. The change of anode voltage does not alter the electron energy, but only its pitch-factor (ratio of its orbital to longitudinal velocity), leading to change of the active and reactive components of conductivity and, hence the change of the operating frequency of the gyrotron.

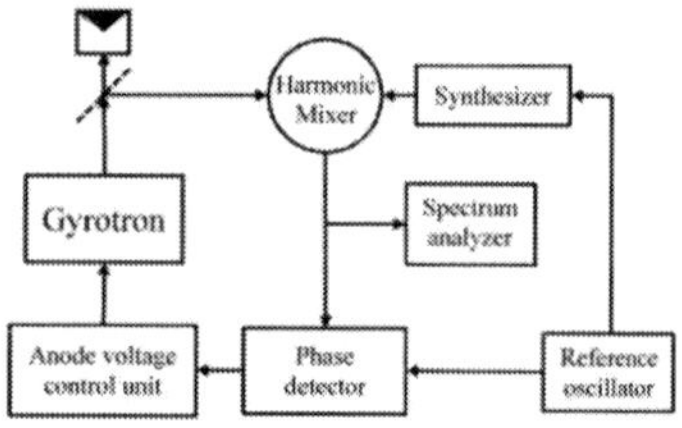

Fig. 1. Block diagram of phase lock loop stabilization system for a gyrotron frequency

The experiment was carried out on the continuous-wave (CW) gyrotron for spectroscopy and various media diagnostics with operating frequency of 263 GHz and output power up to 1 kW in Gaussian beam [6]. For the experiment, the electron beam current was limited at 200 mA and magnetic field was changed for the purpose of lowering output power to about 40 W. The accelerating voltage circuit was modified in order to reduce the high-frequency components of voltage ripple by means of low-pass LC-filter. The fast voltage control unit was installed in the anode voltage circuit and was used as an active element of phase lock control system. This unit allowed variation of anode voltage in range of anode voltage with amplitude up to 1 kV with slopes better than 1 kV/μs. The active element of the unit is the thyratron, that can be controlled by external signal.

The block scheme of the phase lock loop system is presented on the Fig. 1. The output radiation of the gyrotron with frequency 263 GHz is partially transmitted to the harmonic mixer, where it is mixed with the signal from the microwave synthesizer with frequency about 4.8 GHz. The resulting signal on the intermediate frequency of 350 MHz is then fed on the phase detector in order to frequency phase compare it with the signal from reference oscillator. The error signal from phase detector is used as the control signal for anode voltage control unit.

Results of the experiment. Our experimental setup allowed to control frequency of gyrotron radiation with modulation frequency up to 150 kHz. The sensitivity of the gyrotron was 1 MHz per volt of control signal on the anode voltage control unit. The full frequency tuning range was 7 MHz. The width of the frequency spectrum was decreased from 0.5 MHz [6] down to 1 Hz measured at intermediate frequency IF = 350 MHz, which corresponds $\Delta f/f = 3\cdot10^{-12}$ with measurement time of a few seconds (see Fig. 2). The low level of phase noise should be noted.

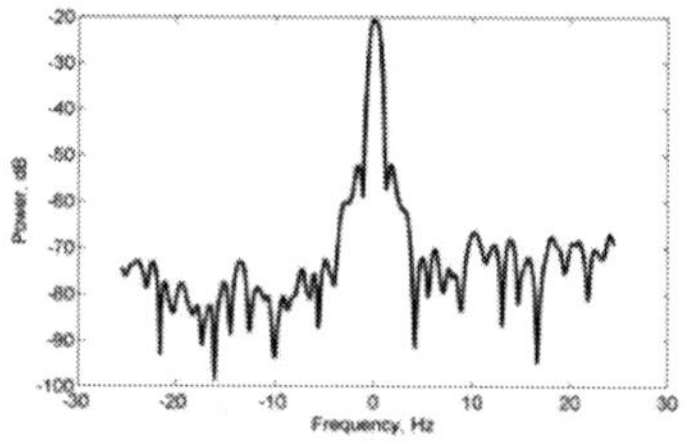

Fig. 2. Observed frequency spectrum of the gyrotron with phase lock loop at the intermediate frequency

The long-term frequency fluctuations were defined by the stability of reference oscillator ($\delta f/f \sim 10^{-9}$ for quartz clock and up to $\delta f/f \sim 10^{-12}$ for rubidium clock). Obtained spectrum width and frequency stability were previously achieved in backward-wave oscillators, utilized in spectroscopy, but with power levels of tens of milliwatts, while stabilization system with anode voltage control in gyrotrons has no apparent limiting factors in the field of gyrotron output power – as it was demonstrated with output power of 40 W.

This work was supported by Russian Scientific Fund (RSCF), grant 14-12-00887.

1. *Emilio A. Nanni, Alexander B. Barnes, Robert G. Griffin, Richard J. Temkin*, THz Dynamic Nuclear Polarization NMR, IEEE Transactions on Terahertz Science and Technology, 2011, V1, Issue: 1

2. *Blank M, Rosay M, Engelke F* Instrumentation for solid-state dynamic nuclear polarization with magic angle spinning NMR J. Magn. Reson., 2016, 264 88

3. *M.A.Koshelev, A.I.Tsvetkov, M.V.Morozkin, M.Yu.Glyavin, M.Yu.Tretyakov*, Molecular gas spectroscopy using radioacoustic detection and high-power coherent subterahertz radiation sources, Journal of Molecular Spectroscopy, 2017, 331, 9–16

4. *Idehara T., Mitsudo S., Ogawa I.*, Development of High-Frequency, Highly Stable Gyrotrons as Millimeter to Submillimeter Wave Radiation Sources, IEEE Transactions on Plasma Science, 2004. V. 32. №3. P. 910-916.

5. *G.Yu. Golubiatnikov, A.F. Krupnov, L.V. Lubyako, A.G. Luchinin, A. B. Pavelyev, M. I. Petelin, A. Fernandez Curto*, Gyrotron frequency control by a phase lock system, Technical Physics Letters, 2006, Volume 32, Issue 8, pp 650–652

6. *Glyavin M Yu et al.*, Experimental tests of a 263 GHz gyrotron for spectroscopic applications and diagnostics of various media Rev. Sci. Instrum., 2015, 86 054705.

EPJ Web of Conferences **149**, 04023 (2017) DOI: 10.1051/epjconf/201714904023
10th International Workshop 2017 "Strong Microwaves and Terahertz Waves: Sources and Applications"

Progress in microwave to sub-THz sources at Strathclyde

A. D. R. Phelps

Physics Department, SUPA, University of Strathclyde, Glasgow G4 0NG, Scotland, UK
a.d.r.phelps@strath.ac.uk

Research progress during recent years at the University of Strathclyde is presented. The main research themes include gyro-devices with particular emphasis on gyro-TWAs [1,2] and gyro-BWOs, and natural sources of radiation such as auroral kilometric radiation (AKR) [3-6], novel Cherenkov mm-wave sources [7,8] and pseudospark-driven sub-THz sources [9-11].

While there has been some theoretical work, the majority of the research has involved either laboratory experiments, or numerical simulations.

Gyro-devices

Most of the work on gyro-devices has involved helical waveguide interaction structures [1,2]. This interest began in the 1990's stimulated by collaborative research with IAP. The early experiments were in the 10GHz frequency range, whereas our more recent experiments have been in the 90 to 100 GHz range. The applications for these mm-wave amplifiers include DNP-NMR, cloud radar and communications, etc.

Gyro-TWA design target	
Frequency band	W-band
Centre frequency	95 GHz
Bandwidth	90 – 100 GHz
Power	~5 kW
Gain	~40 dB
Efficiency	10%
Microwave output	Fundamental Gaussian mode

Fig. 1. Gyro-TWA, 90 – 100 GHz

The gyro-TWA shown in Fig. 1 has produced 3.4kW output power. The limit on the output power is presently the maximum power that is available at the input to the amplifier.

Natural sources of EM radiation

Previous research on laboratory gyro-devices and electron cyclotron masers has provided the background for investigations of the mechanism of some naturally occurring sources of electromagnetic radiation. A research programme stretching over more than 10 years has explored the detailed mechanism of auroral kilometric radiation (AKR) and similar emissions emanating from other astrophysical bodies. Laboratory experiments to simulate the natural phenomena, combined with theoretical analysis and numerical modelling have been compared with natural observations. Using all four approaches it has been possible to understand how AKR arises and the mechanism by which the energy in electron streams can be converted with 1% or 2% efficiency into AKR. The mechanism depends upon the existence of an anisotropy in velocity space. This anisotropy is created when electrons are spiralling down the magnetospheric field lines. The magnetic moment is an adiabatic invariant of the electron motion and as the electrons penetrate further into the increasing magnetic field the electrons increase the perpendicular component of their velocity and decrease the component of their velocity parallel to the magnetic field lines. This results in a horseshoe shaped particle distribution in velocity space. The positive value of the gradient with respect to the perpendicular velocity component of the distribution function drives this kinetic instability that is now known as the "horseshoe instability". The theoretical analysis, the PiC simulations, the laboratory experiment and the space observations all agree.

Cherenkov sources

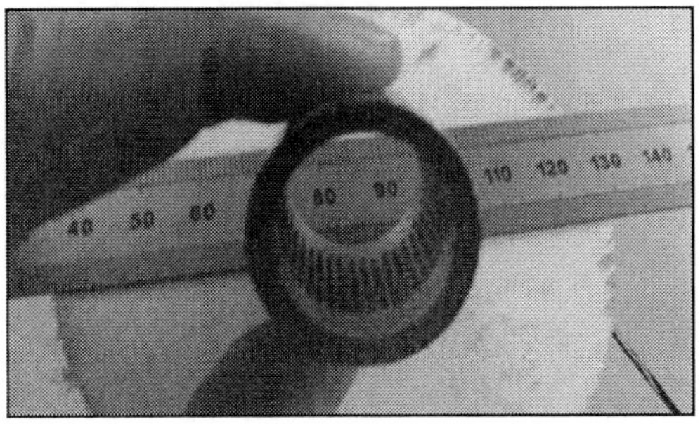

Fig. 2. Cylindrical over-sized Cherenkov structure

One of the aims of the research on Cherenkov sources is to retain reasonably high output power levels as the frequency increases. Traditionally Cherenkov sources have followed the trend that as the frequency increases and the wavelength decreases the interaction structures shrink to retain low order mode operation and the power handling capacity decreases as a result. Methods of selecting and driving modes in highly overmoded structures have been explored. Coupling of modes to form an unique eigenvalue is one of the methods that has been shown to be promising to achieve efficient excitation of a single high power mode in an overmoded structure.

Ginzburg et al. [12] have successfully demonstrated the coherent summation of Cherenkov sources, which provides a pathway to achieving very high intensity fields by combining several sources.

Pseudospark-driven sub-THz sources

Pseudosparks have been known since at least the 1970's and have been developed for several applications including switching. Our interest in pseudosparks is related to their ability to emit intense high brightness electron beams. It has been found that by optimizing the pseudospark geometrical configuration and by employing micro collimating apertures, micro electron beams can be produced that are ideal for driving small structures suitable for creating sub-THz sources.

A W-band (75-110 GHz) Extended Interaction Oscillator (EIO) millimetre wave source was designed and constructed to operate in the 2π mode. With a 35 kV discharge voltage, the EIO successfully produced W-band radiation pulses of 200W peak power and 20ns pulse duration, agreeing with 3D Particle-in-Cell PiC simulations. Simulations show the capability to extend this concept to higher frequencies.

Acknowledgements

The UK Engineering and Physical Research Council (EPSRC) is thanked for supporting this research.

References

1. *Denisov, G. G., Bratman, V. L., Cross, A. W., He, W., Phelps, A. D. R., Ronald, K., Samsonov, S. V., and Whyte, C. G.* Gyrotron traveling wave amplifier with a helical interaction waveguide // Phys. Rev. Lett. 1998. V. 81, No. 25, P. 5680-5683.

2. *Bratman, V. L., Denisov, G. G., Samsonov, S. V., Cross, A. W., Phelps, A. D. R., and Xe, W.* Hgh-efficiency wideband gyro-TWTs and gyro-BWOs with helically corrugated waveguides// Radiophysics and Quantum Electronics. 2007. V. 50. No. 2. P. 95-107.

3. *McConville, S. L., Speirs, D. C., Ronald, K., Phelps, A. D. R., Cross, A. W., Bingham, R., Robertson, C. W., Whyte, C. G., He, W., Gillespie, K. M., Vorgul, I., Cairns, R. A., and Kellett, B. J.* Demonstration of auroral radio emission mechanisms by laboratory experiment // Plasma Phys. Control. Fusion, 2008. V. 50. No. 7. 074010. P.1-13.

4. *Speirs, D. C., Vorgul, I., Ronald, K., Bingham, R., Cairns, R. A., Phelps, A. D. R., Kellett, B. J., Cross, A. W., Whyte, C. G., and Robertson, C.* A laboratory experiment to investigate auroral kilometric radiation emission mechanisms // J. Plasma Phys. 2005. V. 71. No. 5. P. 665-674.

5. *Speirs, D. C., McConville, S. L., Gillespie, K. M., Ronald, K., Phelps, A. D. R., Cross, A. W., Bingham, R., Robertson, C. W., Whyte, C. G., Vorgul, I., Cairns, R. A., and Kellett, B. J.* Numerical simulation of auroral cyclotron maser processes // Plasma Phys. Control. Fusion. 2008. V. 50. No. 7. 074011. P. 1-15.

6. *Speirs, D. C., Bingham, R., Cairns, R. A., Vorgul, I., Kellett, B. J., Phelps, A. D. R., and Ronald, K.* Backward Wave Cyclotron-Maser Emission in the Auroral Magnetosphere // Phys. Rev. Lett.. 2014. V. 113. No. 15. 155002. P. 1-5.

7. *Konoplev, I. V., Fisher, L., Cross, A. W., Phelps, A. D. R., Ronald, K., and Robertson, C. W.* Surface wave Cherenkov maser based on a periodic lattice // Appl. Phys. Lett 2010. V. 96. No. 26. 261101. P.1-3.

8. *Konoplev, I. V., MacLachlan, A. J., Robertson, C. W., Cross, A. W., and Phelps, A. D. R.* Cylindrical periodic surface lattice as a metadielectric: Concept of a surface-field Cherenkov source of coherent radiation // Phys. Rev. A. 2011. V. 84. No. 1. 013826. P. 1-12.

9. *Yin, H., Robb, G. R. M., He, W., Phelps, A. D. R., Cross, A. W., and Ronald, K.* Pseudospark-based electron beam and Cherenkov maser experiments // Phys. Plasmas. 2000. V. 7. No. 12. P. 5195-5205.

10. *Yin, H., Cross, A. W., Phelps, A. D. R., Zhu, D., He, W., and Ronald, K.* Propagation and post-acceleration of a pseudospark-sourced electron beam // J. Appl. Phys. 2002. V. 91. No. 8. P. 5419-5422.

11. *He, W., Zhang, L., Bowes, D., Yin, H., Ronald, K., Phelps, A. D. R., and Cross, A. W.* Generation of broadband terahertz radiation using a backward wave oscillator and pseudospark-sourced electron beam // Appl. Phys. Lett. 2015. V. 107. No. 13. 133501. P.1-4.

12. *Ginzburg, N. S., Cross, A. W., Golovanov, A. A., Mesyats, G. A., Pedos, M. S., Phelps, A. D. R., Romanchenko, I. V., Rostov, V. V., Rukin, S. N., Sharypov, K. A., Shpak, V. G., Shunailov, S. A., Ulmaskulov, M. R., Yalandin, M. I., and Zotova, I. V.* Generation of Electromagnetic Fields of Extremely High Intensity by Coherent Summation of Cherenkov Superradiance Pulses // Phys. Rev. Lett.. 2015. V. 115. No. 11. 114802. P. 1-5.

EPJ Web of Conferences 149, 04024 (2017) DOI: 10.1051/epjconf/201714904024
10th International Workshop 2017 "Strong Microwaves and Terahertz Waves: Sources and Applications"

Code KARAT in simulations of power microwave sources including Cherenkov plasma devices, vircators, orotron, E-field sensor, calorimeter etc.

V.P. Tarakanov[1,2]

[1] JIHT RAS, Moscow, Russia, karat@tarak.msk
[2] National Res.Nuclear Univ.MEPhI, , Moscow, Russia

This paper deals with some last simulations made by electromagnetic PiC code KARAT[1] that is aimed at the solution of 1,2&3D non-stationary electrodynamics problems in a complicated geometry involving beams, plasmas, dielectrics, ionization etc.

A conical vircator

The vircator with conical electrodes for ballistic focusing of the electron beam was offered and investigated [2] including the dynamics of electrons in vircator and the dynamics of the VC as a whole.

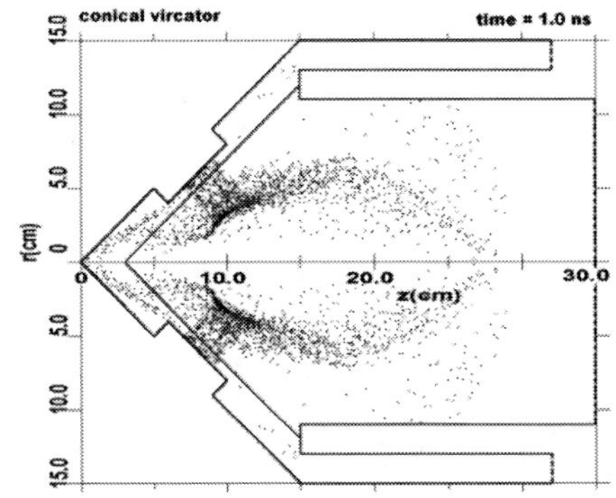

Fig. 1. The evolution of the electron beam and the virtual cathode (VC) in a conical vircator

The VC has the form of curvilinear figure of rotation. It oscillates in both radial and axial directions. The middle VC position depends on the beam current. The generation is resonant at a single frequency, depending on the beam current

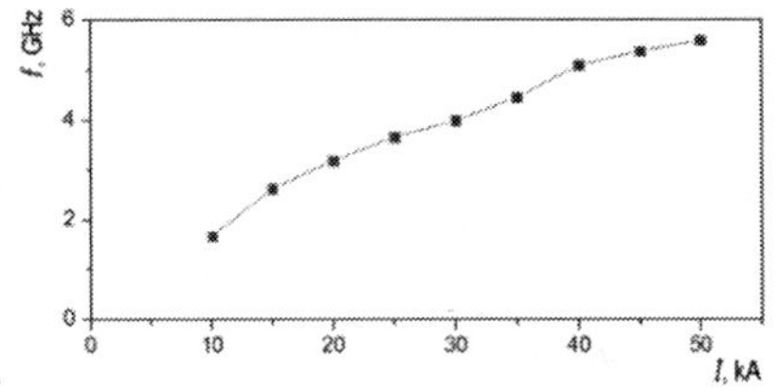

Fig. 2. The dependence of the resonance frequency of the conical vircator generation of the beam current

At the beam current from the range of 10–50 kA the oscillation frequency increases monotonically from 2 to 6 GHz with the beam current increase

Efficient relativistic orotron with mode selection

A high-power V-band relativistic orotron on a TM_{03} oscillation mode has an ovesized ($D/k \approx 2.7$) electrodynamic structure and diffractive output in the TM_{02} mode [3]. Single-mode operation of the oscillator is ensured by the cyclotron selection of the working axisymmetric mode, with simultaneous depression of competing non-axisymmetric modes by means of cutting longitudinal slits in the wall of the SWS. The output power is 350 MW with a power conversion efficiency of 31% when using a 3.6 kA, 310 keV electron beam transported in the 3.9 T magnetic field.

The modeling demonstrated that the simultaneous application of cyclotron and electrodynamic methods of mode selection in the orotron ensures the single-mode operation of the oscillator. The simulated microwave efficiency of the device is close to the theoretical limit. Furthermore, the orotron retains its efficiency 30% even in the case when the energy of the driving electron beam is reduced to 200 keV. Therefore, it seems to be possible to build a compact pulsed source of millimeter-wave radiation with a power of about 10^8 W, based on an orotron and a compact high-current electron accelerator.

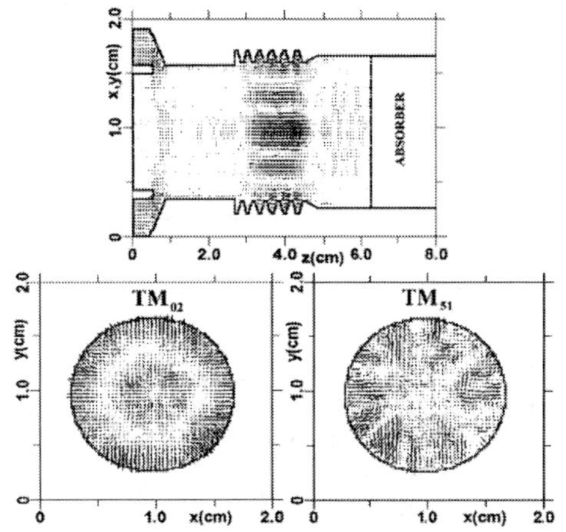

Fig. 3. Snapshots of electric field vectors in 3D modeling: in the axial middle plane of the orotron (*top*), and in a cross section of the output waveguide: in the desired operation regime (*left*) and in the regime with mode competition (*right*)

3D simulations of the orotron employed the Cartesian version of code. In these runs, the grid division was typically 120×120×450 and the number of macro particles was about 600000. The primary goal of the modeling was to observe the composition of competing modes. It is possible to reduce the *Q*-factor of a non-symmetric mode by means of cutting longitudinal slits through the wall of the waveguide (interrupting the azimuthal currents). The number of slits should not equate the azimuthal index of the mode desired to suppress. Indeed, introduction of nine slits in the SWS wall in 3D modeling notably expanded the band of magnetic fields, where TM_{03} oscillation dominates in the orotron.

Simulation of wide-aperture liquid calorimeters

Time domain simulation of diffraction and absorption effects in diagnostics of short high-power microwave pulses using wide-aperture calorimeters [4] was made. To take into account the frequency dispersion of the absorbing mixture, a Lorentz model implemented in the code was applied, based on the presentation of current in the medium by harmonic oscillators.

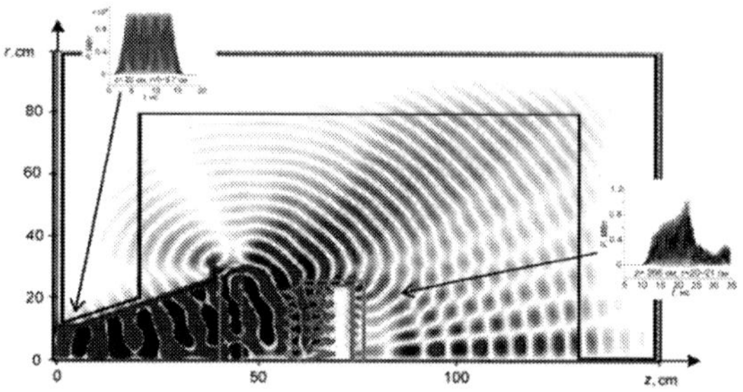

Fig. 4. Simulated B_{theta} field pattern with a fundamental frequency of 3.8 GHz in the vicinity of radiating horn and the load of the calorimeter. Insets show the waveforms of power flux in the neck of the horn and behind the device

The simulation demonstrated that, first, a diffraction field is superimposed on the wave field, which passes through the device with attenuation, while the interference pattern at the front and the decay of the RF pulse is non-stationary. In the field passing through the load, low-frequency components begin to prevail, due to a stronger decay of the high frequency.

UWB strip-line sensor antenna

In designing of the compact semiconductor high power pulse radiator the 3D simulations of the radiator and the compact strip-line sensors were performed [5]. The e/m pulse with the rise time of 50 ps was realized by TEM-horn antennas. The structure of e\m field was measured by various sensors including the new KARAT designed E-field sensor non-symmetrical two-wire type with dielectric stalk with high ε = 16 and small cross-section and 40 cm length.

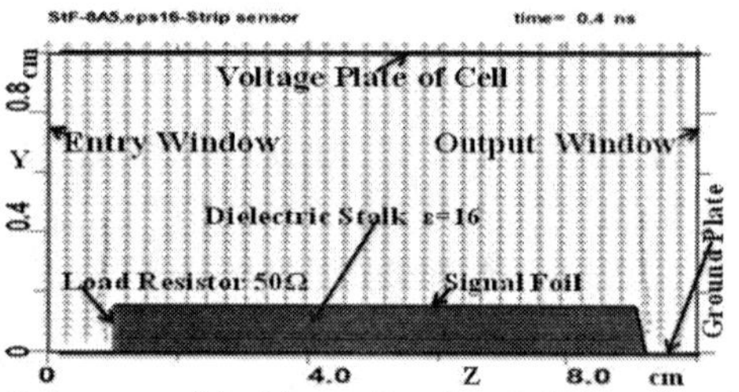

Fig. 5. Sensor model of L_{8CS} = 8 cm length (YZ-section along x = 0.8 cm) in the gauge cell with 1 cm gap. Vertical arrows are $E_0(t)$ – electric field vectors at t = 0.4 ns moment

Simulations of the sensor model gives values by closed to experimental values.

High current relativistic gyrotron

A compact relativistic gyrotron, operating at 3-mm wavelength band was realized with 250 kV, 100 A, 1 μs helical electron beam, forming by a thermionic magnetron-injection electron gun. The device operated in the rotating $TE_{12,5,1}$ mode of a circular waveguide cavity. The mode was transformed into a Gaussian wave beam by a built-in mode converter.

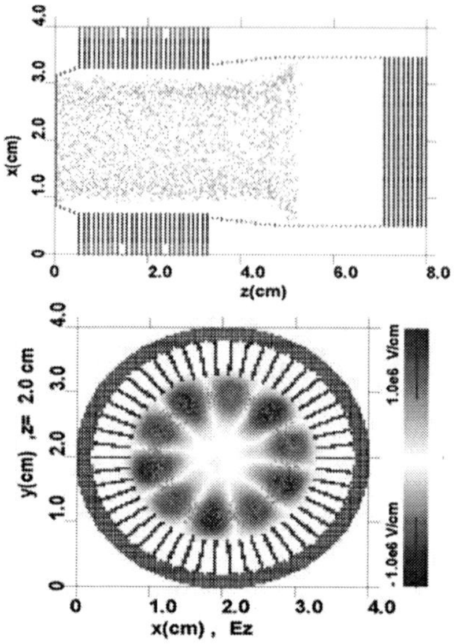

Fig. 6. The gyrotron construction, electron beam and Ez of generated field

Evolution of plasma in plasma electronics devices

Until this work [6] in theoretical investigations parameters of plasmas inside a plasma electronics device were considered constant. We simulated experimental setup were the duration of beam pulse was more hundred of nsec. PiC simulation shows that under conditions when beam current ~ 2 kA, energy ~500 keV, guiding magnetic field ~4 kG, initial Xe plasma density ~ $10^{12}cm^{-3}$ in regime when output microwave power ordered some hundreds kW to time hundred ns kinetic energy of plasma electrons achieves hundred keV. As electrons are frozen in guiding magnetic field then they leave an inner volume alongside of magnetic field and are absorbed on left and right boundaries. A positive potential ordered hundred Volt appears. As plasma ions are not frozen they moves to a side wall and plasma density reduces. As result an optimal parameters of system for amplification or generation of e/m wave are violated and output pulse breaks off.

Work is supported by RAS program "Elemental base and power semiconducting electronics devices" (ОЭ-8, acad. I.V. Grekhov).

References

1. *V.P. Tarakanov.* User's Manual for Code KARAT, BRA Inc., Va., USA, (1992.)
2. *A.E. Dubinov, V.P. Tarakanov* IEEE TRANSACTIONS ON PLASMA SCIENCE, V.44, NO. 8, 2016.
3. *E.M. Totmeninov, I.V. Pegel, and V.P. Tarakanov* Physics of Plasmas **24**, 053106 (2017).
4. *V.P. Tarakanov, A.I. Klimov, I.V. Pegel* (in press).
5. *V. Fedorov* (in press).
6. *P.S. Strelkov, V.P. Tarakanov* (in press).

EPJ Web of Conferences **149**, 04025 (2017) DOI: 10.1051/epjconf/201714904025
10th International Workshop 2017 "Strong Microwaves and Terahertz Waves: Sources and Applications"

Simulations of powerful microwave oscillators with oversized electrodynamics systems

V.Yu. Zaslavsky and N.S. Ginzburg

Institute of Applied Physics of the Russian Academy of Sciences, Nizhny Novgorod, Russia, zas-vladislav@appl.sci-nnov.ru

Introduction

Nowadays, various types of numerical codes are effective tools for analysis and optimization of electron microwave devices. Special codes are widely used to simulate the microwave components, the electron-optical system, and nonlinear dynamics. Nevertheless, up to now most of the simulations related to electron optics and electron-wave interaction were restricted by exploiting 2D self-consistent numerical models. At the same time, the state-of-the-art high-performance computational clusters combined with modern commercial codes look sufficient for 3D physical modeling of vacuum electronic devices with strongly oversized interaction space.

The paper is devoted to review of recent results of the *CST STUDIO SUITE* particle-in-cell (PIC) simulations of broad class of relativistic and sub relativistic microwave oscillators and amplifiers. In this report we consider terahertz-band gyrotrons with conventional cylindrical and planar configurations of the interaction space [1–3]. Here we discuss generators operating at the fundamental harmonic, as well as at the second and third harmonics. For improvement of stability of high cyclotron harmonic operation regime, multiple-beam schemes of gyrotrons are investigated [2]. Studies of relativistic surface wave Cherenkov oscillators with 1D and 2D periodical slow-wave structures were investigated [4, 5]. For all models comparison with experimental data and results of theoretical consideration based on simplified averaged equations are carried out. This analysis demonstrates that at present stage 3D PIC simulations can be considered as an effective tool for design and development of new schemes of powerful microwave generators and amplifiers. In addition, with the use of modern graphics processing units, it is possible to optimize parameters of electron devices for improvement characteristics within a reasonable calculation time.

High cyclotron harmonic operation in the double-beam THz gyrotrons

In this section, we explore a double-beam scheme of a short-wavelength gyrotron [2]. To date, the authors have extended this concept to the development of THz gyrotrons operating in CW regime at the second cyclotron harmonic. A double-beam scheme of a short-wavelength gyrotron with two generating helical electron beams operating at the second cyclotron harmonic is studied using 3D PIC simulations. Analysis shows that the introduction of an additional generating electron beam allows increase drastically the operating current of a second-harmonic gyrotron with the simultaneous suppression of self-excitation of spurious modes at the fundamental harmonic. This PIC code allows integrating directly (without averaging) the Maxwell equations together with the electron motion equations in the real geometry of the interaction space. The simulation parameters are shown in the figure 1. For a single-beam scheme current value of 2 A leads to the simultaneous excitation of the spurious $TE_{1,4}$ mode at the fundamental harmonic (see Fig. 1b). At the same time, single-mode oscillation with the $TE_{8,5}$ mode excited at the second harmonic was obtained in the double-beam scheme for a total current of 2 A (see Fig. 1c). The radiated power of the parasitic modes remains at the noise level. As a result, the radiated power of a double-beam gyrotron exceeds the power of a single-beam gyrotron by a factor of four. The developed concept makes it possible to realize high-power (several hundred watts) single-mode gyrotrons in the 0.7–1.0 THz frequency range.

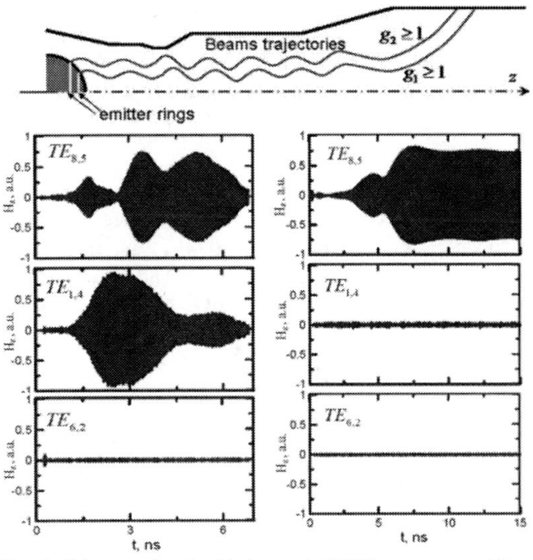

Fig. 1. Scheme of a double-beam 0.78 THz gyrotron with two generating electron beams (top). Temporal evolution of the competing mode amplitudes in the single-beam (left) and double-beam (right) schemes for $I_{total} = 2$ A, $U = 20$ kV, $I_{beam1} = 0.68$ A, $I_{beam2} = 1.32$ A, $g_{1,2} = 1.3$, $R_{beam1} = 0.46$ mm, $R_{beam2} = 0.72$ mm, $R = 3$ mm, $L = 12$ mm

Powerful high-harmonic THz-range planar gyrotron with transverse energy extraction

To increase output power of terahertz gyrotrons to several hundred kilowatts, we suggest using a planar geometry of interaction space with a sheet electron beam and transverse energy extraction [3]. An advantage of this scheme in comparison with conventional cylindrical geometry is the possibility to ensure effective mode selection over the open transverse coordinate in combination with radiation outcoupling that leads to a substantial reduction of Ohmic losses.

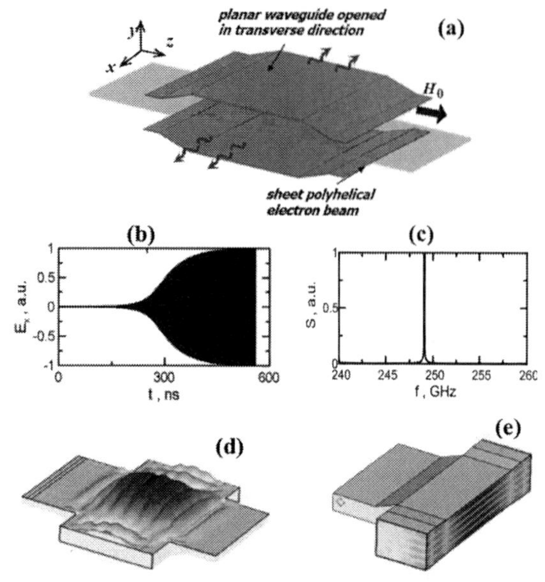

Fig. 2. Results of simulations of planar gyrotron at the third cyclotron harmonic with transverse energy extraction: (*a*) principal scheme planar gyrotron using in the 3D simulations; (*b*) establishment of steady state regime of generation for guide magnetic field of 3.35 T; (*c*) radiation spectrum, (*d*), (*e*) spatial distributions of excited $TE_{1,5}$ mode of open planar waveguide; (*d*) field profile at (x, z) plane with one variation along both coordinates, (*e*) transverse structure at (x, y) plane

Simulations of the nonlinear dynamics of the planar gyrotron at the third cyclotron harmonic were performed for an operating frequency of 250 GHz, an accelerating voltage of 80 kV, a total current of 15 A, a pitch factor of $g = 1$, a distance between plates of 3 mm, and the plate width of 1 cm. At a magnetic field of 3.35 T, the modeling demonstrates establishment of the steady-state regime at the third cyclotron harmonic at the mode with single longitudinal variation and its spatial distribution are shown in Fig. 2. The total efficiency at these parameters was 8%, and the output power ~50 kW.

Oversized Ka-band surface-wave oscillator based on 2D periodical corrugated structure

To study the possibility of achieving a gigawatt power level, we simulate Ka-band surface wave oscillator (SWO) with 2D corrugated structure of cylindrical geometry based on the "SINUS-6" accelerator (IAP RAS) [5]. This accelerator generates an explosive-emission electron beam with particles energy of 500–700 keV, beam current of several kiloamperes and about 20 ns duration in a single-pulse regime.

Figure 3 shows the results of modeling in the frame of 3D PIC code. For a Ka-band oscillator with a sufficiently large oversized waveguide (perimeter of about 16 wavelengths), a steady-state regime settled with a narrow radiation spectrum shown in Fig. 3b. In this regime the output radiation pattern is a superposition of azimuthally-symmetric modes of a regular cylindrical waveguide. At the operating frequency of 36 GHz, with the total beam current of 4 kA and the particles energy of 0.7 MeV, the output power of the simulated oscillator amount to 250 MW with an electron efficiency of about 10%.

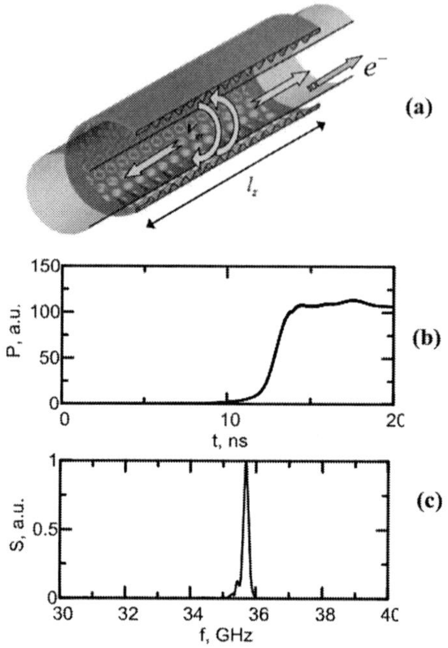

Fig. 3. Results of simulations of oversized Ka-band SWO with 2D corrugated structure based on "SINUS-6" accelerator: (*a*) principal scheme of the oscillator; (*b*) time dependence of the output power and (*c*) radiation spectrum

Acknowledgments

This work is partially supported by the Russian Scientific Foundation (grant #17-19-01605).

References

1. *V.Yu. Zaslavsky, N.S. Ginzburg, M.Yu. Glyavin, et al.* Three-dimensional particle-in-cell modeling of terahertz gyrotrons with cylindrical and planar configurations of the interaction space. // Physics of Plasmas. 2013. V. 20. P. 043103.
2. *N.S. Ginzburg, M.Yu. Glyavin, A.M. Malkin, et al.* Improvement of stability of high cyclotron harmonic operation in the double-beam THz range gyrotrons. // IEEE Transaction on Plasma Science. 2016. V. 44. P. 1303.
3. *N.S. Ginzburg, I.V. Zotova, A.S. Sergeev, et al.* High-Power Terahertz-Range Planar Gyrotrons with Transverse Energy Extraction. // Physical Review Letters. 2012. V. 108. P. 105101.
4. *N.S. Ginzburg, A.M. Malkin, A.S. Sergeev, V.Yu. Zaslavsky.* Quasi-optical theory of relativistic surface-wave oscillators with 1D and 2D periodic planar structures. // Physics of Plasmas. 2013. V. 20, P. 113104.
5. *N.S. Ginzburg, A.M. Malkin, A.S. Sergeev, V.Yu. Zaslavsky.* Oversized co-axial and cylindrical surface-wave oscillators with 2D periodical grating (quasi-optical model). // Journal of Applied Physics. 2013. V. 113. P. 104504.

EPJ Web of Conferences **149**, 04026 (2017) DOI: 10.1051/epjconf/201714904026
10th International Workshop 2017 "Strong Microwaves and Terahertz Waves: Sources and Applications"

W-band 5 MW pulse relativistic gyrotron: development and experimental implementation

Alexander N. Leontyev, E.B. Abubakirov, A.V. Chirkov, G.G. Denisov, Yu.M. Guznov, S.Yu. Kornishin, O.P. Plankin, R.M. Rozental, A.S. Sedov, E.S. Semenov, N.A. Zavolsky, S.A. Zapevalov and V.E. Zapevalov

Institute of Applied Physics of the Russian Academy of Sciences, Nizhny Novgorod, Russia, jamasiro@mail.ru

Powerful sources of the millimeter wavelength range are of considerable interest for a number of applications, such as plasma heating in controlled thermonuclear fusion devices, power supply systems for charged particles linear accelerators with an ultra-high acceleration rate and new radar schemes. The most promising versions of multi-megawatt sources operating in the band seem to be relativistic gyrotrons and gyroklystrons.

A new version of the pulsed relativistic gyrotron operating at the 3-mm wavelength range was developed and tested in experiment. Earlier, significant results were achieved in the realization of centimeter and long-wave millimeter range relativistic gyro-devices in the IAP RAS [1, 2]. These achievements was based on the detailed study [3], which showed the possibility of obtaining high operation efficiency in gyrotrons at any initial electron energy, including the relativistic one. However, the decrease in the working wavelength, while maintaining the energy parameters of the beam and radiation, entails increasing of the cross-section of the electrodynamic system and, accordingly, the transition to higher-order operating modes. Such a transition makes more complex solving of the key problems of the gyrotron development, first of all mode selection end electron beam formation. Also, an important part of the work was the conjugation of a new short-wave device with elements of the high-voltage electron accelerator already in existence.

Scheme of the experiment. The gyrotron is based on the "Saturn-F" electron accelerator [4], which can generate electron beams with a particle energy of up to 500 keV and a current of up to 200 A in pulses with the duration and repetition rate being equal to 1 μs and 10 pps, respectively. Overall view and the schematic diagram and of the relativistic gyrotron is presented in Fig. 1 and Fig. 2, correspondingly. A helical electron beam is formed by a three-electrode magnetron-injection gun. Theoretical analysis and optimization of the gun was undertaken with the ANGEL code [5]. Calculations showed it feasible to obtain an electron beam at 250 kV of the total voltage and 80–100 A of the beam current with the pitch-factor being equal to 1.3 and an acceptable spread of transverse velocities (20–25 %).

The rotating $TE_{12,5}$ mode was chosen as the operating one. Single-mode numerical simulations predicted 7–8 MW of the output power with an efficiency of about 35%.

Earlier relativistic gyrotrons were equipped with axial radiation outputs. That approach results, obviously, in very complicated output field structures, corresponding to the operation mode of the gyrotron, and it also has several other drawbacks. The higher-order mode scatters because of the irregularities of the output waveguide, and it results in an increase in the radiation losses.

The new gyrotron was designed with a built-in mode converter, which provides transformation of the operating mode into the Gaussian wave beam and, simultaneously, separation between the regions of propagation of the RF wave and transportation and collection of the exhausted electron beam (Fig. 3). The converter consists of a radiator of the $TE_{12,5}$ mode into open space, which is formed as a corrugated waveguide with a profiled cut of its edge, feeding a parabolic mirror and two guiding mirrors, and a vacuum boron nitride window. The convertor losses are less than 5%.

Fig. 1. Gyrotron installation overall view

The measurement system consisted of two sensors for determination of the envelope of the microwave signal, a mixer with a local oscillator to observe the spectrum of radiation, and a calorimetric load to measure the total pulse energy. The energy divided by the pulse duration yielded the peak output power.

Operating regimes. The gyrotron demonstrated stable generation when the magnetic field was set in the vicinity of the cyclotron resonance for the operating mode. The maximum output peak power achieved was 5.6 MW with about 20% efficiency at the accelerating voltage of 240 kV, beam current of 100 A and leading magnetic field induction of 4.35 T (Fig. 3). The central frequency was about 94.4 GHz with a spectral width of 3 MHz (Fig. 4).

It was found that the region of the operating magnetic fields was narrower than that predicted theoretically. This seemed to be due to the mode competition, the signs of which were observed in some experiments, where spurious oscillations appeared before and after operation of the working mode (Fig. 5).

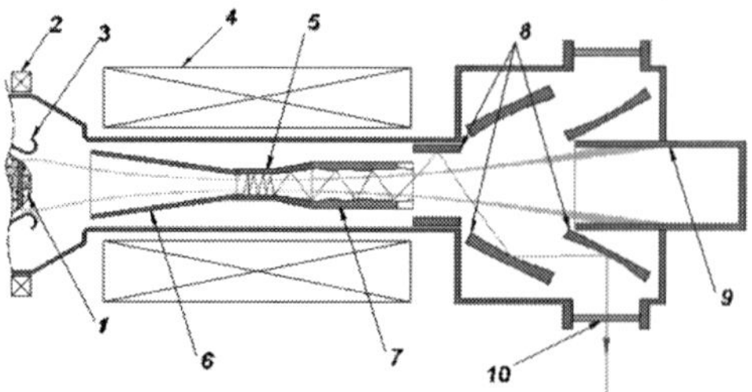

Fig. 2. Schematic diagram of the 94.4 GHz relativistic gyrotron: (1) cathode, (2) cathode coil, (3) anode, (4) cryomagnet, (5) cavity, (6) beam tunnel, (7) working wave radiator, (8) mirrors, (9) collector, (10) output window

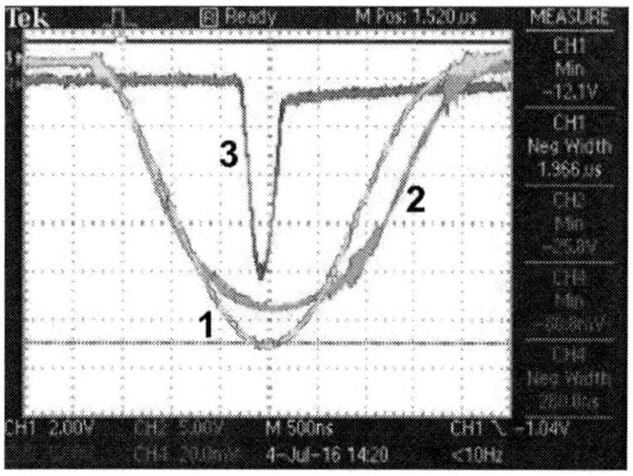

Fig. 3. Oscilloscope traces: (1) waveforms of accelerating voltage (244 kV peak), (2) beam current (103 A peak) and (3) microwave pulse at the regime of maximum output power (5.6 MW). Time scale is 0.5 μs/div

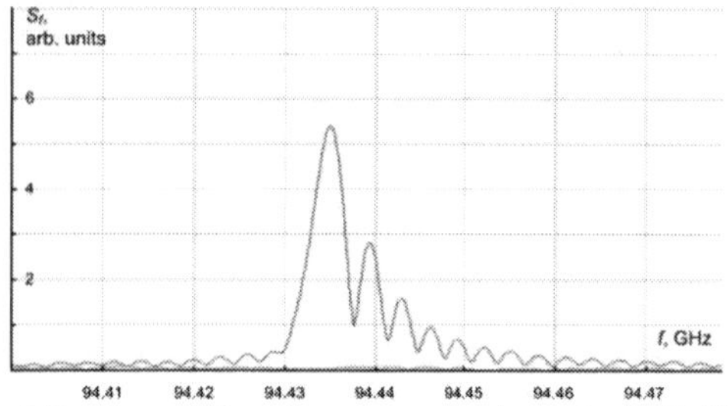

Fig. 4 Spectrum of the microwave pulse shown in Fig.3. High-frequency components correspond to the change of the oscillation frequency during the pulse due to variations in the electron energy

Fine adjustment of the voltage and magnetic fields in the resonator and in the cathode area allowed us to obtain a projected oscillation with a stable character; excitation at frequencies differing from the operating one were suppressed or essentially weakened by the main mode generation.

Conclusion. The possibility of obtaining a multi-megawatt output power in short-wave gyrotrons is demonstrated. A new version of a pulsed relativistic gyrotron has been developed and experimentally implemented in the 3-mm wavelength range. A stable single-mode generation was realized on the $TE_{12.5}$ mode of a circular waveguide with subsequent transformation into a Gaussian wave beam. The output power of the gyrotron exceeds 5 MW in pulses of duration about 300 ns.

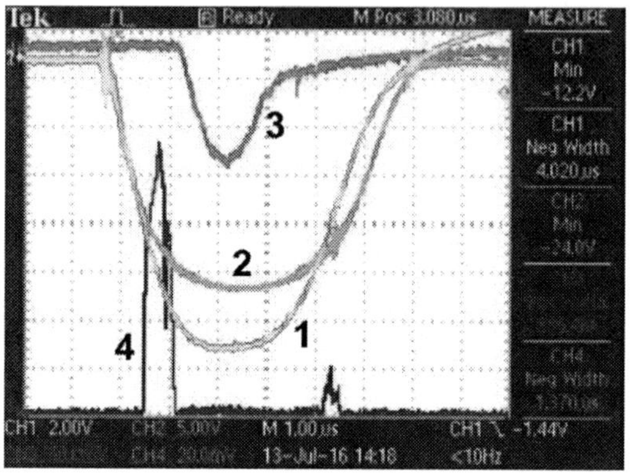

Fig. 5. Waveforms of signals, when parasitic oscillations were detected: accelerating voltage (1); beam current (2); radiation through the main output channel (3) and spurious microwaves registered at the auxiliary vacuum window (4)

References

1. *Zaitsev N.I., Zavolsky N.A., Zapevalov V.E., Ilyakov E.V., Kulagin I.S., Lygin V.K., Moiseev M.A., Nechaev V.E., Petelin M.I., Rozental R.M.* Ten-Megawatt Pulsed Gyrotron with a 1-cm Wavelength and a 50% Efficiency. Radiophysics and Quantum Electronics, 2003, v.46, iss.10, pp.816-819.

2. *Zaitsev N. I., Guznov Yu. M., Kuzikov S. V., Plotkin M. E., Tai E. M., Shevchenko A. S.* Experimental Investigation of the Relativistic Millimeter-Wave Gyroklystron. Radiophysics and Quantum Electronics, 2014, V. 56, Iss. 8-9, pp. 492-496.

3. *Zavolsky N.A., Zapevalov V.E., Moiseev M.A.*, "Efficiency Enhancement of the Relativistic Gyrotron", International Journal of Infrared and Millimeter Waves, Vol. 22, No. 6, 2001, pp.813-833.

4. *Zaitsev N.I., Zapevalov S.A., Ilyakov E.V., Kornishin S.Yu., Kofanov S.V., Kryltsov M.Yu., Kulagin I.S., Lygin V.K., Malygin A.V., Manuilov V.N., Movshevich B.Z., Perminov V.G., Petelin M.I., Fiks A.Sh., Shevchenko A.S., Tsalolikhin V.I., Kladukhin V.V., Krasnykh A.*, "500 keV, 200A microsecond electron accelerator with a repetition rate of 10 Hz", Proceedings of XXI Russian Accelerator Conference, 2008, Russia, Zvenigorod, pp. 339-341.

5. *Plankin O.P., Semenov E.S.* Trajectory Analysis of the Electronic-Optical System of Technological Gyrotron. Vestnik NSU. Series: Physics. 2013. V. 8, iss.2, pp.44-54 (in Russian).

Planar slow-wave structures for miniaturized low-voltage Cherenkov devices

N.M. Ryskin,[1,2] A.I. Benedik,[1] A.G. Rozhnev,[1,2] N.I. Sinitsyn,[2] G.V. Torgashov,[2] R.A. Torgashov[1,2]

[1]Saratov State University, Saratov, Russia, ryskinnm@info.sgu.ru
[2]Saratov Branch, Institute of Radio Engineering and Electronics, Saratov, Russia

1. Introduction

Microfabricated vacuum power amplifiers and oscillators operating at THz frequencies have attracted a great interest. Among them, linear-beam Cherenkov devices such as travelling-wave tubes (TWT) and backward-wave oscillators (BWO) have been actively studied. For miniaturized devices, a possibility of low-voltage operation is of primary importance, since decrease of voltage allows reduce of size and weight of the device. In particular, promising planar slow-wave structures (SWSs) utilizing a metallized meander mi-crostrip line on a dielectric substrate has been proposed [1]. Later, similar SWSs have been considered in [2–4].

In this paper, we present the results of numerical simulation, fabrication, and cold measurement of the microstrip meander-line and interdigital SWSs. Schematics of the SWSs are shown in Fig. 1(a) and Fig. 1(b), respectively. Fig. 1(c) shows qualitative drawing of their dispersion diagrams. One can see that the meander and interdigital SWS are suitable for TWT and BWO operation, respectively. The proposed structures exhibit high slowdown factor that is profitable for low-voltage operation.

2. G-band meander-line SWS

We have developed the meander-line SWS at G-band (180-230 GHz) [5, 6]. For the fabrication, copper films with 1 μm thickness were sputtered over quartz and polycor substrates. After that, the structure was patterned by photolithography. Below the metallic layer, there is a dielectric pedestal, which height g is regulated by the etching time. Dimensions of the SWS are listed in the Table 1.

Basic electromagnetic parameters of the designed meander SWS were calculated. It is assumed that the structure is placed inside a rectangular waveguide, which is not shown in Fig. 1. In Fig. 2(a), dispersion diagram is plotted. One can see that the SWS passband is rather wide. The SWS exhibits normal dispersion with slow-wave factor $c/v_{ph} \sim 7 \div 11$ in the operating frequency range 180-230 GHz. The beam synchronism voltage is as low as 3÷5 kV.

Pierce coupling impedance R_c vs. frequency is shown in Fig. 2(b). In the simulation, it is supposed that a 180×30 μm^2 sheet electron beam propagates at distance h over the metallized strip surface. In Fig. 2(a), the coupling impedance averaged over the beam cross-section is plotted at different values of h. The coupling impedance rapidly decreases with the increase of h, because the electromagnetic field decays exponentially in vertical direction. Nevertheless, $R_c \sim 1 \div 10$ Ohm is attained in the 180-230 GHz frequency range for $h \sim 20 \div 50$ μm.

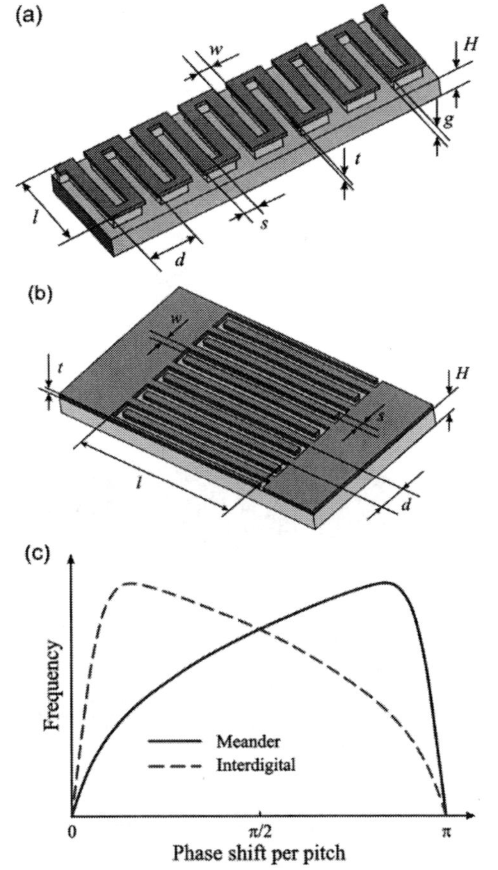

Fig. 1. Schematic diagrams of meander (*a*) and interdigital (b) planar SWSs, and qualitative dispersion diagrams (c)

Table 1

Waveguide cross section, $a\times b$ (mm)	2×1
Substrate thickness H (μm)	500
SWS period, d (μm)	50
Meander width l (μm)	200
Metallized strip width, s (μm)	15
Metallized strip thickness, t (μm)	1

3. V-band meander and interdigital SWSs

Due to the limited capability for cold-test measurements in G-band, we turned to the study of lower-frequency V-band (50-70 GHz) prototype SWSs. Several samples of the meander and interdigital SWS circuits with input/output couplers on quartz and polycor substrate were fabricated.

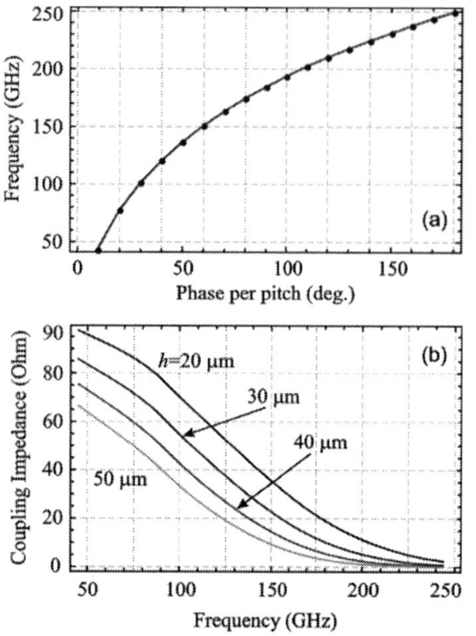

Fig. 2. Dispersion diagram (a) and interaction impedance (b) for the G-band meander-line SWS

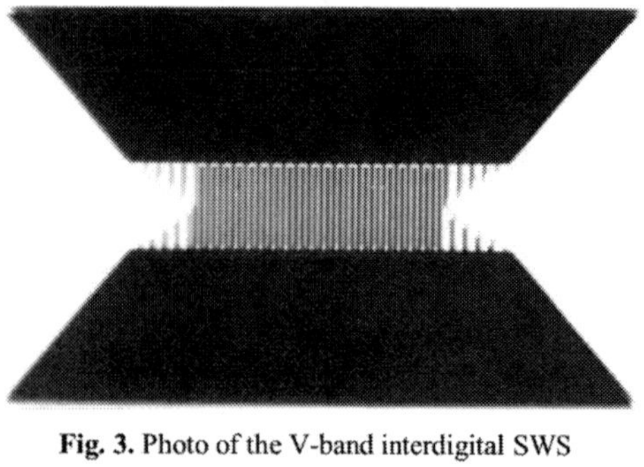

Fig. 3. Photo of the V-band interdigital SWS

In Fig. 3, photo of the interdigital SWS is shown. The SWS consists of 20 periods. Matching sections with linear decrease of pin height are placed at both ends. Dimensions of the SWS are listed in the Table 2. The SWS has anomalous dispersion and exhibits high slow-wave factor $c/v_{ph} \sim 10 \div 30$. The corresponding synchronism voltage is 1÷3 kV. Coupling impedance $R_c \sim 1 \div 10$ Ohm is calculated for a 500×50 μm^2 sheet electron beam propagating at distance h=50-75 μm over the metallic surface.

S-parameters of the circuit were measured using the two-port P2-69 scalar network analyzer in 53.57-78.33 GHz frequency band. For the measurements, the SWSs were placed in the 3.6×1.8 mm^2 waveguide. In the cold-test measurements, VSWR less than 2 and transmission losses less than 5 dB were observed.

The meander SWS exhibits normal dispersion in the 50÷70 GHz frequency range. For the SWS on silica substrate, beam synchronism voltage is 3÷5 kV.

Table 2

Substrate material	Quartz	Polycor
Waveguide cross section (mm)	3.6×1.8	3.6×1.8
SWS period, d (μm)	100	100
Pin length, l (μm)	825	575
Pin width, s (μm)	25	25
Etched strip width, w (μm)	25	25
Metallized strip thickness, t (μm)	1	1
Substrate thickness, h (μm)	500	500

4. Summary

In conclusion, planar SWSs on dielectric substrates possess wide bandwidth and relatively large slow-wave factor. They are suitable for low-voltage millimeter and THz-band Cherenkov devices such as TWT and BWO with sheet electron beam. Several samples of SWSs in G- and V-band were designed, simulated and fabricated. Good transmission characteristics were measured.

This work is supported by the Russian Science Foundation grant No. 17-12-01160.

References

1. *Gulyaev, Y.V., Zhbanov, A.I., Zakharchenko, Y.F., Nefedov, I.S., Sinitsyn, N.I., Torgashov, G.V.* Planar slow wave structures of miniature electrovacuum microwave devices // J. Comm. Technol. Electron., 1994. V. 39, No. 12. P. 2049–2058.

2. *Sengele, S., Jiang, H., Booske, J.H., Kory, C.L., van der Weide, D.W., Ives, R.L.* Microfabrication and characterization of a selectively metallized W-band meander-line TWT circuit // IEEE Trans. Electron Devices. 2009. V. 56, No. 5. P. 730–737.

3. *Shen, F. Wei, Y.-Y., Xu, X., Liu, Y., Yin, H.-R., Gong, Y.-B., Wang W.-X.* 140-GHz V-shaped microstrip meander-line traveling wave tube // J. Electromagn. Waves Appl. 2012. V. 26, No. 1. P. 89–98.

4. *Sumathy, M., Augustin, D., Datta, S.K., Christie, L., Kumar L.* Design and RF characterization of W-band meander-line and folded-waveguide slow-wave structures for TWTs // IEEE Trans. Electron Devices. 2013. V. 60, No. 5. P. 1769–1775.

5. *Benedik, A.I., Rozhnev, A.G., Ryskin, N.M., Torgashov, G.V., Sinitsyn, N.I., Bushuev, N.A., Shalaev P.D.* Study of electrodynamic parameters of the planar meander slow-wave structures for THz band traveling wave tubes // Abstracts of the Sixteenth IEEE International Vacuum Electronics Conference (IVEC2015), Beijing, China, April 27-29, 2015.

6. *Karetnikova, T.A., Benedik, A.I., Rozhnev, A.G., Rys-kin, N.M., Torgashov, G.V., Sinitsyn, N.I., Shalaev P.D.* Development and modeling of G-band vacuum tube power amplifiers with sheet electron beam // 41st International Conference on Infrared, Millimeter, and Terahertz Waves (IRMMW-THz), Copenhagen, Denmark, 2016.

Prospective field emitters for miniature high voltage electronic devices operating at technical vacuum conditions

G.G. Sominskii, V.E. Sezonov, E.P.Taradaev, T.A. Tumareva

Peter the Great Saint Petersburg Polytechnic University, St. Petersburg, Russia, sominski@rphf.spbstu.ru

The use of field emitters is promising for the creation of systems for formation of electron flows in many electronic devices, since they do not require heating and are inertialess. Until recently, the field emitters were used mainly in ultra-high vacuum, low current and low voltage devices. Meanwhile, miniature, but high voltage electronic devices, which it is difficult to use with thermionic cathodes, are attracting more and more interest. Such are, for example, electron-beam microwave devices of short millimeter and sub-millimeter wavelengths, as well as miniature X-ray sources, operating at accelerating voltages of units - tens kV. Emission currents of several tens milliamps at current densities of the order of or more than 100–150 mA/cm^2, are required, as a rule in these applications. Not only the difficulty of sufficiently high currents producing by field emitters, but also their low durability impedes their use in high voltage devices. Basic mechanisms of cathodes destruction are associated with an intense bombardment of their surface by residual gas ions typical for such devices, with the impact of ponderomotive forces, large in strong electric fields, and/or by the thermal effects at the selection of the large emission currents. Using distributed systems as field emitters allows to facilitate the current load on the single emission center, minimize the effect of ponderomotive forces and thermal effects. However usually, ion bombardment restricts the use of field emitters in high-voltage electronic devices.

Multi-tip silicon field emitters (MSE) with new type double-layer metal - fullerene coating were developed and investigated in the laboratory of "High current and microwave electronics" of SPbPU [1, 2]. The main investigations were performed for structures of silicon tips with a height h = 30 μm and tip top radius R = 5–10 nm. The distance L between the tips was set equal to their height. With this relation L and h, there was a partial mutual shielding of the tips [3, 4], which reduced the magnitude of the electric field at their top. However, in the investigated sufficiently high-voltage modes in a diode system with the cathode-anode distance of 1.5 mm at voltages of 5–15 kV, it was possible to create electric fields at the surface of the tips of the order of $(4–6)\cdot 10^7$ V/cm, which were necessary for intense field emission.

The layer of molybdenum with thickness from 5 to 20 nm deposited on the surface of a multi-tip silicon structure increased its conductivity and strength. A thin (from 2 to 10 monolayers) film of fullerene C_{60} molecules was deposited on the molybdenum layer to protect cathode against the destructive effect of ion bombardment. Calculations performed using SRIM program [5] showed that such a thin fullerene film is transparent enough for ions of sufficiently high energy (over 1000 eV). Such fast bombarding cathode ions lose less than 1% of their energy in the film and dissipate the most part of their energy in the molybdenum substrate. On the other hand, for slow molybdenum atoms "knocked out" of the substrate, the thin fullerene coating is practically not transparent. As a result, the created thin fullerene coating minimize ion sputtering of multi-tip cathodes.

The fullerene coating increases the work function of the surface of the tips to approximately 5.3 eV. However, treating a multi-tip cathode with a slow (40 eV) flow of potassium ions makes it possible to reduce the work function of the tip structure surface by 15–20% due to the formation of stable metallo-fullerenes such as endo- (C_{60}@K) and / or exohedrals (K@C_{60}) in a coating.

The operation of cathodes with activated and non-activated coatings was investigated. It was showed, that used activated and non-activated coatings protect emitters from the destructive action of the ion bombardment at energies of bombarding cathode ions up to 15 keV. MSE with area of 0.2–0.3 cm^2 with not activated coating provides the field emission currents of up to about 100 mA at emission current densities up to approximately 0.4–1.0 A/cm^2. The maximal currents of activated cathodes of the approximately same area were smaller and did not exceed 40–50 mA. In our opinion, this is due to the fact that when the coating is activated, some of the potassium atoms that is not captured by the fullerene molecules, intercallate under the fullerene coating and reduce its bond with the substrate.

The created multi-tip emitters are able to provide stable field emission necessary for the functioning of diagnostic gyrotrons of short-wave millimeter and sub-millimeter wavelengths. At present, the concept of creating diagnostic gyrotrons with an operating frequency about 260 GHz with cold multi-tip cathodes has been developed [6] in cooperation with the Institute of Applied Physics RAS. The use of created multi-tip cathodes in portable X-ray sources is also promising [7], as we think.

Along with the investigation of mentioned multi-tip emitters, the possibility of using developed by the authors original multi-layer cathodes in the high-voltage electronic devices is studied in SPbPU [1, 8]. For obtainment of field emission, the fields near the contacts of nano-layers of materials with different work function are used. Performed calculations and experiments have shown that the field emission of the layered cathodes depends on the thickness Δ of the contacted layers, on the difference in the work functions $\Delta e\varphi$ of these layers, and on the number N of pairs of layers. The obtained data indicate that the thickness of the layers with a larger work function should be approximately equal to 2 nm. The optimum thickness of the layers with smaller work function should be approximately 10–15 nm. The cathodes including 10–

20 pairs of ytterbium-carbon layers ($\Delta e\varphi \sim 1.6$ eV) and cathodes from the same number of pairs of hafnium-platinum layers ($\Delta e\varphi \sim 1.8$ eV) were studied in the most detail. The largest field emission currents and current densities were obtained for Hf-Pt cathodes including 20 pairs of layers. Currently, a stable field emission of 2–3 milliamps at average current densities along the surface up to about 200 A/cm^2 was obtained with the cathodes of this type. We hope that there is the fundamental possibility to obtain higher emission currents from the layered cathodes with larger emitting area.

This work was funded by the grant of the Russian Science Foundation №16-12-10010.

References

1. *Sominski, G.G., Sezonov, V.E., Taradaev, E.P., Tumareva, T.A., Zadiranov, Y.M., Kornishin, S.Y., Stepanova, A. N.* Innovative Field Emitters for High-Voltage Electronic Devices // Radiophysics and Quantum Electronics. 2015.V. 58.No.7. P.511-519.

2. *Sominskii, G.G., Tumareva, T.A., Taradaev, E.P., Mishin, M.V., & Stepanova, A.N.* Multitip semiconductor field emitters with new-type bilayer protecting coatings // Technical Physics. 2015. V.60. No. 1.P.133-136.

3. *Eletskii, A.V.* Carbon nanotube-based electron field emitters // Physics-Uspekhi, 2010. V.53. No. 9. P. 863.

4. *Sominski, G.G., Taradaev, E.P.,* Influence of Surface Morphology on The Characteristics of Multi-Tip Field Emitters with Two-Layer Metal - Fullerene Coatings // Proc. of 11th Int. Vacuum Electron Sources Conference. Oktober 18-20. 2016. Seoul, Korea.P.67-68.

5. *Ziegler, James F.* Srim-2003 // Nuclear Instruments and Methods in Physics Research Section B: Beam Interactions with Materials and Atoms. 2004. V. 219–220. P. 1027 – 1036.

6. *Glyavin, M.Y., Manuilov, V.N., Sominskii, G.G., Taradaev, E.P., Tumareva, T.A.* The concept of an electron-optical system with field emitter for a spectroscopic gyrotron. // Infrared Physics & Technology. 2016. V. 78. P. 185-189.

7. *Zhang, M., Naik, R.R., Dai, L. (Eds.).* Carbon Nanomaterials for Biomedical Applications. V. 5. 2015. Springer.

8. *Sominski, G.G., Sezonov, V.E., Tumareva, T.A, Taradaev, E.P.* Field Emitter // RF Patent №118119. Published 10.07.2012.

EPJ Web of Conferences 149, 04029 (2017) DOI: 10.1051/epjconf/201714904029
10th International Workshop 2017 "Strong Microwaves and Terahertz Waves: Sources and Applications"

Synchronization of delay-coupled gyrotron oscillators

A.B. Adilova[1], S.A. Gerasimova[1], N.M. Ryskin[1,2]

[1]Saratov State University, Saratov, Russia, usachevasa@gmail.com
[2] Saratov Branch, Institute of Radio Engineering and Electronics, Saratov, Russia

Introduction

The most important application of high-power continuous-wave (CW) gyrotrons is electron cyclotron plasma heating for nuclear fusion [1]. For such application, a large number of gyrotrons should be used to obtain a very high power. For example, in ITER it is planned to use 26 gyrotrons with 1 MW power level. Providing coherence of radiation of large number of high-power microwave sources is a challenging task. Among the possible solutions there are either injection locking by external master oscillator with stabilized frequency [2, 3], or mutual synchronization of two or more oscillators [4].

In this paper, a model of two delay-coupled gyrotron oscillators is developed and basic regimes of the mutual interaction are studied. Time delay of the coupling signal traveling between two gyrotrons strongly affects the pattern of phase-locking regimes.

Model

In [5], we developed a basic theory of mutual phase locking of two limit-cycle oscillators coupled with delay. The delay in coupling provides essential features of the pattern of synchronization. In particular, it is very sensitive to the phase shift of the coupling signal propagating between two counterparts. However, the analysis presented in [5] is valid only in the case of weak coupling and small frequency mismatch. In this paper, we consider a more general model of two coupled gyrotron oscillators, which are assumed identical except a small frequency mismatch. The model is described by two delay-differential equations (DDE)

$$\begin{aligned}\frac{dA_1}{dt}+i\frac{\Delta}{2}A_1+A_1&=I_s\Phi(A_1)\cdot A_1+\rho e^{-i\psi}A_2(t-\tau),\\ \frac{dA_2}{dt}-i\frac{\Delta}{2}A_2+A_2&=I_s\Phi(A_2)\cdot A_2+\rho e^{-i\psi}A_1(t-\tau).\end{aligned} \quad (1)$$

In (1), $A_{1,2}$ are slow amplitudes, Δ is normalized frequency mismatch, $\Phi=\Phi'+i\Phi''$ is the complex gain factor, I_s is the normalized beam current parameter, ρ is the coupling strength, $\tau=\omega_0 t_d/2Q$ is the normalized delay time t_d, and $\psi=\omega_0 t_d$.

In the quasi-linear theory [6,7], the complex gain factor is obtained as a polynomial expansion $\Phi=\alpha-\beta|A|^2-\gamma|A|^4-\ldots$ as a result of approximate solution of the electron motion equations. However, this approximation is valid only at a slight excess of the start-oscillation current and, consequently at low efficiency. Thus, to extend the range of validity of the model, we obtain Φ by fitting the numerical solution of the equations of time-domain gyrotron theory with fixed Gaussian profile of the RF field [7]. First, oscillation amplitude A and frequency Ω are calculated for an uncoupled gyrotron at different I_s. In a steady-state mode, Φ, A and I_s satisfy the relations

$$1=I_s\Phi'(A),\quad \Omega=I_s\Phi''(A)=\Phi''(A)/\Phi'(A). \quad (2)$$

From (2), one can evaluate Φ' and Φ''.

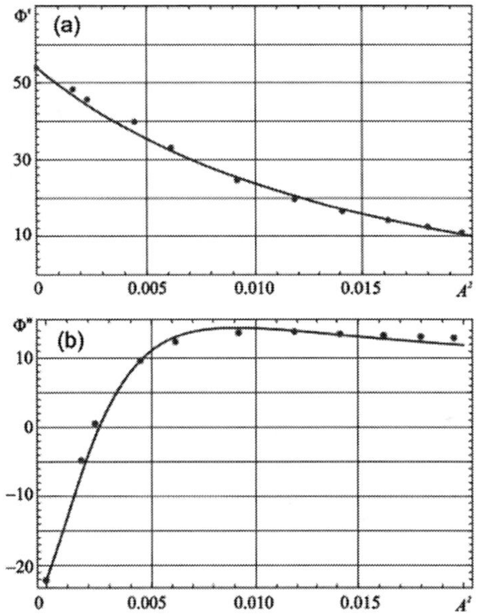

Fig. 1. The real (a) and imaginary (b) parts of the complex gain factor Φ for gyrotron in soft excitation mode

Consider a gyrotron with the following parameters: cyclotron resonance mismatch $\Delta_H=0.4$, and normalized cavity length $\mu=10$. In that case, the gyrotron operates in the soft self-excitation mode. In Fig. 1, the results of numerical simulations shown with circles are compared with fit by formulas

$$\begin{aligned}\Phi'\left(|A|^2\right)&=\frac{\alpha'-\beta'|A|^2}{1+\delta'|A|^2},\\ \Phi''\left(|A|^2\right)&=\frac{\alpha''+\beta''|A|^2+\gamma''|A|^4}{1+\delta''|A|^4},\end{aligned} \quad (3)$$

which provide fairly accurate fitting in a wide range of the amplitudes.

Results

Numerical simulation in a wide range of parameters confirms the basic pattern of synchronization revealed in [5]. In Fig. 3(a), domains of synchronization on the (Δ,ρ) plane are presented for $I_s=0.05$, $\psi=0.4\pi$, $\tau=1.0$. For such a value of I_s, the orbital efficiency of the uncoupled gyrotrons are about 45%. As predicted by the theory developed in [5], there exist in-phase and anti-phase mode of synchronization.

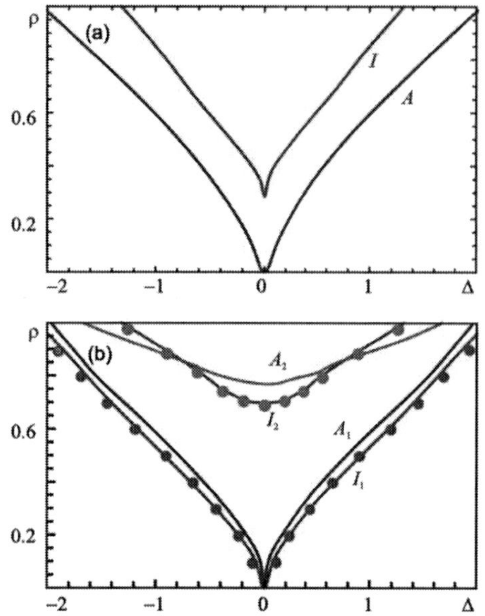

Fig. 2. Domains of stability of the in-phase (I) and anti-phase (A) mode of synchronization on the Δ, ρ plane for $I_s = 0.05$, $\psi = 0.4\pi$, and $\tau = 1.0$ (a) and 5.0 (b)

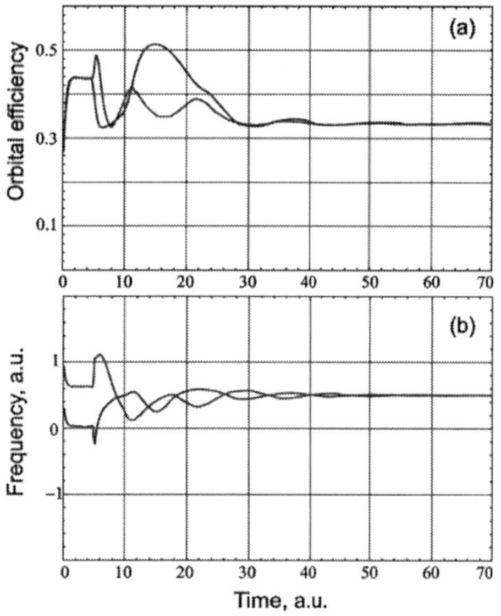

Fig. 3. Time histories of orbital efficiencies (a) and frequencies (b) in the phase-locked mode: $\rho = 0.5$, other parameters same as in Fig. 2

With the increase of the delay, there appear new modes of synchronization, i.e. the multistability becomes more pronounced. Fig. 2(b) illustrates the situation when there exist two in-phase and two anti-phase modes.

In addition, in Fig. 2(b) the stability domains calculated according the time-domain gyrotron theory with fixed Gaussian profile of the RF field [7] are shown with circles. They are in good agreement with the results of simulation for the DDE model (1).

In Fig. 3, typical waveforms of orbital efficiencies and instantaneous frequencies of two oscillators are presented. One can see that at the beginning stage of the transient, the gyrotrons oscillate with nearly same efficiencies but with slightly different frequencies. At $t > \tau$ the signal from the first gyrotron begins to affect the second one and vice versa. When the coupling strength is large enough, mutual synchronization takes place. After a rather short transient time the frequencies become equal, which clearly illustrates the effect of frequency locking.

Summary

Mutual synchronization of two gyrotrons coupled with delay is studied. The model of coupled gyrotrons described by two DDEs (1) is developed. The results of simulation of transition to the mode of synchronization are presented and the conditions of arising of either in-phase or anti-phase synchronization are presented. Qualitative agreement with the theory developed in [5] is observed. The results of simulation using the time-domain gyrotron theory with fixed RF field profile are in good agreement with the DDE model.

This work is supported by the Russian Foundation for Basic Research grant No. 16-32-00124.

References

1. *Thumm, M.* Recent advances in the worldwide fusion gyrotron development // IEEE Trans. Plasma Sci. 2014. V. 42, No. 3. P. 590–599.

2. *Yakunina, K.A., Kuznetsov, A.P., Ryskin, N.M.* Injection locking of an electronic maser in the hard excitation mode // Phys. Plasmas. 2015. V. 22, No 11. 113107.

3. *Bakunin, V.L, Denisov, G.G., Novozhilova, Y.V.* Zones of frequency locking by an external signal in a multimode gyrotron of a megawatt power level // Radiophys. Quantum Electron. 2016. V. 58, No. 12. P. 893–904.

4. *Rozental, R.M., Ginzburg, N.S., Glyavin M.Yu., Sergeev, A.S., Zotova, I.V.* Mutual synchronization of weakly coupled gyrotrons // Phys. Plasmas. 2015. V. 22, No. 9. 093118.

5. *Usacheva S.A., Ryskin N.M.*, Phase locking of two limit cycle oscillators with delay coupling // Chaos. 2014. V. 24, No. 2. 023123.

6. *Nusinovich, G. S.* Mode interaction in gyrotrons // Int. J. Electron. 1981. V. 51, No. 4. P. 457–474.

7. *Nusinovich, G.S.* Introduction to the Physics of Gyrotrons. Johns Hopkins University Press, Baltimore, 2004.

EPJ Web of Conferences **149**, 04030 (2017) DOI: 10.1051/epjconf/201714904030
10th International Workshop 2017 "Strong Microwaves and Terahertz Waves: Sources and Applications"

260 GHz CW gyrotron heating substitution with second-long laser pulses in waveguide semiconductor switches

M.L. Kulygin[1,2], G.G. Denisov[1,2], S.H. Salahetdinov[1], S.V. Shubin[1], E.A. Novikov[1], I.A. Litovsky[3]

[1] Institute of Applied Physics, Nizhny Novgorod, Russia, kmaxim@appl.sci-nnov.ru
[2] GYCOM Ltd., Nizhny Novgorod, Russia
[3] Lobachevski State University, Nizhny Novgorod, Russia

Sub-Terahertz waveguide semiconductor switches driven by laser emission [1] are used to cut a continuous-working (CW) microwave emission to series of wave packets [2, 3]. The main advantage of the switches is low distortion to the phases of the packets' high frequency fillings at the output [4]. The phases are linked to each other since the microwave input emission is coherent, e.g. a gyrotron in a phase stabilization regime. Recent studies achieved nanosecond level of switching performance [5] for 260 GHz frequency band and 532 nm laser emission [6] using a semiconductor plate of plain gallium arsenide.

Up to now, the switches have been tested with a commutated microwave power of up to 50 mW, limited by the backward-wave oscillator (BWO) test source. The maximum possible power is limited by a WR3 waveguide standard safety value; it is about 20 W (about two times lower than the breakdown value). An obvious solution of increasing smoothly the power level from 50 mW to 20 W using modern gyrotrons faces the problem of measuring errors. Since the rated power for the gyrotrons and their measuring calorimeters is about 1 kW, using them at 20 W produces measurement errors of about 100%, and worse than that, for lower powers. There are no precision commercial measurement devices for frequencies around 300 GHz for the moment.

The maximum heating zone is obviously located on the semiconductor plate, not on the metallic switch housing. There are two sources of heating, specifically, laser energy and Joule heating by the commutated microwave signal. So, for the temperature T within the semiconductor we can write:

$$\frac{\partial T}{\partial t} = B\Delta T + J(\vec{r},t) + L(\vec{r},t), \qquad (1)$$

where t is the time, $\vec{r}$ is the position, $B = 0.31$ cm^2/s is the thermal diffusivity for GaAs, Δ is the Laplace operator, $J(\vec{r},t)$ is the Joule heating, and $L(\vec{r},t)$ is the laser heating distribution. However, laser-power precision measurement devices are easily available. So, we can approximate the Joule microwave heating by substituting it with the appropriate laser heating. For infrared lasers [7], the function $L(\vec{r},t)$ from (1) is like $J(\vec{r},t)$, their maxima are at the surface, and they both decay inside the semiconductor. Knowing a part of the dissipated laser power at the semiconductor and another part is reflected, we can convert it to an effective microwave commutated power using the power transmission measurement for a typical switch.

The experiments [7] show the average microwave power transmission coefficient is about –0.4 dB for a wide frequency region. So, the microwave power insertion loss is about 9%. Then, assuming that all the microwave loss is dissipated inside the semiconductor plate, we can say that if the switch withstands a certain laser power, it should withstand a microwave power being 8 times as high.

References

1. *M. Kulygin, G. Denisov, and Vl. Kocharovsky,* J. Infrared Milli Terahz Waves, 2010, **31**(1), 31-40.
2. *M. Kulygin, V. Belousov, et al,* Radiophys. and Quant. Electron., 2014, **57**(7), 568-579.
3. *M. Kulygin, G. Denisov, N. Andreev, et al,* Proc. of 9th Int. W/Shop "Strong Microwave in Plasmas", 2014, Nizhny Novgorod-Perm, Russia, 225-226.
4. *M. Kulygin, S. Salahetdinov, et al,* IEEE COMCAS, 2015, DOI: 10.1109/COMCAS.2015.7360359.
5. *M. Kulygin, G. Denisov, K. Vlasova, et al,* J. Infrared Milli Terahz Waves, 2015, **36**(9), 845-855, DOI: 10.1007/s10762-015-0182-4.
6. *M. Kulygin, G. Denisov, et al,* Rev. Sci. Instr., 2016, **87**(014704), 1-4, DOI: 10.1063/1.4939673.
7. *M. Kulygin, G. Denisov, S. Shubin, et al,* IEEE Trans. THz Sci. Tech., 2017, **7**(2), 225-227, DOI: 10.1109/TTHZ.2017.2647879.

EPJ Web of Conferences **149**, 04031 (2017) DOI: 10.1051/epjconf/201714904031
10th International Workshop 2017 "Strong Microwaves and Terahertz Waves: Sources and Applications"

Theoretical and experimental investigations of oversized Ka-band surface-wave oscillator based on 2D periodical corrugated structure

N.S. Ginzburg, E.V. Ilyakov, I.S. Kulagin, N.Yu. Peskov, A.S. Sergeev and V.Yu. Zaslavsky
Institute of Applied Physics of the Russian Academy of Sciences, Nizhny Novgorod, Russia, peskov@appl.sci-nnov.ru

Introduction

For spatially-extended relativistic electron beams of sheet and tubular geometry the use of a two-dimensional (2D) distributed feedback was proposed for providing spatial coherence of radiation and increase the total radiation power [1]. Such 2D feedback can be realized in planar and co-axial 2D Bragg structures (resonators) having double-periodic corrugations. Experimental studies of FEMs based on novel feedback mechanism are performed in Ka-band (coaxial geometry) at the University of Strathclyde (Glasgow, UK) [2] and in W-band (planar geometry) at the Budker INP RAS (Novosibirsk, Russia) [3] in collaboration with the IAP RAS (N.Novgorod, Russia) and KIT (Karlsruhe, Germany). As a result, narrow-band multi-MW power radiation was obtained.

At present, theoretical and experimental studies of Cherenkov masers with 2D distributed feedback are in progress [4]. Among relativistic masers of such type, surface wave oscillators (SWO) are beneficial due to the large values of the electron-wave coupling impedance. Besides, formation of a surface mode ensures regular field distribution along the coordinate directed perpendicularly to the corrugated surface and, thus, would solve the problem of mode selection over this coordinate. In SWO 2D periodic structure can be exploited both as a slow-wave system and as a high selective Bragg resonator simultaneously [5]. In this case such structure would provide effective mode control over azimuthal coordinate.

Present paper is devoted to theoretical and experimental studies of oversized SWO with 2D-periodical corrugated structure of cylindrical geometry. Simulations of nonlinear dynamics of the oscillator were performed in the frame of quasi-optical approach and using 3D PIC code CST Studio Suite. Experimental studies were performed at IAP RAS based on the thermionic accelerator "Saturn".

Nonlinear dynamics of SWO with 2D periodical grating in the frame of quasi-optical approach

To study nonlinear dynamics of SWO with 2D periodical grating based on oversized cylindrical waveguide we assume that a hollow rectilinear electron beam moves with longitudinal velocity $V_0 = \beta_0 c$ along the 2D grating (Fig. 1)

$$r = \frac{r_1}{4}\left[\cos(\bar{M}\varphi - \bar{h}_z z) + \cos(\bar{M}\varphi + \bar{h}_z z)\right], \qquad (1)$$

which represents a superposition of two helical corrugations with opposite rotations having period d_z and number of turns $\bar{M}$ (variations of the corrugation over azimuth), $\bar{h}_z = 2\pi/d_z$, r_1 is corrugation depth.

We assume that the system parameters satisfy to the Bragg resonance condition

$$\bar{h}_z = \bar{h}_x = \bar{h}, \qquad (2)$$

Fig. 1. Scheme of oversized SWO with 2D corrugated structure

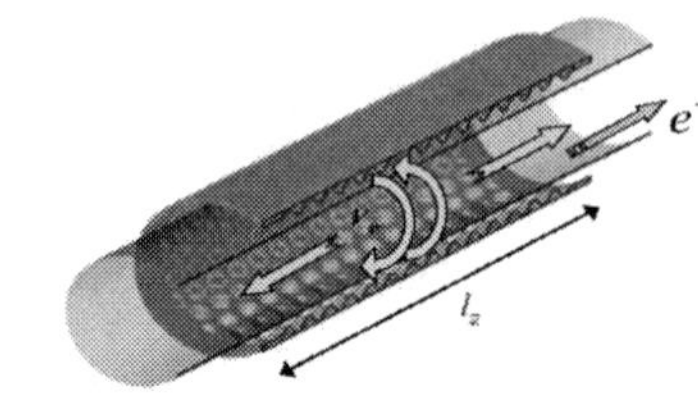

where $\bar{h}_x = \bar{M}/r_0$. If the curvature of the waveguide is small, i.e. its radius r_0 is substantially larger than the wavelength λ, one is able to introduce new coordinate $x = r_0\varphi$ and to use a quasi-planar model. On the 2D corrugated surface (1) under resonance condition (2), coupling and mutual scattering of the four partial quasi-optical wave-beams occurs

$$\vec{H} = \mathrm{Re}\left[\left(\vec{x}_0(C_z^+ e^{-i\bar{h}z} + C_z^- e^{i\bar{h}z}) + \vec{z}_0(C_z^+ e^{-i\bar{h}x} + C_z^- e^{i\bar{h}x}\right)e^{i\bar{\omega}t}\right]. \qquad (3)$$

Two of these beams $C_z^{\pm}$ propagate along the longitudinal $\pm z$ directions and the other two $C_x^{\pm}$ propagate along the transverse (azimuthal) $\pm x$ ($\pm\varphi$) directions. Radial coordinate y is directed normally to the surface of the conductor. Choosing Bragg frequency $\bar{\omega} = \bar{h}c$ as a reference frequency, the excitation of surface waves by the electron beam can be described by the following system of equations [5]

$$i\frac{\partial^2 \hat{C}_z^+}{\partial Y^2} + \frac{\partial \hat{C}_z^+}{\partial Z} + \frac{\partial \hat{C}_z^+}{\partial \tau} = i\hat{\alpha}(\hat{C}_x^+ + \hat{C}_x^-)\delta(Y) - \frac{1}{B_e}\frac{\partial}{\partial Y}JF(Y),$$

$$i\frac{\partial^2 \hat{C}_z^-}{\partial Y^2} - \frac{\partial \hat{C}_z^-}{\partial Z} + \frac{\partial \hat{C}_z^-}{\partial \tau} = i\hat{\alpha}(\hat{C}_x^+ + \hat{C}_x^-)\delta(Y), \qquad (4)$$

$$i\frac{\partial^2 \hat{C}_x^\pm}{\partial Y^2} \pm \frac{\partial \hat{C}_x^\pm}{\partial X} + \frac{\partial \hat{C}_x^\pm}{\partial \tau} = i\hat{\alpha}(\hat{C}_z^+ + \hat{C}_z^-)\delta(Y),$$

where $\delta(Y)$ is the delta function, $F(Y)$ is a function defining the radial current density of the electron beam with an effective thickness of B_e. Amplitude of the RF current

$$J = \frac{1}{\pi}\int_0^{2\pi} e^{-i\theta} d\theta_0$$

can be found from the electron motion equations

$$\left(\frac{\partial}{\partial Z} + \frac{1}{\beta_0}\frac{\partial}{\partial \tau}\right)^2 \theta = \mathrm{Re}\left[\frac{\partial \hat{C}_z^+}{\partial Y} e^{i\theta}\right]. \qquad (5)$$

Here θ is the electron phase with respect to the forward partial wave, Δ is the synchronism detuning on the carrier frequency ($\Delta > 0$), which may vary with changes in the particle energy. Synchronous interaction with a rectilinear electron beam arises due to the formation of an evanescent slow wave. In (4), (5) we used the following normalized variables and parameters: $Z = G\bar{h}z$, $X = G\bar{h}x$, $Y = \sqrt{2G}\bar{h}y$, $\tau = G\bar{\omega}t$, G is the gain parameter, $\hat{\alpha}$ is normalized wave coupling parameter at the corrugation (1). Due to the cylindrical geometry of the cavity the wave-beam amplitudes should satisfy the cyclic boundary conditions

$$\hat{C}^{\pm}_{x,z}(X+L_x;Y;Z)=\hat{C}^{\pm}_{x,z}(X;Y;Z)\,, \qquad (6)$$

where $L_x = 2\pi\bar{h}Gr_0$ is perimeter of the structure. These conditions allow solution of Eqs. (4) to be expanded in a Fourier series

$$\hat{C}^{\pm}_{x,z}(X;Y;Z;\tau)=\sum_{m=-\infty}^{+\infty}\hat{C}^{\pm,m}_{x,z}(Y;Z;\tau)e^{2\pi i m X/L_x}\,, \qquad (7)$$

considering each harmonic as a mode with an azimuthal index m. Boundary conditions for Eqs.(4) at the edges of the corrugated region correspond to the absence of external energy fluxes.

We used Eqs.(4)–(7) to simulate operation of powerful relativistic surface-wave oscillator at the wavelength of 8 mm driven by relativistic electron beam of 1 MeV, linear current density of 100 A/cm and beam radius r_0=2 cm. We consider the 2D corrugation of the period d_z = 7 mm and depth r_1 = 1 mm with the length of corrugated section l_z = 21 cm. These parameters correspond to the normalized values $G\approx 0.01$, $\hat{\alpha}=2.2$, $\Delta\approx 8$, L_z = 1.5, L_x = 1.

Figure 2 shows results of simulations of nonlinear stage of interaction. The temporal dependencies of amplitudes of different azimuthal components of the wave-beam $\hat{C}_z^-$ at various times of accelerating voltage front are presented. With zero rise time of the front we see excitation of a single symmetric mode (Fig. 2a). Amplitudes of others non-symmetric modes are negligibly small. However with the front time of 200 ns (Fig. 2b) the azimuthally non-symmetric mode suppresses symmetric mode and sets on.

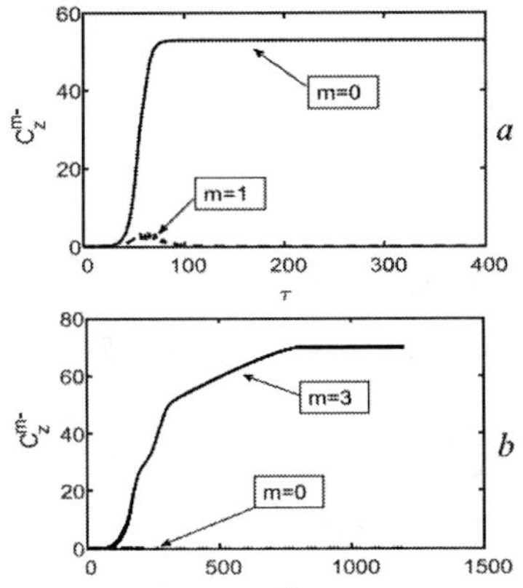

Fig. 2. Temporal dependencies of amplitudes of different azimuthal components of the wave-beam $\hat{C}_z^-$ at duration of the accelerating voltage front of (*a*) 0 and (*b*) 200 ns

Experimental results

Experiments with Ka-band 2D-SWO were carried out based on microsecond accelerator "Saturn" (IAP RAS). Tubular electron beam 100 A / 200–300 keV was formed from a thermionic cathode with a diameter of 5 cm. As a result of the experiments, stable narrow-band generation in strongly oversized system (perimeter-to-wavelength ratio is about 16) using periodical structure with 2D corrugation has been observed in the design region of parameters (Fig. 3). Calorimetric measurement indicates the power level of 1.5–2 MW in up to 200 ns pulses, that corresponds to electron efficiency of ~5%. High stability of the oscillation frequency with respect to variation of the electron beam parameters was demonstrated. In accordance with the results of simulations, frequency stabilization is caused by the presence of azimuthal energy fluxes provided by 2D corrugated structure.

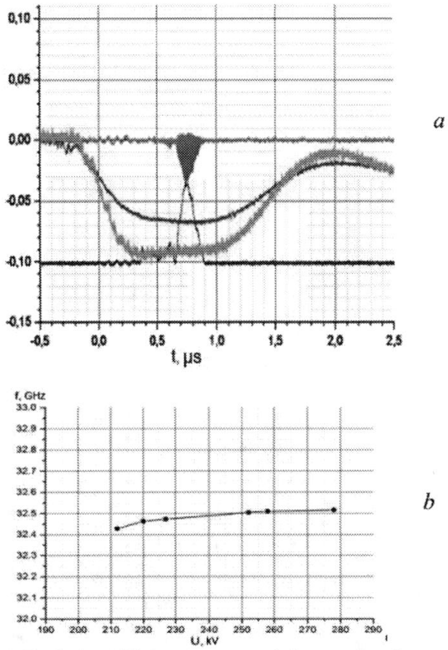

Fig. 3. (*a*) Typical oscilloscope traces of the accelerating voltage (green curve), beam current (red), heterodyne signal (brown) and output RF pulse (blue). (*b*) Dependence of the radiation frequency on the accelerating voltage

Acknowledgments. This work is partially supported by the Russian Foundation for Basic Research (grant #16-02-00890).

References

1. *N.S. Ginzburg, N.Yu. Peskov, A.S. Sergeev* Dynamics of free-electron lasers with two-dimensional distributed feedback // Optics Commun. 1994. V.112. P.151.

2. *I.V. Konoplev, A.W. Cross, A.D.R. Phelps, e.a.* Co-axial Free-Electron Maser based on two-dimensional distributed feedback. // Phys. Rev. E. 2007. V.76. P.056406.

3. *A.V. Arzhannikov, N.S. Ginzburg, P.V. Kalinin, e.a.* Using two-dimensional distributed feedback for synchronization of radiation from two parallel-sheet electron beams in FEM // Phys. Rev. Lett. 2016. V.117. P.114801.

4. *N.S. Ginzburg, E.V. Ilyakov, I.S. Kulagin, e.a.* Synchronization of radiation in an oversized coaxial Ka-band BWO using two-dimensional Bragg structure // Phys. Rev. Accel. Beams. 2015. V.18. P.120701.

5. *N.S. Ginzburg, A.M. Malkin, A.S. Sergeev, V.Yu. Zaslavsky* Oversized co-axial and cylindrical surface-wave oscillators with 2D periodical grating (quasi-optical model) // J. Appl. Phys. 2013. V.113. P.104504.

EPJ Web of Conferences 149, 04032 (2017) DOI: 10.1051/epjconf/201714904032
10th International Workshop 2017 "Strong Microwaves and Terahertz Waves: Sources and Applications"

45GHz/20kW gyrotron setup with automated output power control for ECR ion source

A.I. Tsvetkov[1], A.G. Eremeev[1], V.V. Kholoptsev[1], M.Yu. Shmelev[1], I.V. Plotnikov[1], Yu.V. Bykov[1], E.A. Kopelovich[1], A.Yu. Novikov[1], M.M. Troitskiy[1], M.V. Kuznetsov[1], K.A. Zhurin[1], A.P. Fokin[1], M.V. Morozkin[1], M.Yu. Glyavin[1, 2], M.I. Bakulin[2], G.G. Denisov[1, 2], E.A. Soluyanova[2], E.M. Tai[2]

[1]Institute of Applied Physics RAS, Nizhny Novgorod, Russia, tsvetkov@ipfran.ru
[2]GYCOM Ltd., Nizhny Novgorod, Russia

The 45 GHz / 20 kW automated gyrotron setup described below is intended for use in FECRAL (Fourth generation ECR ion source with Advanced design in Lanzhou, China) superconducting ion source. The FECRAL was proposed by Institute of Modern Physics (Chinese Academy of Science) for using in next generation Heavy Ion Accelerator Facility (HIAF), which requests ion source capable of delivering 50 pμA of $^{238}U^{34+}$ pulsed beam and 25 pμA of $^{238}U^{34+}$ CW beam for injector of a superconducting heavy ion linac [1, 2]. When used as a part of such facility, the microwave source must satisfy the very strong requirements both for output radiation parameters and for possibilities of system integration.

Fig. 1. The gyrotron setup general view

The gyrotron setup developed by IAP RAS jointly with GYCOM Ltd. [3] capable to provide 45 GHz microwave radiation with output power from 1 kW up to 20 kW in both CW and pulse (5–200 ms with up to 10 Hz repetition rate) modes at 45 GHz frequency.

The gyrotron tube is designed to operate with diode-type electron gun (LaB_6 cathode), and using of depressed collector for recuperation of the electron beam residual energy leads to considerable total efficiency increasing. Magnetic system is built on the basis of cryogen-free magnet JMTD-4T140 (Jastec Inc., Japan) with 140 mm warm bore diameter that allows the gyrotron to operate on fundamental harmonic ($TE_{6,3}$ mode).

The system allows triggering operation in pulse mode using external pulse generator as a reference. The transmission line consist of MOU, quasioptical elements, mode converter from the Gaussian beam TEM_{00} to TE_{01} or TE_{11} mode (at the user's option) and vacuum window. The last two mirrors and mode converter locate on the separate platform at the 2 m distance from the gyrotron, which results in 300 kV electrical insulation, required for ECR ion source operation. The efficiency of the quasioptical transmission line is not less than 99% and the measured content of the TE_{01} or TE_{11} modes at the output of the mode converter more than 97% (Fig. 3).

Fig. 2. Part of quasioptical transmission line: mirrors, mode converter from the Gaussian beam to TE_{01} or TE_{11} mode with vacuum window

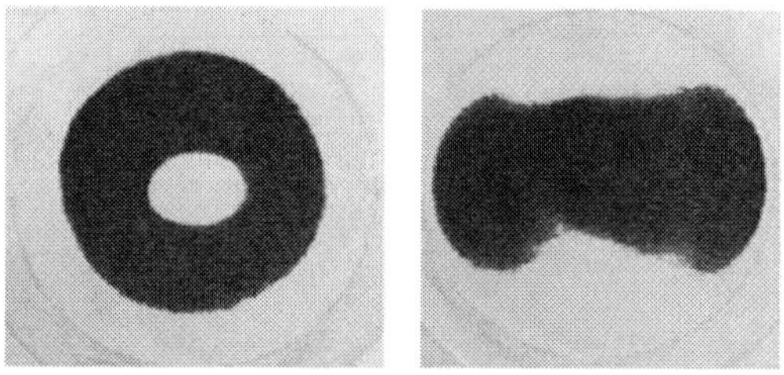

Fig. 3. MW radiation thermal prints at the output of the mode converter: left – TE_{01}, right – TE_{11}

The bi-directional couplers for monitoring and measuring the incident and reflected microwave power are mounted on the MOU mirrors. The attenuated microwave power for the incident and reflected waves is coupled to the rectangular waveguides through a set of the regular located circular holes of small diameter. An additional directional coupler is set in the channel of incident power. It divides the signal to 2 channels, one of them is connected to thermistor MW power meter and another is connected to MW detector. Signal of the reflected microwave power goes to a microwave detector. The microwave detectors and power meter are situated outside the MOU and connected with directional couplers by flexible waveguides. Additional mirror (rotary switch) is situated between the main MOU mirrors and reflects the gyrotron radiation to the calorimetric dummy load.

The set of unique high-voltage power supplies (HVPS) made in IAP RAS provides synchronous operation to provide gyrotron operation at energy recovery regime (depressed collector) and ensure very short radiation power fall time (less than 1 μs) and rise time about 3 μs in pulse operation mode (Fig. 4) in the full range of output power levels. Cathode HVPS provides 25 kV / 2.4 A (negative polarity) output and anode HVPS provide 15 kV / 0.1 A (positive polarity) output in CW and pulse modes. PID-regulation algorithm is used for stabilization of output power level using feedback signal from microwave power meter or semiconductor detector in CW or pulse mode.

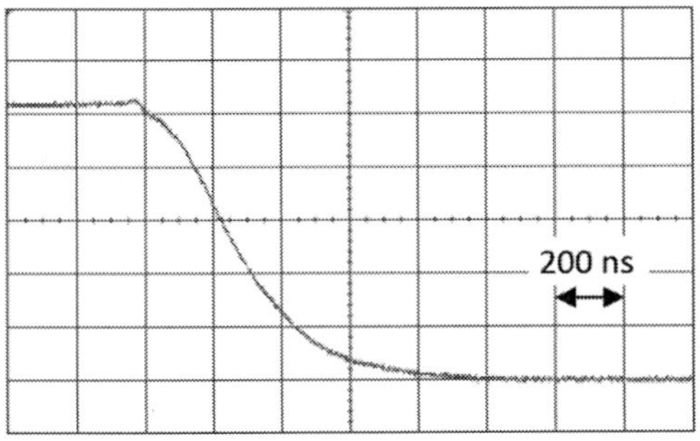

Fig. 4. Oscilloscope trace of the microwave power pulse falling edge

The setup is fully equipped with the automated control, data acquisition and safety interlock systems that guarantee convenient and trouble free remote and local operation. These systems provide starting up the setup and preparing it for operation follows a pre-set algorithm and automatic (full or partial) shutting out the power supplies in the event of failure or in emergency. The interrelation between operations of the setup individual devices is determined by the wiring of the relays without using any central processing unit or software. In order to provide fast switching off in the case of breakdown the signal from ARC-detector goes directly to cathode HVPS. Server software interacts with Data Acquisition System and passes information to the client software for displaying the actual state on the graphical user interface (Fig. 5). Possibility of integration into ion accelerator facility is achieved through the using of EPICS Channel Access network protocol-based interaction with server software.

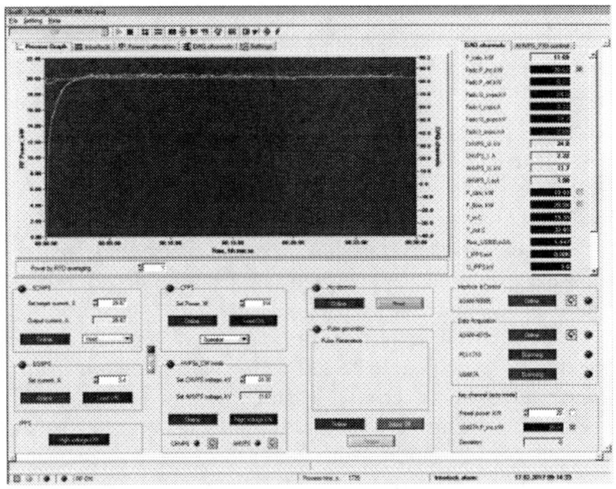

Fig. 5. The Client GUI displaying the results of microwave power control

Acknowledgements

Partially supported by the RF President Council for Grants, grant MK-3452.2017.8.

References

1. *H. Zhao, L. Sun, J. Guo, W. Lu, Y. Yang, X. Zhang,* Book of Abstracts: 16th Int. Conf. on Ion Sources, NY, USA, 2015, ThuPS26.

2. *H. Zhao, L. Sun,* The 22nd Int. Workshop on ECR Ion Sources, Busan, Korea, 2016, MOBO01.

3. *G. Denisov, Yu. Bykov, M. Glyavin* et. al., https://doi.org/10.1109/PLASMA.2016.7533987.

A quasi-optical input for a whispering-gallery-mode gyro-twystron

S. Filchenkov, I. Gachev, M. Goikhman, Yu. Guznov, D. Lukovnikov, A. Tsvetkov
Institute of Applied Physics RAS, Nizhny Novgorod, Russia, tsvetkov@ipfran.ru

Beginning with 2005, the International Science and Technology Center (ISTC) funded a research program (grant No. 3169) for development of Ka-band gyro-klystrons operating at combinations of high order modes. A number of versions differing in RF inputs and in sequences of operating modes have been tested at microsecond pulses [1, 2]. In particular [2], a 35.4 GHz gyro-klystron with the output rotating mode TE73 delivered 15 MW pulses with 33% efficiency and 30 dB gain.

One of further modifications of the existing configuration is planned to operate at sequences of rotating modes with high azimuthal indexes $m >> 1$ and the radial index equal to 1. Operation at such whispering gallery modes would allow to expand the RF amplification band and approximate the gyro-klystron to a gyro-twystron.

In such an amplifier, the electron-beam-modulation section is expedient to be fed with a wave arriving from a waveguide being coaxial relative to the tube axis [3] (Fig. 1). The drive RF signal would be injected to the structure from a horn (Fig. 2). In this case the azimuthal index m_h of the excited coaxial waveguide mode would be excessively high; and to convert this mode into the counter-rotating operating mode with azimuthal index m, the inner wall of the coaxial waveguide would be corrugated with azimuthal index $\overline{m} = |m| + |m_h|$

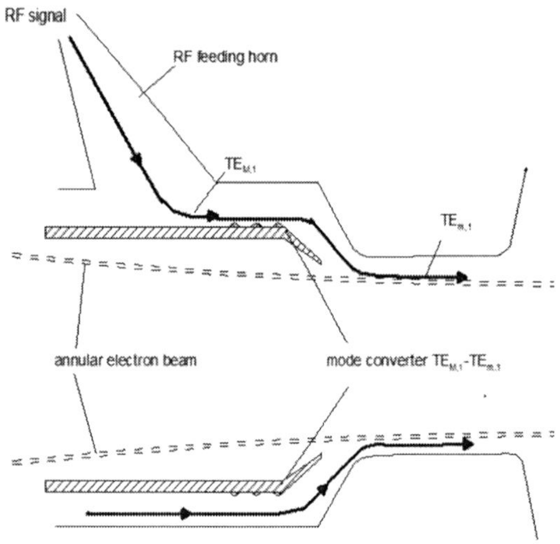

Fig. 1. A scheme of quasi-optical RF input into the electron beam modulating section

Fig. 2. A scheme of matching of a horn to a coaxial wavegude

The above approach was used to preliminary design a 35.4 GHz gyro-twystron to operate at a sequence of TE modes with the common azimuthal index equal to 7. The junction between the feeding coaxial waveguide and the electron beam modulating section (Fig. 3) was optimized to suppress parasitic resonances and minimize parasitic RF losses. A $TE_{15,1}$ mode would be launched from a Fig. 2 horn to the coaxial waveguide and then converted to the $TE_{7,1}$ mode by means of the inner wall corrugation with azimuthal index $\overline{m} = 22$

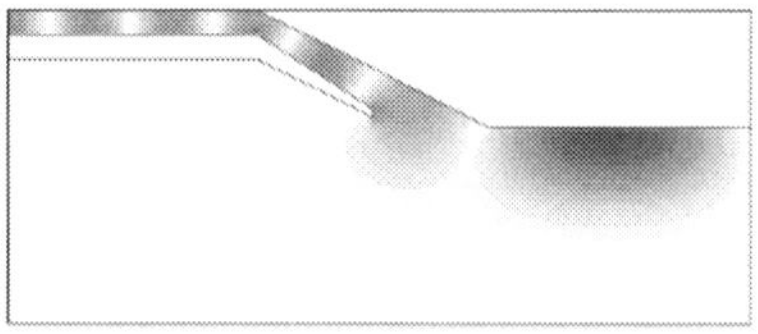

Fig. 3. The coupling of the feeding coaxial waveguide and the electron beam modulating section at the $TE_{7,1}$ mode (numerical simulation)

By analogy with present-day free-running gyro-monotrons, future gyro-twystrons might operate at modes with azimuthal indexes up to 20–30.

Acknowledgements

Authors are grateful to Prof. M. I. Petelin for constructive discussions.

The research was partially supported by RFBR, grant № 16-32-00431.

References

1. *Petelin M., Danilov Yu., Pavelyev V., Zasypkin E., Zaitsev N. A* research to high-order-mode gyroklystrons // Proc of IVEC 2010, 2010, Monterey, California, pp.191-192.

2. *Zaitsev N., Petelin M., Guznov Yu.* // 9th International Workshop Strong Microwaves and Terahertz Waves: Sources and Applications, 2014, p. 174.

3. *Erckmann V., Hirshfield J.L., Kasparek W., Petelin M.I., Salin M.B., Shchegolkov D.Yu., Zaitsev N.I.* // Proc. of IVEC 2008. Monterey. California. USA. 2008. pp. 87–88.

EPJ Web of Conferences **149**, 04034 (2017) DOI: 10.1051/epjconf/201714904034
10th International Workshop 2017 "Strong Microwaves and Terahertz Waves: Sources and Applications"

Development of high-efficient gyrotron based complex for industrial applications

M.D. Proyavin[1], M.Yu. Glyavin[1] and V.N. Manuilov[1,2]

[1]Institute of Applied Physics of Russian academy of Sciences, Nizhny Novgorod, Russia, mdchase@yandex.ru
[2]Lobachevsky State University, Nizhny Novgorod, Russia

Introduction

CW gyrotrons with operating frequencies from 24 to 28 GHz and an output power of several tens of kilowatts are being considered as the next stage in development of well-known technological complexes [1, 2] for the sintering of composite materials, growing CVD diamonds and other advanced technologies. In this regard, the total efficiency of microwave system plays a key role for real industrial employment. The report examines the possibility of reducing power consumption of magnet by the installation of ferromagnetic screens surrounding the main solenoid [3]. The project of 25 kW / CW / 28 GHz gyrotron based system is presented.

Development of the magnetic system

To reduce the power of the main coil it was surrounded by the system of disk screens and inner and outer cylindrical screens (fig. 1 a). As a result, after optimization of the screens shape, it was possible to decrease the value of power consumption of the magnetic system in more than 2–2.5 times (see also fig. 1 b).

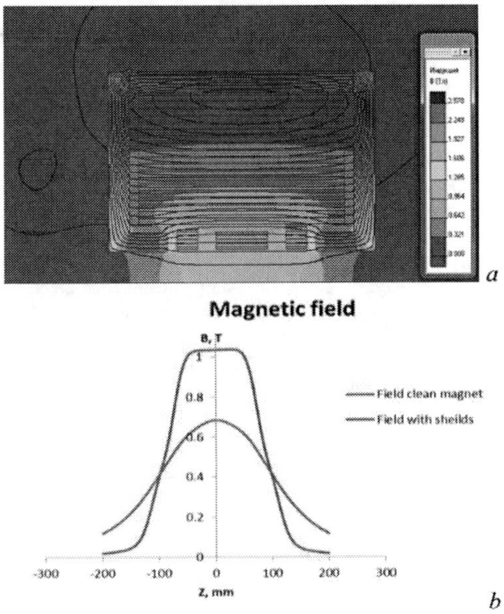

Fig. 1. The scheme of magnetic system with shielded solenoid (a). Magnetic field lines are shown. The axial magnetic field distributions (b) for the cases without (lower curve) and with installed ferromagnetic screens (upper curve) are shown

Small correcting coils placed inside main coil volume allowed to adjust the magnetic field distribution Bz(z) in such a manner as to provide its inhomogeneity less than 1% and reach the operating magnetic field 1.03 T in the cavity.

At the same time ferromagnetic screens substantially increases the inhomogeneity of the magnetic field outside the screens, where the initial formation of the helical electron beam (HEB) takes place. So the coefficient of non-adiabaticity $\varepsilon = h/L_B$ (here h is the step of electron trajectory, L_B is the scale of the magnetic field) exceeds 2 in the formation region (fig. 2). This makes it difficult to use well developed and reliable magnetron-injection gun (MIG) to form HEB with good enough parameters. To smooth the magnetic field distribution the additional cathode coils was installed to the left from the screen region (see fig. 1a). It allowed to form adiabatic distribution of the magnetic field with $\varepsilon < 1$ (fig. 2) and to employ adiabatic electron optical system (MIG) to minimize effects of misalignment, variation of technical parameters, etc.

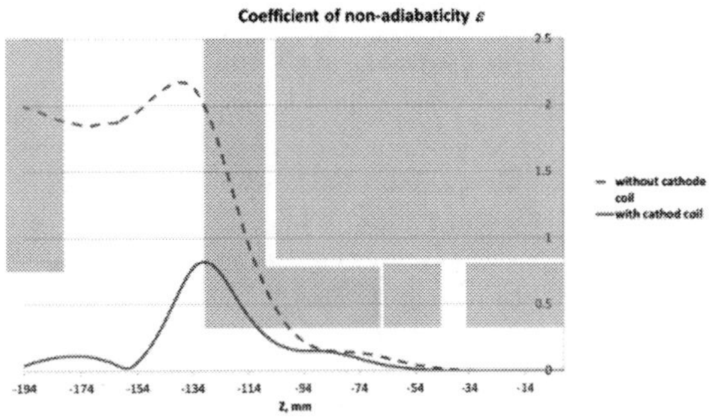

Fig. 2. The dependence of the coefficient of non-adiabaticity ε on the longitudinal coordinate in the optimized magnetic system. $Z = 0$ corresponds to the cavity center

Electron-optical system

Initial form and position of MIG was found on the basis of adiabatic theory of MIGs [4]. Accelerating voltage was chosen 23 kV, maximum operating current 2.4 A, cathode radius 19 mm. To diminish the influence of the space charge forces on the beam quality the angle of emitter cone with respect to the axis was chosen 30 degrees to form laminar beam. Further optimization of the electrodes shape (fig. 3) has shown that HEB parameters (see fig. 4) are acceptable for high efficient (about 50%) microwave generation.

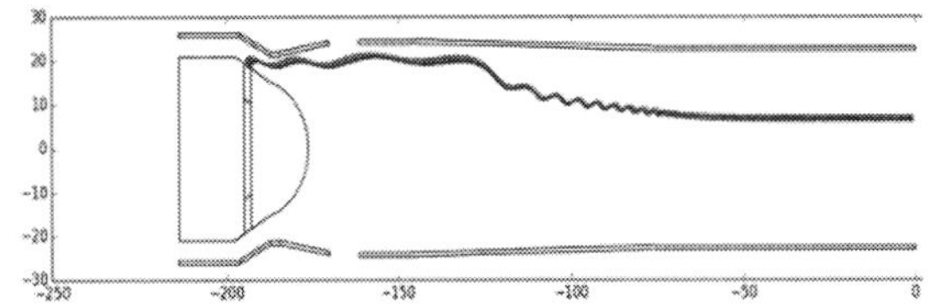

Fig. 3. Optimized geometry of MIG and corresponding electron trajectories

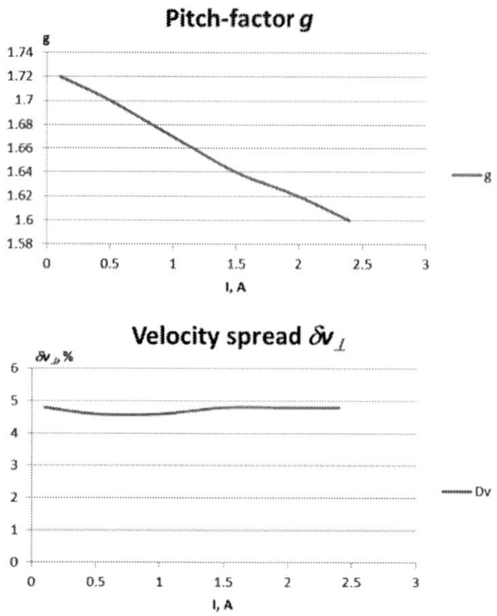

Fig. 4. Dependence of the pitch-factor g and velocity spread on the operating current in the optimized version of MIG

Choice of the operating mode and cavity design

The analysis of the mode spectrum as well as the comparison of value of the coupling factors for some candidates to the operating mode with moderate azimuthal and radial indexes (modes TE_{02}, TE_{22}, TE_{03} TE_{13} and TE_{13} were considered) and the neighboring competiting modes has shown that the most suitable is TE_{13} one. Further preliminary optimization of the output efficiency η of the gyrotron within the model of fixed RF field distribution [6] with taking into account the ohmic losses has shown the possibility to overcome the value $\eta = 50\%$. Final optimization was performed on the basis of more complete model [7], which takes into account the influence of the own beam space charge on the interaction process (see fig. 5). According to the results of numerical simulation the total gyrotron efficiency is 51%, output power 25 kW.

During the optimization the restrictions on the power consumption of the main coil and its length also were considered.

Conclusion

Technological gyrotrons with hot magnets surrounding by ferromagnetic screens allow to increase the operating value of the magnetic field in 1.5–2 times approximately and thus to use the generation on the first cyclotron harmonic in the frequency range 24–28 GHz with power level some tens kW and at the same time to keep the existing and well mastered coil cooling system. This opens up new opportunities in mastering novel promising technologies based on powerful CW radiation sources.

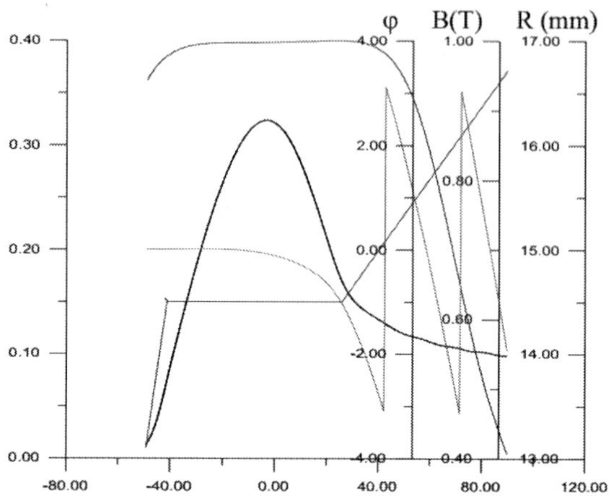

Fig. 5. Distribution of the RF field (black), the cavity profile (red) and the axial magnetic field distribution (blue) along the cavity region

Acknowledgements

The work was supported by the Russian Foundation for Basic Research, project No. 17-02-00183.

References

1. *V.A. Flyagin, A.N.Kuftin, V.K.Lygin, O.V.Malygin, B.V.Raisky, Gh.Tsimring, V.E.Zapevalov.* CW technological gyrotrons in the range 15-50 GHz // 4th Russian-German meeting on ECRH and Gyrotrons, N.Novgotrod, Russia, May 11-17, 1992
2. *Yu. Bykov, A. Eremeev, M. Glyavin, V. Kholoptsev, A. Luchinin, I. Plotnikov, G. Denisov, A. Bogdashev, G. Kalynova, V. Semenov, and N. Zharova.* 24-84 GHz Gyrotron Systems for Technological Microwave Applications // IEEE Transactions on Plasma Science, vol.32, no.1, 2004, pp.67-72
3. *A.N. Kuftin, S.P. Belov.* Magnetic shielded electron optical system of a gyrotron in space charge limited current regime // Applied Physics, 2000, №3, c.76-81.
4. *A.L. Gol'denberg and M.I. Petelin.* The Formation of Helical Electron Beams in an Adiabatic Gun // Radiophys. Quantum Electron., 16(1), 106-111 (1973)
5. *P.V. Krivosheev, V.K. Lygin, V.N. Manuilov, and Sh.E. Tsimring.* Numerical Simulation Models of Focusing Systems of Intense Gyrotron Helical Electron Beams // Int. J. IR MM Waves, 22(8), 1119-1146 (2001)
6. *G.S. Nusinovich.* Introduction to the physics of gyrotrons. Baltimore; London: The Johns Hopkins University Press, 2004, 352 p.
7. *N.S. Ginzburg, G.S. Nusinovich, N.A. Zavolsky.* Theory of non-stationary processes in gyrotrons with low Q resonators // Int. J. Electronics. 1986. Vol. 61, № 6. P. 881.

EPJ Web of Conferences 149, 04035 (2017) DOI: 10.1051/epjconf/201714904035
10th International Workshop 2017 "Strong Microwaves and Terahertz Waves: Sources and Applications"

Non-adiabatic electron-optical system for 170GHz/1MW/CW gyrotron

K.A. Leshcheva[1,2], A.L. Goldenberg[2], M.Y. Glyavin[2] , V.N. Manuilov[1,2]

[1]Lobachevsky State University of Nizhni Novgorodizhny, N. Novgorod, Russia,
e-mail: ksenilescheva@mail.ru
[2]Institute of Applied Physics, Russian Academy of Sciences, N. Novgorod, Russian

Introduction

The classical scheme of a gyrotron's electron-optic system (EOS) employs an adiabatic magnetron injection gun (MIG) [1] to form a helical electron beam (HEB) with the oscillatory energy and the velocity spread of electrons good enough for high efficiency generation of microwaves. From the condition of the stability of the electron beam in the trapped configuration of the magnetic field distribution it follows that the maximum pitch factor increases when the velocity spread goes down. Therefore, the decreasing of the velocity spread is the most important approach for improving the gyrotron's efficiency, especially because the spread tends to grow with increasing of the power and the operating frequency of the device. In the MIGs of millimeter and submillimeter wave tubes the most essential factors contributing to the velocities spread are the presence of the initial thermal velocities of electrons and, more importantly, the emitter's roughness. One of the possible ways to decrease the influence of the mentioned factors is to use a non-adiabatic scheme for HEB formation [2]. In this case a rectilinear beam is formed initially (fig. 1) and, as a result, the influence of these factors becomes small, which makes it possible to significant increase the gyrotron's efficiency.

Non-adiabatic electron gun in the temperature limiting regime

Below a version of a non-adiabatic EOS aimed to be used in a 170GHz/80kV/30A gyrotron is considered. Here the initial transverse velocities of the electrons are formed by injection of a rectilinear electron beam at an angle of Ψ to the magnetic field (fig.1). After that the adiabatic compression of the beam in the increasing magnetic field is applied to achieve the desired value of the pitch-factor. The required value of Ψ is provided by corresponding adjustment of the cathode coils current (fig. 1 and 2). The advantage of the suggested scheme is that the value of the pitch-factor may be varied independently of the beam compression ratio, i.e. the beam position in the cavity.

The optimal shape and position of the cathode coils providing minimal energy consumption for specified value of Ψ is found.

Simple analytical formulas and some qualitative considerations to find the preliminary shape of electrodes and gun position with respect to the main coil are presented. For preliminary estimations of the gun's position, the value of the angle Ψ, the cathode-anode gap and the shape of the electrodes, simple physical model based on the energy conservation law, Bush theorem and transverse adiabatic invariant conservation law is applied. The recommendations on the choice the electrodes' geometries providing close angles of injection into magnetic field of the particles starting from the different positions on emitter are given.

The final gun design (fig. 3) is found on the basis of numerical simulation. The optimization was performed using the software package CST Studio Suite [3] and EPOS [4].

Special attention was paid to avoid the electric breakdown. For this purpose the distances between cathode and first anode, first and second anodes, as well as the radii of the curvature of the corresponding electrodes were optimized (see fig. 4). This ensured the electric field value less than 7.7 kV / mm in any point of the formation system.

The results of numerical simulation showed that the use of this nonadiabatic scheme for formation of a helical electron beam for the 170 GHz / 80 kV / 30 A gyrotron allows one to eliminate the influence of such critical for gyrotron operation factor as the condition of the emitter's surface on the value of the velocity spread and makes it possible to get pitch factors up to 1.6 (instead of 1.2–1.3 for conventional MIGs) at the velocity spread of about 30% (Fig. 5). It makes it possible essential increase efficiency of the device. It is interesting to note that unlike in the MIGs, the pitch-factor increases when the current grows, while the velocity spread changes very smoothly.

Space charge limiting regime of electron gun

The formation of the initially rectilinear beam in the cathode-anode gap allows also to use the beam space charge limiting regime instead of temperature limiting one. It is more convenient for operation due to the improved beam stability and possibility of quick (i.e. inertia-free) control of the beam current.

To ensure the space charge limited current regime, the initial geometry of the electrodes has to be altered because in such regime the strongest space-charge defocusing effect takes place. To compensate the defocusing of the beam, the Pierce like gun shape of electrodes, where the conical surfaces are inclined to the beam at an angle of 67.5°, was used (see Fig. 6).

The simulation results made it possible to find the optimal parameters and geometry of the EOS in the regime of the current limitation by the space charge and to investigate the velocity distribution in the electron beam being formed.

Conclusion

The new version of the non-adiabatic electron-optic system suitable for formation HEBs in high-efficient gyrotrons is considered and optimized. The EOS can be employed both in the temperature limiting and space charge limiting regimes.

Acknowledgements

This work was supported by RFBR grant No 16-02-00674.

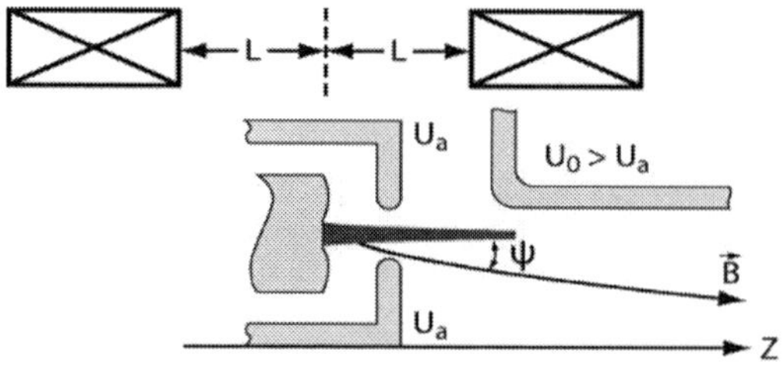

Fig. 1. Principal scheme of the HEB formation region

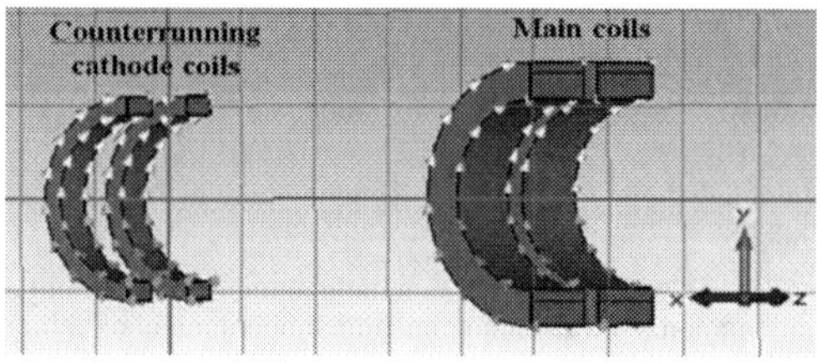

Fig. 2. Principal scheme of the of the gyrotron magnetic system

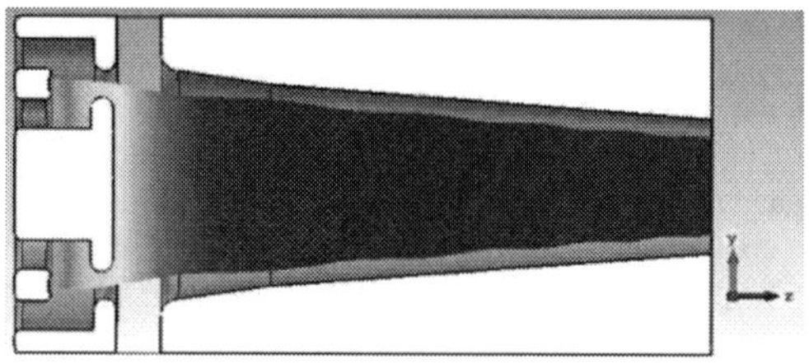

Fig. 3. The configuration of the electrodes in non-adiabatic EOS for 170 GHz/1 MW gyrotron

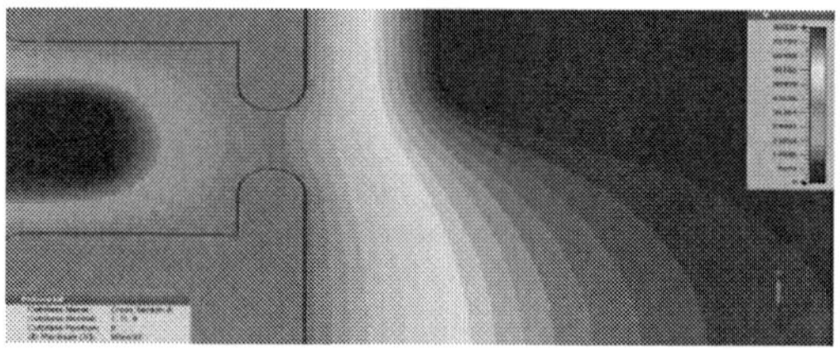

Fig. 4. The map of the equipotential lines in the gun region

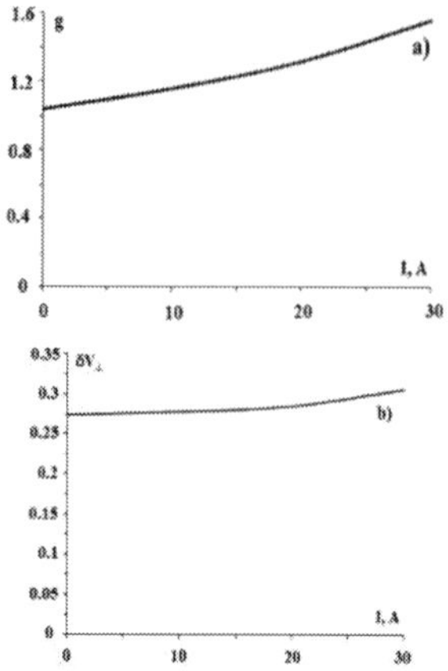

Fig. 5. Dependence of the pitch factor (a) and the velocity spread (b) on the beam current

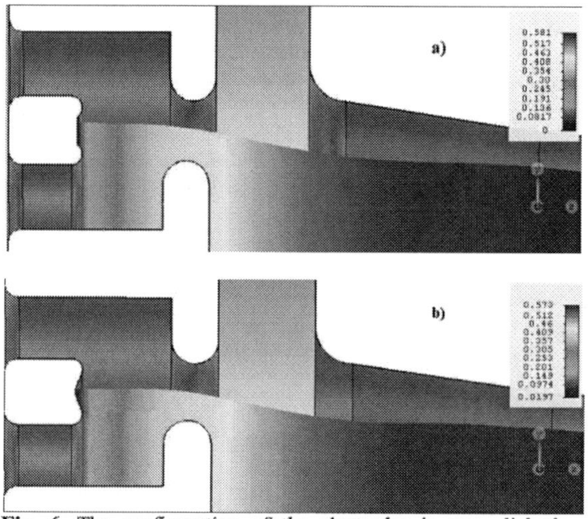

Fig. 6. The configuration of the electrodes in non-adiabatic electron-optical gun for the temperature-limiting emission regime (a) and the space charge limiting regime (b)

References

1. *A.L. Goldenberg, M.I. Petelin.* Formation of helical electron beams in adiabatic gun// Radiophysics and Quantum Electronics, vol.16, No 6, 1973, pp. 141-149.
2. *A.L. Goldenberg, V.N. Manuilov and M.Yu. Glyavin.* Electron-Optical System of a High-Power Gyrotron with Non-adiabatic Electron Gun// Radiophysics and Quantum Electronics. 2005. V. 48. No 6. P. 461-465.
3. CST STUDIO SUITE www.cst.com.
4. *P.V. Krivosheev, V.K. Lygin, V.N. Manuilov, and Sh.E. Tsimring.* Numerical Simulation Models of Focusing Systems of Intense Gyrotron Helical Electron Beams// Int. J. IR MM Waves. 22(8). 1119-1146 (2001).

EPJ Web of Conferences **149**, 04036 (2017) DOI: 10.1051/epjconf/201714904036
10th International Workshop 2017 "Strong Microwaves and Terahertz Waves: Sources and Applications"

Development of field emitter non-adiabatic electron optic system for the spectroscopic 263 GHz/CW gyrotron

V.N. Manuilov[1,3], G.G.Sominskii[2], E.P. Taradaev[2], T.A. Tumareva[2] and M.Yu. Glyavin[1]

[1] Lobachevsky State University, Nizhny Novgorod, Russia, manuilov@rf.unn.ru
[2]Peter the Great St. Petersburg Polytechnic University, St.Petersburg, Russia
[3] Institute of Applied Physics, Nizhny Novgorod, Russia

Introduction

At present, gyrotrons occupy the leading position among high-power microwave radiation sources in the millimeter and submillimeter wavelength ranges. One of the most promising new applications of gyrotrons is the increase of the sensitivity of the spectrometry based on the nuclear magnetic resonance (NMR) and dynamic nuclear polarization (DNP). In such case the power level about some tens watt in the frequency range 200–400 GHz is needed. In spite of successful experiments [1] with first 263 GHz / CW gyrotron, where the output power reached even 1 kW, the traditional scheme of the gyrotron with adiabatic magnetron-injection gun for mentioned above application has some disadvantages. The most important are small sizes of high temperature thermionic cathodes (it causes high sensitivity to thermal deformations), big influence of emitter roughness (lead to big velocity spread) and longtime of the switch on process. To avoid the mentioned above problems it is interesting to examine the quite new scheme of the helical electron beam formation system, based on two principles – to use non-adiabatic electric and magnetic fields [2] and besides, field emitter [3] instead of hot cathode. First approach allows eliminate the influence of the cathode roughness, because the rectilinear electron beam is formed initially; second - to make the device inertia-free and not sensitive to the cathode temperature.

In what follows the feasibility of the described above approach is examined on the example of a gyrotron with an operating frequency of 263 GHz, an accelerating voltage of 15 kV, and an output power about ten watts, which is sufficient for spectroscopy applications.

Gyrotron

To check the feasibility of the described above idea, it was decided to try it using the already existing gyrotron for spectroscopy [8] (f = 263 GHz, power up to 1 kW), with only the beam generation system replaced and the other components (resonator, radiation-to-wave beam converter, collector, and magnetic system) left unchanged. Such an approach reduces the cost of first experiments and checking the principle itself of using field emission in gyrotrons. Owing to this, the main gyrotron parameters, were chosen as in the classical gyrotron implemented before [8].

Field emitter

The field emitter based on the multi-pin silicon structures (fig.1) on a flat substrate was chosen as the source of electrons because the technology of its manufacturing is fairly simple and well mastered [4]. Silicon pins have, as a rule, a low conductivity, but it can be increased using a thin (about 10-20 nm) layer of molybdenum. Fullerene or metallofullerene protective coatings deposited on the molybdenum metallization are effective against the destructive effect of ion bombardment. Estimations show that corresponding cathode can provide currents of up to 40 mA for the emission area S=0.2-0.3 cm^2. Average current densities over the emitter surface reach j~0.15-0.2 A/cm^2. To obtain the required values of the field emission current, approximately N~10^4 pins on the emitting cathode surface must be placed, while the electric field on the emitter surface should exceed 5-5.5 kV/mm. The estimates based on the Fowler-Nordheim formula show that even a field difference by 2-3% changes the emission current density by about 30% and, therefore, makes the space charge field non-uniform in the formation region and consequently may adversely affect the quality of the helical electron beam.

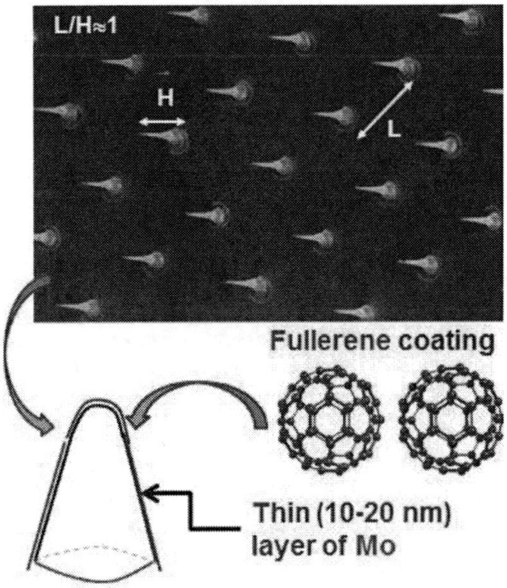

Fig. 1. Typical multi-pin cathode structure. Pin height H = 30 μm. The distance between the pins L = 30 μm. Pin tip diameter D = 36 nm. The scheme of the each pin coating is shown

Electron beam formation system design

The above requirements to the structure and characteristics of the emission surface are fairly rigid and were taken into account in the gyrotron electron optic system (EOS) development.

With the existing technology for manufacturing of multi-pin emitters, which we are going to use in a diagnostic gyrotron, it is nearly impossible to create conic emitters typical of the gyrotrons. However, the multi-pin emitting system in the form of a ring can be mounted on a flat end of a cylindrical cathode. With this cathode configuration, it is virtually excluded to use the well-developed adiabatic magnetron-injection guns (MIGs) [5] for the formation of a helical electron beam (HEB), Therefore, for the HEB formation in a gyrotron with field emitter, it was decided to use a non-adiabatic EOS, the scheme of construction of which is given in detail in [6]. Such an EOS permits one to use butt-end cathodes for the HEB formation. Here, primary gyration energy of the beam is provided by injection of the initially formed rectilinear electron flow obliquely to the magnetic field. The further increase in the oscillatory energy of the particles to the required level is reached by adiabatic compression of the beam in a gradually increasing magnetic field. Note that the scattering field of the main solenoid (fig. 2) has very small angles of inclination of the magnetic field line to the axis, i.e., the field lines are almost parallel to the normal to the cathode, and therefore a pair of counter-running cathode coils located symmetrically with respect to the emitter should be installed before the beam injection obliquely to the magnetic field to ensure the required angle of inclination of the magnetic field to the axis with minimum energy consumption.

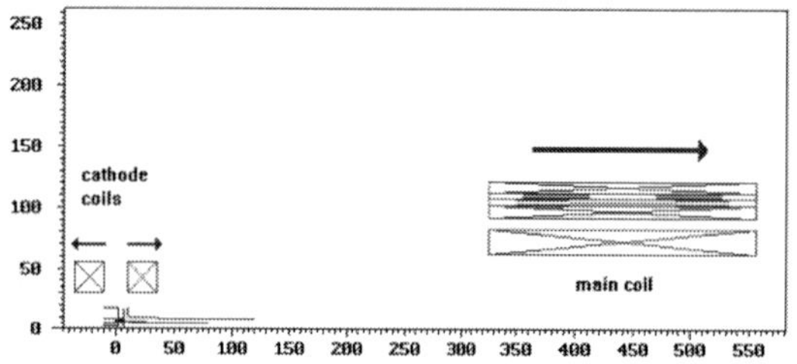

Fig. 2. General view of the system. Arrows show the coil polarities. Left: auxiliary coils at the cathode. Right: the main solenoid. All sizes are given in mm

With the existing technology, it is difficult to manufacture multi-pin ring field emitters with an average radius R greater than 10 mm. On the other hand, the minimal cathode radius is also limited since it is needed to provide the operating beam current about 40–50 mA. Summing up, it can be inferred that the average radius of a ring emitter should satisfy the inequality $4 < R$ [mm] < 10. Therefore during the numerical simulation, the values of R were enumerated with 2-mm step in the mentioned interval.

Numerical optimization of the location, shape, and currents of the counterrunning cathode coils, and the shape of the electrodes was performed using two-dimensional code EPOS [7]. Simulation shows that for optimized version of the electron gun R = 7 mm. In the range of currents 0–50 mA the pitch-factor and velocity spread remains close to 1.2 and 30% respectively that is quite admissible to achieve a typical efficiency of a few percent for such gyrotrons.

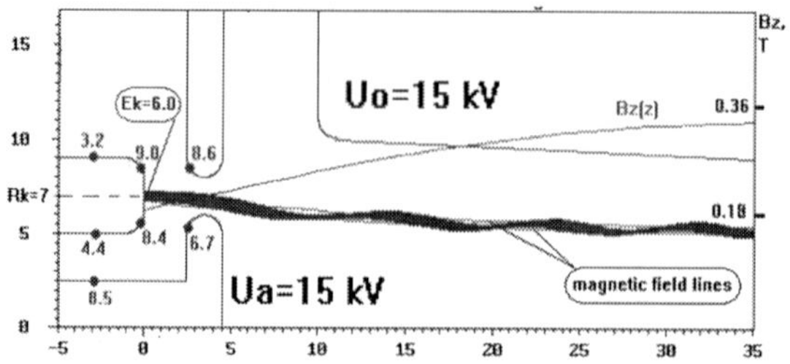

Fig. 3. Scheme of the primary formation of the beam in a non-adiabatic EOS. Distributions of the axial magnetic field B(z) (right axis, data are in T) is presented. Dots and corresponding numbers show the electric field value (kV/mm) at the most critical points. Electric field Ek at the emitter center is also specified. All dimensions are in mm

Conclusion

The concept of an electron-optical system of a sub-terahertz gyrotron with field emitter is described. It is shown that the non-adiabatic scheme of the beam formation with a flat ring multi-pin emitter makes it possible to form the HEB with quality acceptable for operation of modern gyrotrons intended for spectroscopy and diagnostics of various media.

Acknowledgements

The work was supported by the Russian Science Foundation, project No. 16-12-10010.

References

1. *M.Yu. Glyavin et al.* Experimental Tests of 263 GHz Gyrotron for Spectroscopy Applications and Diagnostic of Various Media // Rev. Sci. Instr., 86(5), 054705 (2015).
2. *A.L. Goldenberg, M.Yu. Glyavin, N.A. Zavolsky, V.N. Manuilov* Technological gyrotron with low accelerating voltage.// Radiophys. Quantum Electron., 48 (10-11), 741-747 (2005).
3. *G.G. Sominskii, T.A. Tumareva, E.P. Taradaev, M.V. Mishin, A.N. Stepanova.* Multitip semiconductor field emitters with new-type bilayer protecting coatings. // Tech. Phys., 60(1), 133-136 (2015).
4. *Crystal Wisker Probes.* // Kristallografiya, 51(5), 947–953 (2006).
5. *A.L. Gol'denberg and M.I. Petelin.* The Formation of Helical Electron Beams in an Adiabatic Gun // Radiophys Quantum Electron., 16(1), 106-111 (1973).
6. *A.L. Gol'denberg, M.Yu. Glyavin, K.A. Leshcheva, and V.N. Manuilov.* Nonadiabatic Electorn-Optical System of a Technological Gyrotron // Radiophys. Quantum Electron, 2017, in press.
7. *P.V. Krivosheev, V.K. Lygin, V.N. Manuilov, and Sh.E. Tsimring.* Numerical Simulation Models of Focusing Systems of Intense Gyrotron Helical Electron Beams // Int. J. IR MM Waves, 22(8), 1119-1146 (2001).
8. *M.Yu.Glyavin et al.* Experimental Tests of 263 GHz Gyrotron for Spectroscopy Applications and Diagnostic of Various Media, Rev. Sci. Instr., 86(5), 054705 (2015)

EPJ Web of Conferences **149**, 04037 (2017) DOI: 10.1051/epjconf/201714904037
10th International Workshop 2017 "Strong Microwaves and Terahertz Waves: Sources and Applications"

Influence of weak reflection from a nonresonant load on the operation frequency of the 28 GHz technological gyrotron

A.A. Bogdashov, G.G. Denisov, A.P. Fokin, M.Yu. Glyavin, Yu.V. Novozhilova, A.S. Sedov, A.I. Tsvetkov

Institute of Applied Physics of the Russian Academy of Sciences (IAP RAS) Nizhny Novgorod, Russia, ap.fokin@mail.ru

The frequency of the output radiation of a gyrotron can be unstable because of the fluctuations of the power supply voltage. For several applications of the gyrotrons frequency tunability or frequency stabilization are required. The usual way of frequency variation or stabilization is the control of the accelerating or the modulating voltage [1, 2]; while there exist some other ways to stabilize the frequency of the gyrotron [3] that seem to be promising. In our work we investigate the frequency stabilization of a gyrotron by weak reflection from the nonresonant load. The advantages of his method are the simplicity of reflector manufacturing and absence of complex and expensive fast-acting power supplies.

Experimental setup

In our experiment, we used the continuous wave gyrotron for technological applications with operating frequency 28 GHz at the second cyclotron harmonic. The output power was partially reflected back to the gyrotron cavity by the nonresonant reflector – the ring of the smaller inner diameter, inserted to the output waveguide. The parameters of the ring were selected in such way, that reflection coefficient was 10% for radiation power. Experimental setup allowed the step movement of the reflector in the range of several wavelengths with steps of 2 mm.

For the experiment we limited the beam current at 0.5 A to prevent the excess power reflection back into the gyrotron. The operation point was far from optimal for this gyrotron, and with accelerating voltage of 19.4 kV, the output power is about 2 kW, so the reflector parameters also will not be distorted by thermal losses.

Results of the experiment

Depending on the distance to the reflector, either the frequency stabilization or the increase of the frequency modulation can be achieved. On the Fig. 1. presented the dependence of gyrotron frequency on the magnetic field. In the region of lower magnetic fields (that correspond the optimal region for output power) the presence of the reflector changes the incline, stabilizing the frequency. The spectrum width in the experiment was reduced up to two times and the frequency tenability in range of 3 MHz was achieved.

Along with the change of the dependence of gyrotron frequency on the magnetic field, there is a considerable change of the gyrotron output power with the distance to the reflector.

This can be explained by the change of the Q-factor of the complex resonator, consisting of gyrotron resonator and the waveguide section up to the reflector. The movement of the reflector in the range of wavelength can double the Q-factor, leading to better stability of the frequency.

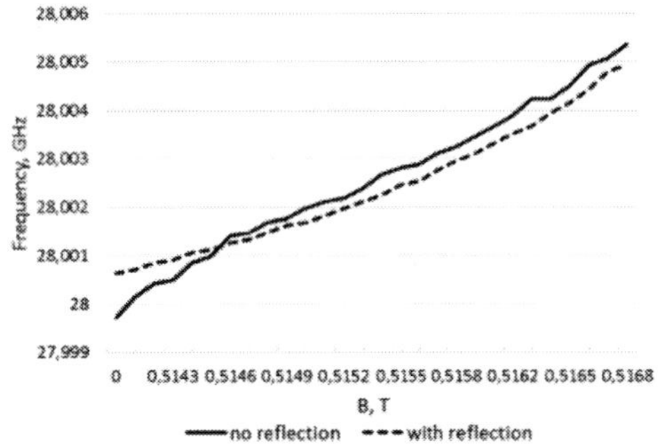

Fig. 1 Experimental dependence of frequency on the magnetic field with and without reflections

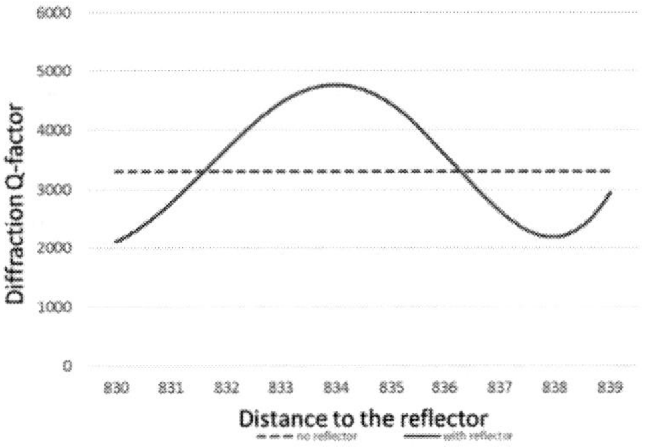

Fig. 2. Calculated dependence of diffractional Q-factor on the distance to the reflector

The movement of the operating point on the interaction length – normalized beam current plane due to the change of Q-factor explains the seeming paradox of gyrotron power increase in presence of noticeable reflection. Such mechanical tunability of frequency and power opens a new way of control over the output parameters of gyrotron radiation.

This work was supported by Council on grants of the President of Russian Federation, project # MK-3452.2017.8

References

1. *Idehara T., Mitsudo S., Ogawa I.,* Development of High-Frequency, Highly Stable Gyrotrons as Millimeter to Submillimeter Wave Radiation Sources // IEEE Transactions on Plasma Science, 2004. V. 32. №3. P. 910-916

2. *G.Yu. Golubiatnikov, et al,* Gyrotron frequency control by a phase lock system // Technical Physics Letters, 2006, Volume 32, Issue 8, pp 650–652

3. *M.Yu. Glyavin, G.G. Denisov, M.L. Kulygin, and Y.V. Novozhilova,* Stabilization of gyrotron frequency by reflection from nonresonant and resonant loads // Tech. Phys. Lett., vol. 41, no. 7, pp 628–631, 2015b.

Non-relativistic hollow electron beam formation for mm-wave BWO

V.L. Bratman[1,2], A.E. Fedotov[1], P.B. Makhalov[1,3], and V.N. Manuilov[1,4]

[1]Institute of Applied Physics of the RAS, Nizhny Novgorod, Russia, fedotov@appl.sci-nnov.ru
[2]Ariel University, Ariel, Israel
[3]KTH University, Stockholm, Sweden
[4]Nizhny Novgorod State University, Nizhny Novgorod, Russia

Slow-wave devices such as traveling-wave tubes (TWTs), backward-wave oscillators (BWOs), and extended-interaction klystrons (EIKs) are widely used as medium-power sources of short-millimeter and subterahertz waves. One of the difficulties that limits the CW or average power of slow-wave devices at short waves is partial interception of the electron beams by fragile slow-wave structures (SWS) causing dangerous heating of the latter. In order to increase the power, the use of spatially distributed electron beams, mostly sheet beams, is widely studied last few years [1, 2]. Another feasible and maybe somewhat simpler configuration is hollow rectilinear thin-wall electron beams interacting with the slow waves of azimuthally-symmetric operating waveguides [3, 4] similar to that of relativistic high-current microwave oscillators. Like in the case of sheet beams, the use of hollow electron beams permits a significant decrease in current density and heating of the microwave structure. To study this option, a W-band BWO with operating voltage of 30–32 kV has been designed in the Institute of Applied Physics. Oscillator operates at the lower axi-symmetric $TM_{0,1}$ mode of microwave structure with a rectangular corrugation with following parameters: the period is 0.8 mm, the inner diameter is 1.8 mm and the outer radius is 2.71 mm. Simulations based on the PIC-code predicts sub-kW level of output power for W-band BWO-oscillator with 1 A current.

The formation of a hollow thin-wall electron beam is a critical point for the further experimental realization of the slow-wave devices discussed above. Since the beam particles travel very close to the corrugated surface, strict requirements on transverse pulsations of electrons should be imposed. Namely, for guiding the beam at a distance of 0.1 mm from the corrugation, parasitic transverse velocities should not exceed 0.05 *c*. To drive the W-band BWO, the 30 kV/1 A hollow electron beam with an outer diameter of 1.6 was designed. The beam is formed in Pierce-like electron gun with thermionic cathode and moderate emission density of 2 A/cm^2. To increase the current density, a 100-fold compression in an increasing magnetic field is used which leads to a tenfold increase in parasitic transverse velocities. The magnetic field is produced by main solenoid and cathode coil with reverse polarity. In the interaction region, the beam is guided in the magnetic field of 0.7 T.

Numerical simulations of the gun were carried out using 2-D code EPOS [5] and 3-D PIC-code CST Studio. In order to diminish the beam edge pulsations, the cathode geometry was optimized to provide very accurate beam formation, see Fig. 1. Simulations predict the beam wall thickness of 0.1 mm in the interaction region and small radii of cyclotron rotations of particles (Fig. 2). The particles at the outer edge of hollow beam have smaller cyclotron radii than at the inner edge, which is favorable for the beam guiding. According to these results, electrons can be guided at the distance of 0.08 mm from SWS wall. In simulations, parasitic cyclotron rotation mainly occurs due to the initial electron transverse velocities arising from the thermal electron motion and emitter surface roughness.

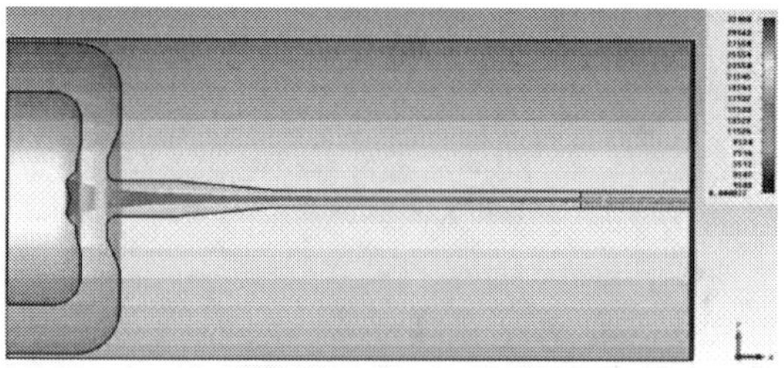

Fig. 1. Electron gun geometry and particles trajectories simulated by CST Studio code

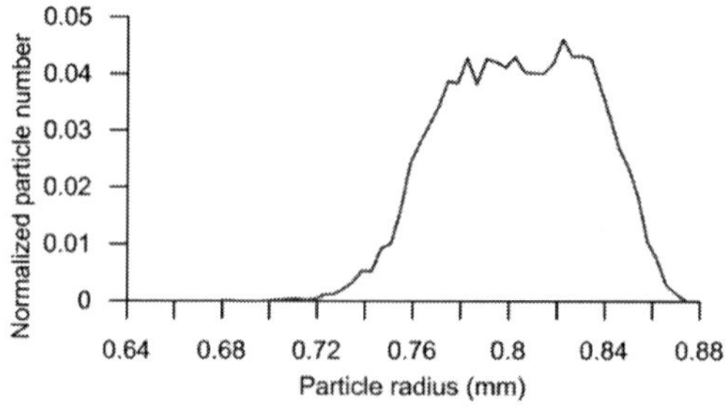

Fig. 2. Simulated distribution of electrons over radius in slow-wave structure

In the preliminary electron-optical experiment, 1-A hollow beam was formed. The outer beam radius could be varied by changing the magnetic compression ratio. For the designed parameters, 0.55 A current was guided through the slow-wave structure with inner radius of 0.9 mm while 0.45 A were intercepted by cutoff narrowing at the cathode side of the structure, see Fig. 3.

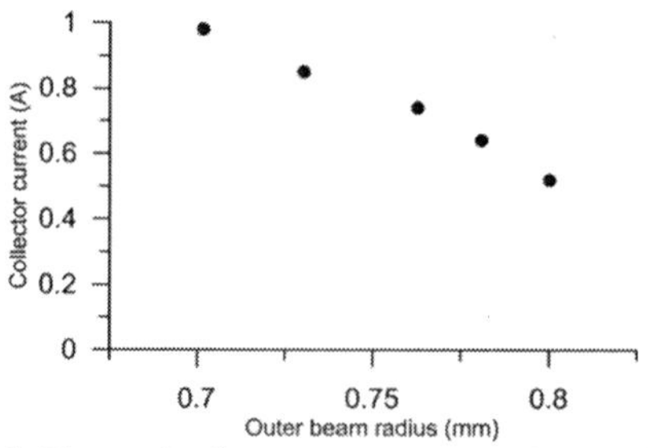

Fig. 3. Measured collector current vs. outer beam radius

Thus, in the experiment, the beam wall thickness of about 0.3 mm was measured while simulations predict beam wall thickness of 0.15 mm. The difference between simulations and experiment can be caused, in principle, by a number of reasons. One of the reasons is the possible misalignment of components of tube and magnetic system. Another probable reason is a cathode surface roughness which could result in an increased velocity spread of electrons. To study this effect, we have analyzed the cathode under microscope, see Fig. 4. The shape of the emitter surface was found numerically using a series of microscope photos made with different distance between the objective and the cathode. Then, this obtained shape was used in direct numerical simulations of the electron motion near the emitter surface made using 3-D PIC-code. Results of the simulations (Fig. 5) predict that emitter surface roughness could not be the main reason for the beam interception. These results are in good agreement with theoretical estimations based on known formulas. Therefore, we conclude that the misalignment of the solenoid and the cathode coil, on the one side, and the anode, cathode, and slow-wave structure, on the other side, is the most probable reason for the beam interception, so efforts will be made to minimize these effects in the following experiments.

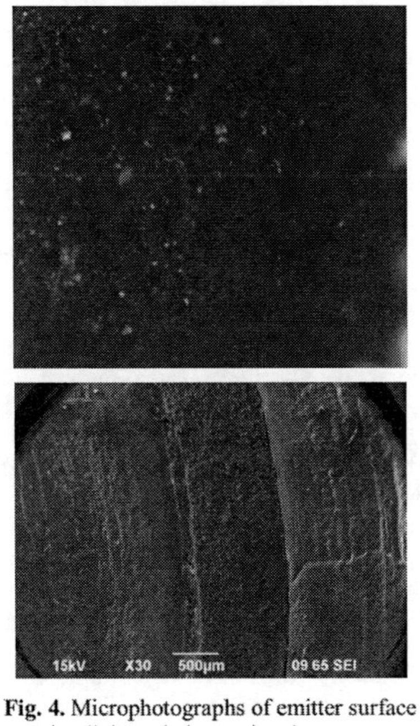

Fig. 4. Microphotographs of emitter surface using light and electronic microscopes

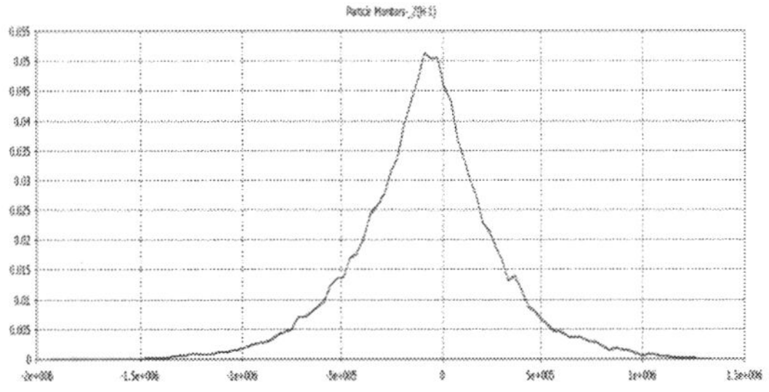

Fig. 5. Simulated distribution of parasitic electron transverse velocities caused by emitter surface roughness

Simulations of electron-wave interaction based on the averaged equations and 3D PIC-code CST predict that BWO with the existing hollow beam can provide output power up to 300 W (Fig. 6) and wideband frequency tuning.

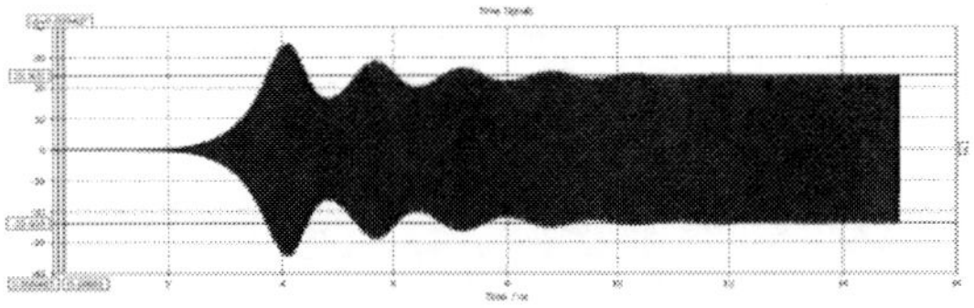

Fig. 6. Simulated output signal for BWO with 0.7 A hollow electron beam with wall thickness of 0.3mm

The work is supported by RFBR, project No. 16-08-01174.

References

1. *Y.-M. Shin, J.-X. Wang, L.R. Barnett, and N.C. Luhmann, Jr.* Particle-In-Cell simulation analysis of a multicavity W-band sheet beam klystron. // IEEE Trans. Electron Dev. 2011. V. 58, No. 1. P. 251-258.

2. *J. Lai, Y. Gong, X. Xu, Y. Wei, Z. Duan, W. Wang, and J. Feng.* W-band 1-kW staggered double-vane traveling-wave tube. IEEE Trans. Electron Dev. 2012. V. 59, No. 2. P. 496-503.

3. *V.L. Bratman, A.E. Fedotov, and P.B. Makhalov.* High-frequency devices with weakly-relativistic hollow thin-wall electron beams. // Physics of Plasmas. 2012. V. 19, No. 2. Art. 020704.

4. *V.L. Bratman, A.E. Fedotov, P.B. Makhalov, and V.N. Manuilov.* Design and numerical analysis of W-band oscillators with hollow electron beam. // IEEE Transactions on Electron Devices. 2014. Vol. 61, No. 6. P. 1795-1799.

5. *P.V. Krivosheev, V.K. Lygin, V.N. Manuilov, and Sh.E. Tsimring.* Numerical simulation models of forming systems of intense gyrotron helical electron beams. // Int. J. Infrared and Millimeter Waves. 2001. V. 22, N. 8. P. 1119-1145.

EPJ Web of Conferences **149**, 04039 (2017) DOI: 10.1051/epjconf/201714904039
10th International Workshop 2017 "Strong Microwaves and Terahertz Waves: Sources and Applications"

Progress in the development of low-voltage gyrotron for integration with NMR spectrometer

V.L. Bratman[1,2], A.E. Fedotov[1], Yu. K. Kalynov[1], V.N. Manuilov[1,3], and I.V. Osharin[1]

[1]Institute of Applied Physics of the RAS, Nizhny Novgorod, Russia, fedotov@appl.sci-nnov.ru
[2]Ariel University, Ariel, Israel
[3]Nizhny Novgorod State University, Nizhny Novgorod, Russia

While high-power, high-voltage gyrotrons for plasma fusion are most important gyrodevices nowadays, compact and less expensive gyrotron tubes with low operating voltage [1–4] could be an attractive opportunity for spectroscopic applications. The project of the medium-power gyrotron which can be integrated with nuclear magnetic resonance (NMR) spectrometer in a single cryomagnet (Fig. 1) is under development in the Institute of Applied Physics [1, 5]. This compact and low-voltage gyrotron ("gyrotrino") is aimed at dynamic nuclear polarization (DNP) technique and should deliver the CW power about of 10 W at the frequency of 263 GHz. The integrated solution [2–3] could eliminate the need for an additional superconducting magnet, result in a shorter terahertz transmission line, and make DNP systems less expensive.

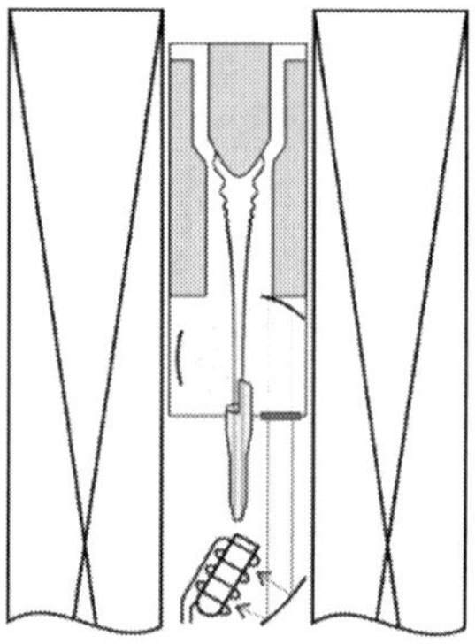

Fig. 1. Layout of the gyrotrino in NMR spectrometer magnet (not in scale)

The integration of the gyrotron with NMR-spectrometer in a single cryomagnet results in a number of specific features of such a THz oscillator, including electron-optical system, beam-wave interaction, and collector-end power output. The very low operating voltage of 1.5 kV results in very small anode-cathode distance in the electron gun and low electrical field at the emitter. To minimize velocity spread, electron gun is designed [6] to form a laminar electron beam, see Fig. 2. As a result, the transverse velocity spread in the electron beam is determined mainly by the initial velocity spread caused by thermal electron velocities and emitter surface roughness; for 5 μm roughness, the calculated spread is about of 30%. The lack of space in the NMR-spectrometer magnet necessitates placing the electron-beam collector in a region of homogeneous magnetic field [5]. Utilization of the electron beam with very high power density of 10 kW/cm^2 is possible due to very thin footprint of the electron beam on the inner collector wall; as it shown in Fig. 3, the heat spreads over all directions, i.e., in the angle range π, unlike the conventional gyrotron collectors, so heat flow density at the outer collector surface is less than 400 W/cm^2.

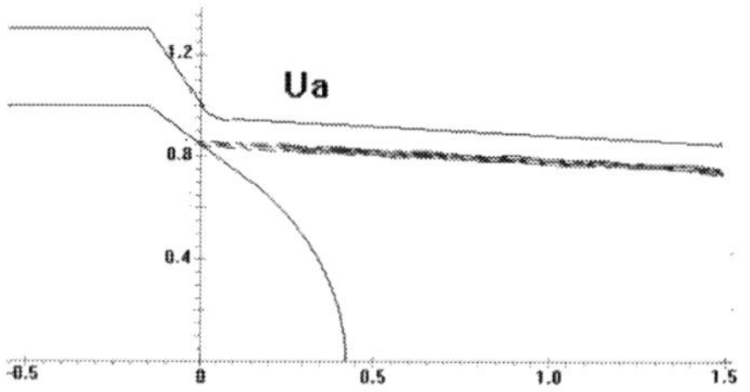

Fig. 2. Designed geometry of a magnetron-injection gun for the gyrotrino and simulated laminar electron beam trajectories in the emitter region. The dimensions are in cm

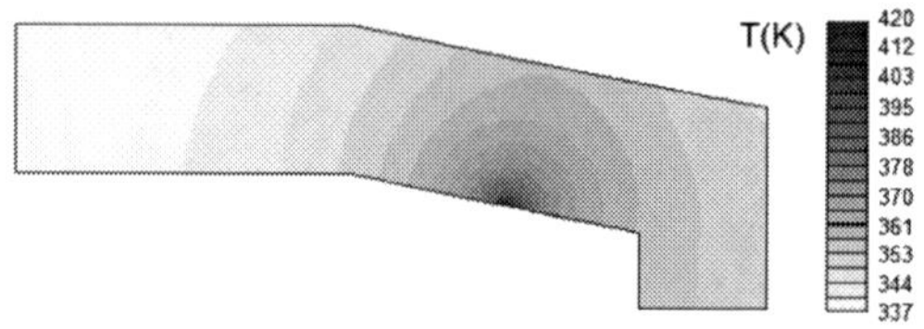

Fig. 3. Temperature distribution in the 2.4-mm-thin copper collector wall. The beam power is 270 W and the beam footprint width is 0.3 mm

The electron-wave interaction for integrated gyrotron configuration also demonstrates a number of specific features caused by lack of space and condition of matching the gyrotron and NMR frequencies. These factors results in very low gyrotron accelerating voltage and short cavity. An alternative way to provide frequency matching for integration of the gyrotron and the spectrometer is substantial modification of the cryomagnet instead of the gyrotron, namely, providing the two distinct uniform-field regions for the gyrotron cavity and for the NMR probe [7].

For gyrotron with low voltage and short cavity, the use of abrupt decrease in cavity diameter instead of smooth cathode narrowing can provide 20 % increase in gyrotron output power (Fig. 4). This result seems counterintuitive, since it is known that smooth cavity up-taper and down-taper typically increase the gyrotron efficiency. The reason for this result is the non-optimal operation regime caused by the limitation on the length for the designed tube. Abrupt termination of the electron-wave interaction cause a decrease in the gyrotron starting current and, therefore, shifts the operation point further from the self-excitation border. In principle, this effect can be used in more common THz gyrotron devices, although it is not so pronounced for long cavities. For 527 GHz second-harmonic gyrotron with operating voltage of 15 kV, simulations predict 5% of power increase due to use of abrupt cutoff narrowing.

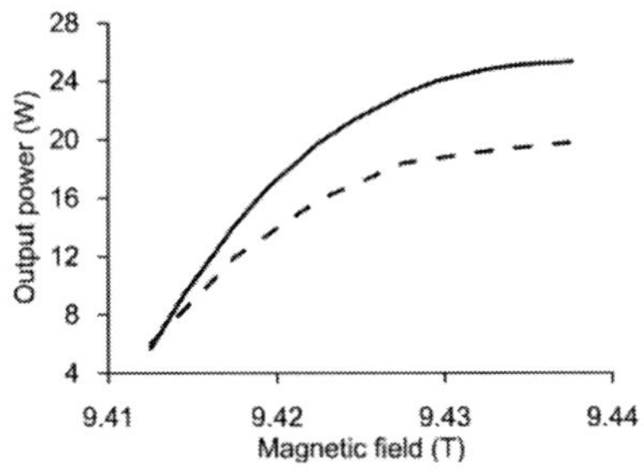

Fig. 4. The output power of 264 GHz low-voltage gyrotron with short cavity and operating mode $TE_{6,2}$ in cases of smooth (dashed line) and abrupt cathode narrowing (solid line). The voltage and current are 1.5 kV and 200 mA

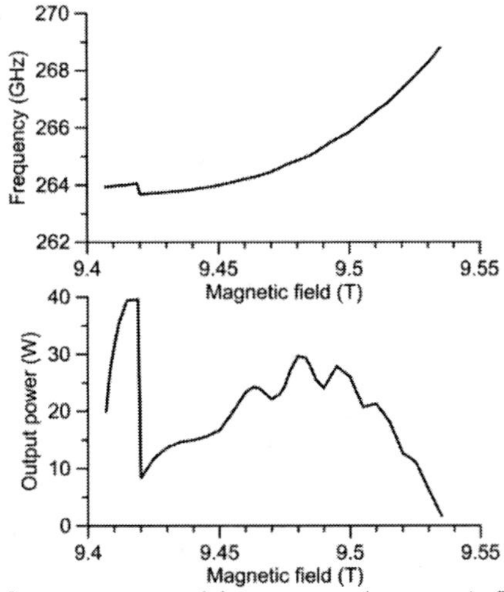

Fig. 5. Gyrotron power and frequency vs. the magnetic field for cavity length of 20 mm with cathode-end power output. The step at the left corresponds to electron synchronism with forward wave (traveling-wave tube regime)

One more surprising result is that the gyrotrino output power at the high axial modes can be significantly higher than at the fundamental mode [8]. This effect is most pronounced at traveling-wave tube regime, see Fig. 5; for a 2-kV gyrotron with beam current of 150 mA, the output power can be as high as 40 W at the second axial mode while only few watts at the fundamental axial mode. This effect can be used to provide smooth frequency tuning in spectroscopy-aimed gyrotrons. The collector-end power output is also favorable [9] for smooth frequency tuning. Simulations show that wider-than-percent FWHM frequency tuning can be achieved in 2-kV device with current of 150 mA.

The possibility of low-voltage gyrotron operation was verified in the demonstration experiment using the gyrotron tube designed for operation voltage of 15 kV. To obtain low-voltage electron beam with acceptable pitch-factor, we applied a higher voltage between cathode and first anode while decreasing the voltage between the cathode and the cavity. The gyrotron operation at the voltages down to 1.5 kV was demonstrated at the frequency of 250 GHz (the operating mode $TE_{5,2}$ at fundamental cyclotron resonance).

The work was supported by the Russian Science Foundation, grant No. 16-12-10445.

References

1. *V.L. Bratman, A.E. Fedotov, Yu.K. Kalynov, P.B. Makhalov, and I.V. Osharin*. Numerical study of a low-voltage gyrotron ("gyrotrino") for DNP/NMR spectroscopy. // IEEE Trans. Plasma Sci. 2017. V. 45, No. 4. P. 644-648.

2. *V.L. Bratman, A.E. Fedotov, Yu.K. Kalynov, and A. Samoson*. THz gyrotron and BWO designed for operation in DNP-NMR spectrometer magnet. // J. Infrared, Millimeter, and THz Waves. 2013. V. 34, No. 12. P. 837-846.

3. *J.R. Sirigiri, T. Maly*, U.S. Patent No. 8,786,284, 22 Jul. 2014.

4. *X.-B. Qi, C.-H. Du, S. Pan, X. Ji, B. Huang, and P.-K. Liu*. Terahertz Broadband-Tunable Minigyrotron With a Pulse Magnet. // IEEE Trans. Electron Dev. 2017. V. 64, No. 2. P. 527-535.

5. *V.L. Bratman, A.E. Fedotov, Yu.K. Kalynov, P.B. Makhalov, V.N. Manuilov*. Project of gyrotron for DNP applications based on NMR magnet. // Proc. of 41st Int. Conf. on Infrared, Millimeter and Terahertz Waves (IRMMW-THz-2016). Copenhagen, Denmark, Sept. 25-30, 2016. H3B.4.

6. *V.L. Bratman, A.E. Fedotov, Yu.K. Kalynov, and V.N. Manuilov*. Electron-Optical System of the Gyrotron Designed for Operation in the DNP-NMR Spectrometer Cryomagnet ('Gyrotrino'). // J. Infrared, Millimeter, and THz Waves. 2017. V. 38 (accepted).

7. *H. Ryan, J. van Bentum, and T. Maly*. A ferromagnetic shim insert for NMR magnets–Towards an integrated gyrotron for DNP-NMR spectroscopy. // J. of Magnetic Resonance. 2017. V. 277. P. 1-7.

8. *A.E. Fedotov, V.L. Bratman, I.V. Bandurkin, A.V. Savilov, I.V. Osharin, Yu.K. Kalynov, N.A. Zavolsky*. Efficient excitation of high axial modes in simulations of low-voltage gyrotron. // Proc. of 18th Int. Vacuum Electronic Conf. (IVEC 2017), London, UK, 24 – 26 April 2017. ID293.

9. *V.E. Zapevalov, A.N. Kuftin, V.N. Manuilov, M.A. Moiseev, A.B. Pavelyev, A.S. Sedov, N.A. Zavolsky*. // Development of 395 GHz gyrotrons for DNP spectroscopy on the basis of expirience of elaboration of a 260 GHz gyrotron. // Proc. of 8th Int. Workshop "Strong Microwaves and Terahertz Waves: Sources and Applications" (SMP-2011), Nizhny Novgorod – St. Petersburg, Russia, July 9-16, 2011. P. 143-144.

EPJ Web of Conferences **149**, 04040 (2017) DOI: 10.1051/epjconf/201714904040
10th International Workshop 2017 "Strong Microwaves and Terahertz Waves: Sources and Applications"

Thermal analysis of gyro-amplifiers with helically corrugated waveguides

Sergey V. Mishakin, Sergey V. Samsonov

Institute of Applied Physics, RAS, Russia, mishakin@ipfran.ru

Introduction

Microwave sources find numerous application in modern physics and technology. In particular, development of wideband amplifiers with high continuous-wave/average output power operating in millimeter and sub-millimeter frequency ranges has drawn a considerable amount of interest in recent years in connection with emerging telecommunication and long-distance high-precision radiofrequency imaging systems [1].

For the mentioned classes of systems, traditional Cherenkov electron devices as well as solid-state devices at frequencies higher than 100 GHz become non-competitive when compared with vacuum electron devices based on electron cyclotron resonance effect. The most prominent class of amplifiers of such type are the gyrotron traveling wave amplifiers – wideband amplifiers in which the electron beam interacts with a non-resonant electromagnetic wave of the interaction circuit.

For a number of years, a variety of gyro-TWTs having the interaction circuit in the form of a waveguide with helical corrugation of its inner surface is being investigated at IAP RAS. A number of Ka-band second-harmonic operating gyro-TWTs of this type have been created during the last two decades and a record-high values of CW / average power have been achieved [2]. Presently, a great deal of effort is directed towards advancing theses devices to higher frequency bands [3, 4].

One of the factors preventing gyro-TWTs from reaching high CW or average output power is the overheating of the interaction circuit caused by Ohmic loss in its walls. The problem becomes increasingly serious for higher operating frequencies. In this work we investigate this problem for a *W*-band high-power gyro-TWT in the cascade of two gyro-TWTs being developed at IAP RAS [4].

Thermal problem statement

The inner surface of a typical interaction circuit (most commonly made of copper) has a form helical waveguide while the outer has a form of a circular cylinder (fig. 1). The inner surface is heated by the dissipation of operating wave's power due to finite conductivity of the circuit's walls.

The steady-state temperature distribution in the bulk of the interaction circuit of the gyro-TWT is governed by stationary heat transfer equation [5]:

$$\nabla(\kappa\nabla T)=0 \qquad (1)$$

where κ – coefficient of thermal conductivity, T is the temperature distribution being sought.

The boundary conditions on the inner surface are that of Neumann type and correspond to setting the value of power being dissipated into the wall

$$-\kappa\nabla T\cdot\mathbf{n}=0.5R_{surface}\cdot\left|\mathbf{j}_{surface}\right|^2 \qquad (2)$$

where $R_{surface}=\sqrt{\pi\cdot f\cdot\rho\cdot\mu_0}$ is reactive part of the surface impedance, f is frequency, ρ – resistivity of the interaction circuit's material, μ_0 – vacuum permeability.

The outer boundary of the interaction circuit is cooled by a stream of water. Mathematically this can be described by the so-called convective boundary condition:

$$-\kappa\nabla T\cdot\mathbf{n}=h_c(T-T_\infty) \qquad (3)$$

where T_∞ is the temperature of the water far away from the surface, h_c – convection coefficient. The convection coefficient depends on the properties of the coolant and the surface.

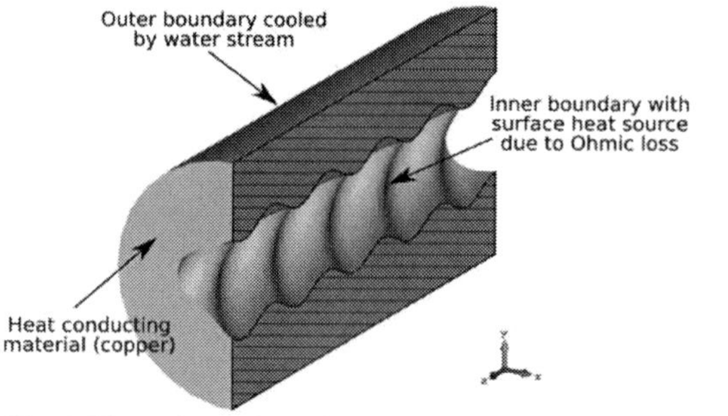

Fig. 1. Illustration to thermal problem

The thermal conductivity, κ, and the electric resistivity, ρ, are in general temperature-dependent, which renders the problem (1)-(3) non-linear. While for copper, the dependence of the thermal conductivity on the temperature for most practical cases may be neglected, the dependence of the ρ on temperature in the operating temperature range is quite strong and cannot be ignored.

Results of simulation

The simulation of thermal distributions in the interaction circuit of the high power gyro-TWT suggested in [4] has been carried out. Since the inner surface of the structure has a complex shape, no analytic method can be employed, instead a full 3D finite element analysis has been performed.

The heat source has been calculated based on the surface current distribution obtained from the particle-in-cell simulation of the gyro-TWT [4]. The current distribution corresponds to maximum output power of about 350 kW (fig. 2). The outer radius of the interaction circuit and the duty cycle were the parameters being investigated. The rest parameters were as follows: k = 0.37 W/ K · mm h_c = 3 W/ K·cm^2, T_∞ = 300 K, f = 95 GHz; the dependence $\rho(T)$ was taken to be linear with the resistivity thermal coefficient of $3.8 \cdot 10^{-3}$ K^{-1}, and the room temperature resistivity of $3.4 \cdot 10^{-8}$ Om·m (resistivity two times higher than that of pure copper). Since the boundary condition (2) non-linearly depends on the temperature, the solution of the problem was carried out in an iterative manner: the surface sources distribution on the current iteration was calculated using the solution from the previous iteration; this procedure was repeated until convergence with some predefined accuracy was achieved.

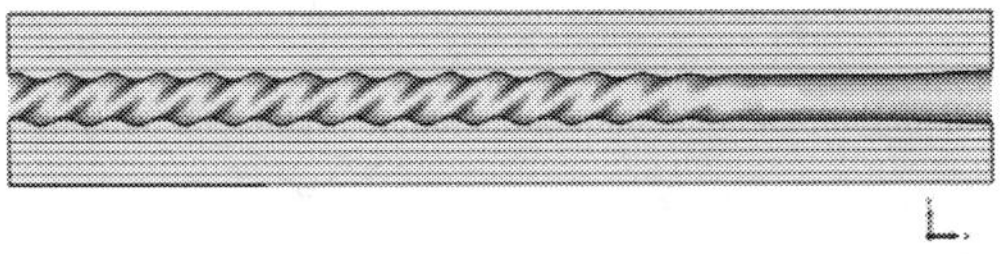

(a)

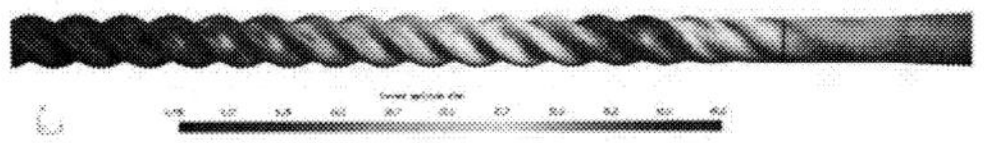

(b)

Fig. 2. (a) Part of the interaction circuit of the gyro-TWT being investigated; (b) distribution of the amplitude of the surface current on the inner surface of the interaction structure in case of maximum output power.

There are two major factor limiting the maximum possible duty cycle (or, alternatively, average power) of the device's interaction circuit: the temperature of the water on the outer surface of the structure should not exceed the boiling point (130 °C for the water stream under pressure of 3 atm.) and the maximum temperature on the inner surface should not exceed some safe value (the value of 200 °C was chosen) to ensure long operating time of the device.

With these limiting values set, the outer radius and the duty cycle could be determined. The simulation shows (fig. 3) that for r_{outer} > 5.5 mm and duty cycle ~ 14% (average power ~ 50 kW) safe operation of the device's interaction circuit is possible.

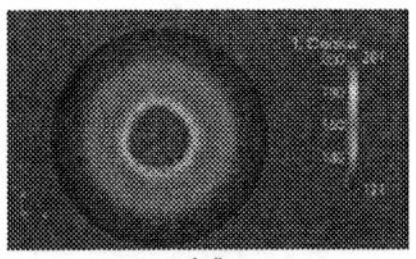

(a)

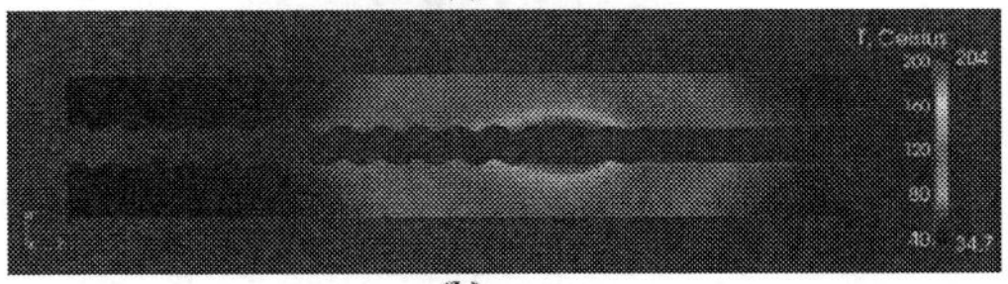

(b)

Fig. 3. Temperature distribution (results of 3D finite element simulation) in the interaction circuit: (a) z cross-section at the point of maximum temperature; (b) - longitudinal distribution.

Acknowledgements

The work was supported by the Russian Science Foundation under project # 16-19-10332.

References

1. *Usoff, J.* Haystack Ultrawideband Satellite Imaging Radar. – Massachusetts Inst of Tech Lexington Lincoln Lab, 2014.
2. *S.V. Samsonov, I.G. Gachev, G.G. Denisov, et al.,* Ka-band Gyrotron Traveling-Wave Tubes with the Highest Continuous-Wave and Average Power // IEEE Transactions on Electron Devices, 2014, vol. 61, no.12, pp.4264-4268.
3. *S.V. Mishakin, S.V. Samsonov, G.G. Denisov,* "A Helical-Waveguide Gyro-TWT at the Third Cyclotron Harmonic", IEEE Trans. Electron Devices, vol. 62, no. 10, pp. 3387–3392, Oct. 2015.
4. *Samsonov, S.V., Bogdashov A.A., Denisov G.G.,et al.,* Cascade of Two W-Band Helical-Waveguide Gyro-TWTs With High Gain and Output Power: Concept and Modeling // IEEE Transactions on Electron Devices. 2017. V. 64, No. 3. P. 1305–1309.
5. *FP Incropera, AS Lavine, TL Bergman, DP DeWitt* Fundamentals of heat and mass transfer. Wiley, 4th edition, 2011.

DOI: 10.1051/epjconf/201714904041

Ultrashort pulse generation based on two coupled helical gyro-TWTs

M.N. Vilkov, N.S. Ginzburg, G.G. Denisov, I.V. Zotova, A.S. Sergeev, A.M. Malkin, S.V. Samsonov and S.V. Mishakin

Institute of Applied Physics RAS, Nizhny Novgorod, Russia, vilkovmn@ipfran.ru

In laser physics, there exists a well-known principle for production of optical ultrashort pulses (USP), based on the effect of passive mode-locking [1], which is achieved by incorporating a saturable absorber into the laser resonator. As it was shown in [2], a similar method of ultrashort pulse generation can be realized in a two-section microwave oscillator consisting of an amplifier and a nonlinear absorber in the feedback loop.

The basic requirement on the amplifying unit is a broad frequency bandwidth that is needed for effective amplification of ultrashort microwave pulses. In the millimeter wave band, it is attractive to utilize a novel type of gyro-TWTs with helically corrugated waveguides operating at harmonics of epy gyrofrequency [3, 4]. Such devices have a number of unique characteristics, including an extremely broad frequency bandwidth up to 20%.

The key problem for development of mode-locked electron oscillators is realization of a nonlinear absorber which is applicable to the microwave frequency band and suitable for operation at the high-power level. In this paper we suggest to use for this purpose the additional section with independent electron beam where interaction also takes place in a helically corrugated waveguide at a harmonic of the gyrofrequency. In such a system saturable absorption realizes for magnetic field slightly exceeding the resonance one. In fact this area corresponds to the so-called Kompfner dip regime, which was originally studied for Cherenkov TWTs [5]. We demonstrate that after proper matching of parameters, the Kompfner dip regime with saturation of absorption can be realized also in a gyro-TWT with a helically corrugated waveguide.

The scheme of the considered microwave USP oscillator with passive mode-locking is presented in Fig. 1. This oscillator includes two gyro-TWT sections with a helical corrugation, one of which is used as an amplifying unit, while another operates in the regime of nonlinear absorption.

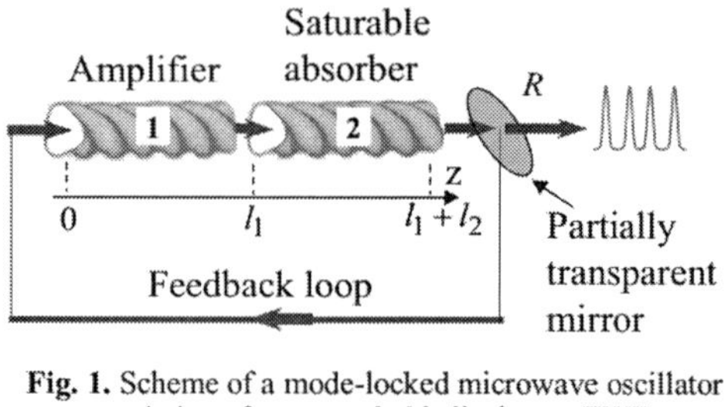

Fig. 1. Scheme of a mode-locked microwave oscillator consisting of two coupled helical gyro-TWTs

Multi-fold helical corrugation provides the coupling of two partial $TE_{m,n}$ modes: a near-cutoff mode A and a far-from-cutoff traveling mode B. Correspondingly, electron-wave interaction in both sections is described by the equations [6]:

$$\begin{aligned}
&\frac{\partial^2 a}{\partial z^2}-2i\frac{\kappa_A}{c}\frac{\partial a}{\partial t}-2\sigma\kappa_A^2 b=\\
&=i\frac{4eI_b}{\pi mc^3}\frac{\kappa_A^2}{\beta_{\|0}\sqrt{N_A}}\frac{s^s}{2^s s!}\int_0^{2\pi} p_+^s d\theta_0,\\
&\frac{\partial b}{\partial z}+\frac{1}{V_{gr}}\frac{\partial b}{\partial t}-i(\bar{h}-h_0)b-i\frac{\kappa_A^2}{h_0}\sigma a=0, \qquad (1)\\
&\frac{\partial p_+}{\partial z}+\frac{1}{V_{\|0}}\frac{\partial p_+}{\partial t}+i\frac{\kappa_A}{2\beta_{\|0}}\frac{p_+}{s}(\Delta+|p_+|^2-\beta_{\perp 0}^2)=\\
&=\frac{s^s}{2^s s!}\frac{\kappa_A}{\beta_{\|0}\sqrt{N_A}}a(p_+^*)^{s-1}.
\end{aligned}$$

Here p_+ is the transverse electron momentum, $a=eA\sqrt{N_A}/mc^2\kappa_A$ and $b=eB\sqrt{N_B}/mc^2\kappa_B$ are the normalized amplitudes of partial waves, $\sigma=(\tilde{r}/2r_0)\left(\nu_B^2-m_A m_B\right)/\sqrt{\left(\nu_A^2-m_A^2\right)\left(\nu_B^2-m_B^2\right)}$ is the wave coupling parameter [3] proportional to the corrugation depth $\tilde{r}$, r_0 is the mean radius of a corrugated waveguide, N_A and N_B are the dimensionless norms of the partial waves, $V_{gr}=h_0 c/\kappa_A$ is the group velocity of the traveling wave B, $V_{\|0}=\beta_{\|0}c$ is the longitudinal velocity of electrons, I_b is the beam current, $\Delta=2\left(\omega_A-s\omega_H\right)/\omega_A$ is the cyclotron resonance detuning, $h_0=h_B\left(\omega_A\right)$ is the longitudinal wave number of the traveling wave B at the frequency ω_A, $\kappa_{A,B}=\nu_{A,B}/r_0$ are transverse wave numbers for partial waves A and B, correspondingly, $\bar{h}=2\pi/d$, d is the corrugation period, s is the electron cyclotron harmonic number.

We assume that the amplifying and absorbing sections are coupled via the traveling partial wave B In accordance with the scheme shown in Fig. 1 boundary conditions can be presented in the form

$$b_2(t,l_1)=b_1(t,l_1),\ b_1(t,0)=Rb_2(t-T,l_1+l_2) \qquad (2)$$

where $R<1$ is the reflection coefficient.

Simulations were performed for parameters the experimentally realized Ka-band helical gyro-TWT [4]: the particles energy of 50–70 keV, the total current of 2–10 A, the pitch-factor of 0.8–1.2. For this amplifier the second helical gyro-TWT section was synthesized (the particles energy is 54 keV, the total current is 6 A, the pitch-factor is 0.53) which operates like a saturable absorber and provides USPs generation. Both TWTs have three-fold helically corrugated waveguides for resonant coupling of the near-cutoff $TE_{2,1}$ and counter-rotating traveling $TE_{1,1}$ mode. The indicated modes are selectively excited by an axis-encircling beam at the second cyclotron harmonic ($s=2$).

In the amplifying section in contrast to the gyro-TWT [4] we choose parameters of corrugation in such a way that the intersection between the beam line and the dispersion curve of the normal mode is provided (Fig. 2a). In this case the wave group velocity substantially differs from the axial velocity of the electrons. As a result due to slippage of em pulse over the electrons it effectively accumulates the energy from different fractions of the electron beam. On the contrary, in the absorbing section, for minimization of the absorber relaxation time the grazing incidence of dispersion characteristics of the electron beam and the normal wave of a corrugated waveguide is beneficial (Fig. 2b).

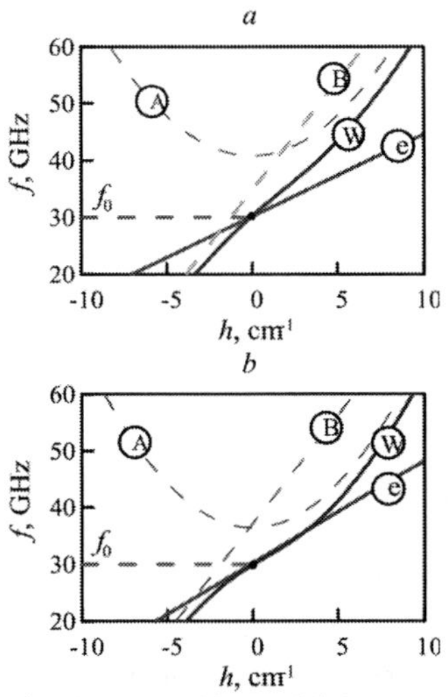

Fig. 2. Dispersion diagrams of the partial (*A,B*), normal (*W*) and electron cyclotron (e) waves in the amplifier (a) and absorbing section (b) for operating frequency 30 GHz

Simulations show that the maximum stability and reproducibility of generated pulses is achieved in a high excitation mode of the considered oscillator, when generation starts after initiation by a sufficiently large initial pulse. In the absence of the source of an external pulse, we can use the startup scenario of USP generation with delayed switching of the absorbing section with respect to the amplifying section. In this case, at the first stage only the amplifier is switched on and, consequently, chaotic generation develops from small initial noises (Fig. 3b). Then the absorbing section is switched on and, as a result, the chaotic regime of generation transforms to the periodical regime of USPs generation (Fig. 3c). For chosen parameters the duration of the generated pulses is 0.2 ns with distance between them of 1 ns. Peak power of pulses achieves 400 kW, that (due to slippage in the amplifying section) exceeds in 2 times the maximum output power level in a helical gyro-TWT [4] with the similar parameters. Currently this type of USP oscillator is under development at the Institute of Applied Physics RAS.

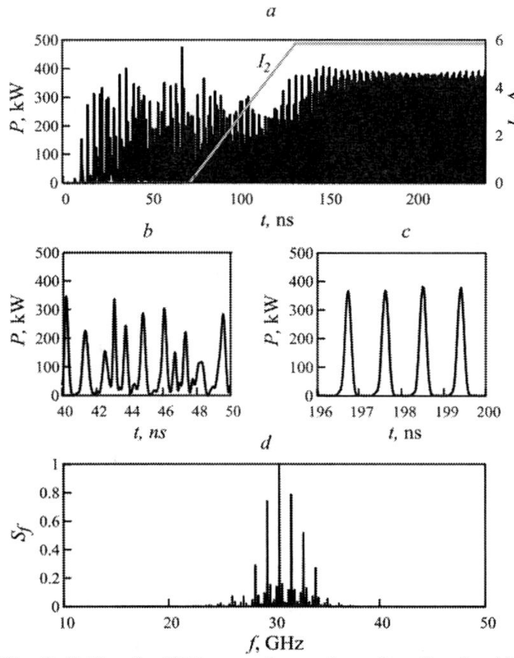

Fig. 3. Setting the USP generation regime when the absorbing section switches on with a delay relative to the amplifying section: temporal dependence of the output power and the current I_2 in the absorber (a), detailed profiles of the generated signal on the stage of developing a chaotic self-modulation (b) and on the stage of the ultrashort pulses generation after the switching of the absorber (c), the radiation spectrum (d)

This work was supported by Russian Science Foundation (RSCF), grant No. 16-42-01078.

References

1. *Haus H.A.* Mode-Locking of Lasers // IEEE J. of Select. Topics in Quant. Electron. 2000. V. 6, No. 6. P. 1173-1185.

2. *Ginzburg N.S., Denisov G.G., Vilkov M.N. et al.* Generation of "gigantic" ultra-short microwave pulses based on passive mode-locking effect in electron oscillators with saturable absorber in the feedback loop // Phys. Plasmas. 2016. V. 23. P. 050702-1-4.

3. *Denisov G.G., Bratman V.L., Cross A.W. et al.* Gyrotron Traveling Wave Amplifier with a Helical Interaction Waveguide // Phys. Rev. Lett. 1998. V. 81, No. 25. P. 5680-5683.

4. *Samsonov S.V., Gachev I.G., Denisov G.G. et al.* Ka-band gyrotron traveling-wave tubes with the highest continuous-wave and average power // IEEE Trans. Electron Devices. 2014. V. 61, No. 12. P. 4264-4267.

5. *Kompfner R.* On the operation of the traveling wave tube at low level // J. Brit. IRE. 1950. V. 10, No 8-9. P. 283-289.

6. *Ginzburg N.S., Zotova I.V., Sergeev A.S., Zaslavsky V.Yu. et al.* Mechanisms of amplification of ultrashort electromagnetic pulses in gyrotron traveling wave tube with helically corrugated waveguide // Phys. Plasmas. 2015. V. 22 P. 113111-1-7.

EPJ Web of Conferences **149**, 04042 (2017) DOI: 10.1051/epjconf/201714904042
10th International Workshop 2017 "Strong Microwaves and Terahertz Waves: Sources and Applications"

Amplification of short-wavelength radiation by relativistic electron beams moving near the impedance surfaces

A.M. Malkin, N.S. Ginzburg, I.V. Zheleznov, and A.S. Sergeev

Institute of Applied Physics RAS, Nizhny Novgorod, Russia, malkin@appl.sci-nnov.ru

Electrodynamical properties of a wide class of surfaces can be described by introduction of an impedance boundary condition on the surface, namely, a linear relation binding the tangential components of the electric and magnetic fields. One of the most well-known is the Leontovich boundary condition valid at the surface of the lossy metal with its skin depth much smaller than the wavelength. Approximate impedance boundary conditions can also be used for description of periodical gratings with a period small in the scale of the wavelength. Impedance surfaces support evanescent waves with a phase velocity that can be smaller than the velocity of light, thus making it possible to provide their interaction with relativistic electron beams. Relativistic TWTs based on such mechanisms operate in centimeter wave bands [1]. In shorter wavelength bands, impedance systems also look viable due to the fact that the excitation of evanescent wave allow solving the transverse mode selection problem inevitable in oversized systems.

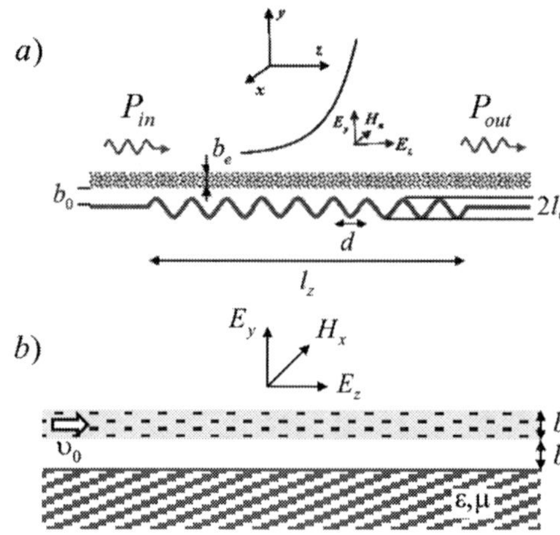

Fig. 1. Scheme of a surface-wave amplifier: an relativistic electron beam interacts synchronously with a decelerated surface wave near the corrugated impedance surface (a); amplification of the field in the beam near a resistive surface (b)

In the paper, we present the results of quasi-optical approach applied to the problem of electron beam excitation of high-frequency surface waves existing near the impedance surfaces. Two types of surfaces are taken into consideration, one of them being the lossy metal surface and the other is the ideal metal corrugation. For the case of shallow sine corrugation we obtain explicitly the expression for impedance based on the coupled wavebeam approach and the surface magnetic current technique [2].

We consider a relativistic electron beam moving in z direction and interacting with a co-moving incident wavebeam near the impedance surface located at $y=0$ with an impedance of χ. The radiation field can be presented via its magnetic component,

$$H_x = \mathrm{Re}\left(H_x^{\omega}(z,y)e^{i(\omega t-kz)}\right),$$

where H_x^{ω} is the slowly varying amplitude.

The beam-wave interaction can be described within the 2D model by the self-consistent set of equations comprising the parabolic equation for the wavebeam supplemented by the equations of electron motion

$$\frac{\partial C}{\partial Z}+i\frac{\partial^2 C}{\partial Y^2}+i\hat{\chi}\delta(Y)C = i\frac{1}{B_e}\frac{\partial}{\partial Y}\left(F(Y)J\right), \quad (1)$$

$$\frac{\partial^2\theta}{\partial Z^2} = \mathrm{Re}\left(i\frac{\partial C}{\partial Y}e^{i\theta} - i\frac{1}{B_e}\left(F(Y)J\right)e^{i\theta}\right), \quad (2)$$

$$\theta\big|_{Z=0} = \theta_0 \in [0,2\pi), \quad \left.\frac{\partial\theta}{\partial Z}\right|_{Z=0} = \Delta, \quad (3)$$

where $J=\pi^{-1}\int_0^{2\pi} e^{-i\theta}d\theta_0$ is the amplitude of electron current. Wave amplitude at the input of the interaction space satisfies the boundary condition $C_{Z=0}=C_0(Y)$, where C_0 is the transverse profile of the incident field. Equations (1)-(3) are put down using the normalized variables $C=\sqrt{2}e\mu H_x^{\omega}/mc^2\gamma_0 kG^{3/2}$, $Z=Gkz$ $Y=\sqrt{2G}ky$, $\hat{\chi}=\sqrt{2/G}\chi$, θ is the electron phase with respect to the wave, $\Delta=(1-\beta_0)/G\beta_0$, $k=\omega/c$, $F(Y)$ is the transverse current distribution function. Normalization parameter (similar to Pierce parameter) has the following form:

$$G=\left(2\sqrt{2}\frac{eI_0}{mc^3}\frac{\mu}{\gamma_0}\lambda\right)^{\frac{2}{3}}.$$

In the case of the sine corrugation on the surface of the ideal metal (Fig. 1a), the impedance is determined by a relation [2]

$$\chi = 2\alpha^2\left(\frac{k}{\sqrt{(k+\bar{h})^2-k^2}}+\frac{k}{\sqrt{(k-\bar{h})^2-k^2}}\right), \quad (4)$$

where $\bar{h}=2\pi/\bar{d}$ is the lattice wavenumber and $\alpha=l_0\bar{h}/8$ is the coupling coefficient proportional to the corrugation depth l_0.

If the impedance boundary conditions are set on the surface of a conductor (Fig.1b), than the impedance is determined by Leontovich formulas as

$$\chi = iw/2, \quad w=\sqrt{\mu/\varepsilon}. \quad (5)$$

Here w is the wave resistance of the material. For the case of impedance corrugation, χ is purely real, while for the lossy case, χ is complex, and there is an energy flow directed inside the metal.

In the case when the corrugated material is non-ideal metal with high conductivity, surface impedance would be a sum of (4) and (5).

For the electron beam with the width of b_e moving at the distance of b_0 from the surface, a linear theory can be constructed and a dispersion equation can be obtained. We have shown that in both cases (***a*** and ***b*** in Fig. 1) only one root of the dispersion equation satisfies the requirements on wave evanescence and energy flow direction. Thus, only one regime of amplification can be realized. In this regime, in the case of corrugated surface, the slow evanescent wave can be amplified using the Cherenkov mechanism of interaction while in the case of lossy surface, the dissipative instability takes place leading to amplification of the spatial charge fields of the beam. This conclusion is illustrated by Fig. 2 where the transverse distribution of electric and magnetic field is presented on the linear stage of amplification for the case of the resistive instability (Fig. 2a,b). If the absolute value of impedance tends to zero or the distance between the beam and the surface increases, the fields tend to the intrinsic field of the beam localized near it. At the same time, for purely real impedance (case *a* in Fig.1), the electric field tends to evanescent profile confined at the corrugation (Fig.2 c,d).

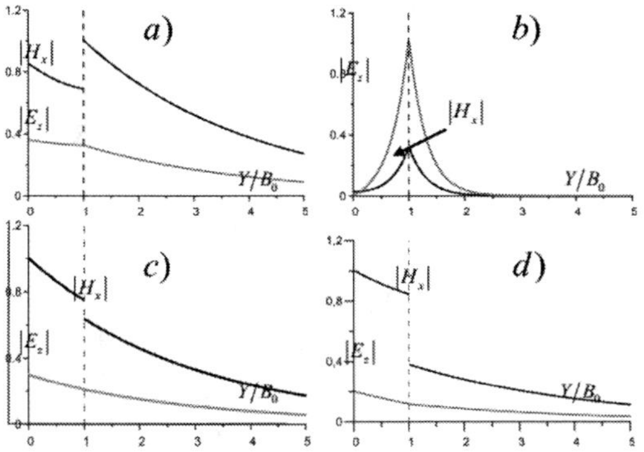

Fig. 2. Transverse distributions of electric and magnetic field amplitudes at the linear stage of instability at $\Delta=9$, $\sqrt{2G}kb_0=0.1$, $b_e=0$, $\hat{\chi}=3-3i$ (a), $\hat{\chi}=0.5(1-i)$ (b), $\hat{\chi}=3$ (c), $\hat{\chi}=2$ (d)

Based on the nonlinear theory given above, let us analyze the possibility of realization of the relativistic sub-mm range amplifier of surface waves based on the corrugated metallic surface operating at 0.9 mm wavelength. We take the electrons energy to be of 1 MeV, linear injection current density of 1.2 kA/cm, corrugation period of 0.25 mm, corrugation depth of 0.136 mm, the distance between the corrugation and the beam of 0.1 mm, and beam thickness 0.2 mm. These values correspond to the following dimensionless parameters: $G=8.8\times10^{-3}$, $\sqrt{2G}kb_0=0.1$, $\sqrt{2G}kb_e=0.2$, $\hat{\chi}=3.5$, $\Delta=7$. For the linear input power density of 50 kW/cm, the output power density of 550 MW/cm (which corresponds to an amplification factor of up to 40 dB and an efficiency of 45%) was reached at the optimal length of about 6 cm. Spatial distribution of the magnetic field amplitude of the amplified field is presented in Fig. 3 together with the dependence of integral radiation power on the z-coordinate.

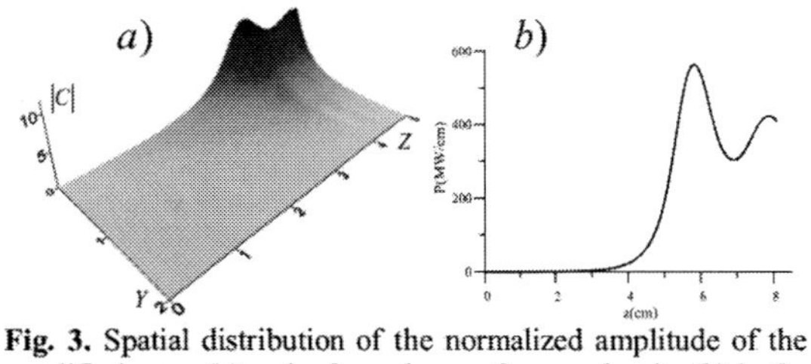

Fig. 3. Spatial distribution of the normalized amplitude of the amplified wave (a) and z dependence of power density (b) in the case of purely real impedance

We also have undertaken simulations of the dissipative amplifier operating in THz frequency range at a wavelength of 0.3 mm. We considered a relativistic electron beam with following parameters: particles energy of 1 MeV, linear current density of 1.3 kA/cm, width of electron beam of 0.1 mm and the distance between the beam and the resistive surface of 0.1 mm. corresponding to the normalized values: $G=4.3\times10^{-3}$, $\sqrt{2G}kb_0=0.2$, $\sqrt{2G}kb_e=0.2$, $\hat{\chi}=3.35-0.3i$, $\hat{\Delta}=14$. The output power density in saturation regime is 10.6 MW/cm with a power gain of up to 28 dB (for a power density of the input signal of about 16 kW/cm). Saturation is reached at a length of about 7 cm. Wave and electron efficiencies are rather small (0.8% and 4%, correspondingly); nevertheless, absolute value of the output power is of interest for THz wavelength range.

This work was supported by RFBR grant No. 16-02-00890.

References

1. *E.B.Abubakirov, A.N.Denisenko, M.I. Fuks, N.G. Kolganov, N.F. Kovalev, M.I. Petelin, A.V. Savelyev, E. Schami-loglu, E.I.Soluyanov, and V.V.Yastrebov,* An X-band gigawatt amplifier // IEEE Trans. on Plasma Sci., 2002, V. 30, N. 3, P. 1041-1051.

2. *N.S. Ginzburg, A.M. Malkin, A.S. Sergeev, V.Yu. Zaslavsky,* Quasi-optical theory of relativistic submillimeter surface-wave oscillators // Appl. Phys. Lett. 99, 121505 (2011).

3. *N.S. Ginzburg, A.M. Malkin, I.V. Zheleznov and A.S. Sergeev,* Evanescent waves propagation along a periodically corrugated surface and their amplification by relativistic electron beam (quasi-optical theory) // Physics of Plasmas 20, 063105 (2013).

EPJ Web of Conferences 149, 04043 (2017) DOI: 10.1051/epjconf/201714904043
10th International Workshop 2017 "Strong Microwaves and Terahertz Waves: Sources and Applications"

Collector system of a gyrotron with magnetically shielded solenoid

M.V. Morozkin[1], M.Yu. Glyavin[1], V.N. Manuilov[1, 2], I.V. Zotova[1] and M.D. Proyavin[1]

[1]Institute of Applied Physics of the Russian Academy of Sciences, N. Novgorod, Russia, morozkin@ipfran.ru
[2]Lobachevsky State University of Nizhny Novgorod, N. Novgorod, Russia

The gyrotron-based complexes developed by IAP RAS are used now for many applications, such as sintering of nanostructured ceramics, metal compacts and metal-ceramic composites, CVD growth of diamond films and disks, and generation of multi-charged ion beams [1–3]. These systems typically operate in 24–28 GHz frequency range with a power of 5–15 kW. As a source of microwave radiation, the systems use CW gyrotrons operating in magnets with oil or water cooling (so called technological gyrotrons). To reduce power consumption of magnets, gyrotrons operate on second cyclotron harmonic.

Nowadays applications of gyrotron-based setups become more demanding in terms of output power. Also new applications appear that require high power, for example fabrication of metal oxides nanopowders [4] or production of oversized diamond disks for vacuum windows.

At present, a powerful gyrotron complex is being developed at the Institute of Applied Physics of the Russian Academy of Sciences, which includes a "warm" solenoid with magnetic shielding [5, 6] (see Fig. 1). Due to ferromagnetic screens, it is possible either to reduce the energy consumption of the solenoid in more than two times, or to increase the frequency of radiation while maintaining the energy of the magnetic system. The main parameters of the gyrotron are:

Frequency f	28 GHz
Cyclotron harmonic s	1
Accelerating voltage U_0	23 kV
Operating current I_0	2.4 A
Operating mode	$TE_{1,3}$
Beam radius in the cavity R_0	6.6 mm

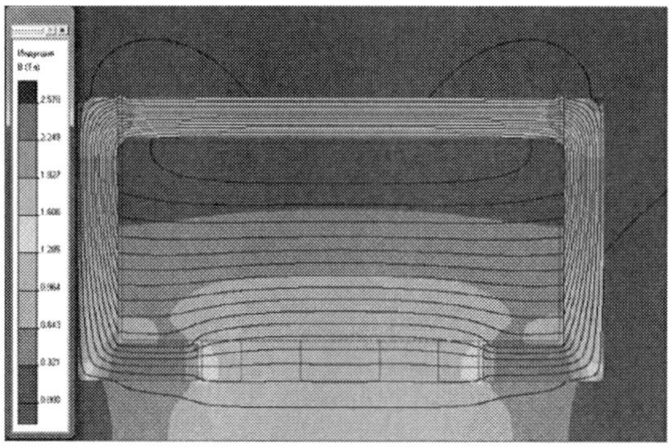

Fig. 1. The scheme of magnetic system with shielded solenoid. Magnetic field lines are shown

At the same time ferromagnetic screens substantially increases the inhomogeneity of the magnetic field outside the screens, in particular in the collector region. So here the coefficient of non-adiabaticity $\varepsilon = h/L_B$ (here h is the step of electron trajectory, L_B is the scale of the magnetic field) exceeds 2 (see Fig. 2).

Due to nonstandard distribution of the magnetic field in such systems and increased energy of an electron beam, collector of a gyrotron with magnetically shielded solenoid would differ significantly from collector of a standard technological gyrotron. The intensity of the magnetic field outside the shielding decays very rapidly; due to this, the motion of the electrons will be nonadiabatic. In this report, a collector design is presented and the simulation of the motion of electrons in the collector region is carried out.

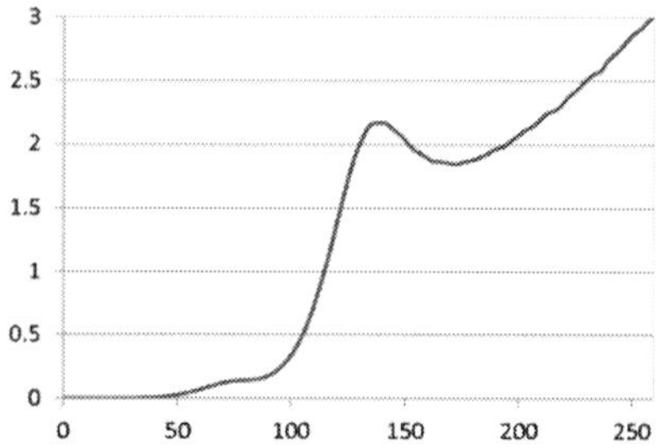

Fig. 2. Dependence of the coefficient of non-adiabaticity ε on the longitudinal coordinate in the optimized magnetic system. $z = 0$ corresponds to the cavity center

The results of the simulation (see Fig. 3) show that, due to the rapid decay of the magnetic field, electrons, falling into the region of the nonadiabatic field, acquire a transverse velocity and twist, starting to move downward along the radius. Then, when the magnetic field almost completely decays, the electron beam continues to fly apart inertially, while the direction of motion of the electrons is determined by the angle of the velocity vector at the exit from the region where the magnetic field is still significant.

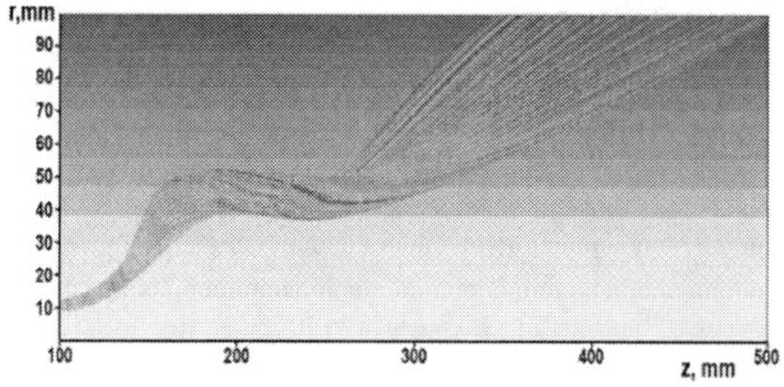

Fig. 3. Trajectories of the electron beam in the collector region (disregarding collector walls)

The inertial expansion of the electron beam provides a large ratio of the beam track length to the radius of the collector Rc and a smooth energy density distribution within the track. However it is possible to use this effect effectively only at large collector radii, which may be relevant for a more powerful (50–100 kW) gyrotron.

For our gyrotron with a power of 20 kW, the most optimal collector radius is 45 mm < *R*c <50 mm, at which part of the electrons will sit on the collector before the moment of twisting, and the part will fly further and descend already at the stage of inertial expansion. In this case, the total length of the electron beam trace will increase by 2–3 times in comparison with the length even at a slightly larger radius (*R*c ~ 60), which will allow achieving low values of dissipated power density without using additional electron beam spreading systems.

Calculations show that the distribution of power density in the region of beam deposition is quite inhomogeneous and represents two peaks spaced in z (see Fig. 4). However, this calculation is made for the worst case in terms of load on the collector, when there is no generation. With the real spent electron beam, the power density distribution can become more homogeneous due to electrons with energies lower than the initial one.

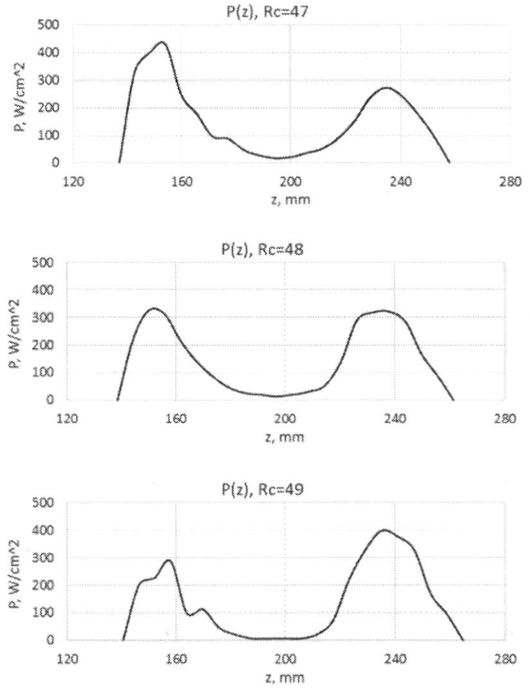

Fig. 4. Dependence of the power density on the collector on the z coordinate for different collector radii (*Rc* = 47, 48 and 49 from top to bottom)

The most optimal position of the collector wall is that at which part of the electrons sit down on the collector at the moment of twist, and the part flies further and settles down already at the stage of inertial expansion. In this case, the effective length of the track of the electron beam increases by 2–3 times, which makes it possible to achieve low values of the power dissipation at a relatively small radius of the collector without using additional systems for spreading the electron beam.

We determined the optimal radius of the collector to minimize the thermal load, and for a 20-kW gyrotron, the maximum power density scattered on the collector was about 300 W/cm^2 (see Fig. 5), which leaves a reserve for increasing the gyrotron power with a shielded warm solenoid up to 40–50 kW.

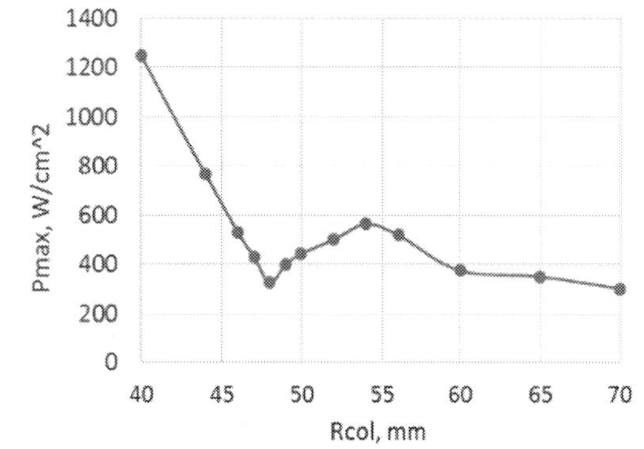

Fig. 5. Dependence of the maximum deposited power on the collector radius

Conclusion

The motion of electrons in the collector region of a gyrotron with a magnetically shielded solenoid was analyzed. A variant of the collector is proposed that uses the motion of electrons in the region of a nonadiabatic magnetic field and makes it possible to obtain a long trace of the deposited electron beam and a low power density on the collector without additional beam spreading systems.

Acknowledgements

This work was supported by the Russian Science Foundation, project No. 14-12-00887.

References

1. *Bykov, Yu., Eremeev, A., Glyavin, M., Kholoptsev, V., Luchinin, A., Plotnikov, I., Denisov, G., Bogdashev, A., Kalynova, G., Semenov, V., Zharova, N.* 24-84-GHz gyrotron systems for technological microwave applications // IEEE Trans. Plasma Sci. 2004, V. 32, No. 1, P. 67-72.
2. *Vikharev, A. L., Gorbachev, A. M., Kozlov A. V., Koldanov, V. A., Litvak, A. G., Ovechkin, N.M., Radishev, D.B., Bykov, Yu.V., Caplan, M.* Diamond films grown by millimeter wave plasma-assisted CVD reactor // Diam. Relat. Mater., 2006, V. 15, No. 4-8, P. 502-507.
3. *Bykov, Yu., Denisov, G., Eremeev, A., Gorbatushkov, V., Kurkin, V., Kalynova, G., Kholoptsev, V., Luchinin, A., Plotnikov, I.* 28 GHz 10 kW gyrotron system for electron cyclotron resonance ion source // Rev. Sci. Instr. 2004, V. 75, P. 1437.
4. *Samokhin, A. V. , Alexeev, N. V., Vodopyanov, A. V., Mansfeld, D. A., Tsvetkov, Yu. V.* Metal Oxide Nanopowder Production by Evaporation-Condensation Using a Focused Microwave Radiation at a Frequency of 24 GHz // J. Nanotechnol. Eng. Med. 2015, V. 6, No. 1, P. 011008.
5. *Kuftin, A.N., Belov, S.P.* Magnetic shielded electron optical system of a gyrotron in space charge limited current regime // Applied Physics 2000, No. 3, P. 76-81.
6. *Glyavin, M. Yu., Manuilov, V. N., Proyavin, M. D.* Electron optics system of continuous wave gyrotron with an operating frequency 24 GHz based on shielded solenoid" // J. Commun. Technol. Electron. 2017, submitted.

EPJ Web of Conferences **149**, 04044 (2017) DOI: 10.1051/epjconf/201714904044
10th International Workshop 2017 "Strong Microwaves and Terahertz Waves: Sources and Applications"

Modulation of microwave radiation in the process of resonant interaction with a counter-propagating rectilinear electron beam

I.V. Zotova[1], N.S. Ginzburg[1], A.S. Sergeev[1], A.E. Fedotov[1], V.Yu. Zaslavsky[1,2]

[1]Institute of Applied Physics RAS, N.Novgorod, Russia, zotova@appl.sci-nnov.ru
[2]N.Novgorod State University, N.Novgorod, Russia

The need for powerful pulsed sources of millimeter and sub-millimeter radiation is motivated by a large number of fundamental problems and practical applications, including diagnostics of plasma, photochemistry, biophysics, new locating systems, and the spectroscopy of various media. Formation of periodic sequences of ultrashort pulses can be achieved based on different methods. For example in [1] the optically controlled switches are realized based on induced photoconductivity effect in semiconductor elements implemented in resonant system. In [2] for the same purpose the method of passive mode locking in electron oscillators with a saturable absorber in the feedback loop is developed.

At the same time, as it was shown in [3], there is an alternative method of periodic ultrashort pulse generation which is based on the cyclotron resonance interaction of power steady-state signal with counter propagating initially rectilinear electron beam. In such system incident stationary signal splits into sequence of short pulses, each of them can be considered as a soliton which propagates in the interaction space without profile change. This paper is devoted to the detail theoretical analysis of described method of short pulses generation with a view to experimental realization of this effect in millimeter waveband.

Basic Equations

Let us consider the resonance interaction of an initially rectilinear annular electron beam guided by a homogeneous magnetic field $\vec{H}=\vec{z}_0 H_0$ with an electromagnetic wave (Fig. 1) in the cylindrical waveguide with a radius R near the cut-off frequency ω_c. Presenting the transverse electric field of the wave in the form $\vec{E}=\mathrm{Re}(A(z,t)\vec{E}_s(r)e^{i\omega_c t-im\phi})$, where $A(z,t)$ is the complex slowly varied amplitude, $\vec{E}_s(\vec{r}_\perp)$ describes the radial structures of the TE_{mn} waveguide mode, we can describe the wave-beam interaction by the set of equation which includes a parabolic equation for the field amplitude combined with averaged electron motion equations:

$$\frac{i\sqrt{G}}{4\beta_{\parallel 0}^2}\frac{\partial^2 a}{\partial Z^2}+\frac{\partial a}{\partial \tau}=p,$$
$$\frac{\partial p}{\partial Z}+\frac{\partial p}{\partial \tau}+ip\left(\Delta+|p|^2\right)=-a. \quad (1)$$

Here the following normalized variables are used:

$$\tau=\frac{\sqrt{G}\omega_c t}{2},\quad Z=\frac{\sqrt{G}\omega_c z}{2V_{\parallel 0}},\quad a=\frac{eAJ_{m-1}(\kappa R_0)}{G^{3/4}mc\omega_c},$$

$$\hat{p}=\frac{p_x+ip_y}{G^{1/4}mc}e^{-i\omega_c t},\quad G=\frac{4eIJ_{m-1}^2(\kappa R_0)}{mc^3\beta_{\parallel 0}(\nu^2-m^2)J_m^2(\nu)},$$

I is the electron current, $\kappa=\omega_c/c$, $J_m(x)$ is the Bessel function, m is the azimuthal index of the operating mode, R_0 is the injection radius, ν is the n-th root of $J'_m(x)=0$, $\Delta=2(\omega_c-\omega_H)/\sqrt{G}\omega_c$ is the initial cyclotron resonance mismatch, $\omega_H=eH_0/mc\gamma$ is the electron gyrofrequency. It should be noted that, in contrast with the gyrotron theory, the transverse velocity of electrons at the input of interaction space is zero, that's why there is no the averaging over cyclotron phases in the right part of parabolic equation in (1). At the cross-section $Z=0$ the standard reflectionless boundary condition for the field amplitude [4] is used. At $Z=L$ (where $L=\sqrt{G}\omega_c l/2V_{\parallel 0}$) where the stationary signal with amplitude a_0 and a normalized frequency $\Omega=2(\omega_0-\omega_c)/\sqrt{G}\omega_c$ enters the system we apply the modified boundary condition [5]:

$$a(L,\tau)+\frac{G^{1/4}}{2\beta_{\parallel 0}\sqrt{\pi i}}\int_0^{\tau}\frac{1}{\sqrt{\tau-\tau'}}\frac{\partial a(L,\tau')}{\partial Z}d\tau'=2a_{in}e^{i\Omega\tau} \quad (2)$$

Simulations show that depending on the parameters several regimes of interaction can be observed in this system including stationary regimes of resonance absorption and non-linear bleaching described in [3,6]. At the same time there is an area of parameters where stationary regimes became unstable replaced by a periodic or chaotic self-modulation.

Periodic regime presented in Fig. 2 realizes for following parameters $L=4$, $a_0=6$, $\Omega_0=0.53$ $\beta_{\parallel 0}=0.28$, $\Delta=-2$ and $G=10^{-4}$. In this case the input stationary signal transforms to a sequences of ultrashort pulses which propagates through the interaction space without transformation of their form, i.e. exhibit behavior typical for solitons. Soliton nature of formed pulses can be demonstrated for $|\Delta|>>1$, when there is a limiting transition to the equations of gyro-BWO. As it was shown analytically in [3] for an initially rectilinear electron beam such equations possess the soluton-like solution for intensity of electromagnetic field.

Results of 3D PIC Simulations

The possibility of modulation of stationary signal in the process of resonance cyclotron interaction with the backward wave is confirmed by direct 3D simulations based on the PIC-code *CST STUDIO SUITE*. Simulations were carried out for operating frequency of 30 GHz and following parameters of an electron beam corresponding to once for planning experiments: electron energy of 15 keV, electron current of 0.75 A, resonance magnetic field of 11.5 kOe. Interaction takes place in a cylindrical waveguide with the radius of 0.3 cm and a length of 12 cm. The input signal with power 30 kW has the transverse structure corresponding to the TE_{11} mode. In Fig.3a the time dependence of an output power is shown. It is seen that after interaction radiation represents the periodical train of short pulses with duration of 3 ns and peak power of 30 kW. The distance between pulses is of 20 ns.

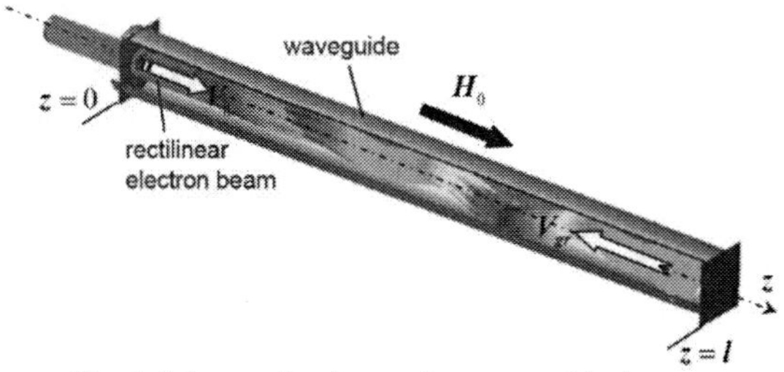

Fig. 1. Scheme of an interaction space with electron trajectories found in PIC simulations

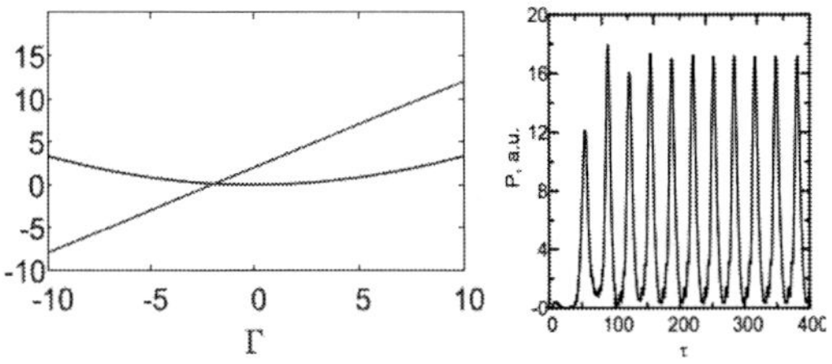

(a) (b)

Fig. 2. Dispersion diagram (a) and formation of periodic sequence of ultrashort pulses (b)

Simulations show that the considered mechanism can be used also for modulation of radiation of short wavelength gyrotrons. In particular, in process of interaction of signal from a 240 GHz/350 W gyrotron [7] with an initially linear annular electron beam having the electron energy 1 keV, current 65 mA and injection radius 0.1 mm the input stationary signal breaks up into solitons with duration of 0.8 ns and peak power of 300 W. Due to a significant part of the energy of incident radiation is absorbed by an electron beam the time averaged output power amounts only 30% from the power of the incident CW signal.

In summary, it was demonstrated that a very simple system consisting of the electromagnetic wave interacting with a counter-propagating beam of initially unexcited oscillators can exhibit a quite complex behavior. The nonlinearity caused by non-isochronous motion of electrons in the magnetic field in combination with time delay effects leads to a nontrivial picture of stationary and non-stationary dynamical regimes. The region of parameters is found in which the incident stationary signal splits into a train of short pulses, and each of them can be interpreted as a soliton.

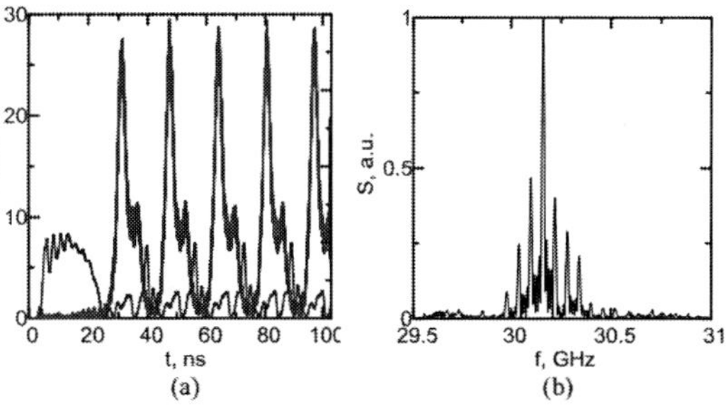

Fig. 3. PIC simulations of solitons formation for an operation frequency of 36 GHz: time dependence of the output power (a) and radiation spectrum (b). The reflected signal is shown by the blue line

This work was supported by the Russian Foundation for Basic Research (RFBR) (No. 15-08-04506).

References

1. *Vikharev, A.A., Denisov, G. G., Kocharovskiy, V.V., Kuzikov, S.V., Parshin, V.V., Peskov, N.Yu., Stepanov, A.N., Sobolev, D.I., Shmelev, M.Yu.* A high-speed quasi-optical wave phase switch based on the induced photoconductivity effect in silicon // Tech.Phys.Lett. 2007. V.33, Iss.9. P.735-739.

2. *Ginzburg, N.S., Denisov, G.G., Vilkov, M.N., Sergeev, A.S., Zotova, I.V., Samsonov, S.V., Mishakin, S.V.* Generation of trains of ultrashort microwave pulses by two coupled helical gyro-TWTs operating in regimes of amplification and nonlinear absorption // Phys. Plasmas. 2017. V. 24, Iss.2. P. 033101.

3. *Zotova, I.V., Ginzburg, N.S., Sergeev, A.S., Kocharovskaya, E.R., Zaslavsky, V.Yu.* Conversion of Electromagnetic Wave into a Periodic Train of Solitons under Cyclotron Resonance Interaction with Backward Beam of Unexcited Oscillators // Phys. Rev. Lett. 2014. V.113 P.143901.

4. *Ginzburg, N.S. , Nusinovich, G.S., Zavolsky, N.A.* Theory of non-stationary processes in gyrotrons with low Q resonators // Int. J. of Electronics 1986. V. 61, P. 881-894.

5. *Ginzburg, N.S., Sergeev, A.S., Zotova, I.V.* Time-domain self-consistent theory of frequency-locking regimes in gyrotrons with low-Q resonators // Phys. Plasmas 2015. V. 22, P. 033101.

6. *Kovalev, N.F., Kolganov, N.G., Palitsin, A.V., Fuchs, M.I.* Proc. of the 4th Int. Workshop "Strong Microwaves: Sources and Applications", N.Novgorod, Russia, 2000, V.2, p. 845.

7. *Venediktov, N.P., Dubrov, V.V., Zapevalov, V.E., Kornishin, S.Yu., Kotov, A.V., Kuftin, A.N, Malygin, O.V., Sedov, A.S., Fiks, A.Sh., Tsalolikhin, V.I.* Experimental tests of a 263 GHz gyrotron for spectroscopic applications and diagnostics of various media // Radiophys. & Quant. Electron. 2010. V. 53, Iss.4 P. 237

EPJ Web of Conferences **149**, 04045 (2017) DOI: 10.1051/epjconf/201714904045
10th International Workshop 2017 "Strong Microwaves and Terahertz Waves: Sources and Applications"

PIC-simulation of efficient cherenkov X-band and V-band HPM sources with moderately relativistic electron beams

V.P. Tarakanov[1,2], E.M. Totmeninov[3], I.V. Pegel[3,4]

[1]Joint Institute for High Temperatures, Russian Academy of Sciences, Moscow, Russia, karat@tarak.msk.su
[2]National Research Nuclear University MEPhI, Moscow, Russia
[3]Institute of High Current Electronics SB RAS, Tomsk, Russia, totm@lfe.hcei.tsc.ru
[4]National Research Tomsk Polytechnic University, Tomsk, Russia

Introduction

Efficient sources of high-power repetitively-pulsed microwave sources with moderately relativistic electron beams have several advantages. Firstly, high-voltage drivers of smaller dimensions and power capacity are required to power these devices, which allows building mobile devices. It is also important that, in the sources operating in repetitively pulsed mode, reducing the energy of electrons in the spent electron beam lowers the thermal load on the collector (and therefore the cooling requirements), and, to an even greater extent, reduces the collector bremsstrahlung and the requirements for X-ray protection.

X-band Twistron with efficiency 54%

A new scheme of high-power microwave oscillator of twistron type [1] using a moderately relativistic high current electron beam is presented (Fig. 1). In this device the reflector position relative to the SWS sets the amplitude and phase relationships between RF current in the electron beam and the fields of main harmonics at the SWS inlet. Next, it is advisable to choose the location of electron beam deposition where the electron bunches reach maximum compactness. It is also appropriate that the slope of the SWS at this location is nearly perpendicular to electron trajectory. In this case, the electric field acting on electrons is represented by z-component only.

In numerical experiment using axisymmetric version of completely electromagnetic PiC-code KARAT [2], a 54% conversion efficiency of electron beam power to electromagnetic radiation was demonstrated. With 230 kV accelerating voltage, 1.7 kA electron beam current, and 1.9 T guiding magnetic field strength, the simulated microwave power was 210 MW at 9.7 GHz (Fig. 2). The "electronic efficiency" of the source reaches 64% (Fig. 1).

With a special phasing, the energy exchange between the beam and the RF field is non-monotonous along the interaction space (Fig. 1). The electron bunch is first boosted by the field of 0th harmonic of forward wave. As a result, it is shifted to the center of decelerating phase of -1st harmonic of the backward wave. The kinetic energy of the bunch is extracted by RF field on a short section (with a length close to corrugation period), and the intensity of the extraction achieves its maximum near the beam dumping location, where the beam coupling to longitudinal RF field is maximized due to steep (normal) tilt of the SWS surface (Fig. 3).

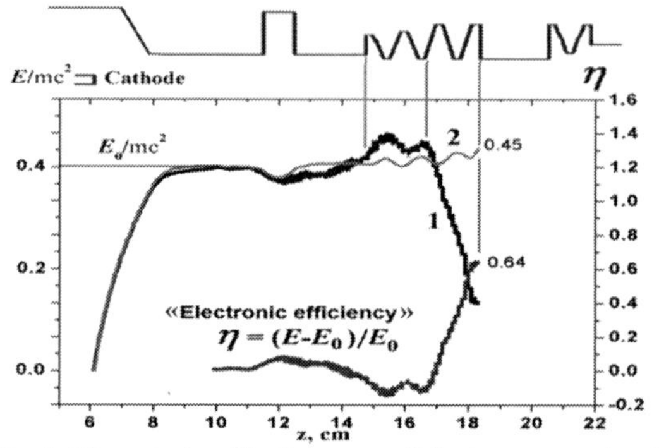

Fig. 1. Dimensionless kinetic energy of electrons versus longitudinal coordinate [(1) saturated steady-state oscillation; (2) prior to oscillation buildup] and the "electronic efficiency" $\eta = (E - E_0)/E_0$ obtained in KARAT simulation

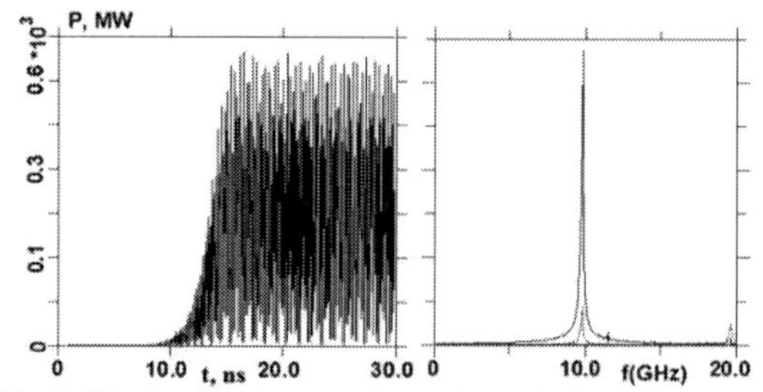

Fig. 2. Simulated non-averaged waveform of microwave power and the radiation spectrum of the oscillator

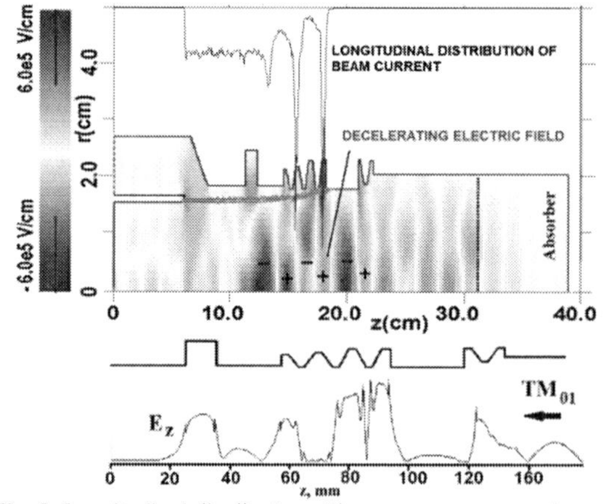

Fig. 3. Longitudinal distribution of beam current, grayscale map of RF electric field strength (z-component) – top; "cold" longitudinal distribution of electric field z-component at 9.7 GHz at the radius of the electron beam – bottom

V-band Orotron with efficiency 34%

A scheme of V-band high power microwave oscillator of orotron type [3] with oversized ($D/\lambda \sim 2.7$) electrodynamic structure is presented. The proposed device operates the axisymmetric TM_{03} mode of a rippled circular waveguide, while the output mode is TM_{02}. The mean radius of the slow-wave structure (SWS) is $R_0 = 6.9$ mm, the length is 18 mm, and the period and amplitude of corrugation are 3.3 mm and 0.6 mm, respectively (Fig. 4). The electron beam radius is $R_{beam} = 6.1$ mm. The plain waveguide section separating a coaxial diode with magnetic insulation and the SWS is supercritical with respect to TM_{03} mode, and the same is true about the waveguide attached downstream of the SWS. That is why radiation generated in the SWS in the form of "trapped" TM_{03} mode can be released only owing to transformation into some mode having smaller radial index and lower critical frequency — in the particular case, the mode TM_{02}.

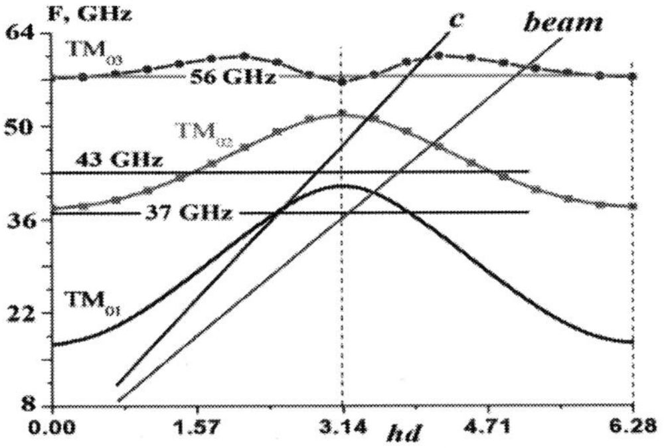

Fig. 4. Dispersion curves of lower axisymmetric electric modes in an infinitely long rippled circular waveguide: "c" marks the speed of light line, and "beam" denotes the electron beam velocity line

Cyclotron mode selection is an efficient method to discriminate between different modes in microwave devices employing tubular electron beams transported in longitudinal magnetic field. The method is based on the phenomenon of cyclotron absorption of electromagnetic waves by magnetized beam electrons. The intensity of absorption for a particular wave depends on its radial structure. Properly choosing the electron beam radius allows efficient absorption of all symmetric waves except the desired working mode.

The simulated output microwave power was 190 MW with 34% efficiency when using 2.5 kA, 220 keV electron beam transported via 2.2 T magnetic field and oscillation frequency 56 GHz (Fig. 5, 6).

Single-mode operation of the oscillator is ensured by using cyclotron mode selection [4] of concurrent modes TM_{01} (37 GHz) and TM_{02} (43 GHz). In the orotron being considered, the resonant magnetic field (energy of electromagnetic wave is converted into kinetic energy of electrons rotating in magnetic field) is about 3.0 T. It is near the field of the selected excitation of TM_{03} mode observed in the simulations. Indeed, for this mode, the squared amplitude of radial electric field (which determines the intensity of cyclotron absorption of any symmetrical TM waves) is proportional to $J_1^2(\nu_{03}R_{beam}/R_0) = 0.019$, which is much smaller than $J_1^2(\nu_{01}R_{beam}/R_0) = 0.324$ (for TM_{01} mode) and $J_1^2(\nu_{02}R_{beam}/R_0) = 0.088$ (for TM_{02} mode).

As for competing non-axisymmetric modes, they can be depressed by interrupting the azimuthal conductivity of the electrodynamic structure (for example, by cutting longitudinal slits through its wall [3]).

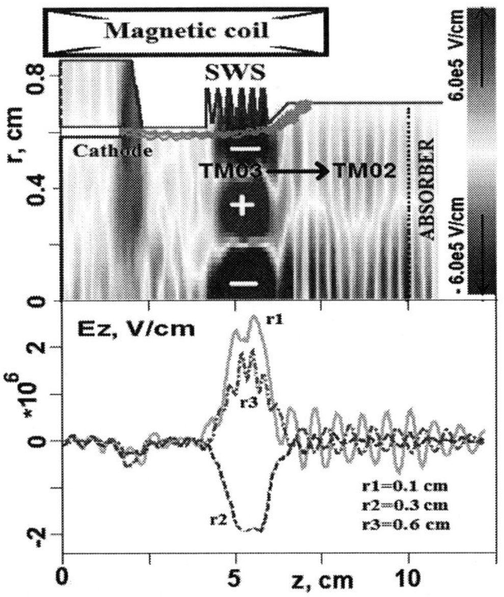

Fig. 5. Snapshot of the driving electron beam and electric field strength (top); longitudinal distributions of the electric field (z-component) at different radii

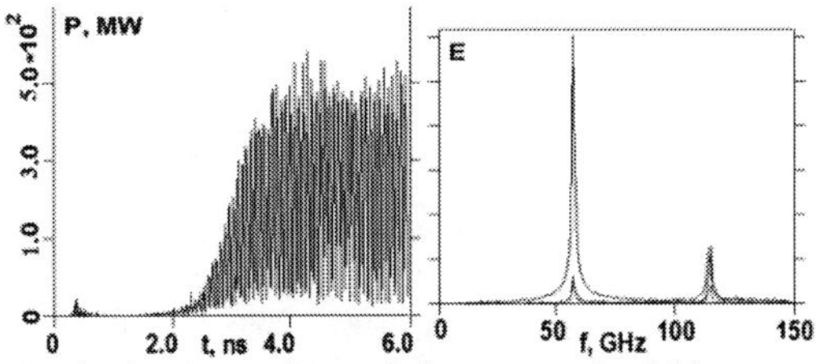

Fig. 6. Simulated waveform of electromagnetic field power flow and spectrum of electric field oscillation in the output waveguide of the orotron

References

1. *Totmeninov E.M., Pegel I.V., Tarakanov V.P.* Highly efficient X-band relativistic twistron // Laser and Particle Beams. 2016. V. 34. No. 4. P. 601-605.

2. *Tarakanov V.P.* Multipurpose electromagnetic code KARAT // *in Mathematical Modeling: Problems and Results*, 2003. Nauka. Russia. Moscow, P. 456-476.

3. *Totmeninov E.M., Pegel I.V., Tarakanov V.P.* Efficient 58-GHz relativistic orotron with combined mode selection // Physics of Plasmas. 2017. V.24. No. 5, 053106.

4. *Abubakirov E.B., Denisenko A.N., Kovalev N.F., Kopelovich E.A., Savel'ev A.V., Soluyanov E.I., Fuks M.I., Yastrebov V.V.* Relativistic backward wave oscillator using a selective mode converter // Tech. Phys, 1999, V. 44. No. 11, P. 1356-1359.

EPJ Web of Conferences **149**, 04046 (2017) DOI: 10.1051/epjconf/201714904046
10th International Workshop 2017 "Strong Microwaves and Terahertz Waves: Sources and Applications"

Time-dependent numerical simulation of diffraction and absorption effects in diagnostics of short high-power microwave pulses using wide-aperture liquid calorimeters

V.P. Tarakanov[1,2], A.I. Klimov[3,4], I.V. Pegel[3,4], P.V. Priputnev[3], and E.M. Totmeninov[3]

[1]Joint Institute for High Temperatures, Russian Academy of Sciences, Moscow, Russia, karat@tarak.msk.su
[2]National Research Nuclear University MEPhI, Moscow, Russia
[3]Institute of High Current Electronics SB RAS, Tomsk, Russia, klimov@lfe.hcei.tsc.ru
[4]National Research Tomsk Polytechnic University, Tomsk, Russia

Introduction

To measure the energy of high-power microwave pulses, calorimeters with disc-shaped wide-aperture absorbing loads filled with a working liquid based on ethanol are used [1, 2]. Their principle of operation is based on an increase in the volume of the liquid due to the absorption of the energy of the microwave pulse. Using a microwave detector recording the envelope of the microwave pulse in conjunction with the calorimeter allows to determine the amplitude of the power in the pulse. The detector with a receiving antenna is usually installed behind the load – for example, to register a signal that has passed with attenuation through the absorbing liquid. In experiments [3, 4] it was found that the measured shape of the microwave signal depends on the registration point and can differ significantly from the observed in the absence of a calorimeter.

To figure out the origins of this dependence, numerical modeling using the axisymmetric (RZ) and planar (XZ) versions of the fully electromagnetic PiC-code KARAT [5] was performed. The parameters of the microwave pulses corresponded to the S-band as it was in experiments [4] using the SINUS-7 accelerator.

To take into account the frequency dispersion of the absorbing mixture, a Lorentz model implemented in the KARAT code was applied, based on the presentation of current in the medium by harmonic oscillators. The number of the oscillators with different eigenfrequencies and Q-factors, and their density, are selected in such a way that the dispersion curves known from the experiment can be approximated with a given accuracy.

The modeling demonstrated that, depending on the position of the registration point, the shape of the recorded pulse undergoes a distortion compared to that measured in the neck of the horn antenna, both because of the diffractive rounding of the calorimeter load by the electromagnetic wave, and because of the effect of frequency-dependent spatial attenuation of the wave passing through the load.

Diffraction and time-dependent interference of microwave field behind the calorimeter

As a rule, a gap is left between the radiating horn antenna and the calorimeter, at least in order to mechanically separate the calorimeter from the microwave source (jitter transmitted from vacuum pumps or shaking from the operation of a pulsed magnetic system leads to a decrease in the sensitivity of the calorimeter and the accuracy of the microwave pulse energy measurement [1]).

The radiation that has passed through this gap, due to the diffraction at the lateral edge of the load of the calorimeter, falls into the region of the geometric shadow behind the calorimeter, where an interference pattern is formed. The amplitude of the diffraction field in this region can substantially exceed the amplitude of the wave transmitted with attenuation through the absorbing liquid. The simulations showed that, not only the amplitude of the detected signal depends on the position of the registration point in this region, but, more importantly, its shape (Fig. 1).

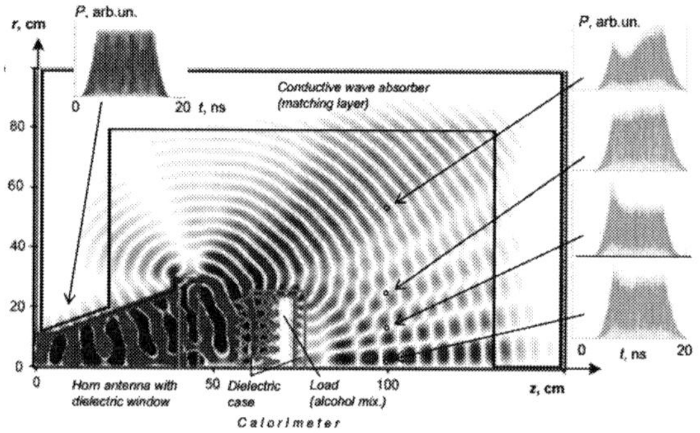

Fig. 1. Simulated B_φ field pattern with a fundamental frequency of 3.8 GHz in the vicinity of radiating horn antenna and the calorimeter load; waveforms of power flux density in the neck of the horn and at several points in a plane behind the calorimeter. The incident pulse has 16 ns full width, with 3 ns rise and fall

Distortion of the shape of the pulse is explained by the fact that the formation of the interference pattern at the front of the incident pulse and its destruction at the decay of the pulse behind the obstacle (the calorimeter) occurs with a delay determined by the difference in the arrival time of the diffraction waves from the edge of the obstacle to the observation point. Thus, distortion of the pulse shape takes place in off-axis regions, where the path difference is not zero. Maximally simplifying the situation, we can, instead of the obstacle in the form of a disk of some diameter, consider a flat strip of the same width, two edges of which will be the sources of cylindrical secondary waves.

Figure 2 shows examples of the envelopes of the "quadratic detector" signals obtained on the basis of the above elementary considerations, with a 2 ns variation in the wave path for the cases of in-phase and anti-phase addition of two secondary waves for a trapezoid initial pulse. As can be seen, the forms of the resulting pulses have the same characteristic features as those observed in the simulation.

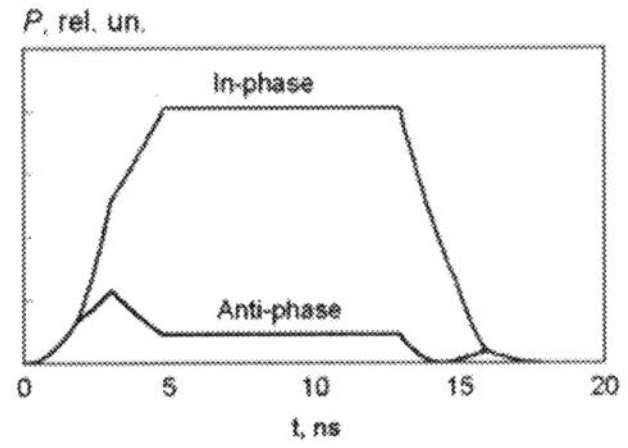

Fig. 2. An example of "quadratic" envelopes with in-phase and antiphase summation of diffraction fields of trapezium pulse duration of 16 ns with fronts of 3 ns at points with a path difference of 2 ns

Frequency-dependent decay of microwave pulses in the absorbing mixture

It is possible to reduce the amplitude of the waves enveloping the calorimeter by means of its closer contact with the radiating antenna or by increasing the diameter of the calorimetric load. However, the pulse, which passes through the absorbing liquid with attenuation, can differ substantially in form from the incident pulse. This is due to frequency dependence of the microwave damping in the absorbing mixture. Indeed, even in the approximation of an absorbing medium with a constant conductivity independent of frequency, the spatial decrement of the wave (the skin depth) is inversely proportional to the square root of the frequency. As a result, if the wave decay at the fundamental frequency is strong (as is done in practice), then low-frequency components dominate in the pulse that reaches the rear side of the load, and its shape may radically differ from the original one.

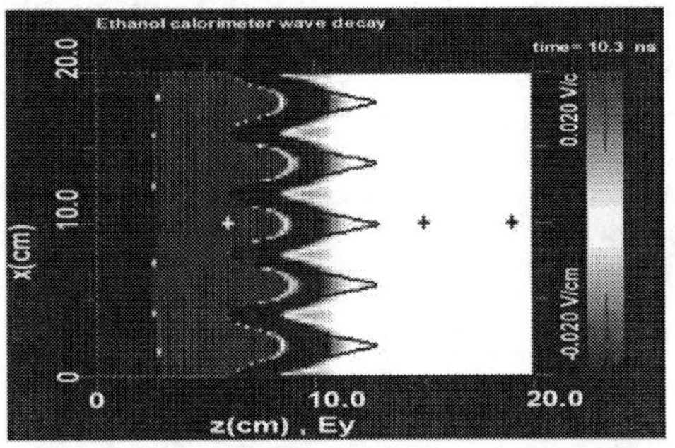

Fig. 3. The pattern of transverse component of the electric field when a plane wave falls on the ribbed surface of the absorbing medium of the calorimeter, simulated in planar (XZ) model. Pluses indicate control points, in which the field waveforms are monitored (Fig. 4)

The simulation confirmed that, with a given thickness of the absorbing medium, most significant distortions are experienced by those pulses that have steep fronts, and, respectively, a strongly "unbalanced" RF oscillation at this point. In this case, the sign of the low-frequency ejection is determined by the initial (or final) phase of the RF oscillation (Fig. 4, a). Pulses in which the RF oscillations build up and decay smoothly, for example, a Gaussian pulse, are transmitted with relatively small low-frequency distortion (Fig. 4, c).

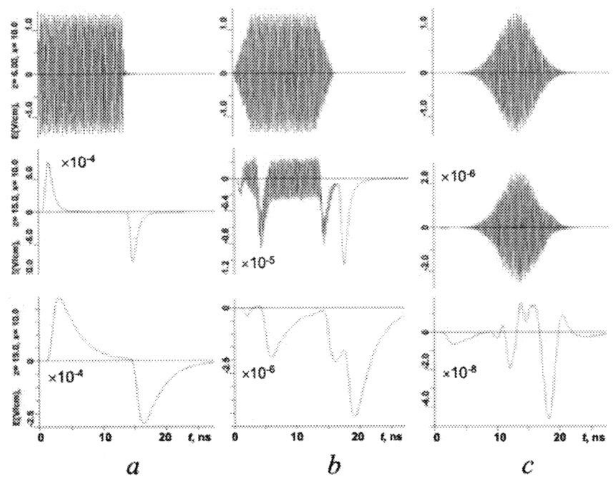

Fig. 4. Simulated waveforms of electric field at control points marked in Fig. 3: outside the absorbing mixture (top) and inside it (middle, bottom), for incident pulses of various shapes: *a* – rectangular pulse containing an integer number of RF periods; *b* – trapezoidal pulse; *c* – pulse with Gaussian enveloping

Conclusion

Thus, the numerical modeling demonstrated that when registering the shape of the envelope of a nanosecond microwave pulse using a detector located behind the absorbing load of a wide-aperture liquid calorimeter, the shape of the detected signal under certain conditions may strongly differ from the pulse emitted by the microwave source. The corresponding limitations must be considered when performing such measurements.

References

1. *Shkvarunets A.G.* A broadband microwave calorimeter of large cross section // Instruments and Experimental Techniques, 1996, V. 39, Issue 4, pp. 535–538.

2. *Vykhodtsev P.V., Elchaninov A.A., Klimov A.I., Kovalchuk O.B., Totmeninov E.M.* Liquid calorimeters for measuring the energy of high-power microwave pulses // Instruments and Experimental Techniques, 2015, V. 58, Issue 4, pp. 510–514.

3. *Tot'meninov E.M., Vykhodtsev P.V., Gunin A.V., Klimov A.I., Rostov V.V.* Increase in the energy efficiency of a pulsed-periodic relativistic backward wave oscillator with a modulating resonant reflector // Technical Physics, 2014, V. 59, Issue 3, pp. 428–433.

4. *Kitsanov S.A., Korovin S.D., Klimov A.I., Rostov V.V., Tot'meninov E.M.* Mechanically tuned relativistic backward wave oscillator // Technical Physics Letters, 2004, V. 30, Issue 8, pp. 619–621.

5. *Tarakanov V.P.* Multipurpose electromagnetic code KARAT // *in Mathematical Modeling: Problems and Results*, 2003. Moscow, Nauka, Russia, pp. 456–476.

EPJ Web of Conferences **149**, 05001 (2017) DOI: 10.1051/epjconf/201714905001
10th International Workshop 2017 "Strong Microwaves and Terahertz Waves: Sources and Applications"

Wave beams with orbital angular momentum: a step towards terahertz

B.A. Knyazev[1,2], V.S. Cherkassky, Yu.Yu. Choporova[1,2], O.E. Kameshkov[1,2], G.N. Kulipanov[1], N.D. Osintseva[1,3], V.S. Pavelyev[4], N.A. Vinokurov[1,2], B.O. Volodkin[4]

[1] Budker Institute of Nuclear Physics SB RAS, Novosibirsk, Russia, b.a.knyazev@inp.nsk.su
[2] Novosibirsk State University, Novosibirsk, Russia
[3] Novosibirsk State Technical University, Novosibirsk, Russia
[4] Samara National Research University, Samara, Russia

Beam formation

The energy and angular momentum are quantum characteristics of photons. The angular momentum is the sum of the spin and orbital angular momenta. In physical optics, the spin momentum is described as the polarization of light beams, and the orbital angular momentum (OAM) is the rotation of beam along the azimuth. Interest in beams with OAM [1, 2], or "vortex beams," appeared shortly after the creation of lasers and increased after the publication of the article [3]. Vortex beams have been formed and investigated in the spectral range from X-ray to radio waves, but only a few publications are devoted to beams with OAM in the terahertz range (see [4] and references therein). In this paper we describe vortex Bessel beams with an average power of 25 W, obtained on the Novosibirsk free-electron laser (NovoFEL) [5].

Vortex beams were formed from the NovoFEL Gaussian beam transformed by binary phase axicons (BPA) with spiral zone structures (Fig. 1, a, b). The zone boundaries are described with the equation $r_\beta = p(l\varphi/2\pi+\beta/2)$, where $p=2\pi/\kappa=3.2$ mm is the period, $\beta=0\ldots(2l-1)$, and κ is the transverse wave number. The relief depth was calculated for a wavelength $\lambda=141$ µm: $H=\lambda_0/2(n-1)=29.1$ µm ($n=3.42$). Beams with the topological charges $l=\pm1$ (Fig. 1, c) were formed with the BPA (a), whereas the BPA (b) formed the beams with intertwined helicoids wavefront ($l=\pm2$) (Fig. 1, d). The average power of the incident Gauss beam was about 100 W. About 50% of this power was reflected from the silicon axicon, and then 41% was diffracted into the converging plane wave, which formed the Bessel beam with a power of 25 W. This value can be increased fourfold by the use an antireflection coating and a telescopic system mounted directly behind the axicon.

Beam characteristics

The radial distribution of the intensities of the obtained beams recorded with a 16.32×12.24 mm^2 microbolometer array [6] moving along the z-axis on a translation stage was in good agreement with the Bessel function (Fig. 2), but the Fourier spectrum of transverse spatial-frequency components of the beams measured at the lens focus consisted, instead of rings observed in the case of ordinary Bessel beams, of segments of $2l$ intertwined arcs (Fig. 2, a, b). Nevertheless, in the rest, the properties of the obtained beams corresponded to the properties of the Bessel beams. In particular, the cross-section of the beams was unchanged over a length of approximately 150 mm ("non-diffraction beams"). The beams are self-reconstructing after passing through obstacles and randomly inhomogeneous media.

The formation of Bessel beams by means of a binary spiral axicon has an advantage in comparison with the use of the combination of spiral phase plate and axisymmetric axicon [7]. In the latter case, the optical system correctly forms a vortex beam only for wavelengths for which the thickness of the spiral plate grows up as $h(\varphi)=\lambda l\varphi/2\pi(n-1)$. On the contrary, a binary phase axicon with spiral zones forms a beam $E(r,\varphi,z)=E_0J_l(\kappa r)\exp[i(k_z z-l\varphi)]$ with a given topological charge for any wavelengths, only its diffraction efficiency changing. The diameters of the beam rings are also independent of the wavelength, which can be useful for many applications.

All these experiments showed that the vortex beams obtained by us possess all the properties of Bessel beams for a certain distance.

Beam diagnostics

The intensity distributions of the beams formed by the axicons are in good agreement with the distributions calculated for the Bessel beams. However, the intensity distributions are identical for both left-handed and right-handed beams and the topological charge cannot be determined from them. The beam helicity parameters were determined in classical diffraction experiments: the interference of vortex and Gaussian beams, rotation of beam obstructed by a half-plane, and the double-slit Young experiment. The Young experiment proved to be the most obvious and simple way to solve this problem.

The diffraction pattern obtained in this experiment for the Bessel beam with the topological charge $l=-2$ is shown in Fig. 3. The fringes are dogleg shaped and can be described with the expression

$$x(y,z)=-a\tan\left[\frac{m\pi}{l}+\frac{2\pi ay}{lz\lambda}+\frac{\pi}{2}\right], m=0\ldots l-1,$$

where $2a$ is the distance between narrow slits.

Beam applications

The advantages of Bessel beams are widely used, and one can assume that they will also find application in the terahertz range. Communication systems and radars are the most obvious uses. Without touching these problems here, we give examples of vortex Bessel beams applied in optical phenomena in which they were not previously investigated.

The first experiment in which the use of vortex beam led to an unusual result was the generation of surface plasmon polaritons by the method of beam diffraction at the edge of metal-dielectric interface ("end-fire coupling"). The plasmons were expected to arise at each intersection of the Bessel beam rings with the surface boundary. However, depending on the direction of the beam twist, plasmons were generated only on one side of the optical axis. This effect, which has not been clearly explained yet, can be used as a new type of binary plasmonic switch.

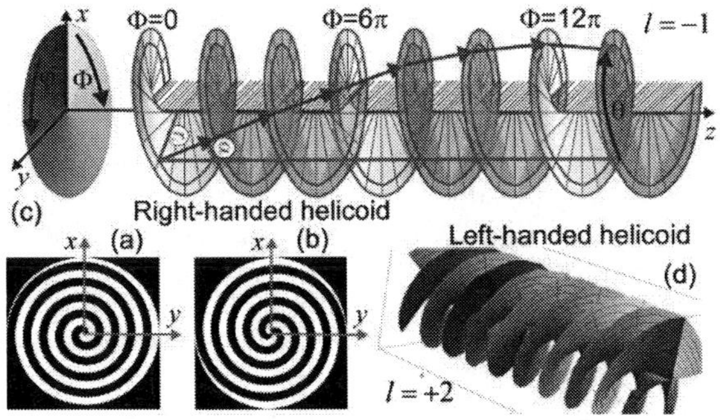

Fig. 1. Binary phase axicons with spiral zones (phase shift by π at $\lambda = 141$ μm) for forming beams with topological charges $l = -1$ (*a*) and $l = +2$ (*b*); (*c*, *d*) helicoidal wavefronts

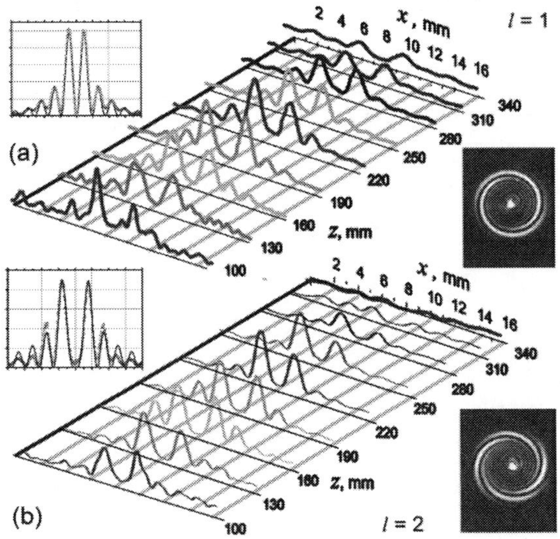

Fig. 2. Cross-sections of beams with topological charges $l = -1$ (*a*) and $l = +2$ (*b*) vs. distance. Left insets: comparison of the Bessel functions with beam cross-sections. Right insets: image of beams in lens focal plane (spatial frequency distribution)

We also investigated for the first time the passage of vortex beams through amplitude and phase periodic gratings, that is, an analog of the Talbot effect. Experiments and numerical calculations have shown that, as in the classical Talbot effect, periodic diffraction patterns are formed in the main and fractional Talbot planes, depending however on a larger number of parameters than the classical effect. The transformation of a large vortex beam into a periodic lattice of small vortex beams was demonstrated.

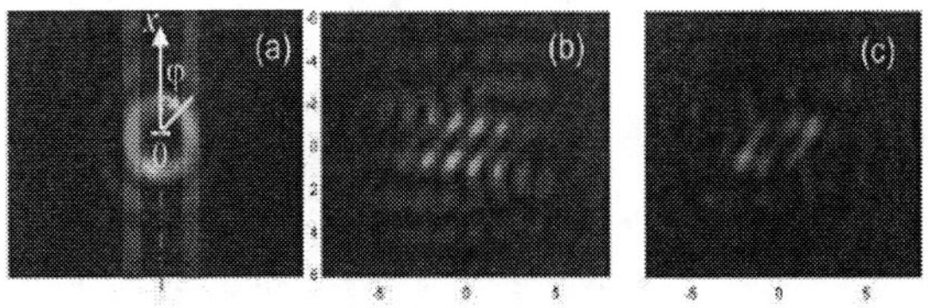

Fig. 3. (*a*) Double slit with spacing of 4.5 mm, illuminated by vortex beam; (*b*) The experimentally observed diffraction pattern 40 mm behind the slits reveals that the beam topological charge is equal to $l = -2$; (*c*) (color online) Simulated diffraction pattern in which colors show phase distribution from 0 (blue) to π (red)

Acknowledgements

Experiments were supported by RFBR grant 15-02-06444. The axicons were fabricated under supported by the Ministry of Education and Science of the Russian Federation (project 16.7894.2017/6.7). The terahertz beamline was designed under support of RSF grant 14-50-00080. The experiments engaged equipment belonging to the Siberian Synchrotron and Terahertz Radiation Center.

References

1. *Baranova, N.B., Zeldovich, B.Ya.* Dislocations of the wave-front surface and zeros of the amplitude // Sov. Phys. JETP 1981. V. 53, P. 925–929.

2. *Bazhenov, V.Yu., Vasnetsov, M.V., Soskin, M.S.* Laser beams with screw dislocations in their wavefronts // JETP Lett. 1990. V. 52, P. 429–431.

3. *Allen, L., Beijersbergen, M.W., Spreeuw, R.J.C., Woerdman, J.P.* Orbital angular momentum of light and the transformation of Laguerre-Gaussian laser modes // Phys. Rev. A. 1992. V. 45, 8185 –8189.

4. *Knyazev, B.A., Choporova, Yu.Yu., Mitkov, M.S., Pavelyev, V.S., Volodkin, B.O.* Generation of terahertz surface plasmon polaritons using nondiffractive Bessel beams with orbital angular momentum // Phys. Rev. Lett. 2015. V. 115, 163901 (5 p.).

5. *Kulipanov, G.N., et al.* Novosibirsk Free Electron Laser—Facility Description and Recent Experiments // IEEE Trans. Terahertz Sci. Technol. 2015. V. 5. P. 798-809.

6. *Dem'yanenko, M.A., Esaev, D.G., Marchishin, I.V., Ovsyuk, V.N., Fomin, B.I., Knyazev, B.A., Gerasimov V.V.*, Application of uncooled microbolometer detector arrays for recording radiation of the terahertz spectral range // 2011. Optoelectron., Instrum. Data Process. V. 47, P. 508–512.

7. *Wei, X., Liu, C., Niu, L., Zhang, Z., Wang, K., Yang, Z., Liu, J.* Generation of arbitrary order Bessel beams via 3D printed axicons at the terahertz frequency range // Appl. Opt. 2015. V. 54. P. 10641-10649.

EPJ Web of Conferences **149**, 05002 (2017) DOI: 10.1051/epjconf/201714905002
10th International Workshop 2017 "Strong Microwaves and Terahertz Waves: Sources and Applications"

Coherent spontaneous THz undulator radiation from dense electron bunches formed in laser-driven photo-injectors

V.L. Bratman[1,2], N. Balal[1], I.V. Bandurkin[2], A. Friedman[1], Yu. Lurie[1], and A.V. Savilov[2]

[1]Ariel University, Ariel, Israel, v_bratman@yahoo.com
[2]Institute of Applied Physics, Russian Academy of Sciences, Nizhny Novgorod, Russia

Introduction

Short dense bunches formed in advanced photo-injectors can be effectively used for generation of powerful coherent spontaneous Doppler-frequency-up-shifted radiation. Several methods have been proposed to weaken longitudinal Coulomb repulsion of electrons in bunches to significantly increase the radiation energy and frequency of such sources. In this presentation, we discuss methods which can be used in the THz source [1] that is now under construction in Ariel University, Israel, and in other relatively compact sources using particle energy of the order of 3–6 MeV. These are pre-modulation and energy chirping of bunches as well as using space charge of the bunch for stabilization of its longitudinal size due to development of Negative Mass Instability (NMI).

Evolution of Modulated Bunches

One method of providing a modulation of bunches is the use of a temporal modulation of laser power at a photo-cathode. Due to fixation of the period of modulation, the electron density can be decreased by enlarging longitudinal or radial sizes of the bunch while maintaining the coherent nature of the process and the magnitude of the radiated energy.

If the bunch shape is close to a thread and its radius is much smaller than the period of modulation in the own reference frame, the Coulomb force decreases many times at the one period; different periods are almost independent and the value of the period stays nearly constant. This may lead to excitation of nonlinear plasma waves with disappearance and reappearance of modulation with narrow spikes at quarter and half of plasma wave period, respectively, (Fig. 1) [2]. The opposite case of quasi-planar disks as well as intermediate case, which are important for powerful generation, will be discussed in this presentation.

In contrast with threads, particle bunching and overtaking are impossible in modulated electron discs and 1D planar layers with a static initial density distribution; in such systems, longitudinal displace-ments of the particles do not generate sufficient restoring forces and plasma oscillations, but only lead to longitudinal expansion with an increasing period of modulation. For example, for the cosine initial distribution in a 1D planar layer (Fig. 2), the density evolves as

$$\rho(x) = \rho_0(x_0)/\left(1+q_0+\varepsilon q_0 \cos hx_0\right),\ \ \rho_0(x_0) = \\ = \rho_0(1+\varepsilon\cos hx_0)$$

where the function $x = x(x_0, t)$ describes motion of an electron with a constant invariant acceleration whose absolute value increases with an initial coordinate x_0, $q_0 = (\omega_b t)^2/2$, and $\omega_b = (4\pi e \rho_0 / m\gamma^3)^{1/2}$ is the electron plasma frequency. A developed 1D theory of evolution of moving layers and disks is confirmed with simulations based on General Particle Tracer (GPT) [3] as well as on original 3D space-frequency WB3D code [4]. The excitation of plasma waves in the treads occurs at constant modulation period and allows excitation of high-frequency radiation harmonics. However, at the same plasma frequency, the amplitude of modulation changes in this case faster than for disks where the period increases. Choosing one or the other bunch shape is mainly determined by the possibilities of specific accele-rators. In both cases, guiding magnetic field is needed for transverse confinement of the electron bunches.

Fig. 1. Excitation of a plasma wave in a thread with 100% cosine density distribution (GPT-simulations; disappearance and reappearance of modulation in accordance with [2])

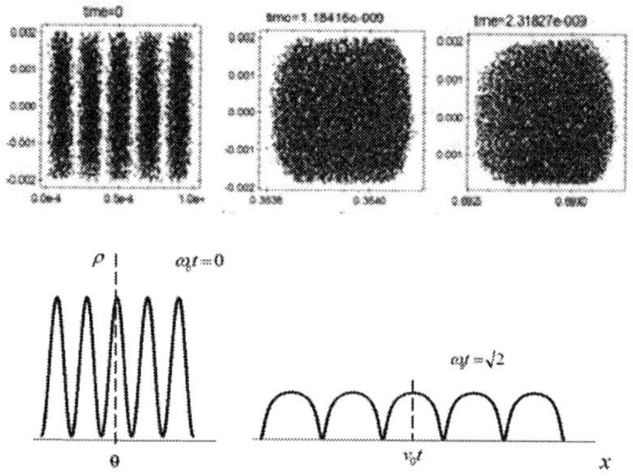

Fig. 2. Expansion of a disk (GPT simulations) and a 1D planar layer with 100% cosine density modulation (analytical result). No plasma wave excitation

Negative Mass Instability in Bunches

Unlike the above situation, the space charge can even maintain the long-living electron "core" in the bunch [5, 6] if conditions for NMI in undulators [7, 8] are fulfilled. It can be achieved in a combined undulator and over-resonance uniform longitudinal magnetic field. In such conditions, increase/decrease in energy of an electron under action of space charge can lead to de-crease/increase in its longitudinal velocity. In this case, development of NMI can lead to formation of a dense "core" with a sufficiently small longitudinal size (Fig. 3) that can coherently radiate high energy at THz frequencies into a narrow vacuum wave-beam or into several dominant modes of a waveguide (Fig. 4).

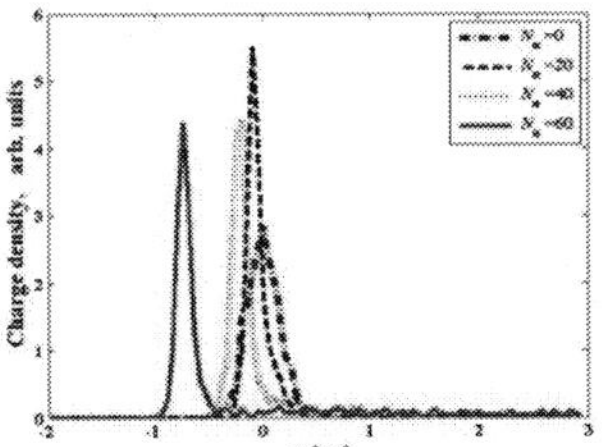

Fig. 3. Bunch density after each 20 undulator periods for charge 0.5 nC. Bunch tail shifts peak of distribution out of center of mass (time is given relative to center)

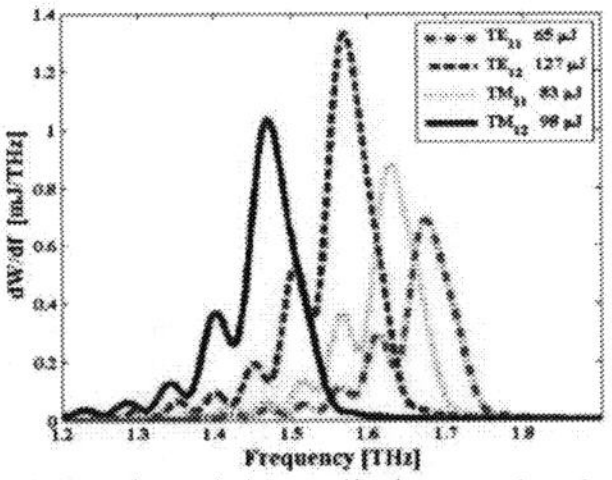

Fig. 4. Spectral density of the radiation at dominant waveguide modes for charge 0.5 nC after 40 undulator periods

It has been demonstrated in detail, both for the bunch moving in a transversely uniform undulator field and radiating in free space based on Lienard-Wichert approach, as well as for a more realistic 3D geometry of undulator field and THz waveguide with WB3D code [7, 8]. Injection of a dense bunch on a stationary helical trajectory with small amplitude of parasitic cyclotron oscillations can be provided if the bunch with a definite chirp of transverse velocities enters first a solenoid with an operating value of the guiding field and after that moves in a section with adiabatically increasing undulator and uniform field.

A strong magnetic field that is necessary in NMI regime can be used for creation of required undulator field. It can be done, for example, using a redistribution of the uniform field of the solenoid by a ferromagnetic (steel) helix (Fig. 5) [5]. For example, a helix with a period of 2.5 cm and an inner diameter of 10 mm wound of a steel wire with a radius 3 mm and mounted into solenoid with a field 75 kOe provides sufficient amplitude of transverse helical field 2 kOe (these parameters are used in Figs. 3, 4, 6, 7).

Simulations demonstrate a possibility for significant efficiency enhancement and narrowing of spectrum for THz undulator radiation from short and very dense electron bunches thanks to both pre-modulation and NMI-stabilization of dense electron bunches. Combination of these two methods allows formation of nearly periodic and long-living sets of dense narrow bunches (Fig. 6) which can be used for more selective excitation of waveguide modes (Fig. 7) and for obtaining radiation at higher undulator harmonics. All these methods open the way for implementation of efficient sources of narrow-band THz pulses with milli-Joule energy, picosecond duration and easy wide-band frequency tuning. Such pulses can be prospective for pump-probe experiments.

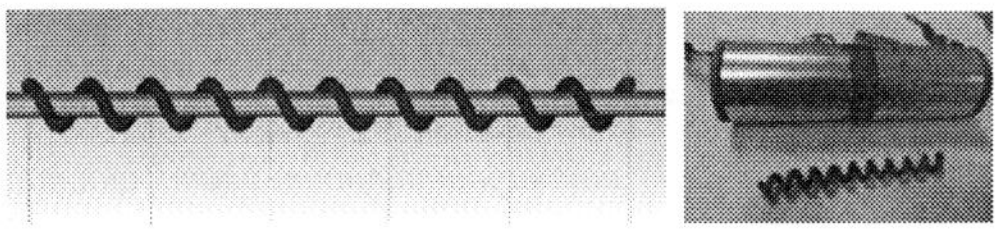

Fig. 5. Helical undulator in the form of steel helix wound on cylindrical copper waveguide and placed inside solenoid with strong guiding field

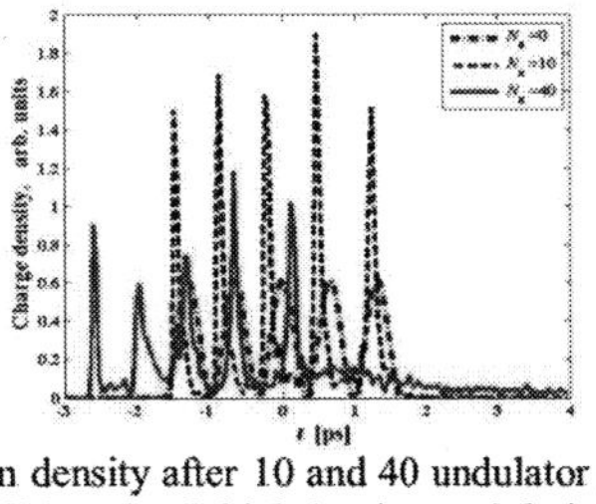

Fig. 6. Electron density after 10 and 40 undulator periods for a thread with 100% cosine initial density modulation and charge 2.5 nC (WB3D-simulations)

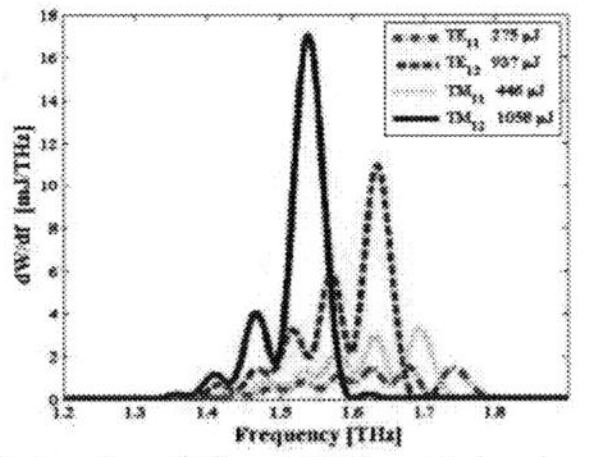

Fig. 7. Spectral density of the radiation at dominant waveguide modes for a thread with 100% cosine initial density modulation, charge 2.5 nC and 40 undulator periods

The work is supported by the Russian Foundation for Basic Research (Project # 16-02-00794) and Israeli Ministry of Science.

References

1. *A. Friedman, N. Balal, E. Dyunin, Yu. Lurie, E. Magori, V.L. Bratman, J. Rosenzweig, H. Lay To, A. Gover.* Configuration and status of the Israeli THz free electron laser.// *Proc. 36th Intern. FEL Conf.*, Basel, 2014, 553.

2. *P. Musumeci, R. K. Li, and A. Marinelli.* Nonlinear Longitudinal Space Charge Oscillations in Relativistic Electron Beams // Phys. Rev. Lett., 106, 184801, 2011.

3. http://www.pulsar.nl/gpt.

4. *Y. Pinhasi, Yu. Lurie, A. Yahalom.* Space-frequency model of ultrawide-band interactions in free-electron lasers// Phys. Rev. E, 71, 036503, 2005.

5. *N. Balal, I. V. Bandurkin, V. L. Bratman, E. Magory, and A. V. Savilov.* Negative-mass mitigation of Coulomb repulsion for terahertz undulator radiation of electron bunches. // *Appl. Phys. Lett.*, 107, 163505, 2015.

6. *Yu. Lurie, V.L. Bratman, and A.V. Savilov.* Energy Enhancement and Spectrum Narrowing in Terahertz Electron Sources due to Negative Mass Instability. // Phys. Rev. AB 19, 050704, 2016.

7. *H.P. Freund and P. Sprangle*, Unstable electrostatic beam modes in free-electron-laser systems. //Phys. Rev. A 28, 1835, 1983.

8. *N.S. Ginzburg and N.Yu. Peskov*, Nonlinear theory of relativistic ubitrons with electron beams formed in adiabatically increasing undulator and uniform longitudinal magnetic fields. //Tech. Phys. 58, 859, 1988.

Terahertz Bessel beams with orbital angular momentum: diffraction and interference

Yu.Yu. Choporova[1,2], B.A. Knyazev[1,2], N.D. Osintseva[1,3], V.S. Pavelyev[4], B.O. Volodkin[4]

[1]Budker Institute of Nuclear Physics, Novosibirsk, Russia, author@instphys.zz
[2]Novosibirsk State University, Novosibirsk, Russia
[3]Novosibirsk State Technical University, Novosibirsk, Russia
[4]Samara National Research University, Samara, Russia

The fact that light beams are able to carry a mechanical angular momentum is well-known. In paraxial beams, angular momentum can be represented as a sum of the spin (SAM) and orbital (OAM) angular momentum. After paper Allen et al. [1], a lot of articles devoted to the generation, study and the use of beams with orbital angular momentum, or "vortex beams", have been published. To date, vortex beams were obtained in the range from radio frequencies [2] to the soft X-ray radiation [3]. Nevertheless, there are only five studies [4–8], in which vortex beams in the terahertz range have been generated.

Recently, using radiation of the Novosibirsk free electron laser [9] and silicon binary spiral phase axicons, terahertz Bessel beams with OAM with topological charges $l = \pm 1$ and $l = \pm 2$ have been generated for the first time [7]. The intensity distributions of the beams formed by the axicons are in good agreement with the distributions calculated for Bessel beams, but they are identical for both left-handed and right-handed helicities. To determine the characteristics of the beams associated with their rotation, we applied classical experiments on diffraction and interference, adapting them to the terahertz range.

A direct method of detecting the rotation of a beam was its diffraction on a half-plane (Fig. 1, a). Similar experiments with Laguerre-Gaussian and Bessel vortex beams were performed in the visible region in [11], where for the Bessel beams only qualitative results were obtained because of the short laser wavelength and the tiny interference pattern.

In Fig. 1, b the diffraction patterns calculated numerically at several distances are shown. In the experiments, a 16.32×12.24 mm^2 microbolometer array (MBA) with pixel size of 0.051 mm was used as a detector for imaging of diffraction pattern. Because of geometrical restrictions, only planes located at a distance z of more than 35 mm could be recorded. The recorded patterns were identical with the calculated ones.

The diffraction patterns shown in Fig. 1 clearly demonstrate the rotation of the beam. According to the theory [12], the rate of change of azimuthal angle of the trajectory of the Poynting vector with z is given by

$$\frac{d\alpha}{dz} = \frac{1}{\rho}\frac{S_\varphi}{S_z} = \frac{l}{k\rho^2}, \qquad (1)$$

where $\rho = (x^2 + y^2)^{1/2}$ and S_φ, S_z are the components of the Poynting vector that is normal to the helical wavefront. Since the speed of the azimuthal rotation is inversely proportional to the square of the radius, the inner ring rotates most rapidly after the obstacle. The Poynting vector makes a complete revolution along the helix at the distance

$$\Delta z_{pitch} = \frac{2\pi k\rho^2}{l}. \qquad (2)$$

From experiments and calculations it follows that one turn of the Poynting vector is 240 mm for the first ring and 340 mm for the second one. The direction of rotation corresponds to the sign of the topological charge.

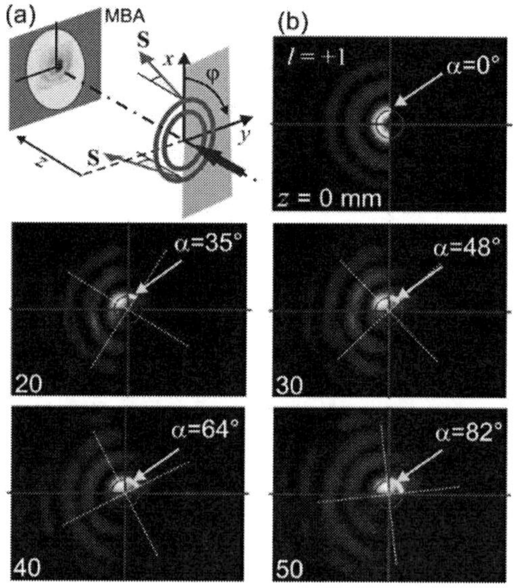

Fig. 1. Diffraction of vortex beams ($\lambda = 141$ μm) on a half-plane: (a) experimental schematic; (b) diffraction patterns for beam with the topological charge $l = +1$ vs. distance D. Radii of first and second beam rings were 0.9 mm and 1.9 mm, respectively

A simpler method for determining the twist parameters of a beam is the double slit Young experiment (Fig. 2, a). To determine all the beam parameters, it requires recording only one diffraction pattern at some distance z from the slits. Because in any plane $z = const$ the beam phase grows up with the azimuthal angle as $\Phi = l\varphi$, interference fringes become distorted. Simple calculations show that the maxima of the intensity of the fringes are described by an equation, the notation in which is clear from Fig. 2

$$x(y) = -a\tan\left[\frac{2\pi ay}{lz\lambda} + \frac{m\pi}{l} + \frac{\pi}{2}\right], \qquad (3)$$

$$m = 0 \ldots l-1.$$

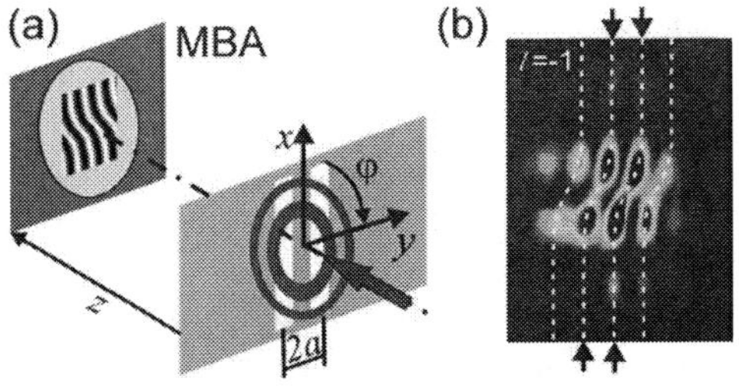

Fig. 2. (a) Schematic of double-slit experiment. (b) Diffraction pattern recorded for vortex beam with topological charge $l = -1$ at $\lambda = 141$ μm

The Bessel beam of a free-electron laser with topological charge -1, expanded five times by a telescopic system, illuminated two slits located at a distance of 4 mm. The dotted curves calculated using Eq. (3) are superimposed on the diffraction pattern (Fig. 2, b), recorded by the microbolometer matrix at a distance of 60 mm. For the beam with a charge equal to -2, the bending of the curves increased, and the shift of the bands from the upper edge to the lower edge of the frame increased by a factor of two. When the direction of rotation of the beam changed, the slope of the bands changed to the opposite.

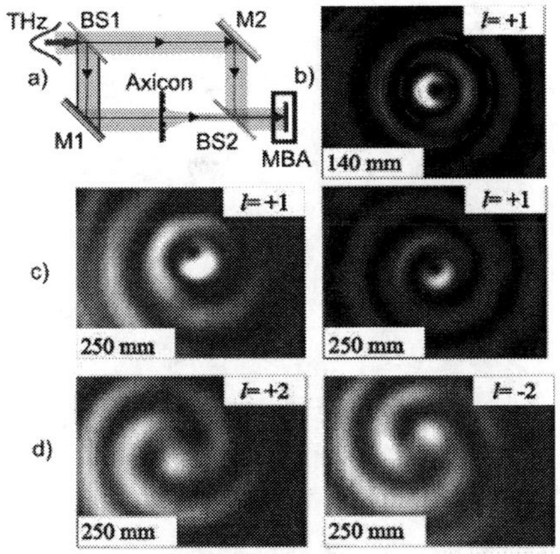

Fig. 3. Interference of vortex and Gaussian beams. (*a*) Mach–Zehnder interferometer; (*b*) interference of Bessel and Gauss beams at different distances from axicon. The dashed rings indicate the maxima of the intensity of the Bessel beam. Left column – experimental data; right column – similations

Bessel beams bounded in the aperture, formally speaking, are not diffractive at a certain distance. Investigation of the interference of the beams obtained with Gaussian wave in the Mach-Zehnder interferometer (Fig. 3, a) made it possible to reveal the details of the decay of the Bessel beam, which begins already in the process of its propagation.

This fact becomes obvious if we observe a change in the interference pattern with the distance (Fig. 3, b–d). The phase of adjacent rings of Bessel beams differs by a value of π, and the phase increases along the rings from 0 to $2\pi l$. The wave front of the Gaussian beam was practically flat. At a distance z = 140 mm, the interference along the rings varies in azimuth l times from constructive to destructive, which indicates that the wave front in this Bessel beam is practically flat. At a distance of 250 mm, the interference pattern changes, forming spirally divergent bands. Since the wavefront of a Gaussian laser beam at such a small distance remains practically unchanged, this indicates the beginning of the transformation of the helicoidal wave front into conically divergent plane waves. The number of spirals and their direction also make it possible to determine the magnitude and sign of the topological charge of the beam.

Acknowledgments

Experiments were supported by RFBR grant 15-02-06444. The axicons were fabricated under supported by the Ministry of Education and Science of the Russian Federation (project 16.7894.2017/6.7). The terahertz beamline was designed under support of RSF grant 14-50-00080. The experiments engaged equipment belonging to the Siberian Synchrotron and Terahertz Radiation Center.

References

1. *Allen, L., Beijersbergen, M.W., Spreeuw, R.J.C., Woerdman, J.P.* Orbital angular momentum of light and the transformation of Laguerre–Gaussian laser modes. Phys. Rev. 1992. V. A 45. P. 8185–8189.
2. *Thide, B., Then, H., Sjoholm, J., Palmer, K., Bergman, J., Carozzi, T.D., Istomin, Y.N., Ibragimov, N.H., Khamitova, R.* Utilization of Photon Orbital Angular Momentum in the Low-Frequency Radio Domain // Phys. Rev. Lett. 2007. V. 99. 087701.
3. *Bahrdt, J., Holldack, K., Kuske, P., Mueller, R., Scheer, M., Schmid, P.* First Observation of Photons Carrying Orbital Angular Momentum in Undulator Radiation // Phys. Rev. Lett. 2013. V. 111. 034801.
4. *He, J., Wang, X., Hu, D., Ye, J., Feng, S., Kan, Q., Zhang, Y.* Opt. Express, 2013. V. 21. P. 20230-20239.
5. *Miyamoto, K., Suizu, K., Akiba, T., Omatsu, T.* Direct observation of the topological charge of a terahertz vortex beam generated by a Tsurupica spiral phase plate // Appl. Phys. Lett. 2014. 104. 261104.
6. *Imai, R., Kanda, N., Higuchi, T., Konishi, K., Kuwata-Gonokami, M.* Generation of broadband terahertz vortex beams // Opt. Lett. 2014. V. 39. P. 3714-3417.
7. *Knyazev, B.A., Choporova, Yu.Yu., Mitkov, M.S., Pavelyev, V.S., Volodkin, B.O.* Generation of Terahertz surface plasmon polaritons using nondiffractive Bessel beams with orbital angular momentum // Phys. Rev. Lett. 2015. V. 115. 163901.
8. *Wei, X., Liu, C., Niu, L., Zhang, Z., Wang, K., Yang, Z., Liu, J.* Generation of arbitrary order Bessel beams via 3D printed axicons at the terahertz frequency range // Appl. Opt. 2015. V. 54. P. 10641-10649.
9. *Kulipanov, G.N., et al.* Novosibirsk Free Electron Laser—Facility Description and Recent Experiments // IEEE Trans. Terahertz Sci. Technol. 2015. V. 5. P. 798-809.
10. *Arlt, J.* Handedness and azimuthal energy flow of optical vortex beams // J. Modern Opt. 2003.V. 50, P. 1573-1580.
11. *Allen, L., Padgett, M.J.* The Poynting vector in Laguerre±Gaussian beams and the interpretation of their angular momentum density // Optics Communs. 2000. V. 184. P. 67-71.
12. *Emile, O., Emile, J.* Young's double-slit interference pattern from a twisted beam. // Appl. Physics B - Laser and Optics, Springer Verlag. 2014. V. 117. P. 487-491.

EPJ Web of Conferences **149**, 05004 (2017) DOI: 10.1051/epjconf/201714905004
10th International Workshop 2017 "Strong Microwaves and Terahertz Waves: Sources and Applications"

Terahertz time-domain measurements by electro-optic crystals with various symmetries

I.E. Ilyakov[1], G.Kh. Kitaeva[2], B.V. Shishkin[1], and R.A. Akhmedzhanov[1]

[1] Institute of Applied Physics RAS, Nizhny Novgorod 603950, Russia, e-mail: igor_ilyakov@mail.ru
[2] M.V. Lomonosov Moscow State University, Faculty of Physics, Moscow 119991, Russia

Terahertz wave radiation (TWR) is attractive due to its applications in spectroscopy, communication technologies, biology, medical science and other fields. One of the popular ways of TWR time-domain measurements is based on its nonlinear-wave interaction with femtosecond laser pulses in electro-optic (EO) crystals [1]. This technique utilizes ellipticity changes of the laser pulse polarization (the Pockels effect) and is widely used because of its simplicity, high sensitivity and broad operating band of EO crystals. In the report, we present our recent theoretical and experimental study [2–6] of alternative EO detection techniques and compare them with the standard one in terms of their sensitivity and compatibility with different crystals.

When linearly polarized laser pulse propagates in isotropic crystal (for example ZnTe or GaP) then in the absence of terahertz electric field its polarization doesn't change. However, in the presence of electric field a refractive index tensor of the crystal is changed and the crystal becomes anisotropic. As a result, orthogonal polarization components of the laser pulse transmit through the matter with different optical densities and the laser pulse polarization becomes elliptic. This polarization ellipticity is measured by an appropriate optical scheme (figure 1, *a*) and terahertz electric field overlapped with the laser pulse is determined. This technique can be used with TWR generated by sources directed by femtosecond laser pulses [1] and from external sources (not synchronized with the laser system) emitting both continuous [7] and pulsed radiation [8]. In spite of high sensitivity of the technique, we have shown that sensitivity of EO measurements can be increased. The sensitivity improvement can be achieved by use of laser pulses with an edge-cut spectrum [4–6]. A new experimental scheme is presented in figure 1, *b*. An edge filter is positioned before the EO crystal to produce a high gradient of the spectral energy density of the laser pulse and the TWR interacted with the laser pulse that had previously passed through an edge filter. It was shown theoretically [5] that changes of polarization ellipticity at the frequencies corresponding to the edge-cut of the laser pulse spectrum could be much higher than those could when laser pulse with Gaussian spectrum is used. To utilize the ellipticity changes at the frequencies corresponding to the edge-cut of the laser pulse spectrum, the laser pulse after the EO crystal can be decomposed in the frequency spectrum by a diffraction grating or can be passed through another edge filter transmitting only the optical waves with frequencies corresponding to the slope of the first filter [5, 6] (figure 1, *b*). The ellipticity changes are measured using an appropriate optical scheme. In works [5,6] the scheme depicted in figure 1, *b* was implemented with GaP crystal. This crystal has no natural birefringence and the new technique demonstrated a much stronger response on the same TWR than the standard one (figure 2, *a*). The method based on ellipticity changes of the laser pulse with an edge-cut spectrum was also successfully used to improve the frequency band of the near-field microscopy technique [9].

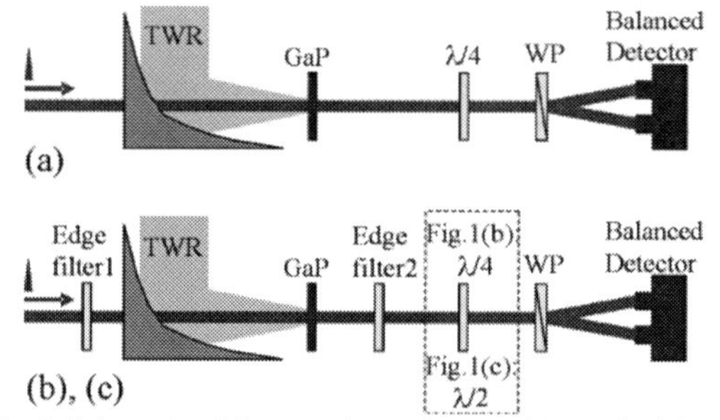

Fig. 1. Schematic of the experimental setup: (a) standard detection scheme; (b), (c) variants of detection schemes with edge filters

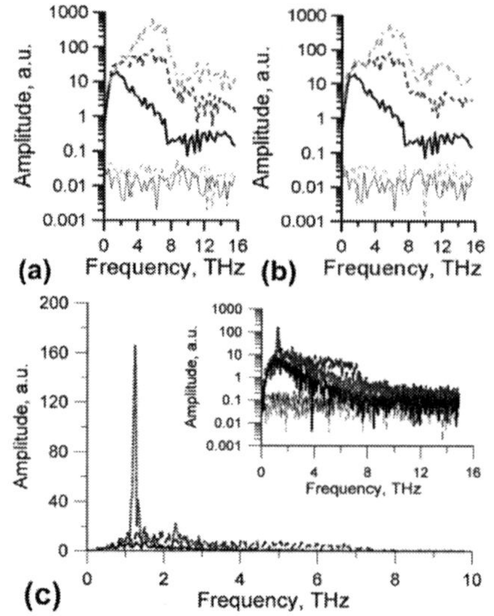

Fig. 2. (a)/(b) Fourier amplitude spectra of time dependences measured by standard measurement scheme (black-solid line) and by the schemes depicted in Figure 1(b)/1(c) with different edge filter 1 orientations (blue-dashed line and green-dashed-dot line). (c) Fourier amplitude spectra (inset – a logarithmic scale) of time dependences measured by standard measurement scheme depicted in Figure 1(a) (black-solid line), by the scheme depicted in Figures 1(c) (blue-dashed line) and its analog with PPLN (red-dotted line)

Further progress in sensitivity, frequency range, compactness and cost-efficiency of the EO detection technique can be achieved by application of EO medium with higher nonlinearity, having new frequencies of optical-TWR synchronism and being applicable with compact low-cost fiber lasers. A number of well-known organic crystals (DSTMS, DAST and OH1) and PPLN crystals previously successfully used for TWR generation are suitable for making progress in these directions. However, these crystals have high natural birefringence and the use of them with ellipsometric detection technique (figures 1, *a* and 1, *b*) is hindered by decoherence of the laser polarization components [10]. Meanwhile, in the last years sensitive schemes for EO detection based on laser pulse energy changes were suggested [2–6]. We demonstrated both experimentally and theoretically that when the laser pulse interacts with TWR in EO crystal not only its polarization ellipticity but also its energy could be changed [2, 3]. These changes depend on the amplitude and phase of the TWR overlapped with the laser pulse and can be used for terahertz time-domain measurements [2, 3]. The method based on energy changes can separately measure the energies of the orthogonal polarization components of the laser pulse and, in contrast to techniques based on ellipticity changes, does not require maintaining coherence between the polarization components. Thus, detection schemes based on this method can potentially be applied to birefringent crystals that are promising for further improvement of the sensitivity of EO time-domain measurements. In our research we applied this technique both with isotropic crystals and with birefringent ones [2–6]. The simplest scheme is based on the total energy changes of the laser pulse. We experimentally demonstrated that it is applicable to terahertz time-domain measurements with the use of birefringent and isotropic EO crystals [2, 3], but for the same TWR electric field and the same isotropic EO crystal (GaP or ZnTe) the magnitude of the registered signal was lower compared to the standard detection technique [3]. To improve the sensitivity of the "energy based" method the laser pulses with the edge-cut spectrum were used. As we have shown theoretically [4, 5] the relative changes of spectral energy density at the frequencies corresponding to the edge-cut of the laser pulse spectrum can be very high also as changes of polarization ellipticity. To utilize them in experiment the laser pulse after the EO crystal can be decomposed in the frequency spectrum by a diffraction grating [4] or can be passed through another edge filter [5] (figure 1, *c*) in the same way as in the case of ellipticity changes (figure 1, *b*). The experimental results obtained with the detection set-up depicted in figure 1, *c* shows much better response for high TWR frequencies comparing with the standard technique figure 2, *b*). To approbate the technique with birefringent crystal we used EO crystal - periodically poled lithium niobate (PPLN) and compared its sensitivity with sensitivities of the detection techniques utilizing a GaP crystal [6]. PPLN crystals demonstrate high optical-TWR conversion efficiency for narrow-band generation at various terahertz frequencies and are applicable to achieve quasi-phase-matching between TWR and the laser pulse radiation for different carrier wavelengths. The experimental results show that the use of PPLN for terahertz time-domain measurements further improves the sensitivity of the detection setup at quasi-phase-matching frequencies of about 1.245 THz (figure 2, *c*) [6]. The proposed technique provides a convenient and effective way of using birefringent crystals in EO time-domain measurements. The use of some other birefringent materials for terahertz time-domain measurements is very interesting in terms of improving the characteristics of the detection technique.

The presented EO detection techniques demonstrate higher sensitivity than the standard one and applicability with greater variety of crystals. The sensitivity of the proposed techniques depends on the parameters of the edge filters and of the optical detectors and their optimization along with the use of other nonlinear materials can lead to further improvement of the sensitivity of EO time-domain measurements.

This work was supported by Russian Foundation for Basic Research (RFBR) Grant Nos. (16-02-01078, 16-02-00258, 16-32-00717).

References

1. *Wu, Q. and Zhang, X.-C.* Free-space electro-optic sampling of terahertz beams // Appl. Phys. Lett. 1995. V. 67. P. 3523–3525.

2. *Kitaeva, G.Kh., Kovalev, S.P., Naumova, I.I., Akhmedzhanov, R.A., Ilyakov, I.E., Shishkin, B.V., and Suvorov E.V.* Quasi-phase-matched probe-energy electro-optic sampling as a method of narrowband terahertz detection // Appl. Phys. Lett. 2010. V. 96. P. 071106.

3. *Ilyakov, I.E., Kitaeva, G.Kh., Shishkin, B.V., and Akhmedzhanov, R.A.* Laser pulse amplitude changes induced by terahertz waves under linear electro-optic effect // Appl. Phys. Lett. 2014. V. 104. P. 151107.

4. *Ilyakov, I.E., Kitaeva, G.Kh., Shishkin, B.V., and Akhmedzhanov, R.A.* Terahertz wave electro-optic measurements with optical spectral filtering // Appl. Phys. Lett. 2015. V. 106. P. 121101.

5. *Ilyakov, I.E., Kitaeva, G.Kh., Shishkin, B.V., and Akhmedzhanov, R.A.* Terahertz time-domain electro-optic measurements by femtosecond laser pulses with an edge-cut spectrum // Opt. Lett. 2016. V. 41. P. 2998–3001.

6. *Ilyakov, I.E., Kitaeva, G.Kh., Shishkin, B.V., and Akhmedzhanov, R.A.* Electro-optic sampling of terahertz waves by laser pulses with an edge-cut spectrum in birefringent crystal // Opt. Lett. 2017. V. 42, No. 9. P. 1704–1707.

7. *Gaal, P., Raschke, M. B., Reimann, K. and Woerner, M.* Measuring optical frequencies in the 0-40 THz range with non-synchronized electro-optic sampling // Nat. Photon. 2007. V. 1. P. 577–580.

8. *Barbieri, S., Ravaro, M., Gellie, P., Santarelli, G., Manquest, C., Sirtori, C., Khanna, S. P., Linfield, E. H. and Davies, A. G.* Coherent sampling of active mode-locked terahertz quantum cascade lasers and frequency synthesis // Nat. Photon. 2011. V. 5. P. 306–313.

9. *Blanchard, F., Tanaka, K.* Improving time and space resolution in electro-optic sampling for near-field terahertz imaging // Opt. Lett. 2016. V. 41. P. 4645–4648.

10. *Han, P.Y., Tani, M., Pan, F., and Zhang, X.-C.* Use of the organic crystal DAST for terahertz beam applications // Opt. Lett. 2000. V. 25. P. 675–677.

EPJ Web of Conferences **149**, 05005 (2017) DOI: 10.1051/epjconf/201714905005
10th International Workshop 2017 "Strong Microwaves and Terahertz Waves: Sources and Applications"

The nonlinearity of the refractive index of optical media in the terahertz spectral range

A.N. Tcypkin[1], S.E. Putilin[1], M.S. Kulya[1], A.A. Drozdov[1], M. Siddiqui[2], S. Choudhary[2], J. Zhao[2], V.G. Bespalov[1], R.W. Boyd[2], X.C. Zhang[1, 2], S.A. Kozlov[1]

[1] ITMO University, Saint-Petersburg, 197101 Russia, kozlov@mail.ifmo.ru
[2] Institute of Optics, University of Rochester, Rochester, NY USA

Introduction

Recently, great progress has been made in the generation of high-intensity terahertz (THz) radiation, which has led to the opportunity to explore non-linear processes in the THz frequency range [1]. A key parameter characterizing the nonlinear response of a material is the non-linear refractive index, denoted as n_2. It has been predicted theoretically [2] that the THz nonlinear response of crystals is extremely large as a consequence of a strong vibrational contribution. An indirect assessment of n_2 for lithium niobate at THz frequencies was reported in [3], leading to a value of approximately 10^{-11}cm^2/W. We find that $n_2 = 4.0 \times 10^{-11}$cm^2/W, which is approximately 3 orders of magnitude larger that the value 2.4×10^{-14} cm^2/W in the NIR spectral range [4].

The present paper is devoted to theoretical analysis and a direct experimental measurement of n_2 for crystalline ZnSe in the THz spectral range. In this paper, we also consider the special aspects of self-action of single-cycle waves in media with given nonlinearities. It is shown that self-focusing efficiency in transparent dispersive medium decreases when the number of oscillations in input radiation is decreased for few-cycle paraxial waves with the same maximum electric field and transverse size, i.e., with the same ratio of the radiation power to the critical self-focusing power.

Experimental setup

The experimental setup is shown in Fig. 1. We used a femtosecond laser system (Regulas 35f1k, Avesta Project, 30 fs duration, pulse energy 2 mJ, repetition rate 1 kHz) as our radiation source. The radiation was split into two beams; one (pump) beam was used for generating the THz radiation and the other (probe) beam was used for monitoring the THz radiation. Generation of the THz radiation took place in the generator (TERA–AX; Avesta Project), which is based on phase-matched optical rectification in a lithium niobate crystal. The intensity of the THz radiation was regulated by attenuating the intensity of the pump beam through use of a polarization attenuator. Temporal (I) and spectral (II) form of the THz pulse generated by TERA–AX are shown on fig 1 (b). The radiation spectrum extended from 0.1 to 2.5 THz.

The THz pulse was focused and collimated using two lenses (*L1* and *L2*) each with a focal length of 12.5 mm. The intensity of the radiation at the waist was 8×10^8 W/cm^2. The sample was translated through the beam waist region of the THz beam through use of a translation stage. The ZnSe sample was 0.3 mm thick and oriented in the 100 direction. Initially, the sample was placed at a distance of 1.5 mm from the waist, which corresponds to the linear propagation regime, and was then scanned through the focal region.

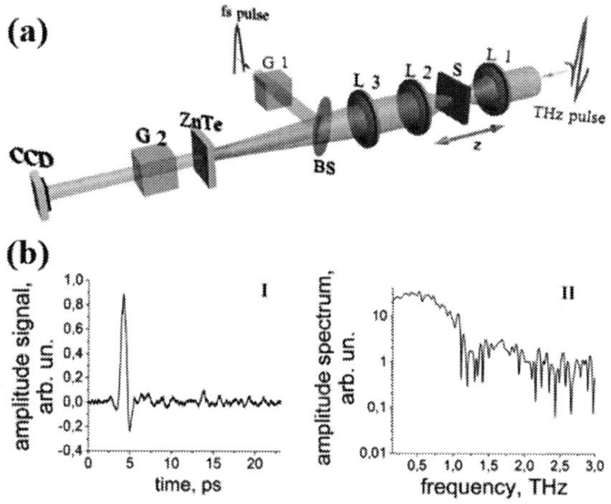

Fig. 1. The experimental setup (a). Temporal (I) and spectral (II) form of the THz pulse generated in the TERA–AX system (b). (*L1, L2, L3* are THz lenses; *S* is the sample; *BS* is the beam splitter; *G1, G2* are polarizers (Glan prism); *CCD* is the camera

We measure the amplitude of the THz radiation by means of the electro-optical effect in a ZnTe crystal that is placed between the crossed polarizers. The THz and fs laser (probe) pulses were made to overlap in space and time at the ZnTe crystal. Initially the THz radiation was overlapped, and the second Glan prism *G2* oriented so that two prisms are in crossed polarization for the probe beam. Next, we opened the THz radiation and using a delay line we found the maximum intensity of the fs probe radiation (corresponding to the maximum of the amplitude module of the THz pulse) which passes through crossed polarizers and fixed delay position and thus THz pulse position regarding the probe pulse.

Then, as in the conventional Z-scan method, we translated the ZnSe crystal along the z axis. In distinction from the classical Z-scan method, in which one measures the change of energy in the central part of the beam, we fix the value of the change of intensity of probe beam at the CCD camera in the area of maximum of the amplitude modulus of the THz pulse. Image processing with CCD-camera is performed as follows: we select an area in the vicinity of the resulting image. We sum the signal from these pixels, and we take this value to represent the amplitude modulus of the THz signal for this particular location in z.

Results

Z-scan traces for different intensities of the THz radiation are presented in Fig. 2. The vertical axis gives the maximum intensity of the probe fs radiation after passing through the crossed polarizers surrounding the electro-optic crystal. As we see from Fig. 2, translation of the ZnSe crystal along the z-axis leads to a variation of the measured intensity that is characteristic of all Z-scan measurements. We thus use the standard formulas [5] for Z-scan measurements to estimate the non-linear refractive index.

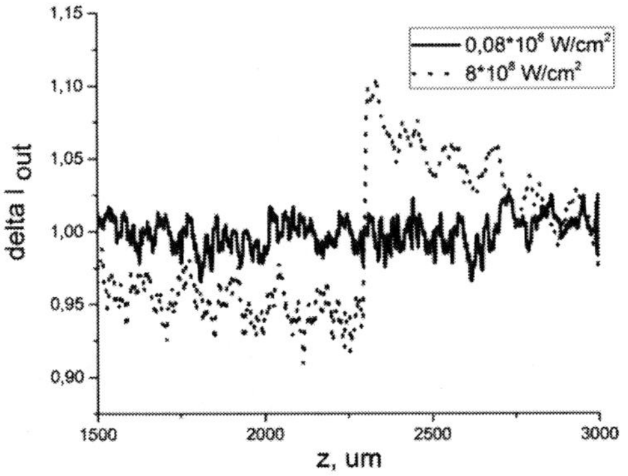

Fig. 2. Z-scan curves for different values of THz radiation intensity

We find that n_2=4.0±2.5×10^{-11}cm^2/W.

Summary

We measure an extremely large value of the n_2 coefficient of crystalline ZnSe in the THz spectral regime. This value is orders of magnitude larger than that of typical materials as measured at optical frequencies. Our results are consistent with the prediction of a recent [2] theoretical model that ascribes the origin of THz nonlinearities in crystals to a vibrational response that is orders of magnitude larger than typical electronic responses.

There is a large error in our measured value of $n_2 = 4.0 \pm 2.5\times10^{-11}$ cm^2/W. We believe that the error is largely methological, resulting for example from shot-to-shot variations in our laser intensity

In evaluating the coefficient of nonlinear refractive index we used the formula for quasi-monochromatic radiation [5], although our THz pulse was only a few cycles in duration. It is known that the nature of nonlinear effects changes substantially with the decrease in the number of oscillations in the pulse [5]. We are presently in the process of investigating the influence of pulse duration on our results. For example, in Fig. 3 you can see the dependence of self-focusing efficiency on number of oscillations N in input THz radiation. Here the case N = 0.3 corresponds initial single-cycle THz radiation.

Nonetheless, our laboratory measurements show that n_2 in the THz spectral range is 3 orders of magnitude larger than the value for the NIR spectral range.

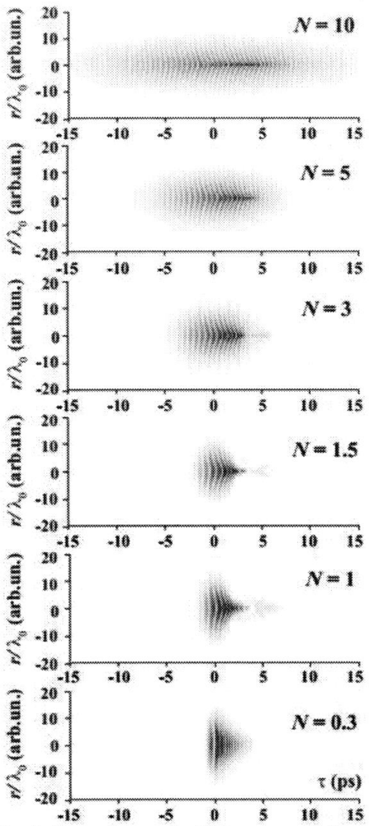

Fig. 3. Dependence of self-focusing efficiency on number of oscillations N in input THz radiation

The results are confirmed by theoretical calculations. A nonlinear response this large holds enormous promise for myriad applications of THz nonlinear optics.

This work was financially supported by the "5-100-2020" Programme of the Russian Federation, Grant No. 074-U01.

References

1. *Nicoletti D., Cavalleri A.* Nonlinear light–matter interaction at terahertz frequencies // Advances in Optics and Photonics. 2016. V. 8, P. 401.

2. *Dolgaleva K., Materikina D.V., Boyd R.W., Kozlov S.A.* Prediction of an extremely large nonlinear refractive index for crystals at terahertz frequencies //Phys Rev. A. 2015. V. 92, P. 023809.

3. *Korpa C.L., Gy T., and J. Hebling J.* Interplay of diffraction and nonlinear effects in the propagation of ultrashort pulses // J. of Phys. B. 2016. V. 49, P. 035401.

4. *Yin M., Li H.P., Tang S.H., Ji W.* Determination of nonlinear absorption and refraction by single Z-scan method //Appl. Phys. B. 2000. V. 70, P. 587–591.

5. *Cheung Y.M., Gayen S.K.* Optical nonlinearities of tea studied by Z-scan and four-wave mixing techniques // J. Opt. Soc. Am. B. 1994. V. 11, P. 636.

6. *Kapoyko Y.A., Drozdov A.A., Kozlov S.A., Zhang Xi-C.,* Evolution of few-cycle pulses in nonlinear dispersive media: Velocity of the center of mass and root-mean-square duration // Phys. Rev. A. 2016. V. 94. P. 033803.

EPJ Web of Conferences **149**, 05006 (2017) DOI: 10.1051/epjconf/201714905006
10th International Workshop 2017 "Strong Microwaves and Terahertz Waves: Sources and Applications"

Study of 0.3–0.8 THz flux generated by magnetized plasma column due to relaxation of high-current REB

A.V. Arzhannikov[1,2], V.V. Annenkov[1,2], A.V. Burdakov[1,3], V.S. Burmasov[1,2], I.A. Ivanov[1,2], A.A. Kasatov[1,2], S.A. Kuznetsov[1,2], M.A. Makarov[1], K.I. Mekler[1], S.V. Polosatkin[1,2], V.V. Postupaev[1,2], A.F. Rovenskikh[1], S. L. Sinitsky[1,2], V.F. Sklyarov[1,2], V.D. Stepanov[1,2], I.V. Timofeev[1,2], and E.P. Volchok[2]

[1]Institute of Nuclear Physics, SB RAS, Novosibirsk, Russia, arzhan1@ngs.ru
[2]Novosibirsk State University, Novosibirsk, Russia
[3]Novosibirsk State Technical University, Novosibirsk, Russia

Studies on submm-wave emission from magnetized plasma due to relaxation of a high current relativistic electron beam (REB) are considerably important since they allow generating the high-power radiation with a prompt tuning frequency [1, 2]. The experimental results were obtained at GOL-PET device (see fig.1). We study the relaxation of the high-power REB (0.8 MeV/ 10kA/ 10 μs) in a plasma column of 2 meters length with the density n_p= (0.5–2) 10^{15} см$^{-3}$ confined by the multiple-mirror magnetic field with the mean value B_{mean} = 4.2 T and the corrugation factor 1.5. The aim of the experiments is to find the conditions for the efficient conversion of plasma waves into electromagnetic ones. The recent experiments have been focused on the study of the properties of 0.2 - 0.8 THz radiation propagating along the axis of the plasma column.

The temporal dynamics of the radiation spectra was measured along the axis by an 8-channel polychromator with semiconductor diodes for the frequency interval from 0.1 up to 0.5 THz (see [2]) and by an additional 2-channel cryogen detection system and single semiconductor diodes covered by narrow band-pass quasi-optical filters for the 0.5–0.9 THz frequency range. During the spectral measurements, the plasma density was measured in eight dots over the diameter of the plasma column via the Thomson scattering system with a 50 ns laser pulse.

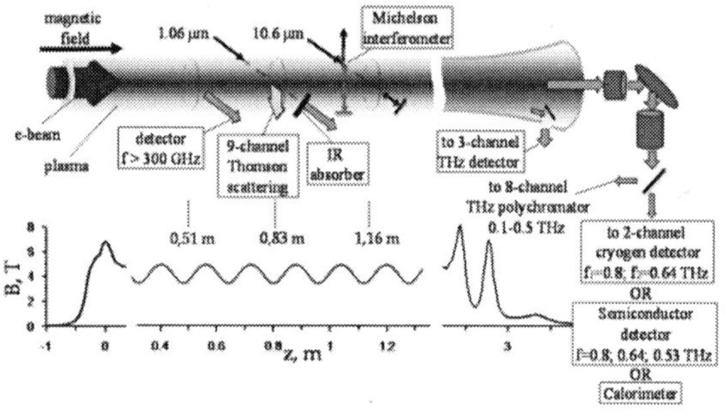

Fig. 1. The layout of the GOL-PET experiments

As an example of the results of such measurements by the Thomson scattering system, the density distribution over the plasma column diameter is presented in fig.2 for the shot t1913.

The temporal evolution of beam and plasma parameters and the signals of the emission for the frequencies 0.26 THz and 0.64 THz are presented in fig. 3 for the same shot.

Measurements of the emission in the longitudinal direction have shown that the spectral power density of this radiation is concentrated in two clear distinct frequency regions as it was demonstrated in the measurements of the emission in transverse direction (see [3]). For the plasma density of 1 10^{15}см$^{-3}$, the first region is located in the frequency interval f_1=0.25–0.35 THz, the second one is in the interval f_2=0.6–0.8 THz. As one can see in fig. 4 presenting the spectrum of the emission for the shot t1913, one of the peaks of the spectral power density is located at the frequency f_1=0.2 THz and the second one at f_1=0.64 THz.

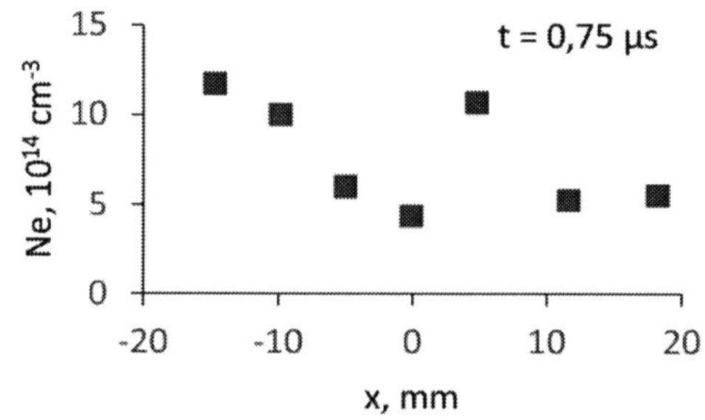

Fig. 2. Plasma density distribution over the column diameter measured in the shot t1913 for the time of the maximal THz emission

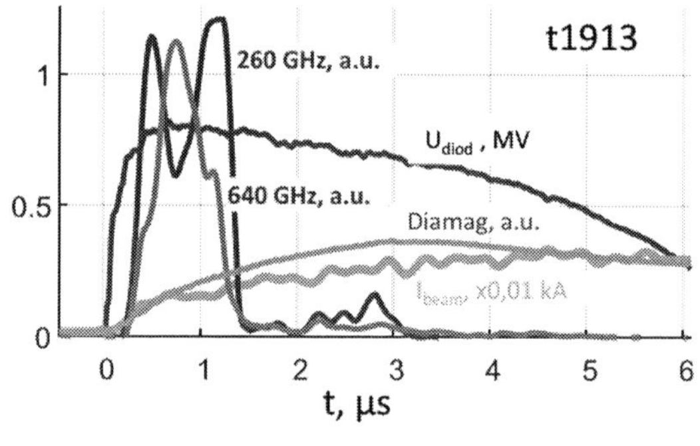

Fig. 3. Temporal evolution of the main beam and plasma parameters and the signals of the THz emission

We have compared the experimental results with the theoretical consideration of the transformation of the plasma oscillations into the electromagnetic radiation for the conditions of our experiments [4-5]. The locations of the frequency ranges with the high level of the spectral power density are associated with the frequency of the upper hybrid mode of plasma oscillations and the doubled value of this frequency.

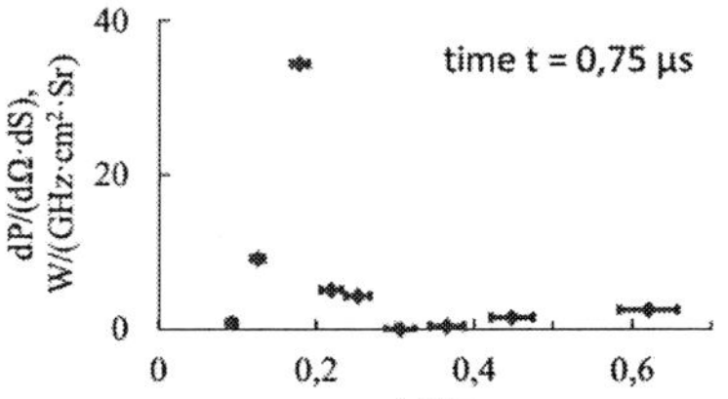

Fig. 4. Spectral density of the radiation measured in shot T1913 at the moment of maximum THz emission

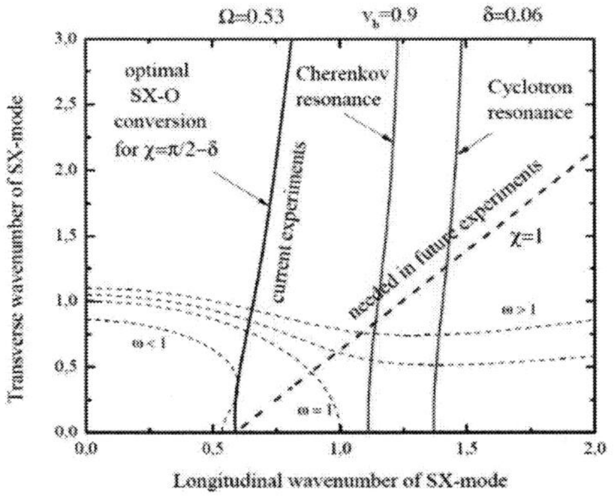

Fig. 5. Lines of the wave numbers of the upper hybrid oscillations in the areas of the resonance with the relativistic electron beam (red lines) and of their effective transformation into the electromagnetic radiation (blue lines)

Taking into account our experimental and theoretical studies on the REB-plasma interaction, we can conclude that the observed emission is explained by the direct transmission of the upper hybrid branch of plasma oscillations to the electromagnetic radiation [4] and merging of two upper hybrid oscillations of the high level plasma turbulence [5]. In current experiments, the line optimal SX-O mode conversion in the wavenumber space is shifted from the area of the intense pumping of the upper hybrid wave by the REB (see fig. 5). The reason for that is almost transverse orientation of magnetic field lines (even in the corrugated regime) to the predominantly radial density gradient. In future experiments, we plan to realize the regime in which the density gradient is directed at an acute angle to the magnetic field. The regime of $\chi=1$ needed to convert the most unstable beam-driven modes is shown in fig.5 by the blue dashed line.

So, the REB-plasma interaction is the sufficient method to generate sub millimeter radiation for frequency range up to 1 THz. This method is based on the intense pumping of the upper hybrid waves by the REB and then the transformation of these waves to the emission through the SX-O mode conversion and the merging of two upper hybrid waves.

The work was supported by Russian Science Foundation in the frame of Project No. 14-12-00610 for experimental and theoretical studies of the spectral composition of the emission and in the frame of Project No. 14-50-00080 for the reconstruction of the GOL-PET device.

References

1. *M.K.A. Thumm, A.V. Arzhannikov, V.T. Astrelin et al.*, Generation of High-Power Sub-THz Waves in Magnetized Turbulent Electron Beam Plasmas // Journal of Infrared, Millimeter and Terahertz Waves. 2014, Vol. 35, Iss. 1, pp. 81-90, DOI 10.1007/s10762-013-9969-3.

2. *A.V. Arzhannikov, A.V. Burdakov, V.S. Burmasov, et al.*, Dynamics and Spectral Composition of Subterahertz Emission From Plasma Column Due to Two-Stream Instability of Strong Relativistic Electron Beam // IEEE Transactions on Terahertz Science and Technology, 2016, Vol. 6, No. 2, P.245.

3. *A.V. Arzhannikov, V.V. Annenkov, A.V. Burdakov, et al.*, Beam-plasma system as a source of powerful submillimeter and terahertz radiation (experimental and theoretical studies) // AIP Conf. Proc., 1771, 070004 (2016); http://dx.doi.org/10.1063/1.4964228.

4. *I.V. Timofeev, V.V. Annenkov, A.V. Arzhannikov*, Regimes of enhanced electromagnetic emission in beam-plasma interactions // Physics of Plasmas, 2015, 22, p.113109, http://dx.doi.org/10.1063/1.4935890.

5. *A.V. Arzhannikov and I. V. Timofeev*. Generation of powerful terahertz emission in a beam-driven strong plasma turbulence // Plasma Phys. Controlled Fusion, 2012, 54, 105004. http://dx.doi.org/10.1088/0741-3335/54/10/105004.

EPJ Web of Conferences **149**, 05007 (2017) DOI: 10.1051/epjconf/201714905007
10th International Workshop 2017 "Strong Microwaves and Terahertz Waves: Sources and Applications"

Instabilities, coherency, and spectra of the NovoFEL radiation

V.V. Kubarev[1,2]

[1]Budker Institute of Nuclear Physics, Novosibirsk, Russia, V.V.Kubarev@inp.nsk.su
[2]Novosibirsk State University, Novosibirsk, Russia

1. Introduction

Novosibirsk free electron laser is user facility consists of three lasers of terahertz (240–90 μm), far infrared (80–30 μm), and infrared (30–5 μm) ranges [1]. The NovoFEL spectrum is one of the main parameter for many user applications. Spatially it is important for ultra-fast high-resolution spectroscopy and radar applications developing on the NovoFEL. The NovoFEL spectra depend on both temporary form of its radiation (continuous trains of 30–150 ps pulses with repetition frequencies of 5.6 MHz) and different radiation instabilities in the device. In the case of full intra-pulse coherency we have narrowest line of width equal to inverse pulse duration (Fourier limit). Coherency between close pulses causes splitting of the spectrum on many spectral components with frequency difference of $c/2L$ (5.6 MHz), where L is optical resonator length. Spectral width of the components is inverse proportional to total coherency length between pulses. On the one hand the coherency must be not less than $L\times Q$-factor of optical resonator due to laser nature of FEL radiation. Moreover in the case of ideal laser coherency length can be much more than the value. But some accumulative phase shift between fields in pulses can take place in conditions of real experiment due to different instabilities. Thus obtaining of new experimental data about fine structure of the NovoFEL radiation is the main subject of this paper.

Other large-scale spectral peculiarities (~ GHz) of the NovoFEL radiation caused by different fast instabilities were investigated early in works [2–7]. We found two types of side-band modes caused by classical FEL instability on trapped electron [3, 5, 7] and by some instability of an unknown nature when the NovoFEL pulses have two-humped form [2, 6, 7]. These mode structures were measured both in a relatively stable regime [2, 3, 5, 7] and a strongly unstable regime [6, 7], where we used our original method of one-pulse spectral measurements. Full stabilization of all side-band instabilities was also demonstrated by using stabilization frequency detuning between electron and light pulses [3, 5–7].

Though the fine spectral structure has to be in all the NovoFEL regimes, its parameters can be strongly different in stable and unstable cases. That is other main task of the investigation.

2. Experimental setup

Two experimental setups we plane to use in the study. The first one is ultra-long resonance waveguide vacuum Fabry-Perot interferometer (FPI) with calculated resolution $(1.5–7)\cdot10^6$ (Fig. 1). It should be noted that its resolution is higher than resolution of the other our spectral devices (Fourier spectrometer, grating spectrometer) by factor of $10^3–10^4$.

If we will find a fine structure with spectral width smaller than 0.5 MHz, than we will use a second setup with spectral resolution of few kHz, based on mixing of the NovoFEL radiation with radiation of ultra-monochromatic terahertz gas laser [8, 9] in high-sensitive fast Schottky diode detector [10].

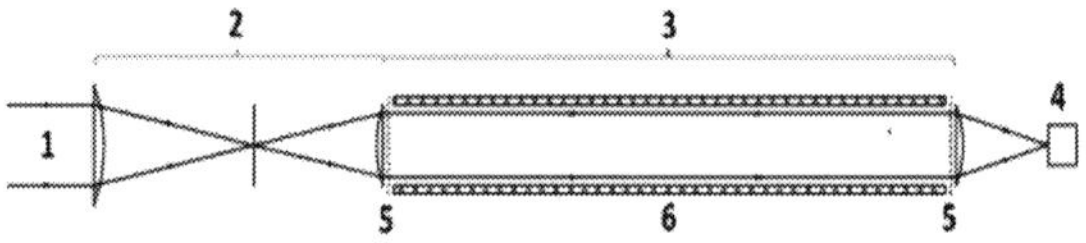

Fig. 1. Optical scheme of installation for measuring of fine spectral structure of the NovoFEL radiation by ultra-long waveguide vacuum Fabry-Perot interferometer: 1 – NovoFEL beam, 2 – beam transformer, 3 – Fabry-Perot interferometer, 4 – lock-in detector, 5 – electroformed mesh, 6 – hollow dielectric waveguide. Length of Fabry-Perot interferometer is 2659 mm (1/10 part of the NovoFEL optical resonator length)

3. Experimental results

Now we obtained only first results with start version of our FPI with nickel meshes. Its calculated spectral resolution was $1.5\cdot10^6$.

Fig. 1 illustrates importance of exact tuning of resonance length of our FPI. This tuning was made by special mechanical system moving one of the mesh mirrors. Fine displacement was carried out by heating of hollow invar rods of the FPI by circulating thermo-stabilized liquid.

Fine structure of the NovoFEL radiation is shown in Fig. 3 and Fig. 4 for resonance (a) and stabilized (b) regimes.

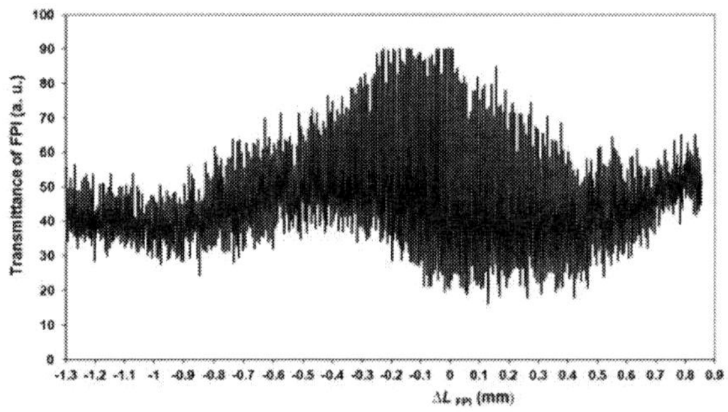

Fig. 2. Transmittance of FPI versus difference of its length to resonance length equal to 1/10 part of NovoFEL optical resonator. Radiation wavelength – 168 μm

For stabilized regimes (b) spectral resolution was practically equal to calculated apparatuses function of our FPI with nickel meshes. Thus the resolution should be increased for correct measurement of real fine structure in the regimes.

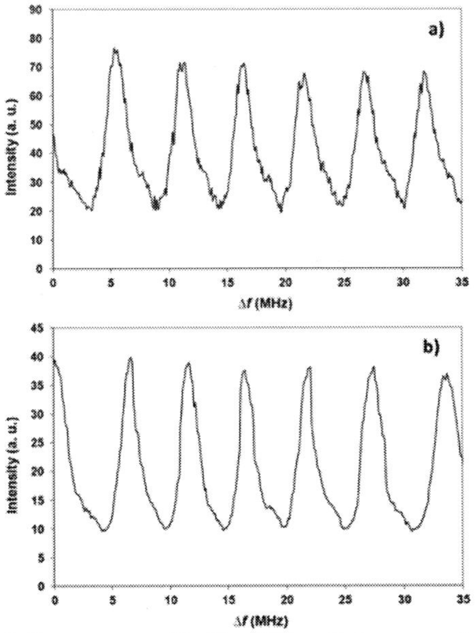

Fig. 3. Fine structure of the NovoFEL radiation at wavelegth 130 μm (2.3 THz) for resonance (a) and stabilized (b) regimes

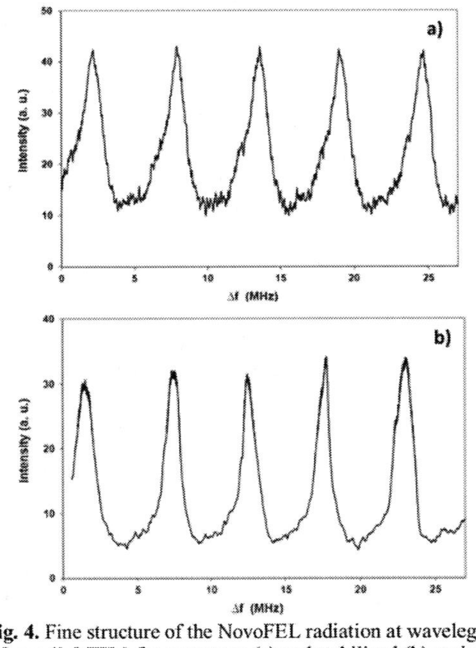

Fig. 4. Fine structure of the NovoFEL radiation at wavelegth 168 μm (1.8 THz) for resonance (a) and stabilized (b) regimes

For resonance regimes (a) we can see some broadening of the mode spectral lines. Therefore real fine structure of NovoFEL radiation is order of apparatus function. We can see also that the broadening look like appearing of weak transverse modes.

We plan to increase spectral resolution of the FPI by 4–5 times making use gold meshes in place of nickel meshes.

References

1. *Kulipanov G.N., Bagryanskaya E.G., Chesnokov E.N., Choporova Yu.Yu., Gerasimov V.V., Getmanov Ya.V., Kiselev S.L., Knyazev B.A., Kubarev V.V., Peltek S.E., Popik V.M., Salikova T.V., Scheglov M.A., Seredniakov S.S., Shevchenko O. A., Skrinsky A. N., Veber S. L., and Vinokurov N. A.* Novosibirsk free electron laser – facility description and recent experiments // IEEE Transaction on Terahertz Science and Technology. 2015. V. 5, No.5. P. 798–809.

2. *Kubarev V.V., Vinokurov N.A., Kotenkov V.V. et al.* Observation of sideband instability in the Novosibirsk terahertz free electron laser // Proc. 31st Int. Conf. IRMMW-THz-2006, Shanhai, China, September 18-22, 2006, P. 415.

3. *Kubarev V.V., Kulipanov G.N., Kolobanov E.I., et al.* Modulation instability, three mode regimes and harmonic generation at the Novosibirsk terahertz free electron laser // Nuclear Instruments & Methods in Nuclear Research, 2009. V. A603. P. 25–27.

4. *Kubarev V.V., Ovchar V.K., and Palagin K.S.* Ultra-fast Terahertz Schottky Diode Detector // Proc. 34th Int. Conf. IRMMW-THz-2009, Busan, Korea, September 21–25, 2009, report 09030439.

5. *Kubarev V.V.* Detailed observation of modulation instability on THz NovoFEL // Proc. 37th Int. Conf. IRMMW-THz-2012, Wollongong, Australia, September 23 – 28, 2012, report Fri-A-4-2.

6. *Kubarev V.V.* Ultrafast high-resolution spectroscopy of separate NovoFEL pulses // Proc. 38th Int. Conf. IRMMW-THz-2013, Mainz, Germany, September 1–6, 2013, report Mo14-4.

7. *Kubarev V.V.* Optical system, diagnostics, and experiments on terahertz and infrared free electron lasers // Doctoral Thesis, Novosibirsk, Russia, 2016, http://www.inp.nsk.su/news/defences.ru.shtml

8. *Kubarev V.V. and Kurenskii E.A.* Ultralow-noise high-power DCN laser with rf pumping // Quantum Electronics 1996. V. 26, No 4. P. 303–306.

9. *Kubarev V.V. and Kurenskii E.A.* Very-low-noise H2O laser with rf pumping // Quantum Electronics 1995. V. 25, No.12. P. 1141–1145.

10. *Kubarev V.V., Kazakevich G.M., Jeong Y.U., Lee B.J.* Quasi-optical highly sensitive Schottky-barrier detector for a wide-band FIR FEL // Nuclear Instruments & Methods for Nuclear Research 2003. V. A 507. P. 523–526.

EPJ Web of Conferences **149**, 05008 (2017) DOI: 10.1051/epjconf/201714905008
10th International Workshop 2017 "Strong Microwaves and Terahertz Waves: Sources and Applications"

Super-radiative self-compression of photo-injector electron bunches and the use of this effect for realization of a THz source based on spontaneous coherent emission from a short dense electron bunch

Andrei Savilov[1,2], I.V. Bandurkin[1], Yu.S. Oparina[1,2]

[1]Institute of Applied Physics, N.Novgorod, Russia, savilov@appl.sci-nnov.ru
[2]Lobachevski State University of Nizhny Novgorod, N.Novgorod, Russia

Laser-driven photo-injectors allow formation of compact and accessible sources of dense electron bunches with a moderate energy of 3–6 MeV, picosecond pulse durations, and charges of up to 1 nC and even greater. These bunches can be used for realization of comparatively simple and compact terahertz sources operating in the regime of spontaneous coherent undulator radiation. This type of radiation is realized, when the effective axial length of bunches is shorter than the radiation wavelength, so that the wave packets emitted by each of the electrons add up basically in phase. The THz source based on the spontaneous emission has a number of advantages as compared to the more traditional free-electron laser (FEL) based on the undulator emission induced due to bunching of a long electron beam by the radiated wave. These are a relatively high efficiency of the emission from the "ready-for-radiation" bunch, a narrow frequency band of the radiated rf signal in a system without a feedback system and an input rf signal, and the phase of the radiated rf signal fixed by the electron bunch phase. These advantages make the THz source based on the spontaneous emission attractive for a number of important applications, including the use of synchronization of such an oscillator with the X-ray FEL in pump-probe experiments.

An example of the THz source based on the spontaneous emission is described in [1, 2]; a 0.3 nC 5 MeV e-bunch with the 0.1 mm initial length propagating in an undulator with a period of 2.5 cm can generate a 20 ps narrow-band rf pulse at a frequency close to 2 THz with a power of tens MW; the efficiency of this process can be over 10 %.

The spontaneous emission occurs while the bunch phase size is smaller than 2π, so that wave packets emitted by each of the electrons add up basically in phase. In the case of the ubitron (as well as Cherenkov) source, this means that the bunch length must stay shorter than the radiation wavelength during the whole radiation region. Therefore, realization of a THz source based on the spontaneous coherent radiation requires (1) creation of short enough (0.3 mm or shorter) dense electron bunches and (2) the presence of a mechanism of stabilization of the phase size of the bunch with respect to the radiated wave. The latter is especially important, taking into account the axial Coulomb repulsion, which is very strong in short dense photo-injector bunches and lead to a significant spreading of the bunch during its motion through a path being as short as few decimeters.

In this work, we propose and study a radiative mechanism for compression and control the axial size of the bunch. This is the axial compression under effect of the spontaneous super-radiation of a long-wavelength wave. In fact, we propose a bicolor (two-wave) rf source operating in the regime of spontaneous coherent undulator radiation (Fig. 1).

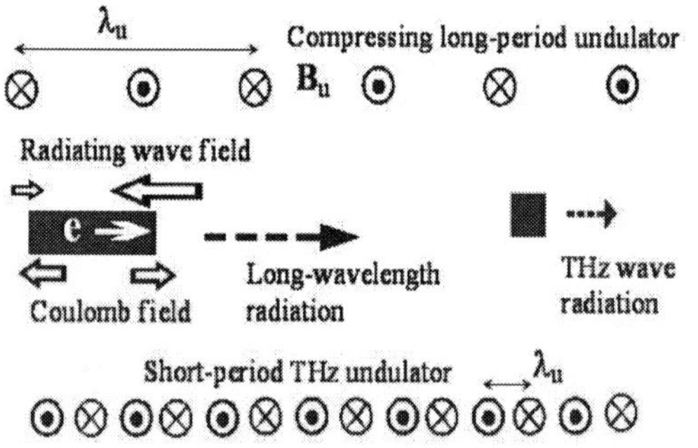

Fig. 1. Bicolor rf source: spontaneous radiation of long-wavelength wave leads to the axial compression of the electron bunch and to the spontaneous excitation of the THz wave

The use of an auxiliary undulator with a relatively long period leads to the super-radiation of a wave with a long wavelength (longer than the initial axial size of the bunch). In certain conditions, the electric field of this auxiliary wave possesses the "proper" phase with respect to the bunch, namely, the bunch front is placed in the maximum of the decelerating phase of this wave, whereas the bunch tail is close to the zero field (Fig. 2). In this situation, the auxiliary long-wavelength wave can provide a significant axial compression of the bunch even in the case of a high charge, when the Coulomb repulsion is strong. The bunch compressed down to a "terahertz" length can radiate the THz signal in the main short-period undulator.

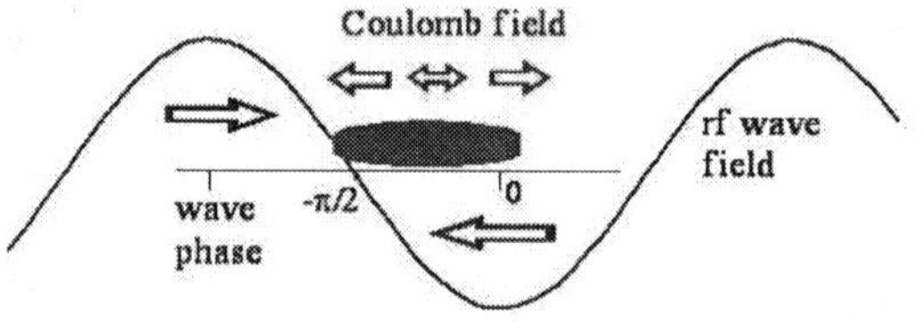

Fig. 2. The bunch phase with respect to the radiated long-wavelength wave

A principal point is the use of the regime of grazing of dispersion characteristics, when the group velocity of the resonant wave coincides with the axial electron velocity, $v_{gr} = v_z$. In this case, the regime of the rf wave emission has super-radiative character; this means that the radiated rf wave does not do not leave the region of the electron bunch. The use of this super-radiation regime provides both a high intensity of the compressing rf field and the "proper" phase of this field (Fig. 3).

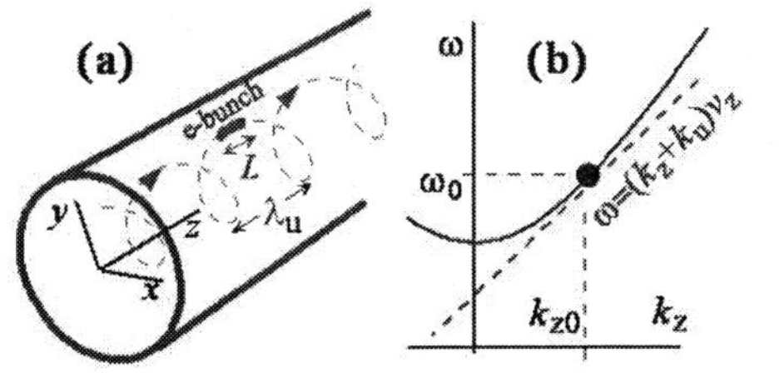

Fig. 3. (a) Electron bunch moving along a helical undulator. (b) Dispersion diagram of the operating waveguide mode

According to a quasi-analytical self-consistent theory describing the motion of a short electron bunch in the radiated wave field and the Coulomb field created by the bunch, the self-compression effect takes place, if the bunch charge is small enough. The permitted charge is determined by the following formula:

$$\frac{I}{I_A}\times\frac{N_w^4}{K_u^8}\approx\left(\frac{R}{L_0}\right)^5\gamma^{-4}.$$

Here, $I = Qc/L_0$ is the electron current, L_0 is the initial bunch length, K_u is the undulator factor, N_w is the norm of the excited transverse waveguide mode, $I_A = mc^3/e$, R is the beam radius, and γ is the initial relativistic Lorants-factor of electrons.

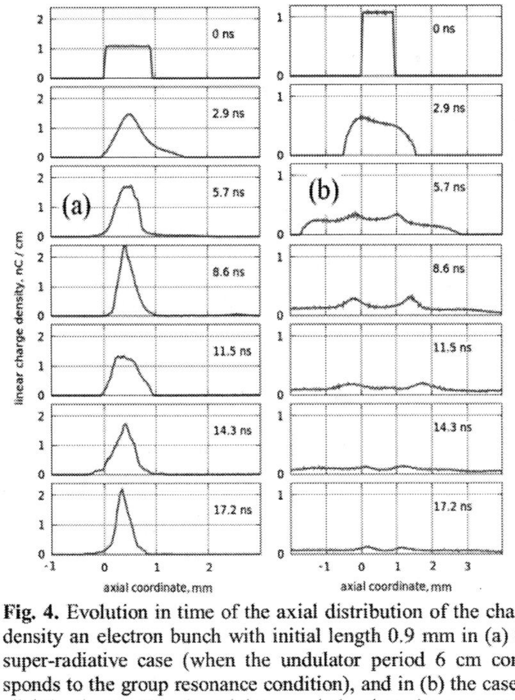

Fig. 4. Evolution in time of the axial distribution of the charge density an electron bunch with initial length 0.9 mm in (a) the super-radiative case (when the undulator period 6 cm corresponds to the group resonance condition), and in (b) the case of slightly shorter (5 cm) undulator period, when the group resonance condition is not fulfilled

Let us consider a bunch with the length L_0 =0.9 mm (3 ps). If the super-radiation of the auxiliary long-wavelength wave in the bicolor source shown in Fig. 1 provides the axial compression of this bunch by a factor of 1/5, then the compressed bunch length is $L = 0.18$ mm, which is enough to provide spontaneous radiation of the operating short-wavelength wave at a frequency of 1 THz. At $R = 1$ mm and $K_u = 0.7$ this leads to a 3 ps bunch charge $Q \approx 170\gamma^{-4}$ nC. For instance, at $\gamma = 7$ (3 MeV) the bunch charge is close to 0.1 nC. In this case, the auxiliary long-wavelength wave frequency $\lambda =$ 1.8 mm, and the proper group velocity of $TE_{1,1}$ waveguide mode ($v_{gr} = v_z$) is provided at the waveguide diameter ≈ 6 mm. The undulator period amounts ≈ 6 cm, and the length of the compressing system amounts 1–2 meters. At the same time, the permitted bunch charge strongly depends on the bunch transverse size, $Q \propto R^5$, so that an increase in the bunch radius up to $R = 1.5$ mm should result in an increase of the permitted charge up to ~ 1 nC.

We examine the superradiation compression by means of direct simulations of the example described above on the basis of the code KARAT. This is a fully electromagnetic code based on Particle-in-Cell method and direct solving Maxwell equations in systems with complex geometry. Figure 4 a illustrates the evolution of the axial distribution of the charge density in the electron bunch in the time in this super-radiative case. Simulations predict axial compression of the electron bunch (with a compression factor of ~1/5) under effect of the radiated wave field. Just for comparison, Fig. b illustrates the same system with one difference, namely, a slightly shorter undulator period ($\lambda_u = 5$ cm), so that the group resonance condition $v_{gr} = v_z$ is not provided, and the process of rf wave emission does not possess super-radiation properties. Due to this fact, the radiated rf wave possesses significantly lower intensity as compared to the super-radiation case shown in Fig. 4 a. In fact, in this case electrons move mainly in the Coulomb electric field of the bunch. This results in a fast destruction of the bunch. This proves that the compression observed in the "proper" case is provided under effect of the rf wave field excited in the regime of super-radiation.

The work is supported by IAP RAS Project 0035-2014-0012, and Russian Foundation for Basic Research Project 16-02-00794 A.

References

1. *N. Balal, I.V. Bandurkin, V.L. Bratman, E. Magory, and A.V. Savilov.* Negative-mass mitigation of Coulomb repulsion for terahertz undulator radiation of electron bunches. // Appl. Phys. Lett. 2015. V. 107, P. 163505.

2. *Y. Lurie, V.L. Bratman, and A.V. Savilov.* Energy enhancement and spectrum narrowing in terahertz electron sources due to negative mass instability // Phys. Rev. AB. 2016. V.19. P. 050704.

Strong terahertz fields: interaction with condensed matter and electron acceleration

Yu.A. Sergeev, I.V. Oladyshkin, S.B. Bodrov, A.I. Korytin, M.D. Tokman, S.V. Kuzikov, A.A. Vikharev, and A.N. Stepanov

Institute of Applied Physics of the Russian Academy of Sciences
46 Ul'yanov Street, 603950 Nizhny Novgorod, Russia, step@ufp.appl.sci-nnov.ru

Recent years witness fast development of the sources of strong single-pulse terahertz radiation, with the record value for terahertz field being about 80 MV/cm [1]. There appeared quite a number of publications where the nonlinear interaction of such strong THz fields with matter was investigated and new effects in intense THz fields were sought for. In this presentation, we will discuss the specific nonlinear properties of graphene in strong electromagnetic fields and propose our approach of using short THz pulses for acceleration of electrons.

The first part of the presentation will be devoted to the discussion of the experimental and theoretical results on optical emission from graphene irradiated by a strong (up to 300 kV/cm) short terahertz pulse. The appearance of emission from graphene was observed in experiment in the 350–600 nm spectral range at high THz fields (Fig. 1). We detected the increase of the optical emission by nearly 3 orders of magnitude, while the THz field increased by a factor of 2 only. The spectrum of the optical emission corresponded to Planck's law for the spectral radiance of a black body. We attribute the optical emission to bias-induced spontaneous emission from energetic charge carriers in the graphene. On the basis of the theoretical analysis, we revealed the importance of the dynamic (ballistic) mechanism of electron-hole pair generation induced by an intense terahertz pulse with subsequent spontaneous radiative recombination.

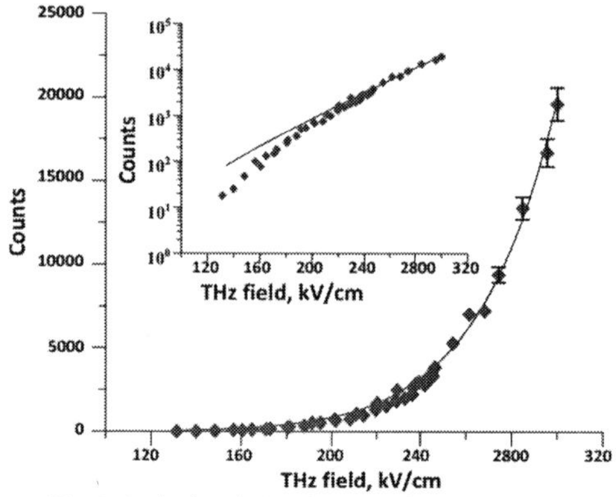

Fig.1. Optical emission from graphene vs THz field. Solid line – theoretical fit

In the second part of the presentation we will discuss our results on the generation of the second optical harmonic (SHG) of laser radiation from graphene in the presence of a high power terahertz pulse. We demonstrated essential amplification of SHG under the action of THz field. Polarization characteristics of SHG specific for graphene will be presented together with the dependence of SHG efficiency on optical and THz fields. Detailed theoretical analysis of von Neumann equation allowed us to propose a new mechanism for SHG in graphene under the combined action of optical and THz fields. This mechanism is caused by broadening of the interband resonance due to quasi-particle acceleration by an intense THz field in the process of interband transition. As a result of the "field" interband resonance broadening, the region of electron-hole generation in k-space becomes asymmetric, which leads to the appearance of necessary anisotropy allowing for SHG in the dipole approximation.

Intense single-cycle THz pulses produced by rectification of laser radiation are suitable for high gradient acceleration of electrons. Naturally, an accelerating structure fed by such pulses should be based on new principles, which would allow propagation of ultra-broadband radiation [2–4]. We propose an accelerating structure (Fig. 2) consisting of a set of waveguides with different adjusted lengths, in which the synchronism of accelerated particles with transversely propagating THz pulse is sustained [3]. A proper mirror shape allows the acceleration of both pre-accelerated particles and particles with zero velocity. The production technology implies a combination of a copper-aluminum "sandwich," electroplating of the reflecting cover mirror, and chemical removal of the aluminum.

For maximizing the acceleration gradient, it is natural to reduce the width of the structure. Unfortunately, this reduction is accompanied by strong diffraction losses due to pulse shape distortion in dispersive waveguides. That is why THz pulse focusing should be implemented in the new structure. The field strength in focus of a parabolic comprise periodic parabolic mirrors with constant focuses fed by a tilted THz pulse. The beam channel in this structure coincides with the focal line. Another problem is the transverse magnetic field of the THz pulse deflecting electrons. There are two solutions of the mentioned problem. First, a THz driven accelerator must have many sections with different directions of magnetic field, which, on the average, do not deflect electrons [3]. The second solution exploits magnetic field distortion (due to dispersion) inside the parabolic mirror. A proper choice of the parabolic mirror focal distance and aperture size allows satisfying the condition of accelerating field maximum and near to zero magnetic field at the same time moment when a bunch goes through the accelerating cell. Preliminary estimates show that for a THz pulse with 80 MV/cm in a 0.1×0.1 mm^2 spot in a structure consisting of 1-cm long parabolic sections, where each section is excited by its own single pulse, the accelerating gradient can be as high as 2.5 GV/m.

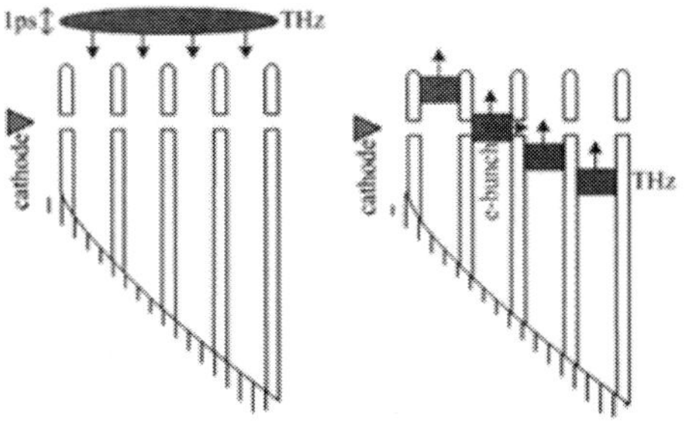

Fig. 2. Sketch of particle acceleration by picosecond THz pulse for two time moments

Another application for intense single-cycle THz pulses is high-brightness electron sources. It was recently proposed to use ~1 picosecond pulse to stimulate field emission in a high-gradient electron gun and simultaneously to provide preliminary acceleration of the electron bunch up to a relativistic energy [4]. A combined THz and RF field helps to overcome beam brightness reduction due to space-charge effects. To achieve high brightness, the beam generated at the cathode is accelerated to a higher energy before the space-charge effects lengthen the bunch. The calculations based on the Fawler-Nordheim equation show that in an accelerating gap fed by a focused mJ THz pulse a subpicosecond electron bunch of about 100 pC charge can be emitted and accelerated so fast that the brightness might be as high as $2\times\times10^{16}$ A/m^2.

Development of electron sources based on THz pulses was supported by the Russian Science Foundation (grant #16-19-10448).

References

1. *C. Vicario, A.V. Ovchinnikov, S.I. Ashitkov, M.B. Agranat, V.E. Fortov, and C.P. Hauri*, Generation of mJ pulses in the 0.1-5 THz gap with electric field exceeding 80 MV/cm // Opt. Lett. 2014. V. 39 (23). P. 6632–6635.

2. *A. Fallahi, M. Fakhari, A. Yahaghi, M. Arrieta, and F.X. Kartner.*Short electron bunch generation using single-cycle ultrafast electron guns // Physical Review Accelerators and Beams 2016. V. 19. P. 081302.

3. *S.V. Kuzikov, A.E. Fedotov, N.Yu. Peskov, A.N. Stepanov, A.A. Vikharev.* Millimeter and submillimeter-wave, high-gradient accelerating structures // AIP Conference Proceedings 2017. V. 1812. P. 060002.

4. *S.V. Kuzikov, S. Shchelkunov, A.A. Vikharev.* Concepts for a Short Wavelength rf Gun // AIP Conference Proceedings 2017. V. 1812. P. 080005.

Bursts of terahertz radiation from relativistic laser-plasma interactions

Yutong Li[1], Guoqian Liao[1], Chun Li[1], Weimin Wang[1], Zhengming Sheng[2,3], Jie Zhang[2]

[1] Institute of Physics, Chinese Academy of Sciences, Beijing 100190, China
[2] Key Laboratory for Laser Plasmas (MoE) and Department of Physics, Shanghai Jiao Tong University, Shanghai 200240, China
[3] SUPA, Department of Physics, University of Strathclyde, Glasgow G4 0NG, UK
ytli@iphy.ac.cn

Terahertz (THz) radiation has attracted increasing attention over last decades for its wide promising applications in many interdisciplinary fields of research. Fundamentally high power THz sources enable one to access to nonlinear THz photonics. Since the plasma is a medium free of a damage threshold, laser-plasma interactions provide a unique opportunity to achieve tabletop high-field THz radiation sources. Recently some experiments and models have been reported on the THz emissions generated from ultra-intense laser interactions with solid targets. In this presentation, we will show generation of coherent THz transition radiation by relativistic laser-accelerated electrons crossing the rear surface of a thin solid target. To verify the generation mechanism, various target parameters and structures are adopted to characterize the THz radiation properties and to distinguish the proposed mechanism from others. The model of transition radiation and particle-in-cell simulations well explain the observations. The presented laser-solid based THz transition radiation provides a new approach toward compact brilliant THz sources.

1. *G. Q. Liao et al.* Plasma Phys. Control. Fusion 59, 014039 (2017).
2. *G. Q. Liao et al.* Phys. Rev. Lett. 116 205003 (2016).

3D terahertz beam profiling from two color laser induced plasma with different focusing

A.A. Ushakov[1,2,3,4], M. Matoba[4], N. Nemoto[4], N. Kanda[5,6], K. Konishi[7], V.A. Andreeva[2,3], N.A. Panov[2,3], D.E. Shipilo[2,3], P.A. Chizhov[1], V.V. Bukin[1], M. Kuwata-Gonokami[4], J. Yumoto[4,7], O.G. Kosareva[2,3], S.V. Garnov[1] and A.B. Savel'ev[2,3]

[1]A.M. Prokhorov General Physics Institute RAS, Moscow, Russia, ushakov.aleksandr@physics.msu.ru
[2]Physics Faculty, M.V. Lomonosov State University, Moscow, Russia
[3]International Laser Center of M.V. Lomonosov Moscow State University, Moscow, Russia
[4]Department of Physics, The University of Tokyo, Tokyo, Japan
[5]RIKEN Center for Advanced Photonics, RIKEN, Tokyo, Japan
[6]Photon Science Center, The University of Tokyo, Tokyo, Japan
[7]Institute for Photon Science and Technology, The University of Tokyo, Tokyo, Japan

Profiling of the Terahertz (THz) beam emerging from the two color laser plasma interaction gain a lot of attention last years, providing insight into the physics of such an interaction and being of crucial importance for future applications. Beam profiling was demonstrated using electro-optical sampling [1], or using 2D THz camera for 3D measurements [2]. Later the beam profiling for different THz spectrum components [3] was demonstrated. However, impact from focusing conditions of the two color laser beam – the key parameter of the THz source in terms of directional output of THz radiation - has not been assessed yet.

In our paper we described the 3D beam profiling by 2D cross-section measurements of THz radiation from the two-color laser induced air plasma by using THz camera [4]. Laser pulses (central wavelength 800 nm, pulse duration 35fs, pulse energy 2.7 mJ, repetition rate 1 kHz) after the BBO crystal (I-type, 400 μm thickness, 13 % conversion efficiency), group velocity dispersion compensator plate and phase plate (half wave plate for fundamental wave and wave plate for second harmonic) were linearly polarized and collinear. Thereafter they were focused by off axis parabolic mirrors with the focal length of 2, 4 or 7.5 inches. For the depicting of plasma source to the THz camera two Tsurupica-lenses with focal length of 10cm were used. 2D THz intensity profile measurements have been done after the two lenses with 1mm step. For each series of experiments band pass THz filters (central frequencies 0.6 THz, 0.8 THz, 1 THz, 1.3 THz, 1.6 THz, 1.9 THz) were used. The sample distributions of THz radiation are presented in figure 1.

We also conduct measurements of distribution of THz field from the two color laser induced plasma by using the interferometric technique [5]. For generation and registration of THz radiation we used laser pulses (central wavelength 800 nm, repetition rate 10 Hz, pulse duration 40 fs, pulse energy 2.5 mJ, Gaussian beam diameter 12 mm FW1/e^2M, horizontal polarisation). The laser radiation was divided into two parts. The main part was used for generation of THz radiation in the source based on the optical breakdown in air induced by two-color femtosecond laser pulses [1]. The other part of laser radiation – the probing pulse – after passing through the variable optical delay line was used to measure the electric field strength, as shown in figure 2.

The principle of THz radiation detection is based on the linear electro-optical effect. The THz pulse and the probing optical pulse (λ = 800 nm) were combined on the ZnTe crystal (10×10×0.5 mm^3, cut <110>). In this case, the polarization of THz radiation was vertical, while the probing pulse was polarized at the angle 45°. As a result of the electro-optic effect in the optically isotropic ZnTe crystal the birefringence was induced, and in the above geometry of the experiment the probing radiation in the crystal was split into two waves (with vertical and horizontal polarizations and equal intensities), for which the difference between the refractive indices was linearly dependent on the electric field strength of the THz pulse.

The nonuniform distribution of the field strength over the crystal cross section caused the nonuniform distribution of the phase difference between the two waves at the output from the crystal. This phase difference was extracted by using an interferometer with polarizer and display of ZnTe crystal surface onto the CCD matrix was realized by telescope.

During the measurements we obtained interferograms in the presence of a THz pulse (signal) and in its absence (reference). To increase the signal-to-noise ratio for each delay time between the optical pulse and the THz one, 50 signal and background records (frames) were acquired. The phase was reconstructed by processing the interferograms by means of the method based on filtration in the Fourier-space [6]. As a result of the processing we obtained a two-dimensional map of the mean phase shift change caused by the presence of the THz pulse field. From the phase shift the THz electric field strength E_{THz} can be extracted [5].

In these experiments the depicting of plasma was realized by using 6cm and 15 cm PTFE Teflon lenses. The spatio-temporal profiles of the THz pulse electric field, obtained in the experiments, are presented in figure 3. The electro-optical crystal was placed in the plane, separated by the distance of 1 cm from the geometric focal point of the PTFE-lens (f=15 cm). Because in this region the focused THz radiation has a spherical wave front, the profiles of the field strength E_{THz} have the shape of rings that correspond to the sections of the spherical wave front of the THz radiation by the planar wave fronts of the probing pulse. With growing delay time (i.e., for later arrival of the probing optical pulse) the ring diameter decreases, collapsing into a spot when the planar front of the probing pulse is tangent to the converging spherical front of the THz pulse.

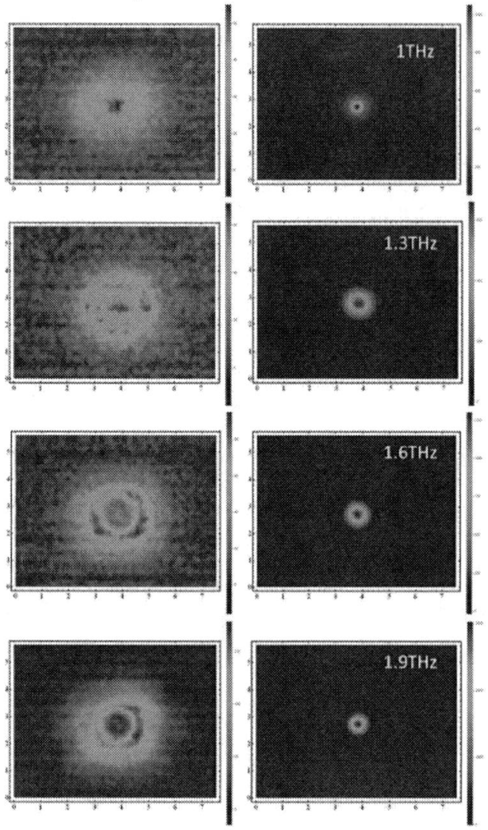

Fig. 1. THz Intensity distribution (at different frequencies) for focused by 4 inches parabola (right) and unfocused (left) THZ beams

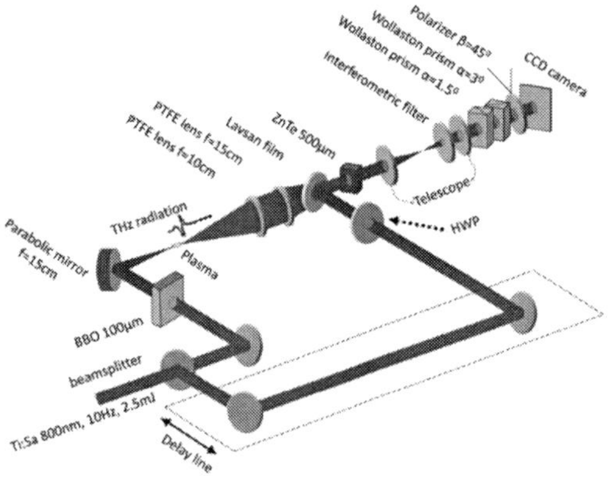

Fig. 2. Scheme of the experimental setup for THz field distribution measurements

Finally we presented theoretical explanation of the data observed using unidirectional pulse propagation equation [7]. To compare with experimental data THz beam propagation through the optical system was also included in the simulations. Simulations have demonstrated that aperture effect of optical elements plays an important role in THz beam profiling and the ring structure is the result of diffraction on the lenses.

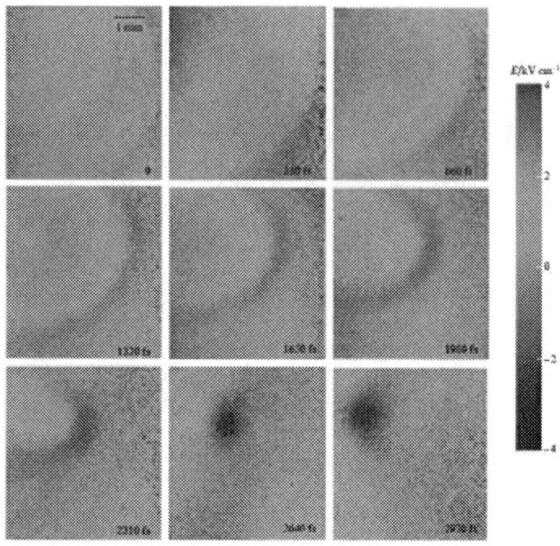

Fig. 3. Images of the spatial distribution of the THz pulse electric field strength at different time delays

This work in part dealing with THz measurements from femtosecond filaments was supported by the Students and Researchers Exchange Program in Sciences (STEPS) and RSF grant #16-42-01060, another part of work was supported by RF President's Grant Council (Support to the Leading Scientific Schools programs, Grant No. NSh-9695.2016.2), Grant of RF President for state support of young Russian scientists (Grant No. MK-8562.2016.2).

References

1. *Zhong H., Karpowich N., Zhang X.C.* Terahertz emission profile from laser induced air plasma. // Appl. Phys. Lett. 2006. V. 88, P. 261103-1–4.
2. *Klarskov P., Strikwerda A.C., Iwaszczuk K., Jepsen P.U.* Experimental three-dimensional beam profiling and modeling of a terahertz beam generated from a two-color air plasma // New Journal of Phys. 2013. V. 15. P. 075012.
3. *Klarskov P., Zalkovskij M., Strikwerda A.C., Jepsen P.U.* Spectrally resolved measurements of the terahertz beam profile generated from a two-color air plasma // Proc. of CLEO:2014. 2014. 6–7.
4. *Nemoto N., Kanda N., Imai R., Konishi K., Miyoshi M., Kurashina S., Sasaki T., Oda N., Kuwata-Gonokami M.* High-Sensitivity and Broadband, Real-Time Terahertz Camera Incorporating a Micro-Bolometer Array With Resonant Cavity Structure // IEEE Trans. Microw. Theory Tech. 2016. V. 6. No. 2. P. 175–182.
5. *Chizhov P.A., Ushakov A.A., Bukin V.V., Garnov S.V.* Measurement of spatio-temporal field distribution of THz pulses in electro-optic crystal by interferometry method // Quantum Electron. 2015. V.45. No. 5. P. 434-436.
6. *Takeda M., Ina H., Kobayashi S.* Fourier-transform method of fringe-pattern analysis for computer-based topography and interferometry // J. Opt. Soc. Am. 1982. V. 72. No. 1. P. 156.
7. *Andreeva V.A., Kosareva O.G., Panov N.A., Shipilo D.E., Solyankin P.M., Esaulkov M.N., Gonzalez De Alaiza Martinez P., Shkurinov A.P., Makarov V.A., Berge L., Chin S.L.,* Ultrabroad Terahertz Spectrum Generation from an Air-Based Filament Plasma // Phys. Rev. Lett. 2016. V. 116, P. 63902.

EPJ Web of Conferences **149**, 05012 (2017) DOI: 10.1051/epjconf/201714905012
10th International Workshop 2017 "Strong Microwaves and Terahertz Waves: Sources and Applications"

Generation of high power terahertz pulses and applications

D.S. Sitnikov, A.V. Ovchinnikov, O.V. Chefonov, and M.B. Agranat

Joint Institute for High Temperatures of the RAS, Moscow, Russia, sitnik.ds@gmail.com

Here we report on our latest results in creating high-power terahertz pulses by means of optical rectification technique and their application in various fields. We investigated terahertz generation in organic crystals DSTMS, DAST and OH1 directly pumped by a Cr: forsterite laser at central wavelength of 1.25 μm. This pump laser technology provides a laser-to-THz energy conversion efficiency higher than 3%. Phase-matching is demonstrated over a broad 0.1–8 THz frequency range. In our simple setup, we achieved hundred μJ pulses in tight focus resulting in electric and magnetic field larger than 10 MV/cm and 3 Tesla [1].

We've applied a novel method for measuring the THz spatial distribution of in the focal plane of a parabolic mirror based on generation of the second-harmonic radiation in the optical region of spectrum in centrosymmetric crystals under action of a powerful terahertz pulse with subsequent image transfer to iCCD camera (Fig. 1).

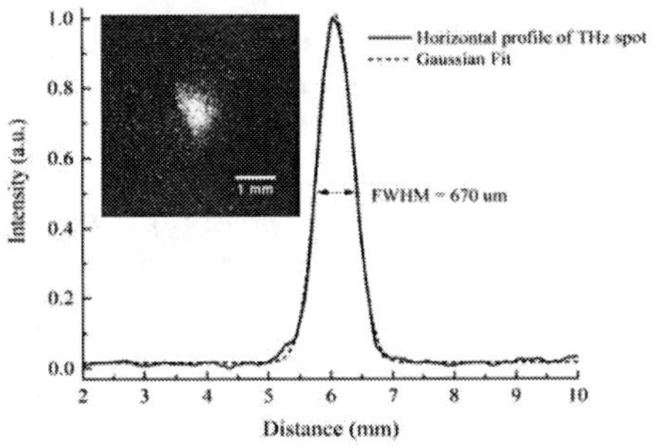

Fig. 1. THz intensity distribution for OH1 crystal

Application of different techniques of spatial profile transformation using interference of two chirped pulses in Mach Zehnder-type interferometer [2] as well as acoustooptical dispersion delay line [3] made it possible to generate THz pulses with a tunable center frequency in the range 0.5–2.5 THz (Fig. 2).

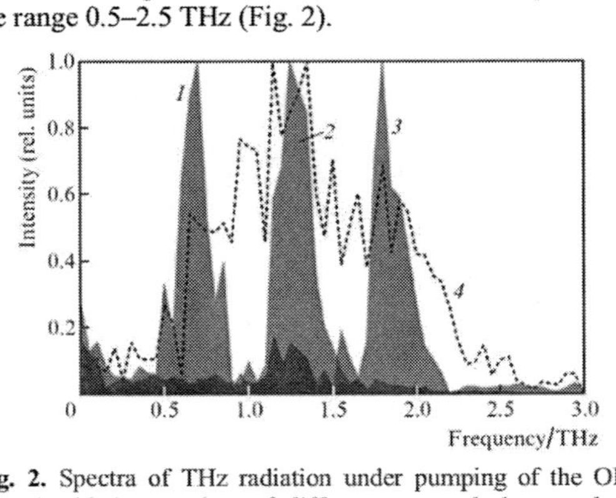

Fig. 2. Spectra of THz radiation under pumping of the OH1 crystal with laser pulses of different temporal shape and the delay between two chirped pulses (*1*) 0.5, (*2*) 1 and (*3*) 1.5 ps; (*4*) is the spectrum in the case of pumping with a single transform-limited pulse

Special attention is devoted to THz electric field-induced second harmonic generation in inorganic ferroelectric and centrosymmetric antiferromagnet as well. Second Harmonic Generation induced by the electric field of a strong nearly single-cycle terahertz pulse with the pick amplitude of 300 kV/cm is studied in a classical inorganic ferroelectric thin film of BaSrTiO3. The dependences of the SHG intensity on the polarization of the incoming light is revealed and interpreted in terms of electric polarization induced in the plane of the film. As the THz pulse pumps the medium in the range of phononic excitations, the induced polarization is explained as a change of the ferrolectric order parameter. It is estimated that under action of the THz pulse the latter acquires an in-plane component up to 3% of the net polarization.

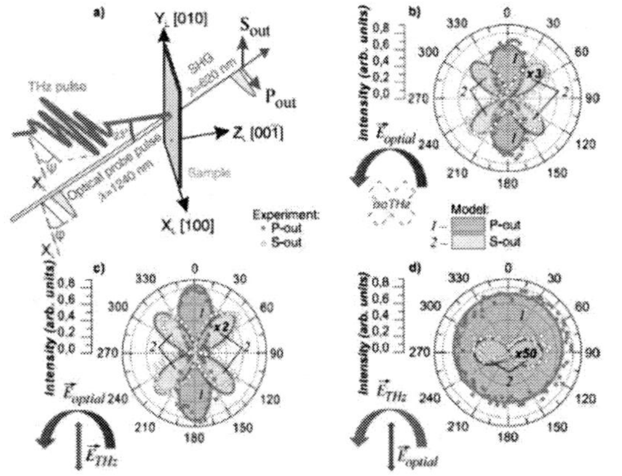

Fig. 3. Experimental geometry and polarization diagrams of the SHG intensity for various experimental geometries. (a) Experimental geometry. The axes of the chosen laboratory frame X_L, Y_L, Z_L correspond to [100], [010] and [001] crystallographic directions, respectively. ϕ – the angle between the electric field of near-infrared probe and the X_L-axis. ψ – the angle between the electric field of the THz pump pulse and the X_L-axis; (b) dependence of the SHG signal on ϕ without any THz pump; (c) the same, when the THz field is applied parallel with respect to the X_L-axis. The polarization of the SHG signal was set either in the P_{out} or S_{out}-state; (d) dependence of the SHG signal on ψ when the probe polarization was set to the P-state ($\phi = 0$). The polarization of the SHG signal was set either in the P_{out} or S_{out}-state. Dots correspond to experimental data and lines are fits. Values for S-state multiplied by 3, 2, and 50 for (b–d) respectively

Hence the data show that the THz electric field clearly affects the process of the second harmonic generation. To reveal ultrafast dynamics of these electric field induced changes, we performed pump-probe measurements of the SHG signal. In particular, the signal was measured as a function of the delay between the THz-pump and near-infrared probe pulses. Fig. 4a shows time-domain trace of the electric field of the THz pulse obtained with the help of electro-optical sampling. The measurements of the SHG signal from the BST film (Fig. 4b) reveal a similar dynamics during the overlap of the probe and THz-pump pulses. It points out that the nonlinear response is proportional to the THz electric field.

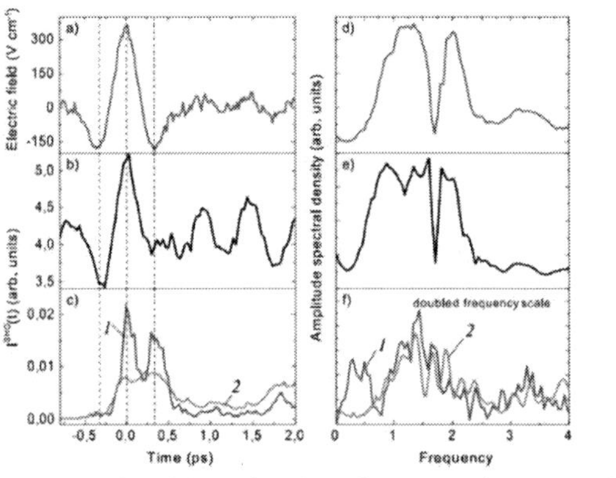

Fig. 4. THz-induced dynamics of nonlinear-optical response of the crystals. (a) Time trace of the electric field of the THz pump pulse; intensities of the SHG-signal: (b) for BST, (c) for STO (line 1) and Si (line 2); (d–f) Fourier transform spectra of the time traces shown in panels (a–c), respectively ((f) plotted in doubled frequency scale)

For comparison, we also measured temporal evolution of the SHG signals from centrosymmetric crystals excited by the intense THz pulse. The SHG transients for $SrTiO_3$ (STO) and Si are shown in Fig. 4c (lines 1 and 2, respectively). First of all, for the both cases of the centrosymmetric crystals the SHG signal of the unperturbed media are zero. Secondly, in the maximum of THz modulated signal the absolute values of the intensity of the SHG from BST is two orders of magnitude higher than those for the centrosymmetric crystals. Thirdly, the shape of the SHG response differs from the shape of the THz pulse.

Optical second harmonic generation at the photon energy of 2 eV in the model centrosymmetric antiferromagnet NiO irradiated with picosecond terahertz pulses (0.4–2.5 THz) at room temperature is detected (Fig. 5). The analysis of experimental results shows that induced optical second harmonic generation at the moment of the impact of a terahertz pulse arises through the electric dipole mechanism of the interaction of the electric field of a pump pulse with the electron subsystem of NiO. Temporal changes in optical second harmonic generation during 7 ps after the action of the pulse are also of an electric dipole origin and are determined by the effects of propagation of the terahertz pulse in a NiO platelet. Coherent oscillations of spins at the antiferromagnetic resonance frequency induced by the magnetic component of the terahertz pulse induce a relatively weak modulation of magnetic dipole optical second harmonic generation.

We have demonstrated that impact of electric component of picosecond single-cycle terahertz pulse (0.4–2.5 THz) leads to dynamic loss of the inversion center in centrosymmetrical (in terms of crystallographic and magnetic symmetry) antiferromagnet NiO.

This is accompanied by enablement for electrodipole SHG. This process takes place at the time of the THz pulse impact and is comparable in intensity to the spontaneous generation of SHG of the magnetodipole type allowed for symmetry at a pulse intensity of I^{THz} = = 0.2 mJ/cm^2 [4].

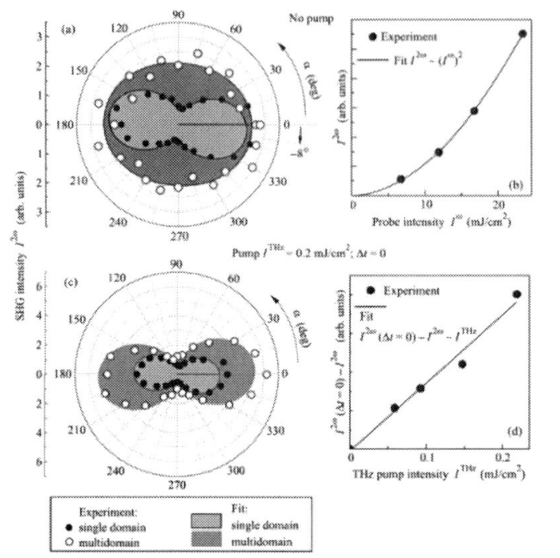

Fig. 5. (a, c) Experimental azimuthal dependencies of SHG intensity in single-domain (•) and multi-domain (○) states obtained in the absence (a) and under action (c) of THz pump pulse (I^{THz} = 0.2 mJ/cm^2) Radiation intensity at the fundamental frequency was I^w = 14 mJ/cm^2. Filled areas are theoretical calculations. (b) Experimental dependence (•) of SHG intensity I^{2w} as a function of pulse intensity at fundamental frequency in the absence of pump pulse and its approximation (curve). (d) Experimental dependence (•) of induced SHG intensity $I^{2w}(\Delta t = 0) - I^{2w}$ as a function of pump pulse intensity I^{THz} and its approximation (curve). Experimental dependencies (c) and (d) are obtained in multi-domain state at analyzer position α = 0

The experiments were conducted using Unique Facility "Terawatt Femtosecond Laser Complex" in the Center for Collective Usage "Femtosecond Laser Complex" of JIHT RAS.

The study was supported by the Russian Science Foundation (project No. 17-19-01261).

References

1. *Vicario, C., Ovchinnikov, A.V., Ashitkov, S.I., Agranat, M.B., Fortov, V.E., and Hauri, C.P.* Generation of 0.9-mJ THz pulses in DSTMS pumped by a Cr:Mg_2SiO_4 laser // Opt. Lett. 2014. V. 39, No. 23, P. 6632–6635.

2. *Vicario, C., Ovchinnikov, A.V., Chefonov, O.V., Shalaby, M., Jazbinsek, M., Hauri, C.P.* Generation of strong-field spectrally-tunable Terahertz pulses // Nature Com. 2017. to be published.

3. *Ovchinnikov, A.V., Chefonov, O.V., Molchanov, V.Ya., Yushkov, K.B., Vicario, C., Hauri, C.* Generation of frequency-tunable pulsed terahertz radiation by a Cr : forsterite laser system with an acoustooptical control of the pulse temporal profile // Quant. Elect. 2016. V. 46, No. 12. P. 1149–1153.

4. *Ovchinnikov, A.V., Chefonov, O.V., Agranat, M.B., Grishunin, K.A., Il'in, N.A., Pisarev, R.V., Kimel, A.V., Kalashnikova, A.M.* Optical second harmonic generation induced by picosecond terahertz pulses in centrosymmetric antiferromagnet NiO // JETP Lett. 2016. V. 104, No. 7, P. 441–448.

EPJ Web of Conferences **149**, 05013 (2017) DOI: 10.1051/epjconf/201714905013
10th International Workshop 2017 "Strong Microwaves and Terahertz Waves: Sources and Applications"

Terahertz spectroscopy for diabetes diagnostics

O.P. Cherkasova[1], M.M. Nazarov[2], A.P. Shkurinov[3, 4]

[1]Institute of Laser Physics of SB RAS, Novosibirsk, Russia, o.p.cherkasova@gmail.com
[2]Kurchatov Institute National Research Center, Moscow, Russia
[3]Crystallography and Photonics Federal Research Center RAS, Moscow, Russia
[4]Lomonosov Moscow State University, Moscow, 119991, Russia

Introduction

Diabetes mellitus is a chronic disease caused by inherited and/or acquired deficiency in production of insulin by the pancreas, or by the ineffectiveness of the produced insulin. Such a deficiency results in increased concentrations of glucose in the blood, which in turn damage many of the body's systems, in particular the blood vessels and nerves. The development of diabetes mellitus leads to violation of carbohydrate, protein, and lipid exchange and is accompanied by a significant increase in the content of glucose, corticosteroid hormones, and some other metabolites in blood. The development of new rapid diagnostic methods for diabetes and its complications is an urgent problem.

Terahertz (THz) radiation has a number of useful features for medical applications. It is sensitive to water concentration and state, it does not ionize biological objects [1] and passes through the skin [2]. A distinctive feature of THz time domain spectroscopy is the possibility of measuring directly the refractive index and absorption coefficient, and hence complex permittivity spectrum of the sample in a single scan and in a broad frequency range [3]. This circumstance makes it possible to obtain a detailed spectral characteristic of a sample during one measurement, due to which rapid and even remote diagnostics is foreseen.

Experimental

Our THz time-domain apparatus was described previously [4, 5]. We used the radiation of a Ti:sapphire laser with a wavelength of 790 nm, a pulse duration of 90 fs with 1 Wt average power. For THz emission, the semiconductor (LT-GaAs) surface was illuminated by laser beam. For THz detection the electro-optical ZnTe crystal of 1 mm thickness was probed by a small, delayed fraction of laser beam. We have used both transmission (for blood plasma) and attenuated total internal reflection (ATR) configurations (for blood plasma and skin). The spectral range of reliable measurements was between 0.05 and 3.2 THz. We analyzed blood plasma of a rat with experimental diabetes. In additional the ATR spectra of human skin on the palm of the hand in vivo were measured at a normal blood glucose concentration (4.2–5.9 mM) and 45 and 90 minutes later after drinking the glucose solution (75 g per 100 ml of water). This is the standard oral glucose tolerance test, which demonstrates an initial spike and rapidly returns to normal levels of blood glucose concentration.

Results

THz spectroscopy of biological samples mainly probes water strong absorption and dispersion in THz ranges. It is known that relaxation time of water molecules involved in hydration of biomolecules are different from those of characterizing free, unbound water molecules, which is reflected in the low-frequency absorption and refraction spectra of biological samples (tissues, blood plasma) [4]. We have previously shown the insertion of glucose into water leads only to an increase of relaxation time τ_1 of the slow Debye process of this solution. The increasing of protein albumin concentration in the solution results in a decrease of the amplitude $\Delta\varepsilon_1$ of the slow Debye relaxation process [5]. This explains observed sensitivity of THz skin reflection spectra and blood plasma properties.

The figure 1 a) shows the averaged absorption coefficient difference for healthy and rat with alloxane diabetes. That is the difference between distilled water and studied solution.

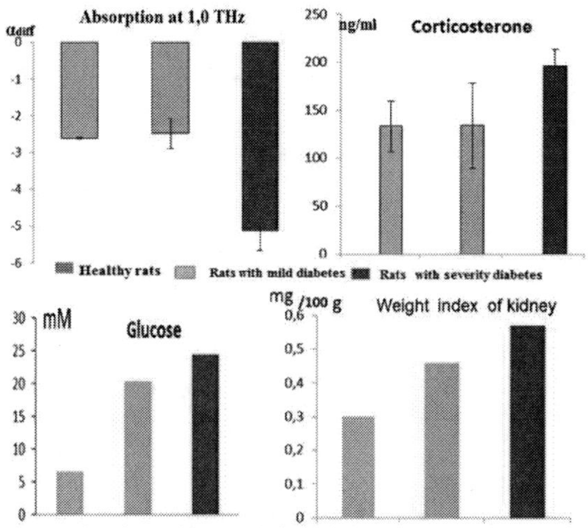

Fig. 1. The absorption coefficient differences at 1 THz, corticosterone and glucose concentration in blood plasma, and weight index of kidney of healthy (blue), mild diabetics (orange) and severity diabetics (red) rats

The absorption coefficient differs in group of rats with severity diabetes by a factor of more than two from the corresponding value in other groups. This fact indicates that some components secreted into the blood plasma under experimental diabetes complications may contribute to the total absorption. The rats with severity diabetes differed by a reliably high level of corticosteroid hormones in blood and adrenal glands, a large weight index of kidney, and a higher level of glucose in blood in comparison with the healthy and mild diabetics rats [6]. The decrease in the amplitudes of absorption spectra of blood plasma of rats with severity diabetes can be considered as an integral estimate of significant biochemical disturbances in Diabetes.

Assuming that the observed spectral changes are due to changes in the state of water in blood plasma, we have selected one of the parameters of the Debye model aqueous solution – $\Delta\varepsilon_l/\tau_1$, leading to the spectral features observed in the experiment. This change in the response of bound water can be the reason of the observed changes in our experiments at increasing glucose concentration in blood. We have demonstrated that when the concentration of glucose in blood rises to 24 mM (in rats with severity diabetes), $\Delta\varepsilon_l/\tau_D$ ratio decreases 1.2 times.

The correlation of the ATR amplitude of human skin and glucose concentration in blood is demonstrated in figure 2. In this case we use ATR amplitude R_{int} integrated over the used frequency range, and time shift ΔT [7]. It can be seen that the variations of the optical characteristics of human skin correlate with the changes in blood glucose level.

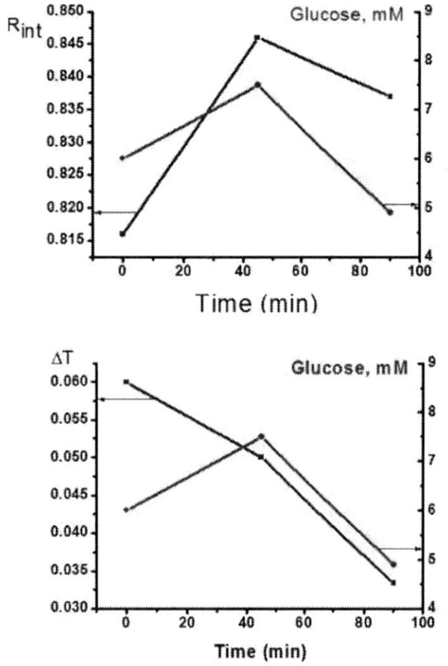

Fig. 2. The ATR amplitude R_{int} and time shift ΔT of human skin and glucose concentration in blood (mM) versus time (min) after glucose intake

The change in the ATR amplitude and phase of skin are determined only by variations of the slow Debye process [7]. When we observe changes of the ATR spectra after ingestion of glucose solution we see the effect that has a high level of blood glucose concentration on tissue and skin.

Conclusion

We showed that diagnostics based on THz time domain spectroscopy allows to make an integral estimation of the content of metabolites in blood plasma and distinguish blood plasma of healthy rats and rats with experimental diabetes of different degrees of severity. To increase diagnostic THz sensitivity we now develop polymer capillary waveguide sensor for such solutions. We expect more sensitivity for refraction changes of the outer media (solution) due to tens of cm interaction length.

The ATR amplitude of human palm skin increased when the glucose concentrations in blood rose above the normal level. The observed change of the spectrum is described with high degree of accuracy by the reduction in the ratio $\Delta\varepsilon_1/\tau_1$ in the Debye model of glucose aqueous solution.

Terahertz Pulsed Spectroscopy can be used as a promising diagnostic method of Diabetes Mellitus.

The work is supported by RFBR grant 16-52-00222.

References

1. *Angeluts, A.A., Gapeyev, A.B., Esaulkov, M.N. et al,* Study of terahertz-radiation-induced DNA damage in human blood leukocytes // Quantum Electron. 2014. V. 44, No. 3. P. 247 - 251.

2. *Cherkasova, O.P., Nazarov, M.M., Berlovskaya, E.E. et al,* Studying human and animal skin optical properties by terahertz time-domain spectroscopy // Bulletin of the Russian Academy of Sciences: Physics. 2016. V. 80. No.4. P. 479-483.

3. *Angeluts, A.A., Balakin, A.V., Evdokimov M.G. et al.,* Characteristic responses of biological and nanoscale systems in the terahertz frequency range // Quantum Electronics. 2014. V. 44. No.7. P. 614 – 632.

4. *Cherkasova, O.P., Nazarov, M.M., Angeluts, A.A., Shkurinov, A.P.* Analysis of blood plasma at terahertz frequencies // Optics and Spectroscopy. 2016. V. 120. No.1. P. 50-57.

5. *Nazarov, M.M., Cherkasova, O.P., Shkurinov. A.P.* Study of the dielectric function of aqueous solutions of glucose and albumin using pulsed THz spectroscopy // Quantum Electronics. 2016. V. 46. No.6. P. 488-495.

6. *Cherkasova, O.P., Nazarov M.M., Smirnova, I.N., Angeluts, A.A., Shkurinov, A.P.* Application of Time-Domain THz Spectroscopy for Studying Blood Plasma of Rats with Experimental Diabetes // Physics of Wave Phenomena. 2014. V. 22. No. 3. P. 185-188.

7. *Cherkasova, O.P., Nazarov, M.M., Shkurinov A.P.* Noninvasive blood glucose monitoring in the terahertz frequency range // Optical and Quantum Electronics. 2016. V. 48. No.3.

EPJ Web of Conferences **149**, 05014 (2017) DOI: 10.1051/epjconf/201714905014
10th International Workshop 2017 "Strong Microwaves and Terahertz Waves: Sources and Applications"

Harmonic terahertz gyrotron with quasi-optical confocal cavity

Wenjie Fu, Xiaotong Guan, Yang Yan, Xiaoyun Li, Yin Huang, and Lin Meng

School of Physical Electronics, University of Electronic Science and Technology of China,
Chengdu, 610054, China, fuwenjie@uestc.edu.cn

To develop terahertz gyrotron with low magnet, harmonic gyrotron is an interesting research area in terahertz gyrotron research area. However, in harmonic gyrotron, mode competition is more terrible than in fundamental gyrotron. Thus, adopting good mode-selectivity structures is a possible approach to improve harmonic numbers. Confocal waveguide is a type of quasi-optical open waveguide, which performs a much lower mode density than conventional cylindrical waveguide, and it is potential to be operated at high order mode with high frequency and high power. The confocal cavity utilized in gyrotron was initially proposed by R. J. Temkin et al. [1, 2], and it has been successfully used in 140 GHz fundamental gyrotron oscillator and travelling wave amplifier in MIT. Due to the good mode selective characteristics, confocal waveguide would greatly suppress the mode competition in harmonic gyrotron. In University of Electronic Science and Technology of China (UESTC), the harmonic terahertz gyrotron with quasi-optical confocal cavity is being carried on, and a 400GHz second harmonic gyro-tron with confocal cavity is designed and experimented.

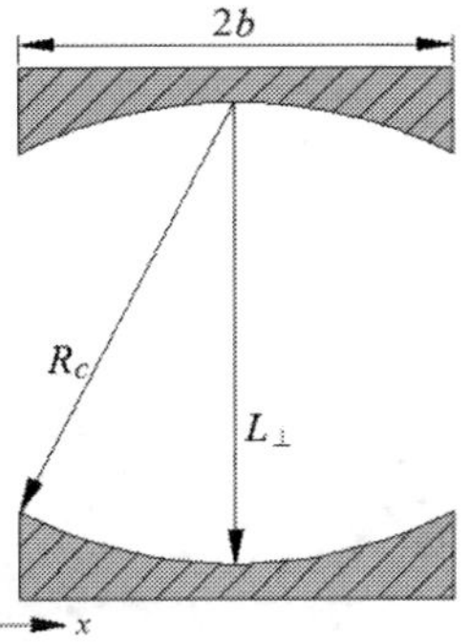

Fig. 1. Transverse geometry of confocal cylindrical waveguide

The cross section of a confocal waveguide is shown as Fig. 1. There are two cylindrical mirrors with each having a curvature radius R_c separated by distance $L_\perp$, which is same length as R_c. Propagation modes in this confocal waveguide are TE and TM modes which are similar to that in cylindrical waveguide. Previous studies show that diffraction losses of TE_{mn} $(m \neq 0)$ modes are greatly larger than that of TE_{0n} modes, and only TE_{0n} modes could be propagated in confocal waveguide. For TE_{0n} mode, eigenvalue of wavenumber is given by

$$k_{\perp,0n} = \left(n + \frac{1}{4}\right)\frac{\pi}{L_\perp} \tag{1}$$

The frequency isolation between neighboring modes in confocal waveguide is $\Delta f = c / 2L_\perp$, which is much higher than in cylinder waveguide. Thus, the quasi-optical confocal waveguide has better mode selection, which greatly reduces the mode density. Equation (1) shows that $TE_{0,s\times n-1}(s)$ mode has the maximum isolation from neighboring fundamental and lower harmonic modes. Therefore, the $TE_{0,s\times n-1}(s)$ mode is potentially to be the operation mode at sth harmonic. In 400 GHz confocal gyrotron design, $TE_{0,11}$ mode is chosen as the operation mode for second harmonic operation, and the eigenvalue characteristics of of $TE_{0,11}$ with potential competition modes are shown in Fig. 2.

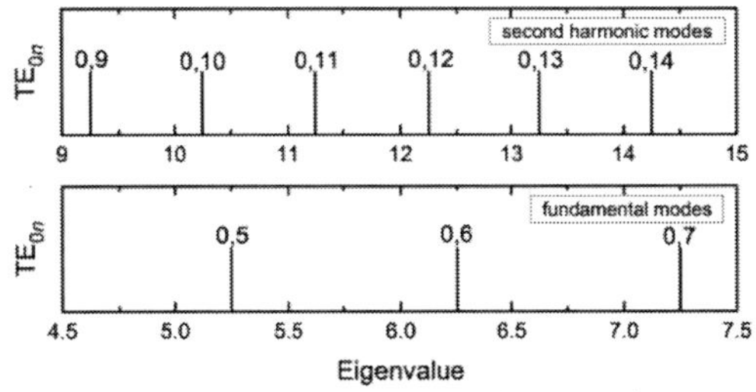

Fig. 2. Mode spectra in confocal waveguide

The confocal waveguide is a quasi-optical structure, some part of power will be diffracted out over the finite mirror aperture $2b$. Thus, different from cylindrical cavity, the confocal cavity contains Ohmic loss, axial diffraction loss and transverse diffraction loss. The total Q factor for TE_{0n} modes in cavity could be expressed as

$$\frac{1}{Q} = \frac{1}{Q_{\perp diff}} + \frac{1}{Q_{z diff}} + \frac{1}{Q_{ohmic}} \tag{2}$$

This denotes that the Q value could be changed by adjusting the mirror aperture. Figure 3(a) displays the relationship between the diffraction loss rate and the mirror aperture. Figure 3(b) plots the calculated total Q factor as a function of mirror aperture $2a$ for different modes. In the design, the mirror aperture is selected to be 3.2 mm in which $TE_{0,11}$ mode has a smaller diffraction loss and a suitable Q factor.

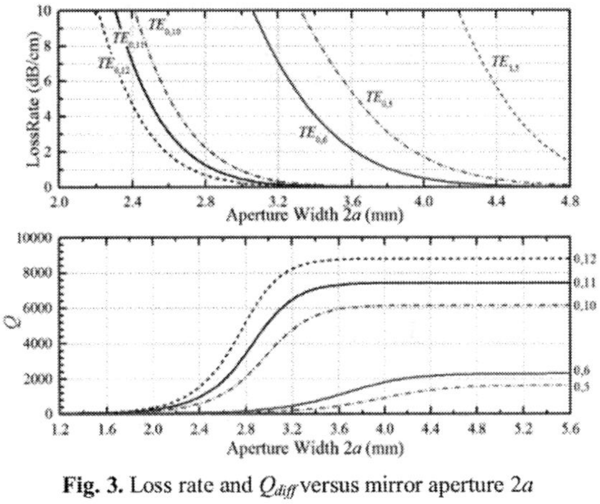

Fig. 3. Loss rate and Q_{diff} versus mirror aperture $2a$

Figure 4 displays the dependence of starting currents I_{st} for different modes and the magnetic field B_0, which calculated by linear theory of gyrotron. The results suggests that quasi-optical cavity performs much fewer competition modes within about 2T magnetic operation range. The potential operation region of $TE_{0,11}$ mode presents a large separation from that of fundamental $TE_{0,5}$ and $TE_{0,6}$ modes. While the operation magnetic field for $TE_{0,5}$-$TE_{0,10}$ and $TE_{0,6}$-$TE_{0,12}$ modes are close to each other, since these mode couples share a similar fundamental frequency of electron cyclotron. It is a good agreement with the above analysis of mode selection. Additionally, the operation regions of second harmonic modes are narrower than that of fundamental cyclotron, which implies that high harmonic can be excited only under strictly conditions.

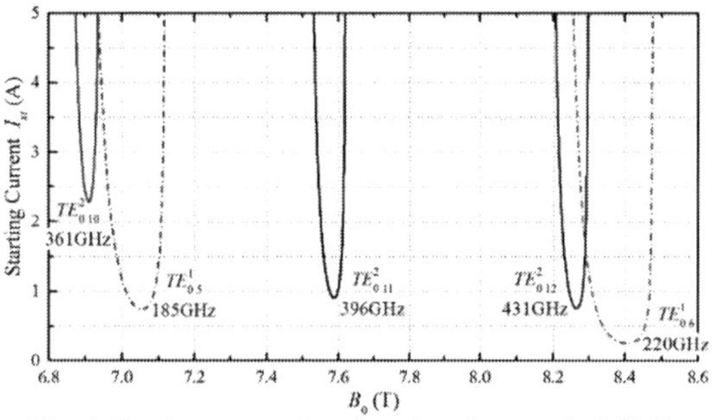

Fig. 4. Starting current I_{st} as function of magnetic field B_0

Fig. 5. Photo of confocal cavity

Fig. 6. Photo of 400GHz second harmonic prototype gyrotron with confocal cavity

To verify theoretical analysis, a 400GHz second harmonic gyrotron with confocal cavity is has been manufactured and assembled in the UESTC. A photo of the confocal cavity are shown in fig. 5, and the photo of assembly prototype tube is presented in fig. 6.

The prototype tube adopts a triode-type magnetron-injection gun (MIG) which developed for 220 GHz gyrotron, and optimized for design parameters. The output from confocal cavity is converted from confocal waveguide to homocentric waveguide by an irregular quasi-optical taper, and transported into cylindrical waveguide.

The initial experiments have been performed based on a 9.2 T superconducting magnet. The accelerating voltage for MIG is provided by a high voltage modulator, which can supply about 120 kV with a pulse width of 100 μs. The output power of gyrotron is measured by a modified laser calorimeter, amplified by a low-noise-amplifier (LNA), and recorded by a Data Acquisition (DAQ) card. The frequency was measured using two VivaTech harmonic mixers. One is used to detect fundamental output, and another is used to detect second harmonic output.

Under the operation point of maximum output power (40 kV, 4.75 A, and 7.51 T), the detected oscilloscope traces are shown in fig. 7(a) which shows the Beam voltage, collector current, IF signal. Figure4 (b) shows the expanded waveform of partial IF signal with a high sample rate of 8 GSa/s. The initial experiment results show that the second harmonic output of $TE_{0,11}$ mode and the fundamental output of $TE_{0,6}$ mode are observed. The frequency of second harmonic output is ~396 GHz with 1.5 kW power. The results demonstrate that confocal cavity has good mode selectivity and would be used to develop high harmonic terahertz gyrotron.

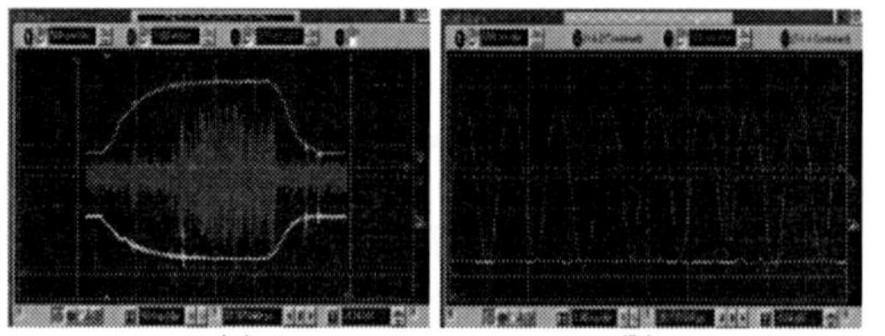

(a) (b)

Fig. 7. Beam voltage, collector current, IF signal envelop and partial expanded IF signal waveform

This work is supported by the National Natural Science Foundation of China under Grant 61401064, and the CAEP THz Science and Technology Foundation under the Grant No. CAEPTHZ201403.

References

1. *Hu, W., Shapiro, A., Kreischer, K. E., Temkin, R. J.* 140-GHz Gyrotron Experiments Based on a Confocal Cavity // IEEE Trans. Plasma Sci. 1998. V. 26, No. 3, PP. 366-374.

2. *Sirigiri, J. R., Shapiro, M. A., Temkin, R. J.* High-Power 140-GHz Quasioptical Gyrotron Traveling-Wave Amplifier // Phys. Rev. Lett. 2003. V. 90, No. 25, P. 258302.

3. *Fu, W., Guan, X., Yan, Y.* High harmonic terahertz confocal gyrotron with nonuniform electron beam // Physics of Plasmas 2016. V. 23, No. 1, P. 013301.

4. *Fu, W., Guan, X., Chen, C., Li, X., Yuan X., Yan, Y.* Design and Experiment of a 220/420-GHz Gyrotron for Nondestructive Evaluation // IEEE Trans. Electron Devices, 2014, V. 61, No. 7, PP. 2531-2537.

Pseudospark excited sub-THz frequency sources

H. Yin[1], L. Zhang[1], Y. Yin[2], J. Zhao[3], G. Shu[1], W. He[1], A. W. Cross[1], A. D. R. Phelps[1]

[1]Department of Physics and SUPA, University of Strathclyde, Glasgow G4 0NG, Scotland, UK
a.d.r.phelps@strath.ac.uk
[2]School of Physical Electronics, University of Electronic Science and Technology of China, Chengdu, 610054, China
[3]High Voltage Division, School of Electrical Engineering, Xi'an Jiaotong University, Xi'an, 710049, China

There are several advantageous features of using a pseudospark (PS) discharge [1] for electron beam production. One of these features is the formation of an ion channel following the pseudospark anode, which enables the beam to propagate and eliminates the need for a guiding magnetic field [2–5]. When a high voltage is applied to the hollow cathode, the electric field across the anode-cathode gap penetrates a short distance into the hollow cathode region due to the small cathode aperture. A PS discharge will occur if the pressure in the system is suitably low (typically 50–500 mTorr) so that the discharge is at the left-hand side (with respect to the minimum) of the Paschen curve. In such a PS discharge condition, the gas breakdown will occur along the longest possible path, allowing a virtual anode to form, extending from the anode into the hollow cathode region. As the virtual anode reaches the cathode surface field-enhanced emission begins to occur. Electrons begin emitting from the cathode surface at an increased rate, augmented by secondary emission and are accelerated toward the aperture by the electric field. Consequentially this rapid increase in electron emission results in a rapid increase in the beam current. As the beam propagates through the anode its front edge ionizes the background gas, forming a plasma channel, while the following beam electrons expel part of the plasma electrons so that an ion-channel is formed, confining the beam and eliminating the need for any external magnetic guide field. A high current density, high brightness electron beam with a sweeping voltage can therefore be generated and propagated by ion channel focusing.

Pseudospark discharges have been explored for various important applications, especially high quality electron beam generation for microwave sources [2, 5, 6] and potential terahertz devices. High frequency sources above 100 GHz are very attractive for a wide range of research and technical applications, including molecular spectroscopy, bio-imaging and security screening. As the frequencies move into the sub-terahertz and terahertz region, the size of device reduces greatly. This brings a challenge with regard to device fabrication. Therefore a compact and simplified structure is desirable, with the pseudospark-sourced electron beam an ideal choice for high power, high frequency sources. This paper presents some experimental results of the electron beam current dependence on the gap separation of a single-gap pseudospark structure. At a certain gap separation, the relationship between the beam current and discharge voltage has been studied [2]. It is found that the electron beam only starts to occur when the charging voltage is above a certain value and increases with the increasing discharge voltage following two tendencies. Under the same discharge voltage, the configuration with the larger electrode gap separation will generate higher electron beam current. Because the energy of the high brightness electrons in the beam produced by a pseudospark is relatively low post acceleration experiments on the hollow cathode beam have also been successfully carried o ut [4, 5].

X-ray emission studies and spatial visualization of the PS beams have been used to improve the characterisation of these beams [7, 8]. Several experiments have successfully produced mm-wave/sub-THz radiation [9, 10]. Zhao et al. [11, 12] have measured the dependence of the PS beam on the gap separation and the beam variability in a post-accelerated PS [13]. Kumar et al. [14] have carried out an analysis of how the PS geometrical factors affect the PS emission. These studies have provided increased insight for the designs of future sub-terahertz and terahertz sources driven by PS electron beams.

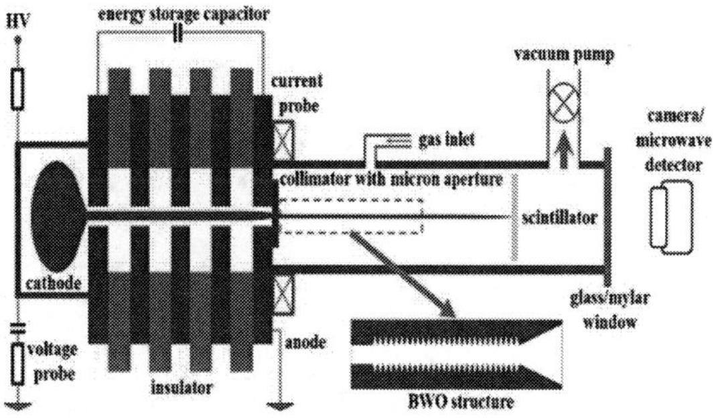

Fig. 1. Experimental configuration producing a micro-electron beam from a PS which is used to excite a 200 GHz BWO structure [9]

Fig. 1 shows the experimental configuration of a multigap PS discharge that produces a micro-electron beam. No magnetic field is needed to transport the electron beam. The rippled wall BWO structure is mounted immediately adjacent to the PS exit micro-aperture, so that the electron beam can pass through the BWO structure. The BWO structure, together with the conical radiation launching horn, was manufactured by high speed grinding of an aluminum former and the subsequent electrodeposition of a 5mm thick layer of copper on the aluminium former, which was later dissolved away in an alkali solution.

An electron beam ~1mm diameter carrying a current of up to 10A and current density of $10^8 Am^{-2}$, with a sweeping voltage of 42 to 25 kV and pulse duration of 25 ns, was generated from the PS discharge. Fig. 2 shows the repeatable time-correlated electron beam voltage, the discharge current and the millimetre wave pulse. The electron beam current has a step of about 5A at the hollow cathode discharge phase and then a peak current of about 10A follows in the conductive phase. The microwave radiation was mainly generated near this first 5 A step, because the correlated beam voltage has stronger coupling with the BWO structure. In the conductive phase, the beam voltage is too small to have efficient beam-wave interaction.

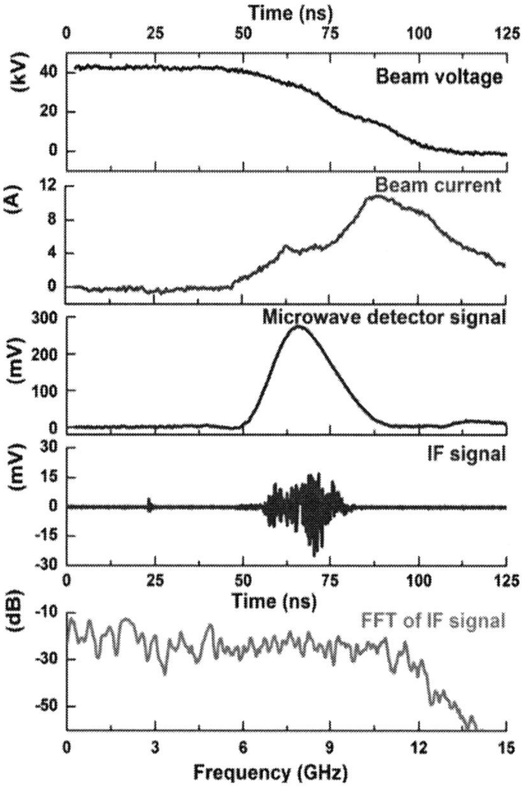

Fig. 2. Time-correlated electron beam voltage, current pulse, the radiation pulse from the 200 GHz BWO, the IF output from a harmonic mixer recorded on a deep memory (20 GHz) single shot digital storage oscilloscope, and FFT result of the intermediate frequency output [9]

The output power was measured using the general antenna theorem with the total power from a launching antenna, calculated by integrating its radiated power density over space. The integration was completed by numerically integrating the normalized mode profile of the launching horn and multiplying by the measured maximum power density. The result of this measurement was that the total power of the BWO in this frequency range was found to be 20 W.

Acknowledgements

The UK Engineering and Physical Research Council (EPSRC) is thanked for supporting this research. David Barclay provided valuable technical assistance.

References

1. *Christiansen, J., and Schultheiss, C.* Production of high current particle beams by low pressure spark discharges // Zeitschrift für Physik A Atoms and Nuclei. 1979. V. 290. No. 1. P. 35-41.

2. *Yin, H., Robb, G.R.M., He, W., Phelps, A.D.R., Cross, A.W. and Ronald, K.* Pseudospark-based electron beam and Cherenkov maser experiments // Phys. Plasmas. 2000. V. 7. No. 12. P. 5195-5205.

3. *Yin, H., He, W., Cross, A.W., Phelps, A.D.R., and Ronald, K.* Single-gap pseudospark discharge experiments // J. Appl. Phys. 2001. V. 90. No. 7. P. 3212-3218.

4. *Yin, H., Cross, A.W., Phelps, A.D.R., Zhu, D., He, W., and Ronald, K.* Propagation and post-acceleration of a pseudospark-sourced electron beam // J. Appl. Phys. 2002. V. 91. No. 8. P. 5419-5422.

5. *Yin, H., Cross, A.W., He, W., Phelps, A.D.R., and Ronald, K.* Pseudospark experiments: Cherenkov interaction and electron beam post-acceleration // IEEE Trans. Plasma Sci. 2004. V. 32. No. 1. P. 233-239.

6. *Yin, H., Cross, A.W., He, W., Phelps, A.D.R., Ronald, K., Bowes, D., and Robertson, C.W.* Millimeter wave generation from a pseudospark-sourced electron beam // Phys. of Plasmas. 2009. V. 16. No. 6. P. 063105.

7. *Bowes, D., Yin, H., He, W., Zhang, L., Cross, A.W., Ronald, K., Phelps, A.D.R., Chen, D., Zhang, P., Chen, X., and Li, D.* X-ray emission as a diagnostic from pseudospark-sourced electron beams // Nuclear Instruments & Methods in Physics Research Section B-Beam Interactions with Materials and Atoms. 2014. V. 335. No. P. 74-77.

8. *Bowes, D., Yin, H.L., He, W.L., Ronald, K., Phelps, A.D. R., Chen, D.F., Zhang, P., Chen, X.D., Li, D.H., and Cross, A.W.* Visualization of a Pseudospark-Sourced Electron Beam // IEEE Trans. Plasma Sci. 2014. V. 42. No. 10. P. 2826-2827.

9. *He, W., Zhang, L., Bowes, D., Yin, H., Ronald, K., Phelps, A.D.R., and Cross, A.W.* Generation of broadband terahertz radiation using a backward wave oscillator and pseudospark-sourced electron beam // Appl. Phys. Lett. 2015. V. 107. No. 13. P. 4.

10. *Shu, G.X., He, W.L., Zhang, L., Yin, H.B., Zhao, J.P., Cross, A.W., and Phelps, A.D.R.* Study of a 0.2-THz Extended Interaction Oscillator Driven by a Pseudospark-Sourced Sheet Electron Beam // IEEE Trans. Electron Devices. 2016. V. 63. No. 12. P. 4955-4960.

11. *Zhao, J., Yin, H., Zhang, L., Shu, G., He, W., Zhang, J., Zhang, Q., Phelps, A.D.R., and Cross, A.W.* Influence of the electrode gap separation on the pseudospark-sourced electron beam generation // Phys. Plasmas. 2016. V. 23. No. 7. P. 6.

12. *Zhao, J., Yin, H., Zhang, L., Shu, G., He, W., Phelps, A.D.R., Cross, A.W., Pang, L., and Zhang, Q.* Study of the beam profile and position instability of a post-accelerated pseudospark-sourced electron beam // Phys. Plasmas. 2017. V. 24. No. 3

13. *Zhao, J., Yin, H., Zhang, L., Shu, G., He, W., Zhang, Q., Phelps, A. D. R., and Cross, A. W.* Advanced post-acceleration methodology for pseudospark-sourced electron beam // Phys. Plasmas. 2017. V. 24. No. 2

14. *Kumar, N., Pal, D.K., Lamba, R.P., Pal, U.N., and Prakash, R.* Analysis of Geometrical Design Parameters for High-Energy and High-Current-Density Pseudospark-Sourced Electron Beam Emission // IEEE Trans. Electron Devices. 2017. V. 64. No. 6. P. 2688 - 2693.

Excitation of THz surface waves in the conductor by a drag current generated by a focused femtosecond pulse

A.A. Frolov[1], S.A. Uryupin[2]

[1]Joint Institute of High Temperature Physics, Russian Academy of Sciences, Moscow, 125412, Russia, frolov@ihed.ras.ru
[2]P.N. Lebedev Physics Institute, Russian Academy of Sciences, Moscow, 119991, Russia, uryupin@sci.lebedev.ru

Surface waves (SW) are of interest because of their unusual physical properties and the possibility of the use of these waves for the diagnostics of surfaces, investigation of the properties of thin films and interfaces between various media, study of spectra of surface excited states, and signal transfer. The linear theory of SW excitation in conductors was developed in many works (see, e.g., [1]). Nonlinear methods of SW excitation are much less studied. At the same time, the experiment reported in [2] indicates the necessity of the development of a nonlinear theory of SW excitation. In application to the action of femtosecond pulses on conductors, we analyzed in [3] the possibility of SW exciting by a normally incident focused pulse. It is shown that the time-dependent ponderomotive force generates SW in a terahertz frequency range. Such a method of excitation can significantly increase the total energy of SW if a cylindrical lens is used to focus laser radiation [4].

In this work, we present the theory of another nonlinear mechanism of SW excitation caused by the generation of the drag current by a laser pulse which is focused by a cylindrical lens and is incident not at a right angle to the surface of a conductor. Since the effective frequencies of electron collisions for typical metals at room and higher temperatures are relatively high, the mechanism described below for the excitation of terahertz surface waves is more efficient than the ponderomotive mechanism.

Let the laser pulse of s-polarized radiation with amplitude of electric field E_{0L}, frequency ω_0 and duration τ be focused by a cylindrical lens in a line of length L_y and the transverse dimension L_x falls on the boundary of the conducting medium at an angle α relative to the normal. The laser field penetrating into conductor generates a slowly varying drag current along the surface of the conductor and provides the ponderomotive action on electrons, which generate low-frequency currents along and across the surface. To determine nonlinear currents $\mathbf{j}(\mathbf{r},t)$, we use the equation

$$\frac{\partial}{\partial t}\mathbf{j}(\mathbf{r},t)+v_s\mathbf{j}(\mathbf{r},t)= =\frac{\omega_p^2}{4\pi e}\left[e\mathbf{E}(\mathbf{r},t)-\nabla W(\mathbf{r},t)+\mathbf{e}_x\frac{2v_h}{c}W(\mathbf{r},t)\sin\alpha\right], \qquad (1)$$

where ω_p -the plasma frequency, $\nu_{h(s)}$–the frequency of electron collisions in the high (low)-frequency field, e is the elementary charge, $W(\mathbf{r},t)$-the ponderomotive potential in the conductor, c-speed of light. Constitutive equation (1) and Maxwell's equations for the electric $\mathbf{E}(\mathbf{r},t)$ and magnetic $\mathbf{B}(\mathbf{r},t)$ fields

$$rot\mathbf{B}=\frac{1}{c}\frac{\partial}{\partial t}\mathbf{D}+\frac{4\pi}{c}\mathbf{j},\ \mathrm{rot}\mathbf{E}=-\frac{1}{c}\frac{\partial}{\partial t}\mathbf{B}, \qquad (2)$$
$$\mathbf{D}(\mathbf{r},t)=\int_{-\infty}^{+\infty}\frac{d\omega}{2\pi}e^{-i\omega t}\varepsilon_0(\omega)\mathbf{E}(\mathbf{r},\omega),$$

where $\mathbf{D}(\mathbf{r}, t)$ is the electric displacement, ε_0 - contribution to the permittivity of bound electrons and lattices, make it possible to describe a low-frequency electromagnetic field in the conductor. In Eq. (1), the term proportional to the gradient of $W(\mathbf{r}, t)$ is due to the ponderomotive action of laser radiation and the term with ν_h corresponds to the effect of the drag current. The effect of the ponderomotive action can be neglected under the conditions

$$\nu_h\tau\gg1,\quad \nu_h\tau\gg(L/L_x)ctg\alpha\ , \qquad (3)$$

where $L=c\tau$ is the longitudinal size of laser pulse. Under conditions (3), the drag current is the main reason for the generation of surface waves. Because of the oblique incidence, a laser pulse generates SW propagating in the direction of the drag current and in the opposite direction, which have different parameters.

Spectral and energy parameters of the surface waves are calculated. The energy of the SW per unit frequency interval $d\omega$ has the form

$$\frac{dW^{\pm}(\Omega,\alpha)}{d\Omega}=\frac{e^2E_{0L}^4L_x^2L^2L_y}{4\sqrt{2}m^2c^4}\frac{(\nu_h\tau)^2}{(\nu_s\tau)^{1/2}}\frac{I^{\pm}(\Omega,\alpha)}{\omega_p^5\tau^5}, \qquad (4)$$

where

$$I^{\pm}(\Omega,\alpha)=\Omega^{5/2}\sin^2\alpha\cos^2\alpha\times \times\exp\left\{-\frac{\Omega^2}{2}\left(1+\frac{L_x^2}{L^2}\frac{1\mp\sin\alpha}{1\pm\sin\alpha}\right)\right\}, \qquad (5)$$

$\Omega=\omega\tau$-dimensionless frequency, m-electron mass, W^+, W^- – energy of the SW in the direction of the drag current and in the opposite direction. The frequency distribution of SW given by Eq. (5) has a maximum at

$$\omega_m^{\pm}=\sqrt{\frac{5}{2}}\frac{1}{\tau}\left(1+\frac{L_x^2}{L^2}\frac{1\mp\sin\alpha}{1\pm\sin\alpha}\right)^{-1/2}. \qquad (6)$$

The maximum in the spectrum for SW, propagating in the direction of the drag current, shifts to the high-frequency region with increasing incidence angle α (Fig. 1). In contrast to this for SW propagating in the direction opposite to the drag current, the maximum in the spectrum shifts to the region of low frequencies with increasing incidence angle α (Fig. 2).

The integration of Eq. (4) with respect to the frequency gives the total energy of surface waves:

$$W^{\pm}(\alpha)=\frac{3}{2^{7/4}}\Gamma\left(\frac{3}{4}\right)\frac{(\nu_h\tau)^2}{(\nu_s\tau)^{1/2}}\frac{e^2E_{0L}^4L_y}{4m^2c^4}\frac{w^{\pm}(\alpha)}{k_p^5L}, \qquad (7)$$

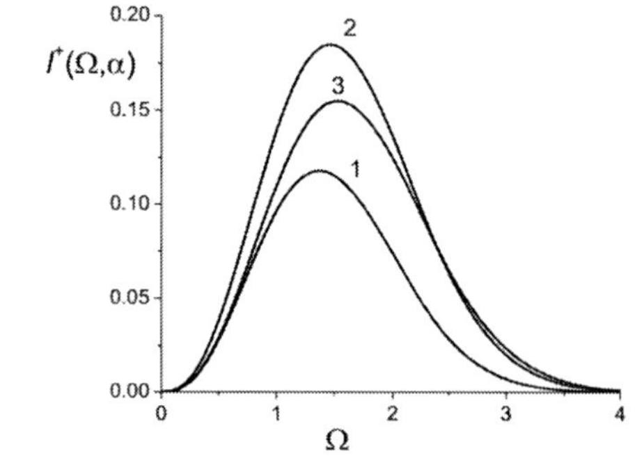

Fig. 1. The dependence of the energy of the SW propagating in the direction of the drag current on the frequency $\omega\tau$ for different angles of incidence α: (*1*) $\alpha = \pi/6$, (*2*) $\alpha = \pi/4$, (*3*) $\alpha = \pi/3$ for $L = L_x$

where $\Gamma(x)$ is the Gamma function, $k_p=\omega_p/c$

$$w^{\pm}(\alpha)=\frac{L_x^2}{4L^2}\sin^2 2\alpha\left(1+\frac{L_x^2}{L^2}\frac{1\mp\sin\alpha}{1\pm\sin\alpha}\right)^{-7/4}. \quad (8)$$

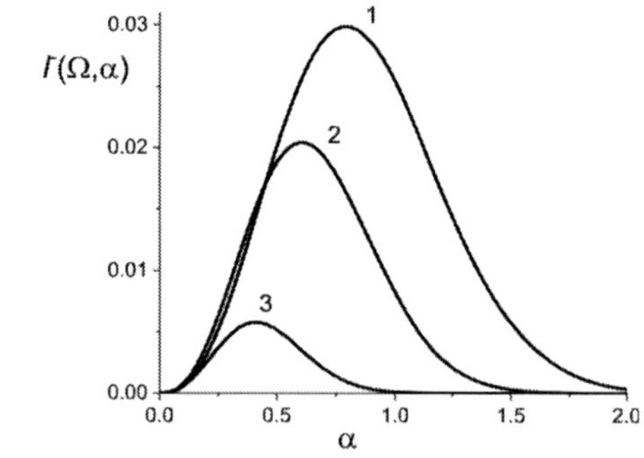

Fig. 2. The dependence of the energy of the SW propagating in the direction opposite to the drag current on the frequency $\omega\tau$ for different angles of incidence α: (*1*) $\alpha = \pi/6$, (*2*) $\alpha = \pi/4$, (*3*) $\alpha = \pi/3$ for $L = L_x$

The dependence of the energy of the SW propagating in the direction of the drag current (8) on the angle α is shown in Fig. 3. In the case of a tightly focused pulse, $L_x << L$, the function $w^{+}(\alpha)$ in Eq. (8) has a maximum of $w^{+} = (L_x/2L)^2$ at $\alpha = \pi/4$. If $L_x = L$, $w^{+}(\alpha)$ has the maximum $w^{+} \approx 0.195$ at $\alpha \approx 49.6°$. Finally, for a poorly focused pulse when $L_x >> L$ $w^{+}(\alpha)$ has a largest value of maximum $w^{+} \approx 1.2$ at $\alpha \approx \pi/2-4L/(L_x\sqrt{3})$. Thus with an increase of the laser pulse transverse size L_x the energy of SW propagating in the direction of the drag current increases.

The dependence of the energy of the SW propagating in the direction opposite to the drag current, on the angle of incidence of the laser pulse α is shown in Fig. 4. As follows from Eq (8) and Fig. 4 with an increase in the transverse size, the energy of the SW propagating in the direction opposite to the drag current decreases.

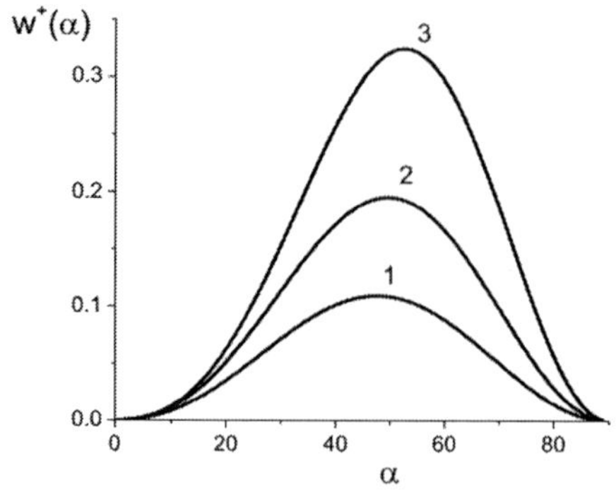

Fig. 3. The dependence of the total energy of the SW propagating in the direction of the drag current on the angle of incidence of the laser pulse α: (*1*) $(L_x/L)^2 = 1/2$, (*2*) $(L_x/L)^2 = 1$, (*3*) $(L_x/L)^2 = 2$

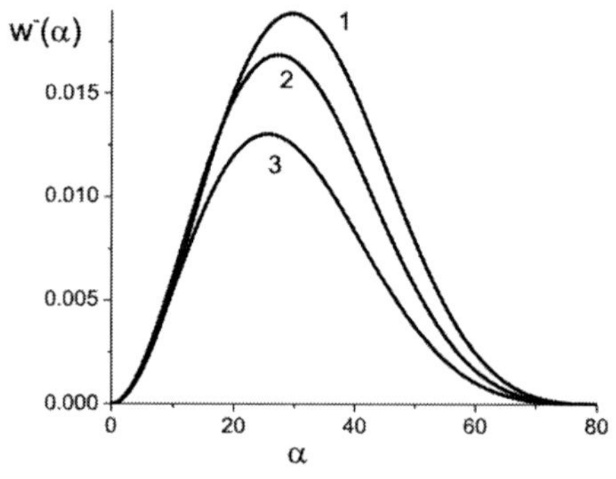

Fig. 4. The dependence of the total energy of the SW propagating in the direction opposite of the drag current on the angle of incidence of the laser pulse α: (*1*) $(L_x/L)^2 = 1/2$, (*2*) $(L_x/L)^2 = 1$, (*3*) $(L_x/L)^2 = 2$

The energy of SW propagating in the direction of the drag current greatly exceeds the SW energy propagating in the opposite direction.

The work supported by RFBR, project N17-02-00648a.

References

1. *Maier S.A.* Plasmonics: Fundamentals and Applications. Springer, Berlin, 2007; R&C Dynamics, Moscow, 2011, P. 22.

2. *Ionin, A.A., Kudryashov, S.I., Seleznev, L.V., Sinitsin, D.V., Emel'yanov, V.I.* Nonlinear regime of the excitation of a surface electromagnetic wave on the silicon surface by an intense femtosecond laser pulse // JETP Lett. 2013. V. 97, No. 3. P. 121–126.

3. *Uryupin, S.A., Frolov, A.A.* Excitation of surface waves by a short laser pulse in a conductor // Quantum Electron. 2013. V. 43. No. 12. P. 1132-1138.

4. *Uryupin, S.A., Frolov, A.A.* Generation of surface waves and low-frequency radiation under exposure of a conductor to a laser pulse focused by a cylindrical lens // Quantum Electron. 2014. V. 44. No. 9. P. 866-872.

EPJ Web of Conferences 149, 05017 (2017) DOI: 10.1051/epjconf/201714905017
10th International Workshop 2017 "Strong Microwaves and Terahertz Waves: Sources and Applications"

Electrodynamic system for two-stage THz-generator on the base of two-channel planar FEM

A.V. Arzhannikov[1,2], N.S. Ginzburg[3], P.V. Kalinin[1,2], N.Yu. Peskov[3], S.E. Sandalov[2], S.L. Sinitsky[1,2], and V.D. Stepanov[1,2]

[1]Budker Institute of Nuclear Physics SB RAS, Novosibirsk, Russia, sinitsky@inp.nsk.su
[2]Novosibirsk State University, Novosibirsk, Russia,
[3]Institute of Applied Physics, RAS, Nizhny Novgorod, Russia

Introduction

Two-dimensional (2-D) distributed feedback gives the opportunity to create generators of a coherent flowof electromagnetic radiation with very wide cross-section. We study such mechanism of spatial synchronization of electromagnetic oscillations for the case of a planar free electron maser (FEM) at the ELMI-device (BINP, Novosibirsk) [1]. To produce high-power THz-radiation few schemes of two-stage generators basing on this mechanism were suggested and investigated [2, 3]. Theory predictsMW-power levelin THz-radiation pulse [3].To get the first experience the simple two-stage scheme similar to our experiments on mm-wave generation in two-channel planar FEM [4] was chosen. The idea is to use single channel of this planar FEM driven by the first sheet beam as a low frequency (LF) section of THz-generator in which the following parameters of the radiation pulses have been achieved: frequency ~75 GHz, spectrum width ~20 MHz, pulse duration ~200 ns and power from single channel ~20 MW. This pulse is planned to be delivered to the high frequency (HF) section by means of Bragg wave deflector. After transformation of the pulse to the cutoff wave in the wave transformer it will be scattered by the second sheet e-beam with the frequency conversion to 1 THz band. The parameters and technical solutions for the elements of electrodynamic system obtained on the base of 3D-modeling and tested in "cold" experiments will be presented in the paper.

Electrodynamic System for Two-Stage Generator

The design of electrodynamic systemincludes two planar channelswith the length of 1.1 m separated by a metal baffle (see Fig. 1). Two sheet e-beams (1 MeV / 1 kA / 5 μs) are transported in a guiding uniform magnetic field $B \leq 1.7$ T inside the lower (cross-section 0.95×9 cm) and the upper (0.6×9 cm) vacuum channels referred as LF- and HF- sections respectively. Besides that, the LF-section is placed in an undulator magnetic field with amplitude $B_\perp \leq 0.17$ T. The electrodynamic system of the LF-section is composed of the two-mirror cavity consisting of 2D and 1D Bragg reflectors connected by a regular waveguide, and the wave deflector to deliver mm-wave to the HF-section. Both 1D and 2D reflectors are formed by two parallel conducting plates with corrugation of their surface in the form of parallel rectangular grooves and "chessboard" corrugation respectively. The 2D-reflector of the cavity has the length $l_{2D} = 20$ cm with its corrugation depth $a_{2D} = 220$ μm and the period of corrugation $d_{2D} = 0.4$ cm. The main Fourier harmonic of the corrugation has the amplitude: $A_{1,1} = 16a_{2D}\left[\cos(h_{2D}z - h_{2D}x) + \cos(h_{2D}z + h_{2D}x)\right]/\pi^2$, where $h_{2D} = 2\pi/d_{2D}$ – x and z components of translation vector. The same parameters for 1D-reflector are: the length $l_{1D} = 20$ cm, depth of corrugation $a_{1D} = 70$ μmand its period $d_{1D} = 0.2$ cm. The amplitude of main Fourier harmonic is: $A_1 = 4a_{1D}\cos(h_{1D}z)/\pi$, where $h_{1D} = 2\pi/d_{1D}$ - translation vector of grating.

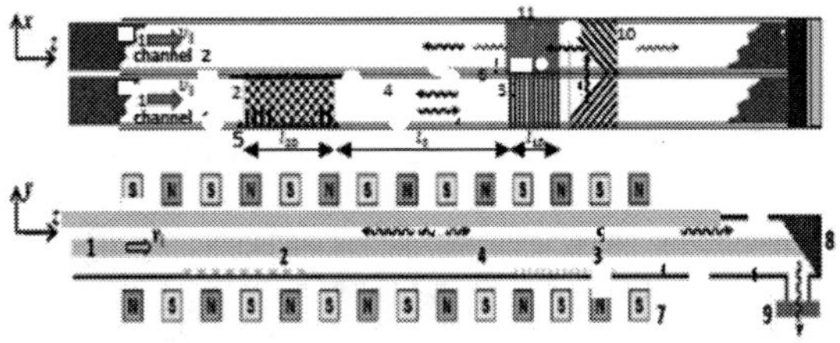

Fig. 1. Layout of the ELMI experiment to study the two-stage scheme for THz generation. 1 – parallel sheet electron beams, 2 – 2D-Bragg reflector, 3 – 1D-reflector, 4 – regular waveguide section, 5 – baffle between channels, 6 – planar undulator (shown in the y-z plane), 7 – beam collector, 8 – output window, 9 – wave deflector,10 – Bragg transformer of travelling to cutoff wave

Wave deflector

The main task for the design of the wave deflector is the achieving the high efficiencyof the radiation transmission from the LF-section to the HF-section and the required wave structure at the exit of the deflector.

To find the optimal configuration of the deflector, two main features of its corrugation were varied in the modeling: the shape of the corrugated region and the corrugation depth. In the result of the numerical solution of the coupled wave equations the optimal expression for the corrugation border was found: $z = 2.69 \cdot \exp(|x|/2.9) - 3.58$ cm, $x = -10 \div 10$ cm, $z = 0 \div 13$ cm. Then the optimal depth of the corrugations was determinedin 3D full electro-dynamic calculations. Fig. 2 shows the results of modeling and measurements for the transformation of H_{10} mode passing in LF-section to the modes of HF-section for the corrugation depth of the deflector 370 μm.

Both 3D calculations and "cold" tests demonstrate the high transmission efficiency (~80%) from the LF-section to the HF-section in the band 74÷76 GHz. As to the mode purity, some parasitic conversion to highest modes increasing with the increase of the corrugation depth was observed.

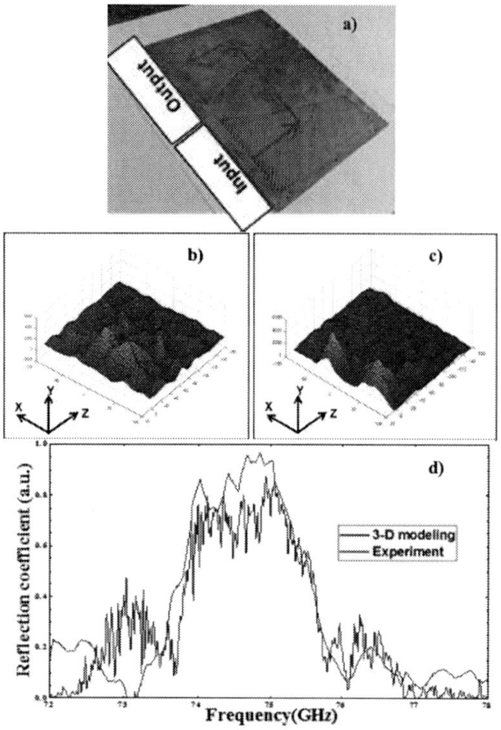

Fig. 2. (a) – photo of the wave deflector plate, (b),(c) – flux density distributions P_x and P_z in the plane (x,z),(d) – transmission efficiency of the deflector from input to outputmeasured in "cold" tests

Wave Transformer

In order to intensify the interaction of the beam with the mm-wave, the transformer of the travelling wave to the cutoff one is planned to be applied in HF-section. Such wave conversion will serve to some decreasing the frequency of scattered radiation to the range 0.6 ÷ 1 THz that is more suitable for further mode selection by means of the cavity with advanced Bragg structures [5]. For this purpose, we developed 1D Bragg structure with one leap phase equal to π in the center of the structure. Such transformer with sizes of its plates $l^{tr} = 19.6\,\text{cm}$, $l_{\perp}^{tr} = 10\,\text{cm}$ and the gap between them $a_0 = 0.6\,\text{cm}$, has rectangular corrugation with the depth $a_{tr} = 70\,\mu\text{m}$ and the period $d_{tr} = 0.4\,\text{cm}$ (see Fig. 3, a).According to 3D-modeling this wave structureprovides substantial increase of the scattering wave amplitude(~10 times) but only in the narrow frequency band 75.05 ÷ 75.12 GHz (see Fig. 3, b).This will make difficultiesfor ingress of the operating mode of LF-section in this band. To overcome them, new Bragg transformer with two leaps of phase is under development now that will allowobtaining more wider band with effective conversion to cutoff wave and larger gain factor.These conditions should provide more effective beam-wave interaction resulting in the increase of the THz-radiation power in the future experiments at the ELMI device.

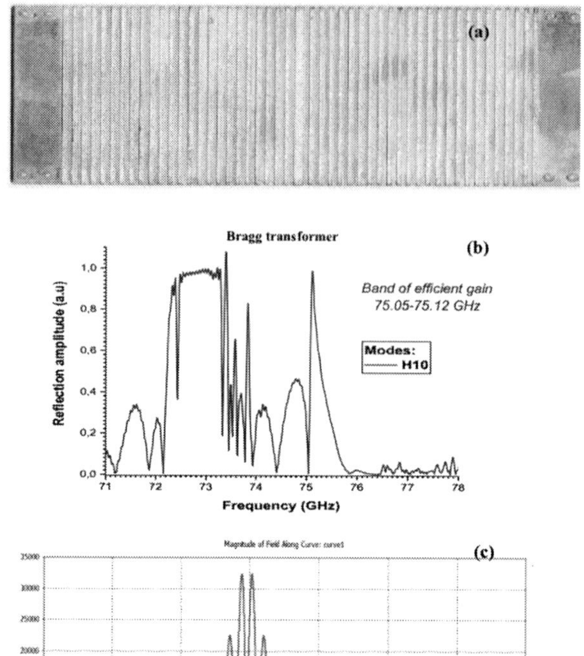

Fig. 3. (a) – photo of the Bragg transformer with aleap phase, (b) – reflection coefficient of Bragg transformer, obtained in 3D modeling, (c) – amplitude amplification

Acknowledgements

The work was partially supported by RSCF Project #14-12-00610 and by the Ministry of Education and Science of the Russian Federation under the State Assignment Contract #3002.

References

1. *A.V. Arzhannikov et al.* Generation of Powerful Narrow Band 75 GHz Radiation in a Free Electron Maser with Two Dimensional Distributed Feedback// Tech. Phys. Lett. 2013. V. 39, No. 9. P. 801–804.
2. *A.V. Arzhannikov et al.*Intercavity scattering scheme for two-stage generationof submillimeter radiation on the base of planar 2DBragg FEM// Proc. of 6 Int. Workshop "Strong microwaves in plasma",N.Novgorod, Russia, 2005. P.228.
3. *A.V. Arzhannikov et al.*A traveling wave ring resonator with Bragg deflectors in a two-stage terahertz free electron laser// Tech. Phys. Lett. 2014. V. 40, No. 9. P. 730-734.
4. *A.V. Arzhannikov et al.* Using Two-Dimensional Distributed Feedback for Synhcronization of Radiation from two Parallel Sheet Electron Beams in Free Electron Maser// Phys. Rev. Lett. 2016. V.117, 114801.
5. *A.V. Arzhannikov et al.* Short-wavelength tunable Bragg reflectors based oncoupling of propagating and cutoff waves: Modeling and experimental studies//Appl. Phys. Lett. 2012. V. 101, 083507.

Terahertz gyrotrons with quasi-regular cavities

I.V. Osharin, I.V. Bandurkin, Yu.K. Kalynov, S.V. Kuzikov, A.V. Savilov

Institute of Applied Physics, Nizhny Novgorod, 603950, Russian Federation, osharin@appl.sci-nnov.ru

The gyrotron [1, 2] is widespread variety of electron cyclotron masers based on the selective excitation of a near-cutoff mode of an open microwave cavity by a beam of electrons rotation in a magnetostatic field. The principal advantages of the gyrotron are high mode selectivity achieved in simple microwave systems and weak sensitivity of the electron-wave interaction to velocity spread of electrons. Since the output frequency of the gyrotron is close to the electron cyclotron frequency or its harmonics, a natural way to decrease the operating magnetic field is the use of operation at the harmonics of the cyclotron frequency. However, at a fixed electron current, a longer operating cavity is required, in general, to provide the start of oscillations at a high cyclotron harmonic. In this situation, the start of the parasitic oscillations of low-frequency waves excited at lower cyclotron harmonics is dangerous.

Thus, a careful mode selection and suppression of the parasitic modes have to be ensured in a gyrotron operating at a high cyclotron harmonic. In [4], a method of selective discrimination of parasitic low-frequency modes in a second cyclotron harmonic gyrotron was proposed, based on the use of special phase correctors located inside the gyrotron resonator. The selection principle is based on the fact that the corrector provides different phase shifts for quasi-critical waves at the fundamental and operating cyclotron harmonics. The method is based on the single-mode approximation, when the corrector is represented by a relatively small and smooth widening of the cavity wall, and the wave phase shift can be calculated in the WKB approximation. Such an approach can be used in sufficiently long and low-mode resonators. However, in the case of a short cavity operating at a high radial mode, this approach can not be realized. This is due to the fact that the restriction on the maximum length of the corrector imposes a restriction on its minimum depth. From a certain set of the resonator parameters, the single-mode approximation in describing the corrector becomes invalid [5, 6].

We propose a modification of the described method of low-frequency parasitic modes discrimination in a gyrotron at the second cyclotron harmonic, which makes it possible to use in relatively short and high-mode resonators. The ultimate goal of the proposed method, as well as in [4], is to realize inside the gyrotron cavity of an element that would noticeably distort the field of parasitic low-frequency modes but would not significantly affect the field of the operating mode at the second cyclotron harmonic. Under certain conditions, a rectangular groove in the resonator wall (Fig. 1) can serve as such an element. Namely, if the radius of the resonator inside this groove is close to the critical one for a mode whose radial index exceeds by one the radial index of the operating mode, and whose azimuthal index coincides with one of the operating mode, then by choosing the width of the groove, it is possible to achieve almost complete absence of scattering of the operating mode on this element.

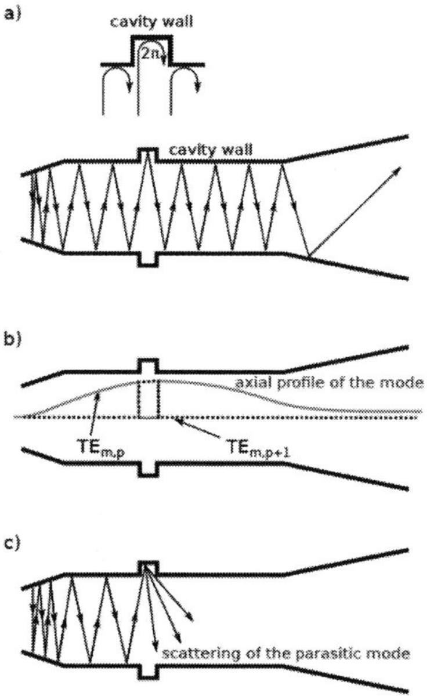

Fig. 1. Explanation of the selection mechanism from the Brillouin concept (a) and cavity with modes transformation point of view (b), as well as illustration of low harmonic mode scattering (c)

To demonstrate the method, we propose to modify the cavity of a 670 GHz gyrotron which presently operates in the Institute of applied physics (IAP) at the fundamental cyclotron resonance at the $TE_{31,8}$ mode [3]. An analysis of the eigenmodes spectrum of this cavity shows that in the operating range of magnetic fields, the resonance is possible at the second cyclotron harmonic at the $TE_{63,15}$ mode at 1.34 THz. Nevertheless, in the traditional cylindrical system the starting current of this mode is slightly higher than the starting current of the low-frequency modes $TE_{30,8}$ and $TE_{31,8}$ excited at the fundamental cyclotron resonance (Fig. 2).

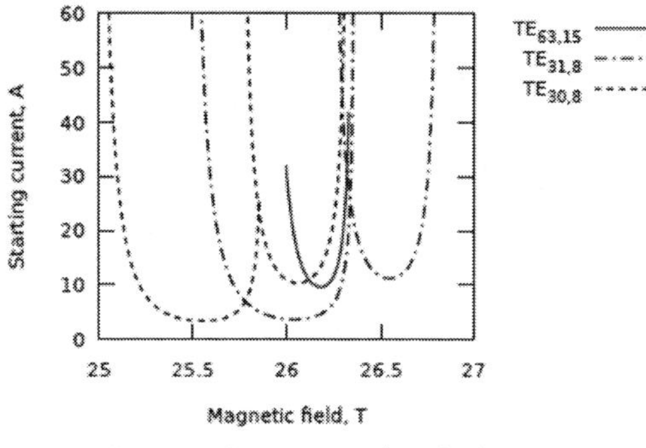

Fig. 2. Starting currents of the fundamental and second cyclotron harmonic oscillations in the regular 670 GHz gyrotron cavity

To discriminate these modes, we propose to apply a groove with depth of about 0.12 mm to the cavity wall having a radius of 4.54 mm. With this depth, the radius of the resonator inside the groove approximately corresponds to the critical radius for the $TE_{63,16}$ mode. The longitudinal position of the groove along the axis of the resonator should divide its length in the ratio 1 to 3 (Fig. 3), which makes it possible to discriminate not only the basic parasitic axial modes, but also modes with two variations of the field along the axis. The eigenmodes of the modified resonator were calculated by solving the three-dimensional Maxwell equations with a given field dependence on the azimuthal coordinate corresponding to the azimuthal index of the mode. The simulation was carried out by the finite element method in the two-dimensional domain (r-z) [5].

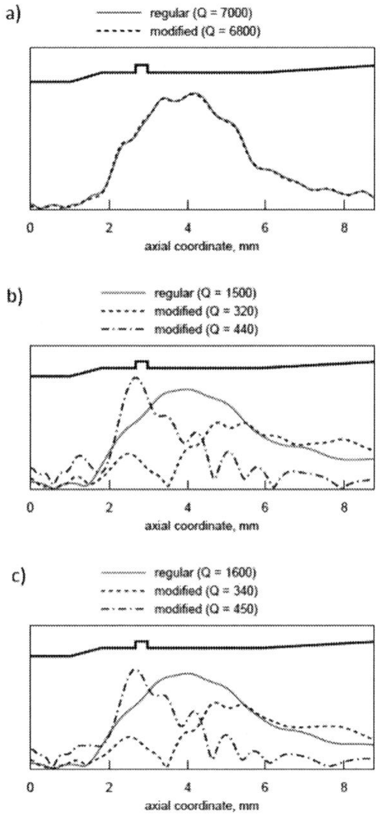

Fig. 3. Results of electrodynamic cavity simulations. Comparison of the axial profiles of the TE63,15 (a), TE30,8 (b) and TE31,8 (c) modes in the vicinity of electron beam for the regular and modified cavity. In the modified cavity, two modes with approximately equal low Q-factors appear in last two cases, which are actually the descendants of the two lowest axial modes of the regular cavity.

By choosing the geometric dimensions of the groove it was possible to achieve its almost complete "invisibility" for the $TE_{63,15}$ mode, keeping both the quality factor of this mode and the longitudinal field profile in the position of the electron beam unchanged (Fig. 3a). Calculation of the eigenmodes of the modified resonator near the resonance at the first cyclotron harmonic, as expected, demonstrated a strong distortion of the longitudinal profiles of these modes with respect to the unperturbed system and a significant decrease in their Q-factors (Fig. 3b,c). Calculation of the starting currents confirmed the effectiveness of discrimination of parasitic modes in the proposed resonator (Fig. 4). The deformation of the low-frequency modes turned out to be so strong that their starting currents became several times greater than the practically unchanged starting current of the operating mode at the second cyclotron harmonic.

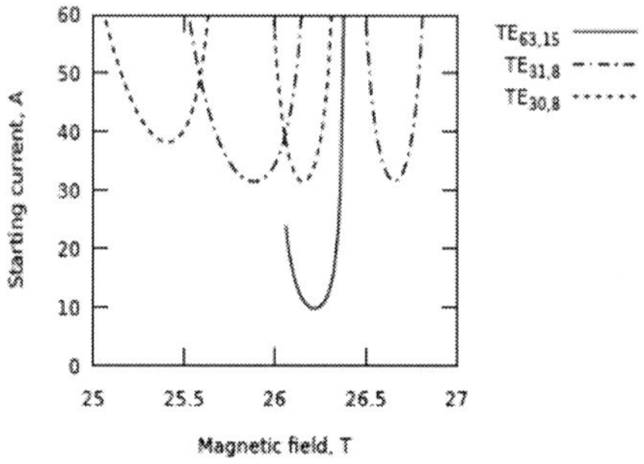

Fig. 4. Starting currents of the fundamental and second cyclotron harmonic oscillations in the modified gyrotron cavity

The proposed method of discrimination of parasitic low-frequency oscillations at the fundamental cyclotron resonance in a gyrotron is the development of the previously proposed methods. According to calculations, it can confidently ensure selective generation at the second cyclotron harmonic in the THz gyrotron. The experimental verification of the proposed scheme should be implemented at the IAP in the near future. In principle, this method can be applied to ensure the generation selectivity at higher than the second cyclotron harmonics as well.

Acknowledgements

The work is supported by the Russian Science Foundation, Project # 17-19-01605

References

1. *A. V. Gaponov, M. I. Petelin,* and *V. K. Yulpatov*, "The induced radiation of excited classical oscillators and its use in high-frequency electronics," Radiophys. Quantum Electron., vol. 10, no. 9, pp. 794–813, 1967.

2. *G. S. Nusinovich*, Introduction to the physics of gyrotrons. The Johns Hopkins University Press, Baltimore-London, 2004.

3. *M. Yu. Glyavin, A. G. Luchinin, G. S. Nusinovich, J. Rodgers, D. G.Kashyn, C. A. Romero-Talamas,* and *R. Pu*, "A 670 ghz gyrotron with record power and efficiency," Appl. Phys. Lett., vol. 101, no. 15, p. 153503, 2012.

4. *Yu. K. Kalynov, I. V. Osharin,* and *A. V. Savilov*, "A method for suppression of spurious fundamental-harmonic waves in gyrotrons operating at the second cyclotron harmonic," Phys. Plasm., vol. 23, no. 5, p. 053116, 2016.

5. *I. V. Bandurkin, Y. K. Kalynov, P. B. Makhalov, I. V. Osharin, A. V. Savilov,* and *I. V. Zheleznov*, "Simulations of sectioned cavity for high-harmonic gyrotron," IEEE Trans. Electron Devices, vol. 64, no. 1, pp. 300–305, 2017.

6. *G. S. Nusinovich* and *O. Dumbrajs*, "Field formation in the interaction space of gyrotrons," J. Infrared Millimeter Terahertz Waves, vol. 37, no. 1, pp. 111–122, 2016.

7. *G. G. Denisov, N. I. Zaitsev, S. V. Kuzikov,* and *M. E. Plotkin*,"Enhancement of cavity selectivity in relativistic gyrotrons operated at axisymmetric modes," Radiophys. Quantum Electron., vol. 51, no. 10, pp. 756–767, 2008.

EPJ Web of Conferences **149**, 05019 (2017)   DOI: 10.1051/epjconf/201714905019
10th International Workshop 2017 "Strong Microwaves and Terahertz Waves: Sources and Applications"

Spontaneous coherent cyclotron THz super-radiation from a short dense photo-injector electron bunch

Yu.S. Oparina and A.V. Savilov

Institute of Applied Physics, N.Novgorod, Russia, savilov@appl.sci-nnov.ru

Laser-driven photo-injectors allow formation of compact and accessible sources of dense electron bunches with a moderate energy of 3-6 MeV, picosecond pulse durations, and charges of up to 1 nC and even greater. These bunches can be used for realization of comparatively simple and compact terahertz sources operating in the regime of spontaneous coherent undulator radiation. This type of radiation is realized, when the effective phase size of the electron bunch with respect to the wave is small enough ($< 2\pi$), so that the wave packets emitted by each of the electrons add up basically in phase. The THz source based on the spontaneous emission has a number of advantages as compared to the more traditional free-electron laser (FEL) based on the undulator emission induced due to bunching of a long electron beam by the radiated wave. These are a relatively high efficiency of the emission from the "ready-for-radiation" bunch, a narrow frequency band of the radiated rf signal in a system without a feedback system and an input rf signal, and the phase of the radiated rf signal fixed by the electron bunch phase. These advantages make the THz source based on the spontaneous emission attractive for a number of important applications, including the use of synchronization of such an oscillator with the X-ray FEL in pump-probe experiments [1-4]. The radiation, which could be characterize as induced radiation, take place, then effective phase size of bunch approximately 2π. In this case electron bunches are formed by spontaneous radiation of bunch edges. The THz source based on the spontaneous emission has a number of advantages. These are a relatively high efficiency of the emission from the "ready-for-radiation" bunch, a narrow frequency band of the radiated rf signal in a system without a feedback system and an input rf signal, and the phase of the radiated rf signal fixed by the electron bunch phase.

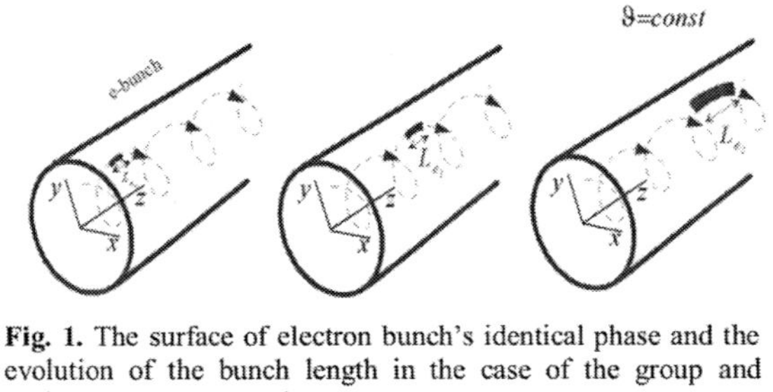

Fig. 1. The surface of electron bunch's identical phase and the evolution of the bunch length in the case of the group and cyclotron resonance regime

However, essential to the realization of a THz source based on the spontaneous coherent radiation is the presence of the mechanism of stabilization of the phase size of the bunch with respect to the radiated wave. Actually, the spontaneous emission occurs while the bunch phase size is smaller than 2π, so that wave packets emitted by each of the electrons add up basically in phase. In the case of the ubitron (as well as Cherenkov) source, this means that the bunch length must stay shorter than the radiation wavelength during the whole radiation region. Therefore, the length of the radiation region of such an rf source is strictly limited by the axial Coulomb repulsion, which is very strong in short dense photo-injector bunches and lead to a significant spreading of the bunch during its motion through a path being as short as few decimeters.

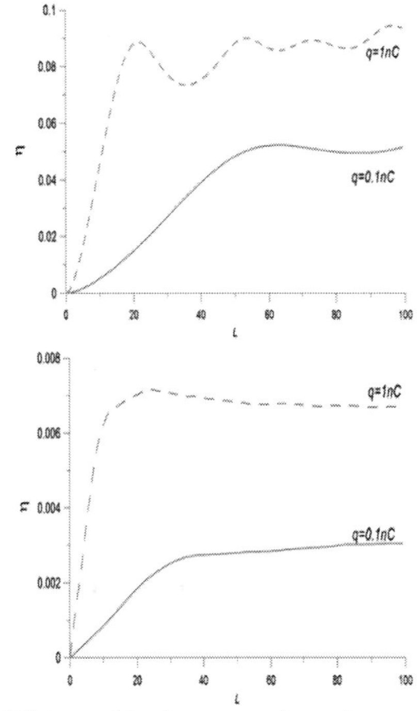

Fig. 2. Efficiency of the electron-wave interaction as a function of cavity length for two different charges of the electron bunch in the case of the group resonance (on the top) and then the group velocity of wave smaller than bunch's velocity (on the bottom)

In this work, we show that this problem is solved automatically in the case, when the cyclotron mechanism of the spontaneous coherent radiation is used. If the group resonance regime is used (the axial electron velocity is close to the group velocity of the radiated wave), then an increase in the length of the bunch (due to the Coulomb repulsion) does not lead (at least in the first approximation) to an increase in the phase size of the bunch (see fig. 1). The change of resonance phase $\vartheta = \omega_0 t - h_0 z - \int \Omega_c / \gamma dt$ (there $\Omega_c = eB_0 / m_e c$ – the cyclotron frequency of electrons, ω_0 – frequency of radiated wave, γ – relativistic Lorentz-factor, h_0 – axial wave number), which is caused by coulomb repulsion, is determined by follow equation

$$\frac{\partial \vartheta}{\partial(\omega_0 t)} \approx -\left(1-\beta_{\parallel 0}^{2}\right)\frac{\beta_{gr}}{\beta_{\parallel 0}}\frac{\Delta\gamma_{jc}}{\gamma_0}+\left(1-\beta_{\parallel 0}\beta_{gr}\right)b\frac{\Delta\gamma_{jc}}{\gamma_0}+\Delta,$$

there $\beta_{\parallel 0} = V_{\parallel 0}/c, V_{\parallel 0}$ – the initial axial velocity of the electron bunch, $\beta_{gr} = V_{gr}/c$, V_{gr} – group wave's velocity, $b = B_0/B_{0res}$ – the factor of the cyclotron mismatch, $\Delta = \left(1-\beta_{gr}\beta_{\parallel 0}\right)\left(1-b\right)$ – the mismatch of the electron-wave resonance. Obviously, the electron cyclotron phase with respect to the radiated wave includes both axial and azimuthal terms, and the changes in these two terms caused by a variation in electron energy due by the Coulomb repulsion have opposite signs, it follows of equation for the electron phase.

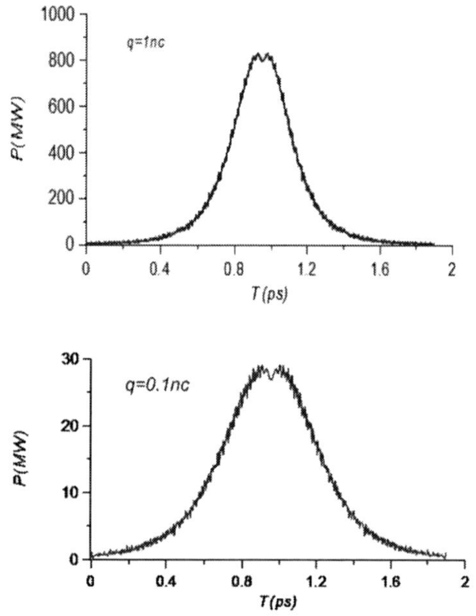

Fig. 3. Radiated pulses in the case of group resonance

The total mutual compensation takes place, when the group resonance regime is provided. It is important, that the group resonance regime is optimal also from the point of view of the maximization of the output rf power (see fig. 2), as in this regime the rf wave emission has the super-radiative character.

We have an opportunity to generate pulses with level of the power of tens-hundreds MW and duration of 1pc, using spontaneous coherent radiation at the frequency 0.5 THz from short bunches with charges 0.1–1 nc (see fig. 3). Also it is possible to realize higher power with the spontaneous radiation of ~10 nC bunches (see fig.4).

The work is supported by IAP RAS Project 0035-2014-0012, and Russian Foundation for Basic Research Project 15-42-02260.

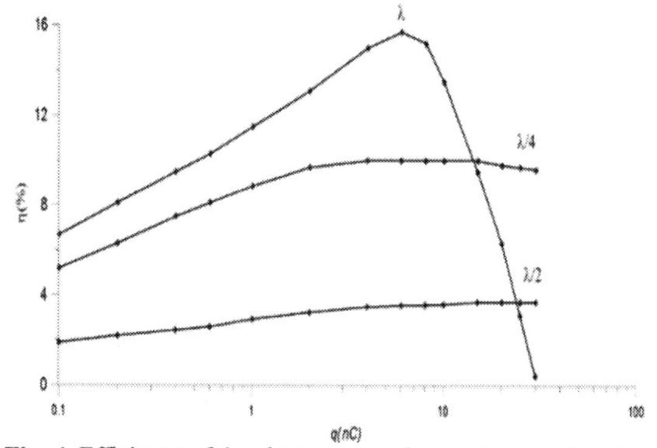

Fig. 4. Efficiency of the electron-wave interaction as a function of the bunch charge for three different bunch lengths

References

1. *E. Chiadroni, A. Bacci, M. Bellaveglia, M. Castellano, G.Di. Pirro el al.* J. of Physics: Conf. Series 359, 012018 (2012).
2. *E.A. Schneidmiller, M.V. Yurkov, M. Krasilnikov, F. Stephan.* Proceedings of SPIE 8778, 877811 (2013).
3. *P. Boonpornprasert, M. Khojoyan, M. Krasilnikov, F. Stephan, B. Marchetti, E. Schneidmiller, M. Yurkov, S. Rimjaem.* Proc. 36th International Free Electron Laser Conference FEL 2014 (Basel, Switzerland), Pages 153-158.
4. *L. Wibmann , M. Emons, M. Kellert, K. Kruse, G. Palmer, M. Pergament, G. Priebe, J. Wang, U. Wegner, and M. J. Lederer.* Proc. 36th International Free Electron Laser Conference FE.

AUTHOR INDEX

AUTHOR INDEX

AUTHOR INDEX

AUTHOR INDEX

AUTHOR INDEX

AUTHOR INDEX